SECOND EDITION

MESHFREE METHODS

Moving Beyond the Finite Element Method

SECOND EDITION

MESHFREE METHODS

Moving Beyond the Finite Element Method

G.R. Liu

CRC Press
Taylor & Francis Group
Boca Raton London New York

CRC Press is an imprint of the
Taylor & Francis Group, an **informa** business

CRC Press
Taylor & Francis Group
6000 Broken Sound Parkway NW, Suite 300
Boca Raton, FL 33487-2742

First issued in paperback 2019

ISBN-13: 978-1-4200-8209-8 (hbk)
ISBN-13: 978-1-138-37270-2 (pbk)

Library of Congress Cataloging-in-Publication Data

Liu, G. R. (Gui-Rong)
    Meshfree methods : moving beyond the finite element method / G.R. Liu. -- 2nd ed.
        p. cm.
    Includes bibliographical references and index.
    ISBN 978-1-4200-8209-8 (alk. paper)
    1. Engineering mathematics. 2. Numerical analysis. I. Title.

TA335.L58 2009
620.001'51--dc22
                                                                                2009013033

Visit the Taylor & Francis Web site at
http://www.taylorandfrancis.com

and the CRC Press Web site at
http://www.crcpress.com

To Zuona,

Yun, Kun, Run,

and my family

for the time, support, and love they gave to me

Contents

Preface

Topics related to modeling and simulation play an increasingly important role in building an advanced engineering system in rapid and cost-effective ways. Since the invention by Sir Thomas Harriet (1560–1621), people have been using the finite difference method (FDM) to perform the task of modeling and simulating engineering systems, in particular to solve partial differential equation systems. The FDM works very well for problems of simple geometry. For the last half-century, we have been using techniques of finite element methods (FEMs) to perform more challenging tasks arising from increasing demands on flexibility, effectiveness, and accuracy in problems with complex geometries. I still remember doing a homework assignment during my university years using the FDM to calculate the temperature distribution in a rectangular plate. This simple problem demonstrated the power of numerical methods and left a profound impression on me. About a year later, I created an FEM program to solve a nonlinear mechanics problem for a frame structural system, as part of my final-year project. Since then, the FEM has been one of my major tools in dealing with many engineering and academic problems. In the last three decades, I have participated in and directed many projects involving engineering problems of very large scales with millions of degrees of freedom (DOFs). I thought, and many of my colleagues agreed, that with the advances made to the FEM and to the computer, there were very few problems left to be resolved. However, I realized that I was wrong, and for a very simple reason. When a class of problems is solved, people simply move on to solve a class of problems that are more complex and to demand results that are more accurate. In reality, problems can only be as complex as we make them to be; hence, we can never claim that problems can be totally resolved. We solve problems that are idealized and simplified by us. Once the simplification is relaxed, new challenges arise. The older methods often cannot meet the demands of new problems of increasing complexity, and newer and more advanced methods are constantly born.

I heard about meshfree methods around 1993, while I was working at Northwestern University, but I was somehow reluctant to move into this new research area probably because I was quite satisfied using FEM techniques. In 1995–1996, I handled a number of practical engineering problems for the defense industry using FEM packages, and encountered difficulties in solving the problems related to mesh distortion. I struggled to use remeshing techniques, but the solution was far from satisfactory. I then began to look for methods that could solve the mesh-distortion problems encountered in my industrial research work. I thus started to learn more about meshfree methods.

I worked alone for about a year, feeling as though I were walking in a maze of this new research area. I wished that I had a book on meshfree methods to guide me. I was excited with the small progress I made, and this motivated me to work hard to write a proposal for a research grant from the NSTB (a research-funding agency of the Singapore government). I was fortunate enough to secure the grant, which enabled me to form a research team at the Centre for Advanced Computations in Engineering Science (ACES) working on element-free methods. The research team at the ACES is still working very hard in the area of meshfree methods. This book will cover many of the research outcomes from this research group.

This book provides a systematic description of meshfree methods, analyzes how they work, and explains how to use and develop a meshfree method, as well as the problems associated with the element-free methods. I experienced difficulties while learning about meshfree methods because I did not have a single book to guide me. I therefore hope my efforts in writing this book will help researchers, engineers, and students who are interested in exploring this field.

Significant advances have been made since the first edition was published in 2002. In this second edition, some of the important developments have been included, especially on the advancement of fundamental theoretical issues.

My work in the area of meshfree methods has been profoundly influenced by the works of Professors Belytschko, Atluri, W.K. Liu, J.S. Chen, late H. Noguchi, and many others working in this area. Without their significant contributions, this book would not have existed.

Many of my colleagues and students have supported and contributed to the writing and preparation of this book. I express my sincere thanks to all of them. Special thanks go to Y.T. Gu, X.L. Chen, L. Liu, V. Tan, L. Yan, K.Y. Yang, M.B. Liu, Y.L. Wu, Z.H. Tu, J.G. Wang, X.M. Huang, Y.G. Wu, Z.P. Wu, K.Y. Dai, and X. Han. For this second edition, special thanks also go to G.Y. Zhang, T. Nguyen-Thoi, Z.Q. Zhang, L. Chen, Y. Li, X.Y. Cui, Q. Tang, and S.C. Wu. Many of these individuals have contributed examples to this book in addition to their hard work in carrying out a number of projects related to the meshfree methods covered in the book.

G. R. Liu

Author

G. R. Liu received his PhD from Tohoku University, Japan, in 1991. He was a postdoctoral fellow at Northwestern University, Evanston, Illinois. He is currently the director of the Centre for Advanced Computations in Engineering Science (ACES), National University of Singapore. He is the president of the Association for Computational Mechanics (Singapore) and an executive council member of the International Associate for Computational Mechanics. He is also a professor at the Department of Mechanical Engineering, National University of Singapore. He has provided consultation services to many national and international organizations. He has authored more than 400 technical publications including more than 300 international journal papers and eight authored books, including two bestsellers: *Meshfree Method: Moving beyond the Finite Element Method* and *Smoothed Particle Hydrodynamics: A Meshfree Particle Methods*. He is the editor-in-chief of the *International Journal of Computational Methods* and an editorial member of five other journals, including the *International Journal for Numerical Methods in Engineering* (*IJNME*). He is the recipient of the Outstanding University Researchers Award, the Defence Technology Prize (national award), the Silver Award at the CrayQuest competition, the Excellent Teachers Award, the Engineering Educator Award, and the APACM Award for Computational Mechanics. His research interests include computational mechanics, meshfree methods, nanoscale computation, micro-biosystem computation, vibration and wave propagation in composites, mechanics of composites and smart materials, inverse problems, and numerical analysis.

1

Preliminaries

In building a modern and advanced engineering system, engineers must undertake a very sophisticated process in modeling, simulation, visualization, analysis, designing, prototyping, testing, fabrication, and construction. The process is illustrated in the flow-chart shown in Figure 1.1. The process is often iterative in nature; that is, some of the procedures are repeated based on the assessment of the results obtained at the current stage to achieve optimal performance.

This book deals with topics related mainly to modeling and simulation, which are underlined in Figure 1.1. The focus will be on mathematical, numerical and computational modeling, and simulation. These topics play an increasingly important role in building advanced engineering systems in rapid and cost-effective ways. Many computational methods and numerical techniques can be employed to deal with these topics. This book mainly focuses on the development and applications of the meshfree methods.

This chapter addresses the *overall* procedures of modeling and simulation using mesh-free methods and the *overall* differences in key numerical techniques between the meshfree methods and other existing methods, especially the well-known and widely used finite element method (FEM) [1–3]. General procedure of meshfree methods is outlined, and common preliminary techniques for different meshfree methods are discussed in detail, including triangulation, determination of local support domains, influence domains, estimation of nodal spacing, and local support node selection techniques.

Some of the discussions may be found too "heavy." Readers are advised to skim through these materials.

1.1 Physical Problems in Engineering

There are a large number of different physical phenomena in engineering systems, so many that it is not possible to model and simulate them all. In fact, only major phenomena, which significantly affect the performance of the system, need to be modeled and simulated to provide a necessary and sufficient in-depth understanding of the system, in order to further improve or optimize the design.

The physical problems covered in this book are in the areas of mechanics for solids, structures, and fluid flows. Classical mathematic models have already been well established for the phenomena in these areas, and different types of differential or partial differential equations (PDEs) that "govern" these phenomena have also been derived. These PDEs are in (obviously) *differential form*, and termed as *strong form* system governing equations. It is required that the PDEs are satisfied at any *point* inside the problem domain: a strong requirement. This kind of physical phenomena can be simulated using proper numerical tools by solving these PDEs with a proper set of boundary and initial conditions. We require, of course, the original physical problem to be *well-posed*.

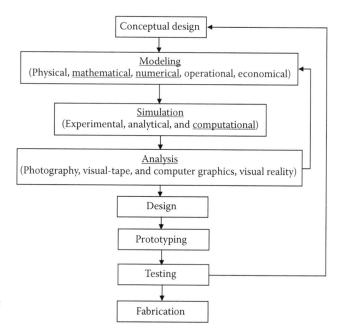

FIGURE 1.1
Processes that lead to building a complicated engineering system.

Remark 1.1: Well-Posed Problems

The original problem setting (PDEs with boundary and initial conditions) must be *well-posed* in the *Hadamard* sense, by which we mean that there *exists* a *unique* solution that depends continuously on the data. Some of the *ill-posed problems* in engineering are treatable using special regularization techniques (resetting, reformulating, adding in new information/assumptions, etc.), but it is beyond the scope of this book. Interested readers are referred to books in the areas of ill-posed problems or inverse problems (see, e.g., [4]). This book deals only numerical methods for solutions to well-posed problems.

If the problem is not well-posed, there is nothing much a numerical method can do, except produce all sort of nonsense numerical numbers or simply breaks down, regardless of how good the method is. For well-posed problems, the currently well-established and often used numerical tools include the conventional FEM [1–3], the finite difference method (FDM) [5], and the finite volume method (FVM) [6]. The meshfree method is one very powerful tool still under rapid development.

1.2 Solid Mechanics: A Fundamental Engineering Problem

We brief now a typical case of engineering problem: solid mechanics is *the default problem* studied in this book, because it is one of the most fundamental problems in all engineering applications. The strong forms for solid mechanics problems are the PDEs defined in the problem domain governing the equilibrium state at any point in linear elastic solids,

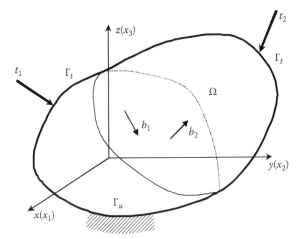

FIGURE 1.2
A constrained 3D solid subjected to external forces.

known as equilibrium equations. Figure 1.2 shows a three-dimensional (3D) solid constrained on a part of the boundary and subjected to body force distributed over the volume and surface forces on another part of the boundary. Figure 1.3 shows a two-dimensional (2D) solid that is very thin in the z-direction, and stress components in the z-direction are all zero, known as *plane stress* problem. Figure 1.4 shows a 2D solid that is very thick in the z-direction and the external forces and constraints are independent of z, resulting in zero strain components in the z-direction, known as *plane strain* problem.

Consider, in general, a d-dimensional solid mechanics problem with a physical domain of $\Omega \in \mathbb{R}^d$ bounded by Γ. The static *equilibrium equation* governing the solid can be written in the differential form:

$$\frac{\partial \sigma_{ij}}{\partial x_j} + b_i = 0, \quad i,j = 1,\ldots,d \text{ in } \Omega \tag{1.1}$$

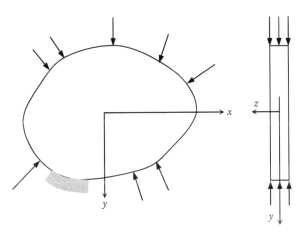

FIGURE 1.3
A 2D plane stress problem.

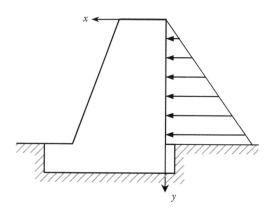

FIGURE 1.4
A 2D plane strain problem.

where

b_i are given external body force

σ_{ij} is the (internal) stress which relates to the strains ε_{ij} via the *constitutive equation* or the generalized Hooke's law:

$$\sigma_{ij} = C_{ijkl}\varepsilon_{kl} \tag{1.2}$$

where C_{ijkl} is the elasticity tensor of the material that is symmetrical:

$$C_{ijkl} = C_{jikl} = C_{ijlk} = C_{klij} \tag{1.3}$$

For isotropic Saint-Venant Kirchhoff elastic materials, we have

$$C_{ijkl} = \lambda\delta_{ji}\delta_{kl} + \mu(\delta_{ik}\delta_{jl} + \delta_{il}\delta_{jk}) \tag{1.4}$$

where λ and μ are the Lame's elastic constants. In the formulations of this book, we consider general anisotropic materials, for which we do not necessarily have Equation 1.4.

The strain tensor ε_{ij} relates to the displacements by the *compatibility equation* (also known as the *kinematic equations*).

$$\varepsilon_{ij} = \frac{1}{2}\left(\frac{\partial u_i}{\partial x_j} + \frac{\partial u_j}{\partial x_i}\right) \tag{1.5}$$

where $u_i, i = 1, \ldots, d$, is the displacement components in the x_i-directions at a point in Ω. Substituting Equations 1.2 and 1.5 into Equation 1.1, we shall have

$$\frac{\partial}{\partial x_j}\left(C_{ijkl}\frac{\partial u_k}{\partial x_l}\right) + b_i = 0, \quad i, j = 1, \ldots, d \text{ in } \Omega \tag{1.6}$$

where the displacement is the primary field variable. In matrix form often used in this book, the equilibrium equation (Equation 1.1) becomes

$$\mathbf{L}_d^{\mathrm{T}}\boldsymbol{\sigma} + \mathbf{b} = 0 \tag{1.7}$$

where the body force vector becomes

$$\mathbf{b} = \left\{ \begin{array}{c} b_1 \\ b_2 \end{array} \right\} \quad \text{for 2D,} \quad \mathbf{b} = \left\{ \begin{array}{c} b_1 \\ b_2 \\ b_3 \end{array} \right\} \quad \text{for 3D} \tag{1.8}$$

The matrix of differential operators $\mathbf{L}_d$ is given by

$$\mathbf{L}_d = \begin{bmatrix} \partial/\partial x_1 & 0 \\ 0 & \partial/\partial x_2 \\ \partial/\partial x_2 & \partial/\partial x_1 \end{bmatrix}_{3 \times 2} \quad \text{for 2D,} \quad \mathbf{L}_d = \begin{bmatrix} \partial/\partial x_1 & 0 & 0 \\ 0 & \partial/\partial x_2 & 0 \\ 0 & 0 & \partial/\partial x_3 \\ 0 & \partial/\partial x_3 & \partial/\partial x_2 \\ \partial/\partial x_3 & 0 & \partial/\partial x_1 \\ \partial/\partial x_2 & \partial/\partial x_1 & 0 \end{bmatrix}_{6 \times 3} \quad \text{for 3D} \tag{1.9}$$

The constitutive equation becomes

$$\boldsymbol{\sigma} = \mathbf{c}\boldsymbol{\varepsilon} \tag{1.10}$$

where $\boldsymbol{\sigma}$ is a vector that collects stress components in the form of

$$\boldsymbol{\sigma} = \left\{ \begin{array}{c} \sigma_{11} \\ \sigma_{22} \\ \sigma_{12} \end{array} \right\} \quad \text{for 2D,} \quad \boldsymbol{\sigma} = \left\{ \begin{array}{c} \sigma_{11} \\ \sigma_{22} \\ \sigma_{33} \\ \sigma_{23} \\ \sigma_{13} \\ \sigma_{12} \end{array} \right\} \quad \text{for 3D} \tag{1.11}$$

and $\boldsymbol{\varepsilon}$ is a vector that collects strain components:

$$\boldsymbol{\varepsilon} = \left\{ \begin{array}{c} \varepsilon_{11} \\ \varepsilon_{22} \\ 2\varepsilon_{12} \end{array} \right\} = \left\{ \begin{array}{c} \varepsilon_{11} \\ \varepsilon_{22} \\ \gamma_{12} \end{array} \right\} \quad \text{for 2D,} \quad \boldsymbol{\varepsilon} = \left\{ \begin{array}{c} \varepsilon_{11} \\ \varepsilon_{22} \\ \varepsilon_{33} \\ 2\varepsilon_{23} \\ 2\varepsilon_{13} \\ 2\varepsilon_{12} \end{array} \right\} = \left\{ \begin{array}{c} \varepsilon_{11} \\ \varepsilon_{22} \\ \varepsilon_{33} \\ \gamma_{23} \\ \gamma_{13} \\ \gamma_{12} \end{array} \right\} \quad \text{for 3D} \tag{1.12}$$

The matrix of material *stiffness* constants $\mathbf{c}$ can be written explicitly in more familiar engineering notations as

$$\mathbf{c} = \begin{bmatrix} c_{11} & c_{12} & c_{13} & c_{14} & c_{15} & c_{16} \\ & c_{22} & c_{23} & c_{24} & c_{25} & c_{26} \\ & & c_{33} & c_{34} & c_{35} & c_{36} \\ & & & c_{44} & c_{45} & c_{46} \\ & sy. & & & c_{55} & c_{56} \\ & & & & & c_{66} \end{bmatrix} \tag{1.13}$$

where "sy" stands for "symmetric."

Note that the $c_{ij} = c_{ji}$ for the symmetry. Thus, there are 21 independent material constants c_{ij}. For isotropic materials, $\mathbf{c}$ can be greatly reduced to

$$\mathbf{c} = \begin{bmatrix} c_{11} & c_{12} & c_{12} & 0 & 0 & 0 \\ & c_{11} & c_{12} & 0 & 0 & 0 \\ & & c_{11} & 0 & 0 & 0 \\ & & & \dfrac{c_{11} - c_{12}}{2} & 0 & 0 \\ & sy. & & & \dfrac{c_{11} - c_{12}}{2} & 0 \\ & & & & & \dfrac{c_{11} - c_{12}}{2} \end{bmatrix} \tag{1.14}$$

where

$$c_{11} = \frac{E(1 - v)}{(1 - 2v)(1 + v)}; \quad c_{12} = \frac{Ev}{(1 - 2v)(1 + v)}; \quad \frac{c_{11} - c_{12}}{2} = \mu = G \tag{1.15}$$

in which E, v, and G are Young's modulus, Poisson's ratio, and shear modulus of the material, respectively. There are only two independent constants among these three constants. The relationship between these three constants is

$$G = \frac{E}{2(1 + v)} \tag{1.16}$$

Given any two of these three constants, the other can then be calculated using the above equation. For 2D plane stress problems we further have

$$\mathbf{c} = \frac{E}{1 - v^2} \begin{bmatrix} 1 & v & 0 \\ v & 1 & 0 \\ 0 & 0 & (1 - v)/2 \end{bmatrix} \quad \text{(Plane stress)} \tag{1.17}$$

For 2D plane strain problems, the matrix of material stiffness constants $\mathbf{c}$ can be obtained by simply replacing E and v, respectively, with $E/(1 - v^2)$ and $v/(1 - v)$, which leads to

$$\mathbf{c} = \frac{E(1 - v)}{(1 + v)(1 - 2v)} \begin{bmatrix} 1 & \dfrac{v}{1 - v} & 0 \\ \dfrac{v}{1 - v} & 1 & 0 \\ 0 & 0 & \dfrac{1 - 2v}{2(1 - v)} \end{bmatrix} \quad \text{(Plane strain)} \tag{1.18}$$

Remark 1.2: Stable Materials

In this book, unless specified, we consider solids and structures made of materials that are physically *stable*: meaning that any amount of strains will result in stresses and hence some positive strain energy. In other words, these material constants are positive definite or the matrix of the material constants $\mathbf{c}$ is symmetric positive definite (SPD).

For stable materials, the stress–strain relation can also be written in the following reverse form:

$$\boldsymbol{\varepsilon} = \mathbf{s}\boldsymbol{\sigma} \tag{1.19}$$

where **s** is a matrix of material *flexibility* constants that can be obtained from the experiments. Because **c** is SPD, **s** must also be SPD, and hence both are invertible. We then have the simple relations:

$$\mathbf{s} = \mathbf{c}^{-1} \quad \text{or} \quad \mathbf{c} = \mathbf{s}^{-1} \tag{1.20}$$

Remark 1.3: Volumetric Locking

The denominator $(1 - 2\nu)$ in Equations 1.15 and 1.18 suggests a possible singularity problem when ν approaches 0.5, which can happen for the so-called incompressible solid materials like rubber. This can have numerical implications for displacement methods, known as *volumetric locking*. Special techniques have been developed in FEM to overcome this numerical problem [1,2]. Techniques in meshfree methods that can deal with volumetric locking problems will be presented in Chapter 8.

The compatibility equation (Equation 1.5) can also be written in the matrix form:

$$\boldsymbol{\varepsilon} = \mathbf{L}_d \mathbf{u} \tag{1.21}$$

where

$$\mathbf{u} = \begin{Bmatrix} u_1 \\ u_2 \end{Bmatrix} \quad \text{for 2D,} \quad \mathbf{u} = \begin{Bmatrix} u_1 \\ u_2 \\ u_3 \end{Bmatrix} \quad \text{for 3D} \tag{1.22}$$

is the displacement vector. Substituting Equation 1.21 into Equations 1.10 and 1.7 gives

$$\mathbf{L}_d^{\mathrm{T}} \mathbf{c} \mathbf{L}_d \mathbf{u} + \mathbf{b} = \mathbf{0} \tag{1.23}$$

The boundary conditions can be of two types: Dirichlet (essential, displacement) boundary condition and Neumann (natural, stress) boundary condition. Let Γ_u denote a part of Γ, on which Dirichlet boundary condition is specified, we then have

$$u_i = u_{\Gamma i}, \quad \text{on } \Gamma_u \in \Gamma \tag{1.24}$$

where $u_{\Gamma i}$ is the specified displacement component on Γ_u. In this book, for simplicity, we consider homogeneous essential boundary condition ($u_{\Gamma i} = 0$) by default. This type of problems is called *force-driving* problem. For nonhomogeneous essential boundary conditions, simple treatments in the standard FEM shall apply. Let Γ_t denotes a part of Γ, on which Neumann boundary condition is satisfied:

$$\sigma_{ij} n_j = t_{\Gamma i}, \quad \text{on } \Gamma_t \in \Gamma \tag{1.25}$$

where
   n_j is the jth component of the unit outward normal
   $t_{\Gamma i}$ is the specified boundary stress on Γ_t

The matrix form of Equation 1.25 is

$$\mathbf{L}_n^T \boldsymbol{\sigma} = \mathbf{t}_\Gamma, \quad \text{on } \Gamma_t \in \Gamma \qquad (1.26)$$

where $\mathbf{L}_n$ is the matrix of the components of the unit outward normal arranged in the form of

$$\mathbf{L}_n = \begin{bmatrix} n_1 & 0 \\ 0 & n_2 \\ n_2 & n_1 \end{bmatrix}_{3\times 2} \quad \text{for 2D,} \quad \mathbf{L}_n = \begin{bmatrix} n_1 & 0 & 0 \\ 0 & n_2 & 0 \\ 0 & 0 & n_3 \\ 0 & n_3 & n_2 \\ n_3 & 0 & n_1 \\ n_2 & n_1 & 0 \end{bmatrix}_{6\times 3} \quad \text{for 3D} \qquad (1.27)$$

Equation 1.1 or 1.23 are the *strong form* system equations governing the mechanics behavior of solids. The *primary* dependent field variables are the displacement functions. They are required to have at least the same order of consistency in the entire problem domain as the order of the differentiations in the PDEs. Such a requirement on consistency for the displacement functions is said *strong*.

1.3 Numerical Techniques: Practical Solution Tools

1.3.1 An Overview

In this section, we provide an overview on the numerical methods/techniques for practical solutions to the above-mentioned problems. The discussion is not meant to be rigorous, but to project a simple and clear picture on this very completed subject to be covered by this entire book.

There are largely two categories of numerical methods for solving these PDEs: direct approach and indirect approach. The direct approach known as strong form methods (such as the FDM and collocation method, e.g., [9]) discretizes and solves the PDEs directly, and the indirect approach known as weak form methods (such as FEM) establishes first an *alternative* weak form system equation that governs the same physical phenomena and then solves it. The weak form equations are usually in an *integral form*, implying that they need to be satisfied only in an integral (averaged) sense: a weak requirement.* In meshfree methods, both strong and weak forms are used. This book focuses on the weak form or weakform-like meshfree methods that are generally more robust, stable, accurate, reliable, efficient, and hence of more practical importance.

The essential idea of a weak form method is to assess a *global behavior* of the entire system and then find a best possible solution to the problem that can strike a *balance* for the system in terms of the global behavior. For a solid mechanics problem, for example, we look at the energy potentials in the entire solid, and try to find a displacement field (solution) that makes the *total* energy potential minimum. The total potential consists of strain energy potential and the work done by all the external forces/loads for a given displacement field.

---

* Terms of strong and weak forms will be discussed in more detail from a different viewpoint in Chapter 5.

At such a minimum (stationary) status, we know (from physics and the well-posedness) that the solid will be stable, and hence in an equilibrium state. How to find out such a displacement solution leading to the minimum potential energy? We just use our usual and most robust engineering practice by "trial and error." We essentially "try" to *create* or *assume* a *set* of all possible and "legal" or "admissible" displacement fields, from which the strain fields can be derived. We then evaluate the energy potentials for each of the displacement fields. Whichever gives the lowest total energy potential is our best possible solution in all these displacement fields we tried. It turns out that this trial and error procedure can be formulated in a systematical manner of a minimization procedure based on the so-called *minimum potential energy principle*, by which a set of discretized system equations can be established and solved routinely. By "admissible" or "legal" we mean that the displacement field has to make physical sense: It has to be *continuous* (in a continuum portion of the solid) and satisfies boundary conditions. Otherwise, it can "upset" the energy potentials: a discontinuous displacement field can result in an energy potential that cannot be properly evaluated! It is clear now that a weak formulation has a very simple physical process. Why is it called weak formulation? This is because the evaluation of energy requires only strains and stresses that can all be obtained by differentiating the displacement function only once, and hence there is no need for the second-order derivatives for the displacement functions as in the strong formulation: a weak requirement.

Now, what has the *functional analysis* in mathematical community got to do with this "engineering" minimum potential energy principle? There is, in fact, a very simple correspondence or equivalence (may not be exactly equal) listed in Table 1.1.

TABLE 1.1

Correspondence of Terminologies Used in the Engineering and Mathematical Communities

Item	Engineering Terminology	Mathematics Terminology
1.	Displacement field (displacement function)	Function
2.	Strain field	Derivatives of function
3.	Set of displacement fields/functions	Function space
4.	A "legal" or "admissible" displacement field	A function from a proper space
5.	A virtual (arbitrary) displacement field	Variation of a function
6.	Energy potential, energy caused by a virtual displacement field	Functional
7.	Work done by the internal stresses in a deformed solid resulted from an imposed virtual displacement field	Bilinear form
8.	Virtual external work, external work	Linear form
9.	Balanced virtual energies	Weak form equation
10.	Minimization	Variation of a functional
11.	"Upset"	Unbounded
12.	Evaluable	Bounded
13.	Minimum potential energy principle	Weak statement
14.	Amplitudes of the displacement field	Norm
15.	(Virtual) strain energy	Seminorm
16.	Displacement boundary	Dirichlet/essential boundary
17.	Force/traction boundary	Neumann/natural boundary

Now, the mathematical (weak) statement for a problem can be generally stated as "the solution to a problem should come from a proper space of functions and satisfying a properly formulated weak equation for all the functions in that space."

Note that the "legality" for functions relates to the weak form equation. If we relax the "laws" for the function, we should establish another proper weak form equation. For example, in the *weakened-weak* (W^2) form we allow discontinuous functions by introducing a new functional or energy potential (see Chapters 3 and 5 for details).

The above "rough" analysis and statements reveals that we need largely three key pieces of numerical techniques in a weak form method:

1. The construction of the shape functions (for creating the displacement field or functions)

2. The integration of over the problem domain (for the evaluation of the energy potentials or functionals)

3. The weak form (equation) used for creating discrete algebraic system equations that can be solved routinely

The procedure in FEM and the meshfree methods based on weak formulation can be outlined in Figure 1.5: They all use these key techniques. The differences between the

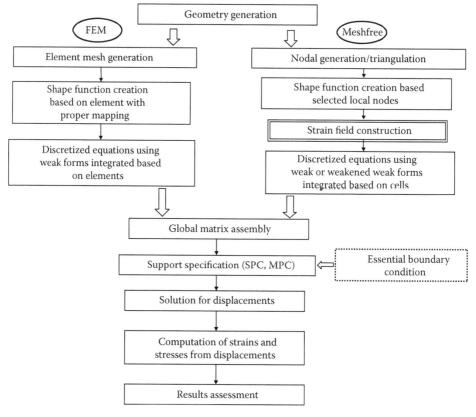

FIGURE 1.5
Flowchart for FEM and meshfree method procedures.

FEM and meshfree methods can be found in these three key numerical techniques, which are further elaborated in the following sections.

1.3.2 Shape Function Construction

In the FEM, the shape function construction is based on the element that acts as a basic building block. Since the shape functions are constructed using individual elements in the *natural* coordinate systems, the shape functions will be the same for the same type of elements. Proper coordinate mapping is needed for complicated domains of general orientations to ensure the compatibility (legality) of the shape functions between the elements. These shape functions are usually predetermined for different types of elements before the finite element analysis starts.

In meshfree methods, however, the shape functions constructed are usually only for the current point or cell of interest using a small number of local *support nodes* selected in the vicinity of the point or cell. The shape function generally changes with the location of the point/cell. The construction of the meshfree shape function is performed *during* the analysis, not *before* the analysis. The procedure to construct the shape functions is very flexible; based on nodes; no coordinate mapping is needed; and the types of meshfree shape functions are much more diversified compared to the FEM. We even allow displacement functions to be discontinuous. Hence, we often need new "legal" frameworks for such cases.

1.3.3 Integration of over the Problem Domain

The integration in the FEM is also, naturally, based on the element in the natural coordinate system. The integral over the domain becomes a simple summation of all the integrals over each of the elements that is performed using a numerical integration scheme, such as the well-known Gauss quadrature.

The integration in meshfree methods is, in general, based on background cells. The cells can be created by various means:

1. Using directly triangular background cells created for the problem domain, which is quite similar as in FEM, and is used in the MFree2D© [7,8] for the element-free Galerkin (EFG) [10] processor.
2. Using node-based smoothing domains created based upon the triangular background cells. This is used in the MFree2D for the node-based point interpolation method (NS-PIM) [11–14] processor.
3. Using edge-based smoothing domains created based upon the triangular background cells. This is used in the MFree2D for the edge-based smoothed point interpolation method (ES-PIM) [15,17] processor.
4. Using triangular subcells that are created by further dividing the background cells, which is used in the cell-based smoothed point interpolation methods (CS-PIMs) [66] and the constructed point interpolation method (SC-PIMs) [18–21,67].

The integral over the problem domain is then a simple summation of all the integrals over each of the integration cells. The means of integration in meshfree methods are also very diversified and can be sophisticated for desired performance and properties.

1.3.4 Use of Weak Forms

Proper principles or weak forms must be followed to establish a set of algebraic equations. These principles differ from types of problems. There are largely four principles:

1. The first principle is based on the variational or energy principles such as the principle of virtual work, Hamilton's principle, the minimum total potential energy principle, and so on. These principles can lead to weak forms of which the Galerkin weak form is the most popular and widely used. Traditional FEM for mechanics problems of solids and structures is founded on these principles, in particular the Galerkin weak form, and they are applicable also to some meshfree methods. Meshfree methods based on weakened-weak formulation use in general the strain-constructed Galerkin or SC-Galerkin weak forms.

2. The second principle is based on the weighted residual methods, and is, in fact, a more general form that can be used for deriving FEM equations both for solids and structures and for fluid flows, as long as the partial differential governing equations are available. In a special setting, it leads to a Galerkin weak form.

3. The third principle is based on the Taylor series, which has led to the formation of the traditional FDM based on essentially regular grids. This book will not discuss further meshfree method based on this formulation. Interested readers may refer to [5]. We will, however, discuss in detail a new type of method called gradient smoothing method (GSM) that works very well for irregular triangular mesh and particularly efficient for fluid flow problems [58–61].

4. The fourth principle is based on the control of conservation laws on each finite volume (element) in the domain. The FVM was established using this approach.

Engineering practice so far has shown that the first two principles are more *often* used for solids and structures, and the other two principles are more *often* used for fluid flow simulations. However, FEM has also been used to develop commercial packages for fluid flow problems, and FDM can be used for solids and structures. The mathematical foundation of all these approaches is the *residual method*. A proper choice of the *test* and *trial* functions in the residual method can lead to a FEM or FDM or FVM formulation.

Meshfree methods can be formulated using all these four principles, and the first two principles are used more often. Formulations based on the first two principles can lead to a *weak form* formulation, and that based on the third principle belongs to *strong form*. The strong form methods are usually less stable and require some special treatments [22,23]. The discretized equation systems derived based on the weak form are usually more stable and can give much more accurate results, because of the well-structured error control measure built in the weak formulations. Therefore, this book mainly covers meshfree methods of weak formulation. Meshfree methods use the standard weak forms such as the Galerkin weak form, and the more general W^2 form of the SC-Galerkin [15] (including the generalized smoothed Galerkin or GS-Galerkin, Galerkin-like weak forms). This is because operations (shape functions, integration, and strain field construction) in meshfree methods are much more diversified. The principles and weak forms for meshfree methods will be discussed in great detail in Chapter 5.

Finally, we note for the discretization in time, the Taylor series is often used in both meshfree and FEM.

1.3.5 On Approximation Techniques: The Characters

In any numerical method, we have to use some basic techniques or tools for the approximation of functions and their derivatives. The basic techniques are (1) interpolation, (2) differentiation, and (3) integration. Understanding the "characters" of these essential techniques is very important in the formulation of any numerical method, because many stability, convergence, and accuracy issues are rooted in the use of these basic techniques. Many years of the author's struggle in using and manipulating these techniques have led to some general understanding on the characters of these techniques, which are summarized in Table 1.2. The interpolation is the most basic tool and is hard to avoid in any numerical method, and hence we have to live with it and use it properly with caution. An improper use of differentiation has been the source of my problems in numerical methods, and hence should be avoided as much as possible. When the interpolation meets with the differentiation, one needs to exercise extra caution. A typical example is the FDM that can be used only with virtually very regular mesh/grids. The integration (with integration by parts) is very useful in creating stable and efficient numerical methods. It is known as a smoothing operator, and it works well with the (lower order) interpolations performed properly. Good examples are the weak and W^2 forms or the carefully designed GSM (Chapters 5, 8, and 9). One should, however, avoid too extensive use of smoothing operations, and ensure capturing sufficient local details. This can be easily done by controlling the size of the area of integration or ensuring the minimum number of smoothing domains. The $\mathbb{G}$ space theory [15] is established with all these careful considerations for unified formulations of a very wide class of stable, convergent, and efficient numerical methods (Chapter 3).

TABLE 1.2

The Characters of Basic Approximation Techniques

Techniques	Application in Numerical Methods	Characters	Remarks/Cautions
Interpolation (extrapolation)	Shape function construction (Chapter 2) Sampling of function values FEM, meshfree, FVM, FDM, etc.	Sensitive to node distribution Source of singularity issues Source of incompatibility	Avoid extrapolation Avoid high order Use RBFs and least squares Use with proper control (Chapter 2)
Differentiation	Evaluation of function derivatives FDM, FEM, meshfree, FVM	Sensitive to node distribution Sensitive to errors in the function Known as "harshening" operator [4] Amplification of local details	Avoid as much as possible Use weak and W^2 forms or GSM (Chapters 5, 8, and 9) Use it to boost sensitivities for inverse problems [4]
Integration (with integration by parts)	Evaluation of "area" behavior of functions, its derivatives and even the entire PDEs FEM, FVM, meshfree (SPH, GSM, PIMs)	Insensitive to mesh distortion Insensitive to error contamination Known as "smoothing" operator [4] Loss of local details	Widely used in weak and W^2 forms or GSM (Chapters 5, 8, and 9) Avoid too extensive use by controlling the size of the integration area

1.4 Defining Meshfree Methods

In the traditional FEM [1,3], the FDM [5], and the FVM [6], the problem spatial domain is *discretized* into meshes. A *mesh* is defined as any of the open spaces or interstices between the strands of a net that is formed by connecting nodes in a predefined manner. In FDM, the meshes used are also often called grids; in the FVM, the meshes are called volumes or cells; and in FEM, the meshes are called elements. The terminologies of grids, volumes, cells, and elements carry sometimes certain physical meanings as they are defined for different physical problems. All these grids, volumes, cells, and elements can be termed *meshes* according to the above definition of mesh. The mesh must be predefined to provide a certain relationship between the nodes, which becomes the building blocks of the formulation procedure of these conventional numerical methods.

By using a properly predefined mesh and by applying a suitable principle, complex, differential, or partial differential governing equations can be discretized to a set of algebraic equations with unknowns of filed variables at the nodes (or the center of the cells) of the mesh. The system of algebraic equations for the whole problem domain can be formed by assembling sets of algebraic equations for all the meshes.

The meshfree method is used to establish a system of algebraic equations for the whole problem domain *without* the use of a predefined mesh, or uses easily generable meshes in a much more flexible or "freer" manner. Meshfree methods essentially use a set of nodes scattered within the problem domain as well as on the boundaries to *represent* the problem domain and its boundaries. The field functions are then approximated locally using these nodes.

There are a number of meshfree methods that use local nodes for field variable approximation, such as the smooth particle hydrodynamics (SPH) [24–28], EFG method [10,29], the meshless local Petrov–Galerkin (MLPG) method [30], reproducing kernel particle method (RKPM) [31], the point interpolation method (PIM) [11–21,32–35], the finite point method [36], the FDM with arbitrary irregular grids [37–39], local point collocation methods [22,23], and so forth.

Because the methodology is still in a rapid development stage, new methods and techniques are constantly proposed. In the recent development, there are considerable works in applying meshfree techniques back to the FEM or FDM settings. These kinds of developments of "merging" or "fusing" different methods are very important in inventing more effective computational methods for more complicated engineering problems. However, these developments make the definition of meshfree methods even more difficult.

In contrast to FEM, the term *element free method* is preferred, and in contrast to FDM, the term *finite difference method using arbitrary or irregular grids* is preferred. Some of the meshfree methods are often termed *meshless methods*. The *ideal* requirement for a "meshless" method is

- No mesh is necessary at all throughout the process of solving the problem of given arbitrary geometry governed by partial differential system equations subject to all kinds of boundary conditions.

The reality is that the meshfree methods developed so far are not entirely "meshless" and fall in one of the following categories:

- Methods that require background cells for the integration of system matrices derived from the weak form over the problem domain. EFG methods may belong to this category. These methods are practical in many ways, as the creation of a background mesh is generally more feasible and can be much more easily automated using a triangular mesh for 2D domains and a tetrahedral mesh for 3D domains. We also use the background triangular mesh to assist in local nodes selection to ensure robustness, reliability, and efficiency. The bottom line is that the mesh used must be easily generable in automatic means, and the numerical operations should be beyond the confinement of the "element" in the mesh.

- Methods that require background cells locally for the integration of system matrices over the problem domain. MLPG methods belong to this category. These methods require only a local mesh and are easier to generate.

- Methods that do not require a mesh at all, but that are less stable and less accurate. Local point collocation methods and FDMs using irregular grids may belong to this category. Selection of nodes based on the type of a physical problem can be important for obtaining stable and accurate results. Automation of nodal selection and improving the stability of the solution are still some of the challenges in these kinds of methods. This type of method has a very significant advantage: It is very easy to implement, because no integration is required. There are, however, vital instability issues that require special treatments [22,23,36].

- Particle methods that require a predefinition of particles for their volumes or masses. The algorithm will then carry out the analyses even if the problem domain undergoes extremely large deformation and separation. SPH methods belong to this category. This type of method suffers from problems in the imposition of boundary conditions. In addition, predefining the particles still technically requires some kind of mesh. More details on SPH can be found in [26] and a recent review article [27]. SPH simulates well the overall behaviors of certain class of problems such as highly nonlinear and momentum-driven problems.

This book uses the term *meshfree* method for the collection of all the different meshfree methods in a loose sense: we permit the use of the mesh as long as the mesh can be automatically generable; numerical operations (integration, interpolation, smoothing, etc.) are beyond the "element" concept; and the solution should not too heavily depend on the quality of the mesh. This loose definition of meshfree method recognizes the reality: (1) Many meshfree methods (often more robust, reliable, and effective ones) do use some kind of mesh, but the mesh is used in much more flexible and "freer" ways; (2) most important motivation of developing meshfree methods was to reduce the reliance on the use of "quality" meshes that are difficult or expensive to create for practical problems of complicated geometries; and (3) if a type of mesh can be created automatically, and it does help in some ways in obtaining better result or in more effective or reliable ways, there is really no reason not to use it.

1.5 Need for Meshfree Methods

FEM is robust and has been thoroughly developed for static and dynamic, linear and nonlinear stress analysis of solids, structures, as well as fluid flows. Most practical engineering problems related to solids and structures are currently solved using a large number of well-developed FEM packages that are now commercially available. However, the following limitations of FEM are becoming increasingly evident:

1. Creation of a quality mesh for the problem domain is a prerequisite in using FEM packages. Usually the analyst spends the majority of his or her time in creating the mesh, and it becomes a major component of the cost of a simulation project because the cost of central processing unit (CPU) time is drastically decreasing. The concern is more the manpower time, and less the computer time. Therefore, ideally the meshing process would be fully performed by the computer without human intervention. However, when the "quality" of the mesh is demanded, the automation of the meshing is very difficult, because meshes that can be easily created are naturally not of good quality.

2. The compatible FEM model is usually "overly stiff," and hence results in a number of issues, such as locking and poor solution in gradient/derivatives. In stress calculations, the stresses obtained using FEM packages are discontinuous and often less accurate. The need for full compatibility in the assumed displacement field in the FEM results in the loss of freedom in the shape function construction.

3. When handling large deformation, considerable accuracy can be lost and the computation can even break down because of element distortions.

4. It is rather difficult to simulate both crack growth with arbitrary and complex paths and phase transformations due to discontinuities that do not coincide with the original nodal lines.

5. It is very difficult to simulate the breakage of material into a large number of fragments as FEM is essentially based on continuum mechanics, in which the elements formulated cannot be broken. The elements can either be totally "eroded" or stay as a whole piece. This usually leads to a misrepresentation of the breakage path. Serious error can occur because the nature of the problem is nonlinear, and therefore the results are highly path dependent.

6. Remesh approaches have been proposed for handling these types of problems in FEM. In the remesh approach, the problem domain is remeshed at steps during the simulation process to prevent the severe distortion of meshes and to allow the nodal lines to remain coincident with the discontinuity boundaries. For this purpose, complex, robust, and adaptive mesh generation processors have to be developed. However, these processors are only workable for 2D problems. There are no reliable processors available for creating quality hexahedral meshes for 3D problems due to technical difficulty.

7. Adaptive processors require "mappings" of field variables between meshes in successive stages in solving the problem. This mapping process often leads to additional computation as well as a degradation of accuracy. In addition, for large 3D problems, the computational cost of remeshing at each step becomes very high, even if an adaptive scheme is available.

8. FDM works very well for a large number of problems, especially for solving fluid dynamics problems. It suffers from a major disadvantage in that it relies on regularly distributed nodes. Therefore, studies have been conducted for a long time to develop methods using irregular grids. Efforts in this direction are still on [39].

9. Last, but surely not least, solutions bound. We know that the fully compatible FEM can produce a "lower bound" to the exact solution. Such a single-sided bound has very limited use in engineering designs. What the engineers really need is "certified" solutions bounded from both sides. However, it is much more difficult to provide an "upper bound" numerical solution for general complicated engineering problems. Meshfree methods can offer such solutions, and we can now bound the solution from both sides using a proper meshfree method together with FEM, as long as a triangular types of meshes can be built for the problem (see Chapter 8). Being able to provide a certified solution has a profound practical significance also in numerical modeling: We know when to stop in refining our models, and need not build an unnecessarily fine model, resulting in substantial savings in resources and manpower as well as the increase of the confidence in the numerical solution obtained.

1.6 The Ideas of Meshfree Methods

A close examination of these difficulties associated with FEM reveals the root of the problem: the heavy and rigid reliance on the use of quality elements that are the building blocks of FEM. A mesh with a predefined connectivity is required to form the elements that are used for both field variable interpolation and energy integration. As long as elements are used in such a rigid manner, the problems mentioned above will not be easy to solve. Therefore, the idea of eliminating or reducing the reliance on the elements and more flexible ways to make use of mesh has evolved naturally. The concept of element-free, meshless, or meshfree method has been proposed, in which the domain of the problem is represented, ideally, only by a set of arbitrarily distributed nodes.

The meshfree methods have shown great potential for solving the difficult problems mentioned above. Adaptive schemes can be easily developed [7,8], as triangular types of mesh that can be much more easily created automatically for complicated 2D and 3D domains, as shown in Figures 1.6 and 1.7. These types of triangular background cells are sufficient for necessary numerical operations in meshfree methods. This provides flexibility in adding or deleting points/nodes whenever and wherever needed. For stress analysis of a solid domain, for example, there are often areas of stress concentration, even singularity. One can relatively freely add nodes in the stress concentration area without worrying too much about their relationship with the other existing nodes. In crack growth problems,

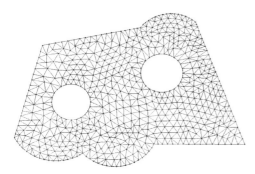

FIGURE 1.6
A triangular mesh of elements or background cells for a complicated 2D domain.

FIGURE 1.7
A tetrahedral mesh of elements or background cells for a complicated 3D domain.

nodes can be easily added around the crack tip to capture the stress concentration with desired accuracy. This nodal refinement can be moved with a propagation crack through background cells associated with the global geometry. Adaptive meshing for a large variety of problems, 2D or 3D, including linear and nonlinear, static and dynamic stress analysis, can be very effectively treated in meshfree methods in a relatively simple manner.

Because there is no need to create a quality mesh, and the nodes can be created by a computer in a much more automated manner, much of the time an engineer would spend on conventional mesh generation can be saved. This can translate to substantial cost and time savings in modeling and simulation projects.

There have been a number of meshfree methods developed thus far. The major features of these methods are listed in Table 1.3 that is definitely not exhaustive. A software package MFree2D with pre- and postprocessor has also been developed [7,8]. Chapter 16 presents MFree2D in detail.

TABLE 1.3

Some Meshfree Methods Using Local Irregular Nodes for Approximation

Method	References	Formulation Procedure	Local Function Approximation	Chapters
Diffuse element method	Nayroles et al. [40]	Galerkin weak form	MLS approximation	6
EFG method	Belytschko et al. [10]	Galerkin weak form	MLS approximation	6, 11, 12, 14 through 16
MLPG method	Atluri and Zhu [30]	Local Petrov–Galerkin	MLS approximation	7
PIMs	Liu et al. [11–21, 32,44–47,66,67]	W^2 forms (GS-Galerkin, SC-Galerkin); local Petrov–Galerkin	Point interpolation using polynomial and radial basis function (RBFs)	8, 10 through 16
SPH	Lucy [24], Gingold and Monaghan [25]	Weakform-like	Integral representation, particle approximation	9
GSM	Liu et al. [58,61]	Weakform-like	Point interpolation and gradient smoothing	9

TABLE 1.3 (continued)

Some Meshfree Methods Using Local Irregular Nodes for Approximation

Method	References	Formulation Procedure	Local Function Approximation	Chapters
Finite point method	Onate et al. [36]	Strong form	MLS approximation	1 (introduction only)
FDM using irregular grids	Liszka and Orkisz [37], Jensen [38], Liu et al. [39]	Strong form	Differential representation (Taylor series) and RBF	1 (introduction only)
Stabilized local collocation method	Liu and Kee [22], Kee et al. [23]	Strong form with regularization	Point interpolation and RBF	1 (introduction only)
RKPM	Liu et al. [31]	Strong or weak form	Integral representation (RKPM)	1 (introduction only)
hp-Clouds	Oden and Abani [41], Armando and Oden [42]	Strong or weak form	Partition of unity, MLS	1 (introduction only)
Partition of unity FEM	Babuska and Melenk [43]	Weak form	Partition of unity, MLS	1 (introduction only)
Meshfree weak–strong form (MWS)	Liu et al. [62,63]	Combined weak and strong forms	Point interpolation RBF, MLS	Not discussed
Boundary node methods (BNM)	Mukherjee and Mukherjee [48,49]	Boundary integral formulation	MLS approximation	13 (brief only)
BPIM	Liu and Gu [53–57]	Boundary integral equation	Point interpolation RBF	13
Combined domain and boundary methods	Liu and Gu [50–53]	Weak form and boundary integral equation	Point interpolation RBF	14

Note: This table is not exhaustive.

1.7 Basic Techniques for Meshfree Methods

We now brief the general procedure and basic steps for meshfree methods. We use solid mechanics problems as an example to describe these basic steps. Common techniques that can be used in different meshfree methods, such as triangulation, determination of local support domains, influence domains, node selection, estimation of nodal spacing, and techniques for local support node selection for field function approximation are presented.

1.7.1 Basic Steps

Step 1: Domain representation/discretization

The geometry of the solid or structure is first reacted in a CAE code or preprocessor, and is triangulated to produce a set of triangular type cells with a set of nodes scattered in the problem domain and its boundary, as described in Section 1.7.2. Boundary conditions and loading conditions are then specified for the model. The density of the nodes depends on the presentation accuracy of the geometry, the accuracy requirement of the solution, and the limits of the computer resources available. The nodal distribution is usually not uniform and a denser distribution of nodes is often used in the area where the displacement gradient is larger. Because adaptive algorithms can be used in meshfree methods, the density is eventually controlled automatically and adaptively in the code of the meshfree method [7,8]. Therefore, we do not have to worry too much about the distribution quality of the initial nodes used in usual situations. In addition, as a meshfree method, it should not demand too much for the pattern of nodal distribution. It should be workable *within reason* for arbitrarily distributed nodes. Because the nodes will carry the values of the field variables in a meshfree formulation, they are often called *field nodes*.

Step 2: Displacement interpolation

The field variable (say, a component of the displacement vector) u at any point at $\mathbf{x} = (x, y, z)$ within the problem domain is approximated or interpolated using the displacements at its nodes within the support domain of the point at $\mathbf{x}$ that is usually a quadrature point, i.e.,

$$u^h(\mathbf{x}) = \sum_{i \in S_n} \phi_i(\mathbf{x}) u_i = \mathbf{\Phi}(\mathbf{x}) \mathbf{d}_s \qquad (1.28)$$

where

S_n is the set of *local* nodes included in a "small local domain" of the point $\mathbf{x}$, such a local domain is called *support domain*, and the set of local nodes are called *support nodes*
u_i is the nodal field variable at the ith node in the support domain
$\mathbf{d}_s$ is the vector that collects all the nodal field variables at these support nodes
$\phi_i(\mathbf{x})$ is the shape function of the ith node created using all the support nodes in the support domain and is often called *nodal shape function*

A support domain of a point $\mathbf{x}$ determines the number of nodes to be used to approximate the function value at $\mathbf{x}$. A support domain can be *weighted* using functions that vanish on the boundary of the support domain, as shown in Figure 1.8. It can have different shapes and its dimension and shape can be different for different points of interest $\mathbf{x}$, as shown in Figure 1.9. The shapes most often used are circular or rectangular, or any shape to include desired supporting nodes.

The concept of support domain works well if the nodal density does not vary too drastically in the problem domain. However, in solving practical problems, such as

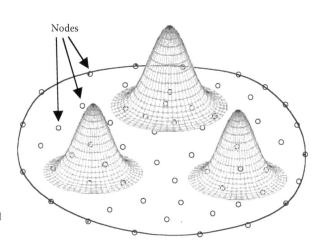

FIGURE 1.8
Domain representation of a 2D domain and nodes in a local weighted support domain.

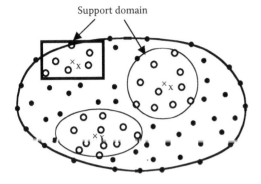

FIGURE 1.9
Support domain determines nodes (marked by o) that are used for approximation or interpolation of field variable at point x. A support domain can have different shapes and can be different from point to point. Most often used shapes are circular or rectangular.

problems with stress singularity, the nodal density can vary drastically. The use of a support domain based on the current point of interest can lead to spatially biased selection of nodes for the construction of shape functions. In extreme situations, all the nodes used could be located on one side only, and the shape functions so constructed can result in serious error, due to extrapolation. To prevent this kind of problem, the concept of influence domain of a node should be used. MFree2D, introduced in Chapter 16, uses the approach of influence domain to select nodes for constructing shape functions. The concept of influence domain is explained in Section 1.7.5. A practical, robust, better controlled, and more efficient way to select node with guaranteed success is the use of T-schemes based on triangular cells to select the field nodes, as detailed in Section 1.7.6. The T-schemes are implemented in the MFree2D, and used in the NS-PIMs [11–14], ES-PIMs [15–17], SC-PIMs [21,67], and CS CS-PIMs [66].

Note also that the interpolation, defined in Equation 1.28, is generally performed for all the components of all the field variables in the same support domain. Taking a 3D solid mechanics problem as an example, the displacement is usually chosen as the field variable, and the displacement should have three components: displacements in the x-, y-, and z-directions. The same shape function is used for all three displacement components in the support domain of the same point. There are, however, situations where different shape functions are used for different field variables. For example, for bending problems of beams, plates, and shells, it is of advantage to use different shape functions, respectively, for deflection and rotation, in overcoming the so-called shear and membrane locking issues (see Chapters 10 through 12).

Step 3: Formation of system equations

The discrete equations of a meshfree method can be formulated using the shape functions and weak or weakened-weak forms. These equations are often written in nodal matrix form and are assembled into the global system matrices for the entire problem domain.

The global system equations are a set of *algebraic equations* for static analysis, *eigenvalue equations* for free-vibration analysis, and *differential equations* with respect to time for general dynamic problems. The procedures for forming system equations are different for different meshfree methods. Hence, we discuss them in later chapters.

Step 4: Solving the global meshfree equations

Solving the set of global meshfree equations, we obtain solutions for different types of problems.

1. *For static problems,* the displacements at all the nodes in the entire problem domain are first obtained. The strain and stress can then be retrieved using strain–displacement relations and constitutive equations. A standard linear algebraic equation solver, such as a Gauss elimination method, LU decomposition method, and iterative methods, can be used.

2. *For free-vibration and buckling problems,* eigenvalues and corresponding eigenvectors can be obtained using the standard eigenvalue equation solvers. Some of the commonly used methods are the following:

   - Jacobi's method
   - Given's method and Householder's method
   - The bisection method (using Sturm sequences)
   - Inverse iteration
   - QR method
   - Subspace iteration
   - Lanczos method

3. *For dynamics problems,* the time history of displacement, velocity, and acceleration are to be obtained. The following standard methods of solving dynamics equation systems can be used:

   - The modal superposition method may be a good choice for vibration types of problems and problems of far field response to low speed impact with many load cases.
   - For problems with a single load or few loads, the *direct integration method* can be used, which uses the FDM for time stepping with implicit and explicit approaches.
   - The implicit method is more efficient for relatively slow phenomena of vibration types of problems.
   - The explicit method is more efficient for very fast phenomena, such as impact and explosion.

For computational fluid dynamics problems, the discretized system equations are basically nonlinear, and one needs an additional iteration loop to obtain the results.

1.7.2 Triangulation

Consider a d-dimensional problem domain of $\Omega \in \mathbb{R}^d$ bounded by Γ that is *Lipschitzian.* By default in this book, we speak of "open" domain that does not include the boundary of the domain. When we refer to a "closed" domain we will specifically use a box: $\boxed{\Omega} = \Omega \cup \Gamma$.

Triangulation is the most flexible way to create background triangular cells for meshfree operations. The process can be almost fully automated for 2D and even 3D domains with complicated geometry. Therefore, it is used in most commercial preprocessors using processes such as the widely used Delaunay triangulation. For one-dimensional (1D) problems, a cell is defined in $\mathbb{R}^1$ and is simply a line segment, for 2D problems it is defined in $\mathbb{R}^2$ and it becomes a triangle, and for 3D problems it is a tetrahedron defined in $\mathbb{R}^3$. The triangulation is nonoverlapping and seamless: $\boxed{\Omega} = \cup_{i=1}^{N_c} \boxed{\Omega_i^c}$ and $\Omega_i^c \cap \Omega_j^c = 0, \forall i \neq j$ where N_c is the total number of cells. In such a triangulation process, N_{cg} straight line segments L_i^c called "edges of the cells" will be produced. These cell edges are the interfaces of the triangular cells.

We assume that a given problem domain can always, in one way or another, be divided into N_c cells with N_n nodes (vertices of the triangles) and N_{cg} edges. In theory, any inner angle θ of the triangles should be strictly larger than 0 and strictly less than $180°$, and in practice we often require $15 < \theta < 120$. We also do not allow any "free" unconnected nodes and any duplicated nodes. If there exists such a node, it should be removed. Under such conditions, we shall have $A_i^c > 0, i = 1, 2, \ldots, N_c$ and $h_i > 0, i = 1, 2, \ldots, N_c$.

A *characteristic dimension* or *length* is then defined. For uniform discretization of isolateral triangles, the edge length of any cell h is the characteristic dimension of cells. In the case of nonuniform discretization of arbitrary triangles, we let

$$h_{\max} = \max_{i=1,\ldots,N_{cg}} (h_i), \quad h_{\min} = \min_{i=1,\ldots,N_{cg}} (h_i) \tag{1.29}$$

and we should assume that the ratio of the smallest and largest cell dimensions is bounded:

$$h_{\max}/h_{\min} = c_{rh} < \infty \tag{1.30}$$

In this case, the largest or the smallest edge length can be used as the *characteristic dimension* of the cells

$$h = h_{\max} \quad \text{or} \quad h = h_{\min} \tag{1.31}$$

The characteristic dimension of cells can also be defined as some kind of averaged cell length, such as

$$h = \sqrt{4A^c/\sqrt{3}} \tag{1.32}$$

where A^c is the "average" area, the triangular cells defined as $A^c = A/N_c$ in which A is the area of the entire problem domain. In Equation 1.32, we assume the cells are all isolateral triangles. Alternatively, we can simply use

$$h = \sqrt{2A^c} \tag{1.33}$$

where we assume the cells are all right-isoceles triangles. Note that Equations 1.31 through 1.33 are "equivalent" in the sense of that "controlling h defined in any of these three ways can put the entire mesh under control." When we say h approaches 0, the dimensions of all the cells in the entire problem domain approach 0. Therefore, in the convergence study (solution approaches exact solution when h approaches 0), any of these equations can be used, and they all should deliver the same *convergence rate* for the same set of meshes used to examine the same problem.

The triangulation can be easily performed using standard algorithms such as the Delaunay algorithm or the advanced front algorithm with proper "cosmetic" treatments to improve mesh quality. Most commercial preprocesses offer such triangulation functions. An example of triangulation of 2D domains to triangular elements/cells is given in Figure 1.6. For 3D domains, the triangulation leads to tetrahedral elements/cells, as shown in Figure 1.7.

Note that this book does not exclude any other types of mesh, as long as it does not give problem in creating such a mesh for the problem to be solved. If other type of mesh is used, we assume that similar controls on mesh as the triangulation defined above must be in place.

1.7.3 Determination of the Dimension of a Support Domain

The accuracy of interpolation depends on the nodes in the support domain of the point of interest (which is often a quadrature point x_Q or the center of integration cells). Therefore, a suitable support domain should be chosen to ensure a proper area of coverage for interpolation. To define the support domain for a point x_Q, the dimension of the support domain d_s is determined by

$$d_s = \alpha_s d_c \tag{1.34}$$

where
   α_s is the dimensionless size of the support domain
   d_c is a characteristic length that relates to the nodal spacing near the point at x_Q

If the nodes are uniformly distributed, d_c is simply the spacing between two neighboring nodes. In the case where the nodes are nonuniformly distributed, d_c can be defined as an "average" nodal spacing in the support domain of x_Q.
   The physical meaning of the dimensionless size of the support domain α_s is the multiple factor of the average nodal spacing. For example, $\alpha_s = 2.1$ means a support domain whose radius is 2.1 times the average nodal spacing. The actual number of nodes, n, can be determined by counting all the nodes in the support domain. The dimensionless size of the support domain α_s should be predetermined by the analyst, usually by carrying out numerical experiments for the same class of problems for which solutions already exist. Generally, an $\alpha_s = 2.0$–3.0 leads to good results. It is, however, not possible to provide a value of α_s for all the problems and node distributions, and the procedure can fail for extremely irregularly distributed nodes. For reliable node selection, the T-schemes detailed in Section 1.7.6 are recommended.

1.7.4 Determination of Local Nodal Spacing

For 1D cases, a simple method of defining an "averaged" nodal spacing is

$$d_c = \frac{D_s}{(n_{D_s} - 1)} \tag{1.35}$$

where
   D_s is an estimated d_s (the estimate does not have to be very accurate but should be known and a reasonably good estimate of d_s)
   n_{D_s} is the number of nodes that are covered by a known domain with the dimension of D_s

By using Equation 1.35, it is very easy to determine the dimension of the support domain d_s for a point at x_Q in a domain with nonuniformly distributed nodes. The procedure is as follows:

1. Estimate d_s for the point at x_Q, which gives D_s.
2. Count nodes that are covered by D_s.

3. Use Equation 1.35 to calculate d_c.

4. Finally, calculate d_s using Equation 1.34 for a given (desired) dimensionless size of support domain α_s.

For 2D cases, a simple method of defining an "average" nodal spacing is

$$d_c = \frac{\sqrt{A_s}}{\sqrt{n_{A_s}} - 1} \tag{1.36}$$

where

A_s is an estimated area that is covered by the support domain of dimension d_s (the estimate does not have to be very accurate but should be known and a reasonably good estimate)

n_{A_s} is the number of nodes that are covered by the estimated domain with the area of A_s

By using Equation 1.36 and the same procedure described for the 1D case, it is very easy to determine the dimension of the support domain d_s for a point at $\mathbf{x}_Q$ in a 2D domain with nonuniformly distributed nodes.

Similarly, for 3D cases, a simple method of defining an "average" nodal spacing is

$$d_c = \frac{\sqrt[3]{V_s}}{\sqrt[3]{n_{V_s}} - 1} \tag{1.37}$$

where

V_s is an estimated volume that is covered by the support domain of dimension d_s

n_{A_s} is the number of nodes that is covered by the estimated domain with the volume of V_s

By using Equation 1.37, and the same procedure described for the 1D case, we can determine the dimension of the support domain d_s for point $\mathbf{x}_Q$ in a 3D domain with nonuniformly distributed nodes.

1.7.5 Concept of the Influence Domain

Note that this book distinguishes between *support domain* and *influence domain*. The support domain is defined as a domain in the vicinity of a *point* of interest $\mathbf{x}_Q$ that can be, but does not have to be, at a node. It is used to include the nodes for shape function construction for $\mathbf{x}_Q$. The extended concept of the support domain means a particular way to select those nodes, not necessarily just by distance. T-schemes detailed in Section 1.7.6 is typical of such an extension.

The influence domain in this book is defined as a domain that a node exerts an influence upon. It goes with a *node*, and an alternative way to select nodes. It works well for very irregularly distributed nodes. Influence domains are defined for each node in the problem domain, and they can be different from node to node to represent the area of influence of the node, as shown in Figure 1.10. Node 1 has an influence radius of r_1, and node 2 has an influence radius of r_2, etc. The node will be involved in the shape function construction for any point that is within its influence domain. For example, in constructing the shape

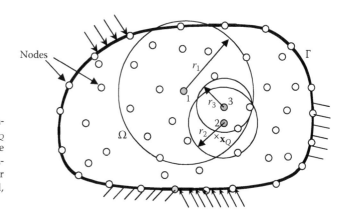

FIGURE 1.10
Influence domains of nodes. In constructing shape functions for point at $\mathbf{x}_Q$ (marked with x), nodes whose influence domains covers x are to be used for construction of shape functions. For example, nodes 1 and 2 are included, but node 3 is not included.

functions for the point marked with **x** at point $\mathbf{x}_Q$ (see Figure 1.10), nodes 1 and 2 will be used, but node 3 will not be used. The fact is that the dimension of the influence domain, which can be different from node to node, allows some nodes to have further influence than others and prevents unbalanced nodal distribution for constructing shape functions. As shown in Figure 1.10, node 1 is included for constructing shape functions for the point at point $\mathbf{x}_Q$, but node 3 is not included, even though node 3 is closer to $\mathbf{x}_Q$ compared with node 1. The dimension of the influence domain can even evolve during the analysis process, as in the SPH methods.

When a triangular background mesh is used, the influence radius of a node can be estimated using Equation 15.1 that is used in the MFree2D for EFG processors.

1.7.6 T-Schemes for Node Selection

Since the background cells are needed for integration for weak or weakened-weak form meshfree methods, background cells are often already made available. Therefore, it is natural to make use of them also for the selection of *supporting nodes* for shape function construction. Background cells of triangular type generated by triangulation detailed in Section 1.7.2 have been found most practical, robust, reliable, and efficient for local supporting node selection. It works particularly well for the family of PIMs. Triangular cell-based node selection schemes are termed as T-schemes, and are listed in Table 1.4.

Figure 1.11 shows the background triangular cells for 2D domain. The node selection is then performed as follows.

T3-Scheme

In the T3-scheme, we simply select three nodes of the home cell of the point of interest. As illustrated in Figure 1.11a, no matter the point of interest **x** located in an interior home cell (cell *i*) or a boundary home cell (cell *j*), only the three nodes of the home cell (i_1-i_3 or j_1-j_3) are selected. T3-scheme is used only for creating linear PIM shape functions by using polynomial basis functions. Note that the linear PIM shape functions so constructed are exactly the same as those in FEM using linear triangular elements. The shape functions can always be constructed (the moment matrix will never be singular), as long as the background cells of triangular type are generated by triangulation detailed in Section 1.7.2.

TABLE 1.4

T-Schemes for Node Selection Based on Triangular Background Cells

Name	Node Selection for Interpolation at Any Point in a Home Cell	Application/Types of Shape Functions
T3-scheme	Three nodes of the home cell	2D domain 3D domain surface PIM
T6/3-scheme	For an interior home cell, three nodes of the home cell and three remote nodes of the three neighboring cells For a boundary home cell, three nodes of the home cell	2D domain 3D domain surface PIM, RPIM
T6-scheme	For an interior home cell, three nodes of the home cell and three remote nodes of the three neighboring cells For a boundary home cell, three nodes of the home cell, two (or one) remote nodes of the neighboring cells plus one (or two) field node which is nearest to the centroid of the home cell	2D domain 3D domain surface PIM, RPIM
T4-scheme	Four nodes of the home tetrahedral cell	3D domain, PIM
T2L-scheme	Nodes of the home cell plus one layer of nodes of the cells connected to the home cell nodes (two layers of nodes are selected)	2D domain 3D domain surface 3D domain RPIM, MLS

Notes: In the definition of types of T-schemes, a *home* cell refers to the cell which hosts the point of interest (usually the quadrature sampling point). An *interior home* cell is a home cell that has no edge on the boundary of the problem domain and a *boundary home* cell is a home cell which has at least one edge on the boundary. A neighboring cell of a home cell refers to the cell which shares one edge with the home cell.

In case of CS-PIMs [66] and SC-PIMs [21,67], the quadrature point x_Q can be on an edge of a triangular cell. In such cases, the T-Schemes are associated with edges, and can have T_2-, T_4-T_2L-Schemes. The details can be found in [66].

T4-Scheme

T4-scheme is the analogy of the T3-scheme, but of node selection for 3D domains with tetrahedral background cells.

T6/3-Scheme

The T6/3-scheme selects six nodes to interpolate a point of interest located in an interior cell and three nodes for those located in boundary home cells. As illustrated in Figure 1.11b, when the point of interest (x) is located in an interior home cell (cell i), we select six nodes: three nodes of the home cell (i_1-i_3) and another three nodes located at the remote vertices of the three neighboring cells (i_4-i_6). When the point of interest at x_Q is located in a boundary home cell (cell j), we select only three nodes of the home cell, i.e., j_1-j_3.

T6/3-scheme was purposely devised for creating high-order PIM shape functions, where quadratic interpolations are performed for the interior home cells and linear interpolations for boundary home cells. This scheme was first used in the NS-PIM [11]. It can not only successfully overcome the singular problem but also improve the efficiency of the method.

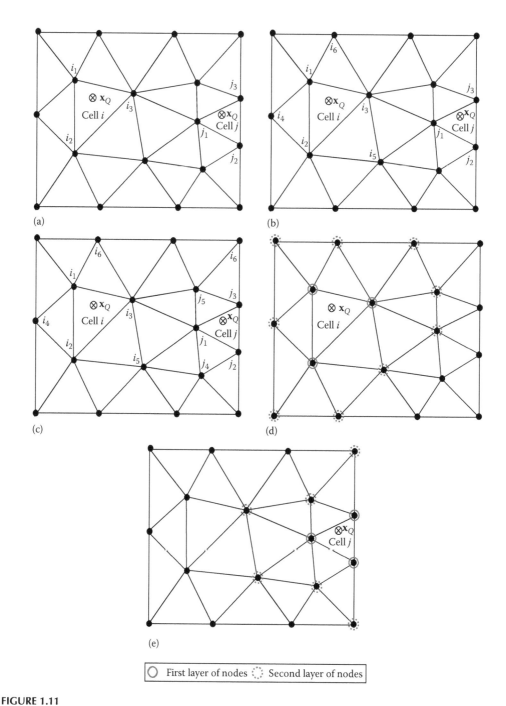

FIGURE 1.11
Background cells of triangles and the selection of support nodes based on these cells. (a) T3-scheme; (b) T6/3-scheme; (c) T6-scheme; (d) T2L-scheme for interior home cell; and (e) T2L-scheme for boundary home cell. Cells i and j are home cells that house the point of interest.

In addition the use of three nodes for boundary home cells insures the pass of the standard patch tests. Using the T6/3-scheme the shape functions can always be constructed, as long as such six nodes can be found for all the interior home cells.

T6-Scheme

Similar to T6/3-scheme, T6-scheme, as shown in Figure 1.11c, also selects six nodes for an interior home cell: three nodes of the home cell and three vertexes at the remote vertices of the three neighboring cells (i_1-i_6 for cell i). However, for a boundary cell (cell j), T6-scheme still selects six nodes: three nodes of the home cell (j_1-j_3), two remote nodes of the neighboring cells (j_4 and j_5), and one field node (j_6) which is nearest to the centroid of the home cell excepting the five nodes that have been selected.

T6-scheme is purposely devised for constructing radial PIM (RPIM) shape functions on considering both accuracy and efficiency. Different from T6/3-scheme, this scheme selects six nodes for all home cells containing the point of interest. The shape functions can always be constructed because the radial moment matrix is always invertible for arbitrary scattered nodes as long as to avoid using some specific shape parameters (see Chapter 2).

T6-scheme can be used for creating RPIM shape functions with linear polynomial basis, and the creation of these shape functions can be very efficient due to the use of very small number of nodes.

T2L-Scheme

T2L-scheme selects two layers of nodes to perform interpolation based on triangular meshes. As shown in Figure 1.11d and e, the first layer of nodes refers the three nodes of the home cell, and the second layer contains those nodes which are directly connected to the three nodes of the first layer.

This scheme usually selects much more nodes than the T6-scheme and leads to more time consumption. The RPIM shape functions can always be constructed because the radial moment matrix is always invertible for arbitrarily scattered nodes. We can use this scheme to create RPIM shape functions with high order of consistence and for extremely irregularly distributed nodes. Such RPIM shape functions can also be used for strong form meshfree method methods where higher order of consistence is required [15,23]. T2L-scheme can also be used for creating moving least squares (MLS) shape functions.

1.8 Outline of the Book

This book provides an introduction to meshfree methods, and their applications to various mechanics problems. This book covers the following types of problems:

- Mechanics for solids (1D, 2D, and 3D)
- Mechanics for structures (beams, plates, and shells)
- Fluid mechanics (fluid flow and hydrodynamics)

The bulk of the material in the book is the result of the intensive research work by G. R. Liu and his research team in the past 12 years. Works of other researchers are also introduced. The significance of this book is as follows:

1. This was the first book published that comprehensively covers meshfree methods.
2. The book covers, in a systematic manner, basic and advanced theories, principles, techniques, and procedures in solving mechanics problems using meshfree methods. It will be very useful for researchers entering this new area of research on meshfree methods, and for professionals and engineers developing computer codes for the next generation of computational methods.
3. Readers will benefit from the research outcome of G. R. Liu's research team and their long-term research projects on meshfree methods founded by the Singapore government and other organizations. Many materials in this book are the results of ongoing projects, and have not been previously published.
4. A large number of examples with illustrations are provided for validating, benchmarking, and demonstrating meshfree methods. These examples can be useful reference materials for other researchers.

The book is written for senior university students, graduate students, researchers, and professionals in engineering and science. Mechanical engineers and practitioners and structural engineers and practitioners will also find the book useful. Knowledge of FEM is not required but would help a great deal in understanding many concepts and procedures of meshfree methods. Basic knowledge of mechanics is also helpful in reading this book smoothly.

The meshfree method is a relatively new area of research. There exist many problems that offer ample opportunities for research to develop the next generation of numerical methods. The method is also in a rapidly developing and growing stage. Different techniques are developed every day. This book addresses some of the current important issues, both positive and negative, related to meshfree methods, which should prove beneficial to researchers, engineers, and students who are interested in venturing into this area of research.

The chapter-by-chapter description of this book is as follows:

Chapter 1: Addresses the background, overall procedures, and common preliminary techniques for different meshfree methods.

Chapter 2: Provides a detailed description of various methods for constructing meshfree shape functions, which is one of the most important issues of meshfree methods. Detailed discussions on the properties of various meshfree shape function and the issues related to use of these shape functions are provided.

Chapter 3: Introduce functions spaces for meshfree methods. The standard spaces that are widely used in FEM and meshfree are first briefed. We then provide a detailed discussion on new $\mathbb{G}$ spaces that are the foundation of the weakened-weak formulation, are particularly useful for both meshfree and FEM settings.

Chapter 4: Introduces some of the techniques used for strain field constructions, which is particularly important for meshfree methods, including strain gradient scaling, strain smoothing, generalized smoothing, point interpolation, and least square projection techniques.

Chapter 5: Introduces the principles and weak forms that will be used for creating discretized system equations, including the standard weak forms such as Galerkin and the weakened-weak forms such as the SC-Galerkin, GS-Galerkin, and Galerkin-like weak forms.

Chapter 6: Introduces the EFG method, one of the widely used meshfree methods. A number of techniques that are used for handling essential boundary conditions are discussed in detail. Issues related to the background integration are also examined.

Chapter 7: Introduces the MLPG method, which requires only local cells of a background mesh for integration. The MLPG method is formulated for both static and dynamic problems. Issues on types of local domains, effects of the dimension of these domains, handling local integration, and procedures in dealing with essential boundary conditions are discussed.

Chapter 8: Introduces the PIMs, which are formulated based on both the GS-Galerkin and SC-Galerkin formulations. Properties of various PIMs, such as the upper bound, free of volumetric locking, superconvergence, and ultra-accuracy are discussed in detail. A comparison study is also presented for PIMs and other methods in terms of efficiency.

Chapter 9: Introduces weakform-like meshfree methods for computational fluid dynamics problems: SPH based on Lagrangian formulation, and GSM based on Eulerian formulation. The SPH works particularly well for highly nonlinear and dynamic momentum-driven problems using particles. The GSM, on the other hand, works well for general fluid dynamics problems using triangular cells.

Chapter 10: Introduces PIM methods developed for analysis of beams governed by the Euler–Bernoulli beam theory. The power of weakened-weak formulation is demonstrated by using linear interpolation to solve fourth differential equations.

Chapter 11: Introduces meshfree methods developed for analysis of plates. Both thin plates governed by the Mindlin plate theory and thick plates governed by the third-order shear deformation theory are used to develop the meshfree methods. Two methods—EFG and PIMs—are formulated. Shear-locking issues for plates and the remedies are also discussed in great detail.

Chapter 12: Introduces meshfree methods developed for analysis of shells. Both thin shells governed by the Kirchhoff–Love theory and thick shells are used to formulate the meshfree methods. Two methods—EFG and PIM—are formulated for static and dynamic problems. Advantages of using meshfree methods for shells are discussed.

Chapter 13: Formulates two boundary-type meshfree methods—boundary point interpolation methods (BPIM) and radial BPIM. Procedures of using the meshfree concept for solving boundary integral equations are provided. A number of examples are presented to demonstrate the advantages of these methods in solving problems of infinite domain.

Chapter 14: Introduces methods that are formulated by coupling domain and boundary types of meshfree methods. Meshfree methods that couple with the traditional FEM and boundary element method (BEM) are also formulated and benchmarked.

Chapter 15: Meshfree methods for adaptive analysis are presented. A number of implementation issues have been discussed for adaptive analysis, such as error estimation and adaptive procedures. The methods are demonstrated using examples of 2D for EFG, 2D and 3D for PIMs.

Chapter 16: Presents MFree2D 2.0, which has been developed by G.R. Liu and coworkers with pre- and postprocessors for adaptive analysis of 2D solids. A number of meshfree solvers are now available in version 2.0. Its functions and usage are introduced.

1.9 Some Notations and Default Conventions

1. We define some of the notations that are often used in this book (Table 1.5).

2. By default in this book, we speak of "open" domain. When a domain is denoted as Ω (bounded by Γ), Ω does not include the boundary Γ. When we refer to a "closed" domain we will specifically use a box: $\boxed{\Omega} = \Omega \cup \Gamma$. We also *in general* require the domain being *Lipschitzian*: it cannot be *singular*. For solids and structures with cracks and sharp corners, special treatments or considerations are needed.

3. The integration used in this book is the sense of *Lebesgue integration*: the value of the integration will not change if we make changes to the value of the integrand only on a set of "zero measures" such as a finite set of points. It is very forgiving of occasional omissions.

4. This book uses both the *matrix* and the *indicial* notations from time to time for more concise presentation. Therefore, we allow the vectors and matrix to have the following forms for easy conversion between these two notations whenever it is needed.

$$
\begin{array}{cc}
\text{Matrix notation} & \text{Indicial notation}
\end{array}
$$

$$
\mathbf{x} = \left\{ \begin{array}{c} x_1 \\ x_2 \\ x_3 \end{array} \right\} = \left\{ \begin{array}{c} x \\ y \\ z \end{array} \right\}, \mathbf{u} = \left\{ \begin{array}{c} u_1 \\ u_2 \\ u_3 \end{array} \right\} = \left\{ \begin{array}{c} u_x \\ u_y \\ u_z \end{array} \right\} = \left\{ \begin{array}{c} u \\ v \\ w \end{array} \right\}
\qquad
\begin{array}{c}
x_i, i = 1,2,3 \\
u_i, i = 1,2,3 \\
b_i, i = 1,2,3 \\
n_i, i = 1,2,3
\end{array}
$$

$$
\mathbf{b} = \left\{ \begin{array}{c} b_1 \\ b_2 \\ b_3 \end{array} \right\} = \left\{ \begin{array}{c} b_x \\ b_y \\ b_z \end{array} \right\}, \left\{ \begin{array}{c} n_1 \\ n_2 \\ n_3 \end{array} \right\} = \left\{ \begin{array}{c} n_x \\ n_y \\ n_z \end{array} \right\}
\tag{1.38}
$$

$$
\begin{aligned}
\boldsymbol{\sigma}^{\mathrm{T}} &= \{ \sigma_{11} \quad \sigma_{22} \quad \sigma_{33} \quad \sigma_{23} \quad \sigma_{13} \quad \sigma_{12} \} \\
&= \{ \sigma_{xx} \quad \sigma_{yy} \quad \sigma_{zz} \quad \sigma_{yz} \quad \sigma_{xz} \quad \sigma_{xy} \} \\
\boldsymbol{\varepsilon}^{\mathrm{T}} &= \{ \varepsilon_{11} \quad \varepsilon_{22} \quad \varepsilon_{33} \quad 2\varepsilon_{23} \quad 2\varepsilon_{13} \quad 2\varepsilon_{12} \} \\
&= \{ \varepsilon_{xx} \quad \varepsilon_{yy} \quad \varepsilon_{zz} \quad \gamma_{yz} \quad \gamma_{xz} \quad \gamma_{xy} \}
\end{aligned}
\qquad
\begin{array}{c}
\sigma_{ij}, i,j = 1,2,3 \\
\varepsilon_{ij}, i,j = 1,2,3
\end{array}
$$

TABLE 1.5

Some Often Used Mathematic Notations

Notations	In Words	Notations	In Words
$\subset$	A subset (or space) of	$\forall$	For all
$\cap$	Union	$\exists$	There exists
$\cup$	Intersection	$\mid$	s.t. (subjected to) or restricted within
$\backslash$	Set minus	$\rightarrow$	Map to
$\oplus$	Assembly (location matched summation)	$\Rightarrow$	If … then
$\Leftrightarrow$	If and only if	$\otimes$	Tensor product

TABLE 1.6

Some Differences between FEM and Meshfree Methods

	Items	FEM	Meshfree Method
1.	Element mesh	Yes	Particles or triangular mesh/cells
2.	Mesh creation and automation	Difficult due to the need for quality elements	Relatively easy and less demand on mesh quality
3.	Mesh automation and adaptive analysis	Difficult for 3D cases	Much easier to perform
4.	Shape function creation	Element based	Node based or cell based
5.	Shape function property	Kronecker delta; valid for all elements of the same type; Compatible	May or may not satisfy Kronecker delta; Different from point to point or cell to cell; May or may not compatible
6.	Strain field construction	Performed some times and element based	A standard and systematic step for some meshfree methods
7.	Integration	Element based	Integration cell based
8.	Discretized system stiffness matrix	Sparse, bonded, symmetrical (for PDEs of symmetric operators)	Sparse, bonded, may or may not be symmetrical depending on the weak form used
9.	Imposition of essential boundary condition	Easy and standard	Special methods may be required; depends on the method used
10.	Computation speed (for same number of nodes)	Fast	0.9–50 times slower compared to the FEM with the same set of nodes depending on the method used
11.	Accuracy (for same number of nodes)	Accurate compared with FDM	Can be much more accurate compared with linear FEM
12.	Computation efficiency	Efficient	0.02–50 times more efficient compared to the FEM depending on the method used
13.	Retrieval of results	Special technique required	Standard routine
14.	Stage of development	Very well developed	Fast developing with promising progress and challenging problems
15.	Commercial software package availability	Many	Very few. Some of the techniques have been implemented in commercial FEM packages

1.10 Remarks

The similarities and differences between FEM and meshfree methods are listed in Table 1.6.

References

1. Zienkiewicz, O. C. and Taylor R. L., *The Finite Element Method*, 5th ed., Butterworth Heimemann, Oxford, 2000.
2. Hughes, T. J. R., *The Finite Element Method: Linear Static and Dynamic Finite Element Analysis*, Prentice-Hall, Englewood Cliffs, NJ, 1987.

3. Liu, G. R. and Quek, S. S., *The Finite Element Method: A Practical Course*, Butterworth Heinemann, Oxford, 2002.

4. Liu, G. R. and Han, X., *Computational Inverse Techniques in Nondestructive Evaluation*, CRC Press, Boca Raton, FL, 2003.

5. Kleiber, M. and Borkowski, A., *Handbook of Computational Solid Mechanics*, Springer-Verlag, Berlin, 1998.

6. LeVeque, R. J., *Finite Volume Methods for Hyperbolic Problems*, Cambridge University Press, New York, 2002.

7. Liu, G. R., Tu, Z. H., and Wu, Y. G., MFree2D$^©$: An adaptive stress analysis package based on meshfree technology, presented at *The Third South Africa Conference on Applied Mechanics*, Durban, South Africa, January 11–13, 2000.

8. Liu, G. R. and Tu, Z. H., An adaptive procedure based on background cells for meshless methods, *Comput. Methods Appl. Mech. Eng.*, 191, 1923–1943, 2002.

9. Liu, G. R. and Gu, Y. T., *An Introduction to Meshfree Method Methods and Their Programming*, Springer, Berlin, 2005.

10. Belytschko, T., Lu, Y. Y., and Gu, L., Element-free Galerkin methods, *Int. J. Numerical Methods Eng.*, 37, 229–256, 1994.

11. Liu, G. R., Zhang, G. Y., Dai, K. Y., Wang, Y. Y., Zhong, Z. H., Li, G. Y., and Han, X., A linearly conforming point interpolation method (LC-PIM) for 2D solid mechanics problems, *Int. J. Comput. Methods*, 2(4), 645–665, 2005.

12. Liu, G. R. and Zhang, G. Y., Upper bound solution to elasticity problems: A unique property of the linearly conforming point interpolation method (LC-PIM). *Int. J. Numerical Methods Eng.*, 74, 1128–1161, 2008.

13. Li, Y., Liu, G. R., Luan, M. T., Dai, K. Y., Zhong, Z. H., Li, G. Y., and Han, X., Contact analysis for solids based on linearly conforming radial point interpolation method, *Comput. Mech.*, 39, 537–554, 2007.

14. Zhang, G. Y., Liu, G. R., Nguyen, T. T., Song, C. X., Han, X., Zhong, Z. H., and Li, G. Y., The upper bound property for solid mechanics of the linearly conforming radial point interpolation method (LC-RPIM), *Int. J. Comput. Methods*, 4(3), 521–541, 2007.

15. Liu, G. R., A $\mathbb{G}$ space theory and weakened weak (W^2) form for a unified formulation of compatible and incompatible methods, Part I: Theory and Part II: Applications to solid mechanics problems, *Int. J. Numerical Methods Eng.*, 2009 (in press).

16. Liu, G. R., A generalized gradient smoothing technique and the smoothed bilinear form for Galerkin formulation of a wide class of computational methods, *Int. J. Comput. Methods*, 5(2), 199–236, 2008.

17. Liu, G. R. and Zhang, G. Y., Edge-based smoothed point interpolation methods, *Int. J. Comput. Methods*, 5(4), 621–646, 2008.

18. Liu, G. R., Xu, X., Zhang, G. R., and Gu, Y. T., An extended Galerkin weak form and a point interpolation method with continous strain field and superconvergence (PIM-CS) using triangular mesh, *Computational Mechancis*, 2009, 43, 651–673, 2009.

19. Liu, G. R., Xu, X., Zhang, G. R., and Nguyen, T. T., A superconvergence point interpolation method (SC-PIM) with piecewise linear strain field using triangular mesh, *Int. J. Numerical Methods Eng.*, 77, 1439–1467, 2009.

20. Xu, X., Liu, G. R., and Zhang, G. Y., A point interpolation method with least square strain field (PIM-LSS) for solution bounds and ultra-accurate solutions using triangular mesh, *Comput. Methods in Appl. Mech. Eng.*, 2009, 198, 1486–1499.

21. Liu, G. R. and Zhang, G. Y., A Strain-constructed point interpolation method (SC-PIM) and strain field construction schemes for mechanics problems of solids and structures using triangular mesh, *Int. J. Solids Struct.* (submitted), 2008.

22. Liu, G. R. and Kee, B. B. T., A stabilized least-squares radial point collocation method (LS-RPCM) for adaptive analysis, *Comput. Methods Appl. Mech. Eng.*, 195, 4843–4861, 2006.

23. Kee, B. B. T., Liu, G. R., and Lu, C., A regularized least-squares radial point collocation method (RLS-RPCM) for adaptive analysis, *Comput. Mech.*, 40, 837–853, 2007.

24. Lucy, L., A numerical approach to testing the fission hypothesis, *Astron. J.*, 82, 1013–1024, 1977.
25. Gingold, R. A. and Monaghan, J. J., Smooth particle hydrodynamics: Theory and applications to non-spherical stars, *Mon. Notice R. Astron. Soc.*, 181, 375–389, 1977.
26. Liu, G. R. and Liu, M. B., *Smoothed Particle Hydrodynamics: A Meshfree Particle Method*, World Scientific, River Edge, NJ, 2003.
27. Liu, M. B., Liu, G. R., and Zong, Z., An overview on smoothed particle hydrodynamics, *Int. J. Comput. Methods*, 5(1), 135–188, 2008.
28. Zhou, C. E., Liu, G. R., and Lou, K. Y., Three-dimensional penetration simulation using smoothed particle hydrodynamics, *Int. J. Comput. Methods*, 4(4), 671–691, 2007.
29. Belytschko, T., Krongauz, Y., Organ, D., Fleming, M., and Krysl, P., Meshless method: An overview and recent developments, *Comput. Methods Appl. Mech. Eng.*, 139, 3–47, 1996.
30. Atluri, S. N. and Zhu, T., A new meshless local Petrov–Galerkin (MLPG) approach in computational mechanics, *Comput. Mech.*, 22, 117–127, 1998.
31. Liu, W. K., Adee, J., and Jun, S., Reproducing kernel and wavelet particle methods for elastic and plastic problems, in *Advanced Computational Methods for Material Modeling*, D. J. Benson, ed., 180/PVP 268 ASME, New Orleans, LA, 1993, pp. 175–190.
32. Liu, G. R. and Gu, Y. T., A point interpolation method, in *Proceedings of the Fourth Asia-Pacific Conference on Computational Mechanics*, Singapore, December 1999, pp. 1009–1014.
33. Liu, G. R. and Gu, Y. T., A point interpolation method for two-dimensional solids, *Int. J. Numerical Methods Eng.*, 50, 937–951, 2001.
34. Liu, G. R. and Gu, Y. T., A matrix triangularization algorithm for point interpolation method, in *Proceedings of the Asia-Pacific Vibration Conference*, W. Bangchun, ed., Hangzhou, China, November 2001, pp. 1151–1154.
35. Liu, G. R., A point assembly method for stress analysis for two-dimensional solids, *Int. J. Solids Struct.*, 39, 261–276, 2002.
36. Onate, E., Idelsohn, S., Zienkiewicz, O. C., and Taylor, R. L., A finite point method in computational mechanics applications to convective transport and fluid flow, *Int. J. Numerical Methods Eng.*, 39, 3839–3866, 1996.
37. Liszka, T. and Orkisz, J., The finite difference method at arbitrary irregular grids and its application in applied mechanics, *Comput. Struct.*, 11, 83–95, 1980.
38. Jensen, P. S., Finite difference techniques for variable grids, *Comput. Struct.*, 2, 17–29, 1980.
39. Liu, G. R., Zhang, J., Li, H., Lam, K. Y., and Kee, B. B. T., Radial point interpolation based finite difference method for mechanics problems, *Int. J. Numerical Methods Eng.*, 68, 728–754, 2006.
40. Nayroles, B., Touzot, G., and Villon, P., Generalizing the finite element method: Diffuse approximation and diffuse elements, *Comput. Mech.*, 10, 307–318, 1992.
41. Oden, J. T. and Abani, P., A Parallel Adaptive Strategy for hp Finite Element Computations, TICAM Report 94-06, University of Texas, Austin, TX, 1994.
42. Armando, D. C. and Oden, J. T., Hp Clouds—A Meshless Method to Solve Boundary Value Problems, TICAM Report 95-05, University of Texas, Austin, TX, 1995.
43. Babuska, I. and Melenk, J. M., The Partition of Unity Finite Element Method, Technical Report Technical Note BN-1185, Institute for Physical Science and Technology, University of Maryland, College Park, MD, April 1995.
44. Liu, G. R. and Gu, Y. T., Vibration analyses of 2-D solids by the local point interpolation method (LPIM), in *Proceedings of the First International Conference on Structural Stability and Dynamics*, Taiwan, China, December 7–9, 2000, pp. 411–416.
45. Liu, G. R. and Gu, Y. T., A local point interpolation method for stress analysis of two-dimensional solids, *Struct. Eng. Mech.*, 11(2), 221–236, 2001.
46. Liu, G. R. and Gu, Y. T., A local radial point interpolation method (LR-PIM) for free vibration analyses of 2-D solids, *J. Sound Vibration*, 246(1), 29–46, 2001.
47. Liu, G. R., Li, Y., Dai, K. Y., Luan, M. T., and Xue, W., A linearly conforming radial point interpolation method for solid mechanics problems, *Int. J. Comput. Methods*, 3, 401–428, 2006.
48. Mukherjee, Y. X. and Mukherjee, S., Boundary node method for potential problems, *Int. J. Numerical Methods Eng.*, 40, 797–815, 1997.

49. Mukherjee, Y. X. and Mukherjee, S., On boundary conditions in the element-free Galerkin method, *Comput. Mech.*, 19, 264–270, 1997.
50. Liu, G. R. and Gu, Y. T., Coupling of element free Galerkin method with boundary point interpolation method, in *Advances in Computational Engineering & Science, ICES'2K*, S. N. Atluri, and F. W. Brust, eds., Los Angeles, CA, August 2000, pp. 1427–1432.
51. Liu, G. R. and Gu, Y. T., Coupling of element free Galerkin and hybrid boundary element methods using modified variational formulation, *Comput. Mech.*, 26(2), 166–173, 2000.
52. Gu, Y. T. and Liu, G. R., A coupled element free Galerkin/boundary element method for stress analysis of two-dimensional solids, *Comput. Methods Appl. Mech. Eng.*, 190/34, 4405–4419, 2001.
53. Gu, Y. T. and Liu, G. R., Hybrid boundary point interpolation methods and their coupling with the element free Galerkin method, *Eng. Anal. Boundary Elem.*, 27(9), 905–917, 2003.
54. Gu, Y. T. and Liu, G. R., A boundary point interpolation method for stress analysis of solids, *Comput. Mech.*, 28, 47–54, 2002.
55. Liu, G. R. and Gu, Y. T., Boundary meshfree methods based on the boundary point interpolation methods, *Eng. Anal. Boundary Elem.*, 28(5), 475–487, 2004.
56. Gu, Y. T. and Liu, G. R., A boundary radial point interpolation method (BRPIM) for 2-D structural analyses, *Struct. Eng. Mech.*, 15(5), 535–550, 2003.
57. Gu, Y. T. and Liu, G. R., A boundary point interpolation method (BPIM) using radial function basis, in *First MIT Conference on Computational Fluid and Solid Mechanics*, MIT, Cambridge, MA, June 2001, pp. 1590–1592.
58. Liu, G. R. and Xu, X. G., A gradient smoothing method (GSM) for fluid dynamics problems, *Int. J. Numerical Methods Fluids*, 56(10), 1101–1133, 2008.
59. Xu, G. X., Liu, G. R., and Lee, K. H., Application of gradient smoothing method (GSM) for steady and unsteady incompressible flow problems using irregular triangles, Submitted to *Int. J. Numerical Methods Fluids*, 2008.
60. Xu, G. X. and Liu, G. R., An adaptive gradient smoothing method (GSM) for fluid dynamics problems, *Int. J. Numerical Methods Fluids*, accepted, 2008.
61. Liu, G. R., Zhang J., and Lam, K. Y., A gradient smoothing method (GSM) with directional correction for solid mechanics problems, *Comput. Mech.*, 41, 457–472, 2008.
62. Liu, G. R. and Gu, Y. T., A meshfree method: Meshfree weak-strong (MWS) form method, for 2-D solids, *Comput. Mech.*, 33(1), 2–14, 2003.
63. Liu, G. R., Wu, Y. L., and Ding, H., Meshfree weak-strong (MWS) form method and its application to incompressible flow problems, *Int. J. Numerical Methods Fluids*, 46, 1025–1047, 2004.
64. Liu, G. R. and Zhang, G. Y., A normed $\mathbb{G}$ space and weakened weak (W^2) formulation of a cell-based smoothed point interpolation method, *Int. J. Comput. Methods*, 6(1), 147–179, 2009.
65. Liu, G. R. and Zhang, G. Y., A novel scheme of strain-constructed point interpolation method for static and dynamic mechanics problems. *International Journal of Applied Mechanics*, 1(1), 233–258, 2009.

2

Meshfree Shape Function Construction

2.1 Basic Issues for Shape Function Construction

2.1.1 Requirements on Shape Functions

Creation of meshfree shape functions is one of the central and most important issues in meshfree methods. Development of more effective methods for constructing shape functions has been one of the most active areas of research in meshfree methods. The challenge is how to efficiently create shape functions of "good" properties using nodes scattered arbitrarily in the problem domain. A good method of shape function construction should satisfy the following requirements:

1. The nodal distribution can be arbitrary within reason, and at least more flexible than that in the finite element method (FEM) (arbitrary nodal distribution).
2. The algorithm must be stable with respect to irregularity (within reason) of the node distribution (stability).
3. The shape functions constructed should possess a certain order of consistency that is the capability to locally reproduce exactly the polynomials of that order (consistency).
4. The support (or influence or smoothing) domain for field variable approximation/ interpolation should be small to include a small number of nodes (compact support).
5. The algorithm should be computationally efficient. It should be of the same order of complexity as that of FEM (efficiency).
6. Ideally, the shape function should possess the Kronecker delta function property (delta function property).
7. Preferably, the nodal shape functions should be compatible throughout the problem domain (compatibility).
8. The set of the nodal shape functions of all the nodes in the problem domain must be linearly independent and hence forms a basis for displacement field construction (linear independence).

Satisfaction of the above requirements facilitates both easy implementation of the meshfree method and accuracy of the numerical solutions. The first requirement is obvious. The second stability requirement should always be checked, because there could be uncertainties caused by the arbitrariness in the distribution of nodes. The consistency requirement is essential for the convergence of the numerical results, when the nodal spacing is reduced.

Therefore, it has to be examined. Satisfaction of the compact condition (requirement 4) leads to a sparse/banded system matrix that can be handled with high computational efficiency in terms of both storage and central processing unit (CPU) time. Requirement 5 prevents unacceptably expensive shape function constructions, because a too costly procedure will eventually become impractical, no matter how good it is.

The consistency and the three requirements will be elaborated in detail in the following sections.

2.1.2 Delta Function Property

The delta function property of shape functions is defined following the Kronecker delta:

$$\phi_i(\mathbf{x}_j) = \begin{cases} 1 & \text{when } i = j \\ 0 & \text{when } i \neq j \end{cases} \tag{2.1}$$

The shape functions that satisfy the delta function property allow easy treatments for essential boundary conditions. Shape functions created using the methods listed in Section 2.1.8 may or may not have the Kronecker delta property. The requirement on delta function property is not rigid because one can use special measures to impose essential boundary conditions, of course, at additional expense. Delta function property is also needed for shape functions to be used to form an *interpolant* that is used in the error estimation procedures.

2.1.3 Consistency

Similar to conventional FEM, a meshfree method must converge, meaning that the numerical solution obtained by the meshfree method must approach the exact solution when the nodal spacing approaches zero. For a meshfree method to converge, the shape functions used have to satisfy a certain degree of consistency. The degree of consistency of shape functions is measured here by "the order of the polynomial functions that the approximation using these shape functions is *capable* to exactly reproduce *locally* (in all the elements or cells that forms the entire problem domain)." If the approximation is capable of producing a constant field function exactly, the approximation is then said to have zero-order consistency, or C^0 consistency. In general, if the approximation can produce a polynomial of up to kth order exactly, the approximation is said to have kth-order consistency, or C^k consistency.

The term of *completeness* means that the approximation of C^k consistency has to be completely consistent for all the lower orders from 0 to $k-1$. In using polynomial shape functions, the C^k completeness is guaranteed by the use of all the polynomial terms completely up to the kth order. In this book, when we require C^k consistency, we imply also all the consistencies from C^0 to C^k. In addition, we require shape functions have at least C^1 consistency.

The requirement on consistency depends on the formulation procedures. For example, in solving any partial differential equation (PDE) based on the Galerkin weak form, there is a minimum consistency requirement for ensuring the convergence of the solution from the discretized equation system. The minimum consistency requirement depends on the order of the PDE. For a PDE of order $2k$, the minimum requirement of the consistency is C^k for Galerkin formulation. This is equivalent to the requirement of representing the polynomial of all orders up to the kth order. An approximation that can exactly represent the polynomial of all the orders up to the kth order can represent any *smooth* function with arbitrary accuracy as the nodal spacing approaches zero [28]. Note the consistency is required only "locally" within the elements, and we often allow lower order of consistency on the interfaces of the elements, as long as the *compatibility* is observed. It is not to confuse the C^k consistency of a function with the $\mathbb{C}^m(\Omega)$ continuity of a function (to be discussed in Chapter 3).

Representing a polynomial exactly is also said to reproduce the polynomial. Therefore, the *reproducing* concept is directly related to the concept of consistency.

2.1.4 Compatibility

The term "compatibility" is very important in weak formulations, and hence we will encounter it multiple times in this volume. It refers to the continuity of the field function over the problem domain. In general, this can occur for any numerical model that uses local approximation. In the FEM settings, for example, incompatibilities can occur on the element interfaces of the elements. For meshfree setting, it could occur at locations when the local support domain updates the support nodes.

The requirement on compatibility depends also on the formulation procedure. In the standard Galerkin weak formulation, it is mandatory. When a strain-constructed Galerkin (SC-Galerkin) weakened-weak (W^2) formulation is used, it is not required. The *global* compatibility is also not necessary if the *local* weighted residual weak form is employed with the $\mathbb{G}$ space theory.

Consistency is usually much easier to achieve. The compatibility is often a "headache" for weak formulations with local approximations. Our hands have been very much tied up, and the numerical maneuver has been confined in a very small "room," until the weakened-weak formulation is theoretically established.

2.1.5 Linear Independence

The linear independence of the nodal shape functions for all the nodes is needed in order to form a *nodal* basis, so that the displacement functions constructed using the nodal basis are linearly independent *necessary* for establishing a stable set of discretized system equations. For a finite model with N_n nodes, we need N_n nodal shape functions for each field variable. The linear independence of these nodal shape functions ϕ_n requires that

$$\sum_{n=1}^{N_n} \alpha_n \phi_n = 0, \quad \Rightarrow \quad \alpha_n = 0, \quad n = 1, 2, \ldots, N_n \tag{2.2}$$

In the FEM, these linearly independent shape functions are created based on elements using mostly polynomial basis functions, and the linear independence is ensured by element-based interpolation, element topology, and properly controlled coordinate mapping. In the meshfree methods no mapping is generally needed, and the linear independence is ensured by the (1) use of proper basis functions and (2) proper local nodes selection with the help of a background cells or proper means to ensure a nonsingular moment matrix.

2.1.6 Basis: An Essential Role of Shape Functions

In any (discrete) numerical method, field functions have to be approximated over the problem domain using a set of nodal values of the functions and the so-called *basis*. Given a linear space S (see Chapter 3 for definition) of dimension N_n, a set of N_n members of functions $\phi_n \in S$, $n = 1, 2, \ldots, N_n$ is a basis for S if and only if $\forall w \in S$, $\exists$ unique $\alpha_n \in \mathbb{R}$, such that

$$w = \sum_{n=1}^{N_n} \alpha_n \phi_n \tag{2.3}$$

Functions ϕ_n in the *basis* is often given in the form of *nodal* shape functions, and hence the basis is also termed as *nodal basis* in the context of FEM and meshfree methods. Equation 2.3 implies that the nodal shape functions must be linearly independent. The dimension of the space S created using ϕ_n $(n = 1, 2, \ldots, N_n)$ is N_n or $\dim(S) = N_n$. For dD solids, the total dimension of the space will be $d \times N_n$, because each node carries d degrees of freedom (DOFs).

2.1.7 Interpolant

Given $w \in S$ where S is a linear space, the *interpolant* $\mathcal{I}_h w$ creates a function that lives in a finite subspace $S_h \subset S$ with N_n dimensions: $\mathcal{I}_h w \in S_h$ where

$$\mathcal{I}_h w(\mathbf{x}) = \sum_{n=1}^{N_n} w(\mathbf{x}_n) \phi_n(\mathbf{x}) \tag{2.4}$$

that satisfies

$$\mathcal{I}_h w(\mathbf{x}_n) = w(\mathbf{x}_n), \quad n = 1, 2, \ldots, N_n \tag{2.5}$$

where
  $\mathbf{x}$ is any point inside the problem domain Ω
  $\mathbf{x}_n$ is the coordinate of the nth node
  ϕ_n is the nodal shape function for the nth node

To satisfy Equation 2.5, we require the shape functions to satisfy the Kronecker delta property Equation 2.1.

Note that a general interpolation defined in Equation 1.28 does not necessarily qualify as an *interpolant* because (1) the function u^h generated by an interpolation using a type of shape functions may not be in a subspace of the function u to be interpolated; or (2) the shape functions may not have the Kronecker delta property. Whether or not a type of shape functions can form an interpolant of a space has implication in the error estimation to relate the interpolation error to the solution error.

2.1.8 Types of Methods for Creating Shape Functions

A number of ways to construct shape functions have been proposed. This book classifies these methods into three major categories:

  1. Integral representation methods, which include
     a. Smoothed particle hydrodynamics (SPH) method
     b. Reproducing kernel particle method (RKPM)
     c. General kernel reproduction (GKR) method

2. Series representation methods, which include
    a. Moving least squares (MLS) methods
        i. MLS approximation
        ii. Modified MLS approximation
    b. Point interpolation methods (PIMs)
        i. Polynomial PIM
        ii. Radial PIM (RPIM)
    c. Partition of unit (PU) methods
        i. Partition of unity finite element (PUFE)
        ii. *hp*-clouds
    d. Least squares (LSs) methods
    e. FEMs
        i. Element-based interpolation with mapping
3. Differential representation methods, which include
        i. Finite difference method (FDM—regular and irregular grids)
        ii. Finite point method (FPM—irregular grids)
4. Gradient smoothing method (GSM)

Figure 2.1 shows the first three methods schematically. Integral representation methods are relatively new, but have found a special place in meshfree methods with the successful development of SPH [1,6]. The function is represented using its information in a local domain (smoothing domain or influence domain) via an integral form, as illustrated in Figure 2.1. Consistency is achieved by properly choosing the weight function.

Series representation methods have a long history of development. They are well developed in FEM, and are very actively studied now in the area of meshfree methods. Consistency is ensured by the use of proper basis functions. The inclusion of special terms in the basis can also improve the accuracy of the results for certain classes of problems. One of the central issues has been the compatibility with this class of methods.

Differential representation methods have also been used for a long time. Convergence of the representation is ensured via the theory of the Taylor series. Differential representation methods are usually used for establishing system equations based on strong formulation, where we may, but usually do not, construct shape functions explicitly.

The GSM is not exactly a method for shape function construction, and hence will not be discussed in detail in this chapter. It is a very powerful technique to approximate the derivatives of the field variable at a point or in a local domain. It is essentially an integral representation method applied to the gradient of field functions, and has various schemes with different types of smoothing domains as presented by Liu et al. [7–10]. When regular nodes are used, some of the GSM schemes become the FDM. The GSM has been applied in strong form formulations for both solid mechanics problems [7] and fluid dynamics problems [8–10], and these methods with proper smoothing domains are found as stable as the weak form methods. It is a weak form-like method and well suited for fluid dynamics problems.

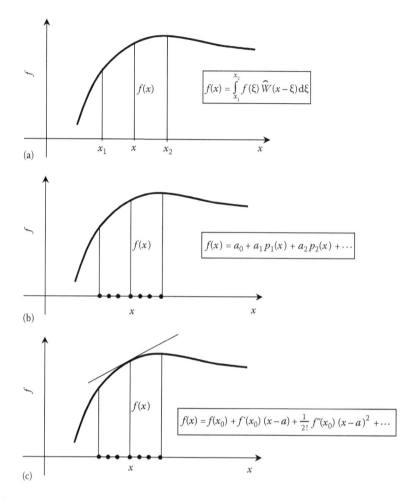

FIGURE 2.1

Methods of function representation at x using the information in the vicinity of x. (a) Integral representation. $\widehat{W}$, weight or smoothing function. (b) Series representation. $p_i(x)$ are basis functions. (c) Differential representation, where derivatives of function are used.

2.2 Smoothed Particle Hydrodynamics Approach

2.2.1 Integral Representation of a Function

The SPH method uses integral representation of a function. Consider an integrable (in the sense of *Lebesgue*) function of $u(\mathbf{x})$ at any point $\mathbf{x} = (x, y, z)$. Its integral representation can be given by

$$u(\mathbf{x}) = \int_{-\infty}^{+\infty} u(\xi)\delta(\mathbf{x} - \xi)\mathrm{d}\xi \qquad (2.6)$$

where $\delta(\mathbf{x})$ is the Dirac delta function. Note that this integral representation of a function is *exact*, if $u(\xi)$ is continuous. However, it is difficult to implement in numerical analyses, where our discretization is always *finite*. In SPH [1–6], $u(\mathbf{x})$ is *approximated* by the following integral form of representation:

$$u^h(\mathbf{x}) = \int\limits_{\Omega_s} u(\xi)\,\widehat{W}(\mathbf{x} - \xi, h)\mathrm{d}\xi \tag{2.7}$$

where
  $u^h(\mathbf{x})$ represents the *approximation* of function $u(\mathbf{x})$
  $\widehat{W}(\mathbf{x} - \xi, h)$ is a kernel or weight or smoothing function
  h is termed the *smoothing length* in SPH

The smoothing length controls the size of the compact *support domain* Ω_s bounded by Γ_s which is often termed the *influence domain* or *smoothing domain* in SPH. The presentation of a function in the integral form of Equation 2.7 can be viewed as an approximation of the integral function representation given in Equation 2.6 over a finite domain.

  In contrast to the differential representation of a function, this approximated integral presentation can be termed as integral representation. The integral representation is also termed as *kernel approximation*. In an integral representation, the weight function is often required to satisfy certain conditions [3,6]:

1. $\widehat{W}(\mathbf{x} - \xi, h) > 0, \forall \xi \in \boxed{\Omega}_s,$ (Positivity)   (2.8)
2. $\widehat{W}(\mathbf{x} - \xi, h) = 0, \forall \xi \notin \boxed{\Omega}_s,$ (Compact)   (2.9)
3. $\int_{\boxed{\Omega}_s} \widehat{W}(\mathbf{x} - \xi, h)\mathrm{d}\xi = 1$ (Unity)   (2.10)
4. $\widehat{W}$ is monotonically decreasing   (Decay)   (2.11)
5. $\widehat{W}(s, h) \rightarrow \delta(s)$ as $h \rightarrow 0,$ (Delta function behavior)   (2.12)

The first *positivity* condition is not necessary mathematically as a function representation requirement, but is important to ensure a stable numerical scheme and a meaningful presentation of some physical phenomena. For example, in fluid dynamics problems, one of the field variables could be the density of the media, which can never be negative. There are different versions of SPH that do not always satisfy this condition, such as the RKPM [11], which ensures higher order reproduction of the function and the derivatives of the function.

  The second condition, *compact*, is important to the SPH method because it enables the approximation to be generated from a local representation of nodes; i.e., $u^h(\mathbf{x})$ will depend only on the values of u at nodes (particles) that are in the smoothing domain in which $\widehat{W}$ is nonzero.

  The third condition, *unity*, assures the zero-order consistency (C^0) of the integral form representation of the continuum function. Note that this does not necessarily guarantee the C^0 consistency of the discrete form of approximation.

  Condition 4 is, again, not a mathematical requirement, but is imposed based on the physical considerations for the SPH method that a force exerted by a particle on another particle decreases with the increase of the distance between the two particles.

  Condition 5 is redundant, as a function that satisfies conditions 1–4 would naturally satisfy condition 5. In addition, the smoothing length h never goes to 0 in practical computation. Condition 5 exists to allow us to observe explicitly that the method is converging to its exact form (Equation 2.6).

In summary, conditions 2 and 3 (compact and unity) are the minimum requirements for constructing a weight function for meshfree methods based on integral representation. When this type of approximation is used in a strong formulation as in SPH, we do not have other means to control the stability and convergence of the solution. Therefore, we often need to impose additional conditions to the weight functions for a viable numerical scheme like SPH that works for certain types of problems.

Remark 2.1: Convergence Property
When W satisfies Equation 2.12 the integral representation of continuous function will be exact at $h \to 0$.

The discretized form of $u^h(\mathbf{x})$ is obtained when nodal quadrature or so-called particle approximation is applied to evaluate the integral in Equation 2.7. The integral is approximated by the summation for all the particles in the support/smoothing domain:

$$u^h(\mathbf{x}) = \sum_{I \in S_n} \widehat{W}(\mathbf{x} - \mathbf{x}_I) u_I \Delta V_I \tag{2.13}$$

where
  S_n is the set of all the nodes (or particles) in the local support domain of $\mathbf{x}$
  ΔV_I represents the volume of particle I

In solving problems of fluid flow, the volumes of particles are treated as field variables, and updated automatically in the solution process. We need, however, an initial definition for these particles that represents the continuum media. The clear advantage of SPH is that, once the initial particles are defined, the subsequent update is handled by the SPH formulation, which can virtually simulate many extreme situations, such as explosion and penetration [6]. Some applications are covered in Chapter 9.

Equation 2.13 can be written in the following form, which is similar (in form) to the finite element formulation:

$$u^h(\mathbf{x}) = \sum_{I \in S_n} \phi_I(\mathbf{x}) u_I \tag{2.14}$$

where $\phi_I(\mathbf{x})$ are the SPH shape functions given by

$$\phi_I(\mathbf{x}) = \widehat{W}(\mathbf{x} - \mathbf{x}_I) \Delta V_I \tag{2.15}$$

Note that, despite the similarity in form, the SPH shape function behaves very differently from the finite element (FE) shape functions [14]. First, Equation 2.15 has low consistency (see Section 2.2.3); second, the SPH shape functions do not satisfy the Kronecker delta function defined by Equation 2.1. From Equation 2.15, it can be seen that the shape function depends only on the weight function (assume the uniform particle distribution). It is very difficult to construct a weight function that satisfies conditions 1–4 and the Kronecker delta function property at the same time.

Because of the lack of the delta function property, we have, in general, $u_I \neq u^h(\mathbf{x}_I)$. Therefore, u_I is termed a *nodal parameter* at node I, which is, in general, not the nodal value of the field variable at the node. The shape functions defined by Equation 2.15 cannot be used to construct an *interpolant*. Equation 2.14 is not an *interpolation* of a function, and it is an *approximation* of a function. Because of this special property of the SPH shape function, the true value of the field variable should be retrieved using Equation 2.14 again, after obtaining the nodal parameters u_I at all the field nodes (particles).

2.2.2 Choice of Weight Function

Weight functions play an important role in meshfree methods. They should be constructed according to the reproducibility requirement. Most meshfree weight functions are bell-shaped. The following is a list of commonly used weight functions.

The cubic spline weight function (W1):

$$\widehat{W}(\mathbf{x} - \mathbf{x}_I) \equiv \widehat{W}(\bar{d}) = \begin{cases} \dfrac{2}{3} - 4\bar{d}^2 + 4\bar{d}^3 & \text{for } \bar{d} \leq \dfrac{1}{2} \\[2mm] \dfrac{4}{3} - 4\bar{d} + 4\bar{d}^2 - \dfrac{4}{3}\bar{d}^3 & \text{for } \dfrac{1}{2} < \bar{d} \leq 1 \\[2mm] 0 & \text{for } \bar{d} > 1 \end{cases} \tag{2.16}$$

The quartic spline weight function (W2):

$$\widehat{W}(\mathbf{x} - \mathbf{x}_I) \equiv \widehat{W}(\bar{d}) = \begin{cases} 1 - 6\bar{d}^2 + 8\bar{d}^3 - 3\bar{d}^4 & \text{for } \bar{d} \leq 1 \\ 0 & \text{for } \bar{d} > 1 \end{cases} \tag{2.17}$$

The exponential weight function (W3):

$$\widehat{W}(\mathbf{x} - \mathbf{x}_I) \equiv \widehat{W}(\bar{d}) = \begin{cases} e^{-(\bar{d}/\alpha)^2} & \bar{d} \leq 1 \\ 0 & \bar{d} > 1 \end{cases} \tag{2.18}$$

where α is constant. We often use $\alpha = 0.3$. In Equations 2.16 through 2.18

$$\bar{d} = \frac{|\mathbf{x} - \mathbf{x}_I|}{d_W} = \frac{d}{d_W} \tag{2.19}$$

where d_W is directly related to the *smoothing length h* that is a characteristic length for the SPH. It defines the dimension of the domain where $\widehat{W} \neq 0$. In general, d_W can be different from point to point.

Following a general procedure for constructing weight (smoothing) functions [11], a relatively new quartic weight (smoothing) function is constructed (W4):

$$\widehat{W}(\mathbf{x} - \mathbf{x}_I) \equiv \widehat{W}(\bar{d}) = \begin{cases} \dfrac{2}{3} - \dfrac{9}{2}\bar{d}^2 + \dfrac{19}{3}\bar{d}^3 - \dfrac{5}{2}\bar{d}^4 & \text{for } \bar{d} \leq 1 \\[2mm] 0 & \text{for } \bar{d} > 1 \end{cases} \tag{2.20}$$

A parabolic weight function also exists but is used less frequently. The formulation is given in Equation 9.30.

Figure 2.2 plots all four weight functions given by Equations 2.16 through 2.18, and 2.20. It can be clearly seen that the quartic weight function (W4) given in Equation 2.20 has a shape

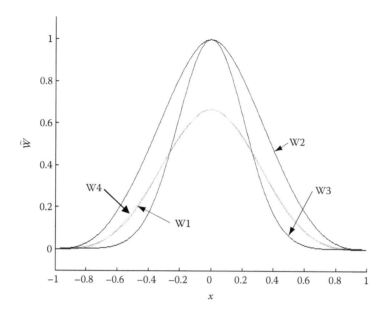

FIGURE 2.2
Weight functions. W1, cubic spline weight function; W2, quartic spline weight function; W3, exponential weight function ($\alpha = 0.3$); and W4, quartic weight function.

very similar to the piecewise cubic spline weight function (W1) given in Equation 2.16, which has been tested and works very well for many applications. The new quartic weight function W4, however, has a simple form of one single piece, and possesses second-order reproducing capacity. Figure 2.3 plots the first derivative of all four weight functions. It is shown that

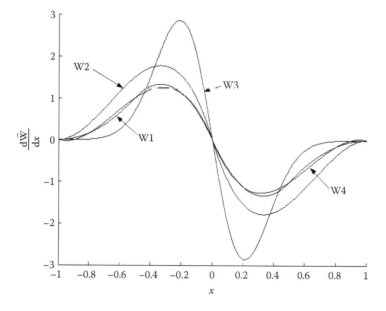

FIGURE 2.3
The first derivative of weight functions. W1, cubic spline weight function; W2, quartic spline weight function; W3, exponential weight function ($\alpha = 0.3$); and W4, quartic weight function.

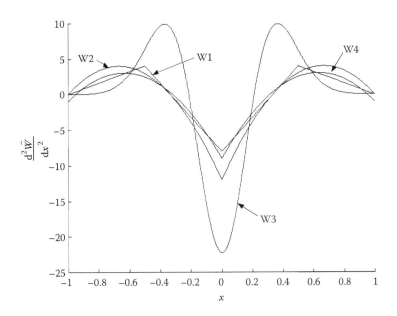

FIGURE 2.4
The second derivative of weight functions. W1, cubic spline weight function; W2, quartic spline weight function; W3, exponential weight function ($\alpha = 0.3$); and W4, quartic weight function.

the first derivatives of all four are smooth. It can be clearly seen that the first derivative of W4 still behaves similarly to that of the piecewise cubic spline function (W1). Figure 2.4 plots the second derivative of all four weight functions and shows that the second derivative of the cubic spline (W1) is no longer smooth. The second derivative of the new quartic weight function (W4) is still smooth but does not precisely equal zero on the boundary. W4 is for readers who prefer the performance of W1 but want a simple one-piece formulation.

Note that the weight functions shown in Equations 2.16 through 2.18, and 2.20 need to be scaled to satisfy the condition of unity defined by Equation 2.10 for problems of different dimensions, if they are used in such integral representation methods as SPH. This is to ensure the consistency of function representation, as is seen in Section 2.2.3. The scaling is immaterial if the shape functions are used in the series representation methods such as MLS to be discussed in Section 2.4.

In SPH methods, the following SPH weight function is often used (for one-dimensional (1D) problems)

$$\widehat{W}(\mathbf{x} - \mathbf{x}_I, h) \equiv \widehat{W}(\bar{d}, h) = \frac{2}{3h} \begin{cases} 1 - \frac{2}{3}\bar{d}^2 + \frac{3}{4}\bar{d}^3 & \text{for } \bar{d} \le 1 \\ \frac{1}{4}(2 - \bar{d})^3 & \text{for } 1 < \bar{d} < 2 \\ 0 & \text{for } \bar{d} \ge 2 \end{cases} \tag{2.21}$$

where $\bar{d} = d/h$ and h is the smoothing length. This SPH weight function is actually exactly the same as the cubic spline function given in Equation 2.16, but different in form and in the dimension of the smoothing domain.

2.2.3 Consistency of SPH Shape Functions

Let us now examine the consistency of the SPH approximation. SPH approximation starts from the integral approximation (Equation 2.7). If we want the approximation to be of C^0 consistency, we need to require it to reproduce a constant c. Assume the field is given by $u(x) = c$. Substituting it into Equation 2.7, we obtain

$$u^h(\mathbf{x}) = \int_{\Omega_s} c\widehat{W}(\mathbf{x} - \xi, h)\mathrm{d}\xi = c \tag{2.22}$$

or

$$\int_{\Omega_s} \widehat{W}(\mathbf{x} - \xi, h)\mathrm{d}\xi = 1 \tag{2.23}$$

This is the condition of unity given in Equation 2.10 that a weight function has to satisfy. It is now clear that the condition of unity (Equation 2.10) ensures the lowest C^0 consistency for the SPH approximation.

Let us examine now whether the SPH approximation possesses C^1 consistency. Assume a linear field given by

$$u(\xi) = c_0 + \mathbf{c}_1^\mathrm{T}\xi \tag{2.24}$$

where $\mathbf{c}_1$ and ξ are, respectively, a constant vector and the Cartesian coordinate vector. For two-dimensional (2D) cases, they should be

$$\mathbf{c}_1 = \begin{Bmatrix} c_{1x} \\ c_{1y} \end{Bmatrix} \tag{2.25}$$

and

$$\xi = \begin{Bmatrix} \xi \\ \eta \end{Bmatrix} \tag{2.26}$$

For three-dimensional (3D) cases, they are

$$\mathbf{c}_1 = \begin{Bmatrix} c_{1x} \\ c_{1y} \\ c_{1z} \end{Bmatrix} \tag{2.27}$$

and

$$\xi = \begin{Bmatrix} \xi \\ \eta \\ \zeta \end{Bmatrix} \tag{2.28}$$

Substituting the above equation into Equation 2.7, we obtain

$$u^h(\mathbf{x}) = \int_{\Omega_s} (c_0 + \mathbf{c}_1^\mathrm{T}\xi)\widehat{W}(\mathbf{x} - \xi, h)\mathrm{d}\xi \tag{2.29}$$

If the approximation possesses C^1 consistency, we should have

$$u^h(\mathbf{x}) = c_0 + \mathbf{c}_1^T \mathbf{x} \tag{2.30}$$

Equating the right-hand sides of the above two equations, we have

$$c_0 + \mathbf{c}_1^T \mathbf{x} = \int_{\Omega_s} (c_0 + \mathbf{c}_1^T \xi)\widehat{W}(\mathbf{x} - \xi, h)d\xi \tag{2.31}$$

or

$$c_0 + \mathbf{c}_1^T \mathbf{x} = c_0 \int_{\Omega_s} \widehat{W}(\mathbf{x} - \xi, h)d\xi + \mathbf{c}_1^T \int_{\Omega_s} \xi\widehat{W}(\mathbf{x} - \xi, h)d\xi \tag{2.32}$$

Using the condition equation (Equation 2.10), the above equation becomes

$$\mathbf{x} = \int_{\Omega_s} \xi\widehat{W}(\mathbf{x} - \xi, h)d\xi \tag{2.33}$$

This gives the condition that the weight function has to satisfy for C^1 consistency. Therefore, in general, SPH does not possess C^1 consistency, if the weight function satisfies only the conditions given in Equations 2.8 through 2.12.

To examine further the conditions required for the weight function to achieve an approximation of C^1 consistency, we first multiply with $\mathbf{x}$ on both sides of Equation 2.23, which gives

$$\mathbf{x} = \int_{\Omega_s} \mathbf{x}\widehat{W}(\mathbf{x} - \xi, h)d\xi \tag{2.34}$$

To show this more clearly, we then subtract Equation 2.33 from Equation 2.34 to obtain

$$0 = \int_{\Omega_s} (\mathbf{x} - \xi)\widehat{W}(\mathbf{x} - \xi, h)d\xi \tag{2.35}$$

The above integral is the first moment of the weight function. Therefore, the condition that the weight function must satisfy for C^1 consistency is that the first moment of the weight function has to vanish. This condition can be satisfied if the weight function is symmetric about the origin. For an infinite problem domain, this symmetric condition is not very difficult to meet, and all the weight functions listed in Section 2.2.2 satisfy this condition. The problem occurs on and near the boundary, where it is not easy to construct a symmetric weight function. Figure 2.5 shows an example of the 1D situation. Figure 2.5a shows a weight function for an interior point, where a symmetric function can be easily defined to have linear consistency. For points near the boundary (Figure 2.5b) and on the boundary (Figure 2.5c and d), it is difficult to maintain the symmetry for linear consistency.

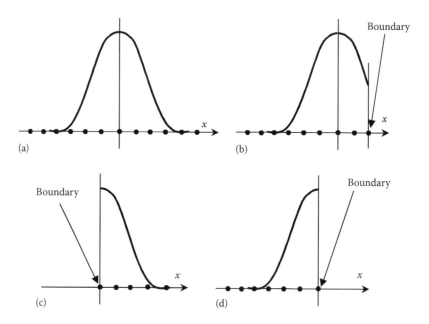

FIGURE 2.5
SPH weight functions for 1D case. (a) For an interior point, the weight function can be symmetric and the first moment vanishes. For a point near the boundary (b), or on the boundaries (c, d), the weight functions are not symmetrical.

Special treatments are required to enforce the linear consistency. This is discussed further in the following section.

Following the same procedure, it is easy to prove that the two quartic weight functions given by Equations 2.17 and 2.20 possess C^2 consistency, when the entire smoothing domain is located within the problem domain [6].

2.3 Reproducing the Kernel Particle Method

Liu et al. [13] have developed a method that ensures the certain degree of consistency of the integral approximation and named it the RKPM. This is achieved by adding a correction function to the kernel in Equation 2.7. This correction function is particularly useful in improving the SPH approximation near the boundaries as well as making it linear or C^1 consistent near the boundary. The integral representation of a function with the correction function can be given by

$$u^h(\mathbf{x}) = \int_{\Omega_s} u(\xi)C(\mathbf{x}, \xi)\widehat{W}(\mathbf{x} - \xi, h)\mathrm{d}\xi \qquad (2.36)$$

where $C(\mathbf{x}, \xi)$ is the correction function. An example of the correction function in 1D is

$$C(\mathbf{x}, \xi) = c_1(\mathbf{x}) + c_2(\mathbf{x})(\xi - \mathbf{x}) \qquad (2.37)$$

where $c_1(\mathbf{x})$ and $c_2(\mathbf{x})$ are coefficients. The coefficients are found by enforcing the corrected kernel to reproduce the function [13]:

$$c_1(\mathbf{x}) = \frac{m_2(\mathbf{x})}{\left(m_0(\mathbf{x})\, m_2(\mathbf{x}) - m_1^2(\mathbf{x})\right)} \tag{2.38}$$

$$c_2(\mathbf{x}) = \frac{-m_1(\mathbf{x})}{\left(m_0(\mathbf{x})\, m_2(\mathbf{x}) - m_1^2(\mathbf{x})\right)} \tag{2.39}$$

where m_0, m_1, and m_2 are the moments of W, defined by

$$m_0(\mathbf{x}) = \int_{\Omega_s} \widehat{W}(\mathbf{x} - \xi)d\xi \tag{2.40}$$

$$m_1(\mathbf{x}) = \int_{\Omega_s} \xi \widehat{W}(\mathbf{x} - \xi)d\xi \tag{2.41}$$

$$m_2(\mathbf{x}) = \int_{\Omega_s} \xi^2 \widehat{W}(\mathbf{x} - \xi)d\xi \tag{2.42}$$

If the integral in Equation 2.36 is discretized, then a function $u(\mathbf{x})$ can be approximated using the surrounding particles:

$$u^h(\mathbf{x}) = \sum_{I \in S_n} C(\mathbf{x}, \mathbf{x}_I)\widehat{W}(\mathbf{x} - \mathbf{x}_I)u_I \Delta V_I = \sum_{I \in S_n} \phi_I(\mathbf{x})u_I \tag{2.43}$$

where $\phi_I(\mathbf{x})$ are the RKPM shape functions given by

$$\phi_I(\mathbf{x}) = C(\mathbf{x}, \mathbf{x}_I)\widehat{W}(\mathbf{x} - \mathbf{x}_I)\Delta V_I \tag{2.44}$$

Note that the corrected weight function may not satisfy the conditions of Equations 2.8 and 2.11. The RKPM method has been applied successfully to solve many problems of solids, structures, acoustics, fluids, etc. Readers are referred to publications by the group led by Liu [12–19].

2.4 Moving Least Squares Approximation

MLS, originated by mathematicians for data fitting and surface construction, is often termed local regression and loss [20,21]. It can be categorized as a method of series representation of functions. A detailed description of the MLS method can be found in a paper by Lancaster and Salkauskas [20]. The MLS method is now a widely used alternative for constructing meshfree shape functions for approximation. Nayroles et al. [22] were the first to use MLS approximation to construct shape functions for their diffuse element method (DEM) for mechanics problems. DEM was modified by Belytschko et al. [23] to become the element-free Galerkin (EFG) method, where the MLS approximation is also employed. The invention of DEM and the advances in EFG have created great impact on the development of meshfree methods. The MLS approximation has two major features that make it popular: (1) the approximated field function is continuous and smooth in the entire problem domain when sufficient nodes are used, which suits well for the constrained Galerkin weak form; and (2) it is capable of producing an approximation with the desired order of consistency, which offers effective ways for field enrichment. The procedure of constructing shape functions for meshfree methods using MLS approximation is detailed in this section.

2.4.1 MLS Procedure

Let $u(\mathbf{x})$ be the function of a field variable defined in the domain Ω. The approximation of $u(\mathbf{x})$ at point $\mathbf{x}$ is denoted as $u^h(\mathbf{x})$. The MLS approximates the field function in the form of series representation:

$$u^h(\mathbf{x}) = \sum_{j}^{m} p_j(\mathbf{x})a_j(\mathbf{x}) \equiv \mathbf{p}^{\mathrm{T}}(\mathbf{x})\mathbf{a}(\mathbf{x}) \tag{2.45}$$

where
    m is the number of terms of monomials (polynomial basis)
    $\mathbf{a}(\mathbf{x})$ is a vector of coefficients given by

$$\mathbf{a}^{\mathrm{T}}(\mathbf{x}) = \{a_0(\mathbf{x}) \quad a_1(\mathbf{x}) \quad \cdots \quad a_m(\mathbf{x})\} \tag{2.46}$$

which are functions of $\mathbf{x}$.

In Equation 2.45, $\mathbf{p}(\mathbf{x})$ is a vector of basis functions that consists most often of monomials of the lowest orders to ensure minimum completeness. Enhancement functions can, however, be added to achieve better efficiency or to produce stress fields of special characteristics, such as singularity at the crack tip and stress discontinuity at interfaces of different types of materials. Here we discuss the use of the pure polynomial basis. In 1D space, a complete polynomial basis of order m is given by

$$\mathbf{p}^{\mathrm{T}}(\mathbf{x}) = \{p_0(x), p_1(x), \ldots, p_m(x)\} = \{1, x, x^2, \ldots, x^m\} \tag{2.47}$$

and in 2D space,

$$\mathbf{p}^{\mathrm{T}}(\mathbf{x}) = \mathbf{p}^{\mathrm{T}}(x, y) = \{1, x, y, xy, x^2, y^2, \ldots, x^m, y^m\} \tag{2.48}$$

In this case, the Pascal triangle shown in Figure 2.6 can be utilized to build $\mathbf{p}^{\mathrm{T}}(\mathbf{x})$, and the number of nodes in the support domain can be chosen accordingly.

In 3D space, we have

$$\mathbf{p}^{\mathrm{T}}(\mathbf{x}) = \mathbf{p}^{\mathrm{T}}(x, y, z) = \{1, x, y, z, xy, yz, zx, x^2, y^2, z^2, \ldots, x^m, y^m, z^m\} \tag{2.49}$$

In this case, the Pascal pyramid shown in Figure 2.7 can be employed to build $\mathbf{p}^{\mathrm{T}}(\mathbf{x})$. The vector of coefficients $\mathbf{a}(\mathbf{x})$ in Equation 2.45 is determined using the function values at a set of nodes that are included in the *support domain* of $\mathbf{x}$. A support domain of a point $\mathbf{x}$ determines the number of nodes that is used locally to approximate the function value at $\mathbf{x}$, as shown in Figure 1.9.

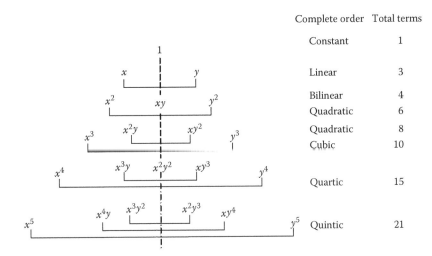

FIGURE 2.6
Pascal triangle of monomials, 2D case.

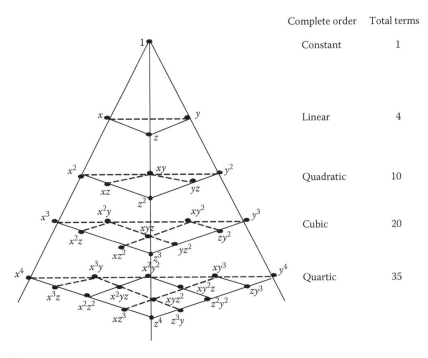

FIGURE 2.7
Pascal pyramid of monomials, 3D case.

Assuming the support domain of $\mathbf{x}$ contains a set of n local nodes $\mathbf{x}_1, \mathbf{x}_2, \ldots, \mathbf{x}_n$, Equation 2.45 is then used to calculate the approximated values of the field function at these nodes:

$$u^h(\mathbf{x}, \mathbf{x}_I) = \mathbf{p}^{\mathrm{T}}(\mathbf{x}_I)\mathbf{a}(\mathbf{x}), \quad I = 1, 2, \ldots, n \tag{2.50}$$

Note that $\mathbf{a}(\mathbf{x})$ here is an arbitrary function of $\mathbf{x}$. A functional of weighted residual is then constructed using the approximated values of the field function and the *nodal parameters,* $u_I = u(\mathbf{x}_I)$:

$$J = \sum_{I}^{n} \widehat{W}(\mathbf{x} - \mathbf{x}_I) \underbrace{\left[u^h(\mathbf{x}, \mathbf{x}_I) - u(\mathbf{x}_I) \right]^2}_{\text{residual}}$$

$$= \sum_{I}^{n} \widehat{W}(\mathbf{x} - \mathbf{x}_I) \left[\mathbf{p}^T(\mathbf{x}_I)\mathbf{a}(\mathbf{x}) - u_I \right]^2 \tag{2.51}$$

where $\widehat{W}(\mathbf{x} - \mathbf{x}_I)$ is a weight function. The nodal parameter u_I of the field variable at node I is shown schematically in Figure 2.8. It is clear that the functional J so constructed is at least semi-symmetric positive definite (semi-SPD).

Note that the weight function used in Equation 2.51 has a different mathematical mission than that used for integral representation methods, such as that in Equation 2.7. The weight function used in Equation 2.51 plays two important roles in constructing globally *continuous* MLS shape functions. The first role is to provide favorable weightings for the residuals at different nodes in the support domain: We usually prefer nodes farther from $\mathbf{x}$ to have small weights, and to give more weighting to the nodes closer to $\mathbf{x}$ where the field variable is to be approximated. The second role is to ensure that nodes leave or enter the support domain in a gradual (smooth) manner when $\mathbf{x}$ moves. The second role of the weight function is very important, because it makes sure that the MLS shape functions constructed satisfy the compatibility condition. Note that the weight function can only play these two roles effectively when sufficient nodes are used: $n \gg m$. Also, the weight function is not responsible for the consistency of the shape functions created.

Theoretically, it can be any function as long as it satisfies the conditions of Equations 2.8, 2.9, and 2.11. Equation 2.9 also ensures a local support feature that leads to a sparse/banded system matrix. Equation 2.11 allows the weight function to play the first role. Any weight functions shown in Equations 2.16 through 2.18 and 2.20 can be and have been used in MLS approximation. The scaling to meet the condition of unity that is required in SPH is not necessary here for MLS approximations.

In the MLS approximation, at an arbitrary point $\mathbf{x}$, $\mathbf{a}(\mathbf{x})$ is chosen to minimize the weighted residual. The minimization condition requires

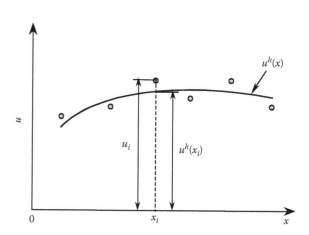

FIGURE 2.8
The approximation function $u^h(x)$ and the nodal parameters u_I in the MLS approximation.

$$\frac{\partial J}{\partial \mathbf{a}} = 0 \tag{2.52}$$

which results in the following linear equation system:

$$\mathbf{A}(\mathbf{x})\mathbf{a}(\mathbf{x}) = \mathbf{B}(\mathbf{x})\mathbf{d}_s \tag{2.53}$$

where $\mathbf{A}$ is called the MLS moment matrix given by

$$\mathbf{A}(\mathbf{x}) = \sum_I^n \widehat{W}_I(\mathbf{x})\mathbf{p}^{\mathrm{T}}(\mathbf{x}_I)\mathbf{p}(\mathbf{x}_I) \tag{2.54}$$

where

$$\widehat{W}_I(\mathbf{x}) \equiv \widehat{W}(\mathbf{x} - \mathbf{x}_I) \tag{2.55}$$

In Equation 2.53, matrix $\mathbf{B}$ has the form of

$$\mathbf{B}(\mathbf{x}) = [\mathbf{B}_1 \ \mathbf{B}_2 \ \cdots \ \mathbf{B}_n] \tag{2.56}$$

$$\mathbf{B}_I = \widehat{W}_I(\mathbf{x})\mathbf{p}(\mathbf{x}_I) \tag{2.57}$$

and $\mathbf{d}_s$ is the vector that collects the (discrete) nodal parameters of the field variables for all the nodes in the support domain:

$$\mathbf{d}_s = \{u_1 \quad u_2 \quad \cdots \quad u_n\}^{\mathrm{T}} \tag{2.58}$$

Assuming that the MLS moment matrix $\mathbf{A}$ is invertible (see Section 2.4.4), Equation 2.53 can then be solved for $\mathbf{a}(\mathbf{x})$:

$$\mathbf{a}(\mathbf{x}) = \mathbf{A}^{-1}(\mathbf{x})\mathbf{B}(\mathbf{x})\mathbf{d}_s \tag{2.59}$$

Substituting the above equation back into Equation 2.50 leads to

$$u^h(\mathbf{x}) = \sum_I^n \sum_j^m p_j(\mathbf{x})\left(\mathbf{A}^{-1}(\mathbf{x})\mathbf{B}(\mathbf{x})\right)_{jI} u_I \tag{2.60}$$

or

$$u^h(\mathbf{x}) = \sum_I^n \phi_I(\mathbf{x})u_I \tag{2.61}$$

where the MLS shape function $\phi(\mathbf{x})$ is defined by

$$\phi_I(\mathbf{x}) = \sum_j^m p_j(\mathbf{x})\left(\mathbf{A}^{-1}(\mathbf{x})\mathbf{B}(\mathbf{x})\right)_{jI} = \mathbf{p}^{\mathrm{t}}\mathbf{A}^{-1}\mathbf{B}_I \tag{2.62}$$

Equation 2.61 can also be written in the following matrix form:

$$u^h(\mathbf{x}) = \boldsymbol{\varphi}(\mathbf{x})\mathbf{d}_s \tag{2.63}$$

where $\boldsymbol{\varphi}(\mathbf{x})$ is the matrix of MLS shape functions corresponding to n nodes in the support domain:

$$\boldsymbol{\varphi}(\mathbf{x}) = \begin{bmatrix} \phi_1(\mathbf{x}) & \phi_2(\mathbf{x}) & \cdots & \phi_n(\mathbf{x}) \end{bmatrix} \tag{2.64}$$

To determine the spatial derivatives of the function of the field variable, which are required for deriving the discretized system equations, it is necessary to obtain the derivatives of the MLS shape functions. For convenience, to obtain the partial derivatives of shape functions, Equation 2.64 is first rewritten as follows using Equation 2.62:

$$\boldsymbol{\varphi}(\mathbf{x}) = \boldsymbol{\gamma}^{\mathrm{T}}(\mathbf{x})\mathbf{B}(\mathbf{x}) \tag{2.65}$$

where $\boldsymbol{\gamma}(\mathbf{x})$ is determined by

$$\mathbf{A}(\mathbf{x})\boldsymbol{\gamma}(\mathbf{x}) = \mathbf{p}(\mathbf{x}) \tag{2.66}$$

The partial derivatives of $\boldsymbol{\gamma}(\mathbf{x})$ can be obtained as follows:

$$\mathbf{A}\boldsymbol{\gamma}_{,i} = \mathbf{p}_{,i} - \mathbf{A}_{,i}\boldsymbol{\gamma} \tag{2.67}$$

$$\mathbf{A}\boldsymbol{\gamma}_{,ij} = \mathbf{p}_{,ij} - (\mathbf{A}_{,i}\boldsymbol{\gamma}_{,j} + \mathbf{A}_{,j}\boldsymbol{\gamma}_{,i} + \mathbf{A}_{,ij}\boldsymbol{\gamma}) \tag{2.68}$$

$$\mathbf{A}\boldsymbol{\gamma}_{,ijk} = \mathbf{p}_{,ijk} - (\mathbf{A}_{,i}\boldsymbol{\gamma}_{,jk} + \mathbf{A}_{,j}\boldsymbol{\gamma}_{,ik} + \mathbf{A}_{,k}\boldsymbol{\gamma}_{,ij} + \mathbf{A}_{,ij}\boldsymbol{\gamma}_{,k} + \mathbf{A}_{,ik}\boldsymbol{\gamma}_{,j} + \mathbf{A}_{,jk}\boldsymbol{\gamma}_{,i} + \mathbf{A}_{,ijk}\boldsymbol{\gamma}) \tag{2.69}$$

where i, j, and k denote coordinates x and y. A comma designates a partial derivative with respect to the indicated spatial variable. The partial derivatives of shape function $\boldsymbol{\Phi}$ can then be obtained as follows:

$$\boldsymbol{\varphi}_{,i} = \boldsymbol{\gamma}_{,i}^{\mathrm{T}}\mathbf{B} + \boldsymbol{\gamma}^{\mathrm{T}}\mathbf{B}_{,i} \tag{2.70}$$

$$\boldsymbol{\varphi}_{,ij} = \boldsymbol{\gamma}_{,ij}^{\mathrm{T}}\mathbf{B} + \boldsymbol{\gamma}_{,i}^{\mathrm{T}}\mathbf{B}_{,j} + \boldsymbol{\gamma}_{,j}^{\mathrm{T}}\mathbf{B}_{,i} + \boldsymbol{\gamma}^{\mathrm{T}}\mathbf{B}_{,ij} \tag{2.71}$$

$$\boldsymbol{\varphi}_{,ijk} = \boldsymbol{\gamma}_{,ijk}^{\mathrm{T}}\mathbf{B} + \boldsymbol{\gamma}_{,ij}^{\mathrm{T}}\mathbf{B}_{,k} + \boldsymbol{\gamma}_{,ik}^{\mathrm{T}}\mathbf{B}_{,j} + \boldsymbol{\gamma}_{,jk}^{\mathrm{T}}\mathbf{B}_{,i} + \boldsymbol{\gamma}_{,i}^{\mathrm{T}}\mathbf{B}_{,jk} + \boldsymbol{\gamma}_{,j}^{\mathrm{T}}\mathbf{B}_{,ik} + \boldsymbol{\gamma}_{,k}^{\mathrm{T}}\mathbf{B}_{,ij} + \boldsymbol{\gamma}^{\mathrm{T}}\mathbf{B}_{,ijk} \tag{2.72}$$

It should be noted that MLS shape functions do not satisfy the Kronecker delta criterion $\phi_I(\mathbf{x}_J) \neq \delta_{IJ}$ that results in $u^h(\mathbf{x}_I) \neq u_I$; i.e., the nodal parameters u_I are not the nodal values of $u^h(\mathbf{x}_I)$. Therefore, they cannot be used to construct an interpolant, but rather approximates of a function. Figure 2.8 gives a 1D example of the MLS approximation. The approximation of the displacement at the Ith node $u^h(\mathbf{x}_I)$ depends not only on the nodal parameter u_I but also on the nodal parameters u_1 through u_n, parameters that correspond to all the nodes within the support domain of node I. This is expressed in the sum given in Equation 2.61. This property makes the imposition of essential boundary conditions more complicated than that in the FEM.

A plot of a typical 1D MLS weight function and shape function is given in Figure 2.9. The shape function is for the node at $x = 0$ and is obtained using five nodes evenly distributed in the support domain of $[-1, 1]$. The quartic spline weight function (W2) is used. It can be seen that the MLS shape function attains a maximum value that is considerably less than 1.

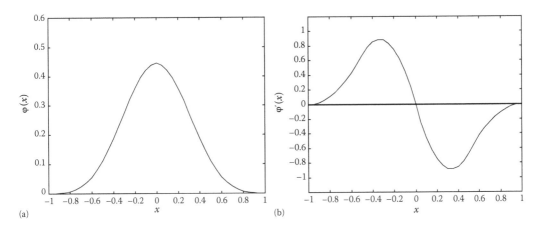

FIGURE 2.9
MLS shape function in 1D space for the node at $x=0$ obtained using five nodes evenly distributed in the support domain of $[-1, 1]$. Quartic spline weight function (W2) is used. (a) MLS shape function; (b) derivative of the shape function. Note that the MLS shape function does not possess the Kronecker delta function property.

For this plot, the quartic weight function (Equation 2.17) is used with $d_W = 0.45\alpha_s$, where $\alpha_s = 2.5$.

Note that the dimension of the support domain d_s in MLS approximation is determined by the dimension of the weight function d_W. Therefore, $d_W = d_s$. The procedure for determining d_s has already been covered in Sections 2.10.2 and 2.10.3. These methods can be used here to determine d_W for both uniformly and nonuniformly distributed nodes in 1D, 2D, and 3D domains.

2.4.2 Consistency of MLS Shape Functions

The consistency of the MLS approximation depends on the complete order of the monomial employed in Equations 2.47 or 2.48. If the complete order of the monomial is k, the MLS shape function will possess C^k consistency. To demonstrate, we follow the argument of [24].

Note that J in Equation 2.51 is positive definite, because the weight function is chosen positive (Equation 2.8). Therefore, its minimum is nonnegative. Consider a field given by

$$u(\mathbf{x}) = \sum_{j}^{k} p_j(\mathbf{x})\alpha_j(\mathbf{x}), \quad k \leq m \tag{2.73}$$

Such a given field can always be written in the form of

$$u(\mathbf{x}) = \sum_{j}^{m} p_j(\mathbf{x})\alpha_j(\mathbf{x}) \tag{2.74}$$

by simply assigning $\alpha_j = 0$ for $j > k$. Then, if we let $\alpha_j(\mathbf{x}) = \alpha_j$, J will vanish and it will necessarily be a minimum, which leads to

$$u^h(\mathbf{x}) = \sum_{j}^{k} p_j(\mathbf{x})\alpha_j(\mathbf{x}) = u(\mathbf{x}) \tag{2.75}$$

This proves that any field given by Equation 2.73 will be exactly represented or reproduced by the MLS approximation. This proof procedure also implies that any function in the basis is reproduced exactly. This feature of MLS approximation is, in fact, very easy to understand by intuition: the MLS approximation seeks a set of coefficient $a(x)$ that can produce a function of $u(x) = \Sigma p_j(x)a_j(x)$ with a minimum distance norm to the actual function. If the actual function is in the basis of $p_j(x)$, MLS approximation will simply produce the basis because the distance norm is 0, which is, of course, the minimum.

The proof of the consistency of MLS approximation is valid for proving another important feature of MLS approximation: Any function that appears in the basis can be reproduced exactly. To have the MLS approximation to exhibit linear consistency, all one need to do is include the constant and linear monomials into the basis. Making use of this feature further, one can develop shape functions for simulating a singular stress field at a crack tip by including singular functions into the basis [25–27]. However, one has to make sure that the weighted moment matrix computed using Equation 2.54 is still invertible, when these additional basis functions are included.

2.4.3 Continuous Moving Least Square Approximation

Belytschko et al. [28] have also shown the relation between the approximations of RKPM and MLS. Their interesting procedure starts with the construction of the continuous form of the MLS approximation. The continuous counterpart of Equation 2.51 can be written as

$$J(\mathbf{x}) = \int_{\Omega_s} \widehat{W}(\mathbf{x} - \xi)\left[u^h(\mathbf{x}, \xi) - u(\xi)\right]^2 d\xi \tag{2.76}$$

where $\widehat{W}(\mathbf{x} - \xi)$ is a weight function of compact support. The approximation of the function has the form:

$$u^h(\mathbf{x}, \xi) = \sum_{i}^{m} p_i(\xi)a_i(\mathbf{x}) \tag{2.77}$$

The condition for minimizing $J(\mathbf{x})$ leads to the following equation for solving $a_j(\mathbf{x})$:

$$\frac{J(\mathbf{x})}{a_i(\mathbf{x})} = 0 \tag{2.78}$$

or

$$2\int_{\Omega_s}\left[\widehat{W}(\mathbf{x} - \xi)\left(\sum_{i}^{m} p_i(\xi)a_i(\mathbf{x}) - u(\xi)\right)\sum_{j}^{m} p_j(\xi)da_j(\mathbf{x})\right]d\xi = 0 \tag{2.79}$$

Because the foregoing equation has to be satisfied for all $da_j(\mathbf{x})$, we obtain the following equation for solving $\alpha_j(\mathbf{x})$:

$$\sum_{j}^{m} \bar{A}_{ij}(\xi)a_j(\mathbf{x}) = \int_{\Omega_s}\left[\widehat{W}(\mathbf{x} - \xi)p_i(\xi)u(\xi)\right]d\xi \tag{2.80}$$

where

$$\bar{A}_{ij}(\mathbf{x}) = \int_{\Omega_s} \left[\widehat{W}(\mathbf{x} - \xi) p_i(\xi) p_j(\xi) \right] d\xi \tag{2.81}$$

which is the continuous counterpart of the discrete moment matrix $\mathbf{A}(\mathbf{x})$ given in Equation 2.54. Solving Equation 2.80 for $a_j(\mathbf{x})$, we have

$$a_j(\mathbf{x}) = \bar{A}_{ij}^{-1}(\mathbf{x}) \int_{\Omega_s} \left[\widehat{W}(\mathbf{x} - \xi) p_i(\xi) u(\xi) \right] d\xi \tag{2.82}$$

Substituting the above equation into Equation 2.77, we obtain

$$u^h(\mathbf{x}, \xi) = p_j(\xi) \bar{A}_{ij}^{-1}(\mathbf{x}) \int_{\Omega_s} \left[\widehat{W}(\mathbf{x} - \xi') p_i(\xi') u(\xi') \right] d\xi' \tag{2.83}$$

Note that in the above equation ξ' is used for the integral variable. The approximation at $\mathbf{x}$ is then obtained by letting $\mathbf{x} = \xi$.

$$u^h(\mathbf{x}) = \int_{\Omega_s} \left[p_j(\mathbf{x}) \bar{A}_{ij}^{-1}(\mathbf{x}) p_i(\xi') \widehat{W}(\mathbf{x} - \xi') u(\xi') \right] d\xi' \tag{2.84}$$

Comparison with Equation 2.7 reveals the similarity between the SPH and MLS approximations. Defining

$$C(\mathbf{x}, \xi') = p_j(\mathbf{x}) \bar{A}_{ij}^{-1}(\mathbf{x}) p_i(\xi') \tag{2.85}$$

produces the additional term for ensuring consistency. This term is similar to the correction function used by Liu et al. [11] for restoring the consistency in SPH. This is not a surprise, because these two methods are essentially the same and the difference is in order of the procedure of the shape function construction: MLS uses first the *consistent basis* to ensure consistency and then using moving weight function to make sure that these consistent shape functions can be successfully produced (invertible moment matrix); while the RKPM constructs the inconsistent shape functions first using SPH weight functions, and then restores consistency later.

2.4.4 Singularity Issues of the MLS Moment Matrix

In obtaining Equation 2.59, we assumed that the MLS moment matrix $\mathbf{A}$ is invertible. In carefully examining $\mathbf{A}$ given in Equations 2.54, it is found that this assumption, however, can fail. It depends on n, the number of the local nodes used in the support domain; m, the number of terms in the polynomial basis $p(\mathbf{x})$; and the locations of the local nodes used.

This is because in all these n vectors $\mathbf{p}^\mathrm{T}(\mathbf{x}_I)$, $I = 1, 2, \ldots, n$ in Equations 2.54, there can be less than m independent vectors. To ensure an invertible $\mathbf{A}$, we have to use a much large number of local nodes, compared to the number of the polynomial bases so that $n \gg m$, hoping there are at least m independent vectors in all these n vectors $\mathbf{p}^\mathrm{T}(\mathbf{x}_I)$, $I = 1, 2, \ldots, n$. This is in fact used in the practice of creating MLS shape functions. It works well for usual situations, but it is not a foolproof: For given n and m, one can design extreme cases of node distribution that can make $\mathbf{A}$ singular. One often needs to adjust the dimension of the support domain to make sure that n is large enough.

On the other hand, however, one does not want to just blindly use a very large n for (1) efficiency concerns in creating MLS shape function; (2) concerns on the sparsity/bandwidth of the system matrix to be created later using the MLS shape functions: large n reduces the sparsity and increases the bandwidth which can be a crucial factor affecting the computational efficiency. When a bandwidth solver is used, the CPU time is approximately proportional to n^2. (3) Concerns on the accuracy of the approximation: overly smoothing can occur when too many nodes are used.

To effectively solve this problem and remove entirely the worries of possible invertible $\mathbf{A}$, the T2L-scheme of node selection based on triangular type mesh (see Section 2.5.4) is recommended, and that is implemented in the MFree2D$^{\copyright}$ for MLS shape function construction with an additional purpose of preventing too biased node selections.

In practice, n is often controlled by concerns of the continuity of the shape functions: Using too small n can lead to incompatibility problems when Galerkin weak form is used. Therefore, one prefers to use more nodes to lessen this concern by sacrificing efficiency. There is no proven theoretical guidance on the optimal n, and it is often relied on in numerical tests to determine a "good" n.

Remark 2.2: Property of MLS Shape Functions
Properly constructed MLS shape functions are compatible, consistent with the order of polynomials included in the formulation. They do not have the Kronecker delta function property. They do not form an interpolant.

2.5 Point Interpolation Method

As the name suggests, PIM obtains its approximation by letting the interpolation function pass through the function values at each scattered node within the support domain. It can be categorized as a series representation method. PIM using polynomial basis functions and local scattered nodes was originally attempted in [29,30]. PIM using radial basis functions (RBFs) and local scattered nodes was suggested in [31,41]. The basic procedure for constructing polynomial PIM shape functions is given as follows.

Consider a function $u(\mathbf{x})$ defined in the problem domain Ω with a number of scattered field nodes. For a point of interest $\mathbf{x}_Q$, the field function $u(\mathbf{x})$ is approximated using the following series representation:

$$u^h(\mathbf{x}) = \sum_{i=1}^{n} B_i(\mathbf{x}) a_i \qquad (2.86)$$

where

$B_i(\mathbf{x})$ is the basis function defined in the Cartesian coordinate space $\mathbf{x}^T = \{x, y, z\}$

n is the number of *support nodes* selected in a local support domain

a_i is the coefficient for the basis function $B_i(\mathbf{x})$

When polynomial basis functions are used, we have

$$u^h(\mathbf{x}) = \sum_{i=1}^{n} p_i(\mathbf{x})a_i = \mathbf{p}^T(\mathbf{x})\mathbf{a} \tag{2.87}$$

where

$p_i(\mathbf{x})$ is the basis function of monomials

a_i is the coefficient for the monomial $p_i(\mathbf{x})$

vector $\mathbf{a}$ has the form

$$\mathbf{a} = \{a_1, a_2, \ldots, a_n\}^T \tag{2.88}$$

Note that the a_i are constants in the vicinity of point of interest $\mathbf{x}_Q$, and are updated only when the support nodes associated with $\mathbf{x}_Q$ are changed. Therefore, in any finite discretization of the problem domain with nonduplicated nodes, $u^h(\mathbf{x})$ is consistent in finite local domains where these support nodes do not change. The order of the consistency depends on the polynomial basis functions used.

The monomial $p_i(\mathbf{x})$ in Equation 2.87 is, in general, chosen in a top-down approach from the Pascal triangle shown in Figures 2.6 and 2.7, so that the basis is complete to a desired order. For 1D problems, we use

$$\mathbf{p}^T(x) = \{1, x, x^2, x^3, x^4, \ldots, x^n\} \tag{2.89}$$

For 2D problems we shall have

$$\mathbf{p}^T(\mathbf{x}) = \mathbf{p}^T(x, y) = \{1, x, y, xy, x^2, y^2, \ldots, x^n, y^n\} \tag{2.90}$$

For 3D problems, we may use

$$\mathbf{p}^T(\mathbf{x}) = \mathbf{p}^T(x, y, z) = \{1, x, y, z, xy, yz, zx, xyz, x^2, y^2, z^2, \ldots, x^n, y^n, z^n\} \tag{2.91}$$

The coefficients a_i in Equation 2.87 can be determined by enforcing Equation 2.87 to be satisfied at the n support nodes. At node i we can have

$$u_i = \mathbf{p}^T(\mathbf{x}_i)\mathbf{a} \quad i = 1, 2, \ldots, n \tag{2.92}$$

where u_i is the nodal value of u at $\mathbf{x} = \mathbf{x}_i$. Equation 2.92 can be rewritten in the following matrix form:

$$\mathbf{d}_s = \mathbf{P}_Q\mathbf{a} \tag{2.93}$$

where $\mathbf{d}_s$ is the vector that collects the values of the field function at all the n nodes:

$$\mathbf{d}_s = \{\, u_1 \quad u_2 \quad \cdots \quad u_n \,\}^{\mathrm{T}} \tag{2.94}$$

and $\mathbf{P}_Q$ is called the *moment matrix* given by

$$\mathbf{P}_Q = \begin{bmatrix} \mathbf{p}^{\mathrm{T}}(\mathbf{x}_1) \\ \mathbf{p}^{\mathrm{T}}(\mathbf{x}_2) \\ \vdots \\ \mathbf{p}^{\mathrm{T}}(\mathbf{x}_n) \end{bmatrix} \tag{2.95}$$

or in detail (for 2D cases):

$$\mathbf{P}_Q = \begin{bmatrix} 1 & x_1 & y_1 & x_1 y_1 & x_1^2 & y_1^2 & x_1^2 y_1 & x_1 y_1^2 & x_1^3 & \cdots \\ 1 & x_2 & y_2 & x_2 y_2 & x_2^2 & y_2^2 & x_2^2 y_2 & x_2 y_2^2 & x_2^3 & \cdots \\ \vdots & \vdots & \vdots & \vdots & \vdots & \vdots & \vdots & \vdots & \vdots & \vdots \\ 1 & x_n & y_n & x_n y_n & x_n^2 & y_n^2 & x_n^2 y_n & x_n y_n^2 & x_n^3 & \cdots \end{bmatrix} \tag{2.96}$$

The moment matrix $\mathbf{P}_Q$ is asymmetric. Assuming that the inverse of the moment matrix $\mathbf{P}_Q$ exists, and using Equation 2.93, we can then have

$$\mathbf{a} = \mathbf{P}_Q^{-1} \mathbf{d}_s \tag{2.97}$$

Substituting Equation 2.97 into Equation 2.87, we obtain

$$u^h(\mathbf{x}) = \sum_{i=1}^{n} \phi_i(\mathbf{x}) u_i \tag{2.98}$$

or in matrix form

$$u^h(\mathbf{x}) = \boldsymbol{\varphi}(\mathbf{x}) \mathbf{d}_s \tag{2.99}$$

where $\boldsymbol{\varphi}(\mathbf{x})$ is a matrix of PIM shape functions $\phi_i(\mathbf{x})$ defined by

$$\boldsymbol{\varphi}(\mathbf{x}) = \mathbf{p}^{\mathrm{T}}(\mathbf{x}) \mathbf{P}_Q^{-1} = [\, \phi_1(\mathbf{x}) \quad \phi_2(\mathbf{x}) \quad \cdots \quad \phi_n(\mathbf{x}) \,] \tag{2.100}$$

Note that it is well possible that the moment matrix $\mathbf{P}_Q$ is singular, which leads to a breakdown of the PIM method, which will be discussed in detail in Section 2.5.4. For now, we assume that the moment matrix is invertible.

Note that derivatives of the PIM shape functions can be obtained very easily when needed, as all the functions involved are polynomials. The *l*th derivative of the shape functions are simply given by

$$\boldsymbol{\varphi}_i^{(l)}(\mathbf{x}) = [\mathbf{p}^{(l)}(\mathbf{x})]^{\mathrm{T}} \mathbf{P}_Q^{-1} \tag{2.101}$$

Note also that no weight function is used in constructing PIM shape functions.

2.5.1 Consistency of the PIM Shape Functions

The consistency of the PIM shape function depends on the complete orders of the monomial $p_i(\mathbf{x})$ used in Equation 2.87, and hence also depends on the number of support nodes. If the complete order of the monomial is n, the shape functions will possess C^n consistency. To demonstrate, we consider a field given by

$$f(\mathbf{x}) = \sum_{j}^{k} p_j(\mathbf{x})\alpha_j, \quad k \leq n \tag{2.102}$$

where $p_j(\mathbf{x})$ are monomials that are included in Equation 2.87. Such a field can always be written using Equation 2.87 using all the basis terms including those in Equation 2.102:

$$f(\mathbf{x}) = \sum_{j}^{n} p_j(\mathbf{x})\alpha_j = \mathbf{p}^{\mathrm{T}}(\mathbf{x})\boldsymbol{\alpha} \tag{2.103}$$

where

$$\boldsymbol{\alpha}^{\mathrm{T}} = [\alpha_1, \alpha_2, \ldots \alpha_k, 0, \ldots, 0] \tag{2.104}$$

Using n nodes in a local support domain of $\mathbf{x}$, we can obtain the vector of nodal function values $\mathbf{d}_s$ as

$$\mathbf{d}_s = \left\{ \begin{array}{c} f_1 \\ f_2 \\ \vdots \\ f_k \\ f_{k+1} \\ \vdots \\ f_n \end{array} \right\} = \underbrace{\left[\begin{array}{cccccc} p_1(x_1) & p_2(x_1) & \cdots & p_k(x_1) & p_{k+1}(x_1) & p_n(x_1) \\ p_1(x_2) & p_2(x_2) & \cdots & p_k(x_2) & p_{k+1}(x_2) & p_n(x_2) \\ \vdots & \vdots & \cdots & \vdots & \vdots & \vdots \\ p_1(x_k) & p_2(x_k) & \cdots & p_k(x_k) & p_{k+1}(x_k) & p_n(x_k) \\ p_1(x_{k+1}) & p_2(x_{k+1}) & \cdots & p_k(x_{k+1}) & p_{k+1}(x_{k+1}) & p_n(x_{k+1}) \\ \vdots & \vdots & \cdots & \vdots & \vdots & \vdots \\ p_1(x_n) & p_2(x_n) & \cdots & p_k(x_n) & p_{k+1}(x_n) & p_n(x_n) \end{array} \right]}_{\mathbf{P}_Q} \left\{ \begin{array}{c} \alpha_1 \\ \alpha_2 \\ \vdots \\ \alpha_k \\ 0 \\ \vdots \\ 0 \\ \boldsymbol{\alpha} \end{array} \right\} = \mathbf{P}_Q \boldsymbol{\alpha}$$

$$\tag{2.105}$$

Substituting Equation 2.105 into Equation 2.99, we have the approximation:

$$u^h(\mathbf{x}) = \mathbf{p}^{\mathrm{T}}(\mathbf{x})\mathbf{P}_Q^{-1}\mathbf{d}_s = \mathbf{p}^{\mathrm{T}}(\mathbf{x})\mathbf{P}_Q^{-1}\mathbf{P}_Q\boldsymbol{\alpha} = \mathbf{p}^{\mathrm{T}}(\mathbf{x})\boldsymbol{\alpha} = \sum_{j}^{k} p_j(\mathbf{x})\alpha_j \tag{2.106}$$

which is exactly Equation 2.102. This shows that PIM shape functions is capable of reproducing exactly any field given by Equation 2.102, as long as the given function is included in the basis functions used to construct these PIM shape functions. This feature of PIM shape function is, in fact, very easy to understand by intuition: Any function given in the form of $f(\mathbf{x}) = \sum_j^k p_j(\mathbf{x})\alpha_j$ can be produced exactly by letting $\alpha_j = \alpha_j$ $(j = 1, 2, \ldots, k)$ and $\alpha_j = 0$ $(j = k+1, \ldots, n)$. This can always be done as long as the moment matrix $\mathbf{P}_Q$ is invertible so as to ensure the uniqueness of the solution for $\mathbf{a}$.

The proof of the consistency of PIM is valid for proving another important feature of PIM: that any function that appears in the basis can be reproduced exactly. This property can be useful for creating fields of special features. For PIM to exhibit linear consistency, all one need to do is to include the constant and linear monomials into the basis. This feature of PIM can be used to compute accurate results for problems by including terms in the basis of PIM that are good approximations of the solution of the problem.

2.5.2 Properties of the PIM Shape Functions

As long as the moment matrix is invertible, the PIM shape functions $\phi_i(\mathbf{x})$ can be uniquely constructed and possess the following characteristics:

1. The shape functions are linearly independent. This is because the polynomial basis functions are linearly independent and $\mathbf{P}_Q^{-1}$ is assumed to exist. The existence of $\mathbf{P}_Q^{-1}$ implies that the shape functions are equivalent to the basis functions (monomials) in function space, as shown in Equation 2.100, and hence are linearly independent.

2. The shape functions possess the Kronecker delta function property, that is,

$$\phi_i(\mathbf{x} = \mathbf{x}_j) = \begin{cases} 1 & i = j, \quad j = 1, 2, \ldots, n \\ 0 & i \neq j, \quad i, j = 1, 2, \ldots, n \end{cases} \tag{2.107}$$

This can be proved easily as follows. Because the PIM shape functions $\phi_i(\mathbf{x})$ are linearly independent, any vector of length n should be uniquely produced by linear combination of these n shape functions. Letting

$$\mathbf{d}_s = \{0, 0, \ldots, u_i, \ldots, 0\}^T \tag{2.108}$$

and substituting the above equation into Equation 2.98, we have at $\mathbf{x} = \mathbf{x}_j$

$$u^h(\mathbf{x}_j) = \sum_1^n \phi_i(\mathbf{x}_j)u_i = \phi_i(\mathbf{x}_j)u_i \tag{2.109}$$

When $i = j$, we obtain

$$u_i = \phi_i(\mathbf{x}_i)u_i \tag{2.110}$$

which leads to

$$\phi_i(\mathbf{x}_i) = 1 \tag{2.111}$$

This proves the first row of Equation 2.107. When $i \neq j$, we have

$$u_j = 0 = \phi_i(\mathbf{x}_j)u_i \tag{2.112}$$

which requires

$$\phi_i(\mathbf{x}_j) = 0 \tag{2.113}$$

This proves that PIM shape functions possess the Kronecker delta function property (Equation 2.107).

3. The shape functions are the partitions of unity

$$\sum_{i=1}^{n} \phi_i(\mathbf{x}) = 1 \tag{2.114}$$

if the constant is included in the basis. This can be proven easily from the reproduction feature of PIM. Letting $u(\mathbf{x}) = c$, where c is a constant, we should have

$$\mathbf{d}_s = c\{1, 1, \ldots, 1\}^{\mathrm{T}} \tag{2.115}$$

Substituting the above equation into Equation 2.98, we obtain

$$u(\mathbf{x}) = c = \sum_{1}^{n} \phi_i(\mathbf{x})u_i = c\sum_{1}^{n} \phi_i(\mathbf{x}) \tag{2.116}$$

which gives Equation 2.114. This shows that the partition of unity of PIM shape functions in the support domain allows a constant field or rigid body movement to be reproduced. Note that Equation 2.114 does not require $0 \leq \phi_i(\mathbf{x}) \leq 1$.

4. The shape functions possess linear reproducing property

$$\sum_{i=1}^{n} \phi_i(\mathbf{x})x_i = \mathbf{x} \tag{2.117}$$

if the first-order monomial is included in the basis. This can be proven easily from the reproduction feature of PIM in exactly the same manner used for proving property 3. Letting $u(\mathbf{x}) = \mathbf{x}$, we should have

$$\mathbf{d}_s = \{\mathbf{x}_1, \mathbf{x}_2, \ldots, \mathbf{x}_n\}^{\mathrm{T}} \tag{2.118}$$

Substituting the above equation into Equation 2.98, we obtain

$$u(\mathbf{x}_j) = \mathbf{x} = \sum_{1}^{n} \phi_i(\mathbf{x}_j)x_i \tag{2.119}$$

which is Equation 2.117.

5. The shape functions are of compact support as long as they are constructed using the nodes in a compact support domain.

6. There is no need for weight functions in constructing PIM shape functions, or the weight function used is a unit (Heaviside type).

7. PIM shape function is not compatible. This is because the bell-shape weight function is not used in constructing the PIM shape function, and the number of nodes is the same as the terms of monomials. The PIM shape function for a node changes suddenly when x_Q moves to a point where the support domain updates its nodes, resulting in discontinuity at that point. Functions created using these PIM shape functions can be discontinuous, and hence lives in a $\mathbb{G}_h^1$ space (see Chapter 3).

8. PIM shape functions can be used to construct an interpolant for functions in $\mathbb{G}_h^1$ spaces. However, it may not be used to construct function in an $\mathbb{H}^1$ space, because $\mathcal{I}_h w$ for a w in an $\mathbb{H}^1$ space may not still be in any of the $\mathbb{H}^1$ subspace due to the incompatibility (see Section 2.1.7 for interpolant and Chapter 3 for space definitions). For linear interpolations using T3-scheme, PIM shape functions can always be created as an interpolant in an $\mathbb{H}^1$ space. For arbitrary polygonal mesh, we can use simple point interpolation techniques given in Refs. [55,56].

Property 2 is important for handling essential boundary conditions. Properties 3 and 4 are essential for a PIM method to pass the standard patch test, which is a conventional test used for decades in FEM for validation of elements. Property 5 leads to sparse/banded discretized system matrices. Property 6 eliminates the question of how to choose a weight function in constructing shape functions. Property 7 implies that a SC-Galerkin weak form (see Chapter 5) should be used for deriving discrete system equations.

The PIM shape function is ideal for meshfree methods in many ways, as it possesses the above excellent properties. Polynomial PIM formulation is the simplest and performs the best. Figure 2.10a shows a PIM shape function in 1D space for a node at $x = 0$ obtained using five nodes evenly distributed in the support domain of $[-1, 1]$. It is clearly seen that the PIM shape function possesses the Kronecker delta function property. That is, $\phi(0.0) = 1.0$, $\phi(-1.0) = \phi(-0.5) = \phi(0.5) = \phi(1.0) = 0.0$. Figure 2.10b plots the first derivative of the shape function.

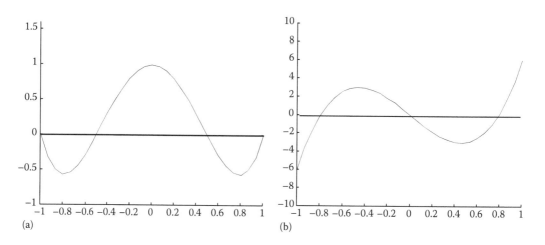

(a)   (b)

FIGURE 2.10
Polynomial PIM shape function in 1D space for the node at $x = 0$ obtained using five nodes evenly distributed in the support domain of $[-1, 1]$. (a) Shape function; (b) derivative of the shape function. Note that the PIM shape function possesses the Kronecker delta function property.

TABLE 2.1

Comparisons between PIM Interpolation and MLS Approximation

	Basis Function	Number of Basis Functions (m) and Number of Nodes (n)	Interpolation Coefficients	Delta Function Property	Compatibility
Point interpolation	Polynomial or radial functions	$m = n$	Constant locally	Yes	No
MLS approximation	Polynomial	$m \neq n$	Function	No	Yes

2.5.3 Differences between the PIM Interpolation and the MLS Approximation

PIM interpolation and MLS approximation are compared in Table 2.1. As the table shows, the main difference between PIM and MLS approximation is that the number of polynomial terms used in PIM is the same as the number of the nodes used in the support domain. The coefficients in PIM are constants, and those in MLS are functions of the coordinates. Most importantly, the PIM shape functions possess the Kronecker delta function property. The Kronecker delta function property allows essential boundary conditions to be easily treated in the same way as in the standard FEM. However, PIM interpolation is not, in general, compatible, while MLS approximation is compatible when sufficient nodes are used. Therefore, PIM shape functions should work with an SC-Galerkin weak form, while the MLS shape functions work with the Galerkin weak form more efficiently. When $m = n$ in MLS approximation, the MLS shape functions become the PIM shape functions: implying trying to reduce the number of local support nodes in a standard Galerkin formulation using MLS shape functions can lead to an incompatible model.

2.5.4 Methods to Avoid a Singular Moment Matrix

As shown above, PIM shape functions possess many excellent properties that are very useful for meshfree methods based on SC-Galerkin weak forms, local weak forms, and strong forms (with a proper regularization technique). However, the process of constructing PIM shape functions can break down as a result of the singularity of the moment matrix $\mathbf{P}_Q$. Figure 2.11 shows a typical example of six nodes in the support domain of a point of interest $\mathbf{x}_Q$. These six nodes sit in two lines parallel to the x axis. When these six nodes are used, the polynomial basis can be of complete second order with respect to both the x and y coordinates:

$$\mathbf{p}^T(\mathbf{x}) = \{1, x, y, xy, x^2, y^2\} \tag{2.120}$$

However, these six nodes, as shown in Figure 2.11a, cannot possibly represent a second-order polynomial in the y direction, as there are only two distinct y coordinate values in all these six nodes. Therefore, the inverse of the moment matrix $\mathbf{P}_Q$ using these six nodes clearly does not exist: it has only a rank of 5.

After selection of nodes and basis functions, the matrix $\mathbf{P}_Q$ is completely determined by the position of the scattered nodes in a given coordinate system; i.e., the existence of the inverse matrix $\mathbf{P}_Q^{-1}$ depends on the node distribution as well as on the coordinate system. Using polynomial basis functions is known to be difficult to guarantee the existence of $\mathbf{P}_Q^{-1}$

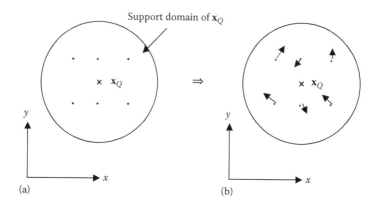

FIGURE 2.11

Node distribution in a domain of support of point x_Q. (a) Six nodes in two parallel lines that lead to a singular moment matrix. (b) Moving nodes by a small distance randomly results in a nonsingular moment matrix.

for a set of arbitrarily scattered nodes. However, the excellent properties of PIM shape functions warrant the effort needed to overcome the singular moment matrix problem. A number of methods for handling the singular moment matrix have been attempted for locally scattered nodes. These methods can be used to obtain an invertible moment matrix and are briefed as follows:

1. The simplest method proposed to obtain a nonsingular moment matrix is to move or shift the nodes in the support domain by a small distance randomly in terms of both direction and the amount of shift, as shown in Figure 2.11b. The method is simple and effective for many problems. However, there is still a chance that the moment matrix will be singular, which may sometimes lead to a badly conditioned $\mathbf{P}_Q$. In addition, there are cases in which we are not allowed to move the nodes.

2. Performing rotational coordinate transformation to produce an invertible moment matrix $\mathbf{P}_Q$. This method makes the use of the fact that the rank of $\mathbf{P}_Q$ depends also on the coordinate system. It is not a full-proof, but has some practical applications. It is quite effective, when a small number of nodes are used. The details are given in Section 2.9.

3. The use of RBFs in constructing PIM shape functions is a method that always works and that guarantees the existence of the inverse of the moment matrix, if the guidelines for choosing shape parameters of the RBFs are followed. The drawback is that it is more expensive as more nodes are required to obtain accurate results comparable with those of polynomial PIM. The details are given in Section 2.6.

4. The use of RBFs with polynomial terms is an approach that restores the polynomial consistency and improves the accuracy of the RPIM, especially for patch tests. It can also reduce the sensitivity of the results on the shape parameters of RBFs. The computational cost is still more expensive as compared with polynomial PIM. The details are given in Section 2.7.

5. The matrix triangularization method is very efficient and works well for many situations, when removal of nodes in the local support domain is permitted. It opens a new window of opportunity to effectively solve the singularity problem of the moment matrix. Details can be found in [47].

6. Triangular-mesh-based node selection scheme (T-scheme). The T-scheme uses a triangular type background mesh to select nodes (see Chapter 1). For 2D domains, we use triangular mesh, and for 3D domains, we use tetrahedron mesh. The T-scheme is practically effective and reliable and is expected to become the major scheme for node selection for meshfree methods for practical problems of complicated geometry with very irregularly distributed nodes. This is because (1) a triangular type background mesh can always be generated easily and automatically for any given set of scattered nodes in 2D and 3D domains; (2) based on a triangular background mesh, some simple node selection schemes can be easily devised to ensure a nonsingular moment matrix; (3) nodes so selected ensure a reasonably good pattern of nodes participation from all directions for even very irregularly distributional nodes (for which the usual selection schemes based on the support domain or influence domains can fail), and hence is most reliable; (4) the number of nodes is very small (much smaller than that using the support domain), and hence very efficient, which is vital for an eventual survival of a numerical method. The T-scheme was used for node selection in the MFree2D code that was launched in 1999 for both EFG and PIMs processors. For all the PIMs discussed in this book, our default choice of node selection is T-schemes, unless specified otherwise.

All the above methods have ample room for improvement, and other better alternatives are also possible.

Remark 2.3: Property of PIM Shape Functions
Successfully constructed PIM shape functions are, in general, incompatible, consistent with the order of polynomials included in the formulation. They have the Kronecker delta function property.

2.6 Radial PIM

2.6.1 Rationale for Radial Basis Functions

The advantage of using a polynomial basis is its simplicity and high accuracy, as will be evident in later chapters. The major drawback of polynomial PIM is that singular moment matrix P_Q may occur and the process breaks down. To create a nonsingular moment matrix, RBFs are introduced in PIM formulation for constructing shape functions using local nodes [31,41]. PIM using RBF is termed RPIM.

2.6.2 PIM Formation Using Radial Basis Functions

In RPIM, we choose radial functions as the basis in Equation 2.86, we thus have

$$u^h(\mathbf{x}) = \sum_{i=1}^{n} R_i(\mathbf{x})a_i = \mathbf{R}^T(\mathbf{x})\mathbf{a} \qquad (2.121)$$

where vector **a** is a vector of unknown constants arranged in the form of Equation 2.88, and R_i is a RBF with r being the distance between point **x** and $\mathbf{x}_i$ defined as

$$r = \begin{cases} \left[(x - x_i)^2 + (y - y_i)^2\right]^{1/2}, & \text{for 2D} \\ \left[(x - x_i)^2 + (y - y_i)^2 + (z - z_i)^2\right]^{1/2}, & \text{for 3D} \end{cases} \tag{2.122}$$

The vector **R** has the form

$$\mathbf{R}^{\mathrm{T}}(\mathbf{x}) = [R_1(\mathbf{x}), R_2(\mathbf{x}), \ldots, R_n(\mathbf{x})] \tag{2.123}$$

There are a number of forms of RBFs used in the mathematics community. Table 2.2 lists the four most often used forms of radial functions with some shape parameters that can be tuned for better performance. A classical form is the multiquadric (MQ) basis proposed by Hardy [32]. This form has been widely used in surface fitting and in constructing approximate solutions for PDEs [33–39]. The MQ basis function shown in Table 2.2 is a general form of the original MQ RBF with arbitrary real shape parameters that was suggested by [31,40,41]. When $q = \pm 0.5$, it reduces to the original MQ RBF proposed by Hardy. When $q = 0.5$, it reduces to the reciprocal MQ RBF. The general form of the MQ radial function has two shape parameters, C and q, which control the shape of the functions. These parameters can be tuned for different problems for better performance. The second form of radial function given in Table 2.2 is called the Gaussian radial function, or EXP, as it is an exponential function of the distance [35]. The EXP RBF has only one shape parameter c, which controls the decay rate of the function. The third radial function in Table 2.2 is called the thin plate spline (TPS) function. The TPS is, in fact, a special case of the MQ radial function. The fourth form of RBF is the logarithmic RBF. This book will use and test the first three forms of radial functions, but will focus more on the first two forms of radial functions (MQ and EXP) and will give preference to the general form of the MQ RBF for reasons to be given in later chapters.

The vectors of coefficients **a** in Equation 2.121 are determined by enforcing interpolation passing through all the n local support nodes selected by means of support domain or a T-scheme. The interpolation at the kth point has the form:

$$u_k = u(x_k, y_k) = \sum_{i=1}^{n} a_i R_i(x_k, y_k) \quad k = 1, 2, \ldots, n \tag{2.124}$$

These n equations given in Equation 2.124 can be written in matrix form:

$$\mathbf{d}_s = \mathbf{R}_Q \mathbf{a} \tag{2.125}$$

TABLE 2.2

Typical RBFs

Item	Name	Expression	Shape Parameters
1	MQ	$R_i(x, y) = (r_i^2 + C^2)^q$	C, q
2	Gaussian (EXP)	$R_i(x, y) = \exp(-c r_i^2)$	c
3	TPS	$R_i(x, y) = r_i^\eta$	η
4	Logarithmic RBF	$R_i(r_i) = r_i^\eta \log r_i$	η

where $\mathbf{d}_s$ is the vector that collects all the field nodal variables at the n local nodes and $\mathbf{R}_Q$ is the moment matrix of RBF:

$$\mathbf{R}_Q = \begin{bmatrix} R_1(r_1) & R_2(r_1) & \cdots & R_n(r_1) \\ R_1(r_2) & R_2(r_2) & \cdots & R_n(r_2) \\ \vdots & \vdots & \ddots & \vdots \\ R_1(r_n) & R_2(r_n) & \cdots & R_n(r_n) \end{bmatrix} \qquad (2.126)$$

where

$$r_k = \begin{cases} \left[(x_k - x_i)^2 + (y_k - y_i)^2\right]^{1/2}, & \text{for 2D} \\ \left[(x_k - x_i)^2 + (y_k - y_i)^2 + (z_k - z_i)^2\right]^{1/2}, & \text{for 3D} \end{cases} \qquad (2.127)$$

Because the distance is directionless, we should have

$$R_i(r_j) = R_j(r_i) \qquad (2.128)$$

Therefore, the moment matrix $\mathbf{R}_Q$ is symmetric. This symmetry property of $\mathbf{R}_Q$ hints that $\mathbf{R}_Q$ will likely be SPD, and hence invertible: It is indeed proven true [35,39,42].* A unique solution for vectors of coefficients $\mathbf{a}$ can then be obtained if the inverse of $\mathbf{R}_Q$ exists:

$$\mathbf{a} = \mathbf{R}_Q^{-1}\mathbf{d}_s \qquad (2.129)$$

Substituting the foregoing equation into Equation 2.121 leads to

$$u^h(\mathbf{x}) = \mathbf{R}^{\mathrm{T}}(\mathbf{x})\mathbf{R}_Q^{-1}\mathbf{d}_s = \boldsymbol{\varphi}(\mathbf{x})\mathbf{d}_s \qquad (2.130)$$

where the matrix of shape functions has the form

$$\begin{aligned} \boldsymbol{\varphi}(\mathbf{x}) &= [R_1(\mathbf{x}), R_2(\mathbf{x}), \dots, R_k(\mathbf{x}), \dots, R_n(\mathbf{x})]\mathbf{R}_Q^{-1} \\ &= [\phi_1(\mathbf{x}), \phi_2(\mathbf{x}), \dots, \phi_k(\mathbf{x}), \dots, \phi_n(\mathbf{x})] \end{aligned} \qquad (2.131)$$

in which $\phi_k(\mathbf{x})$ is the shape function for the kth node given by

$$\phi_k(\mathbf{x}) = \sum_{i=1}^{n} R_i(\mathbf{x})S_{ik}^a \qquad (2.132)$$

where S_{ik}^a is the (i, k) element of matrix $\mathbf{R}_Q^{-1}$, which is a constant matrix for given locations of the n nodes in the support domain.

---

* Following this argument, any distance function can be hopefully used in lieu of the RBFs for scattered point interpolation.

The derivatives of shape functions, when needed, can be easily obtained as

$$\frac{\partial \phi_k}{\partial x} = \sum_{i=1}^{n} \frac{\partial R_i}{\partial x} S_{ik}^a$$
$$\frac{\partial \phi_k}{\partial y} = \sum_{i=1}^{n} \frac{\partial R_i}{\partial y} S_{ik}^a$$

(2.133)

For the MQ basis function shown in Table 2.2, the partial derivatives for the MQ radial functions can be easily obtained using the following simple formulae:

$$\frac{\partial R_i}{\partial x} = 2q\left(r_i^2 + C^2\right)^{q-1}(x - x_i)$$
$$\frac{\partial R_i}{\partial y} = 2q\left(r_i^2 + C^2\right)^{q-1}(y - y_i)$$

(2.134)

For the EXP radial function, the partial derivatives can also be obtained easily as follows:

$$\frac{\partial R_i}{\partial x} = -2cR_i(x, y)(x - x_i)$$
$$\frac{\partial R_i}{\partial y} = -2cR_i(x, y)(y - y_i)$$

(2.135)

2.6.3 Nonsingular Moment Matrix

The only difference between polynomial PIM and RPIM is in the basis functions. Mathematicians have proved that the radial moment matrix $\mathbf{R}_Q$ is always invertible for arbitrary scattered nodes [35,39,42], as long as we avoid using some specific shape parameters, which are known. Therefore, $\mathbf{R}_Q$ can always be symmetric and invertible. The existence of $\mathbf{R}_Q^{-1}$ is the major advantage of using the radial basis over the polynomial basis.

Note that although $\mathbf{R}_Q$ is invertible, but often found bad-conditioned, when too many nodes are used. Fortunately, in creating RPIM shape functions for meshfree method using compact local support domains, only a few local nodes are used that is much smaller compared with those used by [33,34] where all the nodes in the problem domain are used. Therefore, the conditioning in $\mathbf{R}_Q$ in creating RPIM shape functions is much better for the same shape parameters used.

2.6.4 Consistency of RPIM Shape Functions

The RPIM shape function is not consistent with the definition of consistency in this book that is the capability of reproducing polynomials. Mathematicians have found that approximations of any continuous function using RBFs converge. Thus, there is no concern about the convergence issue.

It was found that the use of pure radial functions in the basis of PIM will not pass the standard patch test [41], which has been widely used in the FEM community for testing the performance of finite elements. This is because the radial function cannot produce the linear polynomials exactly, although it can approach polynomials in desired accuracy when the nodes are refined. The consistency of the radial shape functions can be restored by adding polynomial basis functions. For RPIM to pass the patch test, Wang and Liu [41] suggested using radial functions with polynomial terms of up to linear orders so as to construct shape functions with C^1 consistency. Adding polynomial terms to RBFs was also proposed by Powell [35] for function approximation.

2.6.5 Radial Functions with Dimensionless Shape Parameters

The conventional forms of radial functions listed in Table 2.2 have been used by many researchers including the research group of the author. We found that it is very difficult to standardize the shape parameters of the RBFs. We therefore proposed a set of RBFs that has dimensionless parameters by performing some minor modification. Some of the new forms of RBFs are listed in Table 2.3.

The MQ function with dimensionless shape parameters has the form:

$$R_i(x, y) = \left(r_i^2 + (\alpha_C d_c)^2\right)^q \quad \alpha_C \geq 0 \tag{2.136}$$

where
α_C is the dimensionless shape parameter
d_c is the characteristic length that is usually the average nodal spacing for all the n nodes in the support domain

The first- and second-order partial derivatives are obtained as follows:

$$\frac{\partial R_i}{\partial x} = 2q\left(r_i^2 + (\alpha_C d_c)^2\right)^{q-1}(x - x_i) \tag{2.137}$$

$$\frac{\partial R_i}{\partial y} = 2q\left(r_i^2 + (\alpha_C d_c)^2\right)^{q-1}(y - y_i) \tag{2.138}$$

$$R_{i,xx} = 2q\left[r_i^2 + (\alpha_C d_c)^2\right]^{q-1} + 4q(q-1)\left[r_i^2 + (\alpha_C d_c)^2\right]^{q-2}(x - x_i)^2 \tag{2.139}$$

$$R_{i,xy} = 4q(q-1)\left(r_i^2 + (\alpha_C d_c)^2\right)^{q-2}(x - x_i)(y - y_i) \tag{2.140}$$

$$R_{i,yy} = 2q\left[r_i^2 + (\alpha_C d_c)^2\right]^{q-1} + 4q(q-1)\left[r_i^2 + (\alpha_C d_c)^2\right]^{q-2}(y - y_i)^2 \tag{2.141}$$

TABLE 2.3

RBFs with Dimensionless Shape Parameters

Item	Name	Expression[a]	Shape Parameters	Parameter Relations[b]
1	MQs	$R_i(x, y) = \left(r_i^2 + (\alpha_C d_c)^2\right)^q$	$\alpha_C \geq 0, q$	$\alpha_C = C/d_c$ $q = q$
2	Gaussian (EXP)	$R_i(x, y) = \exp\left[-\alpha_c\left(\frac{r_i}{d_c}\right)^2\right]$	α_c	$\alpha_c = c\, d_c$

[a] d_c is a characteristic length that is related to the nodal spacing in the local domain of the point of interest x_Q. d_c is usually the average nodal spacing for all the nodes in the local domain (see Chapter 1 for a method to calculate d_c).

[b] This column gives the relationship between the original parameters and the dimensionless parameters.

The EXP radial function with dimensionless shape parameters can be written as

$$R_i(x, y) = \exp\left(-\alpha_c \left(\frac{r_i}{d_c}\right)^2\right)$$

(2.142)

where α_C is the dimensionless shape parameter. The first- and second-order partial derivatives of the EXP RBFs are obtained as follows:

$$\frac{\partial R_i}{\partial x} = -\frac{2\alpha_c}{d_c^2} R_i(x, y)(x - x_i)$$

(2.143)

$$\frac{\partial R_i}{\partial y} = -\frac{2\alpha_c}{d_c^2} R_i(x, y)(y - y_i)$$

(2.144)

$$R_{i,xx} = \left[-2\left(\frac{\alpha_c}{d_c^2}\right) + 4\left(\frac{\alpha_c}{d_c^2}\right)^2 (x - x_i)^2\right] R_i(x, y)$$

(2.145)

$$R_{i,xy} = 4\left(\frac{\alpha_c}{d_c^2}\right)^2 R_i(x, y)(x - x_i)(y - y_i)$$

(2.146)

$$R_{i,yy} = \left[-2\left(\frac{\alpha_c}{d_c^2}\right) + 4\left(\frac{\alpha_c}{d_c^2}\right)^2 (y - y_i)^2\right] R_i(x, y)$$

(2.147)

Both sets of RBRs listed in Tables 2.2 and 2.3 are used in the example problems in this chapter and later chapters. It is found that the RBFs with dimensionless shape parameters are much easier to use, because good shape parameters found are generally applicable.

2.6.6 On the Range of the Shape Parameters

Note that the shape parameters in RBFs used in a collocation method using all the nodes in the problem domain for solving PDEs (e.g., [33,34]) are usually fixed at discrete values. For example, in the MQ RBF, $q = \pm 0.5$. This maybe because the RBFs with fixed values come from the fundamental solutions of typical PDEs, and hence when (global) collocation methods are used it can produce accurate solution and the convergence property can be proven.

In our local RPIM shape functions, however, the RBFs are merely used as a basis function just like monomials used in the PIM shape functions. These shape functions will be used only as a means of interpolation for field function approximation in the weak and weakened-weak formulations. Whether or not the RBF satisfies a particular PDE is immaterial. This is because (1) we use only local nodes and hence the RBFs with fixed shape parameters may not be able to satisfy the PDE anyway and there is no benefit of accuracy improvement and (2) the stability and convergence of the weak form meshfree methods using local RBFs are controlled also by the weak or weakened-weak forms used. Therefore, we decided to allow the shape parameters to change freely as a real number [41]. The shape parameters can then be tuned to control the incompatibility for better accuracy and performance when a Galerkin weak form is used. Careful investigations on the effects of these shape parameters on the solution accuracy have been conducted and proper guidelines for use of shape parameters for different types of problems have been provided in [41] for Galerkin weak formulations. The reliance of accuracy on the shape parameters can be significantly reduced by adding polynomial basis functions, which are discussed in Section 2.7. Such a reliance is also significantly reduced when a weakened-weak form such as the smoothed Galerkin weak form is applied [43,44]. The major reason for a weak formulation to use RBFs is the invertible moment matrix so that the shape functions can always be constructed.

Remark 2.4: Property of RPIM Shape Functions
Successfully constructed RPIM shape functions are, in general, incompatible, not consist ent, and capable of producing RBFs used in the formulation. They have the Kronecker delta function property.

2.7 Radial PIM with Polynomial Reproduction

2.7.1 Rationale for Polynomials

RPIM with pure radial functions is not (polynomial) consistent and has a problem passing the standard patch tests, meaning that it fails to reconstruct the linear (polynomial) field exactly. The purpose of adding polynomials into the basis functions is to restore the consistency of RPIM shape functions. Adding polynomial terms up to the linear order can ensure the reproduction of the linear field (C^1 consistency) and hence help to pass the standard patch tests. This was our original motivation for adding polynomials to the radial basis for solving solid mechanics problems using local RBFs. Our study later found that, in general, adding polynomials can also improve the accuracy of the results. Another add-itional bonus of this formulation is that we have much more freedom in choosing shape parameters, because the sensitivity of the shape parameters in RBFs on the solution accuracy is reduced.

2.7.2 Formulation of Polynomial Augmented RPIM

By using the n local nodes, RPIM with polynomial basis functions approximates the field variable in the form:

$$u^h(\mathbf{x}) = \sum_{i=1}^{n} R_i(\mathbf{x})a_i + \sum_{j=1}^{m} p_i(\mathbf{x})b_j = \mathbf{R}^{\mathrm{T}}(\mathbf{x})\mathbf{a} + \mathbf{p}^{\mathrm{T}}(\mathbf{x})\mathbf{b} \qquad (2.148)$$

where
  a_i is the coefficient for the radial basis $R_i(\mathbf{x})$ that is listed in Table 2.2
  b_j is the coefficient for the polynomial basis $p_j(\mathbf{x})$ that has the same form as the basis used in polynomial PIM

The number of RBFs n is determined by the number of the nodes in the support domain, and the number of polynomial basis m can be chosen based on the reproduction requirement. We often use a minimum number of terms of polynomial basis and more terms of radial basis ($m < n$) for better stability with respect to the irregularity of node distribution. To pass the patch test for 2D cases, one needs only three terms of polynomial basis.

The vector $\mathbf{a}$ in Equation 2.148 is defined as

$$\mathbf{a}^{\mathrm{T}} = \{a_1 \quad a_2 \quad \cdots \quad a_n\} \tag{2.149}$$

and the vector $\mathbf{b}$ is defined as

$$\mathbf{b}^{\mathrm{T}} = \{b_1 \quad b_2 \quad \cdots \quad b_m\} \tag{2.150}$$

The radial basis vector $\mathbf{R}$ in Equation 2.148 is defined as

$$\mathbf{R}^{\mathrm{T}}(\mathbf{x}) = [R_1(\mathbf{x}), R_2(\mathbf{x}), \ldots, R_n(\mathbf{x})] \tag{2.151}$$

and the polynomial basis vector is written as

$$\mathbf{p}^{\mathrm{T}}(\mathbf{x}) = [p_1(\mathbf{x}), p_2(\mathbf{x}), \ldots, p_m(\mathbf{x})] \tag{2.152}$$

The coefficients a_i and b_j in Equation 2.148 are determined by enforcing that the interpolation passes through all n local nodes. The interpolation at the kth point has the form:

$$u_k = u(x_k, y_k) = \sum_{i=1}^{n} a_i R_i(x_k, y_k) + \sum_{j=1}^{m} b_j p_j(x_k, y_k), \quad k = 1, 2, \ldots, n \tag{2.153}$$

or in matrix form:

$$\mathbf{d}_s = \mathbf{R}_Q \mathbf{a} + \mathbf{P}_m \mathbf{b} \tag{2.154}$$

where $\mathbf{d}_s$ is the vector that collects all the field nodal variables at the n local nodes. The polynomial term has to satisfy an extra requirement that guarantees unique approximation [45] of a function, and the following constraints are usually imposed:

$$\sum_{i=1}^{n} p_j(x_i, y_i) a_i = 0 \quad j = 1, 2, \ldots, m \tag{2.155}$$

or in matrix form:

$$\mathbf{P}_m^{\mathrm{T}} \mathbf{a} = \mathbf{0} \tag{2.156}$$

which is a set of homogeneous equations. Combination of Equations 2.154 and 2.156 gives

$$\begin{bmatrix} \mathbf{R}_Q & \mathbf{P}_m \\ \mathbf{P}_m^{\mathrm{T}} & \mathbf{0} \end{bmatrix} \begin{Bmatrix} \mathbf{a} \\ \mathbf{b} \end{Bmatrix} = \begin{Bmatrix} \mathbf{d}_s \\ \mathbf{0} \end{Bmatrix} \tag{2.157}$$

or

$$G\left\{\begin{array}{c} \mathbf{a} \\ \mathbf{b} \end{array}\right\} = \left\{\begin{array}{c} \mathbf{d}_s \\ 0 \end{array}\right\} \qquad (2.158)$$

The moment matrix corresponding to the radial function basis $\mathbf{R}_Q$ has been given by Equation 2.126, and the moment matrix $\mathbf{P}_m$ is an $n \times m$ matrix given by

$$\mathbf{P}_m = \begin{bmatrix} P_1(x_1, y_1) & P_2(x_1, y_1) & \cdots & P_m(x_1, y_1) \\ P_1(x_2, y_2) & P_2(x_2, y_2) & \cdots & P_m(x_2, y_2) \\ \vdots & \vdots & \vdots & \vdots \\ P_1(x_n, y_n) & P_2(x_n, y_n) & \cdots & P_m(x_n, y_n) \end{bmatrix}_{n \times m} \qquad (2.159)$$

Because matrix $\mathbf{R}_Q$ is symmetric, matrix $\mathbf{G}$ will also be symmetric. A unique solution for vectors of coefficients $\mathbf{a}$ and $\mathbf{b}$ is obtained if the inverse of $\mathbf{G}$ exists:

$$\left\{\begin{array}{c} \mathbf{a} \\ \mathbf{b} \end{array}\right\} = \mathbf{G}^{-1}\left\{\begin{array}{c} \mathbf{d}_s \\ 0 \end{array}\right\} \qquad (2.160)$$

We choose not to directly prove the existence of the inverse of $\mathbf{G}$. Instead, we will try to change the equations in a more efficient form, and then take up the existence issue.

Making use of the fact that Equation 2.156 is homogeneous, Equation 2.157 can be solved in the following more efficient procedure. Starting from Equations 2.154, and using the nonsingular property of matrix $\mathbf{R}_Q$, we have

$$\mathbf{a} = \mathbf{R}_Q^{-1}\mathbf{U}_s - \mathbf{R}_Q^{-1}\mathbf{P}_m\mathbf{b} \qquad (2.161)$$

Substitution of the above expression into Equation 2.156 gives

$$\mathbf{b} = \mathbf{S}_b\mathbf{d}_s \qquad (2.162)$$

where

$$\mathbf{S}_b = \left[\mathbf{P}_m^{\mathrm{T}}\mathbf{R}_Q^{-1}\mathbf{P}_m\right]^{-1}\mathbf{P}_m^{\mathrm{T}}\mathbf{R}_Q^{-1} \qquad (2.163)$$

where $\mathbf{P}_m^{\mathrm{T}}\mathbf{R}_Q^{-1}\mathbf{P}_m$ is termed a transformed moment matrix. Note also that $\mathbf{P}_m^{\mathrm{T}}\mathbf{R}_Q^{-1}$ needs to be computed only once. Substituting Equation 2.162 back into Equation 2.161, we obtain

$$\mathbf{a} = \mathbf{S}_a\mathbf{d}_s \qquad (2.164)$$

where

$$\mathbf{S}_a = \mathbf{R}_Q^{-1}[1 - \mathbf{P}_m\mathbf{S}_b] = \mathbf{R}_Q^{-1} - \mathbf{R}_Q^{-1}\mathbf{P}_m\mathbf{S}_b \qquad (2.165)$$

Note $\mathbf{R}_Q^{-1}\mathbf{P}_m$ can be obtained simply by transposing $\mathbf{P}_m^{\mathrm{T}}\mathbf{R}_Q^{-1}$, which has been already computed.

Finally, the interpolation (Equation 2.148) can be expressed as

$$u(\mathbf{x}) = \left[\mathbf{R}^{\mathrm{T}}(\mathbf{x})\mathbf{S}_a + \mathbf{p}^{\mathrm{T}}(\mathbf{x})\mathbf{S}_b\right]\mathbf{d}_s = \boldsymbol{\varphi}(\mathbf{x})\mathbf{d}_s \qquad (2.166)$$

where the matrix of shape functions $\boldsymbol{\Phi}(\mathbf{x})$ with n shape functions:

$$\boldsymbol{\varphi}(\mathbf{x}) = \left[\mathbf{R}^{\mathrm{T}}(\mathbf{x})\mathbf{S}_a + \mathbf{p}^{\mathrm{T}}(\mathbf{x})\mathbf{S}_b\right] = \left[\phi_1(\mathbf{x}), \quad \phi_2(\mathbf{x}), \quad \cdots \quad \phi_k(\mathbf{x}), \quad \cdots \quad \phi_n(\mathbf{x})\right] \qquad (2.167)$$

in which $\phi_k(\mathbf{x})$ is the shape function for the ith node given by

$$\phi_k(\mathbf{x}) = \sum_{i=1}^{n} R_i(\mathbf{x})S_{ik}^a + \sum_{j=1}^{m} p_j(\mathbf{x})S_{jk}^b \qquad (2.168)$$

where
S_{ik}^a is the (i, k) element of matrix $\mathbf{S}_a$
S_{jk}^b is the (j, k) element of matrix $\mathbf{S}_b$, which are constant matrices for given locations of the
  n nodes in the support domain

The derivatives of shape functions can be easily obtained as

$$\frac{\partial \phi_k}{\partial x} = \sum_{i=1}^{n} \frac{\partial R_i}{\partial x}S_{ik}^a + \sum_{j=1}^{m} \frac{\partial p_j}{\partial x}S_{jk}^b$$

$$\frac{\partial \phi_k}{\partial y} = \sum_{i=1}^{n} \frac{\partial R_i}{\partial y}S_{ik}^a + \sum_{j=1}^{m} \frac{\partial p_j}{\partial y}S_{jk}^b \qquad (2.169)$$

For the MQ basis function shown in Table 2.2, the partial derivatives for the MQ radial functions can be easily obtained using the following simple formulae:

$$\frac{\partial R_i}{\partial x} = 2q(r_i^2 + C^2)^{q-1}(x - x_i)$$

$$\frac{\partial R_i}{\partial y} = 2q(r_i^2 + C^2)^{q-1}(y - y_i) \qquad (2.170)$$

For the EXP radial function, the partial derivatives can also be obtained easily as follows:

$$\frac{\partial R_i}{\partial x} = -2cR_i(x, y)(x - x_i)$$

$$\frac{\partial R_i}{\partial y} = -2cR_i(x, y)(y - y_i) \qquad (2.171)$$

2.7.3 Singularity Issue of the Transformed Moment Matrix

We now examine the positivity of the transformed moment matrix $\mathbf{P}_m^{\mathrm{T}}\mathbf{R}_Q^{-1}\mathbf{P}_m$ in Equation 2.163. Because $\mathbf{R}_Q$ is SPD (symmetric, invertible, with a full rank), the transformed moment matrix $\mathbf{P}_m^{\mathrm{T}}\mathbf{R}_Q^{-1}\mathbf{P}_m$ in Equation 2.163 is at least symmetric. If the columns in $\mathbf{P}_m$

are independent (with a rank of m), we can easily prove that $\mathbf{P}_m^T \mathbf{R}_Q^{-1} \mathbf{P}_m$ is invertible by simply invoking the full rank property of $\mathbf{R}_Q$. To have all the columns of $\mathbf{P}_m$ to be independent could be a problem in theory, but it is very easy to achieve in practice. We simply try to use a minimum number of terms of polynomial bases, so that $n \geq m$. This will ensure that $\mathbf{P}_m$ has a rank of m for *practical* situations. For example, when $m = 3$, all we need is at least there are three support nodes in the support domain that are not in-line.

Of course, if we deliberately arrange all the support nodes in one straight line (for 2D and 3D interpolation), the method will break down. In such an event, all the methods discussed in this chapter will break down. This kind of situation can happen in theory, but it will not happen when the nodes are generated by triangulation defined in Chapter 1 and a T-scheme is used for support node selection. Our purpose of defining the triangulation and T-schemes is to exclude these kinds of extreme situations. When the T6- or T2L-schemes (see Section 1.6.6) are used with triangular background cells, $\mathbf{P}_m^T \mathbf{R}_Q^{-1} \mathbf{P}_m$ will always be invertible at least for RPIM augmented with linear polynomial basis, and hence is recommended. The use of T6-scheme will be more efficient, and T2L-scheme is relatively more robust to extremely irregular distribution nodes.

In a usual situation, it is very rare to have a case where the rank of $\mathbf{P}_m$ is less than m, even when the T-scheme is not used. It is not a guarantee, but it is a workable practical strategy for usual situations. Therefore, $\mathbf{P}_m^T \mathbf{R}_Q^{-1} \mathbf{P}_m$ can be assumed invertible, in usual and in practical cases, when general concept of support domain is used. Note that in the MLS approximation, the singularity issue with the weight moment matrix given in Equation 2.54 is also avoided in using exactly the same strategy of having $n \geq m$.

2.7.4 Examples: RPIM Shape Functions and Shape Parameter Effects

We first give some examples of shape functions constructed using the RBFs. For convenience, notations of MQ-PIM, EXP-PIM, and TPS-PIM refer to RPIM using MQ, EXP, and TPS radial bases, respectively.

Example 2.1: Sample RPIM Shape Functions

Examples of the RPIM shape functions are computed using the formulation given above. The RBFs listed in Table 2.2 are used in the computation. Figure 2.12 shows a typical shape function of MQ-PIM in 1D space. Shape parameters used in MQ-PIM are $C = 1.0$, $q = 0.5$. Five nodes evenly distributed in the support domain of $[-1, 1]$ are used for computing the shape function for the node at $x = 0$. Figure 2.12a shows the shape function, and Figure 2.12b shows the derivative of the shape function.

Figure 2.13 shows a typical shape function of EXP-PIM with a shape parameter of $c = 0.3$. All the other conditions are the same as those used in computing Figure 2.12. Note that these shape functions possess the Kronecker delta function property given in Equation 2.107. As long as the polynomial basis $\{1, x, y\}$ ($m = 3$) is included in the basis, the shape functions so developed also satisfy Equation 2.117.

In using radial functions, one needs to investigate the effects of the shape parameters and to fine-tune these parameters for better performance. This is more important when a weak form is used, because the effect of the incompatibility is mitigated by proper choice of the shape parameters. When weakened-weak form is used, tuning shape parameters are much less important, and a very wide range of parameters can be used. In the following example we examine only how the shape parameters affect the shape functions.

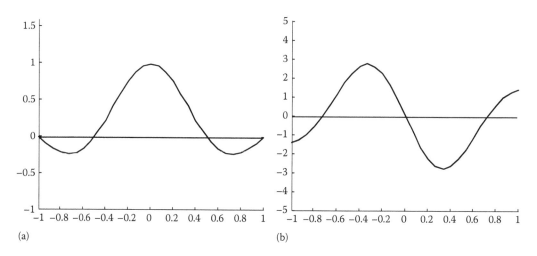

FIGURE 2.12

MQ-PIM ($C=1.0$, $q=0.5$) shape function in 1D space for the node at $x=0$ obtained using five nodes evenly distributed in the support domain of $[-1, 1]$. (a) Shape function; and (b) derivative of the shape function. Note that the RPIM shape function possesses the Kronecker delta function property.

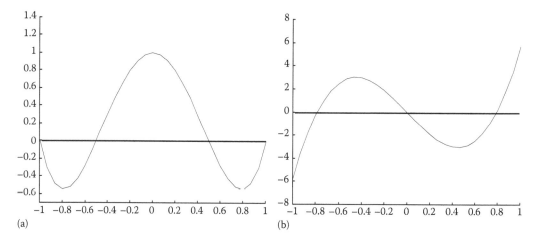

FIGURE 2.13

EXP-PIM ($c=0.3$) shape function in 1D space for the node at $x=0$ obtained using five nodes evenly distributed in the support domain of $[-1, 1]$. (a) Shape function; and (b) derivative of the shape function. Note that the RPIM shape function possesses the Kronecker delta function property.

Example 2.2: Effects of Shape Parameters of RBFs on Shape Function

This example examines how the shape parameters affect the shape functions. To isolate the effects of the shape parameters of the radial functions, polynomial terms are not included in the basis function; that is, pure RPIM shape functions are used. Figure 2.14 shows the shape function of EXP-PIM at the seventh node computed using 16 nodes for a 1D interpolation. Different shape parameters c are used in the computation. Figure 2.14 shows that c affects the decay rate of the

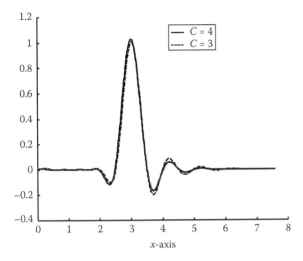

FIGURE 2.14
Effect of shape parameter c on PIM shape functions with EXP radial function basis.

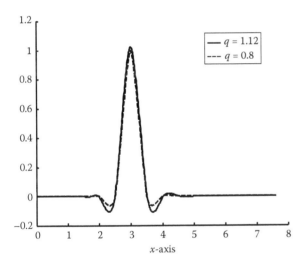

FIGURE 2.15
Effect of shape parameter q on PIM shape functions with MQ RBF ($C = 1.42$).

oscillations behind the first dominant peak. The larger the value of c becomes, the faster the decay. Figure 2.15 shows the PIM shape function at the seventh node computed using 16 nodes for a 1D interpolation using an MQ RBF. The shape parameter C is fixed at $C = 1.42$ and the shape parameter q is investigated for two cases: 0.8 and 1.12. Figure 2.15 shows that q affects the decay rate of the oscillations behind the first dominant peak. The smaller the value of q, the faster the decay.

Remark 2.5: Property of RPIM Shape Functions Augmented with Polynomials
Successfully constructed RPIM shape functions are, in general, incompatible, consistent with the order of polynomials used in the formulation. They have the Kronecker delta function property, and form an interpolant in a $\mathbb{G}_h^1$ space (see Chapter 3).

2.8 Weighted Least Square Approximation

The weighted least square (WLS) approximation is a widely used technique for data fitting. In the WLS, the number of basis, m, is usually predetermined according to the requirements on the consistency for shape functions. Using Equation 2.86, we can approximate a field function $u(\mathbf{x})$ using the polynomial basis as follows:

$$u^h(\mathbf{x}) = \sum_{i=1}^{m} p_i(\mathbf{x})a_i = a_1 + a_2 x + a_3 y + \cdots + a_m p_m(\mathbf{x})$$

$$= \underbrace{\{1 \quad x \quad y \quad \cdots \quad p_m(\mathbf{x})\}}_{\mathbf{p}^T} \underbrace{\left\{ \begin{array}{c} a_1 \\ \vdots \\ a_m \end{array} \right\}}_{\mathbf{a}} = \mathbf{p}^T \mathbf{a} \tag{2.172}$$

where a_i $(i = 1, 2, \ldots, m)$ are the coefficients to be determined, and $\mathbf{p}$ is the vector of basis functions. To determine coefficients $\mathbf{a}$ in Equation 2.172, n $(>m)$ nodes are selected in the local support domain for the approximation. Using Equation 2.172 for all these n nodes, we obtain

$$\mathbf{d}_s = (\mathbf{P}_m)_{(n \times m)} \mathbf{a}_{(m \times 1)} \tag{2.173}$$

The moment matrix, $\mathbf{P}_m$, is

$$\mathbf{P}_m = \begin{bmatrix} 1 & x_1 & y_1 & x_1 y_1 & \cdots & p_m(\mathbf{x}_1) \\ 1 & x_2 & y_2 & x_2 y_2 & \cdots & p_m(\mathbf{x}_2) \\ 1 & x_3 & y_3 & x_3 y_3 & \cdots & p_m(\mathbf{x}_3) \\ \vdots & \vdots & \vdots & \vdots & \ddots & \vdots \\ 1 & x_n & y_n & x_n y_n & \cdots & p_m(\mathbf{x}_n) \end{bmatrix}_{(n \times m)} \tag{2.174}$$

Note that $\mathbf{P}_m$ is not a square matrix because $n > m$. Equation 2.173 is a set of overdetermined system due to $n > m$ meaning that the number of equations is more than the number of unknowns. We can solve Equation 2.173 for $\mathbf{a}$ using the standard WLS method by minimizing the following weighted discrete $\mathbf{L}_2$ norm:

$$J = \sum_{i=1}^{n} \widehat{W}_i \left[u^h(\mathbf{x}_i) - u(\mathbf{x}_i) \right]^2 \tag{2.175}$$

where
     $\widehat{W}_i$ $(i = 1, 2, \ldots, n)$ is the weight coefficient associated with the function value at the ith node in the support domain
     u_i becomes the "nodal parameter" of u at $\mathbf{x} = \mathbf{x}_i$

The stationary condition gives

$$\frac{\partial J}{\partial \mathbf{a}} = 0 \tag{2.176}$$

which leads to the following relation between $\mathbf{a}$ and $\mathbf{d}_s$

$$\mathbf{P}_m^{\mathrm{T}} \widehat{\mathbf{W}} \mathbf{P}_m \mathbf{a} = \mathbf{P}_m^{\mathrm{T}} \widehat{\mathbf{W}} \mathbf{d}_s \tag{2.177}$$

where $\widehat{\mathbf{W}}$ is the diagonal matrix constructed from the weight constants, i.e.,

$$\widehat{\mathbf{W}}_{(n \times n)} = \begin{bmatrix} \widehat{W}_1 & \widehat{W}_2 & \cdots & \widehat{W}_n \end{bmatrix} \tag{2.178}$$

Note that the weights used here are considered as constants (not functions of $\mathbf{x}$) that define the different influences of the nodes in the approximation. The further nodes should have smaller influences while closer nodes have bigger influences, $\widehat{W}_i$ can be computed from any weight function with the bell shape that will be provided in Section 2.2.2. We now let

$$\mathbf{A} = \mathbf{P}_m^{\mathrm{T}} \widehat{\mathbf{W}} \mathbf{P}_m \tag{2.179}$$

$$\mathbf{B} = \mathbf{P}_m^{\mathrm{T}} \widehat{\mathbf{W}} \tag{2.180}$$

Solving Equation 2.177 for $\mathbf{a}$ yields

$$\mathbf{a} = \left(\mathbf{P}_m^{\mathrm{T}} \widehat{\mathbf{W}} \mathbf{P}_m \right)^{-1} \left(\mathbf{P}_m^{\mathrm{T}} \widehat{\mathbf{W}} \right) \mathbf{d}_s \tag{2.181}$$

$$\mathbf{a} = \mathbf{A}^{-1} \mathbf{B} \mathbf{d}_s \tag{2.182}$$

Substituting Equation 2.182 back into Equation 2.172, we have

$$u^h(\mathbf{x}) = \mathbf{p}^{\mathrm{T}} \mathbf{a} = \mathbf{p}^{\mathrm{T}} \mathbf{A}^{-1} \mathbf{B} \mathbf{d}_s = \boldsymbol{\varphi} \mathbf{d}_s \tag{2.183}$$

where the vector of shape functions $\boldsymbol{\varphi}$ is

$$\boldsymbol{\varphi} = \mathbf{p}^{\mathrm{T}} \mathbf{A}^{-1} \mathbf{B} = \{ \phi_1 \quad \phi_2 \quad \cdots \quad \phi_n \} \tag{2.184}$$

where ϕ_i ($i = 1, 2, \ldots, n$) is the WLS shape function corresponding to the ith node in the support domain.

Equation 2.183 is the approximation equation for the WLS. The weight functions used is not "moving" as in the MLS, the shape functions constructed will not be continuous (not in an $\mathbb{H}^1$ space). When a (global) weak form is used, there will be incompatibility issues similar to the PIM shape functions. It should, however, work for the weakened-weak formulations. Because the weighted LSs method is used, the shape functions so constructed do not have the Kronecker delta function property and cautions are needed in imposing essential boundary conditions. However, it is not a big issue in the meshfree methods based on local weak forms, because the *direct interpolation method* can be used to enforce the essential boundary conditions. When we use an identity matrix as the weight matrix, the WLS shape function becomes the LS shape function. The formulation is exactly the same by simply removing $\mathbf{W}$ in the above equations. The WLS (or LS) shape functions can be handily used with a SC-Galerkin weak form; it is also useful in the construction of strain fields.

Remark 2.6: Property of WLS Shape Functions
Successfully constructed WLS shape functions are, in general, incompatible, consistent with the order of polynomials used in the formulation. They do not have the Kronecker delta function property, and do not form an interpolant.

2.9 Polynomial PIM with Rotational Coordinate Transformation

This section introduces a method of (rotational) coordinate transformation to produce an invertible moment matrix $\mathbf{P}_Q$. The method is based on the following realizations: (1) the linear dependence of the columns (or rows) in the moment matrix depends on the coordinates of the nodes used for a fixed set of monomials; (2) such a dependence can be altered through a simple rotational coordinate transformation, and hence a nonsingular moment matrix can be obtained when a proper rotation angle can be found; (3) for a given set of a small number of nodes (except extreme cases), there are only a few rotation angles θ that make the moment matrix singular, and hence for a randomly chosen rotational angle θ, the moment matrix is unlikely to be singular. Therefore, it is generally (not always) workable to produce a nonsingular moment matrix by performing a rotational transformation. The method was originally attempted in [41].

2.9.1 Coordinate Transformation

We first introduce a general coordination transformation between the global coordinate system (x, y) and the local coordinate system (ξ, η), as shown in Figure 2.16. This transformation can be performed using

$$\xi = (x - x_0) \cos \theta + (y - y_0) \sin \theta$$
$$\eta = -(x - x_0) \sin \theta + (y - y_0) \cos \theta \tag{2.185}$$

where (x_0, y_0) is the origin of the local coordinate system (ξ, η), which is defined in a global coordinate system (x, y), and θ is the rotation angle for local coordination system (ξ, η) with respect to the global coordinate system. The inverse transformation is expressed as

$$x = x_0 + (\xi \cos \theta - \eta \sin \theta)$$
$$y = y_0 + (\xi \sin \theta + \eta \cos \theta) \tag{2.186}$$

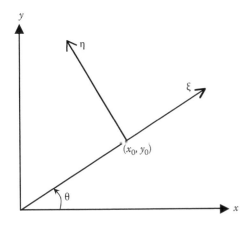

FIGURE 2.16
A local coordinate system (ξ, η) defined in a global coordinate system (x, y). The origin of (ξ, η) is located at (x_0, y_0).

In the local coordinates with a proper rotation angle θ, n polynomial basis terms $p_i(\xi)$ ($i = 0, 1, \ldots, n - 1$) are chosen, and polynomial interpolation is performed to produce a nonsingular moment matrix $\mathbf{P}_0$ for point (0, 0), which corresponds to (x_0, y_0) in the global coordinate system. It is then inverted to obtain $\mathbf{P}_0^{-1}$, and the shape functions can now be computed using (see Equation 2.100)

$$\boldsymbol{\varphi}(\xi) = \mathbf{p}^{\mathrm{T}}(\xi)\mathbf{P}_0^{-1} = [\phi_1(\xi), \phi_2(\xi), \ldots, \phi_n(\xi)] \tag{2.187}$$

Because the coordinate transformation is very simple, and in fact involves only rotation, the derivatives of the shape functions (if needed) can be obtained efficiently using

$$\left\{ \begin{array}{c} \dfrac{\partial \phi_i}{\partial x}\Big|_{(x_0, y_0)} \\[2ex] \dfrac{\partial \phi_i}{\partial y}\Big|_{(x_0, y_0)} \end{array} \right\} = \begin{bmatrix} \cos\theta & -\sin\theta \\ \sin\theta & \cos\theta \end{bmatrix} \left\{ \begin{array}{c} \dfrac{\partial \phi_i}{\partial \xi}\Big|_{(0,0)} \\[2ex] \dfrac{\partial \phi_i}{\partial \eta}\Big|_{(0,0)} \end{array} \right\} \tag{2.188}$$

where

$$\begin{aligned} \dfrac{\partial \phi_i}{\partial \xi}\Big|_{(0,0)} &= (0, 1, 0 \ldots, 0)\mathbf{P}_0^{-1} \\[2ex] \dfrac{\partial \phi_i}{\partial \eta}\Big|_{(0,0)} &= (0, 0, 1, 0 \ldots, 0)\mathbf{P}_0^{-1} \end{aligned} \tag{2.189}$$

2.9.2 Choice of Rotation Angle

The choice of the rotation angle determines the success of the method of coordinate transformation. A study was conducted to reveal the property of the moment matrix $\mathbf{P}_0$ in relation to the rotation angle. We define

$$f(\theta) = |\mathbf{P}_0| \tag{2.190}$$

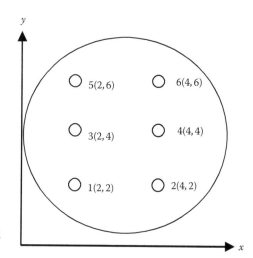

FIGURE 2.17
Six nodes in a support domain distributed in two parallel lines of $x = 2.0$ and 4.0.

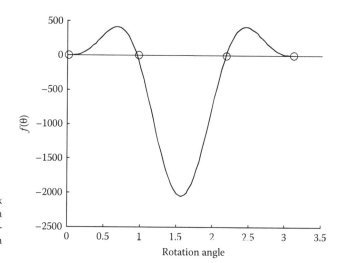

FIGURE 2.18
Determinant of the moment matrix formed in the local coordinate system via rotation. The moment matrix is singular at four rotation angles marked with circles.

where $|\mathbf{P}_0|$ denotes the determinant of $\mathbf{P}_0$. If any θ is not a root of $f(\theta)$, $\mathbf{P}_0^{-1}$ exists and the PIM shape function can be constructed. An example of six nodes, shown in Figure 2.17, is then analyzed. Without coordinate transformation, $\mathbf{P}_0^{-1}$ using these six nodes will not exist. Figure 2.18 shows the function of the determinant of the moment matrix $f(\theta)$ with respect to rotation angle θ. It is seen that the moment matrix is singular only at four rotation angles marked with circles. Using any rotational angle other than these four leads to a nonsingular moment matrix.

Note that how to choose the rotation angle is still an open question. One possible method is to choose a random number. However, this will not give 100% proof for a nonsingular moment matrix for more general cases. Note that the choice of the rotation angle depends on the nodal arrangement in the support domain. It is possible to purposely design a set of nodal arrangement (say all six nodes in one line); no such rotational angle can be found for nonsingular moment matrix. This argument is true for all the methods for shape function construction. When a triangulation defined in Section 1.7.2 is used, there are only a few rotation angles θ that make the moment matrix singular, for a set of nodes selected using some kind of controlling schemes like T-schemes to prevent extreme cases. Hence for a randomly chosen rotational angle θ, the moment matrix is unlikely to be singular, and the method of coordination transform should work. Note that for the special case of using 4 nodes for interpolation as in the CS PIM, there is an "optimal" way to determine θ as shown in [58].

Note that the T-schemes are workable schemes by themselves that do not need this coordinate transformation. We present this coordinate transformation for general cases when T-scheme is not used for constructing PIM shape functions.

2.10 Comparison Study via Examples

Example 2.3: Shape Function Comparison Study (1D Case)

Figure 2.19a shows a comparison of shape functions in 1D space obtained using four different methods: MLS, polynomial PIM, MQ-PIM, and EXP-PIM. The shape functions are obtained using five nodes at $x = -1.0$, -0.5, 0.0, 0.5, and 1.0 in the support domain of $[-1.0, 1.0]$. The shape function shown in Figure 2.19 is for the node at $x = 0.0$. It is noted that the shape functions obtained by the three PIM methods possess the Kronecker delta function property, that is, $\phi(0.0) = 1.0$, $\phi(-1.0) = \phi(-0.5) = \phi(0.5) = \phi(1.0) = 0.0$. The shape function obtained using MLS does not possess the Kronecker delta function property, that is, $\phi(0.0) \neq 1.0$, $\phi(-0.5) = \phi(0.5) \neq 0.0$. The MLS shape vanishes at $x = \pm1.0$, because $x = \pm1.0$ is on the boundary of the support domain. All the PIM shape functions vary more frequently than the MLS shape functions.

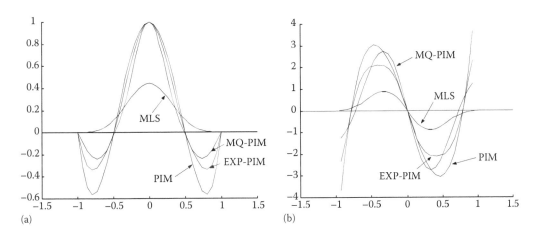

FIGURE 2.19
Comparison of shape functions obtained using four different methods: MLS, polynomial PIM, MQ-PIM, and EXP-PIM. (a) Shape functions; and (b) derivatives of shape functions.

Figure 2.19b shows the first derivatives of shape functions. It is found that the derivatives of the MLS shape function vanish at two boundary points. This is because the quartic weight function is used in the construction of the shape function, whose first derivative also vanishes on the boundary (see Figure 2.4).

Example 2.4: Shape Function Comparison Study (2D Case)

Meshfree shape functions are constructed in a domain of $(x, y) \in [2 \times 2] \times [2 \times 2]$ using 5×5 evenly distributed nodes in the domain. Figure 2.20 shows the shape function and its first derivative with respect to x computed using polynomial PIM. Figure 2.21 shows the shape function and its first derivative with respect to x computed using RPIM (MQ, $C = 1.0$, $q = 0.5$) with linear polynomials. It is shown that the PIM shape functions satisfy the Kronecker delta

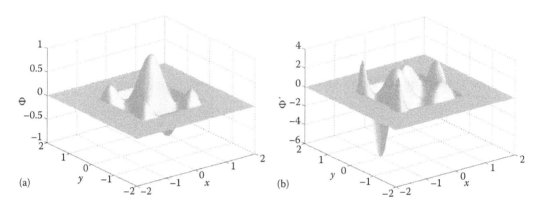

FIGURE 2.20
Polynomial PIM shape function and its derivative with respect to x. (a) PIM shape function; and (b) derivative of the PIM shape function.

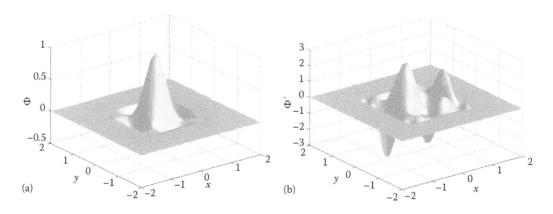

FIGURE 2.21
RPIM (MQ, $C = 1.0$, $q = 0.5$, with polynomial) shape function and its derivative with respect to x. (a) PIM shape function; and (b) derivative of the PIM shape function.

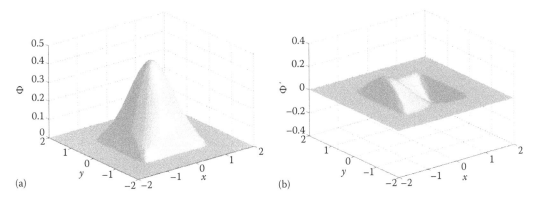

FIGURE 2.22
MLS shape function and its derivative with respect to x. (a) MLS shape function; and (b) derivative of the MLS shape function.

function. Figure 2.22 shows the MLS shape function. It is clear that the MLS shape functions do not satisfy the Kronecker delta function.

Comparing Figures 2.20 and 2.21 with Figure 2.22 reveals that the PIM shape function and its derivatives are much more complex compared with the MLS shape functions. This finding needs to be considered when we perform numerical integrations in computing discrete system equations.

2.11 Compatibility Issues: An Analysis

In using energy principles or the Galerkin weak form the assumed or approximated field function has to be admissible that includes *compatibility* condition and the essential boundary conditions. The compatibility condition requires the field function approximation being continuous in the entire problem domain. In the conventional FEM, the field function approximation in the problem domain is based on the *element* with proper mapping. The continuity of the field function approximation is ensured by either properly choosing shape functions of neighboring elements so that the order of the interpolation on the common boundary of the elements is the same or using so-called multipoint constraint equations (see, e.g., [48]) to enforce the compatibility. These types of elements are called *conforming* elements. It often happens that discontinuity is allowed to formulate so-called nonconforming finite elements. It has also been reported that some of the nonconforming finite elements can perform better than conforming elements.

In meshfree methods, the field function approximation is often based on *moving* support domain or background cells and the nodes used can be beyond the cells. The compatibility of field function approximation using meshfree shape functions may or may not always be satisfied. In using MLS approximation, the compatibility is ensured by choosing weight functions that satisfy Equations 2.8, 2.9, and 2.11 with a sufficient number of nodes. The compatibility of field function approximation using the MLS shape function depends on the weight function used in Equation 2.51. Due to the use of the bell-shaped weight functions, the nodes can enter or leave the support domain in a smooth manner, which ensures continuity and hence the compatibility while the point of interest is moving.

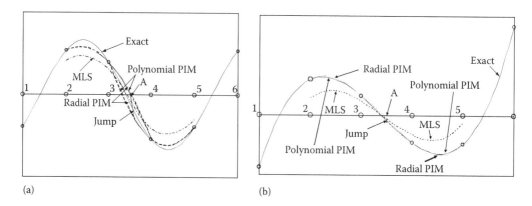

(a)   (b)

FIGURE 2.23

Field function approximation using meshfree shape functions. Functions in the range of [0, 2.5] are approximated using six nodes. MLS, moving polynomial PIM, and moving RPIM are used in function approximation. In MLS approximation the support domain is defined by $d_s = \alpha_s d_i$, where $\alpha_s = 1.9$, d_i is the space between two nodes. In PIMs, the nearest four nodes are used for interpolation (equivalent to $\alpha_s = 2.0$). (a) $f(x) = \sin[(x-0.2)\pi]$; (b) $f(x) = (x-0.2)(x-1.2)(x-2.2)$.

The order of compatibility is determined by the smoothness of the weight function used. From Figures 2.2 through 2.4, it is seen that W1 and W2 provide at least second-order compatibility, as their second derivatives are continuous in the support domain and vanish on the boundary of the support domain. W4 provides first order compatibility, as its first derivatives are continuous in the support domain and vanish on the boundary of the support domain. Although the second derivative of W4 is continuous in the support domain, it does not vanish on the boundary of the support domain (see Figure 2.4). As for W3, its derivatives of all orders are continuous within the support domain, but they are not exactly zero on the boundary of the support domain. Therefore, theoretically, W3 cannot provide compatibility of any order. However, the values of the function and its derivatives are very small on the boundary of the support domain. In practical numerical analyses, W3 provides very high order compatibility with a very small numerical error.

In using PIM shape functions that are constructed based on background cells, the field function approximated could be discontinuous on the cell interfaces, because the nodes used constructing shape functions for one cell can be different from those for the neighboring cells. Therefore, the function approximated using the PIM shape functions can jump over the cell interface. The same situation can happen when moving support domain is used for constructing shape functions. In such a case, the shape functions jump when the nodes in the support domain is updated. Figure 2.23 shows an example of how a field function is approximated using MLS with quartic spline weight function (W2), polynomial PIM, and radial MQ-PIM shape functions, where we consider the approximation of a 1D function using the function values at six nodes. Two functions are considered in this investigation: one is the sine function (Figure 2.23a) that is nonpolynomial, and another is a third-order polynomial function. The following points may be noted from this figure.

1. MLS approximation is continuous, but does not pass through the function values at these nodes.

2. All the PIM approximations pass through the field function values at the nodes due to the delta function property.

3. In approximating the sine functions that are nonpolynomial, the PIM and RPIM approximations are discontinuous at Point A ($x = 1.25$). This happens because, when the function is approximated between Points 2 and A, field nodes 1, 2, 3, 4, and 5 are used, and when the function is approximated between Points A and 5, field nodes 2, 3, 4, 5, and 6 are used for constructing the PIM shape functions. PIM shape functions are incompatible.

Note that for 1D problems, the incompatibility problem can be avoided very easily by simply using PIM shape functions constructed based on background cells. For the case shown in Figure 2.23, for example, one can simply use nodes 2, 3, 4, and 5 to construct PIM shape function for cell of 3–4, and 3, 4, 5, and 6 for cell of 4–5. There will be no discontinuity at node 4, simply because the delta function property of PIM shape functions and the interface of the cell 3–4 and cell 4–5 is just one point. For 2D or 3D domains, however, PIM shape functions based on cells will still be discontinuous, because the interfaces will be line segments (for 2D) and surface segments (3D). Figure 2.24 shows an example for 2D case with triangular background cells, and Figure 2.25 shows that with rectangular background cells. Therefore, when such PIM shape functions are used, the generalized smoothed Galerkin (GS-Galerkin) (see Chapter 5) weak form needs to be used with nodal-based or edge-based smoothing domains. The same analysis applies to RPIM shape functions.

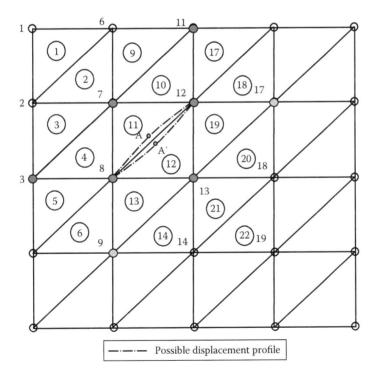

—·—·— Possible displacement profile

FIGURE 2.24
Compatibility of PIM shape functions on the interface of two neighboring background cells. The gap on the interface 8–12 between cells 11 and 12 can occur when six nodes (T6/3-scheme) are used in computing PIM shape functions for these two cells. For background cell 11, nodes 3, 7, 8, 11, 12, and 13 are used, and for cell 12, nodes 7, 8, 9, 12, 13, and 17 are used.

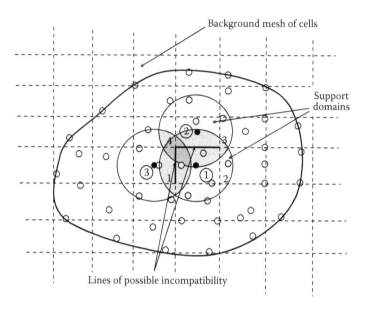

FIGURE 2.25
PIM shape functions construction based on background rectangular cells: Possible incompatibility can occur between the interfaces of neighboring cells.

Note also that when the *local* residual weak form is used to create the discretized system equations, the global compatibility of the trial function is not a requirement. As long as the field approximation is consistent at any point in the quadrature domain, trial function is differentiable and the integrand is integrable. The PIM approximation satisfies all those requirements. Therefore, PIM shape functions can be used with the local residual weak form. The much better choice is a weakened-weak formulation, such as the GS-Galerkin, such implementations will be detailed in Chapter 8.

Table 2.4 lists the features of meshfree shape functions discussed in this chapter.

TABLE 2.4

Features of Meshfree Shape Functions

Shape Functions	Consistency[a]	Global Compatibility	Delta Function Property
SPH	No, on the boundary; yes, in the domain	Yes (for the continuous form of SPH)	No
RKPM	Yes	Yes	No
MLS	Yes	Yes	No
Polynomial PIM	Yes	No	Yes
RPIM	No (but reproduces RBFs)	No	Yes
Radial PIM with polynomial basis	Yes	No	Yes
WLS (LS)	Yes	No	No

[a] Consistency is defined as the ability to produce complete order of polynomials.

2.12 Other Methods

There are a number of other useful methods for constructing shape functions, such as

- Kringing interpolation [51–53]
- Minimum length approximation [54]
- *hp*-clouds method and the partition of unity [49,50]

It has proven that the Kringing interpolation is essentially the same as the RPIM as long as the same basis functions are used [52]. Therefore, properties found for RPIM should apply to Kringing shape functions.

The minimum length approximation is an interesting idea, and may be worth to pursue further. The shape functions created have the delta function property (passing through nodes), but not all the basis terms can be reproduced (do not "pass through" basis terms). This property is in contrast to the MLS shape functions that do not possess the delta function property (do not pass through nodes), but all the basis terms can be reproduced ("pass through" basis terms).

This book will not provide the details on the *hp*-clouds method. This method can be very useful, and readers are referred to [49,50].

Finally, we state that by combining these shape functions with the principles, weak forms, and weakened-weak forms we can develop different types of meshfree methods. Such a combination offers tremendous opportunities for us to develop more effective and robust meshfree methods. Understanding the features of the shape functions in accordance with these weak forms is of importance.

References

1. Lucy, L., A numerical approach to testing the fission hypothesis, *Astronomical J.*, 82, 1013–1024, 1977.
2. Gingold, R. A. and Monaghan, J. J., Smooth particle hydrodynamics: Theory and applications to non-spherical stars, *Mon. Notices R. Astronomical Soc.*, 181, 375–389, 1977.
3. Monaghan, J. J., Why particle methods work, *Siam J. Sci. Stat. Comput.*, 3(4), 423–433, 1982.
4. Liu, M. B., Liu, G. R., and Zong, Z., An overview on smoothed particle hydrodynamics, *Int. J. Comput. Methods*, 5(1), 135–188, 2008.
5. Zhou, C. E., Liu, G. R., and Lou, K. Y., Three-dimensional penetration simulation using smoothed particle hydrodynamics, *Int. J. Comput. Methods*, 4(4), 671–691, 2007.
6. Liu, G. R. and Liu, M. B., *Smoothed Particle Hydrodynamics—A Meshfree Practical Method*, World Scientific, Singapore, 2003.
7. Liu, G. R., Zhang, J., and Lam, K. Y., A gradient smoothing method (GSM) with directional correction for solid mechanics problems, *Comput. Mech.*, 41, 457–472, 2008.
8. Liu, G. R. and Xu, G. X., A gradient smoothing method (GSM) for fluid dynamics problems, *Int. J. Numerical Methods Fluids*, 56(10), 1101–1133, 2008.
9. Xu, G. X., Liu, G. R., and Lee, K. H., Application of gradient smoothing method (GSM) for steady and unsteady incompressible flow problems using irregular triangles, Submitted to *Int. J. Numerical Methods Fluids*, 2008.
10. Xu, G. X. and Liu G. R., An adaptive gradient smoothing method (GSM) for fluid dynamics problems, *Int. J. Numerical Methods Fluids*, accepted, 2008.

11. Liu, G. R., Liu, M. B., and Lam, K. Y., A general approach for constructing smoothing functions for meshfree methods, presented at *Ninth International Conference on Computing in Civil and Building Engineering*, Taipei, China, April 3–5, 2002, pp. 431–436.

12. Liu, W. K., Adee, J., and Jun, S., Reproducing kernel and wavelet particle methods for elastic and plastic problems, in *Advanced Computational Methods for Material Modeling*, Benson, D. J., ed., AMD 180/PVP 268 ASME, New Orleans, LA, 1993, pp. 175–190.

13. Liu, W. K., Jun, S., and Zhang, Y., Reproducing kernel particle methods, *Int. J. Numerical Methods Fluids*, 20, 1081–1106, 1995.

14. Liu, W. K., Chen, Y., Chang, C. T., and Belytschko, T., Advances in multiple scale kernel particle methods, *Comput. Mech.*, 18, 73–111, 1996.

15. Liu, W. K., Jun, S., Sihling, D. T., Chen, Y. J., and Hao, W., Multiresolution reproducing kernel particle method for computational fluid dynamics, *Int. J. Numerical Methods Fluids*, 24, 1–25, 1997.

16. Liu, W. K., Jun, S., Sihling, D. T., Chen, Y. J., and Hao, W., Multiresolution reproducing kernel particle method for computational fluid dynamics, *Int. J. Numerical Methods Fluids*, 24, 1391–1415, 1997.

17. Liu, W. K., Li, S. F., and Belytschko, T., Moving least-square reproducing kernel methods: I Methodology and convergence, *Comput. Methods Appl. Mech. Eng.*, 143, 113–154, 1997.

18. Chen, J. S., Pan, C., Wu, C. T., and Liu, W. K., Reproducing kernel particle methods for large deformation analysis of nonlinear structures, *Comput. Methods Appl. Mech. Eng.*, 139, 195–228, 1996.

19. Uras, R. A., Chang, C. T., Chen, Y., and Liu, W. K., Multiresolution reproducing kernel particle methods in acoustics, *J. Comput. Acoust.*, 5(1), 71–94, 1997.

20. Lancaster, P. and Salkauskas, K., Surfaces generated by moving least squares methods, *Math. Comput.*, 37, 141–158, 1981.

21. Cleveland, W. S., *Visualizing Data*, AT&T Bell Laboratories, Murray Hill, NJ, 1993.

22. Nayroles, B., Touzot, G., and Villon, P., Generalizing the finite element method: Diffuse approximation and diffuse elements, *Comput. Mech.*, 10, 307–318, 1992.

23. Belytschko, T., Lu, Y. Y., and Gu, L., Element-free Galerkin methods, *Int. J. Numerical Methods Eng.*, 37, 229–256, 1994.

24. Krongauz, Y. and Belytschko, T., Enforcement of essential boundary conditions in meshless approximations using finite elements, *Comput. Methods Appl. Mech. Eng.*, 131(1–2), 133–145, 1996.

25. Belytschko, T., Gu, L., and Lu, Y. Y., Fracture and crack growth by element free Galerkin methods, *Model. Simulations Mater. Sci. Eng.*, 2, 519–534, 1994.

26. Belytschko, T., Krongauz, Y., Fleming, M., Organ, D., and Liu, W. K., Smoothing and accelerated computations in the element free Galerkin method, *J. Comput. Appl. Math*, 74, 111–126, 1996.

27. Fleming, M., Chu, Y. A., Moran, B., and Belytschko, T., Enriched element-free Galerkin methods for crack tip fields, *Int. J. Numerical Methods Eng.*, 40, 1483–1504, 1997.

28. Belytschko, T., Krongauz, Y., Organ, D., Fleming, M., and Krysl, P., Meshless method: An overview and recent developments, *Comput. Methods Appl. Mech. Eng.*, 139, 3–47, 1996.

29. Liu, G. R. and Gu, Y. T., A point interpolation method, in *Proceedings of the Fourth Asia-Pacific Conference on Computational Mechanics*, Singapore, December 1999, pp. 1009–1014.

30. Liu, G. R. and Gu, Y. T., A point interpolation method for two-dimensional solids, *Int. J. Numerical Methods Eng.*, 50, 937–951, 2001.

31. Wang, J. G. and Liu, G. R., Radial point interpolation method for elastoplastic problems, in *Proceedings of the First International Conference on Structural Stability and Dynamics*, Taipei, China, December 7–9, 2000, pp. 703–708.

32. Hardy, R. L., Theory and applications of the multiquadrics—Biharmonic method (20 years of discovery 1968–1988), *Comput. Math. Appl.*, 19, 163–208, 1990.

33. Kansa, E. J., A scattered data approximation scheme with application to computational fluid-dynamics—I & II, *Comput. Math. Appl.*, 19, 127–161, 1990.

34. Kansa, E. J., Multiquadrics—A scattered data approximation scheme with applications to computational fluid dynamics, *Comput. Math. Appl.*, 19(8/9), 127–145, 1990.

35. Powell, M. J. D., The theory of radial basis function approximation in 1990, in *Advances in Numerical Analysis*, F. W. Light, ed., Oxford University Press, Oxford, 1992, pp. 303–322.
36. Coleman, C. J., On the use of radial basis functions in the solution of elliptic boundary value problems, *Comput. Mech.*, 17, 418–422, 1996.
37. Sharan, M., Kansa, E. J., and Gupta, S., Application of the multiquadric method for numerical solution of elliptic partial differential equations, *Appl. Math. Comput.*, 84, 275–302, 1997.
38. Fasshauer, G. E., Solving partial differential equations by collocation with radial basis functions, in *Surface Fitting and Multiresolution Methods*, A. L. Mehaute, C. Rabut, and L. L. Schumaker, eds., 1997, pp. 131–138.
39. Wendland, H., Error estimates for interpolation by compactly supported radial basis functions of minimal degree, *J. Approximation Theory*, 93, 258–396, 1998.
40. Wang, J. G. and Liu, G. R., Radial point interpolation method for no-yielding surface models, in *Proceedings of the First MIT Conference on Computational Fluid and Solid Mechanics*, June 12–14, Boston, 2001, pp. 538–540.
41. Wang, J. G. and Liu, G. R., A point interpolation meshless method based on radial basis functions, *Int. J. Numerical Methods Eng.*, 54, 1623–1648, 2002.
42. Schaback, R., Approximation of polynomials by radial basis functions, in *Wavelets, Images and Surface Fitting*, P. J. Laurent, L. Mehaute, and L. L. Schumaker, eds., A. K. Peters Ltd., Wellesley, MA, 1994, pp. 459–466.
43. Liu, G. R., Li, Y., Dai, K. Y., Luan, M. T., and Xue, W., A linearly conforming radial point interpolation method for solid mechanics problems, *Int. J. Comput. Methods*, 3, 401–428, 2006.
44. Li, Y., Liu, G. R., Luan, M. T., Dai, K. Y., Zhong, Z. H., Li, G. Y., and Han, X., Contact analysis for solids based on linearly conforming radial point interpolation method, *Comput. Mech.*, 39, 537–554, 2007.
45. Golberg, M. A., Chen, C. S., and Bowman, H., Some recent results and proposals for the use of radial basis functions in the BEM, *Eng. Anal. Boundary Elem.*, 23, 285–296, 1999.
46. Liu, G. R. and Gu, Y. T., A matrix triangularization algorithm for point interpolation method, in *Proceedings of the Asia-Pacific Vibration Conference*, W. Bangchun, ed., November, Hangzhou, China, 2001, pp. 1151–1154.
47. Liu, G. R. and Gu, Y. T., A matrix triangularization algorithm for the polynomial point interpolation method, *Comput. Methods Appl. Mech. Eng.*, 192(19), 2269–2295, 2003.
48. Liu, G. R. and Quek, S. S., *The Finite Element Method: A Practical Course*. Butterworth Heinemann, Oxford, 2003.
49. Duarte, C. A. and Oden, J. T., An *hp* adaptive method using clouds, *Comput. Methods Appl. Mech. Eng.*, 139, 237–262, 1996.
50. Melenk, J. M. and Babuska, I., The Partition of Unity Finite Element Method: Basic Theory and Applications, TICAM Report 96-01, University of Texas, Austin, TX, 1996.
51. Gu, L., Moving Kringing interpolation and element-free Galerkin method, *Int. J. Numerical Methods Eng.*, 56, 1–11, 2003.
52. Dai, K. Y., Liu, G. R., Lim, K. M., and Gu, Y. T., Comparison between the radial point interpolation and the Kringing based interpolation using in meshfree methods, *Comput. Mech.*, 32, 60–70, 2003.
53. Tongsuk, P. and Kanok, N. W., Further investigation of element-free Galerkin method using moving Kringing interpolation, *Int. J. Comput. Methods*, 1, 345–365, 2004.
54. Liu, G. R., Dai, K. Y., Han, X., and Li, Y., A meshfree minimum length method for 2-D problems, *Comput. Mech.*, 38, 533–550, 2006.
55. Liu, G. R., Dai, K. Y., and Nguyen, T. T., A smoothed finite element method for mechanics problems, *Comput. Mech.*, 39, 859–877, 2007.
56. Dai, K. Y., Liu, G. R., and Nguyen, T. T., An *n*-sided polygonal smoothed finite element method (*n*SFEM) for solid mechanics, *Finite Elem. Anal. Des.*, 43, 847–860, 2007.
57. Liu, G. R. and Zhang, G. Y., A normed $\mathbb{G}$ space and weakened weak (W^2) formulation of a cell-based smoothed point interpolation method, *Int. J. Comput. Methods*, 6, 147–149, 2009.

3

Function Spaces for Meshfree Methods

Chapter 2 has shown various techniques to construct shape functions that can be used to assume functions for field variables such as the displacements in meshfree settings. Because of the diversity, functions assumed can have "unusual" behaviors that can "upset" usual functionals out of "control." Therefore, directly using them in a standard weak formulation can be problematic. To ensure a meshfree method stable and convergent to the exact solution, the functions need to be first "managed" properly, adequate functionals need to be defined/constructed, and then proper formulation procedures have to be used accordingly. Types of weak forms will be discussed in Chapter 4. In this chapter we "manage" these functions by defining different *spaces* that host different types of functions. We start with the standard spaces that are widely used in usual weak formulations such as the finite element method (FEM) [1,2] and are familiar to many. We then provide a detailed discussion on $\mathbb{G}$ spaces that are the foundation of one weakened-weak formulation, are particularly useful for both meshfree and FEM settings, and much less familiar to many.

Terminologies related to spaces are quite difficult for an engineer to comprehend, but for a reasonably accurate and concise presentation of a numerical procedure, it is hard to avoid. The author, as an engineer, attempts to present these related terms in "engineering" languages for easier comprehension. The author has been trying hard to learn and study the formulations in both engineering and mathematics communities, and hope to do something to bridge the gap. If a reader can be patient in reading through this chapter, we hope he/she can find that the gap is getting a little smaller.

In this chapter, we focus only on the minimum necessary terminologies, so as to facilitate our discussions in later chapters without too much confusion and loss of accuracy in presentation. Readers may refer to Tables 1.1 and 1.3, when get lost in the mathematical terminologies and symbols.

3.1 Function Spaces

We first briefly introduce some of the important basic definitions and terminologies. For more intensive and systematic descriptions, readers may refer to [3].

3.1.1 Linear Spaces

A space $\mathbb{S}$ of functions is said to be linear, if the summation of any two member functions from the space still belongs to the same space, and any member function from the space still belongs to the same space after it is scaled. In mathematical expression, a linear space satisfies

$$\begin{array}{lll} \forall w, v \in \mathbb{S}, & (w + v) \in \mathbb{S} & \text{Addition} \\ \forall \alpha \in \mathbb{R} \text{ and } \forall v \in \mathbb{S}, & \alpha v \subset \mathbb{S} & \text{Scaling} \end{array} \tag{3.1}$$

where $\mathbb{R}$ is the real (algebraic) field of all the real numbers.

A group of (assumed) displacement functions for linear elasticity shall form a linear space.

3.1.2 Functionals

A functional J takes a member from a space as input (or argument) and returns a scalar as an output. It is noted as

$$J: \quad \underbrace{\mathbb{S}}_{\text{space offers the input}} \quad \rightarrow \quad \underbrace{\mathbb{R}}_{\text{space hosts the output}} \tag{3.2}$$

meaning that J takes a member function from $\mathbb{S}$ as input, and then produces a real number as output. An energy in a linear elastic solid or structure is a typical functional that takes in a displacement function and returns a (nonnegative) number of energy.

3.1.3 Norms

A functional $\| \cdot \|_{\mathbb{S}}$ from a linear space $\mathbb{S}$ to $\mathbb{R}$ is called a norm *if and only if* it has the following properties:

(a) $\forall v \in \mathbb{R}: \|v\|_{\mathbb{S}} \geq 0, \quad \|v\|_{\mathbb{S}} = 0 \Leftrightarrow v = 0$ Positivity $\qquad$ (3.3)

(b) $\|\alpha v\|_{\mathbb{S}} = |\alpha| \|v\|_{\mathbb{S}}, \quad \forall \alpha \in \mathbb{R}$ $\qquad\qquad$ Scalar multiplication $\qquad$ (3.4)

(c) $\|w + v\|_{\mathbb{S}} \leq \|w\|_{\mathbb{S}} + \|v\|_{\mathbb{S}}, \quad \forall w, v \in \mathbb{S}$ Triangular inequality $\qquad$ (3.5)

where "$\Leftrightarrow$" stands for "if and only if." A norm is used to measure how "big" a function is.

3.1.4 Seminorms

If a functional satisfies conditions (Equations 3.4 and 3.5), and the semipositivity

$$\|v\|_{\mathbb{S}} \geq 0, \quad \forall v \in \mathbb{R} \quad \text{Semipositivity} \tag{3.6}$$

instead of the (full) positivity Equation 3.3, it is called a seminorm and denoted as $| \cdot |_{\mathbb{S}}$. A seminorm is often used to measure the strength of the variation (gradient/derivatives) of a function. For solid mechanics problems, it is used to measure the strength of the strain field resulted from a displacement field. The *strain energy* in a solid or structure is a typical seminorm, and is known as *energy norm*. It is an *overall* or *global* measure on how much is solid being stressed. Because any nonzero displacement field of a rigid motion in a solid does not create any strain or stress in the solid, the strain energy is zero for such a nonzero input. The strain energy is thus semipositive.

3.1.5 Linear Forms

A linear form $L(v)$ is a functional if

$$L: \quad \underbrace{\mathbb{S}}_{\text{space offers the input}} \quad \rightarrow \quad \underbrace{\mathbb{R}}_{\text{space hosts the output}} \tag{3.7}$$

with the following linear property:

$$L(\alpha w + v) = \alpha L(w) + L(v), \quad \forall \alpha \in \mathbb{R}, \quad \forall w, v \in \mathbb{S} \tag{3.8}$$

Work done by external forces over a given displacement field in a linear elastic solid is a typical linear form.

3.1.6 Bilinear Forms

A bilinear form $a(w, v)$ is a functional that takes two inputs of member functions, respectively, from two spaces and returns a scalar as an output:

$$a: \quad \underbrace{\mathbb{S}_1}_{\text{space offers } w} \times \underbrace{\mathbb{S}_2}_{\text{space offers } v} \rightarrow \underbrace{\mathbb{R}}_{\text{space hosts the output}} \tag{3.9}$$

with the following bilinear property:

$$\begin{aligned} a(w, \textbf{v}) \text{ is a linear form in } w \text{ for a fixed } v \\ a(\textbf{w}, v) \text{ is a linear form in } v \text{ for a fixed } w \end{aligned} \tag{3.10}$$

Work done by internal stresses in a deformed solid resulted from a virtual displacement field in a typical bilinear form.

3.1.7 Inner Products

An inner product $(w, v)_\mathbb{S}$ on $\mathbb{S}$ is a scalar valued functional on $\mathbb{S} \times \mathbb{S}$ that satisfies the following conditions:

(a) $(w, v)_\mathbb{S} = (v, w)_\mathbb{S}, \quad \forall w, v \in \mathbb{S}$ Symmetry   (3.11)

(b) $(w, w)_\mathbb{S} \geq 0, \quad \forall w \in \mathbb{S}$ and $(w, w)_\mathbb{S} = 0 \Leftrightarrow w = 0$ Positive definite   (3.12)

(c) $(v + w, u)_\mathbb{S} = (v, u)_\mathbb{S} + (w, u)_\mathbb{S}, \quad \forall v, w \in \mathbb{S}$

      $(\alpha v, w)_\mathbb{S} = \alpha(v, w), \quad \forall v, w \in \mathbb{S}, \quad \forall \alpha \in \mathbb{R}$ Bilinear   (3.13)

A linear space $\mathbb{S}$ on which an inner product can be defined is called an *inner product space*. For such a space, a norm can be associated with the inner product in the form of

$$\|v\|_\mathbb{S} = \sqrt{(v, v)_\mathbb{S}} \tag{3.14}$$

A space of admissible displacement functions is an inner product space.

3.1.8 Cauchy–Schwarz Inequality

For a given inner product space $\mathbb{S}$, we then have the often used Cauchy–Schwarz inequality:

$$(w, v)_\mathbb{S} \leq \|w\|_\mathbb{S} \cdot \|v\|_\mathbb{S} \tag{3.15}$$

Let $a(w, v)$ be a semidefinite bilinear form. The Cauchy–Schwarz inequality can then be expressed in a more general form:

$$|a(w, v)| \leq \sqrt{a(w, w)}\sqrt{a(v, v)} \tag{3.16}$$

The Cauchy–Schwarz inequality is often used to derive bound properties of a bilinear form (strain energy).

3.1.9 General Notation of Derivatives

For the convenience of space definitions, we first define the notation of differentiations:

$$D^\alpha = \frac{\partial^{|\alpha|}}{\partial x_1^{\alpha_1} \cdots \partial x_d^{\alpha_d}} \tag{3.17}$$

where
    d is the dimension of the problem domain
    α is n-tuple of nonnegative integers, $\alpha = (\alpha_1, \ldots, \alpha_d)$
    $|\alpha| = \sum_{i=1}^{d} \alpha_i$

3.2 Useful Spaces in Weak Formulations

In the development of FEM based on weak forms for stable and convergent solutions, we had to be very "choosy": We cannot allow using anyhow created displacement functions. Therefore, we have to classify "qualified" functions in groups called spaces. This section lists some of the often used spaces in FEM formulation, because they are also used in meshfree formulations. We will, however, give only minimum necessary elaboration. Interested readers may refer to [4,5] for more systematic descriptions.

3.2.1 Lebesgue Spaces

The *Lebesgue* space $\mathbb{L}^p(\Omega)$, $p \geq 1$ is defined as

$$\mathbb{L}^p(\Omega) = \left\{ v \Big| \int_\Omega |v|^p \, d\Omega < \infty \right\} \tag{3.18}$$

A Lebesgue space is a *normed* space with the norm defined as

$$\|v\|_{\mathbb{L}^p(\Omega)}^p = \int_\Omega v^p \, d\Omega \tag{3.19}$$

This book uses only the $\mathbb{L}^2$ space that is a Lebesgue space with $p = 2$. It is defined as

$$\mathbb{L}^2(\Omega) = \left\{ v \Big| \int_\Omega |v|^2 \, d\Omega < \infty \right\} \tag{3.20}$$

which means that any function v in $\mathbb{L}^2(\Omega)$ is square integrable (integral is bounded) over Ω. Essentially, it means that the function can be discontinuous, but it has to be bounded in the integral sense defined in Equation 3.20: at least piecewise continuous over the problem domain Ω. The norm of a function in an $\mathbb{L}^2$ space is induced from the following inner product:

$$(w, v)_{\mathbb{L}^2(\Omega)} = \int_\Omega wv \, d\Omega \tag{3.21}$$

The induced $\mathbb{L}^2$ norm of a function v is then defined as

$$\|v\|^2_{\mathbb{L}^2(\Omega)} = \underbrace{\int_\Omega v^2 \, d\Omega}_{(v, v)_{\mathbb{L}^2(\Omega)}} \tag{3.22}$$

3.2.2 Hilbert Spaces

We now note the Hilbert space (for a nonnegative integer m) as

$$\mathbb{H}^m(\Omega) = \{v | D^\alpha v \in \mathbb{L}^2(\Omega), \forall |\alpha| \le m\} \tag{3.23}$$

which hosts all functions whose derivatives up to mth order are all square integrable. The associated inner product is given by

$$(w, v) = \sum_{|\alpha| \le m} \int_\Omega (D^\alpha w) \cdot (D^\alpha v) d\Omega \tag{3.24}$$

and induced (full) norm

$$\|v\|_{\mathbb{H}^m(\Omega)} = \left(\sum_{|\alpha| \le m} \int_\Omega |D^\alpha v|^2 d\Omega \right)^{1/2} \tag{3.25}$$

as well as the seminorm (that includes only the mth derivative):

$$|v|_{\mathbb{H}^m(\Omega)} = \left(\int_\Omega |D^\alpha v|^2 d\Omega \right)^{1/2} \tag{3.26}$$

Essentially, the $\mathbb{H}^m$ full norm measures the first m (including zero) derivatives of v in the $\mathbb{L}^2$ norm.

This book uses most often the $\mathbb{H}^1$ space. It is defined for d-dimensional problem domains as

$$\mathbb{H}^1(\Omega) = \{v | v \in \mathbb{L}^2(\Omega), \ \partial v / \partial x_i \in \mathbb{L}^2(\Omega), \ i = 1, \ldots, d\} \tag{3.27}$$

which means that any function in the space and its first derivatives of the function are all square integrable. Clearly, we see the fact that $\mathbb{H}^1$ space is much smaller than the $\mathbb{L}^2$ space, because the additional conditions are imposed on the derivatives of the functions. The associated inner product is given as

$$(w,v)_{\mathbb{H}^1(\Omega)} = \underbrace{\int_\Omega wv\mathrm{d}\Omega}_{(w,v)_{\mathbb{L}^2(\Omega)}} + \underbrace{\int_\Omega (\nabla w) \cdot (\nabla v)\mathrm{d}\Omega}_{(\nabla w, \nabla v)_{\mathbb{L}^2(\Omega)}} \qquad (3.28)$$

where

$$\nabla v = \left(\frac{\partial v}{\partial x_1}\, \frac{\partial v}{\partial x_2} \right) \qquad (3.29)$$

The full norm of a function v in $\mathbb{H}^1$ is defined as

$$\|v\|^2_{\mathbb{H}^1(\Omega)} = \underbrace{\int_\Omega v^2\mathrm{d}\Omega}_{\|v\|^2_{\mathbb{L}^2(\Omega)}} + \underbrace{\int_\Omega (\nabla v) \cdot (\nabla v)\mathrm{d}\Omega}_{|v|^2_{\mathbb{H}^1(\Omega)}} \qquad (3.30)$$

The $\mathbb{H}^1(\Omega)$ seminorm becomes

$$|v|^2_{\mathbb{H}^1(\Omega)} = \int_\Omega (\nabla v) \cdot (\nabla v)\mathrm{d}\Omega \qquad (3.31)$$

Clearly, we can only expect semipositivity for the seminorm, because a nonzero constant v in $\mathbb{H}^1$ space will produce a zero seminorm.

Functions in $\mathbb{H}^1$ that satisfy the essential (displacement) boundary conditions form a space:

$$\mathbb{H}^1_0(\Omega) = \left\{ v \in \mathbb{H}^1(\Omega) | v = 0 \text{ on } \Gamma_u \right\} \qquad (3.32)$$

where Γ_u stands for the boundary on which the displacement is constrained (essential-boundary). A function in $\mathbb{H}^1_0(\Omega)$ is not "floating": constrained for all possible rigid motions.

Once the functions cannot float, the $\mathbb{H}^1$ seminorm becomes (full) positive, and we have the very important property of $\mathbb{H}^1$ space: the $\mathbb{H}^1$ seminorm is *equivalent* to the $\mathbb{H}^1$ full norm, meaning that there exists a positive constant c_{PF} such that

$$c_{PF}\|w\|^2_{\mathbb{H}^1(\Omega)} \le |w|^2_{\mathbb{H}^1(\Omega)}, \quad \forall w \in \mathbb{H}^1_0 \qquad (3.33)$$

which is known as the Poincare–Friedrichs inequality and c_{PF} is known as the Poincare–Friedrichs constant. The Poincare–Friedrichs inequality is one of the most important inequalities in weak formulation, because it ensures fundamentally the stability of the weak form formulation.

A subspace in $\mathbb{H}^1$ space created using interpolation techniques that ensures compatibility (see Chapter 2) can then be defined as

$$\mathbb{H}_h^1(\Omega) = \left\{ v \in \mathbb{H}^1(\Omega) | v(\mathbf{x}) = \boldsymbol{\varphi}^H(\mathbf{x})\mathbf{d}, \mathbf{d} \in \mathbb{R}^{N_n} \right\} \qquad (3.34)$$

where $\boldsymbol{\varphi}^H(\mathbf{x})$ is the matrix of all the (compatible) nodal shape functions constructed using an FEM model, and can be written as

$$\boldsymbol{\varphi}^H(\mathbf{x}) = \begin{bmatrix} \phi_1^H(\mathbf{x}) & \phi_2^H(\mathbf{x}) & \cdots & \phi_{N_n}^H(\mathbf{x}) \end{bmatrix} \qquad (3.35)$$

Because $\mathbb{H}_h^1(\Omega)$ is a linear space, each of the nodal shape functions $\phi_i^H(\mathbf{x})$ must also be in $\mathbb{H}_h^1(\Omega)$. The linear independence of shape functions $\phi_i^H(\mathbf{x})$, $(i = 1, 2, \ldots, N_n)$ are ensured by a standard FEM procedure (element-based and proper mapping). In Equation 3.34 $\mathbf{d}$ is the vector of all the nodal functions values given in the form:

$$\mathbf{d} = \left\{ v_1 \quad v_2 \quad \cdots \quad v_{N_n} \right\}^{\mathrm{T}} \qquad (3.36)$$

Since the values at each node can change independently, we have $\mathbf{d} \in \mathbb{R}^{N_n}$ where $\mathbb{R}^{N_n}$ stands for a real field of N_n dimensions.

Because $\mathbb{H}_h^1$ is constructed in a discrete form with finite dimensions, it is marked with a subscript "h." Functions in $\mathbb{H}_h^1$ that satisfy the essential (displacement) boundary conditions form a space:

$$\mathbb{H}_{h,0}^1(\Omega) = \left\{ v \in \mathbb{H}_h^1(\Omega) | v = 0 \text{ on } \Gamma_u \right\} \qquad (3.37)$$

A very special subspace in $\mathbb{H}^1$ space created using only linear interpolation with 3-node triangular element mesh (or point interpolation method [PIM] with T3-scheme) can be defined as

$$\mathbb{H}_b^1(\Omega) = \left\{ v \in \mathbb{H}^1(\Omega) | v(\mathbf{x}) = \boldsymbol{\varphi}^b(\mathbf{x})\mathbf{d}, \mathbf{d} \in \mathbb{R}^{N_n} \right\} \qquad (3.38)$$

where $\boldsymbol{\varphi}^b(\mathbf{x})$ is the matrix of all the nodal shape functions written as

$$\boldsymbol{\varphi}^b(\mathbf{x}) = \begin{bmatrix} \phi_1^b(\mathbf{x}) & \phi_2^b(\mathbf{x}) & \cdots & \phi_{N_n}^b(\mathbf{x}) \end{bmatrix} \qquad (3.39)$$

in which $\phi_i^b(\mathbf{x}) \in \mathbb{H}^1(\Omega)$ are compatible and linear. In this book, we often use $\mathbb{H}_b^1(\Omega)$ to create a *base* model that is often called linear FEM or FEM-T3 model. In such a case, the triangular cells defined in Section 1.7.2 become the usual triangular elements. Similarly, functions in $\mathbb{H}_b^1$ that satisfy the essential (displacement) boundary conditions form a space:

$$\mathbb{H}_{b,0}^1(\Omega) = \left\{ v \in \mathbb{H}_b^1(\Omega) | v = 0 \text{ on } \Gamma_u \right\} \qquad (3.40)$$

It is clear that the nodal moving least squares (MLS) shape functions created using sufficient local nodes (Chapter 2) are in a $\mathbb{H}_h^1$ space, and any function created using a linear combination of these shape functions are also in the $\mathbb{H}_h^1$ space. An $\mathbb{H}_h^1$ space is indeed very exclusive, and the methods that can be used to create functions in an $\mathbb{H}_h^1$ space are very much limited: FEM technique and the MLS approximation.

3.2.3 Sobolev Spaces

The Sobolev spaces $\mathbb{W}^{m,p}(\Omega)$ for integer $m \geq 0$ and $p \geq 1$ is defined as

$$\mathbb{W}^{m,p}(\Omega) = \{v | D^{\alpha}v \in \mathbb{L}^p(\Omega), \forall |\alpha| \leq m\} \tag{3.41}$$

with norm

$$\|v\|_{\mathbb{W}^{m,p}(\Omega)} = \left(\sum_{|\alpha| \leq m} \int_{\Omega} |D^{\alpha}v|^p \mathrm{d}\Omega \right)^{1/2} \tag{3.42}$$

Essentially, the $\mathbb{W}^{m,p}$ norm measures the first m derivatives of v in the $\mathbb{L}^p$ norm. We note that $\mathbb{W}^{m,2}(\Omega) = \mathbb{H}^m(\Omega)$. When $p \neq 2$, Sobolev spaces are not Hilbert spaces. We also note $\mathbb{W}^{0,p}(\Omega) = \mathbb{L}^p(\Omega)$.

3.2.4 Spaces of Continuous Functions

The spaces of continuous functions, $\mathbb{C}^m(\Omega)$ for any integer m, is defined as

$$\mathbb{C}^m(\Omega) = \{v | D^{\alpha}v \text{ continuous and bounded over } \Omega, \forall |\alpha| \leq m\} \tag{3.43}$$

For example, $\mathbb{C}^0(\Omega)$ is the space of functions that are continuous over Ω; $\mathbb{C}^{-1}(\Omega)$ is the space of functions whose antiderivatives are continuous over Ω; $\mathbb{C}^1(\Omega)$ is the space of functions whose first derivatives are continuous over Ω; $\mathbb{C}^{\infty}(\Omega)$ is the space of functions of which all derivatives exist and are continuous over Ω.

There is an important theorem called the Sobolev embedding theorem that defines the relationship between the $\mathbb{C}^0(\Omega)$ and Sobolev spaces. For example, it says that for any regular domain $\Omega \subset \mathbb{R}^d$, if $v \in \mathbb{H}^m(\Omega)$, $m > \frac{d}{2}$, then $v \in \mathbb{C}^0(\Omega)$, and

$$\|v\|_{\mathbb{L}^{\infty}(\Omega)} \leq C\|v\|_{\mathbb{H}^m(\Omega)} \tag{3.44}$$

where C is a general constant that does not depend on v. For example, if $d = 1$, $v \in \mathbb{H}^1(\Omega)$ implies that v is also continuous, since $m - 1 > d/2 - 1/2$. However, for $d - 2$ or 3, we can find functions $v \in \mathbb{H}^1(\Omega)$ that are not continuous, because $m = 1 \leq \frac{d}{2} = 1$ (or $= \frac{3}{2}$). This implies that the geometry change has implications to the measures for functions.

3.3 $\mathbb{G}$ Spaces: Definition

$\mathbb{G}$ space theory was introduced recently in [6,7,22,23] as a foundation for a weakened-weak (W^2) formulation. Because it is very new, we provide more details and naturally with many open questions to many of which we do not yet have answers. A $\mathbb{G}$ space is a space of *finite* dimensions for discrete functions created based on background cells in FEM or meshfree setting, over which a set of smoothing domain is also properly created [6]. It allows functions to have certain order of discontinuity in Ω, and hence naturally it is larger than the corresponding $\mathbb{H}$ space. For example, a $\mathbb{G}_h^1$ space allows functions being discontinuous (more precisely, being not squarely integrable). It is therefore larger than the $\mathbb{H}_h^1$ space.

It is not, however, as inclusive as the $\mathbb{L}^2$ space, because the discontinuity for functions in a $\mathbb{G}_h^1$ space is properly controlled, and hence it is smaller than the $\mathbb{L}^2$ space. Since the member function in a $\mathbb{G}_h^1$ space has certain order of discontinuity, we cannot measure the semi-norms in the usual way for functions in an $\mathbb{H}_h^1$ space, because the necessary order of derivative information is not accessible (even the distributional derivatives [4]) anywhere in the domain Ω. The idea put forward in [6] was then to *approximate* the derivative information by introducing the so-called generalized smoothing technique that works for discontinuous functions [7]. To perform such smoothing operations, we need to divide the problem domain into smoothing domains, in a *proper* manner.

3.3.1 Smoothing Domain Creation

Consider a domain Ω bounded by Γ discretized by, for example, triangulation defined in Section 1.7.2 with number of on-overlapping and seamless (NOSL) subdomains cells, such that $\overline{\Omega} = \cup_{i=1}^{N_c} \overline{\Omega_i^c}$ and $\Omega_i^c \cap \Omega_j^c = 0, \forall i \neq j, A_i^c > 0, i = 1, 2, \ldots, N_c$, and $h_i > 0, i = 1, 2, \ldots, N_c$. Such an NOSL cell division results in a set of N_c cells, N_n nodes, and N_{cg} straight-line segments $L_i^c > 0, i = 1, \ldots, N_{cg}$ in between the cells or on Γ. L_i^c is also called cell edge, as shown in Figure 3.1.

We next divide, on top of the triangular cells, the domain Ω into N_s NOSL subdomains called smoothing domains such that $\overline{\Omega} = \cup_{i=1}^{N_s} \overline{\Omega_i^s}$ and $\Omega_i^s \cap \Omega_j^s = 0, \forall i \neq j$. Figure 3.1 shows a typical example of such a division, where node-based smoothing domains are created by sequentially connecting the centroids of the triangular cells with the mid-edge-points on L_i^c of the surrounding triangular cells of a node. These smoothing domains are formed by N_{sg} line segments denoted as $L_i^s > 0, i = 1, \ldots, N_{sg}$, and $A_i^s > 0, i = 1, 2, \ldots, N_s$.

For example, the smoothing domain Ω_n^s for node n is bounded by Γ_n^s that is formed by segments AB, BC, CD, DE, EF, FG, GH, HI, and IJ, as shown in Figure 3.1. In this case, the smoothing domain is supported by five cells. These five cells are called *support cells* for the smoothing domain Ω_n^s. All the nodes of these support cells are called *support nodes*

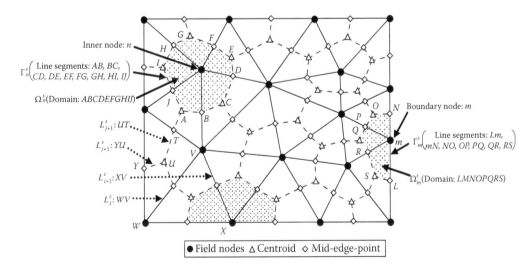

FIGURE 3.1
Background triangular cells and the smoothing domains created by sequentially connecting the centroids with the mid-edge-points of the surrounding triangular cells of a node.

for the smoothing domain. In this particular case, we have six support nodes for smoothing domain Ω_n^s, if the support nodes for each cell is three (linear interpolation). If higher order interpolation is used, the number of support nodes for the support cells will increase, and hence the support nodes for the smoothing domain will increase accordingly. For the node m on the boundary, we have three support cells and five support nodes for the smoothing domain Ω_m^s when the linear interpolation is used.

In the division of Ω into Ω_n^s, the important "no-sharing" rule applies: any Γ_i^s in Ω does not share any *finite* portion of any L_i^c in Ω on which the function is not square integrable. The Γ_i^s can go across L_i^c. Only when the function is continuous on L_i^c, the sharing of Γ_i^s and L_i^c is permitted.

A typical division of domain is given in Figure 3.2 for one-dimensional (1D) domains, where node-based smoothing domains are created upon a set of line background cells. For an interior node, the smoothing domain is supported by two cells and three nodes, and for a boundary node, the smoothing domain is supported by one cell and two nodes. Figure 3.3

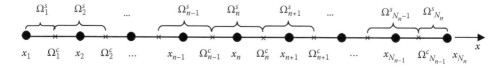

FIGURE 3.2
Background line cells and line smoothing domains created by sequentially connecting the centers of the two neighboring cells.

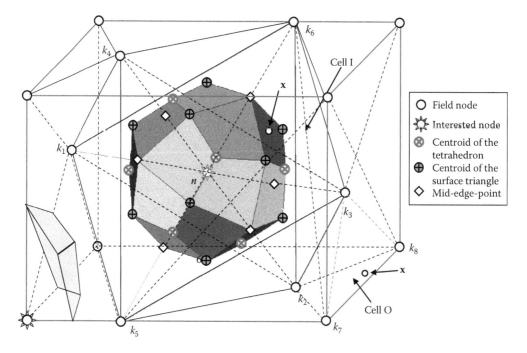

FIGURE 3.3
Background tetrahedral cells and the constructions of node-based smoothing domain for nodes n.

shows an example for 3D domains where node-based smoothing domains are created upon a set of tetrahedral background cells. These types of node-based smoothing domains are intertwined with the cells, and $N_s = N_n$. The node-based smoothing domains were used in the node-based point interpolation method (NS-PIM) [13–16,21] and node-based smoothed finite element method (NS-FEM) [17]. Smoothing domains can also be cell/element based as in the smoothed FEM or smoothed finite element method (SFEM) [10–12], cell-based as in the cell-based smoothed point interpolation method (CS-PIM) [22], edge-based as in the edge-based smoothed finite element method (ES-FEM) [18] and in edge-based point inter-polation method (ES-PIM) [20], as well as face-based as in the face-based smoothed finite element method (FS-FEM) [19].

We note these types of *stationary* smoothing domains are in fact tight-up together with the background cells. When the node (hence the cells) gets refined, the smoothing domains are refined accordingly. In other words, when $h \to 0$, we have $\Omega_k^s \to 0$.

3.3.2 Linearly Independent Smoothing Domains

Linearly independent smoothing domains are defined as a set of smoothing domains that ensure the full rank of the (global) smoothed gradient matrix. The detailed definition and analysis is quite lengthy and can be found in [6]. The essential point to ensure such a linear independence is to have the smoothing domains associated with the nodes, edges, or cells of the mesh. Typical proven divisions that give sufficient number of independent smoothing domains are cell/element based, edge based, as well as face based. The node-based smoothing domains shown in Figure 3.1 through 3.3 have also been proven linearly independent. We will use the node-based and edge-based smoothing domains in Chapter 8.

3.3.3 Minimum Number of Smoothing Domains

To ensure the positivity of the seminorm of a function in a $\mathbb{G}_h^1$ space, we have to use at least the minimum number of smoothing domains N_s, based on the study in [7]. The consideration is that the minimum number of smoothing domains should be at least the same as the number of the unprescribed nodal unknowns N_u, and therefore it depends also on the type of physical problems.

For 1D solid mechanics problem models with one node being fixed, we have immediately $N_s^{\min} = N_u = (N_u - 1)$.

For 2D solid mechanics problem models with n_t (unconstrained) nodes used for displacement field construction, the total number of unknowns in the model should be $N_u = 2n_t$, because one node carries two unknowns (displacement components in x and y directions). On the other hand, the total number of equations that can be sampled from all the smoothing domains should be $3N_n$, because one smoothing domain gives three independent equations for measuring the strain energy norm (each of three strain components produces strain energy independently). Therefore, we must have $N_s^{\min} = 2n_t/3$.

Exactly the same analysis can be done for 3D vector fields of mechanics problems. We now summarize the discussions to Table 3.1.

3.3.4 Integral Representation of Function Derivatives

In Section 2.2 we introduced the integral representation of a function where the function is *approximated* via a convolution performed for the function with a predefined local

TABLE 3.1

Minimum Number of Smoothing Domains $N_s^{\min}$ for Problems with n_t Total Nodal Unknowns

Dimension of the Problem	Minimum Number of Smoothing Domains Vector Field (e.g., Solid Mechanics Problems)
1D	$N_s^{\min} = n_t$
2D	$N_s^{\min} = 2n_t/3$
3D	$N_s^{\min} = 3n_t/6 = n_t/2$

smoothing function. By the same token, the integral representation can also be done for the derivatives of a function:

$$\widehat{\frac{\partial w_l}{\partial x_i}}(\mathbf{x}) = \int_{\Omega_x} \frac{\partial w_l(\boldsymbol{\xi})}{\partial x_i} \widehat{W}(\mathbf{x} - \boldsymbol{\xi}) d\boldsymbol{\xi}, \quad l = 1,2,3; i = 1,2,3 \tag{3.45}$$

where $\widehat{\frac{\partial w_l}{\partial x_i}}(\mathbf{x})$ denotes the smoothed first derivative of w and Ω_x defined for $\mathbf{x}$ is a local smoothing domain bounded by Γ_x, and the smoothing function $\widehat{W}$ is continuously differentiable in Ω_x. When $w_l(\mathbf{x})$ is continuous in Ω_x and hence at least piecewisely differentiable, the usual integration-by-parts can be applied and Equation 3.45 becomes

$$\widehat{\frac{\partial w_l}{\partial x_i}}(\mathbf{x}) = \int_{\Gamma_x} w_l(\boldsymbol{\xi}) n_i \widehat{W}(\mathbf{x} - \boldsymbol{\xi}) d\Gamma - \int_{\Omega_x} w_l(\boldsymbol{\xi}) \frac{\partial \widehat{W}(\mathbf{x} - \boldsymbol{\xi})}{\partial x_i} d\boldsymbol{\xi} \tag{3.46}$$

where n_i is the ith component of the unit outwards normal on Γ_x.

3.3.5 Derivatives Approximation

When $w_l(\mathbf{x})$ is discontinuous in Ω_x, the integration-by-parts is no longer applicable, and hence

$$\widehat{\frac{\partial w_l}{\partial x_i}}(\mathbf{x}) \neq \int_{\Gamma_x} w_l(\boldsymbol{\xi}) n_i \widehat{W}(\mathbf{x} - \boldsymbol{\xi}) d\Gamma - \int_{\Omega_x} w_l(\boldsymbol{\xi}) \frac{\partial \widehat{W}(\mathbf{x} - \boldsymbol{\xi})}{\partial x_i} d\boldsymbol{\xi} \tag{3.47}$$

We, however, still use Equation 3.46 to *approximate* the derivatives of w:

$$\widehat{\frac{\partial w_l}{\partial x_i}}(\mathbf{x}) \approx \int_{\Gamma_x} w_l(\boldsymbol{\xi}) n_i \widehat{W}(\mathbf{x} - \boldsymbol{\xi}) d\Gamma - \int_{\Omega_x} w_l(\boldsymbol{\xi}) \frac{\partial \widehat{W}(\mathbf{x} - \boldsymbol{\xi})}{\partial x_i} d\boldsymbol{\xi} \tag{3.48}$$

This is the generalized gradient (or derivative or strain) smoothing operation [7]. This generalization given in Equation 3.48 is not rigorous in theory, but it is fortunately possible in implementation because no differentiation upon w is required on the right-hand side of Equation 3.48. It was first applied in formulating the node-based smoothed point interpolation methods (NS-PIMs) [13–16] and then in ES-PIM method [6,20] all using incompatible shape functions. This generalization is useful and very important for the establishment of the $\mathbb{G}$ space theory and hence the W^2 formulation that use incompatible functions.

For simplicity, we use in this book the following special smoothing function that is a local constant:

$$\hat{W}(\mathbf{x} - \boldsymbol{\xi}) = \bar{W}(\mathbf{x} - \boldsymbol{\xi}) = \begin{cases} 1/A_{\mathbf{x}} & \boldsymbol{\xi} \in \boxed{\Omega}_{\mathbf{x}} \\ 0 & \boldsymbol{\xi} \notin \boxed{\Omega}_{\mathbf{x}} \end{cases} \tag{3.49}$$

where $A_{\mathbf{x}} = \int_{\Omega_{\mathbf{x}}} d\Omega$ is the area (or volume) of smoothing domain at point $\mathbf{x}$. It is clear that $\bar{W}(\mathbf{x} - \boldsymbol{\xi})$ given above satisfies the conditions of unity, positivity, and decay (defined in Section 2.2). Using Equation 3.49, Equations 3.46 and 3.48 become

$$\overline{\frac{\partial w_l}{\partial x_i}}(\mathbf{x}) = \begin{cases} \frac{1}{A_{\mathbf{x}}} \int_{\Omega_{\mathbf{x}}} \frac{\partial w_l}{\partial x_i} d\Omega = \frac{1}{A_{\mathbf{x}}} \int_{\Gamma_{\mathbf{x}}} w_l(s)n_i ds, & \text{when } w_l(\mathbf{x}) \in \mathbb{C}^0(\Omega_{\mathbf{x}}) \quad \text{(continuous)} \\ \frac{1}{A_{\mathbf{x}}} \int_{\Gamma_{\mathbf{x}}} w_l(s)n_i ds, & \text{when } w_l(\mathbf{x}) \in \mathbb{C}^{-1}(\Omega_{\mathbf{x}}) \quad \text{(discontinuous)} \end{cases} \tag{3.50}$$

Since the smoothing domain used in the above equations changes (or moves) with $\mathbf{x}$, it is termed as *moving* smoothing domain. In many numerical operations we use *stationary* smoothing domains that are fixed in the problem domain. We also require the smoothing domains being created in the way described in Section 3.3.1. In this case, for a smoothing domain Ω_n^s, Equation 3.50 becomes

$$\overline{\frac{\partial w_l}{\partial x_i}}(\mathbf{x}_n) = \begin{cases} \frac{1}{A_n^s} \int_{\Omega_n^s} \frac{\partial w_l}{\partial x_i} d\Omega = \frac{1}{A_n^s} \int_{\Gamma_n^s} w_l(s)n_i ds, & \text{when } w_l(\mathbf{x}) \in \mathbb{C}^0(\Omega_n^s) \quad \text{(continous)} \\ \frac{1}{A_n^s} \int_{\Gamma_n^s} w_l(s)n_i ds, & \text{when } w_l(\mathbf{x}) \in \mathbb{C}^{-1}(\Omega_n^s) \quad \text{(discontinous)} \end{cases} \tag{3.51}$$

where $A_n^s = \int_{\Omega_n^s} d\Omega$. We further assume

$$\overline{\frac{\partial w_l}{\partial x_i}}(\mathbf{x}) = \overline{\frac{\partial w_l}{\partial x_i}}(\mathbf{x}_n) = \underbrace{\frac{1}{A_n^s} \int_{\Gamma_n^s} w_l(s)n_i ds}_{\bar{g}_{l,i}: \text{ constant in } \Omega_n^s}, \qquad \forall \mathbf{x} \in \Omega_n^s \tag{3.52}$$

Therefore the approximated derivative is now constant in Ω_n^s.

The second derivatives of w_l are defined (for 2D cases) as

$$\overline{\frac{\partial^2 w_l}{\partial x_i \partial x_j}} = \underbrace{\frac{1}{2A_n^s} \int_{\Gamma_n^s} \left(\frac{\partial w_l}{\partial x_i} n_j + \frac{\partial w_l}{\partial x_j} n_i \right) ds}_{\bar{g}_{l,ij}: \text{ constant in } \Omega_n^s}, \quad l = 1, 2; \quad i, j = 1, 2 \tag{3.53}$$

Here we assumed the first derivatives for the function exist.

The third derivatives of w_l can always be expressed as the second derivative of a first derivative of w_l, and hence it can be obtained easily using Equation 3.53 by treating the second derivative of w_l as a *new* function. For example,

$$
\overline{\frac{\partial^3 w_l}{\partial x_1^3}} = \overline{\frac{\partial^2}{\partial x_1 \partial x_1}\left(\frac{\partial w_l}{\partial x_1}\right)} = \frac{1}{2 A_n^s} \int_{\Gamma_n^s} \left(\frac{\partial}{\partial x_1}\left(\frac{\partial w_l}{\partial x_1}\right) n_1 + \frac{\partial}{\partial x_1}\left(\frac{\partial w_l}{\partial x_1}\right) n_1\right) ds
$$

$$
= \underbrace{\frac{1}{A_n^s} \int_{\Gamma_n^s} \left(\frac{\partial^2 w_l}{\partial x_1^2} n_1\right) ds}_{\bar{g}_{l,iii}:\ \text{constant in } \Omega_n^s}
\tag{3.54}
$$

In this case we assumed that all the second derivatives for the function exist. The same token can be used for defining any αth smoothed derivatives of w_l, and we should have, in general, a concise form of

$$
\overline{D^\alpha w_l} = \overline{\frac{\partial^{|\alpha|} w_l}{\partial x_1^{\alpha_1} \cdots \partial x_d^{\alpha_d}}}, \quad |\alpha| = \sum_{i=1}^d \alpha_i
\tag{3.55}
$$

as long as all $D^{\alpha-1} w_l$ exist.

We are now ready to define $\mathbb{G}$ spaces.

3.3.6 $\mathbb{G}$ Spaces

The $\mathbb{G}$ spaces are expressed for a nonnegative m:

$$
\mathbb{G}_h^m(\Omega) = \left\{
\begin{array}{l}
v | v(\mathbf{x}) = \boldsymbol{\varphi}(\mathbf{x})\mathbf{d}, \\[4pt]
D^\alpha v \in \mathbb{L}^2(\Omega), \quad \forall \alpha{:}|\alpha| \le (m-1), \\[4pt]
\displaystyle\sum_{n=1}^{N_s} \left(\int_{\Gamma_n^s} (D^\alpha v) n_i ds\right)^2 = 0,\ \forall v \ne 0, \quad i = 1,\ldots,d, \quad \forall \alpha{:}|\alpha| \le (m-1)
\end{array}
\right\} \quad \forall \mathbf{d} \in \mathbb{R}^{N_n}
\tag{3.56}
$$

where $\boldsymbol{\varphi}(\mathbf{x})$ is the matrix of nodal shape functions constructed using a general PIM.

It is observed that the $\mathbb{G}$ space is a set of functions formed using a basis created using a technique presented in Chapter 2. The derivatives of the functions up to the $(m-1)$th orders are square integrable in Ω. Because of the discretized nature of a $\mathbb{G}$ space, it is marked with a subscript "h." The major difference between a $\mathbb{G}$ space and the corresponding *discrete* Hilbert space or $\mathbb{H}_h$ space is that the $\mathbb{H}$ space requires $D^\alpha v \in \mathbf{L}^2(\Omega)$, for $|\alpha| = m$, but in the $\mathbb{G}$ space we require only $D^\alpha v \in \mathbf{L}^2(\Omega)$, for $|\alpha| = (m-1)$. Therefore, the requirement on function is now further weakened upon the already weakened requirement for functions in an $\mathbb{H}$ space, and hence a $\mathbb{G}$ space can be viewed as a space of a set of functions with weakened-weak (W^2) requirements. It is therefore clear now that a $\mathbb{H}_h^m(\Omega)$ is also a $\mathbb{G}_h^m(\Omega)$ space: any function in $\mathbb{H}_h^m(\Omega)$ is surely qualified as a member in $\mathbb{G}_h^m(\Omega)$.

The inner product associated with a $\mathbb{G}$ space is then defined as

$$
(w, v)_{\mathbb{G}^m(\Omega)} = \sum_{|\alpha| \le m-1} \left(\int_\Omega D^\alpha w \cdot D^\alpha v \, d\Omega \right) + \sum_{|\alpha| = m} \sum_{n=1}^{N_s} A_n^s \overline{D^\alpha w} \cdot \overline{D^\alpha v}
\tag{3.57}
$$

The inner product-induced (full) norm is next defined as

$$\|w\|_{\mathbb{G}^m(\Omega)} = \left[\sum_{|\alpha| \leq m-1} \left(\int_\Omega |D^\alpha w|^2 d\Omega \right) + \sum_{|\alpha|=m} \left(\sum_{n=1}^{N_s} A_n^s |\overline{D^\alpha w}|^2 \right) \right]^{1/2} \tag{3.58}$$

The $\mathbb{G}^m(\Omega)$ seminorm is finally defined using only the smoothed αth derivatives:

$$|w|_{\mathbb{G}^m(\Omega)} = \left[\sum_{|\alpha|=m} \left(\sum_{n=1}^{N_s} A_n^s |\overline{D^\alpha w}|^2 \right) \right]^{1/?} \tag{3.59}$$

In this book we use only the $\mathbb{G}_h^1$ spaces, and therefore more details are given in the following sections.

3.3.7 $\mathbb{G}_h^1$ Space

A $\mathbb{G}_h^1$ space can be then defined as

$$\mathbb{G}_h^1(\Omega) = \left\{ \begin{array}{l} v | v(\mathbf{x}) = \boldsymbol{\varphi}(\mathbf{x})\mathbf{d}, \\ v \in \mathbb{L}^2(\Omega), \\ \displaystyle\sum_{n=1}^{N_s} \left(\int_{\Gamma_n^s} v(s)n_i ds \right)^2 > 0, \forall v \neq 0, \quad i=1,\ldots,d, \end{array} \right\} \quad \forall \mathbf{d} \in \mathbb{R}^{N_n} \tag{3.60}$$

where $\mathbf{d} = \{d_1 \quad d_2 \quad \cdots \quad d_{N_n}\}^{\mathrm{T}}$ is the vector of nodal function values, and $\boldsymbol{\varphi}(\mathbf{x})$ is the matrix of nodal shape functions of arbitrary order constructed using any method discussed in Chapter 2, and can be written as

$$\boldsymbol{\varphi}(\mathbf{x}) = \begin{bmatrix} \phi_1(\mathbf{x}) & \phi_2(\mathbf{x}) & \cdots & \phi_{N_n}(\mathbf{x}) \end{bmatrix} \tag{3.61}$$

In creating functions in $\mathbb{G}_h^1$ spaces, we do not restrict the way in which shape functions are created, as long as they satisfy the following conditions:

1. *Linear independency condition*: all these nodal shape functions are linearly independent over Ω and hence are capable to form a basis.

2. *Bound condition*: all the functions constructed using these shape function must be square integrable over the problems domain. This is to ensure the convergence of a numerical model to be created.

3. *Positivity conditions*: there exist a division of Ω_i^s such that $\sum_{n=1}^{N_s} \left(\int_{\Gamma_n^s} v(s)n_i ds \right)^2 > 0$, for all $v \neq 0$, $\forall \mathbf{d} \in \mathbb{R}^{N_n}$, and $i=1,\ldots,d$. This (together with the linearly independent condition) is to ensure the stability of a numerical model to be created.

When PIM or radial point interpolation method (RPIM) shape functions (see Chapter 2) are used, the functions constructed will in general not be continuous over the entire problem domain and hence are not compatible. Such an interpolant is not in an $\mathbb{H}_h^1$ space, but in a $\mathbb{G}_h^1$ space, because all these three conditions can be satisfied, as shown in [6]. Note that since $\phi_i(\mathbf{x})$, $(i=1,2,\ldots,N_n)$ are constructed using nodes selected using a T-scheme (see Section 1.7.6), and at least the three nodes of any home cell are always used. Hence, they always have an equal or higher order compared to the linear FEM base model, and we shall have $\mathbb{H}_b^1 \subseteq \mathbb{G}_h^1$ for all PIM shape functions and RPIM shape functions with linear polynomial basis constructed using a T-scheme.

The major difference between a $\mathbb{G}_h^1$ space and $\mathbb{H}_h^1$ space is that the $\mathbb{H}_h^1$ space requires the first gradient of the function to be square integrable, but in the $\mathbb{G}_h^1$ space we require only the function itself square integrable. Therefore, the requirement on function is now further weakened upon the already weakened requirement for functions in an $\mathbb{H}_h^1$ space, and hence a $\mathbb{G}_h^1$ space can be viewed as space of a set of functions with weakened-weak (W^2) requirements on continuity. In an $\mathbb{H}_h^1$ space, the bound condition is achieved by the imposing the smoothness upon the first derivatives of the function to be square integrable, while in the $\mathbb{G}_h^1$ space, it is controlled by imposing the smoothness only on the function to be square integrable (with a proper construction of smoothing domains). The stability is automatically ensured for functions in an $\mathbb{H}_h^1$ space as long as the smoothness is satisfied, due to the Poincare–Friedrichs inequality. The stability in the $\mathbb{G}_h^1$ space, however, is ensured by imposing the positivity condition.

Because a member in a $\mathbb{G}_h^1$ space is also a member of the $\mathbb{L}^2$ space, therefore, a $\mathbb{G}_h^1$ space is a subspace of $\mathbb{L}^2$ space: $\mathbb{G}_h^1(\Omega) \subset \mathbb{L}^2(\Omega)$.

We note that any function created using as set of shape functions that satisfy the above-mentioned three conditions are also in the $\mathbb{G}_h^1$ space. A $\mathbb{G}_h^1$ space is indeed very accommodating and inclusive, and hence shall have much wider applications.

3.3.8 Normed or Unnormed $\mathbb{G}$ Spaces

$\mathbb{G}$ space is the foundation for the weakened-weak (W^2) formulations, and they can either be normed or unnormed. In this book, we use both normed and unnormed $\mathbb{G}$ spaces for W^2 formulations. Unnormed $\mathbb{G}$ spaces are used for the strain-constructed Galerkin (or SC-Galerkin) models, where admissible conditions for the constructed strains are defined properly in a separated manner. The normed $\mathbb{G}$ space is used for the generalized smoothed Galerkin (or GS-Galerkin) models that are special cases of W^2 formulations. Normed $\mathbb{G}$ spaces require a proper construction of smoothing domains following the rules detailed in Section 3.3.1.

Note that unnormed $\mathbb{G}$ spaces can also be used for establishing strong form meshfree methods by, for example, simple collocation. In such cases, the stability is left "uncontrolled" when the assumed functions are used to create a discrete model, and we need additional procedure such as the regularization techniques to restore the stability [8,9].

For normed $\mathbb{G}_h^1$ spaces, the norms are *induced* from the inner products defined as follows for various cases.

3.3.9 $\mathbb{G}_h^1$ Norms for 1D Scalar Fields

For 1D scalar fields, the associated inner product is given by

$$(w,v)_{\mathbb{G}^1(\Omega)} = \underbrace{\int_\Omega wv\,\mathrm{d}\Omega}_{} + \sum_{n=1}^{N_s} A_n^s \overline{w'} \cdot \overline{v'} = \underbrace{\int_\Omega wv\,\mathrm{d}\Omega}_{(w,v)_{\mathbb{L}^2(\Omega)}} + \underbrace{\sum_{n=1}^{N_s} A_n^s \bar{g}(w)\bar{g}(v)}_{\left(\overline{w'},\overline{v'}\right)_{\mathbb{L}^2(\Omega)}} \qquad (3.62)$$

In Equation 3.62 the (approximated) smoothed derivative for smoothing domain Ω_n^s is denoted as

$$\overline{w}' = \frac{\overline{\partial w}}{\partial x} = \frac{1}{A_n^s}\int_{\Gamma_n^s} w(s)n_x ds = \underbrace{\frac{1}{A_n^s}\left(w_{k+\frac{1}{2}} - w_{k-\frac{1}{2}}\right)}_{=\bar{g},\,\text{constant in }\Omega_n^s} = \bar{g}(w) \tag{3.63}$$

where $\bar{g}(w)$ denotes the smoothed derivatives of w with respect to x, and the smoothing domains Ω_n^s is "centered" at x_n bounded by $x_{n-\frac{1}{2}}$ and $x_{n+\frac{1}{2}}$, as shown in Figure 3.2. Since $\bar{g}(w)$ is constant in Ω_n^s, $\overline{w}'$ will be a piecewise constant function in Ω, and hence is in the $\mathbb{L}^2$ space.

It is easy to show (see Remark 3.4 below) that the $\mathbb{G}_h^1$ inner product defined in Equation 3.62 satisfies the definition given in Section 3.1.7.

The $\mathbb{G}_h^1$ seminorm is next defined as

$$|w|_{\mathbb{G}^1(\Omega)}^2 = \sum_{n=1}^{N_s} A_n^s |\overline{w}'|^2 = \underbrace{\sum_{n=1}^{N_s} A_n^s \bar{g}^2(w)}_{(\overline{w}',\overline{w}')} \tag{3.64}$$

and the $\mathbb{G}_h^1$ full norm becomes

$$\|w\|_{\mathbb{G}^1(\Omega)}^2 = \underbrace{\int_\Omega w^2 d\Omega}_{(w,w)=\|w\|_{L^2}^2} + \underbrace{|w|_{\mathbb{G}^1(\Omega)}^2}_{(\overline{w}',\overline{w}')=|w|_{\mathbb{G}^1(\Omega)}^2} = \|w\|_{L^2}^2 + |w|_{\mathbb{G}^1(\Omega)}^2 \tag{3.65}$$

which is induced from the inner product Equation 3.62.

3.3.10 $\mathbb{G}_h^1$ Norms for 2D Scalar Fields

For 2D scalar fields, the associated inner product is given by

$$(w,v)_{\mathbb{G}^1(\Omega)} = \underbrace{\int_\Omega wv d\Omega}_{(w,v)_{L^2(\Omega)}} + \sum_{n=1}^{N_s} A_n^s \overline{\nabla w}\cdot\overline{\nabla v} = \int_\Omega wv d\Omega + \underbrace{\sum_{n=1}^{N_s} A_n^s(\bar{g}_1(w)\bar{g}_1(v) + \bar{g}_2(w)\bar{g}_2(v))}_{(\overline{\nabla w},\overline{\nabla v})_{L^2(\Omega)}} \tag{3.66}$$

Note that the summation is possible because the division of Ω into Ω_n^s is performed in such a way that the interfaces Γ_i^s of Ω_n^s do not share any *finite* portion of any L_i^c on which the function is not square integrable: no energy loss in the interface of the smoothing domains.

Equation 3.66, the (approximated) smoothed gradient for smoothing domain Ω_n^s is denoted as

$$\overline{\nabla}w = \left(\overline{\frac{\partial w}{\partial x_1}}\ \overline{\frac{\partial w}{\partial x_2}}\right) = \left(\underbrace{\frac{1}{A_n^s}\int_{\Gamma_n^s}w(s)n_1 ds}_{=\bar{g}_1,\text{ constant in }\Omega_n^s}\ \ \underbrace{\frac{1}{A_n^s}\int_{\Gamma_n^s}w(s)n_2 ds}_{=\bar{g}_2,\text{ constant in }\Omega_n^s}\right) = (\bar{g}_1(w)\ \ \bar{g}_2(w))$$

(3.67)

where $\bar{g}_i(w)$ denotes the smoothed derivatives of w with respect to x_i.

The $\mathbb{G}_h^1$ inner product defined in Equation 3.66 is also qualified based on the definition given in Section 3.1.7 for the same reason mentioned in Remark 3.4.

The $\mathbb{G}_h^1$ seminorm is next defined as

$$|w|_{\mathbb{G}^1(\Omega)}^2 = \sum_{n=1}^{N_s}A_n^s|\overline{\nabla}w|^2 = \underbrace{\sum_{n=1}^{N_s}A_n^s\left(\bar{g}_1^2(w)+\bar{g}_2^2(w)\right)}_{(\overline{\nabla}w,\overline{\nabla}w)}$$

(3.68)

and the $\mathbb{G}_h^1$ full norm becomes

$$\|w\|_{\mathbb{G}^1(\Omega)}^2 = \underbrace{\int_\Omega w^2 d\Omega}_{(w,w)=\|w\|_{L^2}^2} + \underbrace{|w|_{\mathbb{G}^1(\Omega)}^2}_{(\overline{\nabla}w,\overline{\nabla}w)} = \|w\|_{L^2}^2 + |w|_{\mathbb{G}^1(\Omega)}^2$$

(3.69)

which is induced from the inner product Equation 3.66. The definitions for 3D scalar fields are a natural extension and hence are omitted here. We now move to the definitions for more complicated vector fields.

3.3.11 $\mathbb{G}_h^1$ Norms for 2D Vector Fields

For vector fields, we need to use vectors of functions. For example, when the function has two components, we should have $\mathbf{w} = \{w_1\ \ w_2\}^T$ where $w_1, w_2 \in \mathbb{G}_h^{1*}$ are two-component functions. In this case, we have the smoothed gradient for smoothing domain Ω_n^s in the following form:

$$\overline{\nabla}\mathbf{w} = \begin{pmatrix}\overline{\dfrac{\partial w_1}{\partial x_1}}\ \overline{\dfrac{\partial w_1}{\partial x_2}}\\[2mm]\overline{\dfrac{\partial w_2}{\partial x_1}}\ \overline{\dfrac{\partial w_2}{\partial x_2}}\end{pmatrix} = \begin{pmatrix}\underbrace{\dfrac{1}{A_n^s}\int_{\Gamma_n^s}w_1(s)n_1 ds}_{=\bar{g}_{11},\text{ constant in }\Omega_s}\ \ \underbrace{\dfrac{1}{A_n^s}\int_{\Gamma_n^s}w_1(s)n_2 ds}_{=\bar{g}_{12},\text{ constant in }\Omega_s}\\[4mm]\underbrace{\dfrac{1}{A_n^s}\int_{\Gamma_n^s}w_2(s)n_1 ds}_{=\bar{g}_{21},\text{ constant in }\Omega_n^s}\ \ \underbrace{\dfrac{1}{A_n^s}\int_{\Gamma_n^s}w_2(s)n_2 ds}_{=\bar{g}_{22},\text{ constant in }\Omega_n^s}\end{pmatrix} = \begin{pmatrix}\bar{g}_{11}(w_1)\ \bar{g}_{12}(w_1)\\\bar{g}_{21}(w_2)\ \bar{g}_{22}(w_2)\end{pmatrix}$$

(3.70)

---

* In this book, when we require a vector is in a space, we require each of the component functions being independently in the space, and the dimension of the space is expanded accordingly.

where $\bar{g}_{ij}(w)$ denotes the smoothed derivatives of w_i with respect to x_j. We notice here that the (smoothed) gradient is now a matrix, and hence there can be many *equivalent* ways to define the associated inner product. In this work, we decide to have the definition associated with the type of physical problems to be studied for convenience of proving necessary theories for that type of the problems. Considering 2D solid mechanics problems, we define the associated inner product in the form:

$$(\mathbf{w}, \mathbf{v})_{\mathbb{G}^1(\Omega)} = \underbrace{\int_\Omega (w_1 v_1 + w_2 v_2) d\Omega}_{(w,v)}$$
$$+ \underbrace{\sum_{n=1}^{N_s} A_n^s \left[\begin{array}{l} \bar{g}_{11}(w_1)\bar{g}_{11}(v_1) + \bar{g}_{22}(w_2)\bar{g}_{22}(v_2) \\ + (\bar{g}_{12}(w_1) + \bar{g}_{21}(w_2))(\bar{g}_{12}(v_1) + \bar{g}_{21}(v_2)) \end{array} \right]}_{(\nabla w, \nabla v)} \quad (3.71)$$

The induced $\mathbb{G}^1(\Omega)$ seminorm is first defined as

$$|\mathbf{w}|^2_{\mathbb{G}^1(\Omega)} = \underbrace{\sum_{n=1}^{N_s} A_n^s \left(\bar{g}_{11}^2(w_1) + \bar{g}_{22}^2(w_2) + (\bar{g}_{12}(w_1) + \bar{g}_{21}(w_2))^2 \right)}_{(\nabla w, \nabla w)} \quad (3.72)$$

It is clear that in our definition of the inner product and hence in the induced seminorm we have intentionally related to the strain components, and hence the $\mathbb{L}^2$ norm of the vector of strains.

The associated $\mathbb{G}_h^1$ full norm can now be defined as

$$\|\mathbf{w}\|^2_{\mathbb{G}^1(\Omega)} = \underbrace{\int_\Omega (w_1^2 + w_2^2) d\Omega}_{(w,w) = \|w\|^2_{L^2}} + \underbrace{|\mathbf{w}|^2_{\mathbb{G}^1(\Omega)}}_{(\nabla w, \nabla w)} = \|\mathbf{w}\|^2_{L^2} + |\mathbf{w}|^2_{\mathbb{G}^1(\Omega)} \quad (3.73)$$

3.3.12 $\mathbb{G}_h^1$ Norms for 3D Vector Fields

For vector fields with three-component functions in three-dimensions, such as the 3D solid mechanics problems, we shall have $\mathbf{w} = \{w_1 \quad w_2 \quad w_3\}^{\mathsf{T}}$ where $w_1, w_2, w_3 \in \mathbb{G}_h^1$. In this case we define, naturally, the associated inner product as

$$(\mathbf{w}, \mathbf{v})_{\mathbb{G}^1(\Omega)} = \int_\Omega (w_1 v_1 + w_2 v_2 + w_3 v_3) d\Omega$$
$$+ \sum_{n=1}^{N_s} A_n^s \left[\begin{array}{l} \bar{g}_{11}(w_1)\bar{g}_{11}(v_1) + \bar{g}_{22}(w_2)\bar{g}_{22}(v_2) + \bar{g}_{33}(w_3)\bar{g}_{3}(v_3) \\ + (\bar{g}_{12}(w_1) + \bar{g}_{21}(w_2))(\bar{g}_{12}(v_1) + \bar{g}_{21}(v_2)) \\ + (\bar{g}_{13}(w_1) + \bar{g}_{31}(w_3))(\bar{g}_{13}(v_1) + \bar{g}_{31}(v_3)) \\ + (\bar{g}_{23}(w_2) + \bar{g}_{32}(w_3))(\bar{g}_{23}(v_2) + \bar{g}_{32}(v_3)) \end{array} \right] \quad (3.74)$$

The associated $\mathbb{G}_h^1$ seminorm is defined as

$$|\mathbf{w}|_{\mathbb{G}^1(\Omega)}^2 = \sum_{n=1}^{N_s} A_n^s \begin{pmatrix} \bar{g}_{11}^2(w_1) + \bar{g}_{22}^2(w_2) + \bar{g}_{33}^2(w_3) \\ + (\bar{g}_{12}(w_1) + \bar{g}_{21}(w_2))^2 \\ + (\bar{g}_{13}(w_1) + \bar{g}_{31}(w_3))^2 \\ + (\bar{g}_{23}(w_2) + \bar{g}_{32}(w_3))^2 \end{pmatrix} \tag{3.75}$$

and the $\mathbb{G}_h^1$ full norm becomes

$$\|\mathbf{w}\|_{\mathbb{G}^1(\Omega)}^2 = \int_\Omega (w_1^2 + w_2^2 + w_3^2)\mathrm{d}\Omega + |\mathbf{w}|_{\mathbb{G}^1(\Omega)}^2 \tag{3.76}$$

We finally define a space for functions that are fixed on the Dirichlet boundaries and hence the functions cannot "float."

$$\mathbb{G}_{h,0}^1 = \{\mathbf{v} \in \mathbb{G}_h^1(\Omega) | \mathbf{v} = 0 \text{ on } \Gamma_u\} \tag{3.77}$$

The $\mathbb{G}_h^1$ spaces defined in this section are "unusual" in two ways: first, we do not use derivatives of functions because we want to accommodate discontinuous functions that can be generated much easily in both the meshfree or finite element settings; second, the Frobenius or trace norms are not used as induced matrix norms, and we intentionally define the inner product in such a way that the inner product-induced norms are related to the $\mathbb{L}^2$ norms of the vector of strains, which facilitates a smoother process in the later part of the derivation of key inequalities.

3.4 $\mathbb{G}_h^1$ Spaces: Basic Properties

Because the normed $\mathbb{G}_h^1$ spaces are defined in the above-mentioned "unusual" manner, we have to show that they possess all the necessary basic properties. Here we discuss only normed $\mathbb{G}_h^1$ spaces.

3.4.1 Linearity

First, a normed $\mathbb{G}_h^1$ space is a *linear* space, meaning that conditions listed in Equation 3.1 can be satisfied. To show this, consider any two functions $w, v \in \mathbb{G}_h^1$. From the definition (Equation 3.60), we shall have

$$\begin{aligned} w(\mathbf{x}) &= \boldsymbol{\varphi}(\mathbf{x})\mathbf{d}_w, \quad \mathbf{d}_w \in \mathbb{R}^{N_n} \\ v(\mathbf{x}) &= \boldsymbol{\varphi}(\mathbf{x})\mathbf{d}_v, \quad \mathbf{d}_v \in \mathbb{R}^{N_n} \end{aligned} \tag{3.78}$$

The addition of w and v becomes

$$w(\mathbf{x}) + v(\mathbf{x}) = \boldsymbol{\varphi}(\mathbf{x}) \underbrace{(\mathbf{d}_w + \mathbf{d}_v)}_{\in \mathbb{R}^{N_n}} \tag{3.79}$$

which must also be in $\mathbb{G}_h^1$, because $\mathbb{R}^{N_n}$ is a linear space that satisfies Equation 3.1: $(\mathbf{d}_w + \mathbf{d}_v) \in \mathbb{R}^{N_n}$. Following exactly the same argument, we shall have that for $\forall w \in \mathbb{G}_h^1$ and $\forall \alpha \in \mathbb{R}, \alpha w \in \mathbb{G}_h^1$. ∎

3.4.2 Positivity

From the definition, for example Equation 3.65, we have

$$\|w\|_{\mathbb{G}^1(\Omega)}^2 = \|w\|_{\mathbb{L}^2}^2 + |w|_{\mathbb{G}^1(\Omega)}^2, \quad \forall w \in \mathbb{G}_h^1 \tag{3.80}$$

Since the $\mathbb{L}^2$ norm is positive and the $\mathbb{G}^1$ seminorm is semipositive, we shall always have

$$\|w\|_{\mathbb{G}^1} > 0, \quad \forall w \in \mathbb{G}_h^1, \quad w \neq 0 \tag{3.81}$$

3.4.3 Scalar Mortification

From the definition, for example Equation 3.62, we have

$$\|\alpha w\|_{\mathbb{G}^1} = |\alpha| \|w\|_{\mathbb{G}^1}, \quad \forall \alpha \in \mathbb{R}, \quad \forall w \in \mathbb{G}_h^1 \tag{3.82}$$

3.4.4 Triangular Inequality

We now prove the *triangular inequality* for $\mathbb{G}_h^1$ norm

$$\|w + v\|_{\mathbb{G}^1} \leq \|w\|_{\mathbb{G}^1} + \|v\|_{\mathbb{G}^1}, \quad \forall w \in \mathbb{G}_h^1, \quad \forall v \in \mathbb{G}_h^1 \tag{3.83}$$

We first prove this for 2D scalar functions:

$$\|w + v\|_{\mathbb{G}^1} = \left[\int_\Omega (w + v)^2 d\Omega + \sum_{n=1}^{N_s} A_n^s \left| \overline{\nabla(w + v)} \right|^2 \right]^{1/2}$$

$$= \left[\int_\Omega w^2 d\Omega + \int_\Omega v^2 d\Omega + 2 \int_\Omega wv d\Omega + \sum_{n=1}^{N_s} A_n^s \left((\bar{g}_1(w) + \bar{g}_1(v))^2 + (\bar{g}_2(w) + \bar{g}_2(v))^2 \right) \right]^{1/2}$$

$$= \left[\int_\Omega w^2 d\Omega + \int_\Omega v^2 d\Omega + 2 \int_\Omega wv d\Omega + \sum_{n=1}^{N_s} A_n^s \left(\begin{array}{c} \bar{g}_1^2(w) + 2\bar{g}_1(w)\bar{g}_1(v) + \bar{g}_1^2(v) \\ + \bar{g}_2^2(w) + 2\bar{g}_2(w)\bar{g}_2(v) + \bar{g}_2^2(v) \end{array} \right) \right]^{1/2}$$

$$\tag{3.84}$$

$$
= \left[\underbrace{\int_\Omega w^2 d\Omega + \sum_{n=1}^{N_s} A_n^s (\bar{g}_1^2(w) + \bar{g}_2^2(w))}_{\|w\|_{G^1}^2} + \underbrace{\int_\Omega v^2 d\Omega + \sum_{n=1}^{N_s} A_n^s (\bar{g}_1^2(v) + \bar{g}_2^2(v))}_{\|v\|_{G^1}^2} \right.
$$

$$
\left. + 2 \left(\underbrace{\int_\Omega wv d\Omega + \sum_{n=1}^{N_s} A_n^s (\bar{g}_1(w)\bar{g}_1(v) + \bar{g}_2(w)\bar{g}_2(v))}_{(w,v)_{G^1} \le \|w\|_{G^1} \|v\|_{G^1}} \right) \right]^{1/2}
$$

$$
\le \left[\|w\|_{G^1}^2 + 2\|w\|_{G^1}\|v\|_{G^1} + \|v\|_{G^1}^2 \right]^{1/2} = \|w\|_{G^1} + \|v\|_{G^1}, \quad \forall w \in \mathbb{G}_h^1, \quad \forall v \in \mathbb{G}_h^1 \qquad (3.85)
$$

In the above proof, we used the Cauchy–Schwarz inequality for our inner product induced norms.

The exact same procedure can applied to prove the triangular inequality for vector functions, but it will be a little lengthy. We prove it here for the 2D case, by examining first the seminorm of the sum of two functions $w, v \in \mathbb{G}^1$ based on the definition of Equation 3.72:

$$
|w + v|_{\mathbb{G}^1(\Omega)}^2 = \sum_{n=1}^{N_s} A_n^s \left(\bar{g}_{11}^2(w_1 + v_1) + \bar{g}_{22}^2(w_2 + v_2) + (\bar{g}_{12}(w_1 + v_1) + \bar{g}_{21}(w_2 + v_2))^2 \right)
$$

$$
= \sum_{n=1}^{N_s} A_n^s \left(\begin{array}{l} \bar{g}_{11}^2(w_1) + \bar{g}_{11}^2(v_1) + 2\bar{g}_{11}(w_1)\bar{g}_{11}(v_1) + \bar{g}_{22}^2(w_2) + \bar{g}_{22}^2(v_2) \\ + 2\bar{g}_{22}(w_2)\bar{g}_{22}(v_2) + ((\bar{g}_{12}(w_1) + \bar{g}_{21}(w_2)) + (\bar{g}_{12}(v_1) + \bar{g}_{21}(v_2)))^2 \end{array} \right)
$$

$$
= \left(\underbrace{\sum_{n=1}^{N_s} A_n^s \left(\bar{g}_{11}^2(w_1) + \bar{g}_{22}^2(w_2) + (\bar{g}_{12}(w_1) + \bar{g}_{21}(w_2))^2 \right)}_{|w|_{\mathbb{G}^1(\Omega)}^2} \right.
$$

$$
\left. + \underbrace{\sum_{n=1}^{N_s} A_n^s \left(\bar{g}_{11}^2(v_1) + \bar{g}_{22}^2(v_2) + (\bar{g}_{12}(v_1) + \bar{g}_{21}(v_2))^2 \right)}_{|v|_{\mathbb{G}^1(\Omega)}^2} \right.
$$

$$
\left. + 2 \sum_{n=1}^{N_s} A_n^s \left(\begin{array}{l} \bar{g}_{11}(w_1)\bar{g}_{11}(v_1) + \bar{g}_{22}(w_2)\bar{g}_{22}(v_2) \\ + (\bar{g}_{12}(w_1) + \bar{g}_{21}(w_2))(\bar{g}_{12}(v_1) + \bar{g}_{21}(v_2)) \end{array} \right) \right) \qquad (3.86)
$$

We then examine the full norm of the sum of two functions $w, v \in \mathbb{G}_h^1$ based on the definition of Equation 3.73:

$$\|w + v\|_{\mathbb{G}^1(\Omega)}^2 = \int_\Omega \left((w_1 + v_1)^2 + (w_2 + v_2)^2 \right) d\Omega + |w + v|_{\mathbb{G}^1(\Omega)}^2$$

$$= \int_\Omega \left(w_1^2 + v_1^2 + w_2^2 + v_2^2 + 2w_1 v_1 + 2w_2 v_2 \right) d\Omega + |w + v|_{\mathbb{G}^1(\Omega)}^2$$

$$= \int_\Omega \left(w_1^2 + v_1^2 \right) d\Omega + \int_\Omega \left(w_2^2 + v_2^2 \right) d\Omega + 2\int_\Omega \left(w_1 v_1 + w_2 v_2 \right) d\Omega + |w + v|_{\mathbb{G}^1(\Omega)}^2 \quad (3.87)$$

Substituting Equation 3.86 into Equation 3.87 gives

$$\|w + v\|_{\mathbb{G}^1}^2 = \int_\Omega \left(w_1^2 + v_1^2 \right) d\Omega + \int_\Omega \left(w_2^2 + v_2^2 \right) d\Omega + 2\int_\Omega \left(w_1 v_1 + w_2 v_2 \right) d\Omega$$

$$+ |w|_{\mathbb{G}^1}^2 + |v|_{\mathbb{G}^1}^2 + 2\sum_{n=1}^{N_s} A_n^s \left(\begin{array}{l} + \bar{g}_{11}(w_1)\bar{g}_{11}(v_1) + \bar{g}_{22}(w_2)\bar{g}_{22}(v_2) \\ + (\bar{g}_{12}(w_1) + \bar{g}_{21}(w_2))(\bar{g}_{12}(v_1) + \bar{g}_{21}(v_2)) \end{array} \right)$$

$$= \|w\|_{\mathbb{G}^1}^2 + \|v\|_{\mathbb{G}^1}^2 + 2 \underbrace{\left(\begin{array}{l} \int_\Omega (w_1 v_1 + w_2 v_2) d\Omega \\ + \sum_{n=1}^{N_s} A_n^s \left(\begin{array}{l} \bar{g}_{11}(w_1)\bar{g}_{11}(v_1) + \bar{g}_{22}(w_2)\bar{g}_{22}(v_2) \\ + (\bar{g}_{12}(w_1) + \bar{g}_{21}(w_2))(\bar{g}_{12}(v_1) + \bar{g}_{21}(v_2)) \end{array} \right) \end{array} \right)}_{(w,v) \le \|w\|_{\mathbb{G}^1(\Omega)} \|v\|_{\mathbb{G}^1(\Omega)}}$$

$$\le \|w\|_{\mathbb{G}^1}^2 + \|v\|_{\mathbb{G}^1}^2 + 2\|w\|_{\mathbb{G}^1}\|v\|_{\mathbb{G}^1} = \left(\|w\|_{\mathbb{G}^1} + \|v\|_{\mathbb{G}^1} \right)^2 \quad (3.88)$$

which is Equation 3.83. Note here we used again the Cauchy–Schwarz inequality.

The other important basic property for a linear space is the completeness. We note here that a $\mathbb{G}_h^1$ space is complete. A detailed discussion on this can be found in [22].

3.4.5 Key Inequalities

For a linear space being useful in constructing stable and convergent numerical methods for well-posed problems, it has to have a set of key inequalities. We disscuss now these key inequalities.

Comparing Equation 3.65 with Equation 3.64, we obtain

$$|w|_{\mathbb{G}^1(\Omega)} \le \|w\|_{\mathbb{G}^1(\Omega)}, \quad \forall w \in \mathbb{G}_h^1 \quad (3.89)$$

meaning that the $\mathbb{G}_h^1$ full norm is always larger than the $\mathbb{G}_h^1$ seminorm.

Remark 3.1: Functions in $\mathbb{G}_h^1$ Space: First Inequality

Functions in a $\mathbb{G}_h^1$ space satisfy the first inequality (Equations 3.90 and 3.91): The full $\mathbb{G}_h^1$ norm of a function is equivalent to the $\mathbb{L}^2$ norm of the nodal values of the function.

$$\|\mathbf{d}\|_{\mathbb{L}^2(\Omega)} \ge c_{dw}^f \|w\|_{\mathbb{G}^1(\Omega)}, \quad \forall w \in \mathbb{G}_h^1 \quad (3.90)$$

or equivalently

$$\|w\|_{\mathbb{G}^1(\Omega)} \geq c_{wd}^f \|\mathbf{d}\|_{\mathbb{L}^2(\Omega)}, \quad \forall w \in \mathbb{G}_h^1 \tag{3.91}$$

where c_{dw}^f and c_{wd}^f are nonzero positive constants independent of w and $\mathbf{d}$. A proof for these inequalities (Equations 3.90 and 3.91) is given in [6].

Remark 3.2: Functions in $\mathbb{L}^2$ Space: Second Inequality
If at least a minimum number of linearly independent smoothing domains are used, we should have the second inequality:

$$|w|_{\mathbb{G}^1(\Omega)} \geq c_{wd}^s \|\mathbf{d}\|_{\mathbb{L}^2(\Omega)}, \quad \forall w \in \mathbb{G}_{h,0}^1(\Omega) \tag{3.92}$$

or equivalently

$$\|\mathbf{d}\|_{\mathbb{L}^2(\Omega)} \geq c_{dw}^s |w|_{\mathbb{G}^1(\Omega)}, \quad \forall w \in \mathbb{G}_{h,0}^1(\Omega) \tag{3.93}$$

where c_{dw}^s and c_{wd}^s are nonzero positive constants independent of w and $\mathbf{d}$. Equations 3.92 and 3.93 mean that the full $\mathbb{G}_h^1$ norm of a function is equivalent to the $\mathbb{L}^2$ norm of the nodal values of the function. A proof for these inequalities (Equations 3.92 and 3.93) has been given in [6].

Remark 3.3: $\mathbb{G}_h^1$ Norm Equivalency: Third Inequality
When a minimum number of independent node-based smoothing domains are used to evaluate the $\mathbb{G}_h^1$ norms, there exists a positive nonzero constant c_G such that

$$c_G \|w\|_{\mathbb{G}^1(\Omega)} \leq |w|_{\mathbb{G}^1(\Omega)}, \quad \forall w \in \mathbb{G}_{h,0}^1 \tag{3.94}$$

meaning that the $\mathbb{G}_h^1$ full norm and the $\mathbb{G}_h^1$ seminorm of any function in a $\mathbb{G}_h^1$ space are equivalent known as the third inequality [6]. It is a generalized Poincare–Friedrichs inequality for $\mathbb{G}_h^1$ spaces. It is fundamentally important for stability of weakened-weak formulation based on $\mathbb{G}$ spaces. A detailed proof on this norm equivalence theorem in $\mathbb{G}$ space theory requires the first and second inequalities, and the details are given in [6] when PIM shape functions are used.

Combining Equations 3.89 and 3.94, when a minimum number of independent node-based smoothing domains are used to evaluate the $\mathbb{G}_h^1$ norms, we arrived at the following chain inequalities:

$$c_G \|w\|_{\mathbb{G}^1(\Omega)} \leq |w|_{\mathbb{G}^1(\Omega)} \leq \|w\|_{\mathbb{G}^1(\Omega)}, \quad \forall w \in \mathbb{G}_{h,0}^1 \tag{3.95}$$

These inequalities are fundamentally important to ensure the stability and convergence of a weakened-weak formulation.

Remark 3.4: $\mathbb{G}_h^1$ Inner Product Space: Cauchy–Schwarz Inequality
The $\mathbb{G}_h^1$ inner product defined in Equation 3.62 is qualified based on the definition given in Section 3.1.7, and hence we shall have the Cauchy–Schwarz inequality:

$$(w, v)_{\mathbb{G}^1(\Omega)} \leq \|w\|_{\mathbb{G}^1(\Omega)} \cdot \|v\|_{\mathbb{G}^1(\Omega)} \tag{3.96}$$

To show this, we first observe the symmetric (Equation 3.11), because swapping places for w and v will not change the value of the inner product. Second, it is positive definite (Equation 3.12), because of the positivity of the $(w, v)_{\mathbb{L}^2(\Omega)}$ and semipositivity of $\left(\overline{w}', \overline{v}'\right)_{\mathbb{L}^2(\Omega)}$. Finally, it is bilinear (Equation 3.13), because of the bilinear property of $(w, v)_{\mathbb{L}^2(\Omega)}$ and $\left(\overline{w}', \overline{v}'\right)_{\mathbb{L}^2(\Omega)}$.

Equation 3.96 is fundamentally important for the continuity of the weakened-weak formulation.

3.4.6 Relationship with Other Spaces

$$\mathbb{H}_h^1 \subset \mathbb{G}_h^1 \tag{3.97}$$

$$\mathbb{G}_h^1(\Omega) \subset \mathbb{L}^2(\Omega) \tag{3.98}$$

3.5 Error Estimation

This section briefs some basic error estimation issues for weak and weakened-weak formulations, for the reference convenience in the discussions in the later chapters. We will focus only on error bounds and convergence rates for methods using linear displacement field, and will not give details of proofs for these inequalities but simply list them with the reference sources. For error measures in $\mathbb{G}$ norms, there are still too many unknown details, and hence we focus only on interpolation error. For simplicity, we consider only 1D problems.

3.5.1 Interpolation Errors

Consider now the error when a "target" function w is to be interpolated using the interpolant defined in Section 2.1.7. We first define such an *interpolation error* for the given target function w as

$$e^I(x) = w(x) - \mathcal{I}_h w(x), \quad \forall x \in \Omega_i^c \tag{3.99}$$

where Ω_i^c is the domain of the ith cell defined in Section 1.7.2. In the standard Galerkin weak formulation such as FEM, we should have the following bounds.

3.5.2 Error in $\mathbb{L}^2$ Norm

For $w \in \mathbb{H}_0^1$ and uniform mesh the interpolation error in $\mathbb{L}^2$ norm [4,5] is

$$|e^l(x)|_{\mathbb{L}^2} = |w - \mathcal{I}_h w|_{\mathbb{L}^2} \leq h^2 \left(\max_{i=1,\ldots,N_c} \max_{\xi \in \Omega_i^c} |w''(\xi)| \right) \qquad (3.100)$$

The proof of this bound in FEM was based on Rolle's theorem. Equation 3.100 gives the h-dependence of the error in $\mathbb{L}^2$ norms in relation to the "strong" norm: the infinity norm of the target function to be approximated by interpolation. In the weak formulation, the h-dependence of the *solution error* in displacement norm is the same as that of the *interpolation error* in $\mathbb{L}^2$ norm [4,5]. Therefore, Equation 3.100 gives the essence of the h-dependence of the solution errors.

3.5.3 Error in $\mathbb{H}^1$ Norm

For $w \in \mathbb{H}_0^1$ and uniform mesh the interpolation error in $\mathbb{H}^1$ norm [4,5] is

$$|e^l(x)|_{\mathbb{H}^1} = |w - \mathcal{I}_h w|_{\mathbb{H}^1} \leq h \left(\max_{i=1,\ldots,N_c} \max_{\xi \in \Omega_i^c} |w''(\xi)| \right) \qquad (3.101)$$

which gives the h-dependence of the error in $\mathbb{H}^1$ norm. In the weak formulation, the h-dependence of the solution error in energy norm is the same as that of the interpolation error in $\mathbb{H}^1$ norm. Therefore, Equation 3.101 reveals the essence of the h-dependence of the solution error in energy norm. We note the following remark without proof.

Remark 3.5: The Rate of Convergence of Linear FEM Solution: A Proven Fact
The theoretical *rate* of convergence of the solution error in $\mathbb{L}^2$ norm (error in displacement norm) for a linear FEM (fully compatible) model is 2.0, and that in $\mathbb{H}^1$ norm (error in strain energy norm) is 1.0.

3.5.4 Error in $\mathbb{G}^1$ Norm

Theorem 3.1: Interpolation Error in $\mathbb{G}_h^1$ Norm

If the target function $w \in \mathbb{G}_{h,0}^1$ and $w|_{\Omega_i^c} \in \mathbb{C}^2(\Omega_i^c)$, $i = 1, \ldots, N_c$, the linear interpolation error satisfies, in general

$$|e^l|_{\mathbb{G}^1(\Omega)} = |w - \mathcal{I}_h w|_{\mathbb{G}^1(\Omega)} \leq h_{\max} \frac{3c_{rh}}{4} \left(\max_{q=1,\ldots,N_e} \max_{\xi \in \Omega_q} |w''(\xi)| \right) \qquad (3.102)$$

where $c_{rh} = h_{\max}/h_{\min}$. In particular, when uniform mesh is used, and when the second derivative of w is constant in the cells sharing the smoothing domains we further have

$$|e^l|_{\mathbb{G}^1(\Omega)} = |w - \mathcal{I}_h w|_{\mathbb{G}^1(\Omega)} = h^{1.5} \frac{1}{4} |w''| \qquad (3.103)$$

This rate of 1.5 is termed as *ideal convergence rate* for a model based on $\mathbb{G}_h^1$ space theory that is achieved for uniform mesh and constant second derivative of w.

Proof Consider a target function $w \in \mathbb{G}_{h,0}^1$ and $w|_{\Omega_i^c} \in \mathbb{C}^2(\Omega_i^c)$, $i = 1, \ldots, N_c$. Using the Taylor's expansion with respect to x_n (see Figure 3.2), there exists a $\xi \in \Omega_n^c$ such that

$$w(x) = w(x_n) + w'(x_n)(x - x_n) + \frac{1}{2}w''(\xi(x))(x - x_n)^2, \quad \forall x \in \Omega_n^c \tag{3.104}$$

In Equation 3.104, we note the fact that ξ is in fact dependent on x. We also note that we do not require the existence of w'' on the boundary of Ω_n^c meaning that the first derivative of w can "jump" there. Using Equation 2.5, the linear interpolant has to pass through two nodes at x_n and x_{n+1}, and hence should be given as

$$\mathcal{I}_h w(x) = w(x_n) + \left[w'(x_n) + \frac{1}{2}w''(\xi(x_{n+1}))h_n \right](x - x_n), \quad \forall x \in \Omega_n^c \tag{3.105}$$

Based on the definition of Equation 3.63, for our 1D problem and the nth smoothing cell, we have

$$\bar{g}(w_n) = \frac{\overline{\partial w}}{\partial x} = \frac{1}{A_n^s}\int_{\Gamma_n^s} w(s)n_1 ds = \frac{1}{A_n^s}\left(w_{n+\frac{1}{2}} - w_{n-\frac{1}{2}}\right) = \frac{2}{h_n + h_{n-1}}\left(w_{n+\frac{1}{2}} - w_{n-\frac{1}{2}}\right) \tag{3.106}$$

where
$$w_n = w(x_n)$$
$$h_0 = h_{N_n} = 0$$

Substitute Equations 3.104 and 3.105 into Equation 3.106, we then have for the nth smoothing cell:

$$\bar{g}(\underbrace{w_n - \mathcal{I}_h w_n}_{e^I}) = \frac{2}{h_n + h_{n-1}} \left\{ \begin{array}{l} w''\left(\xi\left(x_{n+\frac{1}{2}}\right)\right)\left(x_{n+\frac{1}{2}} - x_n\right)^2 - \frac{1}{2}w''(\xi(x_{n+1}))h_n\left(x_{n+\frac{1}{2}} - x_n\right) \\[2mm] -\frac{1}{2}w''\left(\xi\left(x_{n-\frac{1}{2}}\right)\right)\left(x_{n-\frac{1}{2}} - x_{n-1}\right)^2 \\[2mm] +\frac{1}{2}w''(\xi(x_n))h_{n-1}\left(x_{n-\frac{1}{2}} - x_{n-1}\right) \end{array} \right\}$$

$$= \frac{2}{h_n + h_{n-1}} \left\{ \begin{array}{l} \frac{h_n^2}{4}\left[\frac{1}{2}w''\left(\xi\left(x_{n+\frac{1}{2}}\right)\right) - w''(\xi(x_{n+1}))\right] \\[2mm] -\frac{h_{n-1}^2}{4}\left[\frac{1}{2}w''\left(\xi\left(x_{n-\frac{1}{2}}\right)\right) - w''(\xi(x_n))\right] \end{array} \right\} \tag{3.107}$$

Let

$$w''_{max} = \max_{i=1,\ldots,N_e} \max_{\xi \in \Omega_i^c} |w''(\xi)| \tag{3.108}$$

and hence we shall have $w''_{max} \neq 0$, otherwise the second derivative of w will be 0 everywhere in the problem domain, and the interpolation will be exact: no need for error estimation. Using Equation 3.105 into Equation 3.106, Equation 3.107 becomes

$$
\left| \bar{g}(\underbrace{w_n - \mathcal{I}_h w_n}_{e^l}) \right| = \frac{h_{max}^2}{4} \underbrace{\frac{2}{h_n + h_{n-1}}}_{\leq \frac{1}{h_{min}}} |w''_{max}| \underbrace{\left[\begin{array}{c} \underbrace{\frac{1}{2} \frac{h_n^2}{h_{max}^2}}_{\leq 1} \underbrace{\frac{w''\left(\xi\left(x_{n+\frac{1}{2}} \right) \right)}{|w''_{max}|}}_{-1\leq,\leq 1} - \underbrace{\frac{h_n^2}{h_{max}^2}}_{<1} \underbrace{\frac{w''(\xi(x_{n+1}))}{|w''_{max}|}}_{-1\leq,\leq 1} \\ -\underbrace{\frac{1}{2}\frac{h_{n-1}^2}{h_{max}^2}}_{\leq 1} \underbrace{\frac{w''\left(\xi\left(x_{n-\frac{1}{2}} \right) \right)}{|w''_{max}|}}_{-1\leq,\leq 1} + \underbrace{\frac{h_{n-1}^2}{h_{max}^2}}_{\leq 1} \underbrace{\frac{w''(\xi(x_n))}{|w''_{max}|}}_{-1\leq,\leq 1} \end{array} \right]}_{c_{hw} \leq 3}
$$

$$
\leq \frac{3h_{max}^2}{4} \underbrace{\frac{h_{max}}{h_{min}}}_{c_{rh} < \infty} |w''_{max}| = h_{max}\frac{3c_{rh}}{4}|w''_{max}| \tag{3.109}
$$

The error defined in $\mathbb{G}_h^1$ norm in the global problem domain becomes

$$
|w - \mathcal{I}_h w|_{\mathbb{G}}^2 = \sum_{n=1}^{N_s} A_n^s |\bar{g}(w_n - \mathcal{I}_h w_n)|^2 \leq \sum_{n=1}^{N_s} h_{max}\left(h_{max}\frac{3c_{rh}}{4}|w''_{max}| \right)^2
$$

$$
= \underbrace{N_s}_{\leq \frac{1}{h_{max}}} h_{max}\left(h_{max}\frac{3c_{rh}}{4}|w''_{max}| \right)^2 \leq \left(h_{max}\frac{3c_{rh}}{4}|w''_{max}| \right)^2 \tag{3.110}
$$

Therefore, we have Equation 3.102.

In Equation 3.109, we needed to estimate c_{hw} and gave a very sloppy bound of 3. This is in fact a very big overestimate in a usual situation. If a uniform mesh is used, and the target function has a constant second derivative in the cells sharing a smoothing domain, c_{hw} should be 0 for all the inner smoothing domains. In such a situation, Equation 3.109 becomes

$$
|\bar{g}(w_n - \mathcal{I}_h w_n)| = \begin{cases} \frac{h}{4}|w''_{max}|, & n = 1, N_n \\ 0, & s = 2, 3, \ldots, N_n - 1 \end{cases} \tag{3.111}
$$

and Equation 3.110 becomes

$$
|w - \mathcal{I}_h w|_{\mathbb{G}^1}^2 = \sum_{n=1}^{N_s} A_n^s |\bar{g}(w_n - \mathcal{I}_h w_n)|^2
$$

$$
= A_1 |\bar{g}(w_1 - \mathcal{I}_h w_1)|^2 + A_{N_n} |\bar{g}(w_{N_n} - \mathcal{I}_h w_{N_n})|^2
$$

$$
= \frac{h}{2}\left(\frac{h}{4}|w''_{max}| \right)^2 + \frac{h}{2}\left(\frac{h}{4}|w''_{max}| \right)^2 = h\left(\frac{h}{4}|w''_{max}| \right)^2 = \frac{h^3}{4^2}|w''_{max}|^2 \tag{3.112}
$$

which is Equation 3.103. This completes the proof. ∎

Note in Equation 3.103 the bound is sharpest: strict equality. The proof on the above two equations were originally given in [6].

In the W^2 formulation based on $\mathbb{G}$ space theory, we *may* expect the rate of h-dependence of the *solution error* to be the same as that of the *interpolation error*. However, this has not been proven theoretically. We note the following remark of expectation without proof.

Remark 3.6:   The Rate of Convergence of W^2 Solution Based on $\mathbb{G}^1$ Space: An Expectation

The theoretical rate of convergence of the solution error in $\mathbb{L}^2$ norm (error in displacement norm) for W^2 model based on $\mathbb{G}_h^1$ space is expected to be 2.0, and that in $\mathbb{G}_h^1$ norm (error in strain energy norm) is expected to be between 1.0 and the ideal rate of 1.5.

3.5.5 Comparison Errors in $\mathbb{G}^1$ and $\mathbb{H}^1$ Norms

An exact comparison is difficult, due to the space difference between $\mathbb{H}_{h,0}^1$ and $\mathbb{G}_{h,0}^1$ and hence w will be different. But an indicative comparison can be useful. For errors in seminorm measure, Equation 3.101 is quite close to Equation 3.102 for uniform mesh ($c_{rh} = 1.0$): They all give a convergence rate h-dependence of 1.0. Equation 3.103 shows, however, the W^2 formulation *can* provide a convergence rate of 1.5 in seminorm measure for cases of even division of node-based smoothing domains. Compared to Equation 3.101, the convergence *rate* is 50% higher.

Equation 3.103 was obtained under the conditions of (1) uniform division of elements/cells and (2) constant second derivative of w in the cells sharing a smoothing domain. The first condition is essentially the same "symmetrical condition" of smoothing domains for the integral representation to produce the first gradient of a function exactly (see Section 2.2.3). When this condition is satisfied, all the interior smoothing domains become symmetrical, and the smoothing operation will reproduce the first derivative exactly. The only error will be on the boundary where the symmetry condition cannot be satisfied. The second condition of constant second derivative of the target function seems to be very strong, but it can be rather easy to be quite closely satisfied, because all we need is the second derivative of the target function being constant *locally* in the cells sharing a smoothing domain. When the mesh is refined, we can often expect the second derivative being approximately constant locally. Therefore, the rate given in Equation 3.103 can be expected when the mesh is reasonably fine. In practical applications, on the other hand, the first condition is rather very difficult to meet, simply because it is rare to have uniform division of element/cells for practical problems of complicated geometry. However, we can, in fact, expect the smoothing domains to be approximately "symmetric" locally ($h_n \approx h_{n-1}$ for node n for our 1D problems). In such cases, we can still expect Equation 3.103 to hold approximately and hence a convergence rate of 1.5 in $\mathbb{G}_h^1$ seminorm. This has been confirmed in may numerical examples presented in [7,13–20], where numerical rates of about 1.4 were often found. We will also see many examples in Chapter 8.

Let us now further examine for a general problem with reasonably smoothing second derivative. The coefficient c_{hw} in Equation 3.109 becomes

$$
c_{hw} = \left\| \begin{bmatrix} \dfrac{1}{2}\underbrace{\dfrac{h_n^2}{h_{max}^2}}_{\leq 1}\underbrace{\dfrac{w''\left(\xi\left(x_{n+\frac{1}{2}}\right)\right)}{|w''_{max}|}}_{-1\leq,\leq 1} - \underbrace{\dfrac{h_n^2}{h_{max}^2}}_{\leq 1}\underbrace{\dfrac{w''(\xi(x_{n+1}))}{|w''_{max}|}}_{-1\leq,\leq 1} \\[2em] -\dfrac{1}{2}\underbrace{\dfrac{h_{n-1}^2}{h_{max}^2}}_{\leq 1}\underbrace{\dfrac{w''\left(\xi\left(x_{n-\frac{1}{2}}\right)\right)}{|w''_{max}|}}_{-1\leq,\leq 1} + \underbrace{\dfrac{h_{n-1}^2}{h_{max}^2}}_{\leq 1}\underbrace{\dfrac{w''(\xi(x_n))}{|w''_{max}|}}_{-1\leq,\leq 1} \end{bmatrix} \right\|
$$

$$
\approx \frac{1}{2}\frac{1}{h_{max}^2}\underbrace{\frac{|w''(\xi(x_n))|}{|w''_{max}|}}_{\leq 1}|h_n^2 - h_{n-1}^2| = \frac{1}{2}\underbrace{\frac{h_n(h_n + h_{n-1})}{h_{max}^2}}_{\leq 1}\underbrace{\frac{|w''(\xi(x_n))|}{|w''_{max}|}}_{\leq 1}\left| 1 - \underbrace{\frac{h_{n-1}}{h_n}}_{\approx 1}\right|
$$

$$
\leq \left| 1 - \underbrace{\frac{h_{n-1}}{h_n}}_{\approx 1}\right| \tag{3.113}
$$

If the lengths of the two neighboring cells of the interior nodes are different, the contribution of these nodes to the error norm *can* be in control (when mesh is refined), and thus the convergence rate will be about 1.0 and not 1.5. However, c_{hw} in Equation 3.109 can be a very small number. If, for example, there is a 20% length difference in the two neighboring cells of all the interior nodes ($h_{n-1}/h_n = 0.8$), we shall have $c_{hw} \approx |1 - h_{n-1}/h_n| = 0.2$. This means that even the rate of convergence cannot be improved, the results will still be about five times more accurate. This was also observed in the ES-PIM where the edge-based smoothing domains are not quite symmetric, but were often found to be about 2–10 times more accurate in energy norm compared to the FEM using the same mesh [18]. In extreme cases, where all the smoothing domains are not symmetric at all, the results of the W^2 formulation will still be better than the standard weak formulation, as shown in Equations 3.101 and 3.102.

Because a finite $\mathbb{H}_h^1$ space is a subspace of a corresponding $\mathbb{G}_h^1$ space, the convergence rates found in $\mathbb{G}_h^1$ norm measures are applicable to functions in the $\mathbb{H}_h^1$ space. This means that when a weakened-weak formulation is applied to a FEM setting, we will achieve higher convergence rate and/or higher accuracy.

Finally, we note that for the weakened-weak formulation, the relation between the *interpolation error* and the *solution error* is not yet clear. Similar relationships stated in Sections 3.5.2 and 3.5.3 may exist, but are yet to be proved.

3.6 Concluding Remarks

Before closing up this chapter, we mention the following remarks.

The standard weak formulation used in FEM is applicable and has applied to meshfree settings. As long as the meshfree shape functions is created in a proper $\mathbb{H}$ space, the stability and convergence of the meshfree methods is ensured for physically well-posed problems.

Based on the $\mathbb{G}$ space theory presented in this chapter, a weakened-weak formulation is applicable and has applied to meshfree settings as well as FEM settings (because an $\mathbb{H}_h^1$ space is in a $\mathbb{G}_h^1$ space). As long as the meshfree shape functions created is in a $\mathbb{G}_h^1$ space, the stability and convergence of the meshfree methods is ensured by these key inequalities for physically well-posed problems.

With the $\mathbb{G}$ space theory, we can now claim that *as long as a set of linearly independent shape functions that satisfy the bound and positivity conditions can be created, a stable and convergence method can always be established using a weakened-weak formulation.* We do not have to worry about the compatibility issue any more: a peace of mind. A W^2 meshfree method produces symmetric equations systems without any additional degrees of freedoms, much more accurate, higher convergence rate, and hence much more efficient than FEM using the same mesh as demonstrated in [6,7,13–20], and in Chapter 8. Therefore, $\mathbb{G}$ space theory is indeed an important fundamental theory for meshfree, element-based, compatible and incompatible methods.

With these theories on $\mathbb{H}$ and $\mathbb{G}$ spaces, we can now establish weak or weakened-weak statements to formulate meshfree methods for various problems. In these processes, all we need to do now is to form proper bilinear forms for the given type of problems, and then use them in the weak or weakened-weak statement to establish algebraic equations. Details will be given in Chapter 5.

We can also choose to follow physical energy principles (to be presented in Chapter 5) to establish weak or weakened-weak forms for meshfree methods using function spaces defined in this chapter. This physical approach is more preferred by engineers, as we can easily obtain all sorts of weak forms for various types of structures. In addition, we can have more freedom to introduce yet one more step, *strain field construction*, so as to formulate an even wide class of important and efficient meshfree methods with desired properties. The techniques on strain field construction will be detailed in Chapter 4.

References

1. Hughes, T. J. R., *The Finite Element Method*, Prentice-Hall, London, 1987.
2. Liu, G. R. and Quek, S. S., *The Finite Element Method: A Practical Course*, Butterworth Heinemann, Oxford, 2002.
3. Naylor, A. W. and Sell, G. R., *Linear Operator Theory in Engineering and Science*, Springer-Verlag, New York, 1982.
4. Peraire, J., *Lecture Notes on Finite Element Methods for Elliptic Problems*. MIT, Cambridge, MA, 1999.
5. Strang, G. and Fix, G. J., *An Analysis of the Finite Element Method*, Prentice-Hall, Englewood Cliffs, NJ, 1973.
6. Liu, G. R., A $\mathbb{G}$ space theory and a weakened weak (W^2) form for a unified formulation of compatible and incompatible methods, Part I: Theory and Part II: Applications to solid mechanics problems, *Int. J. Numerical Methods Eng.*, 2008 (revised).
7. Liu, G. R., A generalized Gradient smoothing technique and the smoothed bilinear form for Galerkin formulation of a wide class of computational methods, *Int. J. Comput. Methods*, 5(2), 199–236, 2008.
8. Liu, G. R. and Kee, B. B. T., A stabilized least-squares radial point collocation method (LS-RPCM) for adaptive analysis, *Comput. Method Appl. Mech. Eng.*, 195, 4843–4861, 2006.
9. Kee, B. B. T., Liu, G. R., and Lu, C., A regularized least-squares radial point collocation method (RLS-RPCM) for adaptive analysis, *Comput. Mech.*, 40, 837–853, 2007.

10. Liu, G. R., Dai, K. Y., and Nguyen, T. T., A smoothed finite element method for mechanics problems, *Comput. Mech.*, 39, 859–877, 2007.
11. Liu, G. R., Nguyen, T. T., Dai, K. Y., and Lam, K. Y., Theoretical aspects of the smoothed finite element method (SFEM), *Int. J. Numerical Methods Eng.*, 71, 902–930, 2007.
12. Dai, K. Y., Liu, G. R., and Nguyen, T. T., An n-sided polygonal smoothed finite element method (nSFEM) for solid mechanics, *Finite Elem. Anal. Des.*, 43, 847–860, 2007
13. Liu, G. R., Zhang, G. Y., Dai, K. Y., Wang, Y. Y., Zhong, Z. H., Li, G. Y., and Han, X., A linearly conforming point interpolation method (LC-PIM) for 2D solid mechanics problems, *Int. J. Comput. Methods*, 2(4), 645–665, 2005.
14. Liu, G. R. and Zhang, G. Y., Upper bound solution to elasticity problems: A unique property of the linearly conforming point interpolation method (LC-PIM), *Int. J. Numerical Methods Eng.*, 74, 1128–1161, 2008.
15. Li, Y., Liu, G. R., Luan, M. T., Dai, K. Y., Zhong, Z. H., Li, G. Y., and Han, X., Contact analysis for solids based on linearly conforming radial point interpolation method, *Comput. Mech.*, 39, 537–554, 2007.
16. Zhang, G. Y., Liu, G. R., Nguyen, T. T., Song, C. X., Han, X., Zhong, Z. H., and Li, G. Y., The upper bound property for solid mechanics of the linearly conforming radial point interpolation method (LC-RPIM), *Int. J. Comput. Methods*, 4(3), 521–541, 2007.
17. Liu, G. R., Nguyen-Thoi, T., Nguyen-Xuan, H., and Lam, K. Y., A node-based smoothed finite element method (NS-FEM) for upper bound solutions to solid mechanics problems, *Comput. Struct.*, 87, 14–26, 2009.
18. Liu, G. R., Nguyen-Thoi, T., and Lam, K. Y., An edge-based smoothed finite element method (ES-FEM) for static, free and forced vibration analyses in solids, *J. Sound Vibration*, 320, 1100–1130, 2009.
19. Nguyen-Thoi, T., Liu, G. R., Lam, K. Y., and Zhang, G. Y., A face-based smoothed finite element method (FS-FEM) for 3D linear and nonlinear solid mechanics problems using 4-node tetrahedral elements, *Int. J. Numerical Methods Eng.*, 78, 324–353, 2009.
20. Liu, G. R. and Zhang, G. Y., Edge-based smoothed point interpolation method (ES-PIM), *Int. J. Comput. Methods*, 5(4), 621–646, 2008.
21. Zhang, G. Y., Liu, G. R., Wang, Y. Y., Huang, H. T,. Zhong, Z. H., Li, G. Y., and Han, X., A linearly conforming point interpolation method (LC-PIM) for three-dimensional elasticity problems, *Int. J. Numerical Methods Eng.*, 72, 1524–1543, 2007.
22. Liu, G. R. and Zhang, G. Y., A normed $\mathbb{G}$ space and weakened weak (W^2) formulation of a cell-based smoothed point interpolation method, *Int. J. Comput. Methods*, 6(1), 147–179, 2009.
23. Liu, G. R., On $\mathbb{G}$ space theory, *Int. J. Comput. Methods*, 6(2), 1–33, 2009.

4

Strain Field Construction

Novel and more efficient numerical methods can be invented, only if we can always be open-minded. The traditional procedure of three steps used in displacement methods is very simple: (1) domain discretization, (2) displacement field construction (via shape function creation), and (3) weak formulation to derive the discretized algebraic equations system that can be solved using standard routines. The most popular finite element method (FEM) follows this process and its effectiveness has been proven [1,2]. To further advance, in meshfree methods we need to introduce an additional step after step 2: *strain field construction*, so as to separate steps 2 and 3, in order to create a room for improving significantly the solution accuracy, and for solving the compatibility issues in FEM so that incompatible methods can be formulated properly based on the weakened-weak (W^2) formulation (see Chapter 5).

Because techniques for strain field construction are relatively less developed, we are able to introduce only a few techniques, and even for these few there are still many theoretical issues which need to be studied further. The author believes that the potential in this direction of development should not be underestimated, and hence a lot more efforts are needed. The success in this direction of development depends on two major issues: (1) the addition step of strain field construction should be simple, cost-effective, and without introducing addition degrees of freedom to the equations system; (2) it should show sufficient improvement with respect to FEM, and/or offer attractive properties. This chapter discusses some of the techniques used for strain field construction, which focus only on the works for weakened-weak formulations, including the so-called strain-constructed Galerkin (SC-Galerkin) formulations, which are discussed in Chapter 5.

4.1 Why Construct a Strain Field?

Once the displacement in a proper $\mathbb{H}$ space is assumed, the strain field is readily available using simply the strain–displacement relation, known as the compatible strain field. Why we want to construct another strain field? The reasons are described as follows.

In the FEM settings, because the assumed displacement field is compatible over the entire problem domain, and the strain field is obtained using the strain–displacement relation precisely, the Galerkin model or the standard FEM model is said to be fully compatible. Such a standard FEM model is variationally consistent and works well for many practical problems. However, there are three major issues associated with this type of fully compatible Galerkin formulation. The first issue is the well known "overly stiff" phenomenon, which can have possible consequences of (1) the so-called "locking" behavior for many problems, (2) inaccuracy in stress solutions, and (3) poor solutions when using a triangular mesh. The second issue concerns mesh distortion-related problems also, such as significant accuracy loss when the element mesh is heavily distorted. The third issue relates to the difficult task of generation of quality mesh. We, engineers, often prefer using the triangular type of meshes as they can be created much more easily and even automatically for complicated geometries. However, it is well known that a fully compatible FEM model does not like such elements and often gives solutions of very poor accuracy especially for stresses. It demands a good quality mesh of quadrilateral elements.

Many efforts have been made in resolving the overly stiff phenomenon, especially in the area of hybrid or mixed FEM formulations based on two or three field principles [3,4]. Some kind of strain modifications have also been used in FEM settings. However, these efforts are made within the framework of elements. It is clear that for more effective means, more innovative uses of the variational principles or out-of-box approaches beyond the standard variational principles or work beyond the elements (bringing in information from neighboring elements) are necessary.

In meshfree settings, the compatible strain field may not be accessible on the discontinuous lines (even in the distributive sense). We have to construct a new strain field. Even when the compatible strain field is accessible, we often want to construct a better strain field for more accurate solutions having desired properties.

The essential idea in this chapter is to construct a new strain field, hoping that a Galerkin model using the constructed strain field can deliver some good properties. Such a construction can be performed by (1) modifying the compatible strain field obtained from the strain–displacement relation when the compatible strain field is available; and (2) constructing a strain field using only the displacement field without differentiations. The construction works for both meshfree and FEM settings, but the operation is *beyond the element*.

Since the strain field is constructed anew, we can further reduce the requirement on the assumed displacement functions leading to various weakened-weak formulations, for which the SC-Galerkin (Chapter 5) weak forms need to be used. Methods based on an SC-Galerkin weak form are as simple as the standard Galerkin weak form, and very useful for establishing numerical methods, which overcome some of the issues with the standard FEM. They offer many important and attractive properties, such as upper bound, lower bound, superconvergence, free of locking, and ultra-accuracy. In addition, incompatible methods can be effectively formulated with ensured stability and convergence. The idea of strain construction can open a new window of opportunity for a new class of numerical methods outside the box of the Hu–Washizu principle

4.2 Historical Notes

4.2.1 Strain Construction Models of FEM Settings

A typical strain-constructed model (SC-model) is the smoothed FEM (SFEM) [6] formulated with special elements. Although the assumed displacement functions are still in an $\mathbb{H}^1$ space, the expression for the compatible strain field is not generally available. The strain-smoothing operation is then used to construct the *smoothed strain* field (see Remark 4.1 for a precise definition). The SFEM works very effectively for solid mechanics problems including dynamic problems [7,9]; it can produce much more accurate stress solutions and solutions with attractive properties [6]. Because the smoothed strains are obtained via line integrations along the smoothing domain boundary, and the derivatives of shape functions are not used in the formulation, simple point interpolation techniques can be applied to obtain the shape function values needed in the formulation, for which n-sided polygonal elements and very heavily distorted mesh can be used [8]. Detailed theoretical aspects, including stability and convergence of SFEM, can be found in [9]. Since the SFEM uses the smoothed Galerkin weak form and the assumed displacement functions can still be regarded as in an $\mathbb{H}^1$ space, it is variationally consistent.

The second SC-model is the node-based smoothed FEM or NS-FEM [10]. The NS-FEM uses an FEM mesh that is further divided into a set of smoothing domains based on nodes, and the elements and the smoothing domains are overlaid with each other. The strain field is constructed by using strain smoothing over the node-based smoothing domains, and the point interpolation for constructing displacement function values (only on the boundary of the smoothing domain) in an $\mathbb{H}^1$ space. The NS-FEM can have different shapes of elements, including n-sided polygonal elements, using a simple average and point interpolation method for computing the shape function values. It has the properties of upper bound, weak superconvergence, insensitiveness to mesh distortion, and being overly soft. The overly soft behavior leads to spurious modes at a higher energy level for dynamic problems, and hence temporal stabilization techniques are needed. Note that when the linear triangular type of elements are used, the NS-FEM produces the same results as the method of node-based uniform strain elements [16].

The third variationally consistent SC-model is the edge-based smoothed FEM or ES-FEM [11,12]. The ES-FEM is similar as the NS-FEM, and uses an FEM mesh and the point interpolation for displacement function construction. They differ in the division of smoothing domains based on edges of elements. The ES-FEM models are weakly stiff and quite "close-to-exact," and hence have properties of strong superconvergence and ultra-accuracy. No spurious modes are found in ES-FEM models and hence they work very well for both static and dynamic problems.

Note that the NS-FEM and ES-FEM are in fact more like a meshfree method, because their formulations are very much different from FEM. The only one that is in common is that the displacement functions used are still in an $\mathbb{H}^1$ space. The theory, interpolation procedure, integration, solution property, and the use of mesh depart quite a lot from the standard FEM procedure. They are in fact special cases of the node-based point interpolation methods (NS-PIMs) and edge-based point interpolation methods (ES-PIMs) discussed in Chapter 8 based on the generalized smoothed Galerkin (GS-Galerkin) weak form.

We now introduce a variationally inconsistent SC-model, that uses elements: αFEM [17]. The strain field in αFEM is constructed by scaling the gradient of strains with a scaling factor so as to provide some "softness" to the model. The αFEM can give not only much more accurate solutions in stresses but also produce nearly exact solutions in the energy norm for a class of problems. It also offers simple and practical ways to resolve some locking problems. The αFEM is not variationally consistent, and yet it can always produce much better solutions than the FEM that is perfectly variationally consistent! This finding opens an important window for the development of a new class of SC-models via manipulating the compatible strain field obtained directly from the assumed displacements using the strain–displacement relation. We now can commit a variational "crime," as long as we have proper ways to control the assumed strain field so that the solution can be somehow bounded and converges to the exact one [17].

Along with the idea of the αFEM of quadrilateral elements [17], a variationally consistent αFEM (VCαFEM) [29] has also been formulated by scaling only the gradient of strain in the physical coordinates, without scaling the Jacobian matrices. Because the Hellinger–Reissner variational principle is used, it is thus variationally consistent. The VCαFEM can produce both lower and upper bounds to the exact solution in the energy norm for all problems of elasticity by choosing properly the scaling factor α. The important bound property is then used to devise an exact-α approach for ultra-accurate solutions that are very close to the exact solution in the energy norm. Furthermore, VCαFEM performs well for problems with volumetric locking using a stabilization technique [29].

An αFEM using triangular and tetrahedral elements for the exact solution to mechanics problems has also proposed using a partially constructed strain field [25]. Following this line of development, a superconvergent αFEM (SαFEM) using triangular meshes was proposed [30]. A strain field is carefully constructed by combining the compatible strains and the smoothed nodal strains with an adjustable factor α. A Galerkin-like weak form was proposed for SαFEM to establish the discretized equation system, based on the Hellinger–Reissner variational principle. Because of the particular way in which the strain field is constructed, the new weak form is as simple as the Galerkin weak form and the resultant stiffness matrix is symmetric. It was proven theoretically and shown numerically that the results of SαFEM are much more accurate than those of FEM-T3 and even more accurate than those of FEM-Q4 when the same sets of nodes are used. SαFEM can produce both lower and upper bounds to the exact solution in the energy norm for elasticity problems by properly choosing an α. In addition, a preferable α approach has also been devised for SαFEM to produce very accurate and superconvergent solutions for both displacement and energy norms. Furthermore, a model-based selective scheme is proposed to formulate a combined SαFEM/NS-FEM model that handily overcomes the volumetric locking problems. Intensive numerical studies have been conducted to confirm the theory and properties of SαFEM.

Note that all these SC-Galerkin models are in fact quite different from the standard FEM models, but combinations of meshfree and FEM techniques. Such a combination of the standard FEM and meshfree techniques are found indeed very effective and fruitful.

4.2.2 Strain Construction Models of Meshfree Settings

A typical meshfree model is the method of stabilized conforming nodal integration or SCNI proposed by Chen et al. [24]. The spatial instability in the nodal integrated meshfree methods using MLS or RKPM shape functions is resolved effectively using the strain smoothing technique based on the Voronoi nodal domains. It is found that the accuracy in the nodal integrated meshfree methods is considerably improved. Since SCNI uses the smoothed Galerkin weak form and the assumed displacement functions are in an $\mathbb{H}^1$ space, it is variationally consistent.

Based on the strain smoothing technique [24], a generalized strain smoothing technique [23] has been developed for meshfree settings which forms the foundation for a number of SC-models, including NS-PIM* [18–22] and ES-PIM [13]. These methods use a triangular mesh, upon which a set of node- or edge-based smoothing cells are created that are intertwined with the triangular mesh, but allow the use of incompatible shape functions. The GS-Galerkin weak form was established and used in these formulations. These methods are found to have the properties of upper bound, superconvergence, and insensitiveness to mesh distortion. The NS-PIM is found overly soft (with spurious modes), but

---

* NS-PIM was termed initially LC-PIM (linearly conforming point interpolation method) due to the ability to produce at least a linear displacement field exactly. For the same reason, NS-PIM was termed initially also LC-RPIM.

the stiffness of the ES-PIM is very close-to-exact, and hence has properties of strong superconvergence and ultra-accuracy. No spurious modes are found in linear ES-PIM models and hence it works very well for both static and dynamic problems. Note that when linear shape functions are used, the NS-PIM or ES-PIM is the same as the NS-FEM and ES-FEM using linear triangular elements.

Because the constructed strain is assumed constant in the smoothing domains/cells, it is proven that the orthogonal condition is satisfied, and thus these methods are variationally consistent if the assumed displacement functions are in an $\mathbb{H}^1$ space [19]. However, when the assumed displacement functions are in a $\mathbb{G}_h^1$ space, its stability and convergence is ensured by the weakened-weak formulation [37].

Another variationally inconsistent meshfree SC-model is the recently formulated point interpolation method with continuous piecewise linear strain field (PIM-CS) [26]. In PIM-CS, both the displacement field and the strain field are constructed using point interpolation methods, and an SC-Galerkin weak form is used to derive the discrete system equations. Displacement interpolation is performed based on nodal displacements, but strain interpolation uses strains at points obtained using strain smoothing techniques. The points are chosen properly so that the strain field is continuous in the entire problem domain. Furthermore, a superconvergent point interpolation method has been developed for superconvergent solutions as well as solution bounds using the triangular mesh only, by constructing a piecewise constant strain field [27].

The point interpolation method with the strain field constructed using least squares approximation (least square point interpolation method [LS-PIM]) [28] is another example of SC-Galerkin models. This approach is similar to PIM-CS, but with strain fields of different order constructed using the least squares (an orthogonal) projection that satisfies the orthogonal condition. Therefore, LS-PIM is variationally consistent, when linear interpolations for displacement functions (in an $\mathbb{H}^1$ space) are used. When higher order interpolation (quadratic PIM or RPIM) for the displacement field is used, the so-called strain norm equivalent condition and strain convergence conditions are used to ensure the stability and convergence. It is thus another example of a nicely constructed weakened-weak formulation [37]. By controlling the order of the least squares projection for the strain construction, we can build various models with desired properties [28].

These models are often found with strong superconvergence and ultra-accuracy, and can be tuned to produce both lower and upper bounds to the exact solution.

4.3 How to Construct?

4.3.1 Discrete Models: A Base for Strain Construction

Our strain construction techniques operate, in general, on a number of local cells, and beyond the element confinement. To perform a well-controlled construction of a strain field, we first choose a base model that can be any fully compatible FEM model of proven stability and convergence to the exact solution. For easy implementation, robustness and convenience in analysis, and proving important theories and properties of an SC-model, we choose the linear FEM model of triangular elements (FEM-T3) as the base model. We first divide the problem domain Ω with N_c nonoverlapping and seamless (NOSL) triangular back ground cells/elements: $\overline{\Omega} = \bigcup_{i=1}^{N_c} \overline{\Omega}_i^c$ and $\Omega_i^c \cap \Omega_j^c = 0$, $\forall i \neq j$, with a set of N_n

nodes, using the triangulation procedure discussed in Section 1.7.2. Our strain construction is then based on this discrete model of triangular cells.

4.3.2 General Procedure for Strain Construction

The strain field, $\hat{\boldsymbol{\varepsilon}}(\mathbf{u}^h) \in \mathbb{L}^2(\Omega)$, is constructed using the assumed displacement functions, $\mathbf{u}^h \in \mathbb{G}_h^1$.* In general, $\hat{\boldsymbol{\varepsilon}}(\mathbf{u}^h)$ at any point in the problem domain can be given a general form of

$$\hat{\boldsymbol{\varepsilon}}(\mathbf{u}^h) = \begin{cases} P(\tilde{\boldsymbol{\varepsilon}}(\mathbf{u}^h)) = P(\mathbf{L}_d\mathbf{u}^h), & \text{when } \mathbf{u}^h \in \mathbb{H}_h^1(\Omega) \\ B(\mathbf{u}^h), & \text{when } \mathbf{u}^h \in \mathbb{G}_h^1(\Omega) \end{cases} \tag{4.1}$$

where the compatible strain $\tilde{\boldsymbol{\varepsilon}}(\mathbf{u}^h)$ is defined using the strain–displacement relation of linear deformation that can be written in the following matrix form [2]:

$$\tilde{\boldsymbol{\varepsilon}}(\mathbf{u}^h) = \mathbf{L}_d\mathbf{u}^h \tag{4.2}$$

where
  $\mathbf{u}^h$ is the assumed displacement vector
  $\mathbf{L}_d$ is a differential operator matrix given in Equation 1.9

In Equation 4.1, P stands for a general *transformation* operator over space for a given input $\tilde{\boldsymbol{\varepsilon}}(\mathbf{u}^h)$. When, $\mathbf{u}^h \in \mathbb{G}_h^1$, P is not used, because $\tilde{\boldsymbol{\varepsilon}}(\mathbf{u}^h)$ is not generally available: $\mathbf{u}^h$ may be not differentiable on lines/surfaces where the displacement is discontinuous. We need to construct somehow an approximated "strain matrix" B that is also a *transformation* operator over space for a given input $\mathbf{u}^h$.

Because the constructed strain field $\hat{\boldsymbol{\varepsilon}}(\mathbf{u}^h)$ needs only to be in $\mathbb{L}^2(\Omega)$, we can require $\hat{\boldsymbol{\varepsilon}}(\mathbf{u}^h)$ being piecewise continuous in the problem domain Ω. Therefore, in actual practice in a discrete model, the domain Ω in usually divided into a set of quadrature (integration) domains, $\boxed{\Omega} = \bigcup_{i=1}^{N_q} \boxed{\Omega}_i^q$ and the strain construction is done piecewise in each Ω_i^q. In addition, $\hat{\boldsymbol{\varepsilon}}_q(\mathbf{u}^h)$ in Ω_i^q is made continuous (often constant or linear) for convenience and efficiency in strain energy integration. We then try to construct strains at a number of points in domain $\boxed{\Omega}_i^q$. Once the strains at sufficient number of points in $\boxed{\Omega}_i^q$ are obtained, the constructed strain field $\hat{\boldsymbol{\varepsilon}}_q(\mathbf{u}^h)$ in Ω_i^q can be constructed using simply (again) the point interpolation methods (or other approximation methods discussed in Chapter 3), as we do for constructing the displacement field.

After the strain field is constructed, the strain potential energy needs to be evaluated. For any $\mathbf{u}^h \in \mathbb{H}_0^1$, the potential energy for the compatible strain field becomes

$$\tilde{U}_{PE}(\mathbf{u}^h) = \int_{\Omega} \frac{1}{2}\, \tilde{\boldsymbol{\varepsilon}}^{\mathrm{T}}(\mathbf{u}^h)\mathbf{c}\tilde{\boldsymbol{\varepsilon}}(\mathbf{u}^h)\mathrm{d}\Omega \tag{4.3}$$

For any $\mathbf{u}^h \in \mathbb{G}_{h,0}^1$ the strain energy potential for the constructed strain field becomes

$$\hat{U}_{PE}(\mathbf{u}^h) = \int_{\Omega} \frac{1}{2}\, \hat{\boldsymbol{\varepsilon}}^{\mathrm{T}}(\mathbf{u}^h)\mathbf{c}\hat{\boldsymbol{\varepsilon}}(\mathbf{u}^h)\mathrm{d}\Omega \tag{4.4}$$

---

* In this book, when we require a vector is in a space, we require each of the component functions being independently in the space, and the dimension of the space is expanded accordingly.

Both Equations 4.3 and 4.4 are associated to a bilinear form. We further note the following:

- The potential energy computed using the constructed strain field using Equation 4.4 will surely be nonnegative but can be zero depending on how the construction is performed. This means that the resultant stiffness matrix created can be semi-symmetric positive definite (semi-SPD) even for stable materials. The semi-SPD property implies that an SC-model can have *zero energy modes*, and hence can be *spatially* instable. We need to impose some sort of *admissible* conditions upon the constructed strain fields; we cannot expect an anyhow constructed strain field to produce a stable and convergent solution.
- Since the constructed strain field depends entirely on the assumed displacement field, there are no additional unknowns introduced.

4.4 Admissible Conditions for Constructed Strain Fields

The condition of $\hat{\boldsymbol{\varepsilon}} \in \mathbb{L}^2$ is only a minimum condition, ensuring the potential energy defined in Equation 4.4 is bounded, and the stability is left uncontrolled. Here, we present three additional admissible conditions for a constructed strain field for establishing stable and convergent models.

4.4.1 Condition 1: Orthogonal Condition

For any $\mathbf{v} \in \mathbb{H}_{h,0}^1$, when the constructed strain field $\hat{\boldsymbol{\varepsilon}}(\mathbf{v})$ and the compatible strain field $\tilde{\boldsymbol{\varepsilon}}(\mathbf{v}) = \mathbf{L}_d \mathbf{v}$ satisfy the following condition

$$\int_{\Omega} \hat{\boldsymbol{\varepsilon}}^{\mathrm{T}}(\mathbf{v}) \mathbf{c} \hat{\boldsymbol{\varepsilon}}(\mathbf{v}) d\Omega = \int_{\Omega} \hat{\boldsymbol{\varepsilon}}^{\mathrm{T}}(\mathbf{v}) \mathbf{c} (\mathbf{L}_d \mathbf{v}) d\Omega \tag{4.5}$$

a strain-constructed Galerkin model can be derived directly from the single-field Hellinger–Reisnner's principle and hence is variationally consistent (see Chapter 5). Equation 4.5 is the so-called orthogonal condition [3]. This condition is used to construct hybrid FEM models, and is frequently used in meshfree methods to examine whether an SC-model is variationally consistent for assumed displacement functions in an $\mathbb{H}^1$ space.

4.4.2 Condition 2a: Norm Equivalence Condition

For a two-dimensional (2D) and three-dimensional (3D) vector field $\mathbf{v} \in \mathbb{G}_{h,0}^1$ with a vector $\mathbf{U}$ of nodal function values arranged in the form of

$$\mathbf{U} = \left\{ \underbrace{u_{1(1)} \quad u_{2(1)}}_{\text{node}(1)} \quad \underbrace{u_{1(2)} \quad u_{2(2)}}_{\text{node}(2)} \quad \cdots \quad \underbrace{u_{1(n)} \quad u_{2(n)}}_{\text{node}(n)} \quad \cdots \quad \underbrace{u_{1(N_n)} \quad u_{2(N_n)}}_{\text{node}(N_n)} \right\}^{\mathrm{T}}, \quad \text{for 2D}$$

$$\mathbf{U} = \left\{ \underbrace{u_{1(1)} \quad u_{2(1)} \quad u_{3(1)}}_{\text{node}(1)} \quad \underbrace{u_{1(2)} \quad u_{2(2)} \quad u_{3(2)}}_{\text{node}(2)} \quad \cdots \quad \underbrace{u_{1(N_n)} \quad u_{2(N_n)} \quad u_{3(N_n)}}_{\text{node}(N_n)} \right\}^{\mathrm{T}}, \quad \text{for 3D}$$

$$\tag{4.6}$$

the base (discrete and compatible) constant strain field $\tilde{\boldsymbol{\varepsilon}}_b(\mathbf{U})$ is given as

$$\tilde{\boldsymbol{\varepsilon}}_b(\mathbf{U}) = \underbrace{\mathbf{L}_d \boldsymbol{\Phi}_e \mathbf{U}}_{\mathbf{B}_b} = \mathbf{B}_b \mathbf{U} \qquad (4.7)$$

where $\mathbf{B}_b$ is the *global* strain matrix and has entries assembled from the element strain matrix

$$\mathbf{B}_b = \mathbf{L}_d \boldsymbol{\Phi}_e \qquad (4.8)$$

in which the matrix of shape functions have the form

$$\boldsymbol{\Phi}_e = \begin{bmatrix} \underbrace{\phi_{1e}^b \quad 0}_{\text{node 1 element } e} & \underbrace{\phi_{2e}^b \quad 0}_{\text{node 2 element } e} & \underbrace{\phi_{3e}^b \quad 0}_{\text{node 3 element } e} \\ 0 \quad \phi_{1e}^b & 0 \quad \phi_{2e}^b & 0 \quad \phi_{3e}^b \end{bmatrix}, \quad \text{for 2D}$$

$$\boldsymbol{\Phi}_e = \begin{bmatrix} \underbrace{\phi_{1e}^b \quad 0 \quad 0}_{\text{node 1 element } e} & \underbrace{\phi_{2e}^b \quad 0 \quad 0}_{\text{node 2 element } e} & \underbrace{\phi_{3e}^b \quad 0 \quad 0}_{\text{node 3 element } e} & \underbrace{\phi_{4e}^b \quad 0 \quad 0}_{\text{node 4 element } e} \\ 0 \quad \phi_{1e}^b \quad 0 & 0 \quad \phi_{2e}^b \quad 0 & 0 \quad \phi_{3e}^b \quad 0 & 0 \quad \phi_{4e}^b \quad 0 \\ 0 \quad 0 \quad \phi_{1e}^b & 0 \quad 0 \quad \phi_{2e}^b & 0 \quad 0 \quad \phi_{3e}^b & 0 \quad 0 \quad \phi_{4e}^b \end{bmatrix}, \quad \text{for 3D}$$

$$(4.9)$$

The *global* strain matrix $\mathbf{B}_b$ is a very sparse matrix with dimensions of $3N_c \times 2N_n$ for 2D and $6N_c \times 3N_n$ for 3D models. Because all the nodal shape functions are linearly independent, $\mathbf{B}_b$ has $2N_n$ linear independent columns for 2D, and $3N_n$ linear independent columns for 3D models (we assume that the model is properly constrained for all rigid motions, and N_n is the number of unconstrained nodes). The (global) strain vector $\tilde{\boldsymbol{\varepsilon}}_b(\mathbf{U})$ can be written in the form

$$\tilde{\boldsymbol{\varepsilon}}_h(\mathbf{U}) = \left\{ \underbrace{\tilde{\boldsymbol{\varepsilon}}_{b(1)}^{\mathrm{T}}}_{\text{element 1}} \quad \underbrace{\tilde{\boldsymbol{\varepsilon}}_{b(2)}^{\mathrm{T}}}_{\text{element 2}} \qquad \underbrace{\tilde{\boldsymbol{\varepsilon}}_{b(N_c)}^{\mathrm{T}}}_{\text{element } N_c} \right\}^{\mathrm{T}} \qquad (4.10)$$

where N_c is the total number of triangular elements (same as the background cells for the SC-model) for the base model, and

$$\tilde{\boldsymbol{\varepsilon}}_{b(i)} = \{\tilde{\varepsilon}_{11} \quad \tilde{\varepsilon}_{22} \quad 2\tilde{\varepsilon}_{12}\}_{(i)}^{\mathrm{T}}, \quad (i = 1, 2, \dots, N_c) \quad \text{for 2D}$$

$$\tilde{\boldsymbol{\varepsilon}}_{b(i)} = \{\tilde{\varepsilon}_{11} \quad \tilde{\varepsilon}_{22} \quad \tilde{\varepsilon}_{33} \quad 2\tilde{\varepsilon}_{23} \quad 2\tilde{\varepsilon}_{13} \quad 2\tilde{\varepsilon}_{12}\}_{(i)}^{\mathrm{T}}, \quad (i = 1, 2, \dots, N_c) \quad \text{for 3D}$$

$$(4.11)$$

Because linear triangular elements are used for the base model, these strain components $\tilde{\varepsilon}_{ij}$ are constant in any element.

The constructed strain field $\hat{\boldsymbol{\varepsilon}}(\mathbf{U})$ can also be written in a similar form as

$$\hat{\boldsymbol{\varepsilon}}(\mathbf{U}) = \left\{ \underbrace{\hat{\boldsymbol{\varepsilon}}_{(1)}^{\mathrm{T}}}_{\text{cell 1}} \quad \underbrace{\hat{\boldsymbol{\varepsilon}}_{(2)}^{\mathrm{T}}}_{\text{cell 2}} \quad \cdots \quad \underbrace{\hat{\boldsymbol{\varepsilon}}_{(N_q)}^{\mathrm{T}}}_{\text{cell } N_q} \right\}^{\mathrm{T}} \qquad (4.12)$$

where N_q is the total number of the *quadrature or integration cells* for the SC model, and

$$\hat{\boldsymbol{\varepsilon}}_{(i)} = \{\hat{\varepsilon}_{11} \quad \hat{\varepsilon}_{22} \quad 2\hat{\varepsilon}_{12}\}_{(i)}^{\mathrm{T}}, \quad (i = 1, 2, \ldots, N_q) \quad \text{for 2D}$$

$$\hat{\boldsymbol{\varepsilon}}_{(i)} = \{\hat{\varepsilon}_{11} \quad \hat{\varepsilon}_{22} \quad \hat{\varepsilon}_{33} \quad 2\hat{\varepsilon}_{23} \quad 2\hat{\varepsilon}_{13} \quad 2\hat{\varepsilon}_{12}\}_{(i)}^{\mathrm{T}}, \quad (i = 1, 2, \ldots, N_q) \quad \text{for 3D}$$
(4.13)

These constructed strain components $\hat{\varepsilon}_{ij}$ are, in general, not constant in an integration cell.

The constructed strain field $\hat{\boldsymbol{\varepsilon}}(\mathbf{U})$ needs to be *equivalent in a norm* to the base constant strain field $\tilde{\boldsymbol{\varepsilon}}_b(\mathbf{U})$, meaning that there exist nonzero positive constants c_{ac} and c_{ca} such that

$$\|\hat{\boldsymbol{\varepsilon}}(\mathbf{U})\|_{\mathbb{L}^2(\Omega)} \leq c_{ac} \|\tilde{\boldsymbol{\varepsilon}}_b(\mathbf{U})\|_{\mathbb{L}^2(\Omega)} \quad \text{and} \quad \|\tilde{\boldsymbol{\varepsilon}}_b(\mathbf{U})\|_{\mathbb{L}^2(\Omega)} \leq c_{ca} \|\hat{\boldsymbol{\varepsilon}}(\mathbf{U})\|_{\mathbb{L}^2(\Omega)}, \quad \forall \mathbf{U} \in \mathbb{R}_0^{dN_n} \quad (4.14)$$

where $\mathbb{R}_0^{dN_n}$ is a subspace of $\mathbb{R}^{dN_n}$ where the displacements at the nodes on the essential boundary are imposed. Constants c_{ac} and c_{ca} should be independent of $\mathbb{R}_0^{dN_n}$. Since our strain field is defined in a vector form, measuring in $\mathbb{L}^2$ norm is handy and workable for both $\tilde{\boldsymbol{\varepsilon}}_b(\mathbf{U})$ and $\hat{\boldsymbol{\varepsilon}}(\mathbf{U})$ (although they may be different in length). Clearly, to satisfy the conditions in Equation 4.14, the *minimum* requirement should be

$$N_q \geq N_c \quad (4.15)$$

which means that we should sample more locations for the (nonnegative) strain energy in our SC model, as practiced in SFEM [6–9], and proved for general settings based on the argument of ensuring positivity [37].

The norm equivalence ensures only the stability and hence convergence, but we do not know where it converges to. To ensure the solution converges to the exact solution, we need the strain convergence condition.

4.4.3 Condition 2b: Strain Convergence Condition

The convergence condition for the constructed strain is defined as

$$\lim_{\substack{h \to 0 \\ N_n \to \infty}} \hat{\boldsymbol{\varepsilon}}(\mathbf{x}, \mathbf{U}) \to \tilde{\boldsymbol{\varepsilon}}_b(\mathbf{x}, \mathbf{U}), \quad \forall \mathbf{x} \in \Omega, \quad \forall \mathbf{U} \in \mathbb{R}_0^{dN_n} \quad (4.16)$$

When both conditions in Equations 4.14 and 4.16 are satisfied, the SC-Galerkin model is stable and converges to the exact solution of the original strong form, and it is a typical weakened-weak formulation. In this case, it is not variationally consistent in the conventional sense, because $\mathbf{v}$ is not, in general, in an $\mathbb{H}^1$ space.

4.4.4 Condition 3: Zero-Sum Condition

For any $\mathbf{v} \in \mathbb{H}_{h,0}^1$, we construct the strain in the following form:

$$\hat{\boldsymbol{\varepsilon}}(\mathbf{v}) = \tilde{\boldsymbol{\varepsilon}}_b(\mathbf{v}) + \alpha \hat{\boldsymbol{\varepsilon}}_m(\mathbf{v}) \quad (4.17)$$

where
$\tilde{\boldsymbol{\varepsilon}}_b$ is the (compatible) strain field of a base model
$\alpha \in \mathbb{R}$ is a finite adjustable parameter for regularizing the amount of modification
$\hat{\boldsymbol{\varepsilon}}_m$ is the modified portion of the strain field

The *zero-sum condition* is defined as

$$\int_\Omega \hat{\boldsymbol{\varepsilon}}_m(\mathbf{v})d\Omega = 0, \quad \forall \mathbf{v} \in \mathbb{H}_{h,0}^1 \tag{4.18}$$

When both Equations 4.17 and 4.18 can be found, the Galerkin-like weak form is applicable (see Chapter 5), and the formulation is variationally consistent. The zero-sum condition is similar to the orthogonal condition that is used in FEM for the stabilization formulation for quadrilateral elements using reduced integration [14,15].

4.5 Strain Construction Techniques

We now present some of the possible and practical techniques for constructing strain fields. Most of the techniques were proven or tested under specified settings, and hence causations are advised in attempting to extend these techniques to other applications.

4.5.1 At a Glance

Some Schemes for Construction and Modification of the Strain Fields

Item	Schemes	Brief Description	Example/References
1.	Strain gradient scaling	Using a factor to scale-down or -up the gradient of the compatible strain field.	αFEM [17] VCαFEM [29]
2.	Strain smoothing (orthogonal projection)	The compatible strain field is smoothed over smoothing domains that can be element-based, cell-based, node-based, and edge-based. The constructed strain is assumed to be constant in the smoothing domains.	SCNI [5] NS-FEM [10] ES-FEM [11,12]
3.	Generalized smoothing	The strain field is constructed by the generalized smoothing technique over the smoothing domains that can be element-based, cell-based, node-based, and edge-based. The constructed strain is assumed to be constant in the smoothing domains.	SFEM [6,23] NS-PIM [18,19,23] ES-PIM [13,23,37]
4.	Point interpolation method	At points in the problem domain, the strains are assigned with smoothed strains over smoothing domains. The constructed strain field is then constructed via a point interpolation method using these strains at these points.	SC-PIM [23,31] PIM-CS [26,27]
5.	Least squares approximation (orthogonal projection)	At points in the problem domain, the strains are assigned with smoothed strains over smoothing domains. The constructed strain field is then constructed via least squares approximation using these strains at these locations.	LS-PIM [28]

Note: This table may not be exclusive.

4.5.2 Scaling the Strain Gradient

Change the gradient of the compatible strain field obtained from an assumed displacement field in an $\mathbb{H}^1$ space. This was first used in the so-called αFEM presented in [17,29], based on an FEM setting of quadrilateral elements. It was discovered that the compatible strain field in a quadrilateral element can be divided into a constant portion and a portion varying with the coordinates. The gradient of the varying portion can be modified by scaling without affecting the ability to pass the standard patch tests. Therefore, the convenience of the model can be ensured, as long as the stability is maintained. A quadrilateral element was then formulated to obtain the exact or best possible solution for a given problem by scaling the gradient of strains in the natural coordinates and Jacobian matrices with a scaling factor $\alpha \in (-\infty, +\infty)$. It is shown that as long as α is not zero, the stability of the model can be preserved, and will always converge to the exact solution of the original strong form of partial differential equations (PDEs). The method is not necessarily variationally consistent but stable and convergent. The αFEM can produce nearly exact solutions in the strain energy for all overestimation problems, and the best possible solution for underestimation problems. The general procedure of strain gradient scaling works as follows:

For any assumed $\mathbf{u}^h \in \mathbb{H}^1_{h,0}$, the compatible strain field *within an element* can be obtained as

$$\tilde{\boldsymbol{\varepsilon}}(\mathbf{x}) = \mathbf{L}_d \mathbf{u}^h(\mathbf{x}) \tag{4.19}$$

The compatible strain field is then split into two portions:

$$\tilde{\boldsymbol{\varepsilon}}(\mathbf{x}) = \underbrace{\tilde{\boldsymbol{\varepsilon}}_c(\mathbf{x}_c)}_{\tilde{\boldsymbol{\varepsilon}}_c} + \tilde{\boldsymbol{\varepsilon}}_v(\mathbf{x}) \tag{4.20}$$

where
$\tilde{\boldsymbol{\varepsilon}}_c = \tilde{\boldsymbol{\varepsilon}}(\mathbf{x}_c)$ is the constant portion evaluated at the origin of the natural coordinate $\mathbf{x}_c$ defined for the element
$\tilde{\boldsymbol{\varepsilon}}_v(\mathbf{x})$ varies with the coordinate and should satisfy the zero-sum condition in Equation 4.18

The strain field can then be constructed in the following form:

$$\hat{\boldsymbol{\varepsilon}}(\mathbf{x}) = \tilde{\boldsymbol{\varepsilon}}_c + \alpha \tilde{\boldsymbol{\varepsilon}}_v(\mathbf{x}) \tag{4.21}$$

where $\alpha \in (-\infty, +\infty)$ is an adjustable parameter. The stability is ensured by using a small $\alpha \neq 0$. Note $\tilde{\boldsymbol{\varepsilon}}_c$ can be, in general, different from $\tilde{\boldsymbol{\varepsilon}}_b$ in Equation 4.17. It is only the constant portion of the compatible strain field.

The strain gradient scaling technique described above has been tested for quadrilateral elements where the compatible strain field is linear (in the natural coordinate) [25]. Extension of these techniques to other types of elements should be performed with caution, as the proof for general settings for this technique has not yet been determined.

4.5.3 Strain Construction by Smoothing

4.5.3.1 Strain Field Construction by Strain Smoothing for $\mathbf{u}^h \in \mathbb{H}_h^1$

A strain field can be constructed by smoothing the compatible strain field obtained from an assumed displacement field in an $\mathbb{H}^1$ space. Smoothing operation is an often used integral technique to smooth a function. It has a general form of Equation 2.7. It has been used in the nonlocal continuum mechanics [38] to obtain a smoother stress field. The smoothed particle hydrodynamics [36,39–41] uses it for function approximation. The strain smoothing was used as a means of resolving the material instabilities [42] and the special instability in the nodal integrated meshfree methods [5]. It is used later in finite element settings, such as NS-FEM [10] and ES-FEM [11,12], with a number of attractive properties. It is also used with properly constructed smoothing domains for the derivative approximation to establish models of strong formulation for solids [32], compressible fluids [33,35], and incompressible fluids [34].

Since this construction is done by modifying the strain field, it can only be done when the strain field is available, meaning that the assumed displacement there must be at least differentiable in the local smoothing domain Ω_x^s associated with the point of interest $\mathbf{x}$.

4.5.3.2 Strain Field Construction by Generalized Smoothing for $\mathbf{u}^h \in \mathbb{G}_h^1$

When the assumed displacement function is discontinuous, we need to construct a strain field. Based on the existing works on strain smoothing operations, Liu [23] proposed a generalized smoothing technique for functions that are discontinuous, leading to the $\mathbb{G}$ space theory. A strain field can then be constructed using the assumed displacement field in a proper $\mathbb{G}$ space. A GS-Galerkin weak form was then established for models based on elements [6–9,23] and PIM models [13,18,19,23]. The GS-Galerkin is a typical case of the weakened-weak formulation and the stability and convergence have been proven in [37]. Since a function in an $\mathbb{H}_h^1$ space is also in the corresponding $\mathbb{G}_h^1$ space, the GS-Galerkin formulation works also for models with assumed displacement functions in an $\mathbb{H}_h^1$ space. It is variationally consistent if the solution is sought from an $\mathbb{H}_h^1$ space, and is stable and convergent if the solution is sought from a $\mathbb{G}_h^1$ space. The GS-Galerkin offers a unified theoretical foundation for this class of compatible and incompatible methods. The strain modification is a special case of strain construction by generalized smoothing. The strain field constructed and used in the GS-Galerkin is piecewise constant.

The numerical methods developed based on the GS-Galerkin form can possess four major important properties: (1) the stiffness of the discretized model can be reduced compared to compatible FEM models and the exact model, which allows us to obtain upper bound solutions with respect to both the FEM solution and the exact solution; (2) the continuity of the trial and test functions can be reduced as long as they are in a proper $\mathbb{G}$ space, which allows us to use many types of methods to create shape functions for a numerical model; (3) the solution of a numerical method developed using the smoothed bilinear form is less sensitive to the quality of the mesh, and triangular meshes can be well used for problems of complicated domains; and (4) very accurate solutions can be obtained using a triangular mesh.

4.5.3.3 Smoothing over Moving Smoothing Domains

For any assumed displacement $\mathbf{u}^h \in \mathbb{G}_h^1$, the constructed strain field $\hat{\boldsymbol{\varepsilon}}$ is the *smoothed* strain field $\hat{\boldsymbol{\varepsilon}}$ obtained, using the following integral representation:

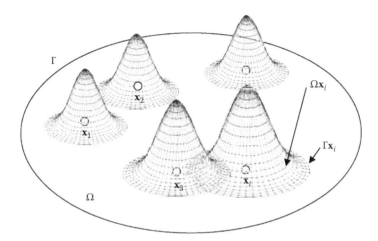

FIGURE 4.1
Moving smoothing domains Ω_x bounded by Γ_x for integral representation of a function at x, over which the smoothing function is defined. The smoothing domain can be different for different x and they can overlap. The smoothing function can also be different for different x.

$$
\hat{\boldsymbol\varepsilon}(x) = \begin{cases} \displaystyle\int_{\Omega_x^s} \underbrace{\tilde{\boldsymbol\varepsilon}(\boldsymbol\xi)}_{L_d(u^h(\boldsymbol\xi))} \, \widehat{\mathbf{W}}(\mathbf{x}-\boldsymbol\xi)\mathrm{d}\boldsymbol\xi = \int_{\Gamma_x^s} \mathbf{L}_n \mathbf{u}^h(\boldsymbol\xi)\widehat{\mathbf{W}}(\mathbf{x}-\boldsymbol\xi)\mathrm{d}\Gamma - \int_{\Omega_x^s} \mathbf{u}^h(\boldsymbol\xi)\mathbf{L}_d^T(\widehat{\mathbf{W}}(\mathbf{x}-\boldsymbol\xi))\mathrm{d}\boldsymbol\xi, \\[4pt] \quad \text{when } \mathbf{u}^h(\boldsymbol\xi) \in \mathbb{C}^0\left(\Omega_x^s\right) \\[10pt] \displaystyle\int_{\Gamma_x^s} \mathbf{L}_n \mathbf{u}^h(\boldsymbol\xi)\widehat{\mathbf{W}}(\mathbf{x}-\boldsymbol\xi)\mathrm{d}\Gamma - \int_{\Omega_x^s} \mathbf{u}^h(\boldsymbol\xi)\mathbf{L}_d^T(\widehat{\mathbf{W}}(\mathbf{x}-\boldsymbol\xi))\mathrm{d}\boldsymbol\xi, \\[4pt] \quad \text{when } \mathbf{u}^h(\boldsymbol\xi) \in \mathbb{C}^{-1}\left(\Omega_x^s\right) \end{cases}
\tag{4.22}
$$

where
L_n is the outward normal matrix given by Equation 1.27
$\mathbf{W}$ is a diagonal matrix of smoothing functions $\widehat{W}$ for the strain components defined in the smoothing domain $\Omega_x^s \subset \Omega$ centered at x and bounded by Γ_x^s, as shown in Figure 4.1

The important "no-sharing" rule applies here: Γ_x^s does not share any *finite* portion of discontinuous lines of the assumed displacement field. As long as this rule is observed, the smoothing domain can be different for different x and they can overlap. Since the smoothing domain Ω_x^s can move with x, this type of smoothing domain is termed the moving smoothing domain.

We require here that $\widehat{W}$ be at least first-order differentiable in (open) Ω_x^s. In the first equation of Equation 4.22 we used the Green's divergence theorem. When $\mathbf{u}^h$ is discontinuous, the Green's divergence theorem is not applicable, and hence the second equation in Equation 4.22 is only an approximation.

Remark 4.1: Smoothed Strain by Boundary Flux Approximation:
An Overall Conservation
The *smoothed strain* defined in Equation 4.22 is a generalized concept. It is not "the strain obtained by smoothing the compatible strain field," because such a compatible strain field

does not, in general, exist! Rigorously speaking, the *smoothed strain* is the outward *flux* of the assumed displacement field through the smoothing domain boundary Γ_x^s. Such a *boundary flux approximation* preserves the overall *conservation* of the strain–displacement relation over the smoothing domain, which is important to ensure the energy balance in energy principles using the (generalized) smoothed strains. In fact, it is the essential reason why the GS-Galerkin weak forms work for all assumed functions in a $\mathbb{G}_h^1$ space.

Observing Equation 4.22, we note that the differentiation on the assumed displacement has now been transferred onto the smoothing functions. Therefore, the continuity requirement on the assumed displacement is reduced by one order.

Generally, the choice of the smoothing function and the smoothing domain will affect how the conservation is achieved internally in the smoothing domain, and hence the property of the weak form. Therefore, how the problem domain is divided into smoothing domains plays an important role. For moving smoothing domains, we have the following remark.

Remark 4.2: Reproducing Property

When $\widehat{W}$ is of the unity property defined in Equation 2.10, and the compatible strain field exists ($\mathbf{u}^h \in \mathbb{H}^1$), the moving-domain integral representation of strain field will be exact at $\mathbf{x}$: $\widehat{\boldsymbol{\varepsilon}} = \tilde{\boldsymbol{\varepsilon}}$. In this case, the GS-Galerkin weak form reduces to the standard Galerkin weak form, and hence is variationally consistent.

Moving smoothing domains can be used in Section 4.5.4 to obtain strains at desired points in the problem domain.

4.5.3.4 Strain Average over Moving Smoothing Domains

If we use the following special smoothing function that is a local constant:

$$\widehat{W}(\mathbf{x} - \boldsymbol{\xi}) = \bar{W}(\mathbf{x} - \boldsymbol{\xi}) = \begin{cases} 1/A_x^s & \boldsymbol{\xi} \in \Omega_x^s \\ 0 & \boldsymbol{\xi} \notin \Omega_x^s \end{cases} \tag{4.23}$$

where $A_x^s = \int_{\Omega_x^s} d\Omega$ is the area (or volume for 3D) of the smoothing domain at the point $\mathbf{x}$. The smoothing function $\bar{W}(\mathbf{x} - \boldsymbol{\xi})$ given above satisfies the conditions of unity. The smoothed strain becomes a smoothed (averaged) strain $\bar{\boldsymbol{\varepsilon}}$ that can be obtained:

$$\bar{\boldsymbol{\varepsilon}}(\mathbf{u}^h(\mathbf{x})) = \begin{cases} \frac{1}{A_x^s} \int\limits_{\Omega_x^s} \tilde{\boldsymbol{\varepsilon}}(\mathbf{u}^h) d\Omega = \frac{1}{A_x^s} \int\limits_{\Omega_x^s} (\mathbf{L}_d \mathbf{u}^h) d\Omega = \frac{1}{A_x^s} \int\limits_{\Gamma_x^s} \mathbf{L}_n \mathbf{u}^h d\Gamma, & \text{when } \mathbf{u}^h(\boldsymbol{\xi}) \in \mathbb{C}^0(\Omega_x^s) \\ \frac{1}{A_x^s} \int\limits_{\Gamma_x^s} \mathbf{L}_n \mathbf{u}^h d\Gamma, & \text{when } \mathbf{u}^h(\boldsymbol{\xi}) \in \mathbb{C}^{-1}(\Omega_x^s) \end{cases}$$

$$\tag{4.24}$$

It can also be obtained by simply dropping the last domain integral term in Equation 4.22 because $\mathbf{L}_d^{\mathrm{T}}(\bar{W}(\mathbf{x} - \boldsymbol{\xi}))$ vanishes for constant smoothing functions.

The potential energy of the smoothed (averaged) strain field becomes

$$\bar{U}_{PE}(\mathbf{u}^h) = \int\limits_\Omega \frac{1}{2} \bar{\boldsymbol{\varepsilon}}^{\mathrm{T}}(\mathbf{u}^h) \mathbf{c} \bar{\boldsymbol{\varepsilon}}(\mathbf{u}^h) d\Omega \tag{4.25}$$

4.5.3.5 Stationary Smoothing Domain Construction

In practical formulation of many numerical methods, we do not allow the smoothing domains to overlap, and the smoothing domains are constructed in a fixed way such that the union of them forms the problem domain $\boxed{\Omega} = \cup_{i=1}^{N_s}\boxed{\Omega}_i^s$, as shown in Figure 4.2, where the problem domain is divided into N_s smoothing domains. The smoothing domain Ω_i^s bounded by Γ_i^s is for the point at $\mathbf{x}_i$. Since the problem domain is divided into a fixed set of the smoothing domain, the smoothing domain is termed the stationary smoothing domain. If the assumed displacement function is discontinuous on segments in Ω, we do not allow Γ_i^s sharing any finite portion of these segments, and Γ_i^s can only go across these segments. It is clear that the strain construction model using the stationary smoothing domain is a typical GS-Galerkin model based on $\mathbb{G}$ space theory (see Chapter 3). In this setting, the smoothing domains are also the integration domains, and hence we have $\Omega_i^s = \Omega_i^q$ and $N_s = N_q$.

The constructed strain in Ω_i^s is assigned to be constant $\widehat{\boldsymbol{\varepsilon}}_i$ and equals the smoothed strain obtained using

$$\widehat{\boldsymbol{\varepsilon}}_i = \widehat{\boldsymbol{\varepsilon}}(\mathbf{x}_i), \quad \forall \mathbf{x} \in \Omega_i^s \tag{4.26}$$

where

$$\widehat{\boldsymbol{\varepsilon}}(\mathbf{x}_i) = \begin{cases} \int\limits_{\Omega_i^s} \underbrace{\widetilde{\boldsymbol{\varepsilon}}(\boldsymbol{\xi})}_{\mathbf{L}_d(\mathbf{u}^h(\boldsymbol{\xi}))} \widehat{\mathbf{W}}(\mathbf{x}-\boldsymbol{\xi})\mathrm{d}\boldsymbol{\xi} = \int\limits_{\Gamma_i^s} \mathbf{L}_n \mathbf{u}^h(\boldsymbol{\xi})\widehat{\mathbf{W}}(\mathbf{x}-\boldsymbol{\xi})\mathrm{d}\Gamma - \int\limits_{\Omega_i^s} \mathbf{u}^h(\boldsymbol{\xi})\mathbf{L}_d^T(\widehat{\mathbf{W}}(\mathbf{x}-\boldsymbol{\xi}))\mathrm{d}\boldsymbol{\xi}, \\ \qquad \text{when } \mathbf{u}^h(\boldsymbol{\xi}) \in \mathbb{C}^0(\Omega_i^s) \\[2ex] \int\limits_{\Gamma_i^s} \mathbf{L}_n \mathbf{u}^h(\boldsymbol{\xi})\widehat{\mathbf{W}}(\mathbf{x}-\boldsymbol{\xi})\mathrm{d}\Gamma - \int\limits_{\Omega_i^s} \mathbf{u}^h(\boldsymbol{\xi})\mathbf{L}_d^T(\widehat{\mathbf{W}}(\mathbf{x}-\boldsymbol{\xi}))\mathrm{d}\boldsymbol{\xi}, \\ \qquad \text{when } \mathbf{u}^h(\boldsymbol{\xi}) \in \mathbb{C}^{-1}(\Omega_i^s) \end{cases} \tag{4.27}$$

where the smoothing function $\widehat{\mathbf{W}}$ satisfies the unity condition defined in Equation 2.10 over the smoothing domain Ω_i^s.

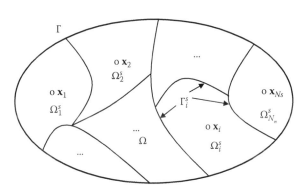

FIGURE 4.2
Division of the problem domain Ω into stationary smoothing domains Ω_i^s bounded by Γ_i^s for the point at $\mathbf{x}_i$. The smoothing domain can be different for different $\mathbf{x}_i$ but they cannot overlap. The smoothing function can also be different for different Ω_i^s.

Remark 4.3: Smoothed Strain: Convergence Property

For any, $\mathbf{u}^h \in \mathbb{H}^1_h$, when Ω^s_i approaches to zero, the smoothing function $\widehat{W}$ approaches the delta function. At such a limit $\hat{\boldsymbol{\varepsilon}} = \tilde{\boldsymbol{\varepsilon}}$, and the constructed strain field is the compatible strain field.

When stationary smoothing domains are used, the minimum number of smoothing domains needs to be determined based on Table 3.1. On violation of the conditions given in Table 3.1, the discretized system equations established using the GS-Gakerkin weak form will be singular, and no unique solution will be obtained. The use of more smoothing domains does not necessarily guarantee a nonsingular set of system equations, because it depends also on how the division of the smoothing domains is performed. Generally, a finer division of smoothing domains leads to a stiffer model (see Theorem 4.3).

When stationary smoothing domains are created based on Section 3.3.1, and the smoothing function $\bar{W}$ is constant within each smoothing domain Ω^s_i:

$$\bar{W}(\mathbf{x}_i - \boldsymbol{\xi}) = \begin{cases} 1/A^s_i & \boldsymbol{\xi} \in \Omega^s_i \\ 0 & \boldsymbol{\xi} \notin \Omega^s_i \end{cases} \tag{4.28}$$

where $A^s_i = \int_{\Omega^s_i} d\Omega$ is the area of the smoothing domain associated with a point at $\mathbf{x}_i$, we have

$$\bar{\boldsymbol{\varepsilon}}(\mathbf{x}_i) = \begin{cases} \frac{1}{A^s_i} \int\limits_{\Omega^s_i} \tilde{\boldsymbol{\varepsilon}}(\mathbf{u}^h) d\Omega = \frac{1}{A^s_i} \int\limits_{\Omega^s_i} (\mathbf{L}_d \mathbf{u}^h) d\Omega = \frac{1}{A^s_i} \int\limits_{\Gamma^s_i} \mathbf{L}_n \mathbf{u}^h d\Gamma, & \text{when } \mathbf{u}^h(\boldsymbol{\xi}) \in \mathbb{C}^0(\Omega_s) \\[2ex] \frac{1}{A^s_i} \int\limits_{\Gamma^s_i} \mathbf{L}_n \mathbf{u}^h d\Gamma, & \text{when } \mathbf{u}^h(\boldsymbol{\xi}) \in \mathbb{C}^{-1}(\Omega_s) \end{cases} \tag{4.29}$$

We further assume that the strain in the entire smoothing domain ω^s_i is constant:

$$\bar{\boldsymbol{\varepsilon}}_i = \bar{\boldsymbol{\varepsilon}}(\mathbf{x}_i), \quad \forall \mathbf{x} \in \Omega^s_i \tag{4.30}$$

When this is done for all the smoothing domains, the strain field is constructed. The discretized form of the potential energy for the smoothed strain becomes

$$\bar{U}^D_{PE} = \int\limits_\Omega \frac{1}{2} \bar{\boldsymbol{\varepsilon}}^T \mathbf{c} \bar{\boldsymbol{\varepsilon}} d\boldsymbol{\xi} = \frac{1}{2} \sum_{i=1}^{N_s} A^s_i \bar{\boldsymbol{\varepsilon}}^T_i \mathbf{c} \bar{\boldsymbol{\varepsilon}}_i \tag{4.31}$$

4.5.3.6 Properties of the Smoothed Strain Field

We now briefly discuss the properties and the strain energy of the smoothed strain field. A more detailed discussion and proofs on these properties can be found in [19–23].

Theorem 4.1: Smoothed Strain: Constructed by Orthogonal Projection

For any $\mathbf{u}^h \in \mathbb{H}^1_h$, the smoothed strain obtained using Equation 4.26 satisfies the orthogonal condition in Equation 4.5, and the smoothed strain field is an orthogonal projection of the compatible strain field onto the space of the smoothed strain fields.

Proof Because the strain field is assumed constant in the stationary smoothing domains, and $\mathbf{u}^h \in \mathbb{H}_h^1$, the integration over the entire problem domain becomes a summation of integrations over each of the smoothing domains. Therefore, the orthogonal condition (Equation 4.5) will be satisfied if the following condition is met for any smoothing domain:

$$\int_{\Omega_i^s} \hat{\boldsymbol{\varepsilon}}_i^T \mathbf{c}(\mathbf{L}_d \mathbf{u}^h) d\Omega = \int_{\Omega_i^s} \hat{\boldsymbol{\varepsilon}}_i^T \mathbf{c}\hat{\boldsymbol{\varepsilon}}_i d\Omega \tag{4.32}$$

Using the fact that the smoothed strain is constant in the smoothing domain, we obtain

$$\int_{\Omega_i^s} \hat{\boldsymbol{\varepsilon}}_i^T \mathbf{c}(\mathbf{L}_d \mathbf{u}^h) d\Omega = \hat{\boldsymbol{\varepsilon}}_i^T \mathbf{c} \underbrace{\int_{\Omega_i^s} (\mathbf{L}_d \mathbf{u}^h) d\Omega}_{A_i^s \hat{\boldsymbol{\varepsilon}}_i} = A_i^s \hat{\boldsymbol{\varepsilon}}_i^T \mathbf{c}\hat{\boldsymbol{\varepsilon}}_i = \int_{\Omega_i^s} \hat{\boldsymbol{\varepsilon}}_i^T \mathbf{c}\hat{\boldsymbol{\varepsilon}}_i d\Omega \tag{4.33}$$

Next, we need to show that Equation 4.26 is a projector, by which we mean that $\mathcal{P}$ defined in Equation 4.1 must be an idempotent:

$$\mathcal{P}^2 = \mathcal{P} \tag{4.34}$$

which physically implies that any projections after the first projection are idle. In the case of the strain smoothing operation, we know that when $\mathbf{u}^h \in \mathbb{H}_h^1$, Equation 4.26 is indeed a projector that satisfies Equation 4.34, because our domains are stationary (fixed), and when it is applied to $\hat{\boldsymbol{\varepsilon}}_i$ again, it is idle: a smoothing to an already smoothed smoothing has no effect.

Now, to show the projector orthogonal, we use Equation 4.32:

$$\int_{\Omega_i^s} \hat{\boldsymbol{\varepsilon}}_i^T \mathbf{c}(\mathbf{L}_d \mathbf{u}^h) d\Omega - \int_{\Omega_i^s} \hat{\boldsymbol{\varepsilon}}_i^T \mathbf{c}\hat{\boldsymbol{\varepsilon}}_i d\Omega = \int_{\Omega_i^s} \hat{\boldsymbol{\varepsilon}}_i^T \mathbf{c}(\tilde{\boldsymbol{\varepsilon}}_i - \hat{\boldsymbol{\varepsilon}}_i) d\Omega = 0 \tag{4.35}$$

which means that the "distance vector" between the constructed and the compatible strain fields is orthogonal (in the energy norm) to the space of the constructed strain fields. This completes the proof. ∎

The discretized form of the potential energy of the smoothed strain becomes

$$\hat{U}_{PE}^D = \int_{\Omega} \frac{1}{2} \hat{\boldsymbol{\varepsilon}}^T \mathbf{c}\hat{\boldsymbol{\varepsilon}} d\boldsymbol{\xi} = \frac{1}{2} \sum_{i=1}^{N_s} A_i^s \hat{\boldsymbol{\varepsilon}}_i^T \mathbf{c}\hat{\boldsymbol{\varepsilon}}_i \tag{4.36}$$

Theorem 4.2: Smoothed Strain: Softening Effects

For any assumed admissible displacement function $\mathbf{u}^h \in \mathbb{H}^1$ and the smoothed strain is obtained using Equation 4.26, we have for solids of stable materials

$$\hat{U}_{PE}^D(\mathbf{u}^h) \leq \tilde{U}_{PE}(\mathbf{u}^h) \tag{4.37}$$

meaning that the potential energy for the constructed strain field is always smaller than that of the compatible strain field, for the same assumed displacement field in a proper Hilbert space.

Proof We first examine the strain energy resulted from the difference of the compatible and smoothed strain fields:

$$\int_\Omega \frac{1}{2}(\boldsymbol{\varepsilon} - \tilde{\boldsymbol{\varepsilon}})^\mathrm{T}\mathbf{c}(\boldsymbol{\varepsilon} - \tilde{\boldsymbol{\varepsilon}})d\Omega$$

$$= \int_\Omega \frac{1}{2}\hat{\boldsymbol{\varepsilon}}^\mathrm{T}\mathbf{c}\hat{\boldsymbol{\varepsilon}}d\Omega - \int_\Omega \frac{1}{2}\hat{\boldsymbol{\varepsilon}}^\mathrm{T}\mathbf{c}\tilde{\boldsymbol{\varepsilon}}d\Omega - \int_\Omega \frac{1}{2}\tilde{\boldsymbol{\varepsilon}}^\mathrm{T}\mathbf{c}\hat{\boldsymbol{\varepsilon}}d\Omega + \int_\Omega \frac{1}{2}\tilde{\boldsymbol{\varepsilon}}^\mathrm{T}\mathbf{c}\tilde{\boldsymbol{\varepsilon}}d\Omega$$

$$= \int_\Omega \frac{1}{2}\hat{\boldsymbol{\varepsilon}}^\mathrm{T}\mathbf{c}\hat{\boldsymbol{\varepsilon}}d\Omega - \int_\Omega \tilde{\boldsymbol{\varepsilon}}^\mathrm{T}\mathbf{c}\hat{\boldsymbol{\varepsilon}}d\Omega + \int_\Omega \frac{1}{2}\tilde{\boldsymbol{\varepsilon}}^\mathrm{T}\mathbf{c}\tilde{\boldsymbol{\varepsilon}}d\Omega \qquad (4.38)$$

where $\tilde{\boldsymbol{\varepsilon}} = \mathbf{L}_d\mathbf{u}^h$ is the compatible strain field. From Theorem 4.1, we know that the orthogonal condition in Equation 4.5 is satisfied, and $\mathbf{c}$ is the SPD for stable materials, and hence, we have

$$\underbrace{\int_\Omega \frac{1}{2}(\boldsymbol{\varepsilon} - \tilde{\boldsymbol{\varepsilon}})^\mathrm{T}\mathbf{c}(\boldsymbol{\varepsilon} - \tilde{\boldsymbol{\varepsilon}})d\Omega}_{\geq 0} = \underbrace{\int_\Omega \frac{1}{2}\tilde{\boldsymbol{\varepsilon}}^\mathrm{T}\mathbf{c}\tilde{\boldsymbol{\varepsilon}}d\Omega}_{\tilde{U}_{PE}(\mathbf{u}^h)} - \underbrace{\int_\omega \frac{1}{2}\hat{\boldsymbol{\varepsilon}}^\mathrm{T}\mathbf{c}\hat{\boldsymbol{\varepsilon}}d\Omega}_{\hat{U}^D_{PE}(\mathbf{u}^h)} \geq 0 \qquad (4.39)$$

which gives Equation 4.37. ∎

Theorem 4.2 implies that a strain constructed model via smoothing is always softer than the compatible model of the same mesh, known as the softening effect discovered in [19,23].

Theorem 4.3: Smoothed Strain: Monotonic Convergence Property

In a given division D_1 of a domain Ω into a set of smoothing domains $\overline{\Omega} = \bigcup_{i=1}^{N_s} \overline{\Omega}_i^s$, if a new division D_2 is created by subdividing a smoothing domain in D_1 into n_{sd} subsmoothing domains, $\overline{\Omega}_i^s = \bigcup_{j=1}^{n_{sd}} \overline{\Omega}_{i,j}^s$, then the following inequality stands:

$$\hat{U}^{D_1}_{PE}(\mathbf{u}^h) \leq \hat{U}^{D_2}_{PE}(\mathbf{u}^h) \qquad (4.40)$$

This implies that the "softening" effect provided by the smoothing operation will be monotonically reduced with the increase of the number of smoothing domains in a nested manner. A simple proof can be given using the triangle inequality of norms: sum of the energy norm of functions is no less than the norm of the summed functions [23].

4.5.4 Strain Construction by Point Interpolation

In this case, the strain field $\hat{\boldsymbol{\varepsilon}}$ is constructed using point interpolation by the following steps:

1. Select a set of points for strain interpolation based on the element mesh (in FEM settings) or the background cells (in meshfree settings). The density of the points should usually be higher than the field nodes to satisfy the norm equivalence condition. We should also try that these points coincide with the nodes so that the strain convergence condition (Equation 4.16) can be easily satisfied. For example, when a triangular mesh is used, we first choose all the field nodes as strain interpolation points, and then add in the middle the points of edges and the centroidals of the triangular cells/elements as the interpolation points. Such a selection of strain interpolation points is used in PIM-CS [26], SαFEM [30], and SC-PIM [27,31].

2. For any assumed admissible displacement $\mathbf{u}^h$, at a strain interpolation point, the strain is assigned as either the compatible strain obtained using Equation 4.2 (if $\mathbf{u}^h \in \mathbb{H}_h^1$) or a smoothed strain of the compatible strain field over a proper local domain for that point using Equation 4.27 or Equation 4.29. For example, when a triangular mesh is used, we choose node-based smoothing domains for points coinciding with nodes, and edge-based smoothing domains for points at the mid edge points. For points at the centroidals, we may simple use the compatible strains, as in PIM-CS [26], SC-PIM [27], and SαFEM [30].

3. A new set of (preferably triangular) cells for the set of strain interpolation points is then created for strain interpolation. This set of cells is used also for integration of the weak form, and is called quadrature or integration cells. For example, when triangular cells are used, the new triangular quadrature cells are hosted by the original triangles, and typically a triangular cell is divided into 1, 3, 4, or 6 integration cells, as in SC-PIM [31] (see Section 8.6).

4. The strain field in each of the quadrature cells is then constructed using point interpolation, based on these strains at the set of points using shape functions created, using the strain interpolation points and any interpolation technique presented in Chapter 3. For example, when a triangular mesh is used, we simply use linear shape functions created for each of the integration cells. In such a case, the integration of the weak form can be performed analytically, using the area coordinates, that is the same as the linear shape functions [2], and no numerical integration is needed. In this case, the strain field within each integration cell is obtained using the linear interpolation:

$$\hat{\boldsymbol{\varepsilon}}(\mathbf{x}) = \sum_{i=1}^{3} \boldsymbol{\Phi}_i(\mathbf{x})\bar{\boldsymbol{\varepsilon}}_i \tag{4.41}$$

where

$\hat{\boldsymbol{\varepsilon}}$ is the vector of constructed strains over the triangular integration cell
$\bar{\boldsymbol{\varepsilon}}$ is the vector of strains at each vertex, which are either the compatible strains $\tilde{\boldsymbol{\varepsilon}}$ or the smoothed strains $\hat{\boldsymbol{\varepsilon}}$ obtained, using a local smoothing domain

In practice, the linear interpolation is often used for the simple reason of easy integration: the strain energy for a triangular integration cell can be obtained exactly so that no numerical integration is needed. In this case, $\mathbf{\Phi}_i$ in Equation 4.41 is a diagonal matrix of linear PIM shape functions that has the following form:

$$\mathbf{\Phi}_i(\mathbf{x}) = \begin{bmatrix} L_i(\mathbf{x}) & 0 & 0 \\ 0 & L_i(\mathbf{x}) & 0 \\ 0 & 0 & L_i(\mathbf{x}) \end{bmatrix} \tag{4.42}$$

where L_i is the area coordinate for node i of the integration cell. Equation 4.41 can be rewritten as

$$\begin{Bmatrix} \hat{\varepsilon}_{11}(\mathbf{x}) \\ \hat{\varepsilon}_{22}(\mathbf{x}) \\ 2\hat{\varepsilon}_{12}(\mathbf{x}) \end{Bmatrix} = \begin{Bmatrix} L_1(\mathbf{x})\breve{\varepsilon}_{11}(\mathbf{x}_1) + L_2(\mathbf{x})\breve{\varepsilon}_{11}(\mathbf{x}_2) + L_3(\mathbf{x})\breve{\varepsilon}_{11}(\mathbf{x}_3) \\ L_1(\mathbf{x})\breve{\varepsilon}_{22}(\mathbf{x}_1) + L_2(\mathbf{x})\breve{\varepsilon}_{22}(\mathbf{x}_2) + L_3(\mathbf{x})\breve{\varepsilon}_{22}(\mathbf{x}_3) \\ 2\left(L_1(\mathbf{x})\breve{\varepsilon}_{12}(\mathbf{x}_1) + L_2(\mathbf{x})\breve{\varepsilon}_{12}(\mathbf{x}_2) + L_3(\mathbf{x})\breve{\varepsilon}_{12}(\mathbf{x}_3)\right) \end{Bmatrix} \tag{4.43}$$

5. The strain energy potential can be now given as

$$\hat{U}_{PE} = \int_\Omega \frac{1}{2} \underbrace{\hat{\varepsilon}^T(\mathbf{x})\mathbf{c}\hat{\varepsilon}(\mathbf{x})}_{finite} \, d\Omega = \frac{1}{2} \sum_{j=1}^{N_q} \int_{\Omega_j^q} \left(\sum_{i=1}^{3} \mathbf{\Phi}_i(\mathbf{x})\breve{\varepsilon}_i\right)^T \mathbf{c} \left(\sum_{i=1}^{3} \mathbf{\Phi}_i(\mathbf{x})\breve{\varepsilon}_i\right) d\Omega$$

$$= \frac{1}{2} \sum_{j=1}^{N_q} \int_{\Omega^q} \left(\begin{array}{c} \breve{\varepsilon}_1^T \mathbf{\Phi}_1^T \mathbf{c}\mathbf{\Phi}_1 \breve{\varepsilon}_1 + \breve{\varepsilon}_2^T \mathbf{\Phi}_2^T \mathbf{c}\mathbf{\Phi}_2 \breve{\varepsilon}_2 + \breve{\varepsilon}_3^T \mathbf{\Phi}_3^T \mathbf{c}\mathbf{\Phi}_3 \breve{\varepsilon}_3 \\ + 2\breve{\varepsilon}_1^T \mathbf{\Phi}_1^T \mathbf{c}\mathbf{\Phi}_2 \breve{\varepsilon}_2 + 2\breve{\varepsilon}_1^T \mathbf{\Phi}_1^T \mathbf{c}\mathbf{\Phi}_3 \breve{\varepsilon}_3 + 2\breve{\varepsilon}_2^T \mathbf{\Phi}_2^T \mathbf{c}\mathbf{\Phi}_3 \breve{\varepsilon}_3 \end{array}\right) d\Omega$$

$$= \frac{1}{2} \sum_{j=1}^{N_q} \left(\begin{array}{c} \breve{\varepsilon}_1^T \mathbf{c}\breve{\varepsilon}_1 \int_{\Omega_q} L_1^2 d\Omega + \breve{\varepsilon}_2^T \mathbf{c}\breve{\varepsilon}_2 \int_{\Omega_q} L_2^2 d\Omega + \breve{\varepsilon}_3^T \mathbf{c}\breve{\varepsilon}_3 \int_{\Omega_q} L_3^2 d\Omega \\ + 2\breve{\varepsilon}_1^T \mathbf{c}\breve{\varepsilon}_2 \int_{\Omega_q} L_1 L_2 d\Omega + 2\breve{\varepsilon}_1^T \mathbf{c}\breve{\varepsilon}_3 \int_{\Omega_q} L_1 L_3 d\Omega + 2\breve{\varepsilon}_2^T \mathbf{c}\breve{\varepsilon}_3 \int_{\Omega_q} L_2 L_3 d\Omega \end{array}\right) \tag{4.44}$$

Using now the Eisenberg–Malvern formula for a triangle with an area A [2].

$$\int_A L_1^p L_2^q L_3^r d\Omega = \frac{p!q!r!}{(p+q+r+2)!} 2A \tag{4.45}$$

which gives

$$\int_{\Omega_j^q} L_1^2 d\Omega = \int_{\Omega_j^q} L_2^2 d\Omega = \int_{\Omega_j^q} L_3^2 d\Omega = \frac{2!}{4!} 2A_j^q = \frac{1}{6} A_j^q$$

$$\int_{\Omega_j^q} L_1 L_2 d\Omega = \int_{\Omega_j^q} L_1 L_3 d\Omega = \int_{\Omega_j^q} L_2 L_3 d\Omega = \frac{1!}{4!} 2A_j^q = \frac{1}{12} A_j^q \tag{4.46}$$

Equation 4.44 becomes

$$\hat{U}_{PE} = \frac{1}{2} \sum_{j=1}^{N_q} \frac{A_j^q}{6} \left(\breve{\boldsymbol{\varepsilon}}_1^T \mathbf{c}\breve{\boldsymbol{\varepsilon}}_1 + \breve{\boldsymbol{\varepsilon}}_2^T \mathbf{c}\breve{\boldsymbol{\varepsilon}}_2 + \breve{\boldsymbol{\varepsilon}}_3^T \mathbf{c}\breve{\boldsymbol{\varepsilon}}_3 + \breve{\boldsymbol{\varepsilon}}_1^T \mathbf{c}\breve{\boldsymbol{\varepsilon}}_2 + \breve{\boldsymbol{\varepsilon}}_1^T \mathbf{c}\breve{\boldsymbol{\varepsilon}}_3 + \breve{\boldsymbol{\varepsilon}}_2^T \mathbf{c}\breve{\boldsymbol{\varepsilon}}_3 \right) \qquad (4.47)$$

It is clear that the strain energy potential for the constructed strain field is the average of the all the six components of energy combinations of strains at these three vertices of the integration cell.

The properties of an SC-model depend on how the strain field is constructed. We often introduce some parameters controlling the constructed strain field for desired properties. The above mentioned steps were used in PIM-CS [26], SC-PIM [27,31], and SαFEM [30].

4.5.5 Strain Construction by Least Squares Approximation

Least squares approximation is a kind of an orthogonal projection, found useful in the strain field construction, and was implemented in LS-PIMs [28]. The constructed strain field can easily satisfy the norm equivalence condition, and as long as the strain convergence condition (Equation 4.16) can be satisfied, we have a convergent SC-model. The procedure for constructing the strain field using least squares approximation is quite similar to the method of interpolation. The four-step procedure given in Section 4.5.4 stands, except to replace the interpolation by a least squares approximation. When the least squares approximation is used, we can have large integration cells, use more strain points for approximation, and have a wider choice of polynomial basis terms for creating models of desired properties. More details are given in [28].

4.6 Concluding Remarks

Remark 4.4: Convergence: Solution to Incompatibility
Strain construction is an effective way to solve the incompatibility problem that we often encounter in the standard FEM formulation and modeling [2]. The essential issue for the incompatibility problem is the possible unbounded potential energy using the incompatible assumed displacement field. Our idea of constructing the strain field offers an effective means to have the strain energy potential always bounded, ensuring the convergence of the model.

Remark 4.5: Stability and Convergence to the Exact Solution
When we open the door to construct the strain field to resolve the convergence issue for general assumed displacement functions in a $\mathbb{G}$ space, possible problems related to instability are also invited, in addition to a possible problem that the solution may not converge to the exact solution. We have given some admissible conditions for the constructed strain field in Section 4.4 to prevent this from happening, but many things are still not clear to the author

at this stage, especially when an assumed displacement function is discontinuous and in an unnormed $\mathbb{G}$ space. Obviously, more thorough studies, and nonstandard theories and principles such as the weakened-weak formulation [37] may be needed. It is quite clear now that as long as a set of linearly independent nodal shape functions that satisfy the bound and positivity conditions can be created, we can always establish a stable and convergence-to-exact-solution model, with proper ways to construct the strain field: the assumed displacement fields do not have to be continuous!

Remark 4.6: Efficiency

To construct the strain field, additional computational efforts or alternative computational treatments are required. From the experience of the author's group, such efforts/treatments do not contribute significantly to the overall computation time. In addition, some of the alternative operations are simpler compared to the models using directly compatible strain fields. For example, in smoothed models, we do not need to compute the derivatives of the shape functions; we do not need to perform domain Gauss integrations; energy integration becomes simple summation; etc. In the strain interpolation models, we do not even need a numerical integration for the weak form, etc. It is the author's opinion that a strain constructed model does not necessarily increase the computation cost compared to the corresponding fully compatible model.

Remark 4.7: Accuracy

A properly constructed SC-model can improve the solution accuracy significantly, as demonstrated in the models mentioned earlier. One can practically establish models that give close-to-exact solutions (in a norm), as shown in PIM-CS [26], SC-PIM [27,31], and SαFEM [30]. This is made possible by using a "knob" α to tune an SC-model for desired properties.

Remark 4.8: Other Properties

Other attractive properties, such as the upper bound, the lower bound, and tight bounds can also be achieved by devising properly a strain construction scheme, especially a model with a knob, as shown in PIM-CS [26], SC-PIM [27], and SαFEM [30]. More discussions are presented in Chapter 8.

References

1. Zienkiewicz, O. C. and Taylor R. L., *The Finite Element Method*, 5th ed., Butterworth Heimemann, Oxford, 2000.
2. Liu, G. R. and Quek, S. S., *The Finite Element Method: A Practical Course*, Butterworth Heinemann, Oxford, 2003.
3. Simo, J. C. and Hughes, T. J. R., *Computational Inelasticity*, Springer-Verlag, New York, 1998.
4. Pian T. H. H. and Wu C. C., *Hybrid and Incompatible Finite Element Methods*, CRC Press, Boca Raton, FL, 2006.
5. Wu, H. C., *Variational Principle in Elasticity and Applications*, Scientific Press, Beijing, China, 1982.

6. Liu, G. R., Dai, K. Y., and Nguyen, T. T., A smoothed finite element method for mechanics problems, *Comput. Mech.*, 39, 859–877, 2007.
7. Dai, K. Y. and Liu, G. R., Free and forced vibration analysis using the smoothed finite element method (SFEM), *J. Sound Vib.*, 301, 803–820, 2007.
8. Dai, K. Y., Liu, G. R., and Nguyen, T. T., An n-sided polygonal smoothed finite element method (nSFEM) for solid mechanics. *Finite Elem. Anal. Des.*, 43, 847–860, 2007.
9. Liu, G. R., Nguyen, T. T., Dai, K. Y., and Lam, K. Y., Theoretical aspects of the smoothed finite element method (SFEM), *Int. J. Numerical Methods Eng.*, 71, 902–930, 2007.
10. Liu, G. R., Nguyen-Thoi, T., Nguyen-Xuan, H., and Lam, K. Y., A node-based smoothed finite element method (NS-FEM) for upper bound solutions to solid mechanics problems, *Comput. Struct.*, 87, 14–26, 2009.
11. Liu, G. R., Nguyen-Thoi, T., and Lam, K. Y., An edge-based smoothed finite element method (ES-FEM) for static, free and forced vibration analyses in solids, *J. Sound Vib.* (in press), 320, 1100–1130, 2009.
12. Nguyen-Thoi, T., Liu, G. R., Lam, K. Y., and Zhang, G. Y., A face-based smoothed finite element method (FS-FEM) for 3D linear and nonlinear solid mechanics problems using 4-node tetrahedral elements, *Int. J. Numer. Methods Eng.* (in press), 2008. doi: 10.1002/nme.2491.
13. Liu, G. R. and Zhang, G. Y., Edge-based smoothed point interpolation methods, *Int. J. Comput. Methods*, 5, 621–646, 2008.
14. Belytschko, T. and Bachrach, W. E., Efficient implementation of quadrilaterals with high coarse-mesh accuracy, *Comput. Methods Appl. Mech. Eng.*, 54, 279–301, 1986.
15. Belytschko, T. and Bindeman, L. P., Assumed strain stabilization of the 4-node quadrilateral with 1-point quadrature for nonlinear problems, *Comput. Methods Appl. Mech. Eng.*, 88, 311–340, 1993.
16. Dohrmann, C. R., Heinstein, M. W., Jung, J., Key, S. W., and Witkowski, W. R., Node-based uniform strain elements for three-node triangular and four-node tetrahedral meshes, *Int. J. Numerical Methods Eng.*, 47, 1549–1568, 2000.
17. Liu, G. R., Nguyen-Thoi, T., and Lam, K. Y., A novel FEM by scaling the gradient of strains with factor α (αFEM), *Comput. Mech.*, 43, 369–391, 2008.
18. Liu, G. R., Zhang, G. Y., Dai, K. Y., Wang, Y. Y., Zhong, Z. H., Li, G. Y., and Han, X., A linearly conforming point interpolation method (LC-PIM) for 2D solid mechanics problems, *Int. J. Comput. Methods*, 2(4), 645–665, 2005.
19. Liu, G. R. and Zhang, G. Y., Upper bound solution to elasticity problems: A unique property of the linearly conforming point interpolation method (LC-PIM), *Int. J. Numerical Methods Eng.*, 2007.
20. Liu, G. R., Zhang, G. Y., Wang, Y. Y., Zhong, Z. H., Li, G. Y., and Han, X., A nodal integration technique for meshfree radial point interpolation method (NI-RPCM), *Int. J. Solids Struct.*, 44, 3840–3860, 2007.
21. Zhang, G. Y., Liu, G. R., Wang, Y. Y., Huang, H. T., Zhong, Z. H., Li, G. Y., and Han, X., A linearly conforming point interpolation method (LC-PIM) for three-dimensional elasticity problems, *Int. J. Numerical Methods Eng.*, 72, 1524–1543, 2007.
22. Zhang, G. Y., Liu, G. R., Nguyen, T. T., Song, C. X., Han, X., Zhong, Z. H., and Li, G. Y., The upper bound property for solid mechanics of the linearly conforming radial point interpolation method (LC-RPIM), *Int. J. Comput. Methods*, 4(3), 521–541, 2007.
23. Liu, G. R., A generalized gradient smoothing technique and the smoothed bilinear form for Galerkin formulation of a wide class of computational methods, *Int. J. Comput. Methods*, 5(2), 199–236, 2008.
24. Chen, J. S., Wu, C. T., and Yoon, S. Y., A stabilized conforming nodal integration for Galerkin mesh-free methods, *Int. J. Numerical Methods Eng.*, 50, 435–466, 2001.
25. Liu, G. R., Nguyen-Thoi, T., and Lam, K. Y., A novel alpha finite element method (αFEM) for exact solution to mechanics problems using triangular and tetrahedral elements, *Comput. Methods Appl. Mech. Eng.*, 197, 3883–3897, 2008.

26. Liu, G. R., Xu, X., Zhang, G. R., and Gu, Y. T., An extended Galerkin weak form and a point interpolation method with continuous strain field and superconvergence (PIM-CS) using triangular mesh, *Comput. Mech.*, 43, 651–673, 2009.

27. Liu, G. R., Xu, X., Zhang, G. Y., and Nguyen-Thoi, T., A superconvergent point interpolation method (SC-PIM) with piecewise linear strain field using triangular mesh, *Int. J. Numerical Methods Eng.* (in press), 2008. doi: 10.1002/nme.2464.

28. Xu, X., Liu, G. R., and Zhang, G. Y., A least square point interpolation method (LS-PIM) for solution bounds and ultra-accurate solutions using triangular mesh, *Comput. Methods Appl. Mech. Eng.* (submitted), 2008.

29. Liu, G. R., Nguyen-Xuan, H., and Nguyen-Thoi, T., A variationally consistent αFEM (VCαFEM) for solution bounds and ultra-accurate solutions, *Int. J. Numerical Methods Eng.* (submitted), 2008.

30. Liu, G. R., Nguyen-Xuan, H., Nguyen-Thoi, T., and Xu, X., A novel weak form and a superconvergent alpha finite element method (SαFEM) for mechanics problems using triangular meshes, *J. Comput. Phys.* (submitted), 2008.

31. Liu, G. R. and Zhang, G. Y., A Strain-constructed point interpolation method (SC-PIM) and strain field construction schemes for mechanics problems of solids and structures using triangular mesh, *Int. J. Solids Struct.* (submitted), 2008.

32. Liu, G. R., Zhang J., and Lam, K. Y., A gradient smoothing method (GSM) with directional correction for solid mechanics problems, *Comput. Mech.*, 41, 457–472, 2008.

33. Liu, G. R. and Xu, X. G., A gradient smoothing method (GSM) for fluid dynamics problems, *Int. J. Numerical Methods Fluids* (accepted), 2008.

34. Xu, G. X., Liu, G. R., and Lee, K. H., Application of gradient smoothing method (GSM) for steady and unsteady incompressible flow problems using irregular triangles, *Int. J. Numerical Methods Fluids* (submitted), 2008.

35. Liu, G. R. and Xu, X. G., An adaptive gradient smoothing method (GSM) for fluid dynamics problems, *Int. J. Numerical Methods Fluids* (accepted), 2008.

36. Monaghan, J. J., Smoothed particle hydrodynamics, *Annu. Rev. Astron. Astrophys.*, 30, 543–574, 1992.

37. Liu, G. R., A $\mathbb{G}$ space theory and weakened weak (W^2) form for a unified formulation of compatible and incompatible methods, Part I Theory and Part II Application to solid mechanics problems, *Int. J. Numerical Methods Eng.*, 2009 (in press).

38. Eringen, Nonlocal polar elastic continua, *Int. J. Eng. Sci.*, 10, 1, 1972.

39. Lucy, L., A numerical approach to testing the fission hypothesis, *Astron. J.*, 82, 1013–1024, 1977.

40. Monaghan, J. J., Why particle methods work, *SIAM J. Sci. Stat. Comput.*, 3(4), 423–433, 1982.

41. Liu, G. R. and Liu, M. B., *Smoothed Particle Hydrodynamics—A Meshfree Practical Method*, World Scientific, Singapore, 2003.

42. Chen, J. S., Wu, C. T., and Belytschko, T., Regularization of material instabilities by meshfree approximations with intrinsic length scales, *Int. J. Numerical Methods Eng.*, 47, 1303–1322, 2000.

5

Weak and Weakened-Weak Formulations

Weak formulations are of fundamental importance for the finite element method (FEM), and naturally, also for meshfree methods. In the FEM, we have basic building blocks of *elements*, and hence, all the numerical operations including function approximation and integration of the weak forms are all naturally based on the elements. In meshfree methods, however, function approximation and integration are virtually independent, and hence it offers many more innovative ways to use and even establish new weak forms such as the weakened-weak (W^2) forms that have unique and important properties, leading to methods that are superior to the standard FEM in many aspects.

We first introduce the weak formulation based on the $\mathbb{H}$-space theory and weakened-weak formulations based on the $\mathbb{G}$-space theory, and discuss some of their important properties. Physical energy principles used for creating weak forms for the FEM and meshfree methods are also outlined in this chapter with emphasis on the novel weakened weak forms. Since both mathematical and physical approaches are used in the literature to establish weak forms, it is sometimes quite confusing to many. This chapter tries to put these two together aiming to show their connections and hence better understand these formulations. Clarity and understanding are often achieved by comparisons.

This book gives high preference to the Galerkin weak form for reasons of simplicity, symmetry, and hence efficiency, which is eventually a crucial factor for any successful numerical method to survive. When we perform all sorts of advances and manipulations, we always try to keep the *form* of Galerkin, even though our formulation has gone far beyond the standard Galerkin weak form, in terms of implementation, solution/function spaces, and properties. This preference is particularly important for partial differential equations (PDEs) of symmetric operators: we want to preserve the symmetry.

We are aware that the contents of this chapter are quite heavy. Readers may skip the proofs, if their interests are on the applications of these formulation procedures.

5.1 Introduction to Strong and Weak Forms

5.1.1 Strong Forms

Strong form equations are those given in the form of PDEs, as briefly discussed in Section 1.2 for solid mechanics problems. The displacement functions are required to have the second order of consistency in the entire problem domain that is the same as the order of the differentiations in PDEs. Such a requirement on consistency for the displacement functions is said *strong*.

Obtaining the *exact* solution for such a strong form system equation is ideal but, unfortunately, it is usually very difficult for practical engineering problems that are often very complicated in both the problem setting and the geometry of the problem domain. We, therefore, search for approximated solutions. The finite difference method (FDM), which uses the finite differential representation (Taylor series) of a function in a local domain, can be used to solve system equations of a strong form to obtain an approximated solution. However, FDM requires essentially a regular mesh of grids, and can usually work only for problems with relatively regular geometry and simple boundary conditions. One of the meshfree methods for solving strong form system equations to obtain an approximate solution is to use arbitrarily distributed grids based on the Taylor series expansions, and least squares (LS) or moving least squares (MLS) approximations. The formulation is very simple, and all one needs to do is to approximate the field functions using a method described in Chapter 2 with sufficient number of local nodes for sufficient consistency so that the function and its derivatives are expressed in terms of the nodal values of the function. Substituting these expressions into the strong form equations directly results in a set of algebraic equations. However, the solution to this set of algebraic equations is often not very stable against the model setting and the node irregularity. The accuracy of the result often depends on boundary condition treatments, node distribution in the problem domain, and the selection of the nodes for the function approximation. This is because this type of direct *collocation* approach has no means to control the stability of the resultant algebraic equations. When the nodal distribution is arbitrary and only local nodes are used for function approximation, practically many things can happen. Therefore, special techniques are needed to stabilize the solution. Some of the detailed discussions on this can be found in [1–3].

Note also that the discretized system equations are generally *asymmetric* for irregularly distributed nodes, even for problems whose PDEs are associated with symmetric operators.

5.1.2 Weak Forms

A weak form, in contrast to a strong form, requires weaker consistency on the assumed field functions. The foundation is on the Sobolev (or $\mathbb{H}$)-space theory (see Chapter 3). The consistency requirement on the assumed functions for field variables is very different from the strong form. For a $2k$th-order differential governing system equation, the strong formulation requires a consistency of the $2k$th order, while the weak formulation requires a consistency of only the kth order. For second-order PDEs for mechanics problems defined in Section 1.2, the assumed displacements can be in an $\mathbb{H}^1$ space and hence need only to be first-order differentiable! This is made possible mathematically by introducing a so-called *test* function into an integral form of a residual formulation to "absorb" one derivative, and "convert" the strong form equations to weak ones. Physically, we create equations that evaluate the energy status for the stressed solid or the structures, instead of the equilibrium status in stresses. When the total energy potential in the solid/structure is at a stationary status, an equilibrium status is observed. When only the strain energy needs to be evaluated, we need only strains and hence the first derivatives of the displacements.

There are basically two major categories of often-used principles for constructing weak forms. They are the weighted residual methods (mathematical approach) and the energy principles (physical approach). The Galerkin formulation can be derived from both the weighted residual methods (by using *test* and *trial* functions from the same space) and the energy principles. It is the most widely used approach for establishing system equations, and is, of course, applicable to derive meshfree equations. Hamilton's principle is often employed to produce approximated system equations for dynamic problems, and is also applicable to meshfree methods. The minimum potential energy principle has been a convenient tool for deriving discrete system equations for FEM and also for many other types of approximation methods. The weighted residual method is a more general and a powerful mathematical tool that can be used for creating discretized system equations for many types of engineering problems. It has been and is still used for developing new meshfree methods. All these approaches are adopted in this book for creating discretized system equations for various types of meshfree methods.

Formulations based on weak forms can produce a stable set of algebraic system equations, because the positivity is ensured in the formulation process using energy (always positive) principles essentially and functions from proper spaces. The discretized system equations are *symmetric* for irregularly distributed nodes for problems of symmetric operators. Therefore, it preserves the symmetry property and hence has good stability, accuracy, and efficiency.

Remark 5.1: Solution Error Control: A Crucial Role of a Weakform
Because of the important stability offered by the weakform, the error in the weakform solution can be properly bounded by the interpolation error (unavoidably) caused during the function approximation using the nodal values for a discretized model. In other words, the role of the weakform is essentially to ensure that the interpolation error will not be uncontrollably amplified in the process of obtaining the numerical solution.

5.1.3 Weakened-Weak Forms

Built upon the weak formulation, weakened-weak (W^2) forms [16,17] have been established. The first W^2 form was the generalized smoothed Galerkin or GS-Galerkin weak form built using the gradient smoothing technique [15] and the generalized gradient smoothing technique [16,17]. The foundation is on the normed $\mathbb{G}$-space theory (see Chapter 3). The second W^2 form was the strain-constructed Galerkin or SC-Galerkin weak form that is a more general W^2 form [17] that uses functions from basically unnormed $\mathbb{G}_h^1$ spaces. The weakened-weak formulations are relatively new but have been used for developing a number of meshfree methods for important properties, including the upper-bound property, ultra-accuracy, and superconvergence [16,17].

The requirement on the assumed functions for field variables is further reduced upon the already reduced weak formulations. For the mechanic cs problem defined in Section 1.2, the assumed displacements can be in a $\mathbb{G}_h^1$ space and hence discontinuous! When the solution is sought from an $\mathbb{H}_h^1$ space (a subspace of a $\mathbb{G}_h^1$ space), the W^2 model becomes softer and the accuracy and the convergence rate of the solution is often found much higher than that of the weak formulation. It offers a systematic and efficient way for both compatible and incompatible displacement methods.

Similar to the weak formulation, the discrete system equations of a weakened-weak formulation are *symmetric* for irregularly distributed nodes in problems of symmetric operators. Therefore, one should always use a weakened-weak formulation for such problems to preserve the symmetry property and hence achieve excellent stability, accuracy, and efficiency.

5.1.4 Weak-Form-Like Formulations

There are also meshfree methods for solving strong form system equations using an integral representation of field variable functions, such as the smooth particle hydrodynamics (SPH) methods and the gradient smoothing method (GSM) [4–6].

The SPH formulation can deal well with dynamic problems of infinite domain, such as problems in astrophysics [7], and it has been found stable for arbitrarily distributed nodes in these types of problems. This is due to the use of the integral representation of field functions, which pass the differentiation operations on the field function to the weight function. Therefore, it reduces the requirement on the order of consistency on the approximated field functions, and is actually somewhat similar to that of weak formulations, and hence is a weak-form-like formulation. The major problem with the SPH methods is the treatment of boundaries of the problem domain and the boundary conditions, when the problem domain is finite. SPH can provide solutions for the overall behavior of a class of highly nonlinear dynamic problems, such as problems that are largely momentum-driven.

The GSM has been found stable for virtually, arbitrarily distributed nodes and for problems with finite domains. This is due to the use of the gradient smoothing technique in proper ways that removes the differentiation upon the field function. Therefore, these *smoothing* operations reduce the requirement on the order of consistency on the approximated field functions, and also are actually quite similar to that of weak formulations. The difference is that the "weakening" operation is implemented in the stage of function approximation, and not in the stage of creating the system equation. The GSM can handle boundary conditions very easily; the solution is as stable as the weak form methods, and works well even for adaptive analysis [6]. However, the discretized system equations are generally *asymmetric*, and hence suit better for problems associated with asymmetric operators, such as fluid dynamics problems [5,6]. In Chapter 9, we use these weak-form-like formulations for problems of fluid flows.

Remark 5.2: Solution Error Control: A Crucial Role of Gradient Smoothing
It is well known that a smoothing operation on a function can "smear" the error of the function [33]. Because of this important error-smearing feature, the error in the solution of a GSM can also be properly bounded by the interpolation error caused during the function approximation using the nodal values for a discretized model. In other words, the gradient smoothing plays essentially a similar role as the weakform in terms of solution error control, and hence such a formulation is termed as weak-form-like formulation. This remark will be observed in Chapter 9 by the fact that only when the gradient smoothing is used for the approximation of all the derivatives that are used in the strong form, and no interpolation is used in the derivative approximation, the GSM scheme performs the best in terms of accuracy.

This chapter discusses all these weak and weakened weak forms beginning with the general weighted residual method.

5.2 Weighted Residual Method

5.2.1 General Form of the Weighted Residual Method

The weighted residual method is a classical, very simple, but a powerful mathematical tool to obtain weak forms of approximated system equations. The concept of the weighted residual method is straightforward and applicable, in principle, to most of the PDEs that govern engineering problems, including mechanics of solids, structures, and fluids.

Let us now consider the general form of the PDEs defined in Ω bounded by Γ that can be rewritten in a concise functional form as

$$\mathbf{F}(\mathbf{u}(x, y, z)) = \mathbf{0} \qquad (5.1)$$

where $\mathbf{u}(x, y, z)$ is the unknown field variable. For example, for solid mechanics problems defined in Section 1.2, we have

$$\mathbf{F}(\mathbf{u}(x, y, z)) = \mathbf{L}_d^{\mathrm{T}} \mathbf{c} \mathbf{L}_d \mathbf{u} + \mathbf{b} - \rho \ddot{\mathbf{u}} \qquad (5.2)$$

In this case, $\mathbf{u}(x, y, z)$ is the displacement field, as discussed in Chapter 1.

In general, it is difficult to obtain the exact solution $\mathbf{u}(x, y, z)$ that satisfies Equation 5.1. We therefore, somehow construct a set of *trial functions* of $\mathbf{u}(x, y, z)$. Any of these trial functions does not, in general, satisfy Equation 5.1, and hence we have a *residual* for a given *trial function* $\mathbf{u}(x, y, z)$:

$$\mathbf{F}(\underbrace{\mathbf{u}(x, y, z)}_{\text{trial function}}) = \underbrace{\mathbf{R}(x, y, z)}_{\text{residual}} \neq \mathbf{0} \qquad (5.3)$$

We then seek for *one* of these trial functions such that the residual becomes zero in a weighted integral sense over the problem domain:

$$\int_{\Omega} \underbrace{\mathbf{W}}_{\text{test functions}} \underbrace{\mathbf{R}(x, y, z)}_{\text{residual}} \, d\Omega = 0 \qquad (5.4)$$

where $\widehat{\mathbf{W}}$ is a vector or a diagonal matrix of the *weight* or *test functions* defined in Ω. For solid mechanics problems, we have

$$\int_{\Omega} \widehat{\mathbf{W}} \underbrace{(\mathbf{L}_d^{\mathrm{T}} \mathbf{c} \mathbf{L}_d \mathbf{u} + \mathbf{b} - \rho \ddot{\mathbf{u}})}_{\text{residual}} \, d\Omega = \int_{\Omega} \widehat{\mathbf{W}} (\mathbf{L}_d^{\mathrm{T}} \mathbf{c} \mathbf{L}_d \mathbf{u}) \, d\Omega + \int_{\Omega} \widehat{\mathbf{W}} \mathbf{b} \, d\Omega - \int_{\Omega} \widehat{\mathbf{W}} \rho \ddot{\mathbf{u}} \, d\Omega = \mathbf{0} \qquad (5.5)$$

We hope that a particular trial function $\mathbf{u}(x, y, z)$ that satisfies the integral form of Equation 5.4 for a set of properly selected $\widehat{\mathbf{W}}$ is a good approximation of the exact solution.

If the weight or test functions are so chosen to be differentiable, we then perform integration-by-parts (or the Gauss divergence theorem) for the first term in Equation 5.5, leading to

$$\int_{\Omega} \left(\mathbf{L}_d \widehat{\mathbf{W}} \right)^{\mathrm{T}} \mathbf{c} (\mathbf{L}_d \mathbf{u}) \, d\Omega - \int_{\Gamma} \left(\mathbf{L}_n \widehat{\mathbf{W}} \right)^{\mathrm{T}} (\mathbf{c} \mathbf{L}_d \mathbf{u}) \, d\Omega - \int_{\Omega} \widehat{\mathbf{W}}^{\mathrm{T}} \mathbf{b} \, d\Omega + \int_{\Omega} \widehat{\mathbf{W}}^{\mathrm{T}} \rho \ddot{\mathbf{u}} \, d\Omega = \mathbf{0} \qquad (5.6)$$

It is clear now we have only first-order derivatives for either the field function $\mathbf{u}$ or test function $\widehat{\mathbf{W}}$, a *weak form*.

This is essentially the idea of the weighted residual approach. It is, indeed, very simple. Either the primitive form of Equation 5.5 or the weak form (Equation 5.6) can be used. The following are well-known versions of weighted residual method.

- Collocation method. When the form of Equation 5.5 is used, the test function needs only to be integrable. The *collocation method* is such an extreme case (not very stable, in general), where the Dirac delta function is used as the test function.
- Subdomain method. The test function is a Heaviside type of a local constant function. In this case both forms of Equations 5.5 and 5.6 are possible.
- Method of moments. Test functions are monomials and both forms of Equations 5.5 and 5.6 are possible.
- LS method. Test functions are the derivatives of the residuals.
- Petrov–Galerkin method. If the weight function constructed is different from the shape functions used, the method is generally termed as the Petrov–Galerkin method.
- Galerkin formulation. If the trial functions are also used as the weight functions in the vector form of $\hat{\mathbf{W}}$, the weighted residual method leads to a *Galerkin formulation*.

Note that in more general formulations, we can include the residuals of the equations of boundary conditions. More details on those methods in meshfree settings can be found in the book [1]. A detailed discussion on the Galerkin formulation is given in Section 5.3.

Although the weighted residual method is simple and, in principle, applicable to most PDEs, different ways of implementation will lead to solutions of different properties and accuracies. The stability and convergence of such a formulation are not generally guaranteed; it offers only the possibilities of solutions but may not be *the solution* to *a problem*. Therefore, a careful construction of numerical models for a type/class of problem is required. Note also that to use the weighted residual method, the strong form of the system equations needs to be known. In general, it is desirable to choose test functions in accordance to the features of the strong form equations.

Note also that if the set of trial functions contains the exact solution, the weighted residual method will produce the exact solutions as long as the weight functions are chosen properly to ensure stability and convergence, and there is no numerical error in the computation. This important feature is useful in testing meshfree methods developed based on the weighted residual method.

5.2.2 Procedure of the Weighted Residual Method

The procedure to create a discrete numerical model for solving a problem using the weighted residual method is as follows:

1. Construct trial functions or shape functions to approximate the field function using the field variables at the nodes in the domain with a certain order of consistency.
2. Construct weight or test functions.
3. Substitute the trial and test functions into Equation 5.5 or Equation 5.6, which leads to a set of differential equations with respect to only time.
4. Solve the set of differential equations with respect to time using standard procedures to obtain the dynamic field.
5. For static problems, step 3 leads to a set of algebraic equations, which can be solved using standard algebraic equation solvers for the static field.

5.2.3 Local Weighted Residual Method

It is important to mention here that the quadrature domain Ω used in Equation 5.5 does not have to be the entire problem domain. When local quadrature domains are used, the weighted residual method can be termed a "local weighted residual method." This local weighted residual method is used (Chapter 7) where the test and trial functions are chosen independently in a meshfree setting.

5.3 A Weak Formulation: Galerkin

5.3.1 Bilinear Form

We first brief the standard weak formulation derived from the weighted residual method. We start with Equation 5.6 without the dynamic term. We choose *test* functions in the vector form $\mathbf{v} = \{v_1, v_2\}^\mathsf{T} \in \mathbb{H}_h^1 \in \mathbb{S} \subset \mathbb{H}_0^1(\Omega);$* we then have

$$\int_\Omega (\mathbf{L}_d\mathbf{v})^\mathsf{T}\mathbf{c}(\mathbf{L}_d\mathbf{u})\mathrm{d}\Omega - \underbrace{\int_{\Gamma_u} (\mathbf{L}_n\mathbf{v})^\mathsf{T}(\mathbf{c}\mathbf{L}_d\mathbf{u})\mathrm{d}\Omega}_{= 0,\, \because\, \mathbf{v}\,=\,0\text{ on }\Gamma_u} - \int_{\Gamma_t} \mathbf{v}^\mathsf{T} \underbrace{\mathbf{L}_n^\mathsf{T}(\mathbf{c}\mathbf{L}_d\mathbf{u})}_{\boldsymbol{\sigma}}\,\mathrm{d}\Omega - \int_\Omega \mathbf{v}^\mathsf{T}\mathbf{b}\mathrm{d}\Omega = 0 \qquad (5.7)$$

$$\underbrace{\phantom{\int_{\Gamma_t} \mathbf{v}^\mathsf{T} \mathbf{L}_n^\mathsf{T}(\mathbf{c}\mathbf{L}_d\mathbf{u})}}_{\mathbf{t}_\Gamma}$$

Because $\mathbf{v} \in \mathbb{H}_{h,0}^1(\Omega)$, we shall have $\mathbf{v}=0$ on Γ_u, and thus the second term in the foregoing equation becomes zero. Using the natural (force) boundary condition on Γ_t, we shall have

$$\underbrace{\int_\Omega (\mathbf{L}_d\mathbf{v})^\mathsf{T}\mathbf{c}(\mathbf{L}_d\mathbf{u})\mathrm{d}\Omega}_{a(\mathbf{v},\mathbf{u})} = \underbrace{\left(\int_{\Gamma_t} \mathbf{v}^\mathsf{T}\mathbf{t}_\Gamma\mathrm{d}\Omega + \int_\Omega \mathbf{v}^\mathsf{T}\mathbf{b}\mathrm{d}\Omega\right)}_{f(\mathbf{v})} \qquad (5.8)$$

We now have the well-known *bilinear form* for the solid mechanics problem defined in Section 1.2:

$$a(\mathbf{w}, \mathbf{v}) = \int_\Omega \underbrace{(\mathbf{L}_d\mathbf{v})^\mathsf{T}}_{\boldsymbol{\varepsilon}^\mathsf{T}(\mathbf{v})}\mathbf{c}\underbrace{(\mathbf{L}_d\mathbf{w})}_{\boldsymbol{\varepsilon}(\mathbf{w})}\mathrm{d}\Omega = \int_\Omega \boldsymbol{\varepsilon}(\mathbf{v})^\mathsf{T}\mathbf{c}\boldsymbol{\varepsilon}(\mathbf{w})\mathrm{d}\Omega \qquad (5.9)$$

The basic properties of the bilinear form are *symmetry, ellipticity* (or coercivity), and *continuity*:

$$a(\mathbf{w}, \mathbf{v}) = a(\mathbf{v}, \mathbf{w}), \quad \forall\mathbf{w} \in \mathbb{H}^1, \quad \forall\mathbf{v} \in \mathbb{H}^1 \quad \text{symmetry} \qquad (5.10a)$$

$$a(\mathbf{v}, \mathbf{v}) \geq C_p\|\mathbf{v}\|_{\mathbb{H}^1(\Omega)}^2, \quad \forall\mathbf{v} \in \mathbb{H}_0^1 \quad \text{ellipticity} \qquad (5.10b)$$

$$a(\mathbf{w}, \mathbf{v}) \leq C_c\|\mathbf{w}\|_{\mathbb{H}^1(\Omega)}^2\|\mathbf{v}\|_{\mathbb{H}^1(\Omega)}^2, \quad \forall\mathbf{w} \in \mathbb{H}^1, \quad \forall\mathbf{v} \in \mathbb{H}^1 \quad \text{continuity} \qquad (5.10c)$$

---

* In this book, when we require a vector is in a space, we require each of the component functions being independently in the space, and the dimension of the space is expanded accordingly.

where C_p and C_c are constants independent of $\mathbf{v}$ and $\mathbf{u}$. The symmetry is obvious; the ellipticity is the consequence of the second Korn's inequality applied to solids of stable materials (see Remark 1.1):

$$C_K \|\mathbf{v}\|_{\mathbb{H}^1(\Omega)} \le \|\boldsymbol{\varepsilon}(\mathbf{v})\|_{L^2}, \quad \forall \mathbf{v} \in \mathbb{H}_0^1 \tag{5.11}$$

where C_K is a general constant independent of $\mathbf{v}$. The second Korn's inequality is in turn rooted at the Poincare–Friedrichs inequality (Equation 3.33). The continuity is resulted from the Cauchy–Schwarz inequality (Equation 3.15) for stable materials.

Using Equation 5.9, we have

$$U_{\text{PE}}(\boldsymbol{\varepsilon}(\mathbf{v})) = \frac{1}{2} \int_\Omega \boldsymbol{\varepsilon}^T(\mathbf{v}) \mathbf{c} \boldsymbol{\varepsilon}(\mathbf{v}) \mathrm{d}\Omega = \frac{1}{2} a(\mathbf{v}, \mathbf{v}) \tag{5.12}$$

where $U_{\text{PE}}(\boldsymbol{\varepsilon}(\mathbf{v}))$ is the (strain) potential energy in the solid for a given displacement field $\mathbf{v} \in \mathbb{H}_0^1$. We see here the relationship between the mathematical term of a bilinear form and the engineering term of strain energy.

The *linear functional* for the solid mechanics problem is defined as

$$f(\mathbf{v}) = \int_{\Gamma_t} \mathbf{v}^T \mathbf{t}_\Gamma \mathrm{d}\Omega + \int_\Omega \mathbf{v}^T \mathbf{b}\, \mathrm{d}\Omega \tag{5.13}$$

This is clearly the work done by the external forces (body force $\mathbf{b}$ and forces on the natural boundary $\mathbf{t}_\Gamma$) under displacement $\mathbf{v}$.

Alternative formulations can be done using indicial notations, which can be found, for example, in [12]. Here we simply list the bilinear and linear forms for later comparison and reference purposes. The well-known *bilinear form* has the form

$$a(w, v) = \int_\Omega \frac{\partial v_i}{\partial x_j} \left(C_{ijkl} \frac{\partial w_k}{\partial x_l} \right) \mathrm{d}\Omega \tag{5.14}$$

where $v, w \in \mathbb{S}$, and the *linear functional* can be written as

$$f(v) = \int_{\Gamma_N} v_i t_i \mathrm{d}\Gamma + \int_\Omega b_i v_i \mathrm{d}\Omega \tag{5.15}$$

Note when the indicial notations are used, v and w will have two components for two-dimensional (2D) and three components for three-dimensional (3D) problems.

5.3.2 Weak Statement

Following Equation 5.7, we now have the *weak statement*: the exact solution of the displacement $\mathbf{u} \in \mathbb{S}$ of the strong form equations given in Section 1.2 satisfies

$$a(\mathbf{u}, \mathbf{v}) = f(\mathbf{v}), \quad \forall \mathbf{v} \in \mathbb{S} \tag{5.16}$$

We now make the following remark for future reference.

Remark 5.3: Galerkin Weak Formulation: Proven Facts
The statement for Equation 5.16 has a *unique* and *stable* solution. This is ensured by the well-known Lax–Milgram theorem due to the *ellipticity* and *continuity* for the bilinear form (Equation 5.10b and c). From Equation 5.8, we observe that we need only the first derivatives for all functions involved in the formulation. This is because a part of the second-order derivatives on **u** has been transferred to the so-called test function **v**. As a result, the continuity requirements on both functions **u** and **v** are *weakened*: they all need to be only first-order differentiable, compared with the requirement of being second-order differentiable in the strong formulation (Equation 5.2). Both functions **u** and **v** live in a proper Hilbert space. Therefore, Equation 5.16 is termed as the *weak* formulation, which is the foundation for the well-known and widely used FEM.

5.3.3 Bilinear Form in an $\mathbb{H}_h^1$ Space: FEM Settings

In practice, it is generally very difficult to solve the governing equations either in strong or weak forms by analytical means for the *exact* solution. We then often resort to numerical methods to obtain *approximate* solutions. The most popular method is the traditional FEM based on the weak formulation where the Galerkin projection is chosen to obtain an approximate solution $\tilde{\mathbf{u}}$ from an $\mathbb{H}_h^1$ space of lower and finite dimension. The FEM formulation can be conveniently done using the weak form statement, as long as we can create a discrete (hence finite) $\mathbb{H}_h^1 \subset \mathbb{H}$ space based on a mesh of elements. It is well known that such an FEM solution is the best (in *a*-norm or energy norm) possible solution in the discrete finite element space $\mathbb{H}_h^1 \subset \mathbb{H}$, such that $\tilde{\mathbf{u}} \to \mathbf{u}$ when $\mathbb{H}_h^1 \to \mathbb{H}$, meaning that the approximate FEM solution approaches the exact solution when the size of the element approaches zero and the dimension of the FEM model $N_n \to \infty$.

The finite element solution $\tilde{\mathbf{u}} \in \mathbb{H}_{h,0}^1 \subset \mathbb{H}_0^1$, as an approximation to the solution to the problem defined in Section 1.2, satisfies the following weak statement

$$a(\tilde{\mathbf{u}}, \mathbf{v}) = f(\mathbf{v}), \quad \forall \mathbf{v} \in \mathbb{H}_{h,0}^1 \tag{5.17}$$

where the displacement field is $\tilde{\mathbf{u}} \in \mathbb{H}_{h,0}^1$.

5.3.4 Discrete System Equations

Using nodal shape functions $\phi_i^H \in \mathbb{H}_h^1$ obtained, based on the elements with proper mapping, the displacement field $\tilde{\mathbf{u}} \in \mathbb{H}_{h,0}^1$ in an element can be obtained for types of elements and expressed in terms of the following interpretation form:

$$\tilde{\mathbf{u}}(\mathbf{x}) = \sum_{i \in S_e} \phi_i^H(\mathbf{x}) \tilde{\mathbf{u}}_i \tag{5.18}$$

where
$\mathbf{x} = \{x_1, x_2\}^T$
S_e is the set of nodes in the element that hosts $\mathbf{x}$
$\tilde{\mathbf{u}}_i$ is a nodal displacement of the element

and

$$\boldsymbol{\phi}_i^H(\mathbf{x}) = \begin{bmatrix} \phi_i^H(\mathbf{x}) & 0 \\ 0 & \phi_i^H(\mathbf{x}) \end{bmatrix} \text{ for 2D,} \quad \boldsymbol{\phi}_i^H(\mathbf{x}) = \begin{bmatrix} \phi_i^H(\mathbf{x}) & 0 & 0 \\ 0 & \phi_i^H(\mathbf{x}) & 0 \\ 0 & 0 & \phi_i^H(\mathbf{x}) \end{bmatrix} \text{ for 3D} \quad (5.19)$$

The nodal shape functions are the Kronecker delta functions: $\phi_i^H(\mathbf{x}_j) = \delta_{ij}$.

We then substitute Equation 5.18 into Equation 5.17, and set $\boldsymbol{\phi}_i^H$, where $i = 1, \dots, N_n$, as the test functions, we have the following discrete set of $d \times N_n$ equations

$$\sum_{j=1}^{N_n} \mathbf{a}\left(\boldsymbol{\phi}_j^H, \boldsymbol{\phi}_i^H\right) \tilde{\mathbf{u}}_j = \mathbf{f}(\boldsymbol{\phi}_i^H), \quad i = 1, \dots, N_n \quad (5.20)$$

where N_n is the total number of (unconstrained) nodes. Equation 5.20 can be written in the matrix form

$$\tilde{\mathbf{K}}\tilde{\mathbf{U}} = \tilde{\mathbf{F}} \quad (5.21)$$

where
 $\tilde{\mathbf{K}}$ is the FEM stiffness matrix with entries of $\tilde{\mathbf{k}}_{ij} = \mathbf{a}\left(\boldsymbol{\phi}_j^H, \boldsymbol{\phi}_i^H\right), 1 \le i, j \le N_n$
 $\tilde{\mathbf{U}}$ is the vector of nodal displacements $\tilde{\mathbf{u}}_i$
 $\tilde{\mathbf{F}}$ is the vector with entries of $\tilde{\mathbf{f}}_i = \mathbf{f}(\boldsymbol{\phi}_i^H)$

Remark 5.4: Full Compatibility
A fully compatible FEM model satisfies three conditions: (1) the strain–displacement relation; (2) the essential (displacement) boundary conditions; and (3) the nodal shape functions ϕ_i^H, are compatible. To ensure the shape functions are compatible in an FEM setting, we practice mainly two tricks: (a) using only nodes of the element to create ϕ_i^H in the natural coordinate system; and (b) use proper mapping for the element to ensure the continuity of ϕ_i^H on the interfaces of the elements. "Of course, we require the integration to be exact."

Remark 5.5: Lower-Bound Property
The strain energy related to the fully compatible FEM solution is a lower bound of the exact strain energy

$$U_{\text{PE}}(\tilde{\boldsymbol{\varepsilon}}) = \frac{1}{2}a(\tilde{\mathbf{u}}, \tilde{\mathbf{u}}) \le \frac{1}{2}a(\mathbf{u}, \mathbf{u}) = U_{\text{PE}}(\boldsymbol{\varepsilon}) \quad (5.22)$$

where
 $\tilde{\boldsymbol{\varepsilon}} = \mathbf{L}_d\tilde{\mathbf{u}}$ are the strains obtained using the FEM displacements $\tilde{\mathbf{u}} \in \mathbb{H}_{h,0}^1 \subset \mathbb{H}^1$
 $\boldsymbol{\varepsilon} = \mathbf{L}_d\mathbf{u}$ is the exact strain obtained using the exact displacements $\mathbf{u} \in \mathbb{H}^1$

For an FEM model, the strain energy can be evaluated using any of the following expressions:

$$U(\tilde{\varepsilon}) = \frac{1}{2} \int_\Omega \tilde{\varepsilon}^T \mathbf{C} \tilde{\varepsilon} d\Omega = \frac{1}{2} a(\tilde{\mathbf{u}}, \tilde{\mathbf{u}}) = \frac{1}{2} \tilde{\mathbf{U}}^T \tilde{\mathbf{K}} \tilde{\mathbf{U}} \tag{5.23}$$

The proof of the lower-bound property can be found in a number of references, for example in variational formulation [12] and in matrix formulation [18] based on the energy principle. The lower-bound property implies the well-known fact that the FEM solution always underestimates the strain potential energy. This is equivalent to saying that the FEM solution overestimates the total potential energy. This property of FEM provides a good global measure of the lower-bound solution with respect to the exact solution. We discuss more in Chapter 8.

Remark 5.6: Monotonic Convergence Property
For a given sequence of n_m *nested* element meshes $M_1, M_2, \ldots, M_{n_m}$, such that the corresponding solution spaces satisfy $\mathbb{H}^1_{M_1} \subset \mathbb{H}^1_{M_2} \cdots \subset \mathbb{H}^1_{M_{n_m}} \subset \mathbb{H}^1$, the following inequalities stand:

$$U(\tilde{\varepsilon}_{M_1}) \leq U(\tilde{\varepsilon}_{M_2}) \leq \cdots \leq U(\tilde{\varepsilon}_{M_{n_m}}) \leq U(\varepsilon) \tag{5.24}$$

where $\tilde{\varepsilon}_{m_i}$ is the FEM-compatible solution of strains obtained using mesh m_i. This property can be shown easily using the arguments given by Oliveira [13].

Remark 5.7: Reproducibility of FEM
If $u \in \mathbb{H}^1_{h,0}$, then a fully compatible FEM model will reproduce the exact solution u.

This property can easily be proven [12,31,32], and can be understood intuitively. Because an FEM model guarantees to produce a unique, stable, and the best possible solution from the given $\mathbb{H}^1_{h,0}$ (in energy norm), if the exact solution is in this space, the FEM model will surely produce it as it must be the best one.

Note that the weak formulation is also applied in meshfree settings, as long as the meshfree shape functions are in a proper $\mathbb{H}$ space. For example, the MLS shape functions can be used in a weak formulation of meshfree methods.

5.4 A Weakened-Weak Formulation: GS-Galerkin

5.4.1 Bilinear Forms in $\mathbb{G}^1_h$ Spaces: A General Setting

For general meshfree and FEM settings, we need to form our bilinear form using functions in a $\mathbb{G}^1_h$ space defined in Chapter 3. The bilinear form in $\mathbb{G}^1_h$ spaces is also called the *(generalized) smoothed* bilinear form because of the use of (generalized) smoothing operations [16].

Consider again a solid mechanics problem governed by strong form PDEs given in Section 1.2. The problem domain Ω is discretized with background cells using the triangulation procedure defined in Section 1.7.2, over which stationary smoothing domains are created following the rules detailed in Section 3.3.1. We require also that at least the minimum number of linearly independent smoothing domains is used (see Section 3.3.3). Following Equation 5.14, the (generalized) smoothed bilinear form $\bar{a}_D(w,v)$ is then defined as

$$\bar{a}_D(w,v) = \sum_{i=1}^{N_s} A_i^s \underbrace{\left(\frac{1}{A_i^s}\int_{\Gamma_i^s} w_i n_j ds\right)}_{\bar{g}_{ij}(w)} C_{ijkl} \underbrace{\left(\frac{1}{A_i^s}\int_{\Gamma_i^s} v_k n_l ds\right)}_{\bar{g}_{kl}(v)} = \sum_{i=1}^{N_s} A_i^s \bar{g}_{ij}(w) C_{ijkl} \bar{g}_{kl}(v) \qquad (5.25)$$

where $w, v \in \mathbb{G}_h^1$. Note that the summation is made possible by the particular way of creating the smoothing domains that ensures the continuity of the functions on Γ_i^s.

In terms of the (generalized) *smoothed strain* defined in Remark 4.1, $\bar{a}_D$ can be written in the matrix form (see Equation 5.9)

$$\bar{a}_D(\mathbf{w},\mathbf{v}) = \sum_{i=1}^{N_s} A_i^s \bar{\boldsymbol{\varepsilon}}_i^{\mathsf{T}}(\mathbf{w}) \mathbf{c} \bar{\boldsymbol{\varepsilon}}_i(\mathbf{v}) \qquad (5.26)$$

where $\bar{\boldsymbol{\varepsilon}}_i$ is the vector of the smoothed strains in the smoothed domain Ω_i^s. For *any* given vector field of displacement $\mathbf{w} = \{\, w_1 \quad w_2 \,\}^{\mathsf{T}}$ with $\mathbf{w} \in \mathbb{G}_h^1$, $\bar{\boldsymbol{\varepsilon}}_i$ can be written as

$$\bar{\boldsymbol{\varepsilon}}_i(\mathbf{w}) = \frac{1}{A_i^s}\int_{\Gamma_i^s} \mathbf{L}_n \mathbf{w}(\mathbf{x}) ds = \{\, \bar{\varepsilon}_{11} \quad \bar{\varepsilon}_{22} \quad 2\bar{\varepsilon}_{12} \,\}_i^{\mathsf{T}}$$

$$-\left\{\, \underbrace{\frac{\overline{\partial w_1}}{\partial x_1}}_{\bar{g}_{11}} \quad \underbrace{\frac{\overline{\partial w_2}}{\partial x_2}}_{\bar{g}_{22}} \quad \left(\underbrace{\frac{\overline{\partial w_1}}{\partial x_2}}_{\bar{g}_{12}} + \underbrace{\frac{\overline{\partial w_2}}{\partial x_1}}_{\bar{g}_{21}}\right) \,\right\}_i^{\mathsf{T}} - \{\, \bar{g}_{11} \quad \bar{g}_{22} \quad (\bar{g}_{12} + \bar{g}_{21}) \,\}_i^{\mathsf{T}} \qquad (5.27)$$

The $\mathbb{L}^2$ norm of the strain vector $\|\bar{\boldsymbol{\varepsilon}}\|_{L^2}^2$ can be written as

$$\|\bar{\boldsymbol{\varepsilon}}\|_{L^2}^2 = \sum_{i=1}^{N_s} A_i^s \left(\bar{\varepsilon}_{11}^2 + \bar{\varepsilon}_{22}^2 + 4\bar{\varepsilon}_{12}^2\right) = \sum_{i=1}^{N_s} A_i^s \left(\bar{g}_{11}^2(w_1) + \bar{g}_{22}^2(w_2) + (\bar{g}_{12}(w_1) + \bar{g}_{21}(w_2))^2\right) \qquad (5.28)$$

It is now clear that the $\mathbb{L}^2$ norm of the strain vector is the same as the $\mathbb{G}_h^1$ seminorm (see Equation 3.72)

$$|\mathbf{w}|_{\mathbb{G}^1(\Omega)} = \|\bar{\boldsymbol{\varepsilon}}\|_{\mathbb{L}^2}, \quad \forall \mathbf{w} \in \mathbb{G}_{h,0}^1 \qquad (5.29)$$

which is useful in proving important properties in the following sections.

Combining Equations 3.94 and 5.29, we obtain

Remark 5.8: Fourth Inequality

$$c_G \|\mathbf{w}\|_{\mathbb{G}^1(\Omega)} \leq \|\bar{\boldsymbol{\varepsilon}}\|_{\mathbb{L}^2}, \quad \forall \mathbf{w} \in \mathbb{G}^1_{h,0} \tag{5.30}$$

which is equivalent to the second Korn's inequality but for functions in an $\mathbb{H}^1_h$ space.
  Combining Equations 3.89 and 5.29, we have the following chain inequality:

$$c_G \|\mathbf{w}\|_{\mathbb{G}^1(\Omega)} \leq \|\bar{\boldsymbol{\varepsilon}}\|_{\mathbb{L}^2} \leq \|\mathbf{w}\|_{\mathbb{G}^1(\Omega)}, \quad \forall \mathbf{w} \in \mathbb{G}^1_{h,0} \tag{5.31}$$

5.4.2 Properties of the Smoothed Bilinear Form

Remark 5.9: Symmetry and Semipositivity
By simple observation, it is clear that the generalized smoothed bilinear form has basic properties of symmetry

$$\bar{a}_D(\mathbf{w}, \mathbf{v}) = \bar{a}_D(\mathbf{w}, \mathbf{v}), \quad \forall \mathbf{w}, \mathbf{v} \in \mathbb{G}^1_h \tag{5.32}$$

for the symmetry property of the elastic material constants, and semipositive definite

$$\bar{a}_D(\mathbf{w}, \mathbf{w}) \geq 0, \quad \forall \mathbf{w} \in \mathbb{G}^1_h \tag{5.33}$$

because of the positivity of the material constants (see Chapter 1).

Remark 5.10: Convergence Property for Functions in an $\mathbb{H}$ Space
For $\mathbf{w}, \mathbf{v} \in \mathbb{H}^1_h$, when $N_s \to \infty$ and all $\Omega^s_k \to 0$, $\bar{W}$ becomes a delta function and the integral representation is exact (see Remark 2.1). At such a limit, $\bar{a}_D(\mathbf{w}, \mathbf{v}) \to a(\mathbf{w}, \mathbf{v})$. Therefore, based on the known property of $a(\mathbf{w}, \mathbf{w})$ (see Equation 5.10b), we should have the *ellipticity* property, meaning there exists a nonzero positive constant c such that

$$\lim_{\substack{N_s \to \infty \\ \text{all } \Omega^s_k \to 0}} \bar{a}_D(\mathbf{w}, \mathbf{w}) \geq c \|\mathbf{w}\|^2_{\mathbb{H}^1(\Omega)}, \quad \forall \mathbf{w} \in \mathbb{H}^1_{h,0} \tag{5.34}$$

The ellipticity ensures the existence and uniqueness of the solution of the W^2 formulation when the smoothing domain is refined. When the $\mathbb{H}^1_h$ space is enriched (element/cell mesh is refined together with the smoothing domains) the solution approaches the exact solution.
  For finite discretization of domains and functions in $\mathbb{G}$ spaces, the smoothed bilinear form has been found to have a number of important properties.

Theorem 5.1: Ellipticity with Respect to the $\mathbb{G}$ Seminorm

For solids of stable materials, there exists a nonzero positive constant c_{aw}^{f} such that

$$\bar{a}_D(\mathbf{w}, \mathbf{w}) \geq c_{aw}^{s} |\mathbf{w}|_{\mathbb{G}^1(\Omega)}^2, \quad \forall \mathbf{w} \in \mathbb{G}_{h,0}^1 \tag{5.35}$$

Proof   Because the material is stable, the matrix of the elastic constants $\mathbf{c}$ is symmetric positive definite (SPD) and can *always* be decomposed into a unitary matrix $\mathbf{V}_m$ of eigenvectors and a diagonal matrix $\boldsymbol{\Lambda}_m$ of all positive eigenvalues:

$$\mathbf{c} = \mathbf{V}_m^T \boldsymbol{\Lambda}_m \mathbf{V}_m \tag{5.36}$$

The proof starts from the definition (Equation 5.26). For $w \in \mathbb{G}_{h,0}^1$, we have

$$
\begin{aligned}
\bar{a}_D(\mathbf{w}, \mathbf{w}) &= \sum_{i=1}^{N_s} A_i^s \bar{\boldsymbol{\varepsilon}}_i^T(\mathbf{w}) \mathbf{c} \bar{\boldsymbol{\varepsilon}}_i(\mathbf{w}) = \sum_{i=1}^{N_s} A_i^s \bar{\boldsymbol{\varepsilon}}_i^T(\mathbf{w}) \underbrace{\mathbf{V}_m^T \boldsymbol{\Lambda}_m \mathbf{V}_m}_{\mathbf{c}} \bar{\boldsymbol{\varepsilon}}_i(\mathbf{w}) \\
&= \sum_{i=1}^{N_s} A_i^s [\mathbf{V}_m \bar{\boldsymbol{\varepsilon}}_i(\mathbf{w})]^T \boldsymbol{\Lambda}_m [\mathbf{V}_m \bar{\boldsymbol{\varepsilon}}_i(\mathbf{w})] \geq \lambda_{\min} \sum_{i=1}^{N_s} A_i^s \bar{\boldsymbol{\varepsilon}}_i^T(\mathbf{w}) \underbrace{\mathbf{V}_m^T \mathbf{V}_m}_{\mathbf{I}} \bar{\boldsymbol{\varepsilon}}_i(\mathbf{w}) \\
&= \lambda_{\min} \|\bar{\boldsymbol{\varepsilon}}(\mathbf{w})\|_{\mathbb{L}^2}^2 = \underbrace{\lambda_{\min}}_{c_{aw}^s} |\mathbf{w}|_{\mathbb{G}^1(\Omega)}^2
\end{aligned}
\tag{5.37}
$$

where $\lambda_{\min}$ is the smallest eigenvalue of $\mathbf{c}$. In the second line of the above equation, we used the facts: (1) the $\mathbb{L}^2$ norm definition by the inner product and (2) the $\mathbb{L}^2$ norm of the strain vector being the same as the seminorm (Equation 5.29). Finally, we have the inequality (Equation 5.35), by letting $c_{aw}^s = \lambda_{\min}$ that is independent of $\mathbf{w}$. ∎

Theorem 5.2: Ellipticity (Coercivity): Fifth Inequality

For solids of stable materials, there exists a nonzero positive constant c_{aw}^{f} such that

$$\bar{a}_D(\mathbf{w}, \mathbf{w}) \geq c_{aw}^{f} \|\mathbf{w}\|_{\mathbb{G}^1(\Omega)}^2, \quad \forall \mathbf{w} \in \mathbb{G}_{h,0}^1 \tag{5.38}$$

which implies the *ellipticity* or *coercivity* of bilinear forms.

Proof   From Equations 3.94 and 5.35, we immediately have

$$\bar{a}_D(\mathbf{w}, \mathbf{w}) \geq c_{aw}^s |\mathbf{w}|_{\mathbb{G}^1}^2 \geq \underbrace{C_{aw}^s C_{\mathbb{G}}}_{c_{aw}^f} \|\mathbf{w}\|_{\mathbb{G}^1}^2 \tag{5.39}$$

which is the fifth inequality (Equation 5.38). ∎

Theorem 5.2 is important because it ensures the existence (and hence the uniqueness), and consequently, the stability of the solution of the W^2 formulation based on the $\mathbb{G}$ space theory.

Theorem 5.3: Continuity: Sixth Inequality

For solids of stable materials, there exists a nonzero positive constant c_{awv}^f such that

$$\bar{a}_D(\mathbf{w}, \mathbf{v}) \le c_{awv}^f \|\mathbf{w}\|_{\mathbb{G}^1(\Omega)} \|\mathbf{v}\|_{\mathbb{G}^1(\Omega)}, \quad \forall \mathbf{w} \in \mathbb{G}_h^1, \quad \forall \mathbf{v} \in \mathbb{G}_h^1 \tag{5.40}$$

Proof We start again from the definition (Equation 5.26). For $\mathbf{w}, \mathbf{v} \in \mathbb{G}_h^1$, we have

$$\bar{a}_D(\mathbf{w}, \mathbf{v}) = \sum_{i=1}^{N_s} A_i^s \bar{\boldsymbol{\varepsilon}}_i^{\mathrm{T}}(\mathbf{w}) \mathbf{c} \bar{\boldsymbol{\varepsilon}}_i(\mathbf{v}) = \sum_{i=1}^{N_s} A_i^s \bar{\boldsymbol{\varepsilon}}_i^{\mathrm{T}}(\mathbf{w}) \underbrace{\mathbf{V}_m^{\mathrm{T}} \Lambda_m \mathbf{V}_m}_{\mathbf{c}} \bar{\boldsymbol{\varepsilon}}_i(\mathbf{v}) \tag{5.41}$$

where the positivity of the material constants is again used to decompose $\mathbf{c}$. Equation 5.41 becomes

$$\bar{a}_D(\mathbf{w}, \mathbf{v}) = \sum_{i=1}^{N_s} A_i^s \bar{\boldsymbol{\varepsilon}}_i^{\mathrm{T}}(\mathbf{w}) \underbrace{\mathbf{V}_m^{\mathrm{T}} \Lambda_m \mathbf{V}_m}_{\mathbf{c}} \bar{\boldsymbol{\varepsilon}}_i(\mathbf{v}) = \sum_{i=1}^{N_s} A_i^s [\mathbf{V}_m \bar{\boldsymbol{\varepsilon}}_i(\mathbf{w})]^{\mathrm{T}} \Lambda_m [\mathbf{V}_m \bar{\boldsymbol{\varepsilon}}_i(\mathbf{v})]$$

$$\le \lambda_{\max} \sum_{i=1}^{N_s} A_i^s \underbrace{\|\mathbf{V}_m \bar{\boldsymbol{\varepsilon}}_i(\mathbf{w})\|_{\mathbb{L}^2(\Omega_i^s)}}_{=\|\bar{\boldsymbol{\varepsilon}}_i(\mathbf{w})\|_{\mathbb{L}^2(\Omega_i^s)}} \underbrace{\|\mathbf{V}_m \bar{\boldsymbol{\varepsilon}}_i(\mathbf{v})\|_{\mathbb{L}^2(\Omega_i^s)}}_{=\|\bar{\boldsymbol{\varepsilon}}_i(\mathbf{v})\|_{\mathbb{L}^2(\Omega_i^2)}} = \lambda_{\max} \sum_{i=1}^{N_s} A_i^s \underbrace{\|\bar{\boldsymbol{\varepsilon}}_i(\mathbf{w})\|_{\mathbb{L}^2(\Omega_i^s)} \|\bar{\boldsymbol{\varepsilon}}_i(\mathbf{v})\|_{\mathbb{L}^2(\Omega_i^s)}}_{=\|\bar{\boldsymbol{\varepsilon}}_i(\mathbf{w})\|_{\mathbb{L}^2(\Omega_i^s)} \|\bar{\boldsymbol{\varepsilon}}_i(\mathbf{v})\|_{\mathbb{L}^2(\Omega_i^s)}}$$

$$= \lambda_{\max} \|\bar{\boldsymbol{\varepsilon}}_i(\mathbf{w})\|_{\mathbb{L}^2(\Omega)} \|\bar{\boldsymbol{\varepsilon}}_i(\mathbf{v})\|_{\mathbb{L}^2(\Omega)} = \lambda_{\max} |\mathbf{w}|_{\mathbb{G}^1(\Omega)} |\mathbf{v}|_{\mathbb{G}^1(\Omega)} \le \underbrace{\lambda_{\max}}_{c_{awv}^f} \|\mathbf{w}\|_{\mathbb{G}^1(\Omega)} \|\mathbf{v}\|_{\mathbb{G}^1(\Omega)}$$

$$= c_{awv}^f \|\mathbf{w}\|_{\mathbb{G}^1(\Omega)} \|\mathbf{v}\|_{\mathbb{G}^1(\Omega)} \tag{5.42}$$

where $\lambda_{\max}$ is the largest eigenvalue of $\mathbf{c}$. In the second line of the above equation, we used these facts: (1) the Cauchy-Schwarz inequality for inner-produced induced $\mathbb{L}^2$ norms, and (2) the $\mathbb{L}^2$ norm preservation property of the unitary matrix. In the third line, we used the fact that the $\mathbb{L}^2$ norm of the strain vector equals the seminorm (see Equation 5.29), and the fact that a seminorm is no larger than a full norm (see Equation 3.89). Finally, we have the sixth inequality (Equation 5.40), by letting $c_{awv}^f = \lambda_{\max}$. ∎

Theorem 5.3 is important because it ensures that our bilinear form is continuous. Together with the ellipticity, it ensures the convergence of the solution of the W^2 formulation based on the $\mathbb{G}$ space theory.

Remark 5.11: Softened Model: Seventh Inequality
For stable materials and any $\mathbf{w} \in \mathbb{H}_h^1$ the smoothed bilinear form is smaller than the bilinear form:

$$\bar{a}_D(\mathbf{w}, \mathbf{w}) \le a(\mathbf{w}, \mathbf{w}), \quad \forall \mathbf{w} \in \mathbb{H}_h^1 \tag{5.43}$$

Remark 5.9 is the same as Theorem 4.2 stated in terms of strain energy, and a proof is given already in Chapter 4. A more general inequality than Equation 5.43 can be found in [16].

Remark 5.9 implies that a model established based on the smoothed bilinear form is "softer" than that of the bilinear form which was discovered in [18].

Remark 5.12: Monotonic Convergence Property: Eighth Inequality
In a given division D_1 of a domain Ω into a set of smoothing domains $\boxed{\Omega} = \cup_{i=1}^{N_s} \boxed{\Omega}_i^s$ where the box stands for an enclosed domain, if a new division D_2 is created by subdividing a smoothing domain in D_1 into n_{sd} subsmoothing domains $\boxed{\Omega}_i^s = \cup_{j=1}^{n_{sd}} \boxed{\Omega}_{ij}^s$, then the following inequality stands:

$$\bar{a}_{D_1}(\mathbf{w}, \mathbf{w}) \leq \bar{a}_{D_2}(\mathbf{w}, \mathbf{w}) \tag{5.44}$$

---

which was found in [18]. Remark 5.10 is the same as the Theorem 4.3 stated in terms of strain energy. It implies that the "softening" effect provided by the W^2 formulation is monotonically reduced with the increase of the number of smoothing domains in a nested manner. One now has a theoretic base to reduce or increase the stiffness or softness of the model.

5.4.3 A Weakened-Weak Form Statement: GS-Galerkin

For solid mechanics problems given in the strong statements in Section 1.2, our W^2 statement becomes: An approximated solution $\bar{\mathbf{u}} \in \mathbb{G}_{h,0}^1$ associated with the strong statements in Section 1.2 satisfies

$$\bar{a}_D(\bar{\mathbf{u}}, \mathbf{v}) = f(\mathbf{v}), \quad \forall \mathbf{v} \in \mathbb{G}_{h,0}^1 \tag{5.45}$$

Note here we make no changes to the linear functional $f(\mathbf{v})$, and Equation 5.13 stands. We now state

---

Theorem 5.4a: W^2 Solution in $\mathbb{H}_h^1$ Spaces: Variationally Consistent

If the solution is sought from an $\mathbb{H}_h^1$ space, the W^2 statement (Equation 5.45) is variationally consistent in the standard weak formulation, and hence the solution is stable, unique, and converges to the exact solution of the strong statement when $h \to 0$ (and hence $\Omega_i^s \to 0$).
   The proof will be discussed based on the energy principle in Theorem 5.4b.

Theorem 5.5a: W^2 Solution in $\mathbb{G}_h^1$ Spaces: Unique Solution

For solids of stable materials, the W^2 statement (Equation 5.45) has a unique solution.

Proof To prove Theorem 5.5a, we need to show: (1) the bilinear form is of ellipticity to ensure the existence of the solution and (2) the bilinear form is continuous to ensure that the solution is bounded. Note the ellipticity has already been given by Theorem 5.2, and the boundedness is given by Theorem 5.3. Based on the Lax–Milgram theorem, the W^2 statement (Equation 5.45) has a unique solution. This completes the proof.   ∎

In Theorem 5.7, we will show (in a more general way) that the unique solution of the W^2 statement will converge to the exact solution of the original strong form (see also, Theorem 5.5b).

Theorem 5.6: Upper Bound to the Galerkin Weak Form Solution

The strain energy of the GS-Galerkin weak form solution $\widehat{\mathbf{u}} \in \mathbb{H}^1_{h,0}$ is no less than that of the Galerkin weak form solution $\tilde{\mathbf{u}} \in \mathbb{H}^1_{h,0}$, when the same mesh is used for creating the numerical model:

$$U(\tilde{\mathbf{u}}) \leq \widehat{U}^D(\widehat{\mathbf{u}}) \tag{5.46}$$

The proof is a little lengthy and can be found in variational formulation [27] and in energy principle formulation [18]. Therefore, it is omitted here. Instead, we provide an intuitive physical explanation on Theorem 5.6: From Remark 5.9, we know that the GS-Galerkin model is always softer than its Galerkin counterpart. Therefore, the displacement field obtained from the GS-Galerkin model should be "larger" and so also the strain field. The strain energy obtained using such a "larger" displacement field should also be larger compared to that of the Galerkin model. In other words, Equation 5.46 is the consequence of the fact that the GS-Galerkin model is always "softer" than the Galerkin model.

Remark 5.13: Upper Bound to the Exact Solution: Special Cases
The strain energy of the GS-Galerkin weak form solution $\widehat{\mathbf{u}}$ is no less than that of the exact solution $\mathbf{u}$, if $\widehat{\mathbf{u}}$ is found from an $\mathbb{H}^1$ space that contains the exact solution:

$$\widehat{U}(\mathbf{u}) \leq \widehat{U}^D(\widehat{\mathbf{u}}) \tag{5.47}$$

This remark can be easily understood intuitively: the $\mathbb{H}^1$ space contains the exact solution, thus, Equation 5.46 becomes Equation 5.47.

Note that when the Galerkin weak form is used to search a solution from a space that contains the exact solution, it will reproduce the exact solution. The GS-Galerkin weak form, however, will not necessarily reproduce the exact solution, if a finite division of the smoothing domain is used. It produces in general a approximated solution that approaches the exact solution when the dimension of the smoothing domains approaches zero.

Note that Remark 5.11 has only a theoretical significance, because it is difficult to assume a space that contains the exact solution, unless the exact solution has a very simple polynomial form. Even this can be done; one can simply use the FEM to reproduce the exact solution, and there is no need for any other form of the solution! Therefore, the following two remarks are more of practical importance.

Remark 5.14: Upper Bound to the Exact Solution: Existence
If a discrete GS-Galerkin model is so large, that the "stiffness" of the model can change reasonably *smoothly* with the change of the number of linearly independent smoothing domains $N_s \geq N_s^{\min}$, the GS-Galerkin model can produce an upper-bound solution by reducing N_s.

Based on the study in [27], to ensure the positivity of the seminorm of a function in $\mathbb{G}_h^1$ space, we have to use at least the minimum number of linearly independent smoothing domains given in Table 3.1. This implies that if the number of smoothing domains N_s used in a GS-Galerkin mode is less than the minimum number given in Table 3.1, the $\mathbb{G}_h^1$ seminorm of a function in $\mathbb{G}_h^1$ space can be zero. In mechanics, this means that the strain energy of such a GS-Galerkin model can be zero: there are finite displacement fields that generate zero energy! In other words, the model has zero stiffness or is extremely soft. This can be understood intuitively using a single degree of freedom (DOF) mechanics model with a stiffness k whose strain energy is given by

$$\underbrace{U_e}_{\text{zero}} = \frac{1}{2} k \underbrace{u^2}_{\text{finite}} \tag{5.48}$$

When u is finite and U_e is zero, k must become zero.

The above analysis implies that we can, in theory, make a GS-Galerkin model as soft as we want to by reducing N_s, and thus always obtain an upper-bound solution, as long as the stiffness of the model changes *smoothly* with the change of N_s.

Remark 5.12 is important because it ensures (with conditions) that a GS-Galerkin model can always be built for an upper bound to the exact solution in the energy norm. This provides a foundation for establishing practical numerical methods for upper-bound solutions.

In actual discrete finite models, however, the stiffness of the model may not change very smoothly with the change of N_s when the number of nodes in the model is too small. For example, for 1D problems (truss) with only one linear element and two nodes of which one is constrained, we have the only choice of using one smoothing domain (using more than one smoothing domains has no effect on the strain energy). Note also that, in practice, the division of smoothing domains is somehow always tied together with the element mesh for (1) convenience in implementation, (2) efficiency in computation, (3) ensuring smoothing domains are created properly, and (4) tighter upper bounds. Therefore, the element mesh is refined, and the smoothing domains are also refined accordingly. In this case, the smoothing effects are reduced with the element mesh refinement. Depending on how they are tied together, the GS-Galerkin model may or may not be able to produce an upper-bound solution.

Remark 5.15: Upper Bound to the Exact Solution: Usual Occurrences
The strain energy of the solution $\bar{\mathbf{u}} \in \mathbb{G}_h^1$ of a sufficiently large GS-Galerkin model is no less than that of the exact solution $\mathbf{u}$, when the smoothing domains are properly chosen for sufficient smoothing effects:

$$U(\mathbf{u}) \leq \bar{U}^D(\bar{\mathbf{u}}) \tag{5.49}$$

A precise proof for Remark 5.13 is difficult due to the difficulty in quantifying the *exceptions*. The inequality (Equation 5.49) is found mostly true when node-based smoothing domains are used as in the node-based point interpolation methods (NS-PIMs). An intuitive explanation and proof by numerical examples can be found in [18–21,27]. A discussion on this is given in Chapter 8. Remark 5.13 is practically important because it implies that a sufficiently large GS-Galerkin model can provide an upper bound to the exact solution in the energy norm by choosing the smoothing domains properly. Here, we emphasis how the smoothing domain is constructed rather than how many smoothing domains are used in an actual GS-Galerkin model.

Remark 5.16: The "Form" of Galerkin
Equation 5.45 has the *form* of Galerkin. The trial and test functions are all from the same $\mathbb{G}$ space that is a much more general setting for solution spaces. The W^2 statement of Equation 5.45 was termed the GS-Galerkin formulation [16]. With the $\mathbb{G}$-space theory established, we now have a very general framework, in a form as simple as the Galerkin form, for establishing a wide class of meshfree methods. We also have plenty of freedom to create smoothing domains and shape functions using FEM settings and meshfree settings. Furthermore, even within the FEM setting, there are now many avenues to explore. Just for an example, we can now conveniently formulate different shapes of elements and even mixed shapes of elements [23–26]. The GS-Galerkin is in fact a special case of SC-Galerkin to be discussed in Section 5.15.

5.4.4 Some Comments

It is seen that the above process for both weak and weakened-weak formulation is quite mathematical. Such a mathematical approach is indeed very handy in proving the theories and properties. It is also convenient for further examination of solution error, convergence rate, solution regularities, and so forth. However, these topics are beyond the scope of this book, hence we will not move any further in that direction.

On the other hand, in such mathematical manipulations one can very easily get lost about the physics of the problem. Hence it is not really to the like of many engineers. In the following we present the *energy principles* to establish weak forms for meshfree methods using function spaces defined in Chapter 3. Using these energy principles, we can easily establish many types of weak forms for different types of structures. In addition, we can quite easily implement the strain construction techniques discussed in Chapter 4 in our formulation. We start with the well-known Hu–Washizu (HW) principle.

5.5 The Hu–Washizu Principle

The HW principle [8] is a very general energy principle with three unknown fields: displacement, strain, and stress. It is hence often called the *three-field principle*. It can be stated as "of all the *admissible* fields of displacement, strain, and stress, the most *accurate** ones are registered at the stationary point of the HW functional." It can be expressed as

$$\Pi_{HW}(\mathbf{u}, \boldsymbol{\varepsilon}, \boldsymbol{\sigma}) = \int_{\Omega} \left[\underbrace{\frac{1}{2}\boldsymbol{\varepsilon}^{\mathrm{T}}\mathbf{c}\boldsymbol{\varepsilon}}_{\substack{\text{Energy potential for the}\\\text{assumed strain field}}} + \underbrace{\boldsymbol{\sigma}^{\mathrm{T}}(\mathbf{L}_d\mathbf{u} - \boldsymbol{\varepsilon})}_{\substack{\text{Energy potential by the assumed stress field}\\\text{over the strain fields difference between}\\\text{the compatible and assumed strain fields}}} \right] d\Omega$$

$$\underbrace{}_{U_{HW}:\ \text{total strain potential}}$$

$$-\underbrace{\left[\underbrace{\int_{\Omega} \mathbf{b}^{\mathrm{T}}\mathbf{u}\,d\Omega}_{\substack{\text{Work done by the}\\\text{body forces}}} + \underbrace{\int_{\Gamma_t} \mathbf{t}_{\Gamma}^{\mathrm{T}}\mathbf{u}\,d\Gamma}_{\substack{\text{Work done by the}\\\text{force on the boundary}}} \right]}_{W_{ext}:\ \text{Work done by the external forces}} - \underbrace{\int_{\Gamma_u} \mathbf{t}^{\mathrm{T}}(\mathbf{u} - \mathbf{u}_{\Gamma})\,d\Gamma}_{\substack{U_{\Gamma_u}:\ \text{Energy potential on the essential}\\\text{boundary by the assumed stress over the}\\\text{displacment difference between the assumed}\\\text{and prescribed displacements}}}$$

= Stationary   (5.50)

where $\Pi_{HW} = U_{HW} - W_{ext} + U_{\Gamma_u}$ is the HW functional in mathematical terminology, and is the *total potential energy* in the system in engineering terminology. It is clear that the HW principle simply counts for all energies and work done by the assumed fields and the given force and boundary conditions. Functional Π_{HW} has three inputs of field functions: strain $\boldsymbol{\varepsilon}$, stress $\boldsymbol{\sigma}$, and displacement $\mathbf{u}$. The admissible conditions are: strain $\boldsymbol{\varepsilon} \in \mathbb{L}^2$, stress $\boldsymbol{\sigma} \in \mathbb{L}^2$, and displacement $\mathbf{u} \in \mathbb{H}_0^1$. Clearly, Equation 5.50 is a weak form requiring only the first-order consistency. The most *accurate* fields will be our *best* possible solution, denoted here as $\mathbf{u}$, $\boldsymbol{\varepsilon}$, and $\boldsymbol{\sigma}$.

The HW principle is powerful in terms of the freedom offered to assume all these three fields. In practice, it is often used to formulate mixed FEM. However, assuming all the three fields in a model results in a large number of unknowns affecting the computational efficiency, which is not usually preferred in practice. In theory, the HW principle forms a foundation for many other practically useful principles.

For future references, we define the HW strain energy potential as

$$U_{HW}(\mathbf{u}, \boldsymbol{\varepsilon}, \boldsymbol{\sigma}) = \int_{\Omega} \left[\frac{1}{2}\boldsymbol{\varepsilon}^{\mathrm{T}}\mathbf{c}\boldsymbol{\varepsilon} + \boldsymbol{\sigma}^{\mathrm{T}}(\mathbf{L}_d\mathbf{u} - \boldsymbol{\varepsilon}) \right] d\Omega \qquad (5.51)$$

which is a functional of functions $\mathbf{u}$, $\boldsymbol{\varepsilon}$, and $\boldsymbol{\sigma}$. The work done by the external forces is defined as

$$W_{ext}(\mathbf{u}) = \int_{\Omega} \mathbf{b}^{\mathrm{T}}\mathbf{u}\,d\Omega + \int_{\Gamma_t} \mathbf{t}_{\Gamma}^{\mathrm{T}}\mathbf{u}\,d\Gamma \qquad (5.52)$$

---

* In energy principles, the accuracy of the solution is usually measured as the error in an energy norm.

which is a linear functional of $\mathbf{u}$. The potential on the constrained (essential) boundary is defined as

$$U_{\Gamma_u}(\mathbf{u}, \boldsymbol{\sigma}) = \int_{\Gamma_u} \mathbf{t}^T(\mathbf{u} - \mathbf{u}_\Gamma)d\Gamma \qquad (5.53)$$

which is a functional of $\mathbf{u}$ and $\boldsymbol{\sigma}$.

The *variational statement* for the HW principle can now be written as

$$\delta\Pi_{HW}(\mathbf{u}, \boldsymbol{\varepsilon}, \boldsymbol{\sigma}) = \delta[U_{HW} - W_{ext} + U_{\Gamma_u}] = 0 \qquad (5.54)$$

which is simply the so-called stationary condition.

5.6 The Hellinger–Reissner Principle

The Hellinger–Reissner principle [9,10] is a principle with two unknown fields: displacement and stress. It is hence often called the two-field principle. The removal of one field in Equation 5.50 is achieved by expressing the strain in terms of stress via the stress–strain relations (the generalized Hooke's law). The Hellinger–Reissner principle can be mathematically expressed as

$$\Pi_{HR}(\mathbf{u}, \boldsymbol{\sigma}) = \underbrace{\int_\Omega \left[-\frac{1}{2}\boldsymbol{\sigma}^T\mathbf{s}\boldsymbol{\sigma} + \boldsymbol{\sigma}^T(\mathbf{L}_d\mathbf{u}) \right]d\Omega}_{U_{HR}} - \underbrace{\left[\int_\Omega \mathbf{b}^T\mathbf{u}d\Omega + \int_{\Gamma_t} \mathbf{t}_t^T\mathbf{u}d\Gamma \right]}_{W_{ext}}$$

$$- \underbrace{\int_{\Gamma_u} \mathbf{t}^T(\mathbf{u} - \mathbf{u}_\Gamma)d\Gamma}_{U_{\Gamma_u}} = \text{Stationary} \qquad (5.55)$$

where the admissible conditions are $\boldsymbol{\sigma} \in \mathbb{L}^2$, $\mathbf{u} \in \mathbb{H}_0^1$, and $\boldsymbol{\varepsilon} = \mathbf{s}\boldsymbol{\sigma}$. For later reference, we define the Hellinger–Reissner potential as

$$U_{HR}(\mathbf{u}, \boldsymbol{\sigma}) = \int_\Omega \left[-\frac{1}{2}\boldsymbol{\sigma}^T\mathbf{s}\boldsymbol{\sigma} + \boldsymbol{\sigma}^T(\mathbf{L}_d\mathbf{u}) \right]d\Omega \qquad (5.56)$$

which is a functional of $\mathbf{u}$ and $\boldsymbol{\sigma}$.

The variational statement for the Hellinger–Reissner principle can now be given as

$$\delta\Pi_{HR}(\mathbf{u}, \boldsymbol{\sigma}) = \delta[U_{HR} - W_{ext} + U_{\Gamma_u}] = 0 \qquad (5.57)$$

5.7 The Modified Hellinger–Reissner Principle

Start again from the three-field principle (Equation 5.50), express the stress in terms of strain via the stress–strain relations; we then have the MHR principle. It is also a principle with two unknown fields: displacement and strain. The MHR principle can be expressed as

$$\Pi_{\text{MHR}}(\boldsymbol{\varepsilon}, \mathbf{u}) = \underbrace{\int_{\Omega} \left[-\frac{1}{2} \boldsymbol{\varepsilon}^{\mathsf{T}} \mathbf{c} \boldsymbol{\varepsilon} + \boldsymbol{\varepsilon}^{\mathsf{T}} \mathbf{c}(\mathbf{L}_d \mathbf{u}) \right] d\Omega}_{U_{\text{MHR}}} - \underbrace{\left[\int_{\Omega} \mathbf{b}^{\mathsf{T}} \mathbf{u} d\Omega + \int_{\Gamma_t} \mathbf{t}_{\Gamma}^{\mathsf{T}} \mathbf{u} d\Gamma \right]}_{W_{\text{ext}}} = \text{Stationary} \quad (5.58)$$

where $\boldsymbol{\varepsilon} \in \mathbb{L}^2$, $\mathbf{u} \in \mathbb{H}_0^1$, and $\boldsymbol{\sigma} = \mathbf{c}\boldsymbol{\varepsilon}$. The MHR potential is defined as

$$U_{\text{MHR}}(\boldsymbol{\varepsilon}, \mathbf{u}) = \int_{\Omega} \left[-\frac{1}{2} \boldsymbol{\varepsilon}^{\mathsf{T}} \mathbf{c} \boldsymbol{\varepsilon} + \boldsymbol{\varepsilon}^{\mathsf{T}} \mathbf{c}(\mathbf{L}_d \mathbf{u}) \right] d\Omega \quad (5.59)$$

which is a functional of $\mathbf{u}$ and $\boldsymbol{\varepsilon}$.

The variational statement for the MHR principle can now be written in the following concise form:

$$\delta \Pi_{\text{MHR}}(\boldsymbol{\varepsilon}, \mathbf{u}) = \delta[U_{\text{MHR}}(\boldsymbol{\varepsilon}, \mathbf{u}) - W_{\text{ext}}(\mathbf{u})] = 0 \quad (5.60)$$

5.8 The Single-Field Hellinger–Reissner Principle

Start now from the MHR principle (Equation 5.58); when the assumed strain field is fully dependent on the assumed displacement, the Hellinger–Reissner functional relies solely on the displacement field:

$$\Pi_{\text{SHR}}(\mathbf{u}) = \underbrace{\int_{\Omega} \left[-\frac{1}{2} \breve{\boldsymbol{\varepsilon}}^{\mathsf{T}}(\mathbf{u}) \mathbf{c} \breve{\boldsymbol{\varepsilon}}(\mathbf{u}) + \breve{\boldsymbol{\varepsilon}}^{\mathsf{T}}(\mathbf{u}) \mathbf{c}(\mathbf{L}_d \mathbf{u}) \right] d\Omega}_{U_{\text{SHR}}} - \underbrace{\left[\int_{\Omega} \mathbf{b}^{\mathsf{T}} \mathbf{u} d\Omega + \int_{\Gamma_t} \mathbf{t}_{\Gamma}^{\mathsf{T}} \mathbf{u} d\Gamma \right]}_{W_{\text{ext}}}$$

$$= \text{Stationary} \quad (5.61)$$

where, $\mathbf{u} \in \mathbb{H}_0^1$, and $\breve{\boldsymbol{\varepsilon}}(\mathbf{u})$ is the strain field that is obtained using the assumed displacement field in some manner. The single-field Hellinger–Reissner (SHR) potential is defined as

$$U_{\text{SHR}}(\mathbf{u}) = \int_{\Omega} \left[-\frac{1}{2} \breve{\boldsymbol{\varepsilon}}^{\mathsf{T}}(\mathbf{u}) \mathbf{c} \breve{\boldsymbol{\varepsilon}}(\mathbf{u}) + \breve{\boldsymbol{\varepsilon}}^{\mathsf{T}}(\mathbf{u}) \mathbf{c}(\mathbf{L}_d \mathbf{u}) \right] d\Omega \quad (5.62)$$

which is a functional of $\mathbf{u}$.

The variational statement for the single-field Hellinger–Reissner principle can now be written as

$$\delta\Pi_{\text{SHR}}(\mathbf{u}) = \delta[U_{\text{SHR}}(\mathbf{u}) - W_{\text{ext}}(\mathbf{u})] = 0 \qquad (5.63)$$

5.9 The Principle of Minimum Complementary Energy

When the assumed stress field satisfies the equilibrium equations, and the prescribed tractions along the boundary of the domain are also satisfied, the Hellinger–Reissner principle reduces to the well-known complementary energy principle. It can be expressed as

$$\Pi_{\text{CE}}(\boldsymbol{\sigma}) = \underbrace{\int_{\Omega} \frac{1}{2}\boldsymbol{\sigma}^{\text{T}}\mathbf{s}\boldsymbol{\sigma}\,\text{d}\Omega}_{U_{\text{CE}}} - \underbrace{\int_{\Gamma_u} \mathbf{t}^{\text{T}}\mathbf{u}_\Gamma\,\text{d}\Gamma}_{W_{\Gamma_u}} = \text{Minimum} \qquad (5.64)$$

The admissible conditions are $\boldsymbol{\sigma} \in \mathbb{H}^{div} = \{\boldsymbol{\tau}|\boldsymbol{\tau} \in \mathbb{L}^2, \mathbf{L}_d^{\text{T}}\boldsymbol{\tau} \in \mathbb{L}^2\}$, $\mathbf{L}_d^{\text{T}}\boldsymbol{\sigma} = \mathbf{b}$, and $\mathbf{n}^{\text{T}}\boldsymbol{\sigma}|_{\Gamma_t} = \mathbf{t}_\Gamma$. We demand for the $\mathbb{H}^{div}$ space, because the stress field assumed has to satisfy the equilibrium equation $\mathbf{L}_d^{\text{T}}\boldsymbol{\sigma} = \mathbf{b}$ where a divergence operator is applied to the stress. The so-called equilibrium methods or the force methods, where we assume the stress field, are established using this principle.

Although in the minimum complementary energy principle, only the stress field is unknown, constructing a stress field that satisfies the equilibrium equations and the prescribed tractions along the boundary is not an easy task for general problems. In addition, computing displacements from given stresses can also be problematic (a kind of inverse problem). This formulation can however provide upper-bound solutions (for homogenous essential conditions).

The stress energy potential is defined as

$$U_{\text{CE}}(\boldsymbol{\sigma}) = \int_{\Omega} \frac{1}{2}\boldsymbol{\sigma}^{\text{T}}\mathbf{s}\boldsymbol{\sigma}\,\text{d}\Omega \qquad (5.65)$$

which is a quadratic functional of $\boldsymbol{\sigma}$.

The variational statement for the principle of minimum complementary energy can now be written as follows:

$$\delta\Pi_{\text{CE}}(\boldsymbol{\sigma}) = \delta[U_{\text{CE}}(\boldsymbol{\sigma}) - W_{\Gamma_u}(\boldsymbol{\sigma})] = 0 \qquad (5.66)$$

which is simply the so-called minimization condition.

5.10 The Principle of Minimum Potential Energy

When the strain field is simply obtained from the assumed displacement via the strain–displacement relation, and the assumed displacement field satisfies the essential boundary conditions where the displacements are prescribed, Equation 5.50 reduces to the well-known principle of minimum potential energy. It can be stated as "of all the *admissible* fields of displacement the most *accurate* one is registered at the minimum point of the functional of total energy potential." Mathematically it can be expressed as

$$
\Pi_{\text{PE}}(\mathbf{u}) = \underbrace{\int_{\Omega} \frac{1}{2} \boldsymbol{\varepsilon}^{\text{T}} \mathbf{c} \boldsymbol{\varepsilon} \, d\Omega}_{U_{\text{PE}}} - \underbrace{\left[\int_{\Omega} \mathbf{b}^{\text{T}} \mathbf{u} \, d\Omega + \int_{\Gamma_t} \mathbf{t}_t^{\text{T}} \mathbf{u} \, d\Gamma \right]}_{W_{\text{ext}}} = \text{Minimum} \qquad (5.67)
$$

where $\Pi_{\text{PE}} = U_{\text{PE}} - W_{\text{ext}}$ is a functional in mathematical terminology and is the total potential energy in engineering terminology. The admissible conditions are $\mathbf{u} \in \mathbb{H}_0^1$ and $\boldsymbol{\varepsilon} = \mathbf{L}_d \mathbf{u}$. We demand for the $\mathbb{H}^1$ space, because the use of the strain–displacement relation, $\boldsymbol{\varepsilon} = \mathbf{L}_d \mathbf{u}$, and the strain energy potential is computed using

$$
U_{\text{PE}}(\boldsymbol{\varepsilon}) = \int_{\Omega} \frac{1}{2} \boldsymbol{\varepsilon}^{\text{T}} \mathbf{c} \boldsymbol{\varepsilon} \, d\Omega \qquad (5.68)
$$

which is a quadratic functional of $\boldsymbol{\varepsilon}$.

The variational statement for the principle of minimum potential energy can now be written as

$$
\delta \Pi_{\text{PE}}(\mathbf{u}) = \delta [U_{\text{PE}}(\mathbf{u}) - W_{\text{ext}}(\mathbf{u})] = 0 \qquad (5.69)
$$

which is the so-called minimization condition. The principle of minimum potential energy is the foundation of the so-called displacement methods where we need to assume only a single field of displacement. The well-known Raleigh–Ritz method that was widely used in early years is essentially the principle of minimum potential energy. The FEM that dominates today's modeling and simulation area follows also this principle. Many meshfree methods can also be established based on this principle.

Remark 5.17: Weak Form Statement vs. Minimization Statement
We now observe the equivalency of the weak form and minimization statements. The minimization process in Equation 5.69 is, in fact, exactly the same as the weak statement given in Equation 5.16:

1. The strain energy potential U_{PE} corresponding to the bilinear form a in the form of Equation 5.12.

2. The external work done W_{ext} defined in Equation 5.52 being the same as the linear form Equation 5.16.

3. The requirement of Equation 5.16 being satisfied for *all* functions in $\mathbb{H}_0^1$ corresponding to the search among all the admissible displacement fields for the one satisfying the stationary (minimization) condition.

4. The condition that $\mathbf{v}$ must be in $\mathbb{H}_0^1$ corresponding to the admissible condition. The admissible condition physically means that the displacement field assumed must satisfy

$$\text{The compatibility conditions within the domain (in-domain compatibility)} \qquad (5.70)$$

$$\text{The essential boundary conditions (on-boundary compatibility)} \qquad (5.71)$$

5. The in-domain compatibility condition (Equation 5.70) implies that the assumed displacement field that lives in the $\mathbb{H}_h^1$ space is at least continuous over the entire problem domain, and the strain field is obtained using the assumed displacement and the strain–displacement relation. The on-boundary compatibility condition (Equation 5.71) requires the assumed displacement field satisfying the prescribed displacement values at any point on the essential boundary, and hence lives in an $\mathbb{H}_{h,0}^1$ space. A displacement method that satisfies conditions (Equations 5.70 and 5.71) together with the kinematical condition (strain–displacement relation) is said *fully compatible*, as stated in Remark 5.2.

6. When the displacement function is not in an $\mathbb{H}_h^1$ space (say, for example, is in a $\mathbb{G}_h^1$ space), the bilinear form a *may* not be bounded, and the weak formulation *can* fail. In engineering terms, when the displacement field has possible gaps or overlaps (discontinuities), the total energy potential cannot be simply evaluated as $\Pi_{\mathrm{PE}} = U_{\mathrm{PE}} - W_{\mathrm{ext}}$, because of the possible energy "leak" on these gaps or overlaps, and hence Equation 5.69 may not be a correct reflection of the system. In such cases, we need to construct a new strain field for proper energy evaluation, leading to *strain-constructed energy principles* (or weakened-weak formulations).

With the presentation of Remark 5.15 and Table 1.1, we no longer need to distinguish so precisely these two formulations and terminologies used in these two formulations. We may use them in a mixed fashion for most convenience in the presentation.

The procedure for solving a problem using the principle of minimum potential energy, for a numerical solution based on a finite discrete model is described as follows:

1. Construct shape functions (see Chapter 2) to approximate the displacement functions $\mathbf{u}^h$ using their values at the nodes in the problem domain. The approximated displacement functions should be in an $\mathbb{H}_{h,0}^1$ space, i.e., they should satisfy the admissible conditions of Equations 5.70 and 5.71.

2. Calculate the potential energy $\Pi_{\mathrm{PE}}(\mathbf{u}^h)$ defined in Equation 5.67 using the assumed displacements $\mathbf{u}^h$ and the strain–displacement relations, which leads to an expression $\Pi_{\mathrm{PE}}(\mathbf{U}^h)$ in terms of nodal displacements $\mathbf{U}^h$. Seek for a set of nodal displacements $\tilde{\mathbf{U}}$ that makes $\Pi_{\mathrm{PE}}(\tilde{\mathbf{U}})$ minimum over all possible nodal displacements, which leads to a set of algebraic equations.

3. Solve the set of algebraic equations for nodal values $\tilde{\mathbf{U}}$.

4. Recover the displacement field $\tilde{\mathbf{u}}$ using again the shape functions and the obtained nodal values. The displacement field $\tilde{\mathbf{u}}$ is the approximated solution for the discrete model.

5. Finally, recover the strains using strain–displacement relations, and the stress field using the generalized Hooke's law.

In the FEM, the shape function is constructed using elements (with proper mapping), and the integration of the energy is also based on these elements. In meshfree methods, MLS shape functions discussed in Chapter 2 based on arbitrarily distributed nodes should be used, and the integration is, in general, based on a set of background cells.

Note that in order to use the energy principles, energy functional has to be obtained, which may not be possible for all problems. On the other hand, using energy principles we do not have to know the strong forms of all the system equations. Our understanding on the energy behavior is sufficient for us to obtain the solution (corresponding to a stable state).

Remark 5.5 has stated that if $\mathbf{u} \in \mathbb{H}_h^1$, then the FEM will reproduce the exact solution $\mathbf{u}$. It is not so difficult to give an "engineering" proof that the principle of minimum potential energy gives the exact solution of the strong form system equations, if the exact solution is in our assumed pool of displacements. Here, we attempt to give such an engineering proof by an energy argument (with some mathematical support). The energy argument is not mathematically rigorous, but should provide the essence on why the principle indeed works.

Remark 5.18: Energy Argument

When a solid of a *stable* material is *constrained* for all the rigid body movements and it is somehow (on surface and/or inside the solid) loaded, a displacement field will be generated over the solid. As a result, there will also be a strain field and hence the strain energy potential is built up that is evaluated as U_{PE} in Equation 5.67. The U_{PE} is a direct result from the work of the loads that is evaluated as W_{ext}. Because W_{ext} is the causal, it should be with a negative sign (when measured as a potential). Now, the total potential becomes $\Pi_{\text{PE}}(\mathbf{u}^h)$ as a quadratic functional of a given displacement field $\mathbf{u}^h$.

Next, we can try to vary $\mathbf{u}^h$: feeding all sorts of $\mathbf{u}^h$ from the space of $\mathbb{H}_0^1$ to $\Pi_{\text{PE}}(\mathbf{u}^h)$. Because our $\mathbf{u}^h$ is all from $\mathbb{H}_0^1$, $\Pi_{\text{PE}}(\mathbf{u}^h)$ will always be bounded (see Chapter 3 for the definition of an $\mathbb{H}^1$ space). This should give us a "hyperparaboloid." Therefore, there must be one (and only one) stationary point. Because the quadratic term U_{PE} in functional $\Pi_{\text{PE}}(\mathbf{u}^h)$ is positive, the stationary point must be a minimum. This proves the existence of the minimum point $\Pi_{\text{PE}}(\mathbf{u}^h)$.

Note U_{PE} (energy) will never be negative. It will not be zero, because we take functions only from $\mathbb{H}_0^1$ where the functions are constraints for rigid motions. Therefore, any loading will surely result in strain and strain energy in our *stable* solid.

At $\Pi_{\text{PE}}(\mathbf{u})$, the solid settles down, because any variation from any direction away from $\mathbf{u}$ results in a higher potential level, and hence it "falls back," which means that the solid is in the equilibrium status with a displacement field $\mathbf{u}$.

The same argument applies to any finite model where we seek for an approximated solution from an $\mathbb{H}_{h,0}^1$ space. In such a case, the principle of minimum potential energy gives the best possible solution given in $\mathbb{H}_{h,0}^1$ in terms of energy error measures. When the mesh is refined, the $\mathbb{H}_{h,0}^1$ space (of continuous functions) gets richer, and the solution (the best in $\mathbb{H}_{h,0}^1$) approaches closer to the exact solution.

Can we take functions from $\mathbb{G}_{h,0}^1$? No, not directly, because U_{PE} may not be bounded. We need to construct a new strain field and use the strain-constructed energy principles or weakened-weak forms, including the SC-Galerkin (Section 5.15), the GS-Galerkin (Section 5.15.1), or the Galerkin-like (Section 5.15.2) weak forms.

5.11 Hamilton's Principle

Hamilton's principle is an energy principle applicable to dynamic problems. It is essentially the same as the minimum potential energy principle, except for the inclusion of the time variation. For dynamic problems, we need to assume the time-history of the displacement field. Therefore, the admissible conditions for the assumed time-history of the displacement field are second-order differentiable with respect to time and satisfy

$$\text{(c) the conditions at the initial } (t_1) \text{ and final } (t_2) \text{ time} \qquad (5.72)$$

in addition to the two conditions given in Equations 5.70 and 5.71. Hamilton's principle can then be stated, "of all the possible admissible time-histories of displacement fields the most *accurate* is registered at the stationary point of the Lagrangian functional." Mathematically, it is stated as

$$L(\mathbf{u}) = \underbrace{\int_{\Omega} \frac{1}{2} \rho \dot{\mathbf{u}}^T \dot{\mathbf{u}} d\Omega}_{T} - \underbrace{\int_{\Omega} \frac{1}{2} \boldsymbol{\varepsilon}^T \mathbf{c} \boldsymbol{\varepsilon} d\Omega}_{U_{\text{PE}}} + \underbrace{\left[\int_{\Omega} \mathbf{b}^T \mathbf{u} d\Omega + \int_{\Gamma_t} \mathbf{t}_{\Gamma}^T \mathbf{u} d\Gamma \right]}_{W_{\text{ext}}} = \text{Stationary} \qquad (5.73)$$

where
  L is known as the Lagrangian functional
  $\mathbf{u} \in \mathbb{H}_0^1$
  $\boldsymbol{\varepsilon} = \mathbf{L}_d \mathbf{u}$

The kinetic energy is defined by

$$T = \int_{\Omega} \frac{1}{2} \rho \dot{\mathbf{u}}^T \dot{\mathbf{u}} d\Omega \qquad (5.74)$$

where the dot stands for differentiation with respect to time. The variational statement becomes

$$\delta \int_{t_1}^{t_2} L = \delta \int_{t_1}^{t_2} [T - U_{\text{PE}} + W_{\text{ext}}] dt = 0 \qquad (5.75)$$

The procedure for solving a problem using Hamilton's principle is described as follows:

1. Construct shape functions to approximate the displacement functions using their values at the nodes in the problem domain. The approximated field function in space should be in an $\mathbb{H}_0^1$ space, thus satisfying the admissible conditions of Equations 5.70 through 5.72.
2. Calculate the kinetic energy T, the strain energy U, and the work done by the external forces W in terms of the approximated displacement functions.

3. Form the Lagrangian functional L using Equation 5.73, seeking for the stationary point, which leads to a set of differential equations with respect only to time.

4. Solve the set of differential equations with respect to time, using standard procedures to obtain the dynamic field.

5.12 Hamilton's Principle with Constraints

In the displacement methods, there are cases often when the approximated displacement functions do not satisfy some of the admissible conditions (Equation 5.70 or 5.71), or parts of the problem domain, including parts of boundaries, discrete curves, and points at locations. For such cases, we need to modify the principles. This section explains a modification of the Hamilton's principle with constraints.

Consider the following given set of k conditions that the approximated field function has to satisfy:

$$\mathbf{C}(\mathbf{u}) = \left\{ \begin{array}{c} C_1(\mathbf{u}) \\ C_2(\mathbf{u}) \\ \vdots \\ C_k(\mathbf{u}) \end{array} \right\} = 0 \tag{5.76}$$

where $\mathbf{C}$ is a given matrix of coefficients.

Our purpose now is to seek the stationary point of the Lagrangian functional subjected to the constraint of Equation 5.76. There are basically two methods often used to modify the functional that accommodates these constraints: the method of Lagrange multipliers and the penalty method.

5.12.1 Method of Lagrange Multipliers

In the method of Lagrange multipliers, the modified Lagrangian is written as follows:

$$\tilde{L} = L + \int_\Omega \boldsymbol{\lambda}^T \mathbf{C}(\mathbf{u}) d\Omega \tag{5.77}$$

where $\boldsymbol{\lambda}$ is a vector of the Lagrange multipliers given by

$$\boldsymbol{\lambda}^T = \{ \lambda_1 \quad \lambda_2 \quad \cdots \quad \lambda_k \} \tag{5.78}$$

These Lagrange multipliers are unknown functions of independent coordinates in the domain Ω. The variational statement of the modified Hamilton's principle then seeks the following stationary condition:

$$\delta \int_{t_1}^{t_2} \tilde{L} \, dt = 0 \tag{5.79}$$

Note that since the Lagrange multipliers are unknown functions, the total number of unknown field functions of the whole system is then increased. In the process of seeking discretized system equations, these Lagrange multipliers must also be approximated in a manner similar to that of the field functions. Therefore, the total number of nodal unknowns in the discretized system equation will also be increased. The method of Lagrange multipliers will, however, rigorously enforce the constraints. The Lagrange multipliers $\boldsymbol{\lambda}$ can be viewed physically as *smart* forces enforcing the constraints. The penalty method, introduced in the following subsection, does not increase the number of unknowns.

5.12.2 Penalty Method

In examining the constraint equations (Equation 5.76) we construct the following functional:

$$\mathbf{C}^{\mathrm{T}}\boldsymbol{\alpha}\mathbf{C} = \alpha_1 C_1^2 + \alpha_2 C_2^2 + \cdots + \alpha_k C_k^2 \tag{5.80}$$

where $\boldsymbol{\alpha}$ is a diagonal matrix given by

$$\boldsymbol{\alpha} = \begin{bmatrix} \alpha_1 & 0 & 0 & 0 \\ 0 & \alpha_2 & 0 & 0 \\ 0 & 0 & \ddots & 0 \\ 0 & 0 & 0 & \alpha_k \end{bmatrix} \tag{5.81}$$

where $\alpha_1, \alpha_2, \ldots, \alpha_k$ are penalty factors. They can be given as functions of the coordinates, but usually they are assigned as positive constant numbers. In any case, $\mathbf{C}^{\mathrm{T}}\boldsymbol{\alpha}\mathbf{C}$ will always be nonnegative, and hence zero is the minimum of the functional $\mathbf{C}^{\mathrm{T}}\boldsymbol{\alpha}\mathbf{C}$. It would be only zero if all the conditions in Equation 5.76 are fully satisfied. Therefore, the following stationary condition of the functional $\mathbf{C}^{\mathrm{T}}\boldsymbol{\alpha}\mathbf{C}$ guarantees the best satisfaction of the constraint equations (Equation 5.76):

$$\delta(\mathbf{C}^{\mathrm{T}}\boldsymbol{\alpha}\mathbf{C}) = 0 \tag{5.82}$$

Performing the variation using the chain rule, we have

$$\delta(\mathbf{C}^{\mathrm{T}}\boldsymbol{\alpha}\mathbf{C}) = 2\mathbf{C}^{\mathrm{T}}\boldsymbol{\alpha}\delta\mathbf{C} = 2\delta\mathbf{C}^{\mathrm{T}}\boldsymbol{\alpha}\mathbf{C} = 0 \tag{5.83}$$

which leads to the following minimization condition:

$$\boldsymbol{\alpha}\mathbf{C}(\mathbf{u}) = \left\{ \begin{array}{c} \alpha_1 C_1(\mathbf{u}) \\ \alpha_2 C_2(\mathbf{u}) \\ \vdots \\ \alpha_k C_k(\mathbf{u}) \end{array} \right\} = 0 \tag{5.84}$$

If $\alpha = 0$, the essential boundary condition is not enforced at all, because any C_i will still satisfy the *i*th equation in Equation 5.84. If α_i goes to infinity, the essential boundary condition is fully enforced, because C_i must be zero in order to satisfy Equation 5.84. The α_i can be viewed physically as *penalty* forces that penalize the dissatisfaction of the constraints.

The above analysis shows that the potential energy related to the penalized constraints should be $\mathbf{C}^T\boldsymbol{\alpha}\mathbf{C}$, and hence the modified Lagrangian is then written as follows:

$$\tilde{L} = L + \frac{1}{2}\int_\Omega \mathbf{C}^T(\mathbf{u})\boldsymbol{\alpha}\mathbf{C}(\mathbf{u})d\Omega \tag{5.85}$$

The $1/2$ serves only to counter the 2 that will be produced in the later variational operation. The important difference between the penalty factor α_i and the Lagrange multiplier λ is that the penalty factor is a given constant (no variation is allowed), whereas the Lagrange multiplier is a variable.

Because α is a known constant, there is no increase in the number of unknowns in the system. The question is how to choose the penalty factor. To impose the constraint fully, the penalty factor must be infinite, which is not possible in practical numerical analysis. Therefore, in the penalty method these constraints cannot be satisfied exactly, but only approximately. In general, the use of a larger penalty factor will enforce the constraint. The problem is that if the penalty factor is too small, the constraints may not be properly enforced, but if it is too large, numerical problems may be encountered. A compromise should be reached. Some kind of formula that is universally applicable would be useful. To find such a formula, one may need to determine the factors that affect the selection of the penalty factor in the actual event of solving the discretized system equations.

Note that use of the penalty method is a routine operation even in the FEM for imposing essential boundary conditions including single-point constraints (SPC) and multi-point constraints (MPC).

5.12.3 Determination of the Penalty Factor

Because discretization errors can be comparable in magnitude to the errors due to the poor satisfaction of the constraint, Zienkiewicz [11] has suggested using the following formula for FEM analysis:

$$\alpha = \text{constant}(1/d)^n \tag{5.86}$$

where
   d is the characteristic length, which can also be the ratio of the element size to the dimension of the problem domain
   n is the order of the elements

In extending this formula to meshfree methods, we suggest that d is the nodal spacing divided by the dimension of the problem domain, and $n = 1$. The constant in Equation 5.86 should relate to the material property of the solid or the structure. It can be 10^{10} times the Young's modulus.

This book prefers the following simple method for determining the penalty factor:

$$\alpha = 1.0 \times 10^{4\sim 13} \times \max \text{ (diagonal elements in the stiffness matrix)} \tag{5.87}$$

In most of the examples reported in later chapters using penalty methods, the foregoing equation is adopted.

It has also been suggested to use

$$\alpha = 1.0 \times 10^{5 \sim 8} \times \text{Young's modulus} \tag{5.88}$$

which works well for some examples.

Note that trials may be needed to choose a proper penalty factor.

5.13 Galerkin Weak Form

5.13.1 Galerkin Weak Form for Static Problems

The Galerkin weak form for mechanics problems of solids and structures can be derived directly from the weak form statement. It can also be easily derived from the principle of the minimum potential energy given in Equation 5.69. The details of Equation 5.69 can be rewritten as

$$\delta \int_\Omega \frac{1}{2}(\boldsymbol{\varepsilon}^T \mathbf{c} \boldsymbol{\varepsilon}) d\Omega - \int_\Omega \delta \mathbf{u}^T \mathbf{b} d\Omega - \int_{\Gamma_t} \delta \mathbf{u}^T \mathbf{t}_\Gamma d\Gamma = 0 \tag{5.89}$$

We see two types of operations in the above equation: variational and integral operations. Changing the order of these operations does not affect the results because these operations operate with respect to different independent variables. The variational operation operates with respect to the displacement field functions, but the integration operation operates with respect to the coordinates of the problem domain. Therefore, it does not matter which one operates first, and hence we can move the variational operation inside the integral in the first term of Equation 5.89:

$$\frac{1}{2} \int_\Omega \delta(\boldsymbol{\varepsilon}^T \mathbf{c} \boldsymbol{\varepsilon}) d\Omega - \int_\Omega \delta \mathbf{u}^T \mathbf{b} d\Omega - \int_{\Gamma_t} \delta \mathbf{u}^T \mathbf{t}_\Gamma d\Gamma = 0 \tag{5.90}$$

The integrand in the first integral term can be written as follows using the chain rule of variation:

$$\delta(\boldsymbol{\varepsilon}^T \mathbf{c} \boldsymbol{\varepsilon}) = \delta \boldsymbol{\varepsilon}^T \mathbf{c} \boldsymbol{\varepsilon} + \boldsymbol{\varepsilon}^T \delta(\mathbf{c} \boldsymbol{\varepsilon}) = \delta \boldsymbol{\varepsilon}^T \mathbf{c} \boldsymbol{\varepsilon} + \boldsymbol{\varepsilon}^T \mathbf{c} \delta \boldsymbol{\varepsilon} \tag{5.91}$$

Because the two terms in the foregoing equation are all scalars, their transposes are still themselves. For the last term in Equation 5.91, we should have

$$(\boldsymbol{\varepsilon}^T \mathbf{c} \delta \boldsymbol{\varepsilon})^T = \delta \boldsymbol{\varepsilon}^T \mathbf{c} \boldsymbol{\varepsilon} \tag{5.92}$$

In the derivation above, we used the symmetry property of the matrix of the material constant **c**. Therefore, Equation 5.91 becomes

$$\delta(\boldsymbol{\varepsilon}^T \mathbf{c} \boldsymbol{\varepsilon}) = 2\delta \boldsymbol{\varepsilon}^T \mathbf{c} \boldsymbol{\varepsilon} \tag{5.93}$$

Substituting Equation 5.93 into Equation 5.90, we arrived at

$$\int_\Omega \delta\boldsymbol{\varepsilon}^T \mathbf{c}\boldsymbol{\varepsilon}\,d\Omega - \left(\int_\Omega \delta\mathbf{u}^T\mathbf{b}\,d\Omega + \int_{\Gamma_t} \delta\mathbf{u}^T\mathbf{t}_\Gamma\,d\Gamma \right) = 0 \qquad (5.94)$$

Equation 5.94 is in fact the well-known and widely used Galerkin weak form. Using the displacement strain relation $\boldsymbol{\varepsilon} = \mathbf{L}_d\mathbf{u}$, Equation 5.94 can be further written explicitly in term of the displacement:

$$\underbrace{\int_\Omega (\mathbf{L}_d\delta\mathbf{u})^T \mathbf{c}(\mathbf{L}_d\mathbf{u})\,d\Omega}_{a(\delta\mathbf{u},\,\mathbf{u})} - \underbrace{\left(\int_\Omega \delta\mathbf{u}^T\mathbf{b}\,d\Omega + \int_{\Gamma_t} \delta\mathbf{u}^T\mathbf{t}_\Gamma\,d\Gamma \right)}_{f(\delta\mathbf{u})} = 0 \qquad (5.95)$$

We now see clearly the relationship between the principle of minimum potential energy and the weak form statement given in Equation 5.16. In Equation 5.16 we require the equation to be satisfied for all $\mathbf{v} \in \mathbb{H}_0^1$, while in Equation 5.95 we require the same for all arbitrary variations of $\mathbf{u}$.

A numerical method based on the Galerkin weak form therefore looks for a displacement field in $\mathbf{u} \in \mathbb{H}_0^1$ that satisfies Equation 5.95 for any arbitrary $\delta\mathbf{u} \in \mathbb{H}_0^1$. Physically, such a displacement field makes the total potential energy in the entire system (solid or structure) minimum, and hence the solid/structure stays stable there.

5.13.2 Galerkin Weak Form for Dynamic Problems

The Galerkin weak form for dynamic problems can be directly derived using Hamilton's principle. Using Equation 5.75, we have

$$\delta \int_{t_1}^{t_2} \left[-\frac{1}{2}\int_\Omega \boldsymbol{\varepsilon}^T\mathbf{c}\boldsymbol{\varepsilon}\,d\Omega + \int_\Omega \mathbf{u}^T\mathbf{b}\,d\Omega + \int_{\Gamma_t} \mathbf{u}^T\mathbf{t}_\Gamma\,d\Gamma + \frac{1}{2}\int_\Omega \rho\dot{\mathbf{u}}^T\dot{\mathbf{u}}\,d\Omega \right] dt = 0 \qquad (5.96)$$

Moving the variation operation into the integral operations, we obtain

$$\int_{t_1}^{t_2} \left[-\frac{1}{2}\int_\Omega \delta(\boldsymbol{\varepsilon}^T\mathbf{c}\boldsymbol{\varepsilon})\,d\Omega + \int_\Omega \delta\mathbf{u}^T\mathbf{b}\,d\Omega + \int_{\Gamma_t} \delta\mathbf{u}^T\mathbf{t}_\Gamma\,d\Gamma + \frac{1}{2}\int_\Omega \delta(\rho\dot{\mathbf{u}}^T\dot{\mathbf{u}})\,d\Omega \right] dt = 0 \qquad (5.97)$$

Using Equation 5.93, we have

$$\int_{t_1}^{t_2} \left[-\int_\Omega \delta\boldsymbol{\varepsilon}^T\mathbf{c}\boldsymbol{\varepsilon}\,d\Omega + \int_\Omega \delta\mathbf{u}^T\mathbf{b}\,d\Omega + \int_{\Gamma_t} \delta\mathbf{u}^T\mathbf{t}_\Gamma\,d\Gamma + \frac{1}{2}\int_\Omega \delta(\rho\dot{\mathbf{u}}^T\dot{\mathbf{u}})\,d\Omega \right] dt = 0 \qquad (5.98)$$

Let us now look at the last term in Equation 5.97, and we see two types of independent integral operations: integral over time and over domain. Changing the order of these two integrals should not affect the results, hence we move the time integration into the spatial integration, and thus obtain

$$\int_{t_1}^{t_2} \left[\frac{1}{2} \int_{\Omega} \delta(\rho \dot{\mathbf{u}}^T \dot{\mathbf{u}}) d\Omega \right] dt = \frac{1}{2} \int_{\Omega} \left[\int_{t_1}^{t_2} \delta(\rho \dot{\mathbf{u}}^T \dot{\mathbf{u}}) dt \right] d\Omega \tag{5.99}$$

Using the chain rule of variation and then the scalar property, the time integration in Equation 5.99 can be changed as

$$\int_{t_1}^{t_2} \delta(\rho \dot{\mathbf{u}}^T \dot{\mathbf{u}}) dt = \rho \int_{t_1}^{t_2} [\delta \dot{\mathbf{u}}^T \dot{\mathbf{u}} + \dot{\mathbf{u}}^T \delta \dot{\mathbf{u}}] dt = 2\rho \int_{t_1}^{t_2} [\delta \dot{\mathbf{u}}^T \dot{\mathbf{u}}] dt \tag{5.100}$$

We again have two types of operations in the above equation: the variational operation and the differential operation. Changing the order of these operations does not affect the results because these operations are operating with respect to different variables. The variational operation operates with respect to the displacement field, but the differential operation operates with respect to time. Therefore, it does not matter which one operates first; we thus have

$$\int_{t_1}^{t_2} [\delta \dot{\mathbf{u}}^T \dot{\mathbf{u}}] dt = \int_{t_1}^{t_2} \left[\frac{d\delta \mathbf{u}^T}{dt} \frac{d\mathbf{u}}{dt} \right] dt \tag{5.101}$$

Integrating by parts with respect to time, we have

$$\int_{t_1}^{t_2} \left[\frac{d\delta \mathbf{u}^T}{dt} \frac{d\mathbf{u}}{dt} \right] dt = \int_{t_1}^{t_2} \left[-\delta \mathbf{u}^T \frac{d^2 \mathbf{u}}{dt^2} \right] dt + \delta \mathbf{u}^T \frac{d\mathbf{u}}{dt} \Big|_{t_1}^{t_2} \tag{5.102}$$

Invoking now the condition given in Equation 5.72, we know that $\mathbf{u}$ has already satisfied the conditions at the initial time (t_1) and the final time (t_2). Therefore, $\delta \mathbf{u}^T$ has to be zero at t_1 and t_2 (no variation can exist for any given constant value). Therefore, the last term in Equation 5.102 vanishes, which gives

$$\int_{t_1}^{t_2} \left[\frac{d\delta \mathbf{u}^T}{dt} \frac{d\mathbf{u}}{dt} \right] dt = \int_{t_1}^{t_2} \left[-\delta \mathbf{u}^T \frac{d^2 \mathbf{u}}{dt^2} \right] dt \tag{5.103}$$

Therefore, Equation 5.99 becomes

$$\int_{t_1}^{t_2} \left[\frac{1}{2} \int_{\Omega} \delta(\rho \dot{\mathbf{u}}^T \dot{\mathbf{u}}) d\Omega \right] dt = \int_{\Omega} \left[\int_{t_1}^{t_2} \left[-\rho \delta \mathbf{u}^T \frac{d^2 \mathbf{u}}{dt^2} \right] dt \right] d\Omega \tag{5.104}$$

We now switch the order of integration to obtain

$$\int_{t_1}^{t_2} \left[\frac{1}{2} \int_{\Omega} \delta(\rho \dot{\mathbf{u}}^{\mathrm{T}} \dot{\mathbf{u}}) d\Omega \right] dt = - \int_{t_1}^{t_2} \left[\int_{\Omega} [\rho \delta \mathbf{u}^{\mathrm{T}} \ddot{\mathbf{u}}] d\Omega \right] dt \qquad (5.105)$$

Equation 5.97 now becomes

$$\int_{t_1}^{t_2} \left[-\int_{\Omega} \delta \boldsymbol{\varepsilon}^{\mathrm{T}} \boldsymbol{\sigma} d\Omega + \int_{\Omega} \delta \mathbf{u}^{\mathrm{T}} \mathbf{b} d\Omega + \int_{\Gamma_t} \delta \mathbf{u}^{\mathrm{T}} \mathbf{t}_{\Gamma} d\Gamma - \int_{\Omega} \rho \delta \mathbf{u}^{\mathrm{T}} \ddot{\mathbf{u}} d\Omega \right] dt = 0 \qquad (5.106)$$

To satisfy the above equation for all possible choices of **u**, the integrand of the time integration has to vanish, which leads to

$$\int_{\Omega} \delta \boldsymbol{\varepsilon}^{\mathrm{T}} \boldsymbol{\sigma} d\Omega - \int_{\Omega} \delta \mathbf{u}^{\mathrm{T}} \mathbf{b} d\Omega - \int_{\Gamma_t} \delta \mathbf{u}^{\mathrm{T}} \mathbf{t}_{\Gamma} d\Gamma + \int_{\Omega} \rho \delta \mathbf{u}^{\mathrm{T}} \ddot{\mathbf{u}} d\Omega = 0 \qquad (5.107)$$

This is the well-known Galerkin weak form for dynamic problems that is widely used for solving mechanics problems of solids and structures. By removing the dynamic term Equation 5.107 reduces to Equation 5.94.

By using the stress–strain relation, and then the strain–displacement relation, Equation 5.107 can be explicitly expressed as follows in terms of the displacement vector **u**:

$$\int_{\Omega} (\mathbf{L}_d \delta \mathbf{u})^{\mathrm{T}} \mathbf{c} (\mathbf{L}_d \mathbf{u}) d\Omega - \int_{\Omega} \delta \mathbf{u}^{\mathrm{T}} \mathbf{b} d\Omega - \int_{\Gamma_t} \delta \mathbf{u}^{\mathrm{T}} \mathbf{t}_{\Gamma} d\Gamma + \int_{\Omega} \rho \delta \mathbf{u}^{\mathrm{T}} \ddot{\mathbf{u}} d\Omega = 0 \qquad (5.108)$$

This is the Galerkin weak form for dynamic problems written in terms of displacement, and therefore it is convenient to use, as the displacement is to be approximated in FEM or meshfree methods. For static problems, Equation 5.108 reduces to Equation 5.95 by dropping the dynamic term.

Remark 5.19: Virtual-Work Argument
Equation 5.107 can also be viewed simply as the principle of virtual work, which states that if a solid is in (dynamically) equilibrium status, the total virtual work performed by all the stresses in the solid and all the external forces applied on the solid should vanish when the solid is subjected to an arbitrary virtual field of displacement. The virtual work can, therefore, be viewed as an alternative statement of the equilibrium equations. In the situation given in Equation 5.107, we can view that the solid is subjected to a virtual displacement of $\delta \mathbf{u} \in \mathbb{H}_0^1$. In Equation 5.107, the first term is the virtual work done by the internal stresses in the problem domain Ω; the second term is the virtual work done by

the external body force; the third term is the virtual work done by the external tractions on the natural boundaries; and the last term is the virtual work done by the inertial forces. Therefore, using the principle of virtual work, we can actually write out Equation 5.107 or Equation 5.94 straightaway. The principle of virtual work is powerful and can, in fact, be used to derive all the weak forms that can be derived from an energy principle. However, the virtual displacement field has to be in a proper $\mathbb{H}^1$ space, so that there are no "gaps," and hence, no "energy leak." If $\delta \mathbf{u} \in \mathbb{G}^1_{h,0}$, we need to evaluate the virtual work done by the internal stresses in a proper manner, for example, by the generalized smoothing operations [15], leading to the GS-Galerkin weak form [16].

Following the virtual work argument, we can now readily and easily write the Galerkin weak form for dynamic problems with damping effects:

$$\int_\Omega (\mathbf{L}_d \delta \mathbf{u})^{\mathrm{T}} \mathbf{c}(\mathbf{L}_d \mathbf{u}) d\Omega - \int_\Omega \delta \mathbf{u}^{\mathrm{T}} (\mathbf{b} - \rho \ddot{\mathbf{u}} - c\dot{\mathbf{u}}) d\Omega - \int_{\Gamma_t} \delta \mathbf{u}^{\mathrm{T}} \mathbf{t}_\Gamma d\Gamma = 0 \qquad (5.109)$$

where we simply treat the inertial and damping forces as body forces and then apply the principle of virtual work.

The Galerkin weak forms are very handy in application to problems of solid mechanics, because one does not need to perform integration-by-parts. The discretized system equations can be derived very easily using approximated displacements. The Galerkin procedure used in meshfree methods is as follows:

1. Approximate the displacement at a point using meshfree shape functions and nodal displacements at the nodes surrounding the point. The approximation should satisfy Equations 5.70 through 5.72.

2. Substitute the approximated displacements into Equation 5.107, and factor out the variations of the nodal displacements for their arbitrariness, leading to a set of differential equations with respect only to time.

3. Solve the set of differential equations with respect to time, using standard procedures to obtain the dynamic field.

4. For static problems, use Equation 5.129 instead, which leads to a set of algebraic equations that can be solved using standard algebraic equation solvers to obtain the static field.

This procedure is applied in the following chapters for mechanics problems of solids and structures solved using meshfree methods based on the Galerkin weak form.

5.14 Galerkin Weak Form with Constraints

For cases when the approximated field function does not satisfy the condition (Equation 5.70 or Equation 5.71) on parts of the problem domain, including parts of the boundaries and discrete points at locations, we should use the Galerkin weak form with constraints. The procedure of obtaining the constrained Galerkin weak form is the same as in Section 5.12. The following presents the final expressions.

5.14.1 Galerkin Weak Form with Lagrange Multipliers

For dynamic problems,

$$
\int_\Omega \delta(\mathbf{L}_d\mathbf{u})^\mathrm{T}\mathbf{c}(\mathbf{L}_d\mathbf{u})\mathrm{d}\Omega - \int_\Omega \delta\mathbf{u}^\mathrm{T}\mathbf{b}\mathrm{d}\Omega - \int_{\Gamma_t} \delta\mathbf{u}^\mathrm{T}\mathbf{t}_\Gamma\mathrm{d}\Gamma
$$

$$
- \int_\Omega \delta\boldsymbol{\lambda}^\mathrm{T}\mathbf{C}(\mathbf{u})\,\mathrm{d}\Omega - \int_\Omega \boldsymbol{\lambda}^\mathrm{T}\delta\mathbf{C}(\mathbf{u})\,\mathrm{d}\Omega + \int_\Omega \delta\mathbf{u}^\mathrm{T}\mathbf{c}\dot{\mathbf{u}}\mathrm{d}\Omega + \int_\Omega \rho\delta\mathbf{u}^\mathrm{T}\ddot{\mathbf{u}}\mathrm{d}\Omega = 0 \qquad (5.110)
$$

For static problems, it simply reduces to

$$
\int_\Omega \delta(\mathbf{L}_d\mathbf{u})^\mathrm{T}\mathbf{c}(\mathbf{L}_d\mathbf{u})\,\mathrm{d}\Omega - \int_\Omega \delta\mathbf{u}^\mathrm{T}\mathbf{b}\mathrm{d}\Omega - \int_{\Gamma_t} 0\mathbf{u}^\mathrm{T}\mathbf{t}_\Gamma\mathrm{d}\Gamma - \int_\Omega \delta\boldsymbol{\lambda}^\mathrm{T}\mathbf{C}(\mathbf{u})\mathrm{d}\Omega - \int_\Omega \boldsymbol{\lambda}^\mathrm{T}\delta\mathbf{C}(\mathbf{u})\mathrm{d}\Omega = 0 \quad (5.111)
$$

When the constrained Galerkin weak form is used, the assumed displacement field may not be in an $\mathbb{H}_0^1$ space, meaning that it may have discontinuities or the essential boundary conditions may not be satisfied at locations. However, proper constraints at those locations need to be established, and used in Equation 5.111. The Lagrange multipliers $\boldsymbol{\lambda}$ can be viewed physically as smart forces that can force the essential boundary conditions.

5.14.2 Galerkin Weak Form with Penalty Factors

For dynamic problems,

$$
\int_\Omega \delta(\mathbf{L}_d\mathbf{u})^\mathrm{T}\mathbf{c}(\mathbf{L}_d\mathbf{u})\,\mathrm{d}\Omega - \int_\Omega \delta\mathbf{u}^\mathrm{T}\mathbf{b}\mathrm{d}\Omega - \int_{\Gamma_t} \delta\mathbf{u}^\mathrm{T}\mathbf{t}_\Gamma\mathrm{d}\Gamma - \int_\Omega \delta\mathbf{C}(\mathbf{u})^\mathrm{T}\boldsymbol{\alpha}\mathbf{C}(\mathbf{u})\,\mathrm{d}\Omega
$$

$$
+ \int_\Omega \delta\mathbf{u}^\mathrm{T}\mathbf{c}\dot{\mathbf{u}}\mathrm{d}\Omega + \int_\Omega \rho\delta\mathbf{u}^\mathrm{T}\ddot{\mathbf{u}}\mathrm{d}\Omega = 0 \qquad (5.112)
$$

For static problems, it simply reduces to

$$
\int_\Omega \delta(\mathbf{L}_d\mathbf{u})^\mathrm{T}\mathbf{c}(\mathbf{L}_d\mathbf{u})\mathrm{d}\Omega - \int_\Omega \delta\mathbf{u}^\mathrm{T}\mathbf{b}\mathrm{d}\Omega - \int_{\Gamma_t} \delta\mathbf{u}^\mathrm{T}\mathbf{t}_\Gamma\mathrm{d}\Gamma - \int_\Omega \delta\mathbf{C}(\mathbf{u})^\mathrm{T}\boldsymbol{\alpha}\mathbf{C}(\mathbf{u})\mathrm{d}\Omega = 0 \qquad (5.113)
$$

The penalty factors $\boldsymbol{\alpha}$ can be viewed physically as penalty forces that penalize the dissatisfaction of the constraints.

5.15 A Weakened-Weak Formulation: SC-Galerkin

The SC-Galerkin weak form is a more general W^2 form that uses functions from basically an unnormed $\mathbb{G}$ space, and has been applied to create SC-PIM models [22]. The stability and convergence of an SC-Galerkin model are controlled by the admissible conditions imposed on the constructed strain fields.

For a vector displacement field $\mathbf{v} \in \mathbb{G}_{h,0}^1$, we first construct a strain field $\hat{\boldsymbol{\varepsilon}}(\mathbf{v})$ in the form (see Chapter 4)

$$\hat{\boldsymbol{\varepsilon}}(\mathbf{u}^h) = \begin{cases} \mathcal{P}(\tilde{\boldsymbol{\varepsilon}}(\mathbf{u}^h)) = \mathcal{P}(\mathbf{L}_d\mathbf{u}^h), & \text{when } \mathbf{u}^h \in \mathbb{H}_h^1(\Omega) \\ \mathcal{B}(\mathbf{u}^h), & \text{when } \mathbf{u}^h \in \mathbb{G}_h^1(\Omega) \end{cases} \tag{5.114}$$

The strain energy potential for the constructed strain field becomes

$$\hat{U}_{\text{PE}}(\mathbf{v}) = \int_\Omega \frac{1}{2}\,\hat{\boldsymbol{\varepsilon}}^\mathsf{T}(\mathbf{v})\mathbf{c}\hat{\boldsymbol{\varepsilon}}(\mathbf{v})\,\mathrm{d}\Omega \tag{5.115}$$

The solution satisfies the SC-Galerkin weak form

$$\int_\Omega \delta\hat{\boldsymbol{\varepsilon}}^\mathsf{T}(\mathbf{u})\mathbf{c}\hat{\boldsymbol{\varepsilon}}(\mathbf{u})\mathrm{d}\Omega - \int_\Omega \delta\mathbf{u}^\mathsf{T}\mathbf{b}\mathrm{d}\Omega - \int_{\Gamma_t} \delta\mathbf{u}^\mathsf{T}\mathbf{t}_\Gamma\mathrm{d}\Gamma = 0 \tag{5.116}$$

where $\mathbf{u} \in \mathbb{G}_{h,0}^1$ and $\hat{\boldsymbol{\varepsilon}} \in \mathbb{L}^2$, and also satisfies (1) the norm equivalence condition (Equation 4.14) and (2) the strain convergence condition (Equation 4.16).

Theorem 5.7: SC-Galerkin: A Stable and Convergent W^2 Formulation

The SC-Galerkin weak form defined in Equation 5.116, using constructed strain fields that satisfy (1) the norm equivalence condition (Equation 4.14) and (2) the strain convergence condition (Equation 4.16), produces a stable and unique solution that converges to the exact solution for stable solids when $h \to 0$ (and hence $\Omega_i^s \to 0$).

Proof For any $\mathbf{v} \in \mathbb{G}_{h,0}^1$ with a vector $\mathbf{U} \in \mathbb{R}_0^{dN_n}$ of nodal unknown function values, because the constructed strain field $\hat{\boldsymbol{\varepsilon}}(\mathbf{U})$ is equivalent in a norm to the base (compatible) constant strain field $\tilde{\boldsymbol{\varepsilon}}_b(\mathbf{U})$, there exists a nonzero positive constant c_{ac} such that

$$\|\hat{\boldsymbol{\varepsilon}}(\mathbf{U})\|_{\mathbb{L}^2(\Omega)} \le c_{ac}\|\tilde{\boldsymbol{\varepsilon}}_b(\mathbf{U})\|_{\mathbb{L}^2(\Omega)}, \quad \forall\mathbf{U} \in \mathbb{R}_0^{dN_n} \tag{5.117}$$

where $\tilde{\boldsymbol{\varepsilon}}_b(\mathbf{U}) = \mathbf{B}_b\mathbf{U}$, and a nonzero positive constant c_{ca} such that

$$\|\tilde{\boldsymbol{\varepsilon}}_b(\mathbf{U})\|_{\mathbb{L}^2(\Omega)} \le c_{ca}\|\hat{\boldsymbol{\varepsilon}}(\mathbf{U})\|_{\mathbb{L}^2(\Omega)}, \quad \forall\mathbf{U} \in \mathbb{R}_0^{dN_n} \tag{5.118}$$

This means that the $\mathbb{L}^2$ norm of the constructed strain field can be bounded from both sides by the norm of the compatible base strain field $\|\tilde{\boldsymbol{\varepsilon}}_b(\mathbf{U})\|_{\mathbb{L}^2}$. Because we choose the linear FEM model with triangular elements with the same nodal unknowns $\mathbf{U} \in \mathbb{R}_0^{dN_n}$ as the base model, all the nodal shape functions of the base model are linearly independent. Since such a base model has been proven stable, the corresponding strain matrix $\mathbf{B}_b$ is a constant matrix, and the columns are all linearly independent. The norm of the strain field of the base model can be expressed as

$$\|\tilde{\boldsymbol{\varepsilon}}_b(\mathbf{U})\|_{\mathbb{L}^2(\Omega)}^2 = (\tilde{\boldsymbol{\varepsilon}}_b(\mathbf{U}))^\mathsf{T}(\tilde{\boldsymbol{\varepsilon}}_b(\mathbf{U})) = \mathbf{U}^\mathsf{T}\underbrace{\mathbf{B}_b^\mathsf{T}\mathbf{B}_b}_{\text{SPD}}\mathbf{U}, \quad \forall\mathbf{U} \in \mathbb{R}_0^{dN_n} \tag{5.119}$$

Due to the linear independent columns in $\mathbf{B}_b$, $\mathbf{B}_b^T\mathbf{B}_b$ must be an SPD matrix, and can always be decomposed into a unitary matrix $\mathbf{V}_b$ of eigenvectors and a diagonal matrix $\mathbf{\Lambda}_b$ of all the positive eigenvalues:

$$\mathbf{B}_b^T\mathbf{B}_b = \mathbf{V}_b^T\mathbf{\Lambda}_b\mathbf{V}_b \tag{5.120}$$

Substituting Equation 5.120 into Equation 5.119, we have

$$\|\tilde{\boldsymbol{\varepsilon}}_b(\mathbf{U})\|_{\mathbb{L}^2(\Omega)}^2 = \mathbf{U}^T \underbrace{\mathbf{V}_b^T\mathbf{\Lambda}_b\mathbf{V}_b}_{\mathbf{B}_b^T\mathbf{B}_b} \mathbf{U} \leq \lambda_{\max}^b \mathbf{U}^T \underbrace{\mathbf{V}_b^T\mathbf{V}_b}_{\mathbf{I}} \mathbf{U} = \lambda_{\max}^b \mathbf{U}^T\mathbf{U} = \lambda_{\max}^b \|\mathbf{U}\|_{\mathbb{L}^2}^2, \quad \forall \mathbf{U} \in \mathbb{R}_0^{dN_n}$$

$$\tag{5.121}$$

where $\lambda_{\max}^b$ is the maximum eigenvalue of $\mathbf{B}_b^T\mathbf{B}_b$, and

$$\|\tilde{\boldsymbol{\varepsilon}}_b(\mathbf{U})\|_{\mathbb{L}^2(\Omega)}^2 = \mathbf{U}^T \underbrace{\mathbf{V}_b^T\mathbf{\Lambda}_b\mathbf{V}_b}_{\mathbf{B}_b^T\mathbf{B}_b} \mathbf{U} \geq \lambda_{\min}^b \mathbf{U}^T \underbrace{\mathbf{V}_b^T\mathbf{V}_b}_{\mathbf{I}} \mathbf{U} = \lambda_{\min}^b \mathbf{U}^T\mathbf{U} = \lambda_{\min}^b \|\mathbf{U}\|_{\mathbb{L}^2}^2, \quad \forall \mathbf{U} \in \mathbb{R}_0^{dN_n}$$

$$\tag{5.122}$$

where $\lambda_{\min}^b$ is the minimum eigenvalue of $\mathbf{B}_b^T\mathbf{B}_b$. This means that the $\mathbb{L}^2$ norm of the constructed strain field will be bounded from both sides by the norm of all $\mathbf{U} \in \mathbb{R}_0^{dN_n}$.

Because the material is stable, we can use Equation 5.36, and the bilinear form $\hat{a}(\mathbf{w}, \mathbf{v})$ for the SC-Galerkin weak form can be given as

$$\begin{aligned}
\hat{a}(\mathbf{w}, \mathbf{v}) &= \int_\Omega \hat{\boldsymbol{\varepsilon}}^T(\mathbf{W}) \underbrace{\mathbf{V}_m^T\mathbf{\Lambda}_m\mathbf{V}_m}_{\mathbf{c}} \hat{\boldsymbol{\varepsilon}}(\mathbf{U})d\Omega = \int_\Omega (\mathbf{V}_m\hat{\boldsymbol{\varepsilon}}(\mathbf{W}))^T\mathbf{\Lambda}_m(\mathbf{V}_m\hat{\boldsymbol{\varepsilon}}(\mathbf{U}))d\Omega \\
&\leq \lambda_{\max}^m \int_\Omega \|\mathbf{V}_m\hat{\boldsymbol{\varepsilon}}(\mathbf{W})\|_{\mathbb{L}^2}\|\mathbf{V}_m\hat{\boldsymbol{\varepsilon}}(\mathbf{W})\|_{\mathbb{L}^2}d\Omega = \lambda_{\max}^m \int_\Omega \|\hat{\boldsymbol{\varepsilon}}(\mathbf{W})\|_{\mathbb{L}^2}\|\hat{\boldsymbol{\varepsilon}}(\mathbf{U})\|_{\mathbb{L}^2} \, d\Omega \\
&= \lambda_{\max}^m \|\hat{\boldsymbol{\varepsilon}}(\mathbf{W})\|_{\mathbb{L}^2(\Omega)}\|\hat{\boldsymbol{\varepsilon}}(\mathbf{U})\|_{\mathbb{L}^2(\Omega)} \leq \underbrace{\lambda_{\max}^m c_{ac}^2}_{C_{SU}} \|\tilde{\boldsymbol{\varepsilon}}_b(\mathbf{W})\|_{\mathbb{L}^2(\Omega)}\|\tilde{\boldsymbol{\varepsilon}}_b(\mathbf{U})\|_{\mathbb{L}^2(\Omega)} \\
&\leq C_{SU}\|\tilde{\boldsymbol{\varepsilon}}_b(\mathbf{W})\|_{\mathbb{L}^2(\Omega)}\|\tilde{\boldsymbol{\varepsilon}}_b(\mathbf{U})\|_{\mathbb{L}^2(\Omega)}, \quad \forall \mathbf{W} \in \mathbb{R}_0^{dN_n}, \quad \forall \mathbf{U} \in \mathbb{R}_0^{dN_n}
\end{aligned} \tag{5.123}$$

where
C_{SU} is a constant independent of $\mathbf{W}$ and $\mathbf{U}$
$\lambda_{\max}^m$ is the maximum eigenvalue of $\mathbf{c}$

In the above derivation, we used the Cauchy–Schwarz inequality in the first line, the $\mathbb{L}^2$ norm conservation property of the unitary matrix in the second line, and Equation 5.117 in the third line. Using Equation 5.121 now, we have

$$\hat{a}(\mathbf{w}, \mathbf{v}) \leq \underbrace{C_{SU}\lambda_{\max}^b}_{C_U} \|\mathbf{W}\|_{\mathbb{L}^2}\|\mathbf{U}\|_{\mathbb{L}^2}, \quad \forall \mathbf{W} \in \mathbb{R}_0^{dN_n}, \quad \forall \mathbf{U} \in \mathbb{R}_0^{dN_n} \tag{5.124}$$

where C_U is a constant independent of $\mathbf{W}$ and $\mathbf{U}$. Equation 5.124 shows the *continuity* of $\hat{a}(\mathbf{w}, \mathbf{v})$.

On the other hand, we also have

$$\hat{a}(\mathbf{v}, \mathbf{v}) = \int_{\Omega} \boldsymbol{\varepsilon}^{\mathrm{T}}(\mathbf{U}) \underbrace{\mathbf{V}_m^{\mathrm{T}} \Lambda_m \mathbf{V}_m}_{\mathbf{c}} \hat{\boldsymbol{\varepsilon}}(\mathbf{U}) d\Omega \geq \lambda_{\min}^m \int_{\Omega} \hat{\boldsymbol{\varepsilon}}^{\mathrm{T}}(\mathbf{U}) \underbrace{\mathbf{V}_m^{\mathrm{T}} \mathbf{V}_m}_{\mathbf{I}} \hat{\boldsymbol{\varepsilon}}(\mathbf{U}) d\Omega$$

$$= \lambda_{\min}^m \|\hat{\boldsymbol{\varepsilon}}^{\mathrm{T}}(\mathbf{U})\|_{\mathbb{L}^2(\Omega)}^2 \geq \underbrace{\frac{\lambda_{\max}^m}{c_{ca}^2}}_{C_{SL}} \|\tilde{\boldsymbol{\varepsilon}}_b(\mathbf{U})\|_{\mathbb{L}^2(\Omega)}^2 = C_{SL} \|\tilde{\boldsymbol{\varepsilon}}_b(\mathbf{U})\|_{\mathbb{L}^2(\Omega)}^2, \quad \forall \mathbf{U} \in \mathbb{R}_0^{dN_n} \quad (5.125)$$

where
   C_{SL} is a constant independent of $\mathbf{U}$
   $\lambda_{\min}^m$ is the maximum eigenvalue of $\mathbf{c}$

In the above derivation, we used Equation 5.118 in the second line. Using Equation 5.122 now, we have

$$\hat{a}(\mathbf{v}, \mathbf{v}) \geq C_{SL} \|\tilde{\boldsymbol{\varepsilon}}_b(\mathbf{U})\|_{\mathbb{L}^2(\Omega)}^2 \geq \underbrace{C_{SL} \lambda_{\min}^b}_{C_L} \|\mathbf{U}\|_{\mathbb{L}^2(\Omega)}^2, \quad \forall \mathbf{U} \in \mathbb{R}_0^{dN_n} \quad (5.126)$$

where C_L is a constant independent of $\mathbf{U}$. Equation 5.124 shows the ellipticity of the SC bilinear form.

With both *continuity* and *ellipticity*, we are sure that the solution will be stable and unique by the well-known Lax–Milgram theorem.

Finally, we need to show that the solution will converge to the exact solution of the strong form. This is ensured by the strain convergence condition given in Equation 4.16. When the mesh is refined, the constructed strain field approaches the strain field of the base model that is guaranteed to converge toward the exact solution. ∎

Remark 5.20: Alternative Base Models
The above proof implies that as long as the base model is stable and convergent, the SC-model will also be stable and convergent, as long as the norm equivalence and strain convergence conditions are satisfied. Therefore, any proven model can in fact be used as a base model to create SC models.

Remark 5.21: Strain-Constructed Minimum Total Potential Energy Principle
The energy principle associated with the SC-Galerkin can be termed as strain-constructed minimum total potential energy principle. Physically, it means that we are looking for an approximated solution that makes strain-constructed total energy potential minimum. Such a constructed strain energy potential should be equivalent to that of a compatible model based on the standard total potential energy principle, and the constructed strain field should converge to that of the compatible model when the model is refined.

Remark 5.22: SC-Galerkin: Variational Consistence

An SC-Galerkin model may or may not be variationally consistent in the usual sense, because it cannot be derived from the HW principle. It is a weakened-weak formulation as the assumed displacement functions are from a proper $\mathbb{G}_h^1$ space, and can be discontinuous. A variationally inconsistent SC-Galerkin model can perform better than variationally consistent ones, which is demonstrated in Chapter 8.

Remark 5.23: SC-Galerkin: Temporal Stability

A spatially stable model (no zero energy modes) does not necessarily ensure a temporal stability. The model can still have *spurious* modes at a higher energy level depending on how the strain field is constructed, which can result in instability in dynamic analysis.*

Remark 5.24: SC-Galerkin: No Additional DOFs

In an SC-Galerkin model, there are no additional unknowns compared to the standard Galerkin model.

5.15.1 GS-Galerkin Weak Form: A Special Case of SC-Galerkin

The GS-Galerkin weak form is a special case of the SC-Galerkin form when the strain is constructed using Equation 4.29 with a minimum number of linearly independent smoothing domains created in the way described in Section 3.3.3:

$$\hat{\boldsymbol{\varepsilon}}(\mathbf{u}) = \bar{\boldsymbol{\varepsilon}}(\mathbf{u}), \quad \mathbf{u} \in \mathbb{G}_{h,0}^1 \tag{5.127}$$

The discretized form of the potential energy for the generalized smoothed strain becomes

$$\bar{U}_{\mathrm{PE}}^D = \int_\Omega \frac{1}{2} \bar{\boldsymbol{\varepsilon}}^{\mathrm{T}} \mathbf{c}\bar{\boldsymbol{\varepsilon}} \, \mathrm{d}\xi = \frac{1}{2} \sum_{i=1}^{N_s} A_i^s \bar{\boldsymbol{\varepsilon}}_i^{\mathrm{T}} \mathbf{c}\bar{\boldsymbol{\varepsilon}}_i \tag{5.128}$$

We then have the following so-called GS-Galerkin weak form:

$$\int_\Omega \delta\bar{\boldsymbol{\varepsilon}}^{\mathrm{T}}(\mathbf{u})\mathbf{c}\bar{\boldsymbol{\varepsilon}}(\mathbf{u})\mathrm{d}\Omega - \int_\Omega \delta\mathbf{u}^{\mathrm{T}}\mathbf{b}\mathrm{d}\Omega - \int_{\Gamma_t} \delta\mathbf{u}^{\mathrm{T}}\mathbf{t}\mathrm{d}\Gamma = 0 \tag{5.129}$$

or

$$\sum_{i=1}^{N_s} A_i^s \delta\bar{\boldsymbol{\varepsilon}}_i^{\mathrm{T}} \mathbf{c}\bar{\boldsymbol{\varepsilon}}_i - \int_\Omega \delta\mathbf{u}^{\mathrm{T}}\mathbf{b}\mathrm{d}\Omega - \int_{\Gamma_t} \delta\mathbf{u}^{\mathrm{T}}\mathbf{t}\mathrm{d}\Gamma = 0 \tag{5.130}$$

---

* In this book, discussion on stability refers to spatial stability, unless specified.

For dynamic problems, we have

$$\int_\Omega \delta\bar{\boldsymbol\varepsilon}^T(\mathbf{u})\mathbf{c}\bar{\boldsymbol\varepsilon}(\mathbf{u})d\Omega - \int_\Omega \delta\mathbf{u}^T\mathbf{b}d\Omega - \int_{\Gamma_t} \delta\mathbf{u}^T\mathbf{t}d\Gamma + \int_\Omega \rho\delta\mathbf{u}^T\ddot{\mathbf{u}}d\Omega = 0 \qquad (5.131)$$

or

$$\sum_{i=1}^{N_s} A_i^s \delta\bar{\boldsymbol\varepsilon}_i^T \mathbf{c}\bar{\boldsymbol\varepsilon}_i - \int_\Omega \delta\mathbf{u}^T\mathbf{b}d\Omega - \int_{\Gamma_t} \delta\mathbf{u}^T\mathbf{t}d\Gamma + \int_\Omega \rho\delta\mathbf{u}^T\ddot{\mathbf{u}}d\Omega = 0 \qquad (5.132)$$

Theorem 5.4b: GS-Galerkin: A Variationally Consistent W^2 Formulation

If $\mathbf{u} \in \mathbb{H}_{h,0}^1$, the GS-Galerkin weak form is variationally consistent; it can be derived from the single-field Hellinger–Reissner principle.

Proof We substitute Equation 5.127 into Equation 5.62 to obtain

$$U_{\mathrm{SHR}}(\mathbf{u}) = \int_\Omega \left[-\frac{1}{2}\bar{\boldsymbol\varepsilon}^T(\mathbf{u})\mathbf{c}\bar{\boldsymbol\varepsilon}(\mathbf{u}) + \bar{\boldsymbol\varepsilon}^T(\mathbf{u})\mathbf{c}(\mathbf{L}_d\mathbf{u}) \right]d\Omega \qquad (5.133)$$

From Theorem 4.1, we now see that $\bar{\boldsymbol\varepsilon}$ satisfies the orthogonal condition (Equation 4.5), and hence Equation 5.133 becomes

$$U_{\mathrm{SHR}}(\mathbf{u}) = \int_\Omega \frac{1}{2}\bar{\boldsymbol\varepsilon}^T(\mathbf{u})\mathbf{c}\bar{\boldsymbol\varepsilon}(\mathbf{u})\,d\Omega \qquad (5.134)$$

which means that the single-field Hellinger–Reissner principle will lead to the GS-Galerkin weak form. ∎

Proving a GS-Galerkin is a workable model can also be done in a more simple, intuitive way. When $\mathbf{u} \in \mathbb{H}_{h,0}^1$, the displacement field is the same as in the standard Galerkin weak form. When the smoothing domains are created in the way described in Section 3.3.1, they are tied together with the element or the background cell mesh: when $h \to 0$, we have $\Omega_i^s \to 0$. Using the reproducing property (Remark 4.2), we know that $\bar{\boldsymbol\varepsilon}(\mathbf{u}) \to \tilde{\boldsymbol\varepsilon}(\mathbf{u}) \equiv \mathbf{L}_d\mathbf{u}$. This means that the GS-Galerkin model approaches the standard Galerkin model.

Theorem 5.5b: GS-Galerkin: A W^2 Formulation in a $\mathbb{G}_h^1$ Space

If $\mathbf{u} \in \mathbb{G}_{h,0}^1$, the GS-Galerkin weak form produces a stable and unique solution that convergences to the exact solution.

Proof When $\mathbf{u} \in \mathbb{G}_{h,0}^1$, Theorem 5.5b is an extended statement of Theorem 5.5a and has already been proven mathematically. Here, we again attempt to give an engineering proof by an energy argument (with some mathematical support), similar to what we have done for Remark 5.16. The energy argument is not mathematically rigorous, but should provide the essence on why the principle indeed works.

For a given $\mathbf{v} \in \mathbb{G}_{h,0}^1$ the strain energy potential is evaluated based on smoothing domains Ω_i^s. On the interface of the smoothing domains Γ_i^s these displacements are continuous (except at a few omissible points that the Lebesgue integration "forgives"). Therefore, there is no energy loss on Γ_i^s. Inside a smoothing domain Ω_i^s, the strain is measured as the generalized smoothed strain based on the given displacements $\mathbf{v}$ on Γ_i^s of the smoothing domain. Such a smoothed strain is overall conserving (see Remark 4.1). It is surely bounded because the given $\mathbf{v}$ on Γ_i^s is bounded and hence $\bar{\boldsymbol{\varepsilon}}(\mathbf{v})$ is obtained using Equation 4.29. This ensures the *continuity* of $\bar{U}_{\mathrm{PE}}^D(\mathbf{v})$: for any finite input of $\mathbf{v} \in \mathbb{G}_{h,0}^1$ there is a finite output of $\bar{U}_{\mathrm{PE}}^D(\mathbf{v})$.

The next thing that we may worry is the possibility of $\bar{U}_{\mathrm{PE}}^D = 0$ for some given nonzero $\mathbf{v} \in \mathbb{G}_{h,0}^1$. For $\bar{U}_{\mathrm{PE}}^D = 0$ to happen, $\bar{\boldsymbol{\varepsilon}}(\mathbf{v})$ in all the smoothing domains Ω_i^s has to be zero (see Equation 5.128). Now, because the smoothing domains are created in the way described in Section 3.3.1, and a minimum number of linearly independent smoothing domains are used (Section 3.3.3), to have $\bar{\boldsymbol{\varepsilon}}(\mathbf{v}) = 0$ in all the smoothing domains will surely lead to a zero $\mathbf{U}$ (the nodal value of $\mathbf{v}$). Since our $\mathbb{G}_h^1$ space is constructed using a basis of linearly independent shape functions, $\mathbf{v}$ has to be zero everywhere in Ω if $\mathbf{U} = 0$. This contradicts our earlier nonzero $\mathbf{v}$ assumption. Therefore, for any nonzero $\mathbf{v} \in \mathbb{G}_{h,0}^1$, we always have $\bar{U}_{\mathrm{PE}}^D > 0$. This implies that $\bar{U}_{\mathrm{PE}}^D(\mathbf{v})$ is an SPD (the symmetry is obvious from Equation 5.128).

With both *continuity* and the SPD property, we are sure that the solution will be stable, unique, and convergent by the well-known Lax–Milgram theorem.

Finally, we need to be sure that the solution will converge to the exact solution of the strong form, for which we use the same argument given in the proof of Theorem 5.7, because GS-Galerkin is a special case of SC-Galerkin. ∎

An Alternative Proof for Theorems 5.4 and 5.5

Using Theorem 5.7, we can quickly give an intuitive proof for Theorems 5.4 and 5.5. All we need to show is that the smoothed strain field using stationary smoothing domains satisfies (1) the norm equivalence condition (Equation 4.14) and (2) strain convergence condition (Equation 4.16).

First, from the fourth inequality (Equation 5.31), we have for all $\mathbf{w} \in \mathbb{G}_{h,0}^1$,

$$c_G \|\mathbf{w}\|_{\mathbb{G}^1(\Omega)} \le \|\bar{\boldsymbol{\varepsilon}}\|_{\mathbb{L}^2} \le \|\mathbf{w}\|_{\mathbb{G}^1(\Omega)} \tag{5.135}$$

meaning that $\|\bar{\boldsymbol{\varepsilon}}\|_{\mathbb{L}^2}$ and $\|\mathbf{w}\|_{\mathbb{G}^1(\Omega)}$ are equivalent. Using the second inequality (Equations 3.92 and 3.93), we then have

$$c_G c_{wd}^f \|\mathbf{U}\|_{\mathbb{L}^2(\Omega)} \le \|\bar{\boldsymbol{\varepsilon}}(\mathbf{U})\|_{\mathbb{L}^2} \le \frac{1}{c_{dw}^f} \|\mathbf{U}\|_{\mathbb{L}^2(\Omega)}, \quad \forall \mathbf{U} \in \mathbb{R}_0^{dN_n} \tag{5.136}$$

Now, since $\phi_i(\mathbf{x}), (i = 1, 2, \ldots, N_n)$ used to form $w \in \mathbb{G}_{h,0}^1$ are constructed using nodes selected using a T-scheme (see Section 1.7.6), they are always of an equal or higher order compared to the linear FEM, and we shall have $\mathbb{H}_b^1 \subseteq \mathbb{G}_h^1$ (see Equation 3.97) for point interpolation method (PIM) shape functions (including radial point interpolation method [RPIM] shape functions with linear polynomial basis). Therefore, we shall also have

$$c_G c_{wd}^f \|\mathbf{U}\|_{\mathbb{L}^2(\Omega)} \le \|\tilde{\boldsymbol{\varepsilon}}_b\|_{\mathbb{L}^2} \le \frac{1}{c_{dw}^f} \|\mathbf{U}\|_{\mathbb{L}^2(\Omega)}, \quad \forall \mathbf{U} \in \mathbb{R}_0^{dN_n} \tag{5.137}$$

Equations 5.136 and 5.137 show clearly that $\|\bar{\boldsymbol{\varepsilon}}(\mathbf{U})\|_{\mathbb{L}^2}$ and $\|\tilde{\boldsymbol{\varepsilon}}_b(\mathbf{U})\|_{\mathbb{L}^2}$ are equivalent, and therefore $\hat{\boldsymbol{\varepsilon}}(\mathbf{U}) = \bar{\boldsymbol{\varepsilon}}(\mathbf{U})$ satisfies Equations 5.117 and 5.118, and hence the norm equivalence condition (Equation 4.14).

Next, we need to show the convergence condition (Equation 4.16) for the smoothed strain:

$$\lim_{\substack{h \to 0 \\ N_n \to \infty}} \bar{\boldsymbol{\varepsilon}}(\mathbf{x}, \mathbf{U}) \to \tilde{\boldsymbol{\varepsilon}}_b(\mathbf{x}, \mathbf{U}), \quad \forall \mathbf{x} \in \Omega, \quad \forall \mathbf{U} \in \mathbb{R}_0^{dN_n} \tag{5.138}$$

Using Equation 4.29, for any $\mathbf{U} \in \mathbb{R}_0^{dN_n}$, at any point $\mathbf{x}$ in Ω_i^s, we shall have

$$\bar{\boldsymbol{\varepsilon}}(\mathbf{x}) - \tilde{\boldsymbol{\varepsilon}}_b(\mathbf{x}) = \frac{1}{A_i^s} \int_{\Gamma_i^s} \mathbf{L}_n \mathbf{u}_P(\mathbf{x}) d\Gamma - \frac{1}{A_i^s} \int_{\Gamma_i^s} \mathbf{L}_n \mathbf{u}_b(\mathbf{x}) d\Gamma = \frac{1}{A_i^s} \int_{\Gamma_i^s} \mathbf{L}_n (\mathbf{u}_P(\mathbf{x}) - \mathbf{u}_b(\mathbf{x})) d\Gamma \tag{5.139}$$

where
  $\mathbf{u}_P \in \mathbb{G}_0^1$ denotes the displacement functions constructed using PIM shape functions based on a T-scheme
  $\mathbf{u}_b \in \mathbb{H}_0^1$ denotes the displacement functions constructed using a linear FEM shape function based on the same set of triangular mesh

Because of the way that the smoothing domain is constructed (Section 3.3.3), $\mathbf{u}_P$ is always continuous at any $\mathbf{x}$ on Γ_i^s. We know that $\mathbf{u}_b$ is linear with respect to $\mathbf{x}$, and $\mathbf{u}_P$ is linear or of a higher order with respect to $\mathbf{x}$ in the local coordinate system defined for Ω_i^s. If $\mathbf{u}_P$ is linear, we immediately have $\bar{\boldsymbol{\varepsilon}}(\mathbf{x}) = \tilde{\boldsymbol{\varepsilon}}_b(\mathbf{x})$ which completes the proof. If $\mathbf{u}_P$ is of a higher order, we invoke the consistency property of PIM shape functions (Section 2.5), and thus the linear field $\mathbf{u}_b$ is reproduced. Therefore the difference between $\mathbf{u}_P(\mathbf{x})$ and $\mathbf{u}_b(\mathbf{x})$ should be at least of an order 2 with respect to $\mathbf{x}$:

$$\mathbf{u}_P(\mathbf{x}) - \mathbf{u}_b(\mathbf{x}) = \mathbf{O}(\mathbf{x}^2) \tag{5.140}$$

At the limit of $h \to 0$ and $N_n \to \infty$, we have $\Omega_i^s \to 0$ and $\mathbf{x} \to 0$, leading to $\mathbf{u}_P(\mathbf{x}) \to \mathbf{u}_b(\mathbf{x})$ and finally $\bar{\boldsymbol{\varepsilon}}(\mathbf{x}) \to \tilde{\boldsymbol{\varepsilon}}_b(\mathbf{x})$. This completes our proof. ∎

5.15.2 Galerkin-Like Weak Form

Galerkin-like weak form was formulated first in [14]. It is a special form resulted from the single-field Hellinger–Reissner principle, where the constructed strain can be written in the following form:

$$\hat{\boldsymbol{\varepsilon}}(\mathbf{u}) = \tilde{\boldsymbol{\varepsilon}}_b(\mathbf{u}) + \alpha \hat{\boldsymbol{\varepsilon}}_m(\mathbf{u}) \tag{5.141}$$

where
  $\mathbf{u} \in \mathbb{H}_h^1$, $\tilde{\boldsymbol{\varepsilon}}_b(\mathbf{u})$ is a (compatible) strain field of a base model
  $\alpha \in (-\infty, +\infty)$ is an adjustable parameter for regularizing the amount of modification
  $\hat{\boldsymbol{\varepsilon}}_m(\mathbf{u})$ is the modified portion of the strain field that satisfies the *zero-sum condition* defined in Section 4.4

The weak form is written as

$$\int_\Omega \delta \, (\hat{\boldsymbol{\varepsilon}}_b(\mathbf{u}) + \alpha\hat{\boldsymbol{\varepsilon}}_m(\mathbf{u}))^{\mathsf{T}} \mathbf{c}(\hat{\boldsymbol{\varepsilon}}_b(\mathbf{u}) - \alpha\hat{\boldsymbol{\varepsilon}}_m(\mathbf{u})) \mathrm{d}\Omega - \int_\Omega \delta\mathbf{u}^{\mathsf{T}}\mathbf{b}\mathrm{d}\Omega - \int_{\Gamma_t} \delta\mathbf{u}^{\mathsf{T}}\mathbf{t}\,\mathrm{d}\Gamma = 0 \qquad (5.142)$$

where the countersigns for the modified portion of the strain field play an important role in softening the model, which offers a possibility to create a model as soft as desired for upper-bound solutions. The variationally consistent αFEM (VCαFEM) [14] and the super-convergent αFEM (SαFEM) [30] are typical models created using the Galerkin-like weak form.

Theorem 5.8: Galerkin-Like: Variational Consistence

Equation 5.142 is variationally consistent and it can be derived from the single-field Hellinger–Reissner principle, as proven in [30]. It is as simple as and has all the required properties (symmetry, positivity, etc.) of the standard Galerkin weak form.

Remark 5.25: Galerkin-Like: Explicit Softening Effect
Equation 5.142 provides an important softening effect induced by the constructed strain, and the mount of the softening effect can be adjusted by tuning the "knob" α. Therefore, models can be created for both lower- and upper-bound solutions, and can also be tuned for close-to-exact or superconvergent solutions [30].

5.16 Parameterized Mixed Weak Form

We now move one more step further to present another very useful and important weak form that is, in general, also not variationally consistent: the parameterized mixed weak form. It has the following form:

$$\delta \left[\alpha \underbrace{\int_\Omega \frac{1}{2}\hat{\boldsymbol{\varepsilon}}^{\mathsf{T}}(\mathbf{u})\mathbf{c}\hat{\boldsymbol{\varepsilon}}(\mathbf{u})\mathrm{d}\Omega}_{\hat{U}_{\mathrm{PE}}} + (1-\alpha) \underbrace{\int_\Omega \left(-\frac{1}{2}\hat{\boldsymbol{\varepsilon}}^{\mathsf{T}}(\mathbf{u})\mathbf{c}\hat{\boldsymbol{\varepsilon}}(\mathbf{u}) + \hat{\boldsymbol{\varepsilon}}^{\mathsf{T}}(\mathbf{u})\mathbf{c}(\mathbf{L}_d\mathbf{u}) \right) \mathrm{d}\Omega}_{\hat{U}_{\mathrm{MHR}}} \right]$$

$$- \underbrace{\int_\Omega \delta\mathbf{u}^{\mathsf{T}}\mathbf{b}\mathrm{d}\Omega - \int_{\Gamma_t} \delta\mathbf{u}^{\mathsf{T}}\mathbf{t}\mathrm{d}\Gamma}_{W_{\mathrm{ext}}} = 0 \qquad (5.143)$$

where $\mathbf{u} \in \mathbb{H}_h^1$, α is an adjustable parameter that is finite real in general: $\alpha \in (-\infty, +\infty)$, and $\hat{\boldsymbol{\varepsilon}}$ is a constructed strain field using techniques discussed in Section 4.5. $\hat{U}_{PE}$ is the strain energy used in the variationally inconsistent SC-Galerkin weak form, and is given by

$$\hat{U}_{\mathrm{PE}} = \int_{\Omega} \frac{1}{2}\hat{\boldsymbol{\varepsilon}}^{\mathrm{T}}(\mathbf{u})\mathbf{c}\hat{\boldsymbol{\varepsilon}}(\mathbf{u})\mathrm{d}\Omega \tag{5.144}$$

$\hat{U}_{\mathrm{MHR}}$ is the strain energy used in the MHR variational principle, and is given by

$$\hat{U}_{\mathrm{MHR}} = \int_{\Omega} \left(-\frac{1}{2}\hat{\boldsymbol{\varepsilon}}^{\mathrm{T}}(\mathbf{u})\mathbf{c}\hat{\boldsymbol{\varepsilon}}(\mathbf{u}) + \hat{\boldsymbol{\varepsilon}}^{\mathrm{T}}(\mathbf{u})\mathbf{c}(\mathbf{L}_d\mathbf{u}) \right)\mathrm{d}\Omega \tag{5.145}$$

It is clear that when $\alpha = 1$, the mixed weak form becomes the variationally inconsistent SC-Galerkin weak form, and when $\alpha = 0$, it becomes the single-field Hellinger–Reissner weak form that is variationally consistent.

Remark 5.26: Parameterized Mixed Weak Form: Variational Consistence
The parameterized mixed weak form Equation 5.143 is not, in general, variationally consistent; it approaches to being variationally consistent when the mesh is refined. However, it can produce stable and convergent models.

This can be understood intuitively: if the solutions of both the SC-Galerkin model and the MHR model are all bounded and convergent, the solution of a model that is a linear combination of these two models will surely be bounded and hence convergent! It simply has no chance to go unbounded.

Use of the parameterized mixed weak form offers a very effective way to establish methods of good properties: both lower and upper bounds, superconvergence, ultra-accuracy in both displacement and stress, and good performance for triangular types of mesh. There is also a bonus: by adjusting α, one can have chances to obtain numerical solutions that are very close to the exact solution!

Application of the parameterized mixed weak form has produced a point interpolation method with a continuous strain field (PIM-CS) [28].

Finally, we state that the treatment of the constraint equations for the SC-Galerkin weak forms and the parameterized mixed weak form is largely the same as in the Galerkin weak form discussed in Section 5.13, and hence requires no further discussion.

5.17 Concluding Remarks

We finally make the following four remarks:

- The procedures used in applying the weak forms in meshfree methods can be different from those in FEM, because of the difference in the forms of the shape functions. The integration domains are often not element-based. The integration domains may no longer be the union of the element, and they may even overlap depending on the meshfree method used.

- The constrained principles or weak forms are used often for imposing additional constraints and links, and enforcing compatibilities (see, e.g., [29]). The same is applicable to meshfree methods.

- There still are a number theoretical issues related to W^2 formulation, which have not yet been resolved, such as the $\mathbb{G}$ dual spaces, regularity of the solution, allowable linear functionals, convergence rate of the W^2 solution, more sophisticated smoothing functions, etc. In other words, we only know it works very well, but do not yet know much of how well and to what extent. Help from mathematicians in this kind of theoretical work is needed.

- The extension of the Galerkin-like formulation to functions in $\mathbb{G}$ spaces is possible, but no much detailed work has been done so far. The same goes for the parameterized mixed weak forms.

In the following chapters, we use the Hamilton principle, the Galerkin weak form, the GS-Galerkin weak form, the more general SC-Galerkin weak form, and the local Petrov–Galerkin method to formulate various meshfree methods for mechanics problems of solids and structures.

References

1. Liu, G. R. and Gu, Y. T., *An Introduction to Meshfree Methods and Their Programming*, Springer, Dordrecht, the Netherlands, 2005.
2. Liu, G. R. and Kee, B. B. T., A stabilized least-squares radial point collocation method (LS-RPCM) for adaptive analysis, *Comput. Method Appl. Mech. Eng.*, 195, 4843–4861, 2006.
3. Kee, B. B. T., Liu, G. R., and Lu, C., A regularized least-squares radial point collocation method (RLS-RPCM) for adaptive analysis, *Comput. Mech.*, 40, 837–853, 2007.
4. Liu, G. R., Zhang J., and Lam, K. Y., A gradient smoothing method (GSM) with directional correction for solid mechanics problems, *Comput. Mech.*, 41, 457–472, 2008.
5. Liu, G. R. and Xu, X. G., A gradient smoothing method (GSM) for fluid dynamics problems, *Int. J. Numerical Methods Fluids*, 56(10), 1101–1133, 2008.
6. Liu, G. R. and Xu, X. G., An adaptive gradient smoothing method (GSM) for fluid dynamics problems, *Int. J. Numerical Methods Fluids*, 2008 (accepted).
7. Monaghan, J. J., Smoothed particle hydrodynamics, *Annu. Rev. Astronomy Astrophys.*, 30, 543–574, 1992.
8. Wu, H. C., *Variational Principle in Elasticity and Applications*, Scientific Press, Beijing, China, 1982.
9. Hellinger, E., Der allgermeine Ansatz der Mechanik der Kontinun, *Encycl. Math. Wiss.*, 4(Part 4), 602, 1914.
10. Reissner, E., On a variational theorem in elasticity, *J. Math. Phys.*, 29, 90, 1950.
11. Zienkiewicz, O. C. and Taylor, R. L., *The Finite Element Method*, 5th ed., Butterworth Heinemann, Oxford, 2000.
12. Hughes, T. J. R., *The Finite Element Method: Linear Static and Dynamic Finite Element Analysis*, Prentice-Hall, Englewood Cliffs, NJ, 1987.
13. Oliveira Eduardo, R. and De Arantes, E., Theoretical foundations of the finite element method, *Int. J. Solids Struct.*, 4, 929–952, 1968.
14. Liu, G. R., Nguyen-Xuan, H., and Nguyen-Thoi, T., A variationally consistent αFEM (VCαFEM) for solution bounds and ultra-accurate solutions, *Int. J. Numerical Methods Eng.*, 2008 (revised).
15. Chen, J. S., Wu, C. T., and Yoon, S. Y., A stabilized conforming nodal integration for Galerkin mesh-free methods, *Int. J. Numerical Methods Eng.*, 50, 435–466, 2001.

16. Liu, G. R., A generalized gradient smoothing technique and the smoothed bilinear form for Galerkin formulation of a wide class of computational methods, *Int. J. Comput. Methods*, 5(2), 199–236, 2008.
17. Liu, G. R., A $\mathbb{G}$ space theory and weakened weak (W^2) form for a unified formulation of compatible and incompatible displacement methods, *Int. J. Numerical Methods Eng.*, 2009 (in press).
18. Liu, G. R. and Zhang, G. Y., Upper bound solution to elasticity problems: A unique property of the linearly conforming point interpolation method (LC-PIM), *Int. J. Numerical Methods Eng.*, 2007.
19. Liu, G. R., Zhang, G. Y., Wang, Y. Y., Zhong, Z. H., Li, G. Y., and Han, X., A nodal integration technique for meshfree radial point interpolation method (NI-RPCM), *Int. J. Solids Struct.*, 44, 3040–3060, 2007.
20. Zhang, G. Y., Liu, G. R., Wang, Y. Y., Huang, H. T,. Zhong, Z. H., Li, G. Y., and Han, X., A linearly conforming point interpolation method (LC-PIM) for three-dimensional elasticity problems, *Int. J. Numerical Methods Eng.*, 72, 1524–1543, 2007.
21. Zhang, G. Y., Liu, G. R., Nguyen, T. T., Song, C. X., Han, X., Zhong, Z. H., and Li, G. Y., The upper bound property for solid mechanics of the linearly conforming radial point interpolation method (LC-RPIM), *Int. J. Comput. Methods*, 4(3), 521–541, 2007.
22. Liu, G. R. and Zhang, G. Y., A strain-constructed point interpolation method (SC-PIM) and strain field construction schemes for mechanics problems of solids and structures using triangular mesh, *Int. J. Solids Struct.*, 2008 (submitted).
23. Dai, K. Y., Liu, G. R., and Nguyen, T. T., An n-sided polygonal smoothed finite element method (nSFEM) for solid mechanics, *Finite Elem. Anal. Des.*, 43, 847–860, 2007.
24. Nguyen-Thoi, T., Liu, G. R., Nguyen-Xuan, H., and Lam, K. Y., An n-sided polygonal edge-based smoothed finite element method (nES-FEM) for solid mechanics, *Finite elements in analysis and design.* (submitted 2008).
25. Liu, G. R., Nguyen, T. T., Dai, K. Y., and Lam, K. Y., Theoretical aspects of the smoothed finite element method (SFEM), *Int. J. Numerical Methods Eng.*, 71, 902–930, 2007.
26. Liu, G. R., Nguyen-Thoi, T., Nguyen-Xuan, H., and Lam, K. Y., A node-based smoothed finite element method (NS-FEM) for upper bound solutions to solid mechanics problems, *Comput. Struct.*, 87, 14–26, 2009.
27. Liu, G. R., A generalized gradient smoothing technique and the smoothed bilinear form for Galerkin formulation of a wide class of computational methods, *Int. J. Comput. Methods*, 5(2), 199–236, 2008.
28. Liu, G. R., Xu, X., Zhang, G. Y., and Gu, Y. T., An extended Galerkin weak form and a point interpolation method with continous strain field and super convergence (PIM-CS)-using triangular mesh, *Computational Mechanics*, 2009, 43, 651–673, 2009.
29. Liu, G. R. and Quek, S. S., *The Finite Element Method: A Practical Course*, Butterworth Heinemann, Oxford, 2003.
30. Liu, G. R., Nguyen-Xuan, H., Nguyen, T. T., and Xu, X., A novel weak form and a super-convergent alpha finite element method (SαFEM) for mechanics problems using triangular meshes, *J. Comput. Phys.*, 228, 3911–4302, 2009.
31. Peraire, J., *Lecture Notes on Finite Element Methods for Elliptic Problems*, MIT, Cambridge, MA, 1999.
32. Strang, G. and Fix, G. J., *An Analysis of the Finite Element Method*, Prentice-Hall, Englewood Cliffs, NJ, 1973.
33. Liu, G. R. and Han, X., *Computational Inverse Techniques in Nondestructive Evaluation*, CRC Press, Boca Raton, FL, 2003.

6

Element-Free Galerkin Method

The element-free Galerkin (EFG) method is developed by Belytschko et al. [1] based on the diffuse elements method (DEM) originated by Nayroles et al. [2]. The major features of the DEM and the EFG method are as follows:

1. Moving least squares (MLS) approximation is employed for the construction of shape functions.
2. Galerkin weak form with constraints is employed to develop the discrete system equations.
3. Background cells are required to perform the numerical integration for computing system matrices.

This chapter introduces the EFG method with detailed formulations, procedure, applications, and some discussions. Technical issues, especially related to background cell integration, will also be examined. A typical benchmark problem of a rectangular cantilever is considered to illustrate the relationship between the density of the field nodes and the density of the global background mesh, as well as the number of integration sampling points. The findings and remarks are applicable to any meshfree method that requires background integration for strain energy.

Note that the EFG is a standard weak formulation that is variationally consistent due to the use of compatible MLS shape functions and the Galerkin approach with constraints to impose the essential boundary conditions. This fact will be evidenced by the examples of patch tests.

6.1 EFG Formulation with Lagrange Multipliers

6.1.1 Formulation

A two-dimensional (2D) linear solid mechanics problem is used to present the procedure of the EFG method in formulating discrete system equations. The partial differential equation and boundary conditions for a 2D solid mechanics problem can be written in the form:

$$\mathbf{L}_d^T \boldsymbol{\sigma} + \mathbf{b} = 0 \quad \text{Equilibrium equation in problem domain} \tag{6.1}$$

$$\mathbf{u} = \mathbf{u}_\Gamma \quad \text{on essential boundary } \Gamma_u \tag{6.2}$$

$$\mathbf{L}_n^T \boldsymbol{\sigma} = \mathbf{t}_\Gamma \quad \text{on natural boundary } \Gamma_t \tag{6.3}$$

where
$\mathbf{L}_d$ is the matrix differential operator defined by Equation 1.9
$\boldsymbol{\sigma}$ is the vector of the stress components defined by Equation 1.11

u is the vector of displacement components given by Equation 1.22
b is the body force vector containing force components
$\mathbf{t}_\Gamma$ is the prescribed traction on the natural (stress) boundaries
$\mathbf{u}_\Gamma$ is the prescribed displacement on the essential (displacement) boundaries
$\mathbf{L}_n$ is the matrix of the components of unit outward normal on the natural boundary
   defined by Equation 1.27

In the EFG method, the problem domain Ω is discretized by a set of nodes scattered in the problem domain and on the boundaries of the domain. The MLS approximation procedure described in Section 2.4 is then used to approximate the displacement field **u** at a point of interest within the problem domain using the *nodal parameters* of displacement at the nodes in the *support domain* of the point. The concept of the *nodal displacement parameter* comes from the fact that the displacement vector obtained by solving the discretized EFG system equation is not the actual displacement at the nodes. This is due to the lack of Kronecker delta function properties in the MLS shape function. We will pursue this issue in more detail later.

As discussed in Chapter 2 the MLS approximation is both consistent and compatible when sufficient number of local nodes is used; the Galerkin procedure can then be used to establish a set of discretized system equations for solving the displacement parameters. The displacement at any point (including the nodes) is retrieved using MLS approximation and the computed nodal displacement parameters. The strains at any point are also retrieved using the derivatives of the MLS shape functions and the nodal displacement parameters.

For the solid mechanics problem stated by Equations 6.1 through 6.3, the Galerkin weak form with Lagrange multipliers for constraints can be given by

$$\int_\Omega \delta(\mathbf{L}_d\mathbf{u})^\mathrm{T}(\mathbf{c}\mathbf{L}_d\mathbf{u})\mathrm{d}\Omega - \int_\Omega \delta\mathbf{u}^\mathrm{T}\mathbf{b}\mathrm{d}\Omega - \int_{\Gamma_t} \delta\mathbf{u}^\mathrm{T}\mathbf{t}_\Gamma\mathrm{d}\Gamma - \int_{\Gamma_u} \delta\boldsymbol{\lambda}^\mathrm{T}(\mathbf{u} - \mathbf{u}_\Gamma)\mathrm{d}\Gamma - \int_{\Gamma_u} \delta\mathbf{u}^\mathrm{T}\boldsymbol{\lambda}\,\mathrm{d}\Gamma = 0 \quad (6.4)$$

where $\mathbf{u} \in \mathbb{H}^1$. Equation 6.4 is formed using Equation 5.111, by changing the area integrals for the constraint-related two terms into curve integrals because the constraints (essential boundary conditions) given in Equation 6.2 are defined only on the boundary. The last two terms in Equation 6.4 are produced by the method of Lagrange multipliers for handling essential boundary conditions for cases when $\mathbf{u} - \mathbf{u}_\Gamma \neq 0$. The Lagrange multipliers $\boldsymbol{\lambda}$ here can be viewed physically as *smart forces* that can force $\mathbf{u} - \mathbf{u}_\Gamma = 0$. If the trial function **u** can be chosen so that $\mathbf{u} - \mathbf{u}_\Gamma = 0$, the last two terms will vanish.

The MLS approximation is now used to express both the trial and test functions at any point of interest **x** using the nodes in the support domain of the point **x**. For displacement component u, we have

$$u^h(\mathbf{x}) = \sum_{I \in S_n} \phi_I^H(\mathbf{x})u_I \qquad (6.5)$$

where S_n is the set of the nodes in the support domain of the point **x** for constructing the MLS shape function $\phi_I^H(\mathbf{x}) \in \mathbb{H}_h^1(\Omega)$. The procedure for constructing $\phi_I^H(\mathbf{x})$ is detailed in Section 2.4, and the formulation of $\phi_I^H(\mathbf{x})$ is given in Equation 2.62.

For displacement component v, we should also have

$$v^h(\mathbf{x}) = \sum_{I \in S_n} \phi_I^H(\mathbf{x})v_I \qquad (6.6)$$

Combining Equations 6.5 and 6.6, we obtain

$$\mathbf{u}^h = \begin{Bmatrix} u \\ v \end{Bmatrix}^h = \sum_{I \in S_n} \underbrace{\begin{bmatrix} \phi_I^H & 0 \\ 0 & \phi_I^H \end{bmatrix}}_{\phi_I^H} \underbrace{\begin{Bmatrix} u_I \\ v_I \end{Bmatrix}}_{\mathbf{u}_I} = \sum_{I \in S_n} \mathbf{\Phi}_I^H \mathbf{u}_I \tag{6.7}$$

where $\mathbf{\Phi}_I^H$ is the matrix of shape functions.

By using Equation 6.7, the compatible strain $\mathbf{L}_d \mathbf{u}^h$ becomes

$$\mathbf{L}_d \mathbf{u}^h = \mathbf{L}_d \sum_{I \in S_n} \mathbf{\Phi}_I^H \mathbf{u}_I = \sum_{I \in S_n} \mathbf{L}_d \mathbf{\Phi}_I^H \mathbf{u}_I = \sum_{I \in S_n} \begin{bmatrix} \dfrac{\partial}{\partial x} & 0 \\ 0 & \dfrac{\partial}{\partial y} \\ \dfrac{\partial}{\partial y} & \dfrac{\partial}{\partial x} \end{bmatrix} \begin{bmatrix} \phi_I^H & 0 \\ 0 & \phi_I^H \end{bmatrix} \mathbf{u}_I$$

$$= \sum_{I \in S_n} \underbrace{\begin{bmatrix} \phi_{I,x}^H & 0 \\ 0 & \phi_{I,y}^H \\ \phi_{I,y}^H & \phi_{I,x}^H \end{bmatrix}}_{\mathbf{B}_I} \mathbf{u}_I = \sum_{I \in S_n} \mathbf{B}_I \mathbf{u}_I \tag{6.8}$$

where

$\phi_{I,x}^H$ and $\phi_{I,y}^H$ represent the derivatives of the MLS shape function with respect to x and y, respectively

$\mathbf{B}_I$ is the *strain matrix* for node I

The use of MLS shape functions in Equations 6.5 and 6.6 leads to $\mathbf{u} - \mathbf{u}_\Gamma \neq 0$ on the essential boundary, and the last two terms in Equation 6.4 are hence needed. The following equations explain why, using detailed formulation.

As described in Chapter 2, the use of MLS approximation produces shape functions $\phi_I^H(\mathbf{x})$ that do not possess the Kronecker delta function property, i.e.,

$$\phi_I^H(\mathbf{x}_J) \neq \delta_{IJ} \tag{6.9}$$

This feature of the MLS shape function results in

$$u^h(\mathbf{x}_I) = \sum_{I \in S_n} \phi_I^H(\mathbf{x}_I) \quad u_I \neq u_I \tag{6.10}$$

$$v^h(\mathbf{x}_I) = \sum_{I \in S_n} \phi_I^H(\mathbf{x}_I) \quad v_I \neq v_I \tag{6.11}$$

This implies that the essential boundary condition (Equation 6.2) cannot be exactly satisfied via enforcing

$$\mathbf{u}_I = \mathbf{u}_{\Gamma I} \quad \text{for node } I \text{ on } \Gamma_u \tag{6.12}$$

because what we really need to enforce is

$$\mathbf{u}(\mathbf{x}_I) = \mathbf{u}_{\Gamma I} \quad \text{for node } I \text{ on } \Gamma_u \tag{6.13}$$

Therefore, the fourth and fifth terms in Equation 6.4 are required to enforce the boundary conditions at the essential boundary Γ_u.

The Lagrange multiplier λ in Equation 6.4 is an unknown function of the coordinates, which needs also to be treated as a field variable and interpolated using the nodes on the essential boundaries to obtain a set of discrete system equations, i.e.,

$$\lambda(\mathbf{x}) = \sum_{I \in S_\lambda} N_I(s)\lambda_I \quad \mathbf{x} \in \Gamma_u \tag{6.14}$$

where
   S_λ is the set of nodes used for this interpolation
   s is the curvilinear coordinate along the essential boundary
   λ_I is the Lagrange multiplier at node I on the essential boundary
   $N_I(s)$ can be a Lagrange interpolant used in the conventional finite element method (FEM)

The Lagrange interpolant of order r can be given in a general form of

$$N_k^r(s) = \frac{(s - s_0)(s - s_1) \cdots (s - s_{k-1})(s - s_{k+1}) \cdots (s - s_r)}{(s_k - s_0)(s_k - s_1) \cdots (s_k - s_{k-1})(s_k - s_{k+1}) \cdots (s_k - s_r)} \tag{6.15}$$

If we choose a first-order Lagrange interpolant, we have $r = 1$ and the Lagrange interpolants at point $s = s_0$ and $s = s_1$ will be

$$N_0(s) = \frac{(s - s_1)}{(s_0 - s_1)}, \quad N_1(s) = \frac{(s - s_0)}{(s_1 - s_0)} \tag{6.16}$$

In this case, the Lagrange multiplier at s is interpolated using two bound nodes of s. A higher order Lagrange interpolant can also be used with more nodes on the boundary. From Equation 6.14, the variation of Lagrange multiplier, $\delta\lambda$, can be obtained as

$$\delta\lambda(\mathbf{x}) = \sum_{I \in S_\lambda} N_I(s)\delta\lambda_I \quad \mathbf{x} \in \Gamma_u \tag{6.17}$$

The vector of Lagrange multipliers in Equation 6.4 can be written in the matrix form:

$$\lambda = \sum_{I \in S_\lambda} \underbrace{\begin{bmatrix} N_I & \\ & N_I \end{bmatrix}}_{\mathbf{N}_I} \underbrace{\begin{Bmatrix} \lambda_{uI} \\ \lambda_{vI} \end{Bmatrix}}_{\lambda_I} = \sum_{I \in S_\lambda} \mathbf{N}_I \lambda_I \tag{6.18}$$

where N_I is the Lagrange interpolant for node I on the essential boundary.

Substituting Equations 6.7 and 6.8 into Equation 6.4, we have

$$
\int_\Omega \delta \left(\sum_{I \in S_n} \mathbf{B}_I \mathbf{u}_I \right)^{\mathrm{T}} \left(\mathbf{c} \sum_{J \in S_n} \mathbf{B}_J \mathbf{u}_J \right) d\Omega - \int_\Omega \delta \left(\sum_{I \in S_n} \mathbf{\Phi}_I^H \mathbf{u}_I \right)^{\mathrm{T}} \mathbf{b} \, d\Omega - \int_{\Gamma_t} \delta \left(\sum_{I \in S_n} \mathbf{\Phi}_I^H \mathbf{u}_I \right)^{\mathrm{T}} \mathbf{t}_\Gamma d\Gamma
$$

$$
- \int_{\Gamma_u} \delta \boldsymbol{\lambda}^{\mathrm{T}} \left(\left(\sum_{I \in S_n} \mathbf{\Phi}_I^H \mathbf{u}_I \right) - \mathbf{u}_\Gamma \right) d\Gamma - \int_{\Gamma_u} \delta \left(\sum_{I \in S_n} \mathbf{\Phi}_I^H \mathbf{u}_I \right)^{\mathrm{T}} \boldsymbol{\lambda} d\Gamma = 0 \tag{6.19}
$$

Notice that we use a different summation index for the term in the second bracket in the first integral term, to distinguish it from that in the first bracket. First, let us look at the first term in the above equation.

$$
\int_\Omega \delta \left(\sum_{I \in S_n} \mathbf{B}_I \mathbf{u}_I \right)^{\mathrm{T}} \left(\mathbf{c} \sum_{J \in S_n} \mathbf{B}_J \mathbf{u}_J \right) d\Omega = \int_\Omega \delta \left(\sum_{I \in S_n} \mathbf{u}_I^{\mathrm{T}} \mathbf{B}_I^{\mathrm{T}} \right) \left(\mathbf{c} \sum_{J \in S_n} \mathbf{B}_J \mathbf{u}_J \right) d\Omega \tag{6.20}
$$

Note that the summation, variation, and integration operate on different variables, and therefore they are exchangeable (see the arguments given in Chapter 5.). Hence, we can have

$$
\int_\Omega \delta \left(\sum_{I \in S_n} \mathbf{u}_I^{\mathrm{T}} \mathbf{B}_I^{\mathrm{T}} \right) \left(\mathbf{c} \sum_{J \in S_n} \mathbf{B}_J \mathbf{u}_J \right) d\Omega = \sum_{I \in S_n} \sum_{J \in S_n} \delta \mathbf{u}_I^{\mathrm{T}} \underbrace{\int_\Omega \mathbf{B}_I^{\mathrm{T}} \mathbf{c} \mathbf{B}_J d\Omega}_{\mathbf{K}_{IJ}} \mathbf{u}_J
$$

$$
= \sum_I^{n_t} \sum_J^{n_t} \delta \mathbf{u}_I^{\mathrm{T}} \mathbf{K}_{IJ} \mathbf{u}_J \tag{6.21}
$$

In the above equation, we made two changes. The first is the substitution of $\mathbf{K}_{IJ}$, which is a 2×2 matrix called as *nodal stiffness matrix* in this book. Note that the integration is over the entire problem domain, and therefore the union of all the sets S_n becomes the total number of nodes in the problem domain. Therefore, the summation limits have to be changed to n_t, which is the total number of nodes in the entire problem domain. Note also that the last summation in Equation 6.21 is a matrix assembly. To view this, we expand the summation and then group the components into matrix form as follows:

$$
\sum_I^{n_t} \sum_J^{n_t} \delta \mathbf{u}_I^{\mathrm{T}} \mathbf{K}_{IJ} \mathbf{u}_J = \delta \mathbf{u}_1^{\mathrm{T}} \mathbf{K}_{11} \mathbf{u}_1 + \delta \mathbf{u}_1^{\mathrm{T}} \mathbf{K}_{12} \mathbf{u}_2 + \cdots + \delta \mathbf{u}_1^{\mathrm{T}} \mathbf{K}_{1n_t} \mathbf{u}_{n_t}
$$

$$
+ \delta \mathbf{u}_2^{\mathrm{T}} \mathbf{K}_{21} \mathbf{u}_1 + \delta \mathbf{u}_2^{\mathrm{T}} \mathbf{K}_{22} \mathbf{u}_2 + \cdots + \delta \mathbf{u}_2^{\mathrm{T}} \mathbf{K}_{2n_t} \mathbf{u}_{n_t}
$$

$$
\vdots
$$

$$
+ \delta \mathbf{u}_{n_t}^{\mathrm{T}} \mathbf{K}_{n_t 1} \mathbf{u}_1 + \delta \mathbf{u}_{n_t}^{\mathrm{T}} \mathbf{K}_{n_t 2} \mathbf{u}_2 + \cdots + \delta \mathbf{u}_{n_t}^{\mathrm{T}} \mathbf{K}_{n_t n_t} \mathbf{u}_{n_t}
$$

$$
= \delta \mathbf{U}^{\mathrm{T}} \mathbf{K} \mathbf{U} \tag{6.22}
$$

where $\mathbf{K}$ is the global stiffness matrix assembled using the nodal stiffness matrices in the form:

$$\mathbf{K} = \begin{bmatrix} \mathbf{K}_{11} & \mathbf{K}_{12} & \cdots & \mathbf{K}_{1n_t} \\ \mathbf{K}_{21} & \mathbf{K}_{22} & \cdots & \mathbf{K}_{2n_t} \\ \vdots & \vdots & \ddots & \vdots \\ \mathbf{K}_{n_t 1} & \mathbf{K}_{n_t 2} & \cdots & \mathbf{K}_{n_t n_t} \end{bmatrix} \tag{6.23}$$

The dimension matrix $\mathbf{K}$ should be $(2n_t) \times (2n_t)$, because $\mathbf{K}_{IJ}$ is 2×2.

Vector $\mathbf{U}$ is the *global displacement parameter vector* that collects the nodal parameter vectors of displacement at all nodes in the entire problem domain, which has the form:

$$\mathbf{U} = \begin{Bmatrix} \mathbf{u}_1 \\ \mathbf{u}_2 \\ \vdots \\ \mathbf{u}_{n_t} \end{Bmatrix} \tag{6.24}$$

where $\mathbf{u}_I$ is the *nodal displacement parameter vector* at node I, i.e.,

$$\mathbf{u}_I = \begin{Bmatrix} u_I \\ v_I \end{Bmatrix} \tag{6.25}$$

The length of vector $\mathbf{U}$ should be $(2n_t)$.

Next, let us examine the second term in Equation 6.19.

$$\int_\Omega \delta \left(\sum_{I \in S_n} \mathbf{\Phi}_I^H \mathbf{u}_I \right)^\mathrm{T} \mathbf{b} \, d\Omega = \sum_{I \in S_n} \delta \mathbf{u}_I^\mathrm{T} \underbrace{\int_\Omega \left(\mathbf{\Phi}_I^H \right)^\mathrm{T} \mathbf{b} \, d\Omega}_{\mathbf{f}_I} = \sum_I^{n_t} \delta \mathbf{u}_I^\mathrm{T} \mathbf{f}_I \tag{6.26}$$

In the above equation, we made two changes. The first is a substitution of

$$\mathbf{f}_I = \int_\Omega \left(\mathbf{\Phi}_I^H \right)^\mathrm{T} \mathbf{b} \, d\Omega \tag{6.27}$$

where $\mathbf{f}_I$ is called the *nodal force vector*. Note again that the integration is over the entire problem domain, and therefore the summations have to be changed for all the nodes in the problem domain, which is the second change in Equation 6.26. The last summation in Equation 6.26 can be expanded and then form a product of matrices as follows:

$$\sum_I^{n_t} \delta \mathbf{u}_I^\mathrm{T} \mathbf{f}_I = \delta \mathbf{u}_1^\mathrm{T} \mathbf{f}_1 + \delta \mathbf{u}_2^\mathrm{T} \mathbf{f}_2 + \cdots + \delta \mathbf{u}_{n_t}^\mathrm{T} \mathbf{f}_{n_t} = \delta \mathbf{U}^\mathrm{T} \mathbf{F} \tag{6.28}$$

Vector **F** in Equation 6.28 is the *global force vector*, which collects force vectors at all the nodes in the problem domain and has the form:

$$\mathbf{F} = \begin{Bmatrix} \mathbf{f}_1 \\ \mathbf{f}_2 \\ \vdots \\ \mathbf{f}_{n_t} \end{Bmatrix} \tag{6.29}$$

where $\mathbf{f}_I$ is the nodal force vector at node I calculated by Equation 6.27, and consists of two components arranged as

$$\mathbf{f}_I = \begin{Bmatrix} f_{xI} \\ f_{yI} \end{Bmatrix} \tag{6.30}$$

where f_{xI} and f_{yI} are two components of nodal force in the x- and y-directions. The length of vector **F** should be $(2n_t)$.

The treatment for the third term in Equation 6.19 is exactly the same as that for the second term, except that the body force vector is replaced by the traction vector on the natural boundary, and the area integration is accordingly changed to the curve integration. Therefore, the additional nodal force vector can be given as

$$\mathbf{f}_I = \int_{\Gamma_t} \left(\mathbf{\Phi}_I^H \right)^{\mathrm{T}} \mathbf{t}_\Gamma d\Gamma \tag{6.31}$$

The force vector, therefore, receives contributions from both the external body force and the external force applied on the natural boundaries.

Before we examine the fifth term, let us look at the last term in Equation 6.19.

$$\int_{\Gamma_u} \delta \left(\sum_{I \in S_n} \mathbf{\Phi}_I^H \mathbf{u}_I \right)^{\mathrm{T}} \lambda d\Gamma = \int_{\Gamma_u} \delta \left(\sum_{I \in S_n} \mathbf{\Phi}_I^H \mathbf{u}_I \right)^{\mathrm{T}} \left(\sum_{J \in S_\lambda} \mathbf{N}_J \lambda_J \right) d\Gamma$$

$$= \sum_{I \in S_n} \sum_{J \in S_\lambda} \delta \mathbf{u}_I^{\mathrm{T}} \underbrace{\int_{\Gamma_u} \left(\mathbf{\Phi}_I^H \right)^{\mathrm{T}} \mathbf{N}_J d\Gamma}_{-\mathbf{G}_{IJ}} \lambda_J$$

$$= -\sum_I^{n_t} \sum_J^{n_{\lambda t}} \delta \mathbf{u}_I^{\mathrm{T}} \mathbf{G}_{IJ} \lambda_J = -\delta \mathbf{U}^{\mathrm{T}} \mathbf{G} \lambda \tag{6.32}$$

where
  $n_{\lambda t}$ is the total number of nodes on the essential boundary
  G is also a global matrix formed by assembling its *nodal matrix* $\mathbf{G}_{IJ}$

Note that the dimension of matrix $\mathbf{G}_{IJ}$ is also 2×2, but it concerns only the nodes on the essential boundaries. The dimension of matrix **G** should be $(2n_t) \times (2n_{\lambda t})$.

Finally, let us examine the fourth term in Equation 6.19. Using Equations 6.7 and 6.18, we have

$$
\int_{\Gamma_u} \delta\boldsymbol{\lambda}^{\mathrm{T}}\left(\left(\sum_{J\in S_n}\boldsymbol{\Phi}_J^H\mathbf{u}_J\right)-\mathbf{u}_\gamma\right)\mathrm{d}\Gamma = \int_{\Gamma_u}\delta\left(\sum_{I\in S_\lambda}\mathbf{N}_I\boldsymbol{\lambda}_I\right)^{\mathrm{T}}\sum_{J\in S_n}\boldsymbol{\Phi}_J^H\mathbf{u}_J\mathrm{d}\Gamma - \int_{\Gamma_u}\delta\left(\sum_{I\in S_\lambda}\mathbf{N}_I\boldsymbol{\lambda}_I\right)^{\mathrm{T}}\mathbf{u}_\Gamma\mathrm{d}\Gamma
$$

$$
= \sum_{I\in S_\lambda}\sum_{J\in S_n}\delta\boldsymbol{\lambda}_I^{\mathrm{T}}\underbrace{\int_{\Gamma_u}\mathbf{N}_I^{\mathrm{T}}\boldsymbol{\Phi}_J^H\mathrm{d}\Gamma}_{-\mathbf{G}_{IJ}}\mathbf{u}_J - \sum_{I\in S_\lambda}\delta\boldsymbol{\lambda}_I^{\mathrm{T}}\underbrace{\int_{\Gamma_u}\mathbf{N}_I^{\mathrm{T}}\mathbf{u}_\Gamma\mathrm{d}\Gamma}_{-\mathbf{q}_I}
$$

$$
= -\sum_I^{n_{\lambda t}}\sum_J^{n_t}\delta\boldsymbol{\lambda}_I^{\mathrm{T}}\mathbf{G}_{IJ}^{\mathrm{T}}\mathbf{u}_J + \sum_I^{n_{\lambda t}}\delta\boldsymbol{\lambda}_I^{\mathrm{T}}\mathbf{q}_I
$$

$$
= -\delta\boldsymbol{\lambda}^{\mathrm{T}}\mathbf{G}^{\mathrm{T}}\mathbf{U} + \delta\boldsymbol{\lambda}^{\mathrm{T}}\mathbf{q} \tag{6.33}
$$

where matrix $\mathbf{G}$ is defined by Equations 6.32 and 6.40. The vector $\mathbf{q}$ in Equation 6.33 is of the length ($2n_{\lambda t}$), and is assembled using nodal vector $\mathbf{q}_I$.

Finally, using Equations 6.22, 6.26, 6.28, 6.32, and 6.33, we obtain

$$
\underbrace{\int_\Omega\delta\left(\sum_{I\in S_n}\mathbf{B}_I\mathbf{u}_I\right)^{\mathrm{T}}\left(\mathbf{c}\sum_{J\in S_n}\mathbf{B}_J\mathbf{u}_J\right)\mathrm{d}\Omega}_{\delta\mathbf{U}^{\mathrm{T}}\mathbf{K}\mathbf{U}} - \underbrace{\int_\Omega\delta\left(\sum_{I\in S_n}\boldsymbol{\Phi}_I^H\mathbf{u}_I\right)^{\mathrm{T}}\mathbf{b}\mathrm{d}\Omega - \int_{\Gamma_t}\delta\left(\sum_{I\in S_n}\boldsymbol{\Phi}_I^H\mathbf{u}_I\right)^{\mathrm{T}}\mathbf{t}_\Gamma\mathrm{d}\Gamma}_{\delta\mathbf{U}^{\mathrm{T}}\mathbf{F}}
$$

$$
\underbrace{-\int_{\Gamma_u}\delta\boldsymbol{\lambda}^{\mathrm{T}}\left(\left(\sum_{I\in S_n}\boldsymbol{\Phi}_I^H\mathbf{u}_I\right)-\mathbf{u}_\Gamma\right)\mathrm{d}\Gamma}_{\delta\boldsymbol{\lambda}^{\mathrm{T}}[\mathbf{G}^{\mathrm{T}}\mathbf{U}-\mathbf{q}]} \underbrace{-\int_{\Gamma_u}\delta\left(\sum_{I\in S_n}\boldsymbol{\Phi}_I^H\mathbf{u}_I\right)^{\mathrm{T}}\boldsymbol{\lambda}\,\mathrm{d}\Gamma}_{\delta\mathbf{U}^{\mathrm{T}}\mathbf{G}\boldsymbol{\lambda}} = 0 \tag{6.34}
$$

which is

$$
\delta\mathbf{U}^{\mathrm{T}}[\mathbf{K}\mathbf{U} + \mathbf{G}\boldsymbol{\lambda} - \mathbf{F}] + \delta\boldsymbol{\lambda}^{\mathrm{T}}[\mathbf{G}^{\mathrm{T}}\mathbf{U} - \mathbf{q}] = 0 \tag{6.35}
$$

Because $\delta\mathbf{U}$ and $\delta\boldsymbol{\lambda}$ are arbitrary, the above equation can be satisfied only if

$$
\begin{aligned}
\mathbf{K}\mathbf{U} + \mathbf{G}\boldsymbol{\lambda} - \mathbf{F} &= 0 \\
\mathbf{G}^{\mathrm{T}}\mathbf{U} - \mathbf{q} &= 0
\end{aligned} \tag{6.36}
$$

The above two equations can be written in the following matrix form:

$$
\begin{bmatrix} \mathbf{K} & \mathbf{G} \\ \mathbf{G}^{\mathrm{T}} & 0 \end{bmatrix}\begin{Bmatrix} \mathbf{U} \\ \boldsymbol{\lambda} \end{Bmatrix} = \begin{Bmatrix} \mathbf{F} \\ \mathbf{q} \end{Bmatrix} \tag{6.37}
$$

This is the final discrete system equation for the entire problem domain. We now summarize all the nodal matrices and vectors that form Equation 6.37 for easy reference later.

$$
\mathbf{K}_{IJ} = \int_\Omega\mathbf{B}_I^{\mathrm{T}}\mathbf{c}\mathbf{B}_J\mathrm{d}\Omega \tag{6.38}
$$

$$\mathbf{B}_I = \mathbf{L}\boldsymbol{\phi}_I^H = \begin{bmatrix} \phi_{I,x}^H & 0 \\ 0 & \phi_{I,y}^H \\ \phi_{I,y}^H & \phi_{I,x}^H \end{bmatrix} \tag{6.39}$$

$$\mathbf{G}_{IJ} = -\int_{\Gamma_u} \mathbf{N}_I^{\mathrm{T}} \boldsymbol{\Phi}_J^H \mathrm{d}\Gamma \tag{6.40}$$

$$\boldsymbol{\Phi}_I^H = \begin{bmatrix} \phi_I^H & 0 \\ 0 & \phi_I^H \end{bmatrix} \tag{6.41}$$

$$\mathbf{N}_I = \begin{bmatrix} N_I & 0 \\ 0 & N_I \end{bmatrix} \tag{6.42}$$

$$\mathbf{F}_I = \int_{\Omega} \left(\boldsymbol{\Phi}_I^H\right)^{\mathrm{T}} \mathbf{b} \mathrm{d}\Omega + \int_{\Gamma_t} \left(\boldsymbol{\Phi}_I^H\right)^{\mathrm{T}} \mathbf{t}_\Gamma \mathrm{d}\Gamma \tag{6.43}$$

$$\mathbf{q}_I = -\int_{\Gamma_u} \mathbf{N}_I^{\mathrm{T}} \mathbf{u}_\Gamma \mathrm{d}\Gamma \tag{6.44}$$

The *nodal stiffness matrix* $\mathbf{K}_{IJ}$ defined in Equation 6.38 is the basic component for assembling the global stiffness matrix of EFG.

By using the symmetric property of $\mathbf{c}$ matrix, it is obvious that

$$[\mathbf{K}_{IJ}]^{\mathrm{T}} = \int_{\Omega} \left[\mathbf{B}_I^{\mathrm{T}} \mathbf{c} \mathbf{B}_J\right]^{\mathrm{T}} \mathrm{d}\Omega = \int_{\Omega} \left[\mathbf{B}_J^{\mathrm{T}} \mathbf{c}^{\mathrm{T}} \mathbf{B}_I\right] \mathrm{d}\Omega = \int_{\Omega} \left[\mathbf{B}_J^{\mathrm{T}} \mathbf{c} \mathbf{B}_I\right] \mathrm{d}\Omega = \mathbf{K}_{JI} \tag{6.45}$$

Therefore $\mathbf{K}$ given in Equation 6.23 is symmetric.

From Equation 6.38, it is shown that the nodal stiffness matrix $\mathbf{K}_{IJ}$ contains the stiffness coefficients between nodes I and J evaluated at a point in the problem domain. It is a function of coordinates, and needs to be integrated over the entire problem domain. It needs to be assembled to the global stiffness matrix $\mathbf{K}$, as long as the nodes I and J are covered by the support domain of at least one quadrature point. If nodes I and J are far apart and they do not share the same support domain of any quadrature point, $\mathbf{K}_{IJ}$ vanishes, and thus there is no need to compute. Therefore, as long as the support domain is compact and does not cover too wide the problem domain, many $\mathbf{K}_{IJ}$ in Equation 6.23 will be zero matrix, and the global stiffness matrix $\mathbf{K}$ will be sparse. If the nodes are properly numbered, $\mathbf{K}$ will also be banded. We discuss briefly in Chapter 15 how to reduce the bandwidth of $\mathbf{K}$ by optimizing the numbering of the nodes.

From Equations 6.31, 6.38, 6.40, and 6.44 it is evident that there is a need to perform the integration over the problem domain and the curve integrations for both natural and essential boundaries. These integrations have to be carried out via numerical quadrature techniques, and the Gauss quadrature scheme is most often used. In using the numerical quadrature scheme, a background mesh of cells is required for the integration. The background mesh is similar to the mesh used in FEM and no overlap or gap is permitted. It is used merely for the integration of the system matrices, and not for field variable interpolation. In principle, the background mesh can be totally independent of the arrangement of nodes. The only consideration in designing the cells of the background mesh is to ensure the accuracy of integration for the system matrices.

The matrices in Equation 6.37 can be much larger than the stiffness matrix in FEM, because of the presence of matrix **G** produced by the essential boundary conditions. Depending on the number of nodes on the essential boundaries in relation to the total number of nodes in the problem domain, solution efficiency can be reduced significantly. From Equation 6.37, it can also be clearly seen that the matrix of the enlarged system is still symmetric but no longer positive definite, which further reduces computational efficiency in solving the system equations.

In the later sections of this chapter, we introduce alternative methods for enforcing essential boundary conditions, which lead to system equations with matrices of the same size as in conventional FEM; the system matrix will also remain positive definite.

6.1.2 EFG Procedure

The solution procedure of the EFG method is similar to that in FEM. The geometry of the problem domain is firstly modeled, and a set of nodes is generated to discretize the problem domain, as shown in Figure 6.1. The system matrices are assembled via two loops. The outer loop is for all the cells of the background mesh, and the inner loop is for all the Gauss quadrature points in a cell. The flowchart of the algorithm for stress analysis using the EFG method is presented in Figure 6.2.

6.1.3 Background Integration

This section deals with the domain integration issues in meshfree methods. The discussion here on the integration issues is based on the work by Liu and Yan [3].

In either FEM or EFG, numerical integration is a time-consuming process required for the computation of the stiffness matrix that is established based on the weak forms. The problem domain is divided into cells to carry out numerical integrations. In FEM, the integration cells are the same as the elements. The density of the element mesh controls the accuracy of the field approximation, and the number of integration points in each

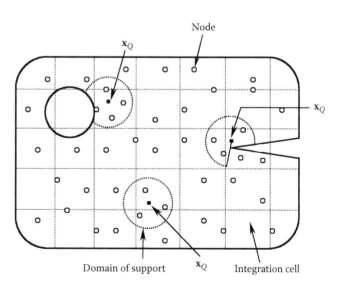

FIGURE 6.1
Meshfree model for EFG method with background mesh of cells for integration.

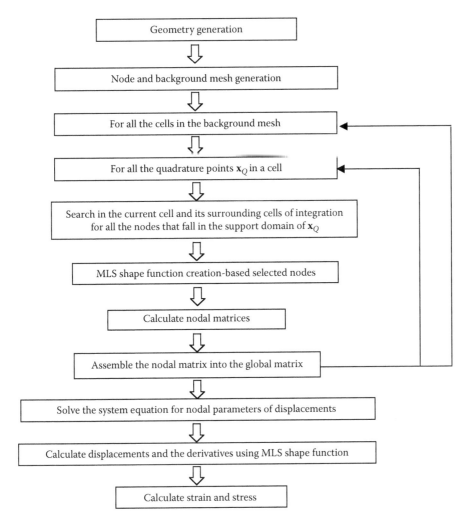

FIGURE 6.2
Flowchart of EFG method.

element controls the accuracy of the integration. In EFG, however, the background mesh is required only in performing the integration of computing the stiffness matrix. Therefore, a background mesh with proper density needs to be designed to evaluate the integrals for the desired accuracy. This can only be done after performing a detailed investigation to reveal the relationship between the density of the field nodes and the density of the background mesh. The first thing that needs to be addressed is the minimum number of integration points when numerical integration is adopted.

Zienkiewicz [4] has shown for FEM that if the number of independent relations provided by all integration points is less than the number of unknowns (displacements at all points in the element), the stiffness matrix **K** must be singular. This concept should also be applicable, in principle, to EFG. For a 2D problem, the number of unknown variables N_u should be

$$N_u = 2 \times n_t - n_f \tag{6.46}$$

where n_t and n_f are the node numbers in domain Ω and the number of constrained degrees of freedoms, respectively.

In evaluating the integrand at each quadrature (integration) point, three independent strain relations are used. Therefore, the number of independent equations used in all the quadrature points, N_Q, is

$$N_Q = 3 \times n_Q \tag{6.47}$$

where n_Q is the number of total quadrature points in domain Ω. Therefore, N_Q must be larger than N_u to avoid the singularity in the solution, and the minimum number of quadrature points must be greater than $N_u/3$. In other words, the total number of quadrature points n_Q should be at least two-thirds of the total number of unconstrained field nodes in the problem domain, i.e.,

$$N_Q > N_u \approx 2n_t \quad \text{or} \quad n_Q > \frac{2}{3}n_t \quad \text{for 2D problems} \tag{6.48}$$

Note also that this rule is only a necessary requirement, not necessarily a sufficient requirement. In addition, this requirement is only about the singularity aspect of **K**. The accuracy in the evaluation of **K** is a separate matter that requires more detailed analysis. The following section presents an analysis on effects of the number of quadrature points on the accuracy of the solution, using benchmark problems with known analytic solutions.

6.1.4 Numerical Examples

An EFG code has been developed based on the formulations provided above. Linear basis functions are employed in analyzing the following examples. The examples presented in this section are mainly designed for testing and benchmarking the EFG methods. In computing the system equations, the Gauss integration scheme is used in the same way as in the FEM.

Example 6.1: Patch Test

For a numerical method to work for solid mechanics problems, the sufficient condition is to pass the standard patch test, which has been used frequently in developing finite elements. The first numerical example, therefore, performs the standard patch test using the EFG method. The basic requirements for a patch are that the patch must have at least one interior node and that a linear displacement is imposed on all the edges of the patch in the absence of body force. Satisfaction of the patch test then requires that the displacement at any interior node should be given by the same linear function and that the strain and stress should be constant in the entire patch. In our patch test, a square patch of dimension $L_x = 2$ by $L_y = 2$ is used. The displacements are prescribed on all outside boundaries by a linear function of x and y:

$$\mathbf{u}_\Gamma(x, y) = \begin{Bmatrix} x \\ y \end{Bmatrix} \tag{6.49}$$

The patch is represented using a set of scattered nodes with some nodes in the interior of the patch. Both the regular and irregular nodal arrangements shown in Figure 6.3 are used for the test.

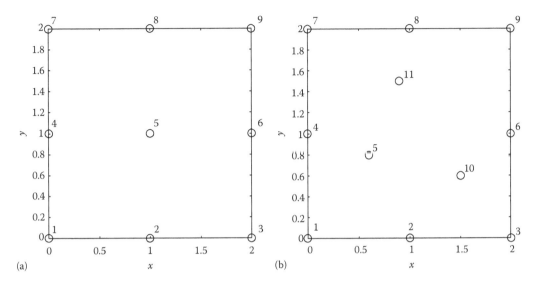

FIGURE 6.3
Nodal arrangement for patches: (a) regular nodal arrangement; and (b) irregular nodal arrangement.

The patch with regular nodal arrangement has eight boundary nodes and one interior node, as shown in Figure 6.3a, and the patch with irregular nodal arrangement has eight boundary nodes and three arbitrarily distributed interior nodes, as shown in Figure 6.3b. The numerical results show that the linear displacement field and the constant strain field have been reproduced within the patch to machine accuracy, as long as the numerical integration is accurate. This confirms that the EFG method "exactly" passes the patch test for both meshes, when an accurate numerical integration is performed. The issue of accurate integration is discussed in great detail in the next example.

Without the use of Lagrange multipliers, patch tests will fail, as reported in [1]. The test was performed using the patch shown in Figure 6.3a for different locations of node 5. The relative errors of stresses are listed in Table 6.1. When node 5 is located at the center of the patch, there is no error in the results. This is, however, a very special case. When node 5 moves away from the center of the patch, error appears. When node 5 is far from the center, meaning that the patch is highly irregular, the error is as large as almost 200%. This test clearly shows that the results can be very erroneous, if Lagrange multipliers are not used in enforcing the displacement (essential) boundary conditions.

TABLE 6.1

Relative Error of Stresses at Node 5 in Patch Shown in Figure 6.3a

Coordinate of Node 5 (x, y)	Error in σ_{xx} (%)	Error in σ_{yy} (%)	Error in σ_{xy} (%)
(1.0, 1.0)	0.00	0.00	0.00
(1.1, 1.1)	−0.98	−0.77	−0.86
(1.9, 1.8)	194.59	142.13	164.46
(1.9, 0.1)	134.95	127.23	−130.43

Note: Lagrange multiplier is not used for imposing essential boundary conditions on the patch edges.

Remark 6.1: Requirements for Galerkin Methods to Pass the Patch Test

The requirements for all the methods based on the standard Galerkin weak form to pass the patch test are as follows:

1. The shape functions are of at least linear consistency (see Chapter 2). This implies that the shape function is of linear field reproduction.

2. The field function approximation using the shape functions must be compatible.

3. The essential boundary conditions (displacement constraints on the boundary of the patch) have to be accurately imposed.

In addition, we require accurate numerical operations, such as integration, to form system equations in the process of testing.

MLS shape functions can satisfy the first requirement very easily as long as linear polynomial functions are included in the basis for constructing the shape functions. The second can also be satisfied if sufficient numbers of nodes in the support domains are used. To satisfy the third condition, the constrained Galerkin form is required for constructing the system equations. We therefore conclude that the EFG method formulated in this chapter can pass the standard patch test, if sufficient numbers of nodes in the support domains are used. In its standard and accurate numerical implementation, the EFG is fully compatible, and it should provide the lower bound of the solution (for force driving problems), and the displacement should converge to the exact solution from below when the nodal spacing approaches zero (Remark 5.2).

Example 6.2: Rectangular Cantilever: A Study on Numerical Integration

Numerical study is conducted for a rectangular cantilever, which is often used for benchmarking numerical methods because the analytical solution for this problem is known. Our purpose here is to investigate issues related to background integration in the EFG method. There are a number of factors affecting the accuracy of the numerical results of the EFG method. These factors include the number of field nodes n, the background mesh density, and the order of Gauss integration or the number of Gauss sampling points. To provide a quantitative indication of how these factors affect the accuracy of results, a rectangular cantilever subjected to a load at the free end, as shown in Figure 6.4, is analyzed in detail using our in-house EFG code. The exact solution of this problem is available as follows [5].

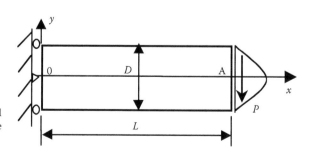

FIGURE 6.4
Rectangular cantilever loaded with an external force P distributed in a parabolic fashion at the end of the cantilever.

The displacement in the x-direction is

$$u_x = -\frac{Py}{6EI}\left[(6L - 3x)x + (2 + \nu)\left[y^2 - \frac{D^2}{4}\right]\right] \qquad (6.50)$$

where the moment of inertia I for a cantilever with rectangular cross section and unit thickness is given by

$$I - \frac{D^3}{12} \qquad (6.51)$$

The displacement in the y-direction is

$$u_y = \frac{P}{6EI}\left[3\nu y^2(L - x) + (4 + 5\nu)\frac{D^2x}{4} + (3L - x)x^2\right] \qquad (6.52)$$

The normal stress on the cross section of the cantilever is

$$\sigma_x = -\frac{P(L - x)y}{I} \qquad (6.53)$$

The normal stress in the y-direction is

$$\sigma_y = 0 \qquad (6.54)$$

The shear stress on the cross section of the cantilever is

$$\tau_{xy} = \frac{P}{2I}\left[\frac{D^2}{4} - y^2\right] \qquad (6.55)$$

In this example, the parameters for this rectangular cantilever are taken as follows:

Loading: $P = -1000$ N
Young's modulus: $E = 3 \times 10^7$ N/m^2
Poisson's ratio: $\nu = 0.3$
Height of the cantilever: $D = 12$ m
Length of the cantilever: $L = 48$ m

The force P is distributed in the form of parabola at the right end of the cantilever:

$$t_{xy} = \frac{P}{2I}\left[\frac{D^2}{4} - y^2\right] \qquad (6.56)$$

Strain energy error e is employed as an indicator of accuracy of the EFG numerical results:

$$e_e = \left\{\frac{1}{2}\int_\Omega \left(\varepsilon^{num} - \varepsilon^{exact}\right)^T D\left(\varepsilon^{num} - \varepsilon^{exact}\right)d\Omega\right\}^{\frac{1}{2}} \qquad (6.57)$$

At the left boundary $(x = 0)$ the displacements are prescribed using the analytical solutions (Equations 6.50 and 6.52):

$$u_x = -\frac{P(2 + v)y}{6EI}\left[y^2 - \frac{D^2}{4}\right]$$ (6.58)

$$u_y = \frac{PvL}{2EI}y^2$$ (6.59)

On the right boundary $(x = L)$, the applied external traction force is computed from the analytical solution in Equation 6.56.

For convenience of analysis, uniformly distributed nodes and background integration cells, as schematically shown in Figure 6.5, are used in the computation. N_x and N_y are the number of nodes with respect to the x- and y-directions. The density of the background mesh of cells is defined by $n_x \times n_y$, where n_x and n_y are, respectively, the number of background cells along the x- and y-directions. Table 6.2 summarizes the results of the error in energy norm defined by Equation 6.57 obtained by the EFG method using different numbers of Gauss integration points and different densities of background mesh of cells. In Table 6.2, n_g is the number of Gauss sampling points within a cell. The total number of quadrature points can be calculated using

$$n_Q = n_g \times n_x \times n_y$$ (6.60)

The total number of independent equations used in all the quadrature points, N_Q, can then be calculated using Equation 6.47, i.e.,

$$N_Q = 3 \times n_Q = 3 \times (n_g \times n_x \times n_y)$$ (6.61)

It is confirmed from Table 6.2 that, when $N_Q < N_u$, no stable solutions are obtained and even $N_Q > N_u$ does not necessarily guarantee an accurate or even a stable solution. It can also be seen from Table 6.2 that an acceptably accurate result can be obtained using

$$N_Q > 4N_u \sim 5N_u \cong 9n_t$$ (6.62)

or

$$n_Q > 3n_t$$ (6.63)

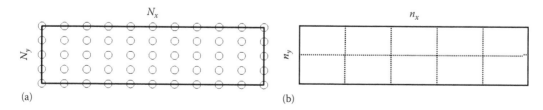

(a)   (b)

FIGURE 6.5
EFG model for the rectangular cantilever: (a) nodal arrangement; and (b) background mesh.

TABLE 6.2

Strain Energy Error ($\times 10^{-2}$) Resulting from Using Different Number of Gauss Sampling Points and Number of Background Mesh ($n_t = 55$, $N_u = 100$)

Gauss Points (n_g)	Background Mesh ($n_x \times n_y$)				
	1×1	2×1	4×1	8×2	12×3
2×2	$e = \infty$	$e = \infty$	$e = \infty$	$e = 8.10$	$e = 3.12$
	$N_Q = 12$	$N_Q = 24$	$N_Q = 48$	$N_Q = 192$	$N_Q = 432$
3×3	$e = \infty$	$e = \infty$	$e = \infty$	$e = 4.01$	$e = 2.95$
	$N_Q = 27$	$N_Q = 54$	$N_Q = 108$	$N_Q = 432$	$N_Q = 972$
4×4	$e = \infty$	$e = \infty$	$e = 4.37$	$e = 3.62$	$e = 2.90$
	$N_Q = 48$	$N_Q = 96$	$N_Q = 192$	$N_Q = 768$	$N_Q = 1728$
5×5	$e = \infty$	$e = 58.2$	$e = 4.63$	$e = 2.90$	$e = 2.89$
	$N_Q = 75$	$N_Q = 150$	$N_Q = 300$	$N_Q = 1200$	$N_Q = 2700$
6×6	$e = \infty$	$e = 4.74$	$e = 3.74$	$e = 2.89$	$e = 2.89$
	$N_Q = 108$	$N_Q = 216$	$N_Q = 432$	$N_Q = 1728$	$N_Q = 3888$
7×7	$e = \infty$	$e = 4.92$	$e = 2.96$	$e = 2.89$	$e = 2.89$
	$N_Q = 147$	$N_Q = 294$	$N_Q = 588$	$N_Q = 2352$	$N_Q = 5292$
8×8	$e = \infty$	$e = 3.70$	$e = 2.99$	$e = 2.89$	$e = 2.89$
	$N_Q = 192$	$N_Q = 384$	$N_Q = 768$	$N_Q = 3072$	$N_Q = 6912$
9×9	$e = \infty$	$e = 6.55$	$e = 2.93$	$e = 2.89$	$e = 2.89$
	$N_Q = 243$	$N_Q = 486$	$N_Q = 972$	$N_Q = 3888$	$N_Q = 8748$
10×10	$e = 41.5$	$e = 9.52$	$e = 2.90$	$e = 2.89$	$e = 2.89$
	$N_Q = 300$	$N_Q = 600$	$N_Q = 1200$	$N_Q = 4800$	$N_Q = 10800$

Note: $N_Q = 3 \times n_Q = 3 \times (n_g \times n_x \times n_y)$.

Therefore, we may conclude as a rough rule-of-thumb that a sufficient number of all the Gauss points should be at least about three times the total number of nodes. This finding is more or less inline with those in FEM using quadrilateral elements [6].

Efforts have been made to develop the EFG method without using a background mesh. Some of the so-called nodal integration approaches carry out the integration using only the field nodes without the use of additional Gauss points, to avoid using a background mesh of cells. In such cases, $n_Q \approx n_t$, which satisfies the minimum requirement of Equation 6.48. However, the above important finding of the $n_Q > 3n_t$ rule implies that any attempt at such a nodal integration scheme may suffer significant loss in accuracy and even instability, unless special measures, such as the use of stabilization terms [7–9], can be taken to prevent that from happening. Note that the requirement of $n_Q > 2n_t/3$ is only the minimum requirement for a *nonsingular* system matrix; it does not guarantee the accuracy of the solution.

Figure 6.6 plots the exact and numerical solutions of EFG for the deflection of the cantilever along the x-axis. The plot shows the excellent agreement between the exact solution and the numerical results for all the background meshes used. This fact reveals that the displacement is less sensitive to the background integration. A very coarse mesh can yield good displacement results. Figure 6.7 shows the distribution of stress σ_{xx} on the cross section of $x = L/2$ of the rectangular cantilever. Errors in stress between the exact solution and the numerical results are evident. This fact implies that the stresses that are obtained using the derivatives of the displacement field are very sensitive to the way the integration is performed. A much finer mesh and more Gauss points have to be used for an

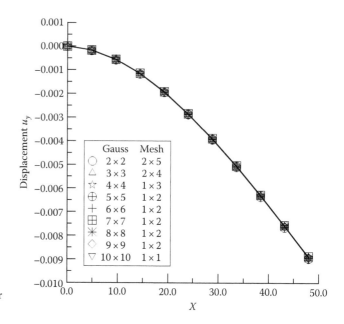

FIGURE 6.6
Deflection of the rectangular cantilever along $y = 0$.

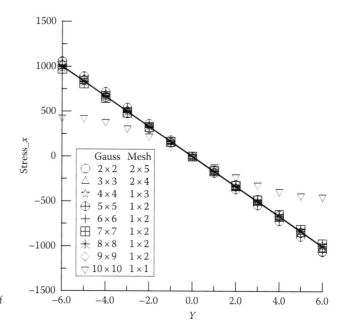

FIGURE 6.7
Distribution of stress σ_{xx} on the section of $x = L/2$ of the rectangular cantilever.

accurate stress field. Figures 6.8 and 6.9 show, respectively, the stress components σ_{yy} and σ_{xy}. It is clearly shown again that the stresses are very sensitive to the cell density of the background integration, especially the shear stress σ_{xy}.

The total number of Gauss points depends on both the density of the background cell and the number of Gauss points used in each cell, and these two have to be balanced. A finer background integral mesh can improve the accuracy, but there is a limit. On the

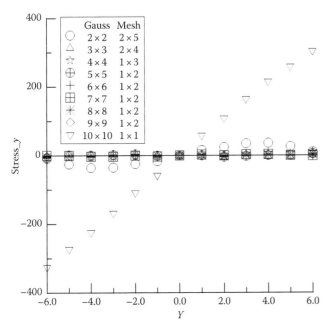

FIGURE 6.8
Distribution of stress σ_{yy} on the section of $x = L/2$ of the rectangular cantilever.

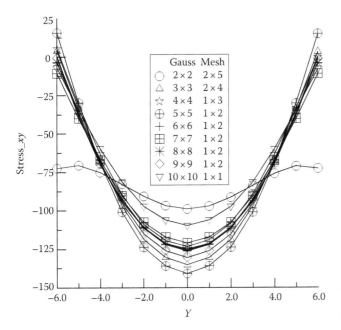

FIGURE 6.9
Distribution of stress τ_{xy} on the section of $x = L/2$ of the rectangular cantilever.

other hand, more Gauss integration points give, in general, higher accuracy, but the background mesh should not be too coarse. By using different densities of background mesh and different numbers of Gauss sampling points, the displacement field and stress field are computed using EFG, and the accuracy is investigated. From Figures 6.7 through 6.9, it can be observed that the cell density of the background mesh and the number of the

Gauss points have to be balanced in order to obtain good results. Too fine a mesh without enough Gauss points or too many Gauss points with too coarse a mesh will not give accurate results. One should always avoid biases to either the number of Gauss points or the number of cells of the background mesh. The balance point should, in general, depend on the complexity of the field to be analyzed, and the basis used in the MLS approximation. Our study for solid mechanics problems has found that, when a linear basis function is used, the proper number of Gauss sampling points should be between $n_g = 2 \times 2$ and $n_g = 6 \times 6$. Once the number of the Gauss points is chosen, the density of the background mesh should then be calculated using the guideline of $n_Q > 3n_t$.

To investigate how the node number affects the accuracy of the result, the strain energy error is calculated for the rectangular cantilever using different nodal densities along the x- and y-directions. The background mesh is fixed at $n_x \times n_y = 12 \times 3$, and Gauss points of $n_g = 4 \times 4$ are used for the integration. Results are presented in Figure 6.10. It is clearly seen that increasing the number of nodes in the domain can improve the accuracy. For this particular problem of rectangular cantilever, increasing the nodes along the y-direction improves the accuracy more efficiently than increasing the nodes along the x-direction. Generally, when N_x is greater than eight, the results are sufficiently accurate. Further study on the effects of Gauss integration points is conducted by changing n_g from 2×2 to 10×10, and no significant change in strain energy error is observed.

Because both nodal density and background mesh density affect the accuracy of the stress field, the ratio of quadrature points to field nodes is defined as

$$\alpha_n = \frac{n_Q}{n_t} \tag{6.64}$$

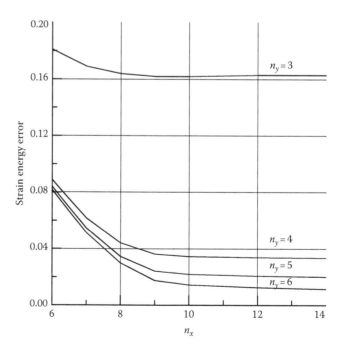

FIGURE 6.10
Effects of the nodal density on the strain energy error.

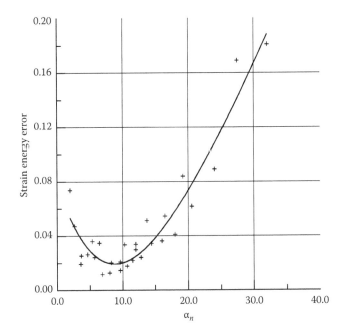

FIGURE 6.11
Relation between strain energy error and the ratio of quadrature points to field nodes α_n.

Further investigation has been conducted on the relationship between strain energy error and the ratio of quadrature points to nodes, α_n. The results are summarized in Figure 6.11. It is clearly shown that when α_n is around 7.0–9.0, the result obtained is most accurate. A reasonable result can be obtained using $\alpha_n > 3$. This confirms the finding from Table 6.2.

In the application of EFG for practical problems, the density of the field nodes should be determined by the gradient of the field variables. For most practical problems, the field nodes are not evenly distributed. The sampling points should also be distributed accordingly in an uneven manner with α_n around 3.0–9.0.

Note that the Gauss point number suggested by Belytschko et al. [1] is $n_Q = n_t \left(\sqrt{n_c} + 2 \right)^2$, where n_c is the number of nodes in a cell. When $n_c = 1$, we have $\alpha_n = 9$. For $n_c \geq 2$, this suggestion demands more Gauss points.

6.1.5 Concluding Remarks

The EFG method is based on global variational formulation—Galerkin variational principle. Therefore, although an element mesh is not required for field variable approximation over the problem domain, a global background mesh is still required to evaluate the integrals for calculating stiffness and mass matrices. Our numerical examination of the relationship between the density of field nodes and background mesh for 2D stress analysis problems shows

a. The minimum number of integration points must be greater than two-thirds of the total number of the unfixed field nodes, i.e., $n_Q > 2n_t/3$. This requirement of $n_Q > 2n_t/3$ is only the minimum requirement to ensure a *nonsingular* system matrix; it does not guarantee the accuracy of the solution.

b. The ratio of the integration points to the field nodes, α_n, is around 3–9, and economic results with acceptable accuracy can be obtained using $\alpha_n = 3$. This means that a sufficient number of integration points should be about three times the number of the field nodes.

c. The displacement field can be obtained rather accurately with a coarse background mesh of integration cells, whereas a finer background mesh is necessary for computing the stress field.

d. Accuracy of the stress field can be improved efficiently by increasing the density of the field nodes, together with sufficient density of the background mesh.

e. Finally, from the experience of the author's research group, triangular type of back ground mesh is most flexible, robust, and efficient for problems with complicated domain. We prefer also to have the field nodes coincide with the vertices of the background triangular mesh, so that the nodal density and density of integral sampling can be naturally tied together. In addition, the background mesh can also be utilized in the selection of nodes for MLS shape function construction using the T2L-Scheme (see Section 1.7.6). This approach is used in MFree2D$^{\copyright}$.

6.2 EFG with Penalty Method

As described in the previous chapters, the use of MLS approximation produces shape functions that do not possess the Kronecker delta function property, i.e., $\phi_I^H(\mathbf{x}_J) \neq \delta_{IJ}$. This leads to $u^h(\mathbf{x}_J) = \sum_I^n \phi_I^H(\mathbf{x}_J)u_I \neq u_J$, which implies that one cannot impose essential boundary conditions in the same way as in conventional FEM. In the previous chapter, the essential boundary conditions are imposed by introducing Lagrange multipliers in the weak form. This method of Lagrange multipliers results in an enlarged nonpositive system matrix, as shown in Equation 6.37. The bandedness of the system matrix is also distorted. Therefore, it requires much more computational cost in solving such system equations as Equation 6.37.

In this section, an alternative method—the penalty method—is introduced for the imposition of essential boundary conditions. The use of the penalty method produces equation systems of the same dimensions that conventional FEM produces with the same number of nodes, and the modified stiffness matrix is still positive definite. The problem with the penalty method lies in choosing a penalty factor that can be used universally for all problems. The penalty method has been used by many researchers; this section follows the formulation reported in [10].

6.2.1 Formulation

The penalty method has been frequently used in conventional FEM for enforcing single or multipoint constraints (MPCs) [4]. In the EFG method, the essential boundary conditions needed to be enforced have the form

$$\sum_I^n \phi_I^H(\mathbf{x})\mathbf{u}_I = \mathbf{u}_\Gamma(\mathbf{x}) \quad \text{on } \Gamma_u \tag{6.65}$$

where $\mathbf{u}_\Gamma(\mathbf{x})$ is the prescribed displacement on the essential boundary. Note that Equation 6.65 is nothing but the continuous form of the so-called MPC equations that are used very often in FEM analyses. Therefore, the penalty method can, of course, be applied to impose the essential boundary conditions in the meshfree methods that use MLS shape functions for field variable approximation.

This section presents the formulation of the penalty method to impose essential boundary conditions in the EFG method. System equations for boundary value problems of both homogeneous and inhomogeneous materials are derived. The discrete system matrices derived using the penalty method from the constrained Galerkin weak form are positive definite (unless the essential boundary condition is improperly set) and banded, and the treatment of boundary conditions is as simple as it is in conventional FEM. Numerical examples are presented to demonstrate the procedure of enforcing the essential boundary conditions.

The present approach is also applied to the problems with continuity conditions on the interfaces of multimaterial bodies, such as composite materials. Numerical examples demonstrate that the EFG method with penalty method is applicable in handling problems for composite materials, where the continuity conditions between different types of materials need to be enforced.

6.2.2 Penalty Method for Essential Boundary Conditions

Consider again the problem stated in Equations 6.1 and 6.2. Instead of using Lagrange multipliers, we introduce a penalty factor to penalize the difference between the displacement of MLS approximation and the prescribed displacement on the essential boundary. The constrained Galerkin weak form using the penalty method can then be proposed as follows:

$$\int_\Omega \delta(\mathbf{L}_d\mathbf{u})^\mathrm{T}\mathbf{c}(\mathbf{L}_d\mathbf{u})\mathrm{d}\Omega - \int_\Omega \delta\mathbf{u}^\mathrm{T}\cdot\mathbf{b}\mathrm{d}\Omega - \int_{\Gamma_t} \delta\mathbf{u}^\mathrm{T}\cdot\mathbf{t}_\Gamma\mathrm{d}\Gamma$$
$$- \delta\int_{\Gamma_u} \frac{1}{2}(\mathbf{u}-\mathbf{u}_\Gamma)^\mathrm{T}\cdot\boldsymbol{\alpha}\cdot(\mathbf{u}-\mathbf{u}_\Gamma)\,\mathrm{d}\Gamma = 0 \tag{6.66}$$

Equation 6.66 is formed from Equation 5.113 by changing the area integrals for the constraint-related terms into curve integrals because the constraints (essential boundary conditions) given in Equation 6.2 are defined only on the boundary. Note that the difference between Equation 6.66 and Equation 6.4 is that the fourth and fifth terms in Equation 6.4 are replaced by the fourth term in Equation 6.66, where $\boldsymbol{\alpha} = [\alpha_1, \alpha_2, \dots, \alpha_k]^\mathrm{T}$ is a diagonal matrix of penalty factors, where $k=2$ for 2D cases and $k=3$ for 3D cases. The penalty factors α_i ($i=1,\dots, k$) can be a function of coordinates and they can be different from each other, but in practice we often assign them an identical constant of a large positive number, which can be chosen by following the method described in Section 5.12.2.

Substituting the expression of the MLS approximation for the displacement of Equation 6.7 into the weak form of Equation 6.66, and after similar manipulations given in Section 6.1.1, we can arrive at the final system equation of

$$[\mathbf{K} + \mathbf{K}^\alpha]\mathbf{U} = \mathbf{F} + \mathbf{F}^\alpha \tag{6.67}$$

where

> **F** is the global external force vector assembled using the nodal force matrix defined by Equation 6.31
>
> **K** is the global stiffness matrix assembled using the nodal stiffness matrix given in Equation 6.38

The additional matrix $\mathbf{K}^\alpha$ is the global penalty matrix assembled using the nodal matrix defined by

$$\mathbf{K}_{IJ}^\alpha = \int_{\Gamma_u} \mathbf{\Phi}_I^{\mathrm{T}} \boldsymbol{\alpha} \, \mathbf{\Phi}_J \mathrm{d}\Gamma \tag{6.68}$$

where $\mathbf{\Phi}_I^H$ is the matrix of MLS shape functions given by Equation 6.41.

The force vector $\mathbf{F}^\alpha$ is caused by the essential boundary condition, and its nodal vector has the form

$$\mathbf{F}_I^\alpha = \int_{\Gamma_u} \left(\mathbf{\Phi}_I^H\right)^{\mathrm{T}} \boldsymbol{\alpha} \mathbf{u}_\Gamma \mathrm{d}\Gamma \tag{6.69}$$

Note that the integration is performed along the essential boundary, and hence matrix $\mathbf{K}^\alpha$ will have entries only for the nodes near the essential boundaries Γ_u, which are covered by the support domains of all the quadrature points on Γ_u.

Comparing Equation 6.67 with Equation 6.37, the advantages of the penalty method are obvious:

- The dimension and positive definite property of the matrix are preserved, as long as the penalty factors chosen are positive.
- The symmetry and the bandedness of the system matrix are preserved.

These advantages make the penalty method much more efficient and hence much more attractive compared with the Lagrange multipliers method. Detailed studies on implementation of the penalty method and computation of actual application problems have indicated the following disadvantages of the penalty method compared with the Lagrange multipliers method:

- It is necessary to choose penalty factors that are universally applicable for all kinds of problems. One hopes to use as large as possible penalty factors, but too large penalty factors often result in the augmentation of condition number in the system matrix leading to possible numerical problems, as we have experienced in the imposition of multipoint boundary condition in FEM.
- The results obtained are, in general, less accurate, compared with the method of Lagrange multipliers.
- An essential boundary condition can never be precisely imposed. It is imposed only approximately.

Despite these disadvantages, the penalty method is often more favorable for many researchers. It is also implemented in MFree2D (see Chapter 16). Below is another application of the penalty method.

6.2.3 Penalty Method for Continuity Conditions

The domain of a problem can be composed of subdomains with different materials. The treatment of material discontinuity is straightforward in the conventional FEM, because we have elements to use. All one need do is have the element edges coincide with the interface of different materials. The properties of the material are used for elements that are located in the corresponding subdomain of the material. There is no special treatment required.

In meshfree methods, however, there is no mesh of elements, and hence the material interface cannot be defined based on elements. When EFG is used, all the stress components will be continuous around the interface, which is not the case physically. Therefore, special treatment is therefore needed. Cordes and Moran [11] dealt with material discontinuity problems using the method of Lagrange multipliers. The conditions on the material interfaces were treated as a special essential boundary condition, and the approaches in the original EFG procedure discussed in Section 6.1 could then be followed to handle the material discontinuity. Krongauz and Belytschko [12] have proposed a method to model material discontinuities in the EFG method by adding special shape functions that contain discontinuities in derivatives.

The penalty method introduced in this section can be an alternative for dealing with the material discontinuity problem. The detailed procedure was suggested by Liu and Yan [3], and is reported in detail in Yang's master's thesis. This section details the penalty method for handling the material discontinuity problem.

MPCs enforced using the penalty method are often used in FEM for modeling different kinds of connections between two subdomains of a structural system. The penalty method is also applicable for modeling similar situations in mechanics problems (see, e.g., [6]).

The penalty method is presented in this section to model two subdomains of different materials connected in a prescribed manner. A perfect connection is considered, but the approach is applicable for all kinds of connections, including partial connections.

Consider first an inhomogeneous medium consisting of two homogeneous bodies. On the boundary of the two homogeneous bodies, an interface is first defined by a set of nodes that belong to both materials. We then impose a *nonpenetration rule* for the influence domains of the nodes. The nonpenetration rule states that points contained in material 1 can only be influenced by the nodes in material 1 plus the interface nodes; points contained in material 2 can only be influenced by the nodes contained in material 2 plus the interface nodes. Our following treatment is based on this nonpenetration rule of influence domains. This rule confines the influence domain of a node within the subdomain of the material of the node.

Figure 6.12 illustrates the determination of the domains of influence for the nodes in problem domains of homogeneous and inhomogeneous materials. In Figure 6.12, circular domains of influence are employed. For the homogeneous case (Figure 6.12a), point a is contained in the influence domains of both nodes 4 and 5. Therefore, nodes 4 and 5 are considered as the neighbors of point a. Similarly, point b has neighbors of nodes 3 and 5, and point c has neighbors of nodes 1 and 2. However, as the interface exists in the inhomogeneous materials in Figure 6.12b, the neighbors of each of the points a, b, and c may change, due to the blockage of the material interface. The influence domain for node 4 is the same as in homogeneous materials because the influence domain of node 4 does not intersect the interface, and therefore, point a is still within its influence. The influence domain for node 5 is also the same as in homogeneous materials because it lies on the interface of both materials. Therefore, point a is still within the influence domain of node 5. The neighbors of point b still include nodes 3 and 5 since each pertains to material 1.

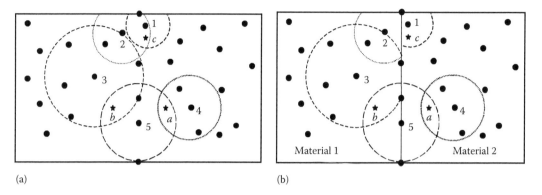

(a)   (b)

FIGURE 6.12
Determination of domains of influence. (a) Domains of influence in a homogeneous body; and (b) domains of influence in an inhomogeneous body.

However, point c is not included in the support domain of node 2 due to the nonpenetration of the influence domain of node 2. In this case, point c has only one neighbor, that is, node 1.

The connection of the subdomains is achieved by enforcing conditions of continuity. On the interface of the two materials Γ_m, we enforce the following constraint:

$$\mathbf{u}_\Gamma^+ = \mathbf{u}_\Gamma^- \tag{6.70}$$

where $\mathbf{u}_\Gamma^+$ and $\mathbf{u}_\Gamma^-$ correspond to the displacement in the two materials but on the interface of Ω^+ and Ω^-, respectively. This constraint is then imposed using the penalty method in the constrained Galerkin weak form, i.e.,

$$\int_\Omega \delta(\mathbf{L}_d\mathbf{u})^\mathrm{T}(\mathbf{c}\mathbf{L}_d\mathbf{u})\mathrm{d}\Omega - \int_\Omega \delta\mathbf{u}^\mathrm{T}\mathbf{b}\mathrm{d}\Omega - \int_{\Gamma_t} \delta\mathbf{u}^\mathrm{T}\mathbf{t}_\Gamma\mathrm{d}\Gamma$$

$$- \delta\frac{1}{2}\int_{\Gamma_u} (\mathbf{u} - \mathbf{u}_\Gamma)^\mathrm{T}\boldsymbol{\alpha}(\mathbf{u} - \mathbf{u}_\Gamma)\mathrm{d}\Gamma - \delta\frac{1}{2}\int_{\Gamma_m} (\mathbf{u}_\Gamma^+ - \mathbf{u}_\Gamma^-)^\mathrm{T}\boldsymbol{\beta}(\mathbf{u}_\Gamma^+ - \mathbf{u}_\Gamma^-)\mathrm{d}\Gamma = 0 \tag{6.71}$$

Note that the difference between Equations 6.71 and 6.66 is the additional term in Equation 6.71, where $\boldsymbol{\beta}$ is a diagonal matrix of penalty factors that have the same form of α, but the values in $\boldsymbol{\beta}$ may be different. The approximate values of $\mathbf{u}_\Gamma^{+h}$ and $\mathbf{u}_\Gamma^{-h}$ are expressed using MLS approximation (see Equation 6.7).

$$\mathbf{u}_\Gamma^{+h}(\mathbf{x}) = \sum_{I\in S_{n+}} \boldsymbol{\Phi}_I^+(\mathbf{x})\mathbf{u}_I^+ \tag{6.72}$$

and

$$\mathbf{u}_\Gamma^{-h}(\mathbf{x}) = \sum_{I\in S_{n-}} \boldsymbol{\Phi}_I^-(\mathbf{x})\mathbf{u}_I^- \tag{6.73}$$

where S_{n^+} and S_{n^-} are the set of nodes that have influences on $\mathbf{x}$ in Ω^+ and Ω^-, respectively, and $\mathbf{\Phi}_I^+$ and $\mathbf{\Phi}_I^-$ are the matrices of MLS shape functions created using these nodes.

Substituting Equations 6.5, 6.72, and 6.73 into Equation 6.71 leads to a set of discrete system equations:

$$[\mathbf{K} + \mathbf{K}^\alpha + \mathbf{K}^\beta]\, \mathbf{U} = \mathbf{F} + \mathbf{F}^\alpha \tag{6.74}$$

where

$$\mathbf{K}_{IJ}^\beta = \int\limits_{\Gamma_m} \left[\mathbf{\Phi}_I^+ - \mathbf{\Phi}_I^-\right]^T \beta \left[\mathbf{\Phi}_J^+ - \mathbf{\Phi}_J^-\right] d\Gamma \tag{6.75}$$

Matrix $\mathbf{K}^\beta$ is the stiffness matrix for connecting the two different materials. Note that the integration is performed along the interfaces of two materials, and hence matrix $\mathbf{K}^\beta$ will have entries for the nodes near (not only on) the interfaces, which have influence on the quadrature points on the interface between different materials.

6.2.4 Numerical Examples

6.2.4.1 Numerical Examples for Treating Essential Boundary Conditions

Example 6.3: Patch Test

The first numerical example is the standard patch test. The same patch tests conducted in Example 6.1 are repeated here using the penalty method for imposing the linear displacement on the boundaries of the patches shown in Figure 6.3. The EFG method with penalty method exactly passes the test for both kinds of meshes to machine accuracy. In both cases, the maximum errors in the displacement are of order 10^{-13}; the stresses remain the same in the patch and the maximum errors are of order 10^{-11}. The displacements for the regular and irregular nodal arrangements are given in Tables 6.3 and 6.4.

TABLE 6.3

Coordinates and Displacements Solved for the Patch Test with Regular Nodal Arrangement Using EFG with the Penalty Method

Nodes	Coordinates (x, y)	Displacements Solved (u_x, u_y)
1	(0, 0)	(0.00000000000000, 0.00000000000000)
2	(1, 0)	(1.00000000000000, 0.00000000000000)
3	(2, 0)	(1.99999999999999, 0.00000000000000)
4	(0, 1)	(0.00000000000000, 1.00000000000000)
5	(1, 1)	(1.00000000000007, 0.99999999999996)
6	(2, 1)	(2.00000000000000, 1.00000000000000)
7	(0, 2)	(0.00000000000000, 2.00000000000000)
8	(1, 2)	(1.00000000000000, 2.00000000000000)
9	(2, 2)	(2.00000000000000, 2.00000000000000)

TABLE 6.4

Coordinates and Displacements Solved for the Patch Test with Irregular
Nodal Arrangement Using EFG with the Penalty Method

Nodes	Coordinates (x, y)	Displacements Solved (u_x, u_y)
1	$(0, 0)$	$(0.00000000000000, -0.00000000000001)$
2	$(1, 0)$	$(1.00000000000000, 0.00000000000000)$
3	$(2, 0)$	$(2.00000000000000, 0.00000000000000)$
4	$(0, 1)$	$(0.00000000000000, 1.00000000000000)$
5	$(0.6, 0.8)$	$(0.59999999999995, 0.80000000000009)$
6	$(1, 2)$	$(1.00000000000000, 2.00000000000000)$
7	$(0, 2)$	$(0.00000000000000, 2.00000000000000)$
8	$(1, 2)$	$(1.00000000000000, 2.00000000000000)$
9	$(2, 2)$	$(2.00000000000000, 2.00000000000000)$
10	$(1.5, 0.6)$	$(1.49999999999997, 0.60000000000001)$
11	$(0.9, 1.5)$	$(0.90000000000000, 1.50000000000004)$

Example 6.4: Rectangular Cantilever

Consider a cantilever of characteristic length L and height D subjected to a parabolic traction at
the free end, as shown in Figure 6.4, which was examined in Example 6.2. The cantilever is of unit
thickness and the plane stress state is considered. The exact solution is given by Timoshenko and
Goodier [5], and is listed in Equations 6.50 through 6.55. The parameters used in this section are
as follows:

Loading: $P = 100$ N
Young's modulus: $E = 3 \times 10^6$ N/m^2
Poisson's ratio: $\nu = 0.3$
Height of the cantilever: $D = 1.2$ m
Length of the cantilever: $L = 4.8$ m

Both a regular and irregular arrangement of nodes and a regular background mesh of cells are
used for numerical integrations to calculate the system equations, as shown in Figure 6.13. In each
integration cell, a 4×4 Gauss quadrature scheme is used to evaluate the stiffness matrix. The
linear basis function and cubic spline weight function are used in the MLS approximation. The
dimension of the support domain α_s is chosen to be 3.5 so that the domain of support of each
quadrature point contains at least 40 nodes to avoid the singularity of the moment matrix in
constructing MLS shape functions.

Figure 6.14 plots the analytical solution based on 2D elasticity and the numerical solution
using the present EFG method for the deflection of the cantilever along the x-axis. The plot
shows excellent agreement between the analytical and present numerical results for both
regular and irregular nodal arrangements.

Figures 6.15 and 6.16 illustrate the comparisons between the stresses calculated using the
analytical solution and the EFG with penalty method. The normal stress σ_x at the section of
$x = L/2$ is shown in Figure 6.15, and the shear stress τ_{xy} is shown in Figure 6.16. A very
good agreement is observed. It should be noted that the accuracy of the shear stress in the
case of the irregular nodal arrangement is lower than that in the regular arrangement.

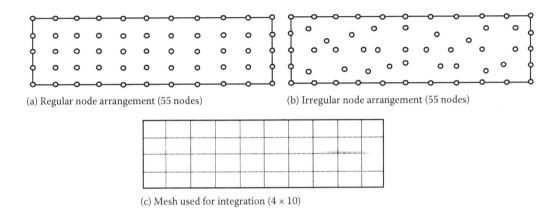

(a) Regular node arrangement (55 nodes)   (b) Irregular node arrangement (55 nodes)

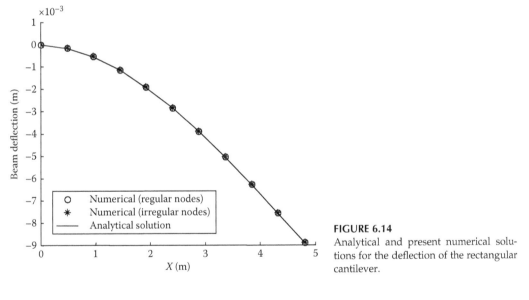

(c) Mesh used for integration (4 × 10)

FIGURE 6.13
Nodal arrangements and background mesh for the rectangular cantilever. (a) Regular node arrangement;
(b) irregular node arrangement; and (c) mesh used for integration.

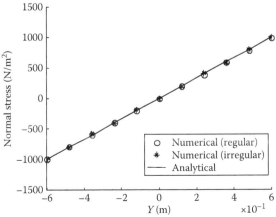

FIGURE 6.14
Analytical and present numerical solutions for the deflection of the rectangular cantilever.

FIGURE 6.15
Analytical and present numerical solutions for the normal stress at the section $x = L/2$ of the rectangular cantilever.

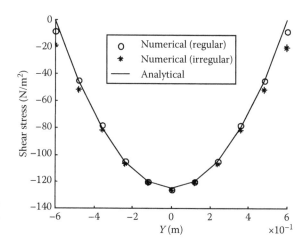

FIGURE 6.16
Analytical and numerical solutions for the shear stress at the section $x = L/2$ of the rectangular cantilever.

Table 6.5 compares the numerical result for the vertical displacement at point A on the cantilever (see Figure 6.4) with the exact vertical displacement given in Equation 6.52. The calculation was performed for models with 10, 18, 55, and 189 nodes. This table shows that the numerical result approaches the exact solutions as the number of the nodes increases.

Figure 6.17 is a plot of the rate of convergence in L_2 energy error for the cantilever problem. The rate of convergence in energy is calculated using Equation 6.57. The value h

TABLE 6.5

Comparison of Vertical Deflection u_y(m) at End of Cantilever

Number of Nodes	Exact	EFG (Penalty)	Error (%)
10	−0.0089	−0.008099	9
18	−0.0089	−0.008511	4.4
55	−0.0089	−0.008883	0.2
189	−0.0089	−0.008898	0.02

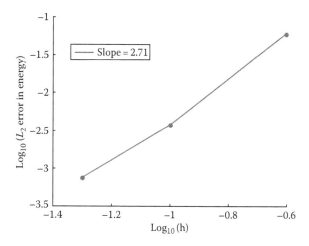

FIGURE 6.17
Rate of convergence in energy error tested on the rectangular cantilever.

TABLE 6.6

Computational Time Using EFG with Different Methods
for Imposing Essential Boundary Conditions

Nodes	CPU Time (s)	
	EFG (Lagrange Multiplier)	EFG (Penalty)
55	1.1	0.6
189	35.4	3.5
561	115.2	13.8

was chosen to be the horizontal nodal spacing in the model. The dimensionless size of the support domain is $\alpha_s = 3.5$. The cubic spline weight function in Equation 2.16 is used. The slope of the line plotted in Figure 6.17 is approximately 2.71, which is greater than the theoretical rate of 2.0 for linear finite elements (Remark 3.5).

Computational time

Because the penalty method does not increase the size of the stiffness matrix, and the stiffness matrix is still banded, computational efficiency can be improved greatly compared with the use of Lagrange multipliers in EFG. Table 6.6 compares the central processing unit (CPU) time of the penalty method vs. the method of Lagrange multipliers used in EFG for the rectangular cantilever. The computation is performed on the same HP workstation using the same half-bandwidth technique to store the system matrices and solve the system equations. It can be seen that the penalty method is much faster than the method of Lagrange multipliers, especially for large numbers of field nodes. Note that if a special solver designed for Equation 6.37 is used, the efficiency for the method of Lagrange multipliers can be improved.

6.2.4.2 Numerical Examples for Treating Continuity Conditions

Example 6.5: Rectangular Cantilever Composed of Two Parts

To further examine the present method, it is applied to the same cantilever but it is now treated as two "different" parts connected at boundary Γ_m, as shown in Figure 6.18. We assume that these two parts have the same material properties; therefore, this cantilever can be regarded as a homogeneous cantilever and the analytical solutions can still be used to check our numerical approach for the interface treatment. In the numerical analysis using EFG with penalty method, we still view the cantilever as two different parts and do not allow the influence domains in both

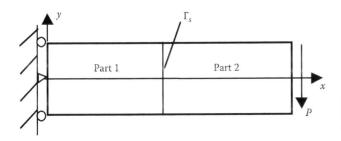

FIGURE 6.18
Rectangular cantilever made of two parts of different materials.

subdomains to go across the interface. The penalty method is applied on Γ_m to enforce the connectivity of these two parts.

The parameters of the cantilever in this case are the same as those in Example 6.4. Figure 6.19 shows the comparison between the analytical and numerical results for the deflection of the cantilever. The solution for the normal stress σ_x on the cantilever sections at the upper surface of the cantilever is shown in Figure 6.20. These numerical solutions also exhibit a good agreement between the meshfree and the analytical results, confirming that the interface treatment works well.

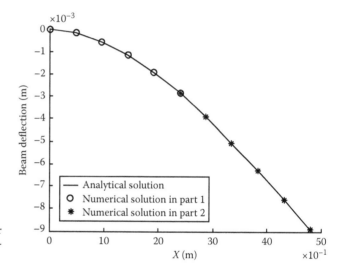

FIGURE 6.19
Analytical and numerical solutions for the deflection of the rectangular cantilever modeled as two connected parts.

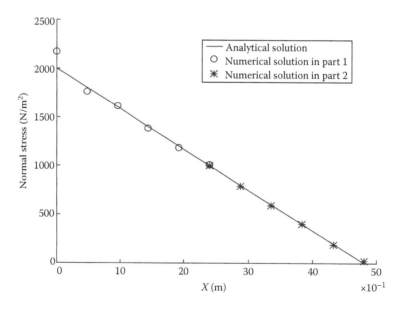

FIGURE 6.20
Analytical and numerical solutions for the normal stress at the upper surface ($y = D/2$) of the cantilever treated as two connected parts.

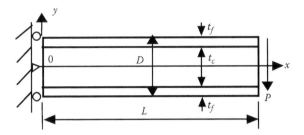

FIGURE 6.21
Sandwich composite cantilever.

Example 6.6: Sandwich Composite Cantilever

In this example, a sandwich composite cantilever consisting of three layers of two materials, shown in Figure 6.21, is simulated. The two upper and lower surface layers are the same material thickness t_f and the thickness of the core material is denoted by t_c. The surface layer is stiffer than the core material, and all three layers are assumed to be perfectly connected together. The connection is enforced using the penalty method formulated in the above section. The parameters for this example are as follows:

Loading: $P = 100$ N

Young's modulus for the material of two surface layers: 1.67×10^9 N/m^2

Young's modulus for the core material: 1.67×10^8 N/m^2

Poisson's ratios for two materials: $\nu = 0.3$

Thickness of the two surface layers: $t_f = 3$

Thickness of the core layer: $t_c = 6$

Height of the cantilever: $D = 1.2$ m

Length of the cantilever: $L = 4.8$ m

The stresses calculated from the present method and PATRAN/FEA are compared in Figures 6.22 and 6.23. The normal stress σ_x at the section of $x = L/2$ is shown in Figure 6.22, and the shear stress τ_{xy} is shown in Figure 6.23. A very good agreement is observed. The discontinuity of the normal stress at the interface is clearly captured.

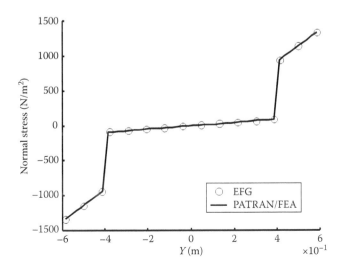

FIGURE 6.22
Numerical solutions for the normal stress at the section $x = L/2$ of the sandwich composite cantilever.

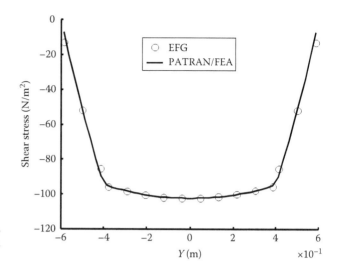

FIGURE 6.23
Numerical solutions for the shear stress at the section $x = L/2$ of the sandwich composite cantilever.

6.2.5 Concluding Remarks

In this section, the penalty method is used to impose the essential boundary and continuity conditions in the EFG method that uses MLS shape functions. It overcomes the drawbacks in the method of Lagrange multipliers. The main advantage of the penalty method is that it leads to a positive definite and banded stiffness matrix. The stiffness matrix also has a smaller dimension than those using Lagrangian multipliers, which improves computational efficiency. Numerical examples have demonstrated the good performance of the penalty method.

The penalty method was also applied for the treatment of problems with material discontinuity. However, in treating the material discontinuity, the FEM is more straightforward.

6.3 Summary

This chapter presents the EFG method. Works by Nayroles and coworkers and Belytschko and coworkers have, in fact, offered a direction in the development of meshfree methods. Following their work, a large number of researchers have also contributed significantly to the development of the EFG method. It is so far one of the most widely used meshfree methods. It was included in the software package, MFree2D showcased in 1999. The MFree2D is, to the best knowledge of the author, the first software package on meshfree methods that is fully packaged with pre- and postprocessors that is capable of carrying out adaptive analysis automatically.

The extended finite element method (XFEM) developed by Belytschko and coworkers has further enhanced the capability in dealing with more complicated problems such as the crack propagation problems. Readers are referred to publications by Belytschko and coworkers for the recent developments.

References

1. Belytschko, T., Lu, Y. Y., and Gu, L., Element-free Galerkin methods, *Int. J. Numerical Methods Eng.*, 37, 229–256, 1994.
2. Nayroles, B., Touzot, G., and Villon, P., Generalizing the finite element method: Diffuse approximation and diffuse elements, *Comput. Mech.*, 10, 307–318, 1992.
3. Liu, G. R. and Yan, L., A study on numerical integrations in element free methods, in *Proceedings of APCOM'99*, Singapore, 1999, pp. 979–984.
4. Zienkiewicz, O. C. and Taylor R. L., *The Finite Element Method*, 5th ed., Butterworth Heinemann, Oxford, 2000.
5. Timoshenko, S. P. and Goodier, J. N., *Theory of Elasticity*, 3rd ed., McGraw-Hill, New York, 1970.
6. Liu, G. R. and Quek, S. S., *The Finite Element Method: A Practical Course*, Butterworth Heinemann, Oxford, 2003.
7. Beissel, S. and Belytschko, T., Nodal integration of the element-free Galerkin method, *Comput. Methods Appl. Mech. Eng.*, 139, 49–74, 1996.
8. Chen, J. S., Wu, C. T., Yoon, S., and You, Y., A stabilized conforming nodal integration for Galerkin mesh-free methods, *Int. J. Numerical Methods Eng.*, 50, 435–466, 2001.
9. Liu, G. R., Zhang, G. Y., Wang, Y. Y., Zhong, Z. H., Li, G. Y., and Han X., A nodal integration technique for meshfree radial point interpolation method (NI-RPCM), *Int. J. Solids Struct.*, 44, 3840–3860, 2007.
10. Liu, G. R. and Yang, K. Y., A penalty method for enforce essential boundary conditions in element free Galerkin method, in *Proceedings of the 3rd HPC Asia'98*, Singapore, 1998, pp. 715–721.
11. Cordes, L. W. and Moran, B., Treatment of material discontinuity in the element-free Galerkin method, *Comput. Methods Appl. Mech. Eng.*, 139, 75–89, 1996.
12. Krongauz, Y. and Belytschko, T., EFG approximation with discontinuous derivatives, *Int. J. Numerical Methods Eng.*, 41(7), 1215–1233, 1998.

7

Meshless Local Petrov–Galerkin Method

The element-free Galerkin (EFG) method requires a mesh of background cells for integration in computing the system matrices. The reason behind the need for background cells for integration is the use of the Galerkin weak form for generating the discrete system equations. Is it possible not to use the weak form? The answer is yes: meshfree methods that operate on strong forms, such as the irregular finite difference method [1,2], finite point method [3], and local point collocation methods [4–8] have been developed. However, these kinds of methods are generally not very stable against node irregularities, and the results obtained can be less accurate. Efforts are still being made to stabilize these methods, especially in the direction of using local radial functions with properly devised regularization techniques [7,8].

In using the weighted residual method, if we try to satisfy the equation point-by-point using information in a local domain of the point as we do in the point collocation methods, the integration form can then be implemented locally by carrying out numerical integration over the local domain. The meshless local Petrov–Galerkin (MLPG) method originated by Atluri and Zhu [9] uses the so-called local weak form of the Petrov–Galerkin residual formulation. MLPG has been fine-tuned, improved, and extended over the years [10–17]. This chapter details the MLPG method for two-dimensional (2D) solid mechanics problems.

In the MLPG implementation, moving least squares (MLS) approximation is employed for constructing shape functions. Therefore, similar to the EFG method, there is an issue of imposition of essential boundary conditions. The original MLPG proposed in [10,11] uses the penalty method. In the formulation in [13], a method called direct interpolation is used. This chapter formulates both methods, in addition to the orthogonal transformation method for free-vibration problems.

A number of benchmark examples are presented to illustrate the procedure and effectiveness of the MLPG method. The effects of different parameters including the dimensions of different domains of MLPG on the accuracy of the results are also investigated via these examples.

Although the node-by-node procedure in MLPG is quite similar to that of the collocation method, the MLPG is more stable against nodal irregularity due to the use of a local weak form of locally integrated weighted residuals. Due to the use of the MLS shape functions in the local Petrov–Galerkin formulation, the MLPG can reproduce the polynomials that are included in the basis of MLS shape functions. This fact will be evidenced in the examples of patch tests.

7.1 MLPG Formulation

We consider again a 2D solid mechanics problem, as shown in Figure 7.1, for illustrating the procedure for formulating the MLPG method. The problem domain is denoted

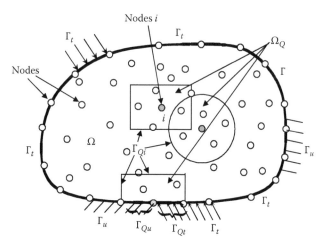

FIGURE 7.1
Domains and their boundaries. Problem domain Ω with essential (displacement) boundary Γ_u, natural (force or free) boundary Γ_t, quadrature domain of Ω_Q with the interior boundary Γ_{Qi} that is located within the problem domain, the essential boundary Γ_{Qu} that intersects with Γ_u, and the natural boundary Γ_{Qt} that intersects with Γ_t.

by Ω, which is bounded by boundaries including essential (displacement) boundary Γ_u and natural (force or free) boundary Γ_t. The strong form of the problem has been given in Equations 6.1 through 6.3.

7.1.1 The Idea of MLPG

In the MLPG method, the problem domain is represented by a set of arbitrarily distributed nodes, as shown in Figure 7.1. The weighted residual method is used to create the discrete system equations. The weighted residual method is, of course, in integral form, and a background mesh of cells is still required for the integration. The major idea in MLPG is that the implementation of the integral form of the weighted residual method is confined to a very small local subdomain of a node. This means that the weak form is satisfied at each node in the problem domain in a local integral sense. Therefore, the weak form is integrated over a "local quadrature domain" that is independent of other domains of other nodes. This is made possible by the use of the Petrov–Galerkin formulation, in which one has the freedom to choose the weight and trial functions independently. If the Galerkin formulation is used, one has to use the weight and trial functions from the same space, which presents difficulties in confining the integration to localized domains of desired simple shapes.

Because MLPG requires integrations only over localized quadrature domains, what we need now is only a background mesh of cells for the local quadrature. The quadrature domain can be arbitrary in theory, but very simple regularly shaped subdomains, such as circles and rectangles for 2D problems and bricks and spheres for 3D problems, are often used for ease of implementation (see Figure 7.1 for the 2D case). Because of the simplicity of the quadrature domain, the creation of the local background mesh of cells is easier to perform compared to the creation of the background mesh for the entire problem domain.

Note that the shape and dimensions of the quadrature domains do not have to be the same for all the nodes. Therefore, one is free to choose a proper shape for the local quadrature domain based on the local situation. This feature is important when the local quadrature domain encounters the global boundary of the problem. In addition, as long as the support domain is compact, MLPG will produce sparse system matrices. The major drawback of MLPG is the asymmetry of the system matrices due to the use of the Petrov–Galerkin formulation. The implications are the less efficiency in computation, the convergence and bound properties are less certain compared to the Galerkin formulations. Therefore, MLPG may suit better for fluid dynamic problems. Another drawback of MLPG is that the local background integration can be very tricky due to the complexity of the integrand produced by the Petrov–Galerkin approach, especially for quadrature domains that intersect with the boundary of the problem domain with complicated geometry.

7.1.2 Formulation of MLPG

Following the formulation of [9], we use the strong form of Equations 6.1 through 6.3 in indicial notations:

$$\sigma_{ij,j} + b_i = 0 \tag{7.1}$$

The boundary conditions are

$$\text{Essential boundary condition: } u_i = u_{\Gamma i} \quad \text{on } \Gamma_u \tag{7.2}$$

$$\text{Natural boundary condition: } \sigma_{ij} n_j = t_{\Gamma i} \quad \text{on } \Gamma_t \tag{7.3}$$

where
$i, j = 1, \ldots, d$
n_j is the jth component of the unit outward normal vector on the boundary

For node I, the local weighted residual method can be stated as (see Equation 5.4)

$$\int_{\Omega_Q} (\sigma_{ij,j} + b_i) \widehat{W}_I d\Omega - \alpha \int_{\Gamma_{Qu}} (u_i - u_{\Gamma i}) \widehat{W}_I d\Gamma = 0 \tag{7.4}$$

where $\widehat{W}$ is the weight or test function defined for the node, and we use the same weight function for all the d equations. Here we require $\widehat{W} \in \mathbb{C}^0(\Omega_Q)$. Ω_Q is the domain of quadrature (integration) for node I, Γ_{Qu} is the part of the essential boundary that intersects with the quadrature domain Ω_Q (see Figure 7.1), and α is the penalty factor that we have seen in Chapter 6. Here we use the same penalty factor for all the displacement constraint equations (essential boundary conditions) [9]. The first term in Equation 7.4 is for the equilibrium (in locally weighted average sense) requirement at node I, and the second term is only for the case when the essential boundary of the problem domain is part of the boundary of the local quadrature domain Ω_Q. If Ω_Q does not intersect with the essential boundary of the problem domain, the second term should be dropped.

Using the divergence theorem, we obtain

$$\int_{\Gamma_Q} \sigma_{ij} n_j \widehat{W}_I d\Gamma - \int_{\Omega_Q} \sigma_{ij} \widehat{W}_{I,j} d\Omega + \int_{\Omega_Q} b_i \widehat{W}_I d\Omega - \alpha \int_{\Gamma_{Qu}} (u_i - u_{\Gamma i}) \widehat{W}_I d\Gamma = 0 \tag{7.5}$$

where
$\Gamma_Q = \Gamma_{Q0} \cup \Gamma_{Qu} \cup \Gamma_{Qt}$
Γ_{Q0} is the internal boundary of the quadrature domain
Γ_{Qt} is the part of the natural boundary that intersects with the quadrature domain
Γ_{Qu} is the part of the essential boundary that intersects with the quadrature domain

When the quadrature domain is located entirely within the global domain, Γ_{Qt} and Γ_{Qu} vanish and $\Gamma_Q = \Gamma_{Q0}$. Unlike the Galerkin method, the Petrov–Galerkin method chooses the trial and test functions from different spaces. The weight function $\widehat{W}$ is purposely selected in such a way that it vanishes on Γ_{Q0} (in this case, we cannot use Heaviside weight functions). Note that the weight functions mentioned in Chapter 2, for example, the cubic or quartic spline, can be chosen to equal zero along the boundary of the internal quadrature domains; hence, they can be used as the weight functions for MLPG.

Equation 7.5 shows that the differentiation on the stresses is now "transferred" to the weight function. This reduces the consistency requirement when we approximate the trial displacement function that is used for obtaining the stresses.

Using a weight function $\widehat{W}$ that vanishes on Γ_{Q0}, we can then change the expression of Equation 7.5 to

$$\int_{\Omega_Q} \sigma_{ij} \widehat{W}_{I,j} d\Omega + \alpha \int_{\Gamma_{Qu}} u_i \widehat{W}_I d\Gamma - \int_{\Gamma_{Qu}} \sigma_{ij} n_j \widehat{W}_I d\Gamma = \int_{\Gamma_{Qt}} t_{\Gamma i} \widehat{W}_I d\Gamma + \alpha \int_{\Gamma_{Qu}} u_{\Gamma i} \widehat{W}_I d\Gamma + \int_{\Omega_Q} b_i \widehat{W}_I d\Omega$$

$$(7.6)$$

which is the local Petrov–Galerkin weak form. Here we require $u_i \in \mathbb{C}^0(\Omega_Q)$. When the quadrature domain is located entirely in the domain, integrals related to Γ_{Qu} and Γ_{Qt} vanish, and the Petrov–Galerkin form can be simplified as

$$\int_{\Omega_Q} \sigma_{ij} \widehat{W}_{I,j} d\Omega = \int_{\Omega_Q} b_i \widehat{W}_I d\Omega \qquad (7.7)$$

Equation 7.7 is used to establish the discrete equations for all the nodes whose quadrature domain falls entirely within the problem domain. Equation 7.6 is used to establish the discrete equations for all the boundary nodes or the nodes whose quadrature domain intersects with the problem boundary.

Using Equation 7.6 or 7.7 and integrating over the quadrature domain, leads to discretized system equations for each node in the problem domain. This gives a set of algebraic equations for each node. By assembling all these sets of equations, a set of discretized system equations for the entire problem domain can then be obtained. Note that MLPG establishes algebraic equations based on nodes in the problem domain, which is in fact very similar to the collocation or the finite difference procedure. It will be shown later that this feature can be used for imposition of the essential boundary conditions.

Equation 7.6 or 7.7 suggests that instead of solving the strong form of the system equation given in Equations 7.1 and 7.2, a "relaxed" weak form with integration over a small local quadrature domain is employed. This integration operation can "smear" out the numerical error and, therefore, make the discrete equation system more stable and accurate compared to the meshfree procedures that operate directly on the strong forms of system equations. MLPG guarantees satisfaction of the equilibrium equation at a node in an integral sense over a quadrature domain, but it does not ensure satisfaction of the system equation of strong form exactly at the node. The size of the quadrature domain determines the "relaxing" extent to the strong form differential equation. It will be shown later in the example problems that the quadrature domain needs to have sufficient dimension to produce an accurate and stable solution, and that too large a support domain does not necessarily provide significantly better results.

In MLPG, Equation 7.6 or 7.7 is to be satisfied for all the local quadrature domains for each and every node in the entire problem domain, including the boundaries. This implies that the equilibrium equation and the boundary conditions are satisfied node by node in a weak sense of the local weighted residual.

Note that if the delta function is used as the weight function, the method becomes a collocation method that is known to be unstable. This fact implies also when the local smoothing domain is too small the MLPG can become unstable.

Following the MLS approximation procedure, one can generate the shape function for each node using the nodes in support domain Ω_s of a point (not necessarily a node). The procedure is exactly the same as that in the EFG method, and for 2D problems, we shall have

$$\mathbf{u}^h = \left\{ \begin{array}{c} u_1 \\ u_2 \end{array} \right\}^h = \sum_{i \in S_n} \underbrace{\left[\begin{array}{cc} \phi_i^H & 0 \\ 0 & \phi_i^H \end{array} \right]}_{\boldsymbol{\Phi}_i} \underbrace{\left\{ \begin{array}{c} u_1 \\ u_2 \end{array} \right\}_i}_{\mathbf{u}_i} = \sum_{i \in S_n} \boldsymbol{\Phi}_i^H \mathbf{u}_i \tag{7.8}$$

where

S_n denotes the set of the nodes in the support domain Ω_s of point $\mathbf{x}_Q$

$\boldsymbol{\Phi}_i^H$ is the matrix of shape functions given by

$$\boldsymbol{\Phi}_i^H = \left[\begin{array}{cc} \phi_i^H & 0 \\ 0 & \phi_i^H \end{array} \right] \tag{7.9}$$

in which ϕ_i^H is the MLS shape function for node i that is created using nodes in the support domain Ω_s of point $\mathbf{x}_Q$. In Equation 7.8, $\mathbf{u}_i$ is the nodal displacement for node i:

$$\mathbf{u}_i = \left\{ \begin{array}{c} u_1 \\ u_2 \end{array} \right\}_i \tag{7.10}$$

After using the divergence theorem that leads to Equation 7.6, we have the discrete system equations in matrix form:

$$\int_{\Omega_Q} \widehat{\mathbf{V}}_I^{\mathrm{T}} \boldsymbol{\sigma} d\Omega + \alpha \int_{\Gamma_{Qu}} \widehat{\mathbf{W}}_I \mathbf{u} d\Gamma - \int_{\Gamma_{Qu}} \widehat{\mathbf{W}}_I \mathbf{t} d\Gamma = \int_{\Gamma_{Qt}} \widehat{\mathbf{W}}_I \mathbf{t}_\Gamma d\Gamma + \alpha \int_{\Gamma_{Qu}} \widehat{\mathbf{W}}_I \mathbf{u}_\Gamma d\Gamma + \int_{\Omega_Q} \widehat{\mathbf{W}}_I \mathbf{b} d\Omega \tag{7.11}$$

where $\widehat{\mathbf{V}}_I$ is a matrix that collects the derivatives of the weight functions in Equation 7.6, which has the form:

$$\widehat{\mathbf{V}}_I = \left[\begin{array}{cc} \widehat{W}_{I,x} & 0 \\ 0 & \widehat{W}_{I,y} \\ \widehat{W}_{I,y} & \widehat{W}_{I,x} \end{array} \right] \tag{7.12}$$

It is in fact the "strain" matrix caused by the weight (test) functions $\widehat{\mathbf{W}}$. In Equation 7.11, $\boldsymbol{\sigma}$ denotes the stress vector defined as

$$\boldsymbol{\sigma} = \mathbf{c}\boldsymbol{\varepsilon} = \mathbf{c}\mathbf{L}_n\mathbf{u}^h = \mathbf{c}\begin{bmatrix} \dfrac{\partial}{\partial x} & 0 \\ 0 & \dfrac{\partial}{\partial y} \\ \dfrac{\partial}{\partial y} & \dfrac{\partial}{\partial x} \end{bmatrix}\sum_{j\in S_n}\boldsymbol{\Phi}_j^H\mathbf{u}_j = \mathbf{c}\sum_{j\in S_n}\mathbf{B}_j\mathbf{u}_j \tag{7.13}$$

where

$$\mathbf{B}_j = \begin{bmatrix} \phi_{j,x}^H & 0 \\ 0 & \phi_{j,y}^H \\ \phi_{j,y}^H & \phi_{j,x}^H \end{bmatrix} \tag{7.14}$$

$\widehat{\mathbf{W}}$ is a matrix of weight functions given by

$$\widehat{\mathbf{W}}_I = \begin{bmatrix} \widehat{W}_I & 0 \\ 0 & \widehat{W}_I \end{bmatrix} \tag{7.15}$$

The tractions $\mathbf{t}$ of a point $\mathbf{x}$ can be written as

$$\mathbf{t} = \underbrace{\begin{bmatrix} n_x & 0 & n_y \\ 0 & n_y & n_x \end{bmatrix}}_{\mathbf{L}_n^T}\boldsymbol{\sigma} = \mathbf{L}_n^T\mathbf{c}\sum_{j\in S_n}\mathbf{B}_j\mathbf{u}_j \tag{7.16}$$

in which (n_x, n_y) is the unit outward normal vector on the boundary:

$$\mathbf{L}_n^T = \begin{bmatrix} n_x & 0 & n_y \\ 0 & n_y & n_x \end{bmatrix} \tag{7.17}$$

Substitution of Equations 7.8 and 7.13 through 7.16 into Equation 7.11 leads to the following discrete systems of linear equations for the Ith node:

$$\int_{\Omega_Q}\widehat{\mathbf{V}}_I^T\mathbf{c}\sum_{j\in S_n}\mathbf{B}_j\mathbf{u}_j d\Omega + \alpha\int_{\Gamma_Q}\widehat{\mathbf{W}}_I\sum_{j\in S_n}\boldsymbol{\phi}_j^H\mathbf{u}_j d\Gamma - \int_{\Gamma_Q}\widehat{\mathbf{W}}_I\mathbf{L}_n^T\mathbf{c}\sum_{j\in S_n}\mathbf{B}_j\mathbf{u}_j d\Gamma$$

$$= \int_{\Gamma_Q}\widehat{\mathbf{W}}_I\mathbf{t}_\Gamma d\Gamma + \alpha\int_{\Gamma_Q}\widehat{\mathbf{W}}_I\mathbf{u}_\Gamma d\Gamma + \int_{\Omega_Q}\widehat{\mathbf{W}}_I\mathbf{b}d\Omega \tag{7.18}$$

The matrix form of Equation 7.18 can be assembled as

$$\sum_{j\in S_n}\mathbf{K}_{Ij}\mathbf{u}_j = \mathbf{f}_I \tag{7.19}$$

where $\mathbf{K}_{Ij}$ is a 2×2 matrix called a *nodal stiffness matrix*, given by

$$\mathbf{K}_{Ij} = \int_{\Omega_Q} \hat{\mathbf{V}}_I^T \mathbf{c} \mathbf{B}_j d\Omega + \alpha \int_{\Gamma_Q} \hat{\mathbf{W}}_I \boldsymbol{\Phi}_j^H d\Gamma - \int_{\Gamma_Q} \hat{\mathbf{W}}_I \mathbf{L}_n^T \mathbf{c} \mathbf{B}_j d\Gamma \qquad (7.20)$$

and $\mathbf{f}_I$ is the *nodal force vector* with contributions from body forces applied in the problem domain, tractions applied on the natural boundary, as well as the penalty force terms.

$$\mathbf{f}_I = \int_{\Omega_Q} \hat{\mathbf{W}}_I \mathbf{b} d\Omega + \int_{\Gamma_{Qt}} \hat{\mathbf{W}}_I \mathbf{t}_\Gamma d\Gamma + \alpha \int_{\Gamma_{Qu}} \hat{\mathbf{W}}_I \mathbf{u}_\Gamma d\Gamma \qquad (7.21)$$

Equation 7.19 presents two linear equations for node I. Using Equation 7.19 for all n_t nodes in the entire problem domain, two independent linear equations can be obtained for each node. Assemble all these $2n_t$ equations to obtain the final global system equations of

$$\mathbf{K}_{2n_t \times 2n_t} \mathbf{U}_{2n_t \times 1} = \mathbf{F}_{2n_t \times 1} \qquad (7.22)$$

It can be easily seen that the system stiffness matrix $\mathbf{K}$ in the MLPG method is banded as long as the support domain is compact, but it is usually asymmetric.

7.1.3 Types of Domains

A number of subdomains are involved in practical implementation of the MLPG method. Each subdomain carries a different meaning, and some are similar to the domains used in the EFG method. The names of these domains are in fact very confusing in the current literature. Considering the terminology in the EFG method, we suggest the following systems of names for these subdomains.

All the subdomains are schematically drawn in Figure 7.2. The quadrature domain Ω_Q of node i at $\mathbf{x}_i$ is a domain for the integration in Equation 7.4. The weighted domain Ω_W is the domain where the weight (test) function is nonzero; i.e., $\hat{W}_I \neq 0$. Theoretically, the quadrature domain Ω_Q and the weighted domain Ω_W do not have to be the same, and Ω_W can be

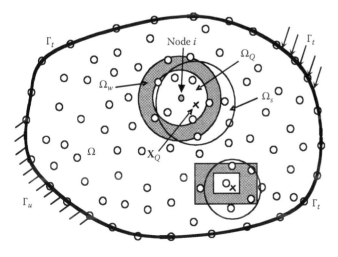

FIGURE 7.2
For node i, there are a number of subdomains: weighted domain Ω_W of a node at $\mathbf{x}_i$ is a domain in which $W_I \neq 0$; quadrature domain Ω_Q is in Ω_W and often $\Omega_Q = \Omega_W$; and the support domain Ω_s for a quadrature point $\mathbf{x}_Q$. (From Gu, Y.T. and Liu, G.R., *Comput. Mech.*, 27, 188, 2001. With permission.)

larger than Ω_Q. However, in practice we often use the same for both, i.e., $\Omega_Q = \Omega_W$, so that the curve integration along the interior boundary of Ω_Q will vanish, which simplifies the formulation and computation. Therefore, we always assume in this book that the quadrature domain is the weighted domain, unless specifically mentioned. The quadrature (or weight) domain can be theoretically arbitrary in shape. A circle or rectangular support domain is often used in practice for convenience.

For the local integration over Ω_Q, the Gauss quadrature may be used in the MLPG method. Therefore, for a mesh of local integration cells over its quadrature domain Ω_Q, a node is needed to employ the Gauss quadrature scheme. Because the quadrature domain is chosen to be simple, the creation of the local integral cells is not difficult. For each quadrature point $\mathbf{x}_Q$ in a cell, MLS interpolation is performed to compute the shape function and to obtain the integrand. A subdomain is then needed to choose the nodes for constructing the shape function. This subdomain noted as Ω_s carries exactly the same physical meaning of the support domain defined in Chapter 1. The support domain Ω_s is independent of the quadrature domain Ω_Q (or Ω_W).

The dimensions of these different domains will, of course, affect the results. These effects are addressed in later sections of this chapter via numerical examples.

7.1.4 Procedures for Essential Boundary Conditions

Enforcement of essential boundary conditions by the penalty method involves the choice of penalty factor α. If α is chosen improperly, instability or erroneous results will sometimes occur. Alternatively, methods using an orthogonal transformation technique [11,12] have been proposed for imposition of essential boundary conditions. This section introduces a *method of direction interpolation* for the imposition of essential boundary conditions, which makes use of the special feature of the MLPG. This method was used in [13] to simplify the MLPG formulation.

As discussed above, the MLPG method establishes equations node by node, which makes it possible to use different sets of equations for the interior and boundary nodes. For node J located on the essential boundary, one can enforce the boundary condition using the equation of MLS approximation in a collocation manner, i.e.,

$$u_J^h(\mathbf{x}) = \sum_{i \in S_n} \phi_i(\mathbf{x}) u_i = u_{\Gamma J} \qquad (7.23)$$

where $u_{\Gamma J}$ is the specified displacement at node J on the essential boundary. The foregoing equation is basically a linear algebraic equation for node J on the essential boundary. Therefore, for all the nodes on the essential boundary, there is no need to establish Equation 7.19. The essential boundary condition of Equation 7.23 is directly assembled into the global system equation. This treatment of the essential boundary condition is straightforward and very effective. It simplifies significantly the procedure of imposing essential boundary conditions, and the essential boundary conditions are satisfied exactly. Moreover, computation for all the nodes on the essential boundary has been simplified. This simple treatment is made possible because the MLPG method establishes discrete equations node by node.

Note also that this direct approach of imposing essential boundary conditions destroys the symmetry of the stiffness matrix. Fortunately, this does not create additional problems, because the stiffness matrix created using MLPG is not symmetric originally. If it were possible to apply this method in the EFG method, which produces symmetric matrices, it would not be used, because one probably loses more efficiency when the symmetry of the matrix is destroyed.

7.1.5 Numerical Investigation

Equations 7.20 and 7.21 require integration over the local quadrature domain and on the boundary that intersects with the quadrature domain. The integration has to be carried out via numerical quadrature schemes. In practice, the quadrature domain often needs to be further divided into cells, and the Gauss quadrature scheme is often used to evaluate the integration for each cell. Therefore, there will be a number of issues involved in the process, such as the number of cells and the number of the Gauss points to be used.

There exist, in general, difficulties in obtaining the exact (to machine accuracy) numerical integration in meshfree methods [11,13]. Insufficiently accurate numerical integration may cause deterioration in the numerical solution. In MLPG, the integration difficulty is more severe, because of the complexity of the integrand that results from the Petrov–Galerkin formulation. First, the shape functions constructed using MLS approximation have a complex feature, the shape functions have different forms in each small integration region, and the derivatives of the shape functions might have oscillations. Second, the overlapping of interpolation domains makes the integrand in the overlapping domain very complicated. To improve the accuracy of the numerical integration, the quadrature domain Ω_Q should be divided into small, regular partitions. In each small partition, more Gauss quadrature points should be used [11].

In this section, several numerical examples are employed to illustrate the implementation issues in the MLPG method using MLS approximation. The work was originally performed by Liu and Yan [13]. Rectangular quadrature domains Ω_Q are used, and the dimension of the quadrature domain for node I is defined as

$$(\alpha_Q \cdot d_{xI}) \times (\alpha_Q \cdot d_{yI}) \tag{7.24}$$

where
α_Q is the dimensionless size of the rectangular quadrature domains
d_{xI} is the average nodal spacing in the horizontal direction between two neighboring nodes in the vicinity of node I
d_{yI} is that in the vertical direction

The support domain Ω_s used for constructing MLS shape functions is also a rectangle. The tensor product weight function for 2D problems is given by

$$\widehat{W}(\mathbf{x} - \mathbf{x}_I) = \widehat{W}(r_x) \cdot \widehat{W}(r_y) = \widehat{W}_x \cdot \widehat{W}_y \tag{7.25}$$

where $\widehat{W}(r_x)$ and $\widehat{W}(r_y)$ are any of the weight functions listed in Chapter 2 where $\bar{d}$ is replaced by r_x and r_y, which are given by

$$r_x = \frac{\|x - x_I\|}{x_{\max}} \tag{7.26}$$

$$r_y = \frac{\|y - y_I\|}{y_{\max}} \tag{7.27}$$

where x_max and y_max are, respectively, the dimensions of the rectangle in the x and y directions given by

$$x_\text{max} = \alpha_s d_{xI} \tag{7.28}$$

$$y_\text{max} = \alpha_s d_{yI} \tag{7.29}$$

where α_s is the dimensionless size of the support domain for computing the MLS shape functions.

The quadrature domain Ω_Q for constructing the weight (test) function is also a rectangle, and the weight function given in Chapter 2 may be used, where $\bar{d}$ is also replaced by r_x and r_y, which are defined by Equations 7.26 and 7.27. However, x_max and y_max are given by

$$x_\text{max} = \alpha_Q d_{xI} \tag{7.30}$$

$$y_\text{max} = \alpha_Q d_{yI} \tag{7.31}$$

where α_Q is the dimensionless size of the quadrature or weight domain. Note that in our implementation of MLPG, the weighted domain Ω_W coincides with the domain of quadrature Ω_Q; hence, the dimension of the quadrature domain is also the same as that of the weighted domain.

The quadrature domain is divided evenly by $n_c \times n_c$ cells, and 4×4 Gauss sampling points are used for each cell. To assess the accuracy, the relative error is defined as

$$r_e = \frac{\|\varepsilon^\text{exact} - \varepsilon^\text{num}\|}{\|\varepsilon^\text{exact}\|} \tag{7.32}$$

where the energy norm is defined as

$$\|\varepsilon\| = \left(\frac{1}{2} \int_\Omega \boldsymbol{\varepsilon}^\text{T} \mathbf{c} \boldsymbol{\varepsilon} \, d\Omega \right)^{\frac{1}{2}} \tag{7.33}$$

7.1.6 Examples

An MLPG code has been developed based on the above-mentioned formulation, and is used to conduct the following investigations. The direct approach is used for the imposition of essential boundary conditions. In the examples presented in this section, a rectangular support domain is used, and the dimension of the support domain is fixed at $\alpha_s = 3.5$, unless specified otherwise.

Example 7.1: Patch Test

Consider a standard patch test in a domain of dimension $[0, 2] \times [0, 2]$ with a linear displacement applied along its boundary: $u_x = x$, $u_y = y$. Satisfaction of the patch test requires that the value of u_x, u_y at any interior node be given by the same linear displacement function and that the derivative of the displacement be constant.

Three patterns of nodal arrangement shown in Figure 7.3 are considered: (a) 9 nodes with regular arrangement, (b) 9 nodes with a randomly distributed internal node, and (c) 25 nodes

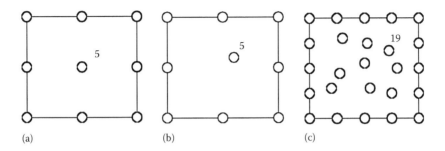

FIGURE 7.3
Nodal arrangement in patches for standard patch test: (a) 9 regular node patch; (b) 9 irregular node patch; and (c) 25 irregular node patch.

with irregular arrangement. The computational results have confirmed that MLPG can pass all the patch tests exactly (to machine accuracy) for both cubic and quartic spline weight functions with the given linear displacement boundary, as long as the integration is carried out accurately. Issues related to integration are discussed in detail in Example 7.2.

Example 7.2: High-Order Patch Test

A high-order patch of 3×6 shown in Figure 7.4 is used to study the effect of the order of polynomial basis used in MLS approximation and the quadrature domain on the numerical results of MLPG. The dimensionless material properties for the patch are as follows:

Young's modulus (dimensionless): $E = 1$

Poisson's ratio: $\nu = 0.25$

The following two cases are examined.

CASE 1

A uniform axial stress with unit intensity is applied on the right end. The exact solution for this problem should be

$$
\begin{aligned}
u_x &= x \\
u_y &= -y/4
\end{aligned}
\tag{7.34}
$$

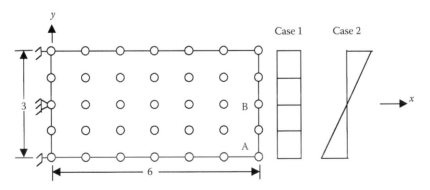

FIGURE 7.4
Nodal arrangement for high-order patch test.

TABLE 7.1

Displacement at the Right End of the High-Order Patch

	u_x at Point A			u_y at Point B		
α_Q	Exact	Numerical	Error (%)	Exact	Numerical	Error (%)
1.0	−6.0	−6.418	6.97	−12.0	−13.65	13.8
1.5	−6.0	−5.942	−0.97	−12.0	−11.84	−1.33
2.0	−6.0	−5.959	−0.68	−12.0	−11.86	−1.17

Note: Number of nodes in the entire domain n_t: $5 \times 7 = 35$; Gauss points: 4×4. α_Q, dimensionless size of the quadrature domain.

CASE 2

A linearly variable normal stress is applied on the right end.
    The exact solution for this problem should be

$$u_x = 2xy/3$$
$$u_y = -(x^2 + y^2/4)/3 \tag{7.35}$$

The linear basis is first used in MLS for creation of the shape functions. It is found that the MLPG can produce an exact solution to machine accuracy for case 1 and hence pass the patch test exactly. For case 2, however, it fails. Table 7.1 shows the relative error of displacement u_x and u_y at point A when the linear basis and quartic spline are used for case 2. It is shown that the MLPG results converge as the dimension of the quadrature domain α_Q increases, although it cannot produce the exact solution. Note that the exact solution of the displacement field for case 2 is quadratic, as shown in Equation 7.35. For the MLS shape function to reproduce the quadratic displacement field exactly, quadratic polynomial basis functions have to be used (see consistency issues with MLS discussed in Chapter 2). The patch is then attempted again using the quadratic basis, which confirms that MLPG has passed the patch test for case 2 as well.

Remark 7.1: Requirements for Local Petrov–Galerkin Methods to Pass the Patch Test

1. The shape functions are of at least linear consistency (see Chapter 2). This implies that the shape function is capable of producing at least linear field.

2. The essential boundary conditions (displacement constraints on the boundary of the patch) have to be satisfied accurately.

In addition, we naturally require sufficiently large local quadrature domains with proper weight functions and accurate numerical operations, such as integration to form system equations.

Example 7.3: Rectangular Cantilever

The rectangular cantilever described in Chapter 6 (Example 6.2) is tested again using the MLPG code. The cantilever is schematically drawn in Figure 6.4. The parameters for this example are as follows:

    Loading: $P = -1000$ N
    Young's modulus for the material: $E = 3.0 \times 10^7$ N/m^2

Poisson's ratio for two materials: $\nu = 0.3$

Height of the cantilever: $D = 12.0$ m

Length of the cantilever: $L = 48.0$ m

Three patterns with $55(5 \times 11)$, $189(9 \times 21)$, and $697(17 \times 41)$ regularly distributed nodes are employed to study the convergence of the MLPG method. Both linear basis and quartic spline are used. The results are presented for different numbers of integration cells n_c in each local quadrature domain Ω_Q (which is the same as the weighted domain Ω_W) and different dimensions of the quadrature domain.

Figures 7.5 and 7.6 plot the relationship between the relative error in strain defined in Equation 7.32 and the cell number n_c. Results in these two figures clearly show that accuracy can be improved significantly by increasing the number of cells for local integration. These findings suggest the importance of subdivisions of the quadrature domain. Both Figures 7.5 and 7.6 also indicate an important fact: too large ($\alpha_Q = 3.0$) or too small ($\alpha_Q = 1.0$) a quadrature domain will give less accurate results. When the quadrature domain is too small, the area for the "smear" error is not large enough, and when the quadrature is too large, the error in the numerical integration will affect the accuracy of the results. Both Figures 7.5 and 7.6 show that a quadrature domain of $\alpha_Q = 1.5$ gives the most accurate results. This suggests that the dimension of the rectangular quadrature domain should be about 1.5 times the local nodal distance. This rule of 1.5 times nodal distance is widely used in meshfree methods based on local weak forms.

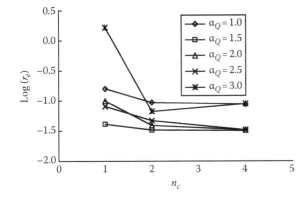

FIGURE 7.5
Convergence in terms of relative error in strains computed in the rectangular cantilever using MLPG with MLS approximation (n_c = number of subdivision of the rectangular quadrature domain; α_Q = the dimension parameter of the rectangular quadrature domain; domain of support: $\alpha_s = 3.5$; total number of nodes: $n_t = 55$).

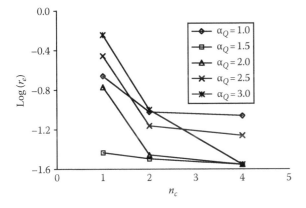

FIGURE 7.6
Same as Figure 7.5, but the number of field nodes $n_t = 189$.

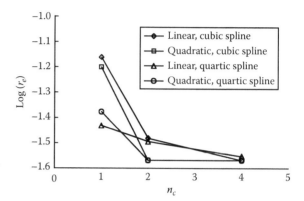

FIGURE 7.7

Convergence in terms of relative error in strains computed in the rectangular cantilever using MLPG with MLS approximation. Effects of spline weight functions ($\alpha_Q = 1.5$; $n_t = 189$; $\alpha_s = 3.5$).

The effects of the spline and basis functions on the relative error have also been investigated. The results are shown in Figure 7.7. It is found that there is no significant difference in accuracy using cubic and quartic spline weighted functions for a sufficient number of subdivisions of the quadrature domain. Using a quadratic basis function can somewhat increase the accuracy of the results, but it is not a clear indication.

Irregularity of the nodal arrangement is also investigated. The irregularity of nodes is created by changing the coordinates of the interior nodes in the cantilever in the following manner:

$$x_I = x_I \pm c_n d_{xI}$$
$$y_I = y_I \pm c_n d_{yI} \qquad (7.36)$$

where c_n is the parameter that controls the irregularity of the nodes. For $c_n = 0.4$, some of the internal nodes are moved up to $0.4 d_{xI}$ in the horizontal direction and $0.4 d_{yI}$ in the vertical direction from its regular position. Parameter c_n is allowed to vary randomly in the range of 0.0–0.4. Figure 7.8 shows a typical irregular nodal arrangement. Table 7.2 shows the relative error defined in Equation 7.32 obtained using irregular node arrangements of different c_n values. It is seen that the irregularity of nodes has very little effect on the accuracy of the results. This fact reveals a very important feature of MLPG: it is stable for irregular nodal arrangements. The results based on $c_n = 0.4$ are obtained and plotted in Figure 7.9a for deflection of the cantilever, and in Figure 7.9b for the shear stress distribution at the central section of $x = 24$ m. Those results confirm that the effect of the nodal irregularity is very small.

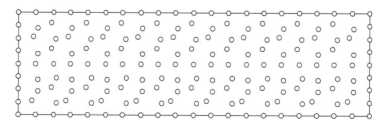

FIGURE 7.8

Irregular nodal arrangement for the rectangular cantilever (189 nodes).

TABLE 7.2

Relative Error for Irregular Node Arrangement

c_n	0.0	0.1	0.2	0.3	0.4
$r_e \ (\times 10^{-2})$	2.77	2.81	2.84	2.87	2.89

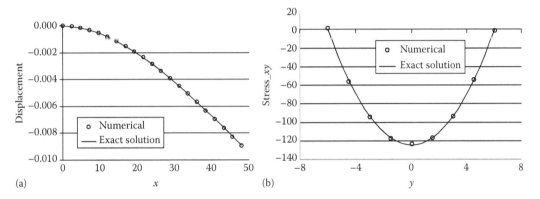

(a)

(b)

FIGURE 7.9

Results for the cantilever computed using MLPG with 189 irregular nodes. Comparison with the exact solution. (a) Deflection; (b) shear stress distribution at central section.

Example 7.4: Square Panel with a Circular Hole

A square panel of 2D solid with a circular hole subjected to a unidirectional tensile load in the x-direction is considered, as shown in Figure 7.10a. The plane stress condition is assumed. Due to symmetry, only the upper right quarter of the square panel is modeled, as shown in Figure 7.10b. Corresponding symmetric boundary conditions are applied on $x = 0$ and $y = 0$, i.e.,

$$u_x = 0, \quad \sigma_{xy} = 0 \quad \text{when } x = 0 \tag{7.37}$$

and

$$u_y = 0, \quad \sigma_{xy} = 0 \quad \text{when } y = 0 \tag{7.38}$$

The boundary condition at the right edge is

$$\sigma_{xx} = p, \quad \sigma_{yy} = \sigma_{xy} = 0 \quad \text{when } x = 5 \tag{7.39}$$

and the boundary condition at the upper edge is

$$\sigma_{xx} = 0, \quad \sigma_{yy} = \sigma_{xy} = 0 \quad \text{when } y = 5 \tag{7.40}$$

The parameters are listed as follows:

Loading: $p = 1$ N/m
Young's modulus: $E = 1.0 \times 10^3$ N/m^2
Poisson's ratio: $\nu = 0.3$
Diameter of the hole: $a = 1.0$ m
Length of the panel: $b = 5$ m

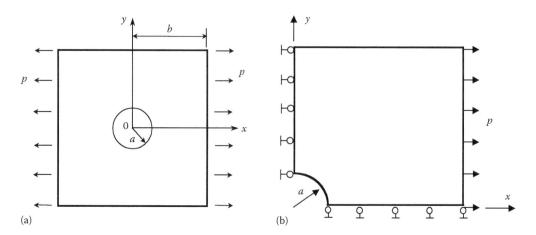

FIGURE 7.10
Square panel with a hole subjected to a tensile load in the horizontal direction. (a) Problem setting; (b) a quarter model.

The analytical solution of displacement and the stress fields within the panel are listed below in the polar coordinates of (r, θ).

The displacement in the radial direction is given by

$$u_r = \frac{p}{4\mu} \left\{ r \left[\frac{\kappa - 1}{2} + \cos 2\theta \right] + \frac{a^2}{r} [1 + (1 + \kappa) \cos 2\theta] - \frac{a^4}{r^3} \cos 2\theta \right\} \qquad (7.41)$$

and the displacement in the tangent direction can be calculated using

$$u_\theta = \frac{p}{4\mu} \left[(1 - \kappa) \frac{a^2}{r} - r - \frac{a^4}{r^3} \right] \sin 2\theta \qquad (7.42)$$

where

$$\mu = \frac{E}{2(1 + v)} \qquad \kappa = \begin{cases} 3 - 4v & \text{plane strain} \\ \dfrac{3 - v}{1 + v} & \text{plane stress} \end{cases} \qquad (7.43)$$

The normal stress in the x-direction can be obtained using

$$\sigma_x(x, y) = p \left\{ 1 - \frac{a^2}{r^2} \left\{ \frac{3}{2} \cos 2\theta + \cos 4\theta \right\} + \frac{3a^4}{2r^4} \cos 4\theta \right\} \qquad (7.44)$$

The normal stress in the y-direction is

$$\sigma_y(x, y) = -\frac{a^2 p}{r^2} \left\{ \frac{1}{2} \cos 2\theta - \cos 4\theta \right\} - \frac{3a^4}{2r^4} \cos 4\theta s \qquad (7.45)$$

and the shear stress is given by

$$\sigma_{xy}(x, y) = -\frac{a^2 p}{r^2} \left\{ \frac{1}{2} \sin 2\theta + \sin 4\theta \right\} + \frac{3a^4}{2r^4} \sin 4\theta s \qquad (7.46)$$

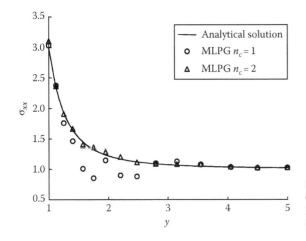

FIGURE 7.11
Comparison between the exact solution and MLPG with MLS approximation for σ_{xx} at $x = 0$ ($\alpha_s = 3.5$, $\alpha_Q = 3.0$).

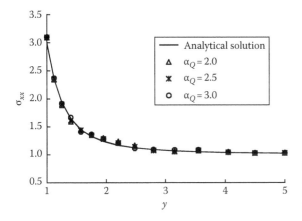

FIGURE 7.12
Comparison between the exact solution and MLPG with MLS approximation for σ_{xx} at $x = 0$ ($N = 165$, $\alpha_s = 3.5$, $n_c = 2$).

where (r, θ) are the polar coordinates and θ is measured counterclockwise from the positive x axis. When the condition $b/a = 5$ is satisfied, the solution of a finite square panel should be very close to that of an infinite 2D solid. Therefore, the analytical results given in Equations 7.44 through 7.46 are employed as the reference results for comparison.

A total of 165 nodes are used to represent the domain. The stress components σ_{xx} obtained at $x = 0$ for $\alpha_s = 3.0$ are compared with the exact solution in Figure 7.11. Two types of cells of $n_c = 1$ and $n_c = 2$ are used. Figure 7.11 shows that finer cells ($n_c = 2$) give a more accurate result than coarse cells ($n_c = 1$). The results suggest again the importance of subdivision in the local quadrature domain. Figure 7.12 plots the results for different sizes of support domain α_Q, which implies a stable result with different α_Q.

Example 7.5: Half-Plane Problem

Stress analysis is carried out for a half plane of elastic solid subjected to a concentrated force, as shown in Figure 7.13. The results are compared with those obtained using the finite element method (FEM) at the section S–S' for the same distribution of nodes. We present the results of comparison only for stress, as it is much more critical than displacement. Figure 7.14 shows the distribution of the normal stress σ_x and the shear stress τ_{xy} along section S–S'. The results are

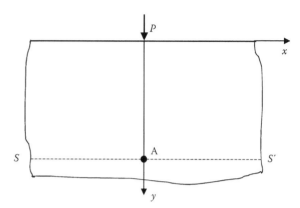

FIGURE 7.13
Half plane of elastic solid subjected to a vertical
concentrated force P.

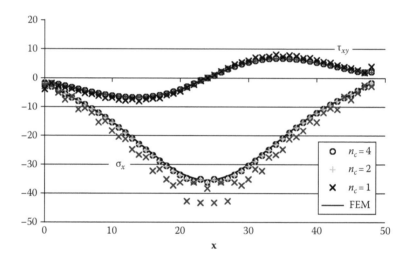

FIGURE 7.14
Stress distribution at section of $S–S'$ ($\alpha_Q = 1.5$).

TABLE 7.3

Comparison of Normal Stress at Point A with FEM
Result ($= 35.4$)

Number of Subdivision of Quadrature Domain	Dimension of Quadrature Domain (α_Q)		
	1.0	2.0	3.0
1×1	46.0	—	—
2×2	44.1	36.7	—
4×4	44.2	36.7	35.7

obtained using different integral cells n_c and for given $\alpha_Q = 1.5$. Table 7.3 lists the normal stress at point A compared with those obtained using FEM. The results indicate that a sufficiently large dimension of quadrature domain together with a corresponding sufficient number of subdivisions of the quadrature domain for integration is necessary to obtain an accurate solution. Subdivision of the quadrature domain, however, leads to additional computation work.

An acceptable combination of the number of subdivisions and the dimension of the quadrature domain is quite complicated, not very clear at this stage, and needs to be further investigated. The above examples seem to suggest that four subdivisions ($n_c = 2$) work well for rectangular quadrature domains of $\alpha_Q = 2.0$. It has been found for these examples that $n_c = \alpha_Q$ seems to be a necessary condition to obtain a reasonably accurate result. Because we find no reason to use a quadrature domain larger than $\alpha_Q = 2.0$, four subdivisions should be a good and economic choice for normal cases. If, for any reason, a large quadrature domain has to be used, our suggestion would be

$$n_c = \text{round-up-to-the-nearest-even-number} \, (\alpha_Q) \tag{7.47}$$

7.2 MLPG for Dynamic Problems

7.2.1 Statement of the Problem

Vibration analysis for structures is a very important field in computational mechanics. These dynamics problems are classically described by a linear partial differential equation associated with a set of boundary conditions and initial conditions. Exact analyses of these types of boundary and initial value problems are usually very difficult. Analytical solutions are available for very few problems with very simple geometry and boundary and initial conditions. Therefore, numerical techniques with different discretization schemes, such as FEM, have been widely used in solving practical vibration problems in science and engineering.

Examples presented in previous sections have demonstrated that the MLPG method works well for static mechanics problems. It is, therefore, a natural extension to develop further MLPG for dynamic mechanics problems of 2D solids for both free-vibration and forced-vibration analyses [16].

This section introduces their formulation. First, the local weak forms are presented using the weighted residual method locally and the strong form of partial differential dynamic system equations. MLS approximation is used to obtain the MLS shape functions, which are fed to the local Petrov–Galerkin formulation to derive a set of discretized dynamic system equations. In free-vibration analysis, the essential boundary conditions are formulated separately using the method of direct interpolation. The boundary conditions are then imposed utilizing orthogonal transform techniques to modify the discretized unconstrained dynamic system equations to obtain the eigenvalue equation. Frequencies and eigenmodes of free vibration are obtained by solving the eigenvalue equation. In forced-vibration analysis, the penalty method is used to implement the essential conditions. Both the explicit time integration method (the central difference method [CDM]) and the implicit time integration method (the Newmark method) are used to solve the forced vibration system equations.

Programs of the MLPG method have been developed, and a number of numerical examples of free-vibration and forced-vibration analyses are presented to demonstrate the convergence, validity, and efficiency of the present methods. Some important parameters on the performance of the present MLPG method are also investigated in great detail.

The strong form of the initial/boundary value problem for small displacement elastodynamics can be given as follows (see Chapter 3):

$$\sigma_{ij,j} + b_i = \rho \ddot{u}_i + \eta_c \dot{u}_i \tag{7.48}$$

where
  ρ is the mass density
  η_c is the damping coefficient
  u_i is the displacement
  $\ddot{u}_i = \partial^2 u_i/\partial t^2$ is the acceleration
  $\dot{u}_i = \partial u_i/\partial t$ is the velocity
  σ_{ij} is the stress tensor
  b_i is the body force tensor
  $()_{,j}$ denotes the operation of $\partial/\partial x_j$

The auxiliary conditions are given as follows:

$$\text{Naural boundary condition: } \sigma_{ij}n_j = t_{\Gamma i} \quad \text{on } \Gamma_t \tag{7.49}$$

$$\text{Essential boundary condition: } u_i = u_{\Gamma i} \quad \text{on } \Gamma_u \tag{7.50}$$

$$\text{Displacement initial condition: } u_i(\mathbf{x}, t_0) = \tilde{u}_i(\mathbf{x}) \quad \mathbf{x} \in \Omega \tag{7.51}$$

$$\text{Velocity initial condition: } \dot{u}_i(\mathbf{x}, t_0) = \tilde{\dot{u}}_i(\mathbf{x}) \quad \mathbf{x} \in \Omega \tag{7.52}$$

in which the $u_{\Gamma i}$, $t_{\Gamma i}$, $\tilde{u}_0$, and $\tilde{\dot{u}}_i$ denote the prescribed displacements, tractions, initial displacements, and velocities, respectively. Note that the differences between the dynamic system equations for static and dynamic problems are (1) the inertial and damping terms in the equilibrium equations and (2) the additional equations of initial conditions.

7.2.2 Free-Vibration Analysis

The governing equation for an undamped free vibration can be written as follows:

$$\sigma_{ij,j} = m\ddot{u} \tag{7.53}$$

The boundary condition for the free vibration is reduced to only the essential boundary condition, Equation 7.50. In free-vibration analysis, the system is assumed to undergo harmonic motion, and the displacement $\mathbf{u}(\mathbf{x}, t)$ can be written in the form:

$$\mathbf{u}(\mathbf{x}, t) = \mathbf{u}(\mathbf{x})\sin(\omega t + \varphi) \tag{7.54}$$

where ω is the frequency of free vibration. Substituting Equation 7.54 into Equation 7.53 leads to the following equations of equilibrium for free vibration:

$$\sigma_{ij,j} + \omega^2 \rho u_i = 0 \tag{7.55}$$

It should be noted that the stresses, σ, and displacements, $\mathbf{u}$, in Equation 7.55 are only functions of coordinate $\mathbf{x}$ for a given frequency ω.

A local weak form of Equation 7.55, over a local quadrature domain Ω_Q bounded by Γ_Q, can be obtained using the weighted residual method with integration over Ω_Q for a node in the problem domain.

$$\int_{\Omega_Q} \widehat{W}\left(\sigma_{ij,j} + \omega^2 \rho u_i\right) d\Omega = 0 \tag{7.56}$$

where $\widehat{W}$ is the weight (or test) function, and here we use the same weight function for all the equations and nodes.

The first term on the left-hand side of Equation 7.56 can be integrated by parts to become

$$\int_{\Gamma_Q} \widehat{W}\sigma_{ij}n_j d\Gamma - \int_{\Omega_Q} \left(\widehat{W}_{,j}\sigma_{ij} - \widehat{W}\omega^2 m u_i \right) d\Omega = 0 \tag{7.57}$$

The choice of Ω_Q is the same as that discussed in Section 7.1.2 for static problems. As shown in Figure 7.1, the boundary Γ_Q for the support domain Ω_Q is usually composed of three parts: the internal boundary Γ_{Q0} and the boundaries Γ_{Qu} and Γ_{Qt}, over which the essential and natural boundary conditions are specified. Imposing the natural boundary condition and noting that $\sigma_{ij}n_j = \partial u/\partial n \equiv t_i$ in Equation 7.57, and the fact that $\widehat{W}(\mathbf{x}) = 0$ on Γ_{Q0}, we obtain

$$\int_{\Gamma_{Qu}} \widehat{W}t_i d\Gamma + \int_{\Gamma_{Qt}} \widehat{W}t_{\Gamma i} d\Gamma - \int_{\Omega_Q} \left(\widehat{W}_{,j}\sigma_{ij} - \widehat{W}\omega^2 m u_i \right) d\Omega = 0 \tag{7.58}$$

For a support domain located entirely within the global domain, there is no intersection between Γ_Q and the global boundary Γ, and $\Gamma_{Q0} = \Gamma_Q$. This leads to the vanishing of integrals over Γ_{Qu} and Γ_{Qt}. Also note that for free vibration, we should have $t_{\Gamma i} = 0$ on Γ_t; the integrals over Γ_{Qt} vanish for all nodes in the free-vibration analysis. Equation 7.58 can be further expressed as follows. For nodes whose support domains do not intersect with the problem boundary:

$$\int_{\Omega_Q} \left(\widehat{W}_{,j}\sigma_{ij} - \widehat{W}\omega^2 \rho u_i \right) d\Omega = 0 \tag{7.59}$$

For nodes whose support domains intersect with the problem boundary:

$$\int_{\Gamma_{Qu}} \widehat{W}t_i d\Gamma - \int_{\Omega_Q} \left(\widehat{W}_{,j}\sigma_{ij} - \widehat{W}\omega^2 \rho u_i \right) d\Omega = 0 \tag{7.60}$$

MLS approximation (Equation 7.8) is then used to approximate the field variables at any point in the support domain Ω_s of a node. Substituting Equation 7.8 and a weight function into the local weak form Equation 7.59 or 7.60 for each and every node in the problem domain leads to the following discrete system equations. The procedure is exactly the same as in Section 7.2.1, except that the inertial term needs to be treated, which leads to

$$\mathbf{KU} - \omega^2 \mathbf{MU} = 0 \tag{7.61}$$

where the global stiffness matrix $\mathbf{K}$ is assembled using the nodal stiffness matrix $\mathbf{K}_{IJ}$. For nodes whose quadrature domains intersect with the problem boundary, we have

$$\mathbf{K}_{IJ} = \int_{\Omega_Q} \widehat{\mathbf{V}}_I^T \mathbf{cB}_J d\Omega - \int_{\Gamma_{Qu}} \widehat{\mathbf{W}}_I \mathbf{L}_n^T \mathbf{cB}_J d\Gamma \tag{7.62}$$

where $\widehat{\mathbf{V}}_I$, $\mathbf{B}_I$, $\widehat{\mathbf{W}}_I$, and $\mathbf{L}_n$ are defined, respectively, by Equations 7.12, 7.14, 7.15, and 7.17. For nodes whose quadrature domains do not intersect with the problem boundary, the nodal stiffness matrix is simplified as

$$\mathbf{K}_{IJ} = \int_{\Omega_Q} \widehat{\mathbf{V}}_I^{\mathrm{T}} \mathbf{c} \mathbf{B}_J \mathrm{d}\Omega \tag{7.63}$$

The "mass" matrix $\mathbf{M}$ is obtained using

$$\mathbf{M}_{IJ} = \int_{\Omega_Q} \rho \widehat{\mathbf{W}}_I \mathbf{\Phi}_J \mathrm{d}\Omega \tag{7.64}$$

where $\mathbf{\Phi}_J$ is a matrix of the MLS shape function for node J, given by Equation 7.9.

For free-vibration analysis, Equation 7.61 can also be written in the following form of eigenvalue equation:

$$(\mathbf{K} - \lambda \mathbf{M})\mathbf{q} = 0 \tag{7.65}$$

where

$\mathbf{q}$ is the eigenvector
λ is termed an eigenvalue that relates to the natural frequency in the form

$$\lambda = \omega^2 \tag{7.66}$$

Equation 7.65 is the unconstrained eigenvalue equation that contains the rigid movement of the solid. To determine the frequencies, ω, and free-vibration modes for a constrained system, it is necessary to impose the essential boundary condition defined by Equation 7.50.

7.2.3 Imposition of Essential Boundary Conditions for Free Vibrations

In the MLPG method that uses MLS shape functions, special effort has to be made to enforce essential boundary conditions for dynamic problems, because the shape functions constructed by MLS approximation lack the delta function property. In previous sections we have shown the penalty method [9] and the direct collocation method [13] for problems of static stress analyses. Here, we introduce the method using orthogonal transform techniques [11,12,18,19] to establish a system equation of constrained free vibration.

Note the fact that for free-vibration analysis, the essential boundary conditions are always homogeneous, meaning that $\mathbf{u}_{\Gamma i} = 0$ in Equation 7.50. Substituting Equation 7.8 into Equation 7.50, we find a set of algebraic linear equations of constraints

$$\mathbf{C}\mathbf{q} = 0 \tag{7.67}$$

where $\mathbf{C}$ is a flat matrix of $n_c \times N_t$ with many zero elements, n_c is the total number of constrained degrees of freedom, N_t is the total number of degrees of freedom of the entire system, and n_r is the rank of $\mathbf{C}$, namely, the number of independent constraints. If the domain is represented by N_n nodes, $N_t = 2N_n$ for 2D solid mechanics problems. Using singular value decomposition [20], matrix $\mathbf{C}$ can be decomposed as

$$\mathbf{C} = \mathbf{U}_{n_c \times n_c} \begin{bmatrix} \boldsymbol{\sigma}_{n_r \times n_r} & 0 \\ 0 & 0 \end{bmatrix}_{n_c \times N_t} \mathbf{V}^{\mathrm{T}}_{N_t \times N_t} \tag{7.68}$$

where
 $\mathbf{U}$ and $\mathbf{V}$ are orthogonal matrices
 Σ is a diagonal matrix with singular values of $\mathbf{C}$ for its diagonal terms

The matrix $\mathbf{V}$ can be written as

$$\mathbf{V}^{\mathrm{T}} = \left\{ \mathbf{V}_{N_t \times n_r}, \mathbf{V}_{N_t \times (N_t - n_r)} \right\}^{\mathrm{T}} \tag{7.69}$$

Performing coordinate transformation of

$$\mathbf{q} = \mathbf{V}_{N_t \times (N_t - n_r)} \tilde{\mathbf{q}} \tag{7.70}$$

the change of coordinates satisfies the constraint equation of Equation 7.67. Substituting Equation 7.71 into Equation 7.65 leads to

$$(\tilde{\mathbf{K}} - \omega^2 \tilde{\mathbf{M}}) \tilde{\mathbf{q}} = 0 \tag{7.71}$$

where the condensed stiffness matrix $\tilde{\mathbf{K}}$ given by

$$\tilde{\mathbf{K}}_{(N_t - n_r) \times (N_t - n_r)} = \mathbf{V}^{\mathrm{T}}_{(N_t - n_r) \times N_t} \mathbf{K}_{N_t \times N_t} \mathbf{V}_{N_t \times (N_t - n_r)} \tag{7.72}$$

and the condensed stiffness mass matrix $\tilde{\mathbf{M}}$ becomes

$$\tilde{\mathbf{M}}_{(N_t - n_r) \times (N_t - n_r)} = \mathbf{V}^{\mathrm{T}}_{(N_t - n_r) \times N_t} \mathbf{M}_{N_t \times N_t} \mathbf{V}_{N_t \times (N_t - n_r)} \tag{7.73}$$

Equation 7.72 is the eigenvalue equation for free vibration of a constrained solid.

It can be easily seen that the stiffness matrix $\mathbf{K}$ and the mass matrix $\mathbf{M}$ developed using the Petrov–Galerkin approach will be asymmetric. They will be banded as long as the support domain is compact.

A numerical integration is needed to evaluate the integration for computing both the stiffness and mass matrices, and Gauss quadrature can be used. Here we investigate also the effects of the dimensions of three local domains shown in Figure 7.2. In computing the stiffness matrix, it should be noted that for nodes whose quadrature domain intersects with the boundary of the problem domain, Equation 7.62 should be used. For interior nodes whose quadrature domains do not intersect with the boundary of the problem domain, Equation 7.63 should be used. In computing the mass matrix, Equation 7.64 is used for all the nodes.

Because the problem domains in the following examples are rectangles, rectangular local domains defined in Section 7.1.5 are used.

7.2.4 Numerical Examples

The MLPG method is used for free-vibration analysis of structures made of 2D solids.

Example 7.6: Rectangular Cantilever

The MLPG method is applied to analyze the free vibration of a rectangular cantilever, as shown in Figure 6.4. A plane stress problem is considered. The parameters for this example are as follows:

Young's modulus for the material: $E = 2.1 \times 10^4$ kgf/mm^2

Poisson's ratio for two materials: $\nu = 0.3$

Mass density: $\rho = 8.0 \times 10^{-10}$ kgfs^2/mm^4

Thickness of the cantilever: $t = 1$ mm

Height of the cantilever: $D = 10$ mm

Length of the cantilever: $L = 100$ mm

The problem has been analyzed in [21] using the node-by-node meshless (NBNM) method. The cantilever is first represented using a number of field nodes. Figure 7.15 shows two kinds of nodal arrangements: coarse (63 nodes) and fine (306 nodes). The effects of dimensions of the quadrature domain are investigated using different α_Q defined in Equations 7.30 and 7.31. It has been found that $\alpha_Q = 1.5$ to 2.5 can yield almost identical results in free-vibration analysis. This finding agrees well with that for static analyses presented in the previous section. From the static problems, we found that the quadrature domain used should be as small as possible, to reduce the burden in numerical integration. Therefore, $\alpha_Q = 1.5$ is used in the following free-vibration analysis.

Frequency results of these two nodal arrangements obtained by MLPG are listed in Table 7.4. The results obtained by the FEM software ABAQUS and the NBNM method [21] are also listed in the table. The mesh used in the FEM mode has the same nodal arrangement. From this table, one can observe that the results by the present MLPG method are in a good agreement with those obtained using FEM and the NBNM method. The convergence of the present method is also demonstrated in this table. As the number of nodes increases, results obtained by the present MLPG approach the FEM results (if we consider the FEM results as a reference). The lowest 10 vibration modes obtained by the MLPG method are plotted in Figure 7.16. Comparison of the FEM results with [21] results reveals that they are almost identical.

For cantilevers with small slenderness ratios, the Euler–Bernoulli beam theory can be applied to obtain an analytical solution for their natural frequencies. The slenderness of a cantilever is expressed by the slenderness ratio, r/L, where $r = \sqrt{I/A}$ is the radius of gyration of the cross section, I is the moment of inertia of the cross section of the cantilever, and L is the length of the cantilever. To further benchmark the MLPG code developed, cantilevers with two slenderness ratios, $r/L = 0.029$ ($L = 100$, $D = 10$, $t = 1.0$) and 0.144 ($L = 100$, $D = 50$, $t = 1.0$), are analyzed.

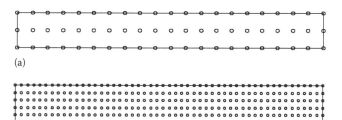

(a)

FIGURE 7.15
Nodal arrangement: (a) 63 nodes; (b) 306 nodes. (From Gu, Y.T. and Liu, G.R., *Comput. Mech.*, 27, 188, 2001. With permission.)

(b)

TABLE 7.4

Natural Frequencies (Hz) of a Rectangular Cantilever Computed Using Meshfree and FEM Models with Different Nodes ($\alpha_Q = 1.5$, $\alpha_s = 3.5$ for MLPG)

Mode	Coarse Model (63 Nodes)			Fine Model (306 Nodes)		
	MLPG	[21]	FEM (ABAQUS)	MLPG	[21]	FEM (ABAQUS)
1	919.47	926.10	870	824.44	844.19	830
2	5,732.42	5,484.11	5,199	5070.32	5,051.21	4979
3	12,983.25	12,831.88	12,830	12,894.73	12,827.60	12,826
4	14,808.64	14,201.32	13,640	13,188.12	13,258.21	13,111
5	26,681.81	25,290.04	24,685	24,044.43	23,992.82	23,818
6	38,961.74	37,350.18	37,477	36,596.15	36,432.15	36,308
7	40,216.58	38,320.59	38,378	38,723.90	38,436.43	38,436
8	55,060.24	50,818.64	51,322	50,389.01	49,937.19	49,958
9	64,738.59	63,283.70	63,584	64,413.89	63,901.16	63,917
10	68,681.87	63,994.48	65,731	64,937.83	64,085.90	64,348

Source: Gu, Y.T. and Liu, G.R., *Comput. Mech.*, 27, 188, 2001. With permission.

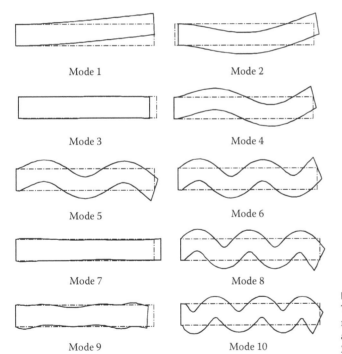

Mode 1 Mode 2

Mode 3 Mode 4

Mode 5 Mode 6

Mode 7 Mode 8

Mode 9 Mode 10

FIGURE 7.16
The lowest 10 vibration modes of the rectangular cantilever. (From Gu, Y.T. and Liu, G.R., *Comput. Mech.*, 27, 188, 2001. With permission.)

The frequency results are listed in Table 7.5. For a cantilever of small slenderness ratio, one expects a very good prediction from the Euler–Bernoulli beam theory. Comparison with the Euler–Bernoulli beam results reveals that, as the slenderness ratio r/L decreases, the natural frequencies of this 2D cantilever approach the values for an Euler–Bernoulli model. For the slender case of $r/L = 0.029$, the Euler–Bernoulli solution can be considered very close to the exact solution, and should be used as the reference. Table 7.5 shows that MLPG gives more accurate results

TABLE 7.5

Natural Frequencies (Hz) of a Rectangular Cantilever with Different Slenderness

Modes	$r/L = 0.144$			$r/L = 0.029$		
	MLPG	FEM (ABAQUS)	Euler Beam	MLPG	FEM (ABAQUS)	Euler Beam
1	3,565.81	3,546.1	4,138.23	824.44	830.19	827.65
Error with Euler beam (%)	−13.83	−14.31	—	−0.39	0.31	—
2	13,025.06	12,864	25,933.86	5070.32	4979	5186.77
Error with Euler beam (%)	18.56	20.6	—	−2.24	−4.01	—

Source: Gu, Y.T. and Liu, G.R., *Comput. Mech.*, 27, 188, 2001. With permission.
Note: 306 nodes are used in MLPG and FEM; $\alpha_Q = 1.5$, $n_s = 3.5$ are used for MLPG.

compared with finite element results. It is well known that the fundamental (first) frequency of a 2D solid obtained by FEM should be larger than the exact value, meaning that the FEM results always give the upper bound of the exact results in terms of eigenvalues, and approach the exact results from the top. The MLPG result, however, does not guarantee an upper-bound solution. It approaches the exact solution from both sides. This is caused by the use of the local residual weak formulation.

Example 7.7: Cantilever with Variable Cross Section

In this example, the present MLPG method is used in the free-vibration analysis of a cantilever with varying cross section, as shown in Figure 7.17. Results are obtained for the following parameters:

Young's modulus for the material: $E = 3.0 \times 10^7 \text{ N/m}^2$

Poisson's ratio for two materials: $\nu = 0.3$

Mass density: $\rho = 1 \text{ kg/m}^3$

Thickness of the cantilever: $t = 1 \text{ m}$

Length of the cantilever: $L = 10 \text{ m}$

Height of the cantilever: $D(0) = 5 \text{ m}$ and $D(10) = 3 \text{ m}$

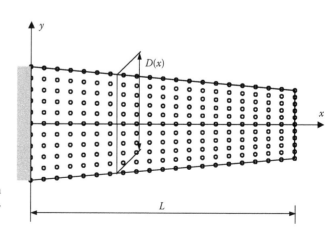

FIGURE 7.17
Cantilever of varying cross section. (From Gu, Y.T. and Liu, G.R., *Comput. Mech.*, 27, 188, 2001. With permission.)

TABLE 7.6

Natural Frequencies of the Cantilever of Varying Cross Section
($\alpha_Q = 1.5$, $\alpha_s = 3.5$ for MLPG)

	ω (rad/s)				
Modes	1	2	3	4	5
MLPG method	263.21	923.03	953.45	1855.14	2589.78
FEM (ABAQUS)	262.09	918.93	951.86	1850.92	2578.63

Source. Gu, Y.T. and Liu, G.R., *Comput. Mech.*, **27**, 188, 2001. With permission.

The nodal arrangement is shown in Figure 7.17. Results obtained by the MLPG method and the FEM software ABAQUS are listed in Table 7.6 for comparison. Results obtained by these two methods are in a very good agreement.

Example 7.8: Shear Wall

MLPG is employed for free-vibration analysis of a shear wall with four openings, as shown in Figure 7.18. The shear wall is fully clamped on the bottom, and all the rest of the boundaries are free of external forces. The problem is considered a plane stress problem and a unit thickness is used. This problem has been studied using the boundary element method (BEM) by some researchers [23] with parameters of $E = 1000$, $v = 0.2$, and $\rho = 1.0$. A total of 574 nodes are used to represent the problem domain shown in Figure 7.18. The problem is analyzed using MLPG and compared with the BEM and the FEM software ABAQUS. The natural frequencies of the lowest eight vibration modes are summarized and listed in Table 7.7. Results obtained by BEM and FEM are listed in the same table. Results obtained by the MLPG method are in a very good agreement with those obtained using BEM and FEM.

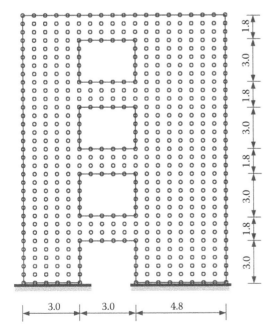

FIGURE 7.18
Shear wall with four openings (dimensions: m). (From Gu, Y.T. and Liu, G.R., *Comput. Mech.*, 27, 188, 2001. With permission.)

TABLE 7.7

Natural Frequencies of a Shear Wall ($\alpha_Q = 1.5$, $\alpha_s = 3.5$ for MLPG)

Mode	MLPG Method	FEM (ABAQUS)	BEM [23]
		ω (rad/s)	
1	2.069	2.073	2.079
2	7.154	7.096	7.181
3	7.742	7.625	7.644
4	12.163	11.938	11.833
5	15.587	15.341	15.947
6	18.731	18.345	18.644
7	20.573	19.876	20.268
8	23.081	22.210	22.765

Source: Gu, Y.T. and Liu, G.R., *Comput. Mech.*, 27, 188, 2001. With permission.

7.2.5 Forced-Vibration Analysis

The strong form of governing equation for forced vibration of 2D solids is given by Equation 7.48. The boundary conditions and initial conditions are given in Equations 7.49 through 7.52. The penalty method is used to enforce the essential boundary conditions. A local weak form of the partial differential Equation 7.48, over a local quadrature domain Ω_Q bounded by Γ_Q, can be obtained using the weighted residual method locally

$$\int_{\Omega_Q} \widehat{W}(\sigma_{ij,j} + b_i - m\ddot{u}_i - \eta_c \dot{u}_i)\mathrm{d}\Omega - \alpha \int_{\Gamma_{Qu}} \widehat{W}(u_i - u_{\Gamma i})\mathrm{d}\Gamma = 0 \tag{7.74}$$

The first term on the left-hand side of Equation 7.75 can be integrated by parts. Using the natural boundary condition defined by Equation 7.49, we obtain

$$\int_{\Omega_Q}(\widehat{W}m\ddot{u}_i+\widehat{W}\eta_c\dot{u}_i+\widehat{W}_{,j}\sigma_{ij})\mathrm{d}x - \int_{\Gamma_{Qu}}\widehat{W}t_i\mathrm{d}\Gamma+\alpha\int_{\Gamma_{Qu}}\widehat{W}u_i\mathrm{d}\Gamma=\int_{\Gamma_{Qt}}\widehat{W}t_{\Gamma i}\mathrm{d}\Gamma+\alpha\int_{\Gamma_{Qu}}\widehat{W}u_{\Gamma i}\mathrm{d}\Gamma+\int_{\Omega_Q}\widehat{W}b_i\mathrm{d}\Omega \tag{7.75}$$

In the forced-vibration analysis, u is a function of both the spatial coordinates and time. MLS approximation over the spatial domain is performed, and Equation 7.8 is rewritten as

$$\mathbf{u}^h(\mathbf{x}, t) = \sum_I^n \Phi_I(\mathbf{x})\mathbf{u}_I(t) \tag{7.76}$$

Substituting Equation 7.77 into the local weak form Equation 7.76 for all nodes leads to the following set of discrete equations:

$$\mathbf{M}\ddot{\mathbf{U}}(t) + \mathbf{C}\dot{\mathbf{U}}(t) + \mathbf{K}\mathbf{U}(t) = \mathbf{F}(t) \tag{7.77}$$

where the global mass matrix $\mathbf{M}$ is given by Equation 7.64. The global stiffness matrix $\mathbf{K}$ is obtained by assembling the nodal stiffness matrix defined by

$$\mathbf{K}_{IJ} = \int_{\Omega_Q} \hat{\mathbf{V}}_I^T \mathbf{c} \mathbf{B}_J d\Omega - \int_{\Gamma_{Qu}} \hat{\mathbf{W}}_I \mathbf{L}_n^T \mathbf{c} \mathbf{B}_J d\Gamma + \alpha \int_{\Gamma_{Qu}} \hat{\mathbf{W}}_I \Phi_J d\Gamma \qquad (7.78)$$

where $\hat{\mathbf{V}}_I$, $\mathbf{B}_I$, $\hat{\mathbf{W}}_I$ and $\mathbf{L}_n^T$ are defined, respectively, by Equations 7.12, 7.14, 7.15, and 7.17 and Φ_I is a matrix of the MLS shape function for node I given by Equation 7.9.

The damping matrix $\mathbf{C}$ is obtained using

$$\mathbf{C}_{IJ} = \int_{\Omega_Q} \eta_c \hat{\mathbf{W}}_I \Phi_J d\Omega \qquad (7.79)$$

and the force vector $\mathbf{f}$ is defined as

$$\mathbf{f}_I(t) = \int_{\Gamma_{Qt}} \hat{\mathbf{W}}_I \mathbf{t}_\Gamma(t) d\Gamma + \alpha \int_{\Gamma_{Qu}} \hat{\mathbf{W}}_I \mathbf{u}_\Gamma d\Gamma + \int_{\Omega_Q} \hat{\mathbf{W}}_I \mathbf{b}(t) d\Omega \qquad (7.80)$$

Again, Equation 7.78 will be reduced to Equation 7.63 for interior nodes whose boundary of quadrature domain does not intersect with the essential boundary.

7.2.6 Direct Analysis of Forced Vibrations

The procedure of solving the discrete dynamic equation (Equation 7.78) is very much the same as that in standard FEM. There are two major approaches to solve Equation 7.78. One is the modal analysis approach, in which the natural frequencies and the vibration modes obtained in Section 7.2.2 are used to transform Equation 7.78 into a set of decoupled differential equations of second order with respect to time. These second-order differential equations can then be solved simply using the standard approach. The second approach is the methods of direct integration operating on Equation 7.78. The direct integration methods are utilized in this section. Several direct integration methods have been developed to solve the dynamic equation set similar to Equation 7.78, such as CDM and the Newmark method (see, e.g., [24,25]). Both the central difference and the Newmark methods are introduced here in a concise and easy-to-understand manner.

7.2.6.1 The Central Difference Method

The CDM consists of expressing the velocity and acceleration at time t in terms of the displacement at time $t - \Delta t$, t, and $t + \Delta t$ using central finite difference formulation:

$$\ddot{\mathbf{u}}(t) = \frac{1}{\Delta t^2}(\mathbf{u}(t - \Delta t) - 2\mathbf{u}(t) + \mathbf{u}(t + \Delta t)) \qquad (7.81)$$

$$\dot{\mathbf{u}}(t) = \frac{1}{2\Delta t}(-\mathbf{u}(t - \Delta t) + \mathbf{u}(t + \Delta t)) \qquad (7.82)$$

where Δt is a time step. The response at time $t + \Delta t$ is obtained by evaluating the equation of motion at time t. CDM is, therefore, an explicit method and is widely used in finite element packages for transient analysis.

CDM is conditionally stable, meaning that the solution is stable when the time step is sufficiently small. In FEM, the critical time step is calculated based on the size of the smallest element. The principle to be followed in calculating the critical time step is that the critical time should be smaller than the time the fastest wave propagates across the element. In the meshfree method, there is no element, and there is a need for a new formula to compute the critical time. The critical time step for CDM can be obtained from the maximum frequencies based on the dispersion relation [22]:

$$\Delta t^{\mathrm{crit}} = \max_i \frac{2}{\omega_i} \left(\sqrt{\xi_i^2 + 1} - \xi_i \right) \tag{7.83}$$

where
   ω_i is the frequency
   ξ_i is the fraction of critical damping in this mode

For nonuniform arrangements of the nodes, the critical time step can be obtained by the eigenvalue inequality. The formula is

$$\Delta t^{\mathrm{crit}} = \min \frac{2}{(\max_Q \lambda^Q_{\mathrm{max}})^{1/2}} \tag{7.84}$$

where λ^Q_{max} is the maximum eigenvalue at the quadrature point $\mathbf{x}_Q$. The value of λ^Q_{max} depends on the size of local integration cells and the size of the interpolation domain.

7.2.6.2 The Newmark Method

The Newmark method is a generalization of the linear acceleration method. This latter method assumes that the acceleration varies linearly within the time interval of $(t, t + \Delta t)$. This gives

$$\ddot{\mathbf{u}} = \ddot{\mathbf{u}}_t + \frac{1}{\Delta t} (\ddot{\mathbf{u}}_{t+\Delta t} - \ddot{\mathbf{u}}_t)\tau \tag{7.85}$$

where $0 \leq \tau \leq \Delta t$, and

$$\dot{\mathbf{u}}_{t+\Delta t} = \dot{\mathbf{u}}_t + [(1 - \delta)\ddot{\mathbf{u}}_t + \delta\ddot{\mathbf{u}}_{t+\Delta t}]\Delta t \tag{7.86}$$

$$\mathbf{u}_{t+\Delta t} = \mathbf{u}_t + \dot{\mathbf{u}}\Delta t + \left[\left(\frac{1}{2} - \beta \right)\ddot{\mathbf{u}}_t + \beta\ddot{\mathbf{u}}_{t+\Delta t} \right]\Delta t^2 \tag{7.87}$$

The response at time $t + \Delta t$ is obtained by evaluating the equation of motion at time $t + \Delta t$. The Newmark method is, therefore, an implicit method.

The Newmark method is unconditionally stable, meaning that the solution will always be stable regardless of the time step used, provided

$$\delta \geq 0.5 \quad \text{and} \quad \beta \geq \frac{1}{4}(\delta + 0.5)^2 \tag{7.88}$$

It has been found that $\delta = 0.5$ and $\beta = 0.25$ lead to acceptable results for most problems. Therefore, $\delta = 0.5$ and $\beta = 0.25$ are always used in this section for all the example problems.

Note that although the Newmark method is unconditionally stable and any time step used will produce stable results, the accuracy of the results is not guaranteed. A sufficiently small time step still must be used for accurate results.

7.2.7 Numerical Examples

Example 7.9: Rectangular Cantilever

For forced-vibration analysis, a rectangular cantilever shown in Figure 6.4 is first examined using MLPG for benchmarking purposes. A plane stress problem is considered, and a unit thickness is used. The parameters used for this example are as follows:

Young's modulus for the material: $E = 3.0 = 10^7$ N/m^2

Poisson's ratio for two materials: $v = 0.3$

Mass density: $\rho = 1$ kg/m^3

Length of the cantilever: $L = 48$ m

Height of the cantilever: $D = 12$ m

External excitation load: $P = -1000g(t)$

where $g(t)$ is a function of time. The external force is applied downward and distributed in a parabolic fashion at the right end of the cantilever. A total of 55 uniformly distributed nodes are used, as shown in Figure 7.19, to represent the problem domain. Displacements and stresses for all nodes are obtained. Detailed results of vertical displacement, u_y, on the middle node, A, of the free end of the cantilever are presented. For comparison, solutions for this problem are also obtained using the finite element software ABAQUS Explicit.

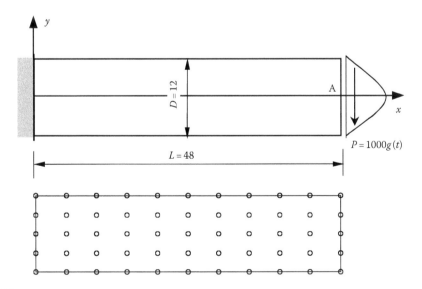

FIGURE 7.19
Configuration and nodal arrangement for the rectangular cantilever fixed at the left end and subjected to a dynamic force at the right end of the cantilever. (From Gu, Y.T. and Liu, G.R., *Comput. Mech.*, 27, 188, 2001. With permission.)

Example 7.9a: Simple Harmonic Loading

Consider first an external load of sinusoidal time function, i.e.,

$$g(t) = \sin(\omega_f t) \tag{7.89}$$

where ω_f is the frequency of the dynamic load, and $\omega_f = 27$ is used in this example. First, the effects of dimension parameter α_Q of the quadrature domain on the performance of the method for dynamic problems are investigated. Using Equation 7.85, the critical time is calculated to have $\Delta t^{crit} \approx 1 \times 10^{-3}$.

The results of $\alpha_Q = 0.5$, 1.0, 1.5, and 2.0 are computed using the MLPG code. The displacements u_y at point A are plotted in Figures 7.20 and 7.21. These figures show that the results will be unstable for both CDM and the Newmark method when $\alpha_Q \leq 1.0$. Increasing α_Q is crucial to improve the accuracy and the stability for both CDM and the Newmark method. However, if the quadrature domain is too large, more subcells are needed to obtain accurate integrations, which will be computationally more expensive. Our study has found that $\alpha_Q = 1.5$–2.5 works for most problems of transient analyses. This finding is the same as that found for static and free-vibration analyses. In the following transient analyses $\alpha_Q = 1.5$ is employed.

To investigate the property of two different direct time integration methods, CDM and the Newmark method, results of different time steps are obtained and plotted in Figure 7.22. It can be found that for $\Delta t = 1 \times 10^{-4}$ both methods obtain results in a very good agreement with FEM. When $\Delta t > \Delta t^{crit}$ ($\Delta t^{crit} \approx 1 \times 10^{-3}$ according to Equation 7.85), the results obtained using CDM becomes unstable. However, the Newmark method is always stable for any time step used. It has been confirmed numerically that CDM is a conditionally

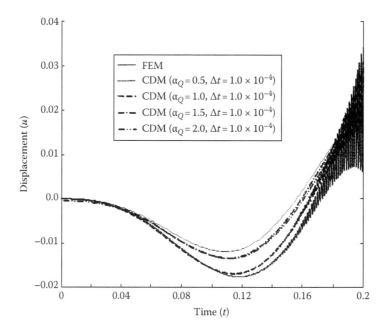

FIGURE 7.20
Displacement in the y-direction at point A using CDM ($g(t) = \sin(\omega t)$). Results are unstable when $\alpha_Q \leq 1.0$. (From Gu, Y.T. and Liu, G.R., *Comput. Mech.*, 27, 188, 2001. With permission.)

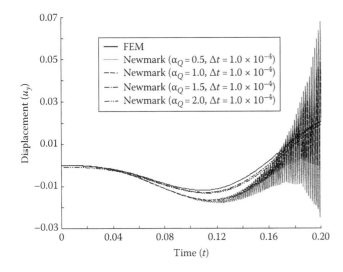

FIGURE 7.21
Displacement in the y-direction at point A using the Newmark method ($\delta = 0.5$ and $\beta = 0.25$, with $g(t) = \sin(\omega t)$). Results are unstable when $\alpha_Q \leq 1.0$. (From Gu, Y.T. and Liu, G.R., *Comput. Mech.*, 27, 188, 2001. With permission.)

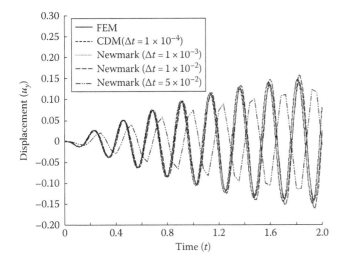

FIGURE 7.22
Displacement in the y-direction at the point A ($g(t) = \sin(\omega t)$). (From Gu, Y.T. and Liu, G.R., *Comput. Mech.*, 27, 188, 2001. With permission.)

stable method and that the Newmark method is an unconditionally stable method. A larger time step can be used in the Newmark method. A time step as large as $\Delta t = 1 \times 10^{-2}$ has been used, and very good results have been obtained using the Newmark method. However, it should be noted that the computational error would increase with the increase of time step in the Newmark method. For this example, it was found that the accuracy of the Newmark method would become unacceptable when the time step is too large, such as $\Delta t = 5 \times 10^{-2}$.

Many time steps are calculated to check the stability of the presented MLPG formulation. The Newmark method with $\Delta t = 5 \times 10^{-3}$ is used, and the damping coefficient, $c = 0.4$, is considered. Results for up to 20 s (about 100 natural vibration periods) are plotted in Figure 7.23. It can be found that a very stable result is obtained. After a long period of time, the forced vibration under the action of the sinusoidal dynamic loading becomes a steady sinusoidal vibration with the frequency of the external excitation ω_f. From the vibration theory [24], a resonance will occur when $\omega_f = \omega_i$, where ω_i is the ith natural frequency.

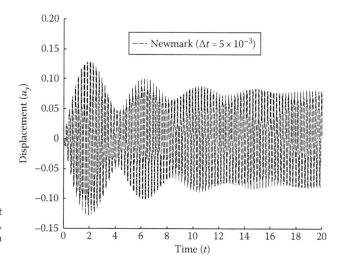

FIGURE 7.23
Displacement in the y-direction at point A ($g(t) = \sin(\omega t)$). (From Gu, Y.T. and Liu, G.R., *Comput. Mech.*, 27, 188, 2001. With permission.)

Figure 7.23 shows that the amplitude of vibration is very large (i.e., about 15 times the static displacement) because $\omega_f \approx \omega_i$. In addition, a beat vibration with the period T_b occurs when $\omega_f \approx \omega_1$. T_b can be obtained from Figure 7.23 as $T_b \approx 4.3$. From vibration theory, $T = 2/|\omega_f - \omega_1|$, the first natural frequency of the system can be found as $\omega_1 = 28.3$, which is nearly the same as the result obtained in the free-vibration analysis by FEM, $\omega_1^{FEM} = 28$.

Example 7.9b: Transient Loading

The transient response of a cantilever subjected to a suddenly loaded and suddenly vanished force $P = 1000g(t)$ is considered. The time function $g(t)$ is shown in Figure 7.24. The present MLPG method is used to obtain the transient response with and without damping. The Newmark method is utilized in this analysis. The result for a damping coefficient of $\eta_c = 0$ is plotted in Figure 7.25. For comparison, the result obtained by the finite element software ABAQUS/Explicit is shown in the same figure. Results obtained by the present MLPG method are in a very good agreement with those obtained using FEM. Many time steps are calculated to check the stability of the presented MLPG formulation. The result for a damping coefficient $\eta_c = 0.4$ is plotted in Figure 7.26; which shows that the response declines with time because of damping. A very stable result is again obtained.

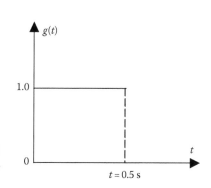

FIGURE 7.24
Rectangular pulse as the time function of the external force $g(t)$. (From Gu, Y.T. and Liu, G.R., *Comput. Mech.*, 27, 188, 2001. With permission.)

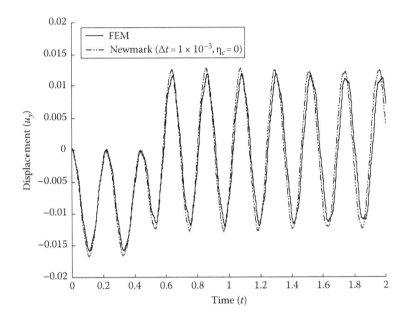

FIGURE 7.25

Transient displacement in the y-direction at point A using the Newmark method ($\delta = 0.5$ and $\beta = 0.25$, $\acute{\eta}_c$ is the damping coefficient). (From Gu, Y.T. and Liu, G.R., *Comput. Mech.*, 27, 188, 2001. With permission.)

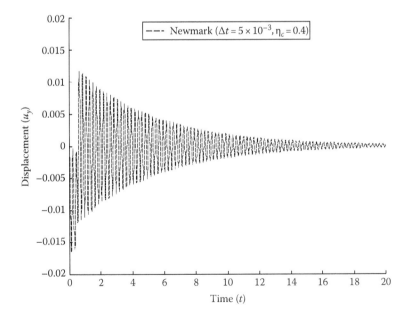

FIGURE 7.26

Transient displacement in the y-direction at point A using the Newmark method ($\delta = 0.5$ and $\beta = 0.25$, $\acute{\eta}_c$ is the damping coefficient). (From Gu, Y.T. and Liu, G.R., *Comput. Mech.*, 27, 188, 2001. With permission.)

7.3 Concluding Remarks

MLPG formulations for static, free-vibration, and forced-vibration analyses of 2D solids have been presented in this chapter. Some important parameters affecting the performance of the method have been investigated; the important findings are summarized as follows:

1. MLPG works well for static and dynamic analysis of 2D solids.

2. The dimension of the quadrature domain is very important to the accuracy as well as the stability of the results. Larger α_Q will in general give more accurate and stable results, if the numerical integration can be carried out accurately. However, too large an α_Q often leads to difficulties in accurate numerical integration. A choice of $\alpha_Q = 1.5$ to 2.5 works for most problems, and $\alpha_Q = 1.5$ is recommended as an economic choice.

3. The dimension parameter of the support domain is also very important to the accuracy as well as the stability of the results. A choice of $\alpha_s = 2.5$ to 3.5 is good for most problems.

4. For the numerical integration in the quadrature domain, four subdivisions (2×2) works well for rectangular quadrature domains of $\alpha_Q = 2.0$.

5. MLPG is not as efficient as FEM in terms of computation time, because the system matrices produced are symmetric. For governing equations of symmetric operators, the Galerkin type of weak forms is clearly preferred. The process of computing the MLS shape functions and their derivatives is more expensive than the FEM shape functions, which are usually given analytically. Improvement on these two issues for MLPG is very important.

6. One of the difficulties in MLPG is the integration for nodes near the boundaries of the problem domain, because the local quadrature domains for these nodes may intersect with the global boundary of the problem domain and create local quadrature domains of complex geometry. One method for solving this problem might be the use of a triangular mesh. We have also tried a simple trick, using very small regular quadrature domains for these nodes so that the boundary of their quadrature domains just touches but does not intersect with the global boundary. This simple trick works for some problems we have studied; however, we also found that the accuracy of the results for many problems could be affected by using this trick.

The MLPG method is one of the widely used meshfree methods. Improvements and advancements have been made during the past years by many, especially by Atluri's group. Readers are referred to publications by Atluri and coworkers for the recent developments.

References

1. Liszka, T. and Orkisz, J., The finite difference method at arbitrary irregular grids and its application in applied mechanics, *Comput. Struct.*, 11, 83–95, 1980.
2. Jensen, P. S., Finite difference techniques for variable grids, *Comput. Struct.*, 2, 17–29, 1980.

3. Onate, E., Idelsohn, S., Zienkiewicz, O. C., and Taylor, R. L., A finite point method in computational mechanics applications to convective transport and fluid flow, *Int. J. Numer. Methods Eng.*, 39, 3839–3866, 1996.

4. Cheng, M. and Liu, G. R., A finite point method for analysis of fluid flow, in *Proceedings of the 4th International Asia-Pacific Conference on Computational Mechanics*, Singapore, December 1999, pp. 1015–1020.

5. Xu, X. G. and Liu, G. R., A local-function approximation method for simulating two-dimensional incompressible flow, in *Proceedings of the 4th International Asia-Pacific Conference on Computational Mechanics*, Singapore, December 1999, pp. 1021–1026.

6. Song, B., Liu, G. R., Xu, D., and Pan, L. S., Application of finite point method to fluid flow and heat transfer, in *Proceedings of the 4th Asia-Pacific Conference on Computational Mechanics*, Singapore, December 1999, pp. 1091–1096.

7. Liu, G. R. and Kee, B. B. T., A stabilized least-squares radial point collocation method (LS-RPCM) for adaptive analysis, *Comput. Method Appl. Mech. Eng.*, 195, 4843–4861, 2006.

8. Kee, B. B. T., Liu, G. R., and Lu, C., A regularized least-squares radial point collocation method (RLS-RPCM) for adaptive analysis, *Comput. Mech.*, 40, 837–853, 2007.

9. Atluri, S. N. and Zhu, T., A new meshless local Petrov–Galerkin (MLPG) approach in computational mechanics, *Comput. Mech.*, 22, 117–127, 1998.

10. Atluri, S. N., Cho, J. Y., and Kim, H. G., Analysis of thin beams, using the meshless local Petrov–Galerkin method, with generalized moving least squares interpolations, *Comput. Mech.*, 24, 334–347, 1999.

11. Atluri, S. N., Kim, H. G., and Cho, J. Y., A critical assessment of the truly meshless local Petrov–Galerkin (MLPG), and local boundary integral equation (LBIE) methods, *Comput. Mech.*, 24, 348–372, 1999.

12. Ouatouati, A. E. and Johnson, D. A., A new approach for numerical modal analysis using the element free method, *Int. J. Numer. Methods Eng.*, 46, 1–27, 1999.

13. Liu, G. R. and Yan, L., A modified meshless local Petrov–Galerkin method for solid mechanics, in *Advances in Computational Engineering and Sciences*, N. K. Atluri, and F. W. Brust, eds., Tech Science Press, Palmdale, CA, 2000, pp. 1374–1379.

14. Liu, G. R. and Gu, Y. T., Meshless local Petrov–Galerkin (MLPG) method in combination with finite element and boundary element approaches, *Comput. Mech.*, 26, 536–546, 2000.

15. Liu, G. R. and Gu, Y. T., On formulation and application of local point interpolation methods for computational mechanics, in *Proceedings of the First Asia-Pacific Congress on Computational Mechanics*, Sydney, Australia, November 20–23, 2001, pp. 97–106 (invited paper).

16. Gu, Y. T. and Liu, G. R., A meshless local Petrov–Galerkin (MLPG) method for free and forced vibration analyses for solids, *Comput. Mech.*, 27(3), 188–198, 2001.

17. Gu, Y. T. and Liu, G. R., A meshless local Petrov–Galerkin (MLPG) formulation for static and free vibration analyses of thin plates, *Comput. Model. Eng. Sci.*, 2(4), 463–476, 2001.

18. Liu, G. R. and Chen, X. L., Static buckling of composite laminates using EFG method, in *Proceedings of the 1st International Conference on Structural Stability and Dynamics*, Taipei, China, December 7–9, 2000, pp. 321–326.

19. Liu, G. R. and Chen, X. L., A mesh-free method for static and free vibration analyses of thin plates of complicated shape, *J. Sound Vibration*, 241(5), 839–855, 2001.

20. Strang, G., *Linear Algebra and Its Application*, Academic Press, New York, 1976.

21. Nagashima, T., Node-by-node meshless approach and its application to structural analyses, *Int. J. Numer. Methods Eng.*, 46, 341–385, 1999.

22. Belytschko, T., Guo, Y., Liu, W. K., and Xiao, S. P., A unified stability of meshless particle methods, *Int. J. Numer. Methods Eng.*, 48, 1359–1400, 2000.

23. Brebbia, C. A., Telles, J. C., and Wrobel, L. C., *Boundary Element Techniques*, Springer Verlag, Berlin, 1984.

24. Petyt, M., *Introduction to Finite Element Vibration Analysis*, Cambridge University Press, Cambridge, U.K., 1990.

25. Liu, G. R. and Quek, S. S., *The Finite Element Method: A Practical Course*, Butterworth Heinemann, Oxford, 2003.

8

Point Interpolation Methods

In Chapters 6 and 7, we introduced the element-free Galerkin (EFG) method and the meshless local Petrov–Galerkin (MLPG) method. Both methods use moving least squares (MLS) approximation for constructing shape functions, and hence they are accompanied by issues related to essential boundary conditions. We also discussed a number of ways to tackle these issues, which require extra efforts both in formulation and computation.

The point interpolation method (PIM) was proposed by Liu et al. [1–6] to replace MLS approximation for creating shape functions in meshfree settings. The major advantages of the PIM, as shown in Chapter 2, are the excellent accuracy in function fitting and the Kronecker delta function property, which allows simple imposition of essential boundary conditions as in the standard finite element method (FEM) (e.g., [51]). In the lengthy process of developing PIM, the battle has been on two fronts. The first front was on how to overcome the problem related to the singular moment matrix using local irregularly distributed nodes. The second one was on how to create weak forms that can always ensure stable and convergent solutions. On the first battle front, two significant advances have been made over the past years, after multiple attempts. The first is the use of RPIM shape functions allowing the use of virtually randomly distributed nodes [7]. The second approach is to use T-Schemes to create polynomial or radial PIM shape functions efficiently with a small number of local nodes selected, based on triangular cells.

The second battle is essentially the restoration of the conformability of the PIM methods caused by the incompatibility of the PIM (or RPIM) shape functions. This incompatibility problem has been a very difficult one to overcome, and lots of efforts have been made during the past years. It is now well resolved by the use of weakened-weak (W^2) formulations based on the $\mathbb{G}$ space theory [27,28,62]: the generalized smoothed Galerkin (GS-Galerkin) weak form, and the strain-constructed Galerkin (SC-Galerkin) weak forms. The W^2 formulations not only solves the compatibility problem effectively, but also offers a variety of ways to implement PIM models with excellent properties (upper bound, lower bound, superconvergence, free of locking, works well with triangular types of mesh, etc.). This chapter details a number of PIMs based on W^2 formulations for stress analysis of solids. For convenience and easy reference, we list various PIM methods in Table 8.1.

8.1 Node-Based Smoothed Point Interpolation Method

The node-based smoothed point interpolation method (NS-PIM) is a typical GS-Galerkin model based on the normed $\mathbb{G}$ space theory. Therefore, theories discussed in Chapter 3 related to $\mathbb{G}$ spaces apply. This section details the NS-PIM formulations for mechanics problems for two-dimensional (2D) and three-dimensional (3D) solids, based on these theories. We first focus our discussion on the 2D case because it is much easier to describe and follow. We can then extend it to the 3D case by simply highlighting the differences between 2D and 3D cases.

TABLE 8.1

Versions of Point Interpolation Methods (Triangular Background Cells, T-Scheme)

Abbreviation	Full Name	Formulation	Features
NS-PIM (2D and 3D)	Node-based smoothed point interpolation method	GS-Galerkin Polynomial PIM shape functions Smoothing operation based on nodes	Linearly conforming Volumetric locking free Upper bound Superconvergence
NS-RPIM (2D and 3D)	Node-based smoothed radial point interpolation method	GS-Galerkin RPIM shape functions Smoothing operation based on nodes	Linearly conforming Volumetric locking free[a] Upper bound Superconvergence
ES-PIM (2D) FS-PIM (3D)	Edge-based (Face-based) smoothed point interpolation method	GS-Galerkin Polynomial PIM shape functions Smoothing operation based on the edges (faces) of the cells	Linearly conforming Ultra-accuracy Very efficient Superconvergence
ES-RPIM (2D) FS-RPIM (3D)	Edge-based (Face-based) smoothed radial point interpolation method	GS-Galerkin RPIM shape functions Smoothing operation based on the edges (faces) of the cells	Linearly conforming Ultra-accuracy Superconvergence
CS-PIM (2D)	Cell-based smoothed point interpolation method	GS-Galerkin PIM shape functions Smoothing operation based on the cells or subdivided cells	Linearly conforming Ultra-accuracy Superconvergence
CS-RPIM (2D)	Cell-based smoothed point interpolation method	GS-Galerkin RPIM shape functions Smoothing operation based on the cells or subdivided cells	Linearly conforming Ultra-accuracy Superconvergence
SC-PIM (2D) SC-PIM (3D)[a]	Strain-constructed point interpolation method	SC-Galerkin Polynomial PIM shape functions Strain construction based on cells using (node- or edge-based) smoothed strains	Linearly conforming Ultra-accuracy Lower and upper bounds Superconvergence

[a] Yet to confirm or under development.

The NS-PIM [3] was formulated using the PIM shape functions created with T-Schemes (Chapter 1) and the GS-Galerkin weak form (Chapter 5) that allows discontinuous functions. It was termed as linearly conforming PIM (LC-PIM) initially because it is at least linearly conforming. We now term it as NS-PIM because the smoothing operation is nodal based. It is more convenient for the presentation and in distinguishing from other models of PIMs, many of which are linearly conforming but use different types of smoothing domains for different properties.

NS-PIM possesses the following novel features: (1) the T-Schemes based on triangular background cells are used for node selection, which overcomes the singular moment matrix issue, and ensures the efficiency in computing PIM shape functions; (2) shape functions generated using polynomial basis functions and simple interpolation ensure that the PIM shape functions possesses at least linearly consistent and delta function property, which facilitates easy implementation of essential boundary conditions; and (3) the use of the weakened-weak form allows the use of incompatible displacement functions. Due to these excellent features, the present NS-PIM is easy to implement, guarantees stability and faster convergence, can produce upper-bound solution with respect to the exact solution, and computationally as efficient as the FEM using the same mesh.

8.1.1 Domain Discretization and Node Selection

In a PIM, the problem domain is first represented by properly scattered points as in any meshfree method. It is then discretized by triangulation detailed in Section 1.7.2 with N_e triangular background cells and N_n nodes. A group of field nodes can then be selected for constructing shape functions based on cells, using T3-Scheme for linear NS-PIM and T6/3-Scheme for quadratic NS-PIM.

The T3-Scheme is the simplest and leads to a linear NS-PIM that uses the same shape functions as the FEM using linear triangular elements; however, all the other numerical operations and the solution for NS-PIM-T3 will be very much different from that of FEM-T3 even when the same mesh is used. The FEM-T3 is known to behave "overly stiff" and this produces a lower-bound solution (for "force-driving" problems). The NS-PIM model is much "softer" than the FEM model and can produce an upper-bound solution. The NS-PIM is often found "overly soft," which will be discussed in detail in the Section 8.2.

Once the nodes are selected, techniques detailed in Section 2.5 are then used to compute the PIM shape functions. Note that in the NS-PIM formulation, the derivatives of the shape functions are not required as will be shown in the formulation below.

8.1.2 Construction of Node-Based Smoothing Domains

On top of the triangular cells created via triangulation, the domain is divided into N_s smoothing domains associated with nodes following the rules given in Section 3.3.1. Each smoothing domain contains a node and covers portions of elements sharing the node. In this case, $N_s = N_n$, the number of the smoothing domain is the same as that of the nodes. The basic rule in the construction is that the boundaries of Ω_i^s should not share any *finite* portion of the interfaces of the triangular cells, if the assumed displacements are incompatible there; they come across each other (sharing only points). The smoothing domains are also used as quadrature cells for the integration, leading to a simple summation over the smoothing domains.

8.1.2.1 Equally Shared Smoothing Domain

In the NS-PIM, the GS-Galerkin weak form that allows incompatible assumed displacement functions will be used to create the discretized system equations. In this section, we use the so-called node-based equally shared smoothing domains created based on the triangular type background cells, as shown in Figure 3.1. The problem domain Ω is divided into N_s smoothing domains each of which contains one node, for example Ω_i^s bounded by Γ_i^s contains node i, as shown in Figure 8.1. The subdomain Ω_i^s is constructed by connecting sequentially the mid-edge-point to the centroids of the triangles. The union of all Ω_i^s forms exactly Ω: no overlap or gap is allowed. Because the area of a triangular cell is equally shared by its three vertices and the length of the edge of a cell is also equally shared by the two vertices of the edge, the smoothing domain created is termed as equally shared smoothing domain.

8.1.2.2 Voronoi Smoothing Domains

In the NS-PIM, we can also use an alternative node-based smoothing domain called the Voronoi smoothing domain that was used in [10] for stabilizing the nodal integrated

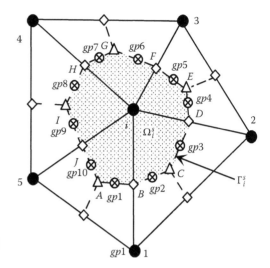

FIGURE 8.1
An equally shared smoothing domain Ω_i^s bounded by Γ_i^s
for node i.

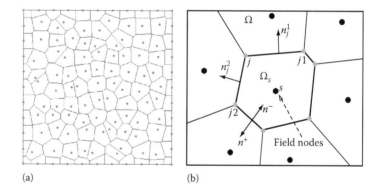

(a)   (b)

FIGURE 8.2
Voronoi smoothing domain in a 2D domain. (a) Voronoi smoothing domains, and (b) one Voronoi smoothing
domain.

meshfree method. The Voronoi smoothing domain is constructed in the same manner
but using the standard algorithm for Voronoi diagram construction, which results in
smoothing domains of different shapes. A typical Voronoi smoothing domain is shown
in Figure 8.2. It is clear that the Voronoi smoothing domain is convex, and the equally
shared smoothing domain is generally not. Note whether or not the smoothing domain
being convex is immaterial, as the formulation of NS-PIM or node-based smoothed radial
point interpolation method (NS-RPIM) does not require the convexity. The author's group
uses mainly the equally shared smoothing domain, but we will use the Voronoi smoothing
domain in Section 8.2, just to demonstrate this alternative.

The equally shared smoothing domain is preferred in NS-PIM because of the use of the
PIM shape functions that are constructed based on the triangular cells, especially when
linear interpolation is used one does not even need to construct the shape functions
explicitly; and the displacements at any point on Γ_n^s can be interpolated using nodal
displacements very easily by simple inspection. Therefore, the equally shared smoothing
domain is used in this book by default, unless specified otherwise.

8.1.3 Formulation for the NS-PIM

Consider a 2D problem of solid mechanics in domain Ω bounded by Γ. The strong form of system equation is given by Equations 6.1 through 6.3. Because PIM shape functions are used, we use the following GS-Galerkin weak form that allows incompatible functions (see Equation 5.130).

$$\sum_{i=1}^{N_s} A_i^s (\delta \bar{\boldsymbol{\varepsilon}}_i)^{\mathrm{T}} \mathbf{c} \bar{\boldsymbol{\varepsilon}}_i - \int_\Omega \delta \mathbf{u}^{\mathrm{T}} \mathbf{b} \, \mathrm{d}\Omega - \int_{\Gamma_t} \delta \mathbf{u}^{\mathrm{T}} \mathbf{t} \, \mathrm{d}\Gamma = 0 \tag{8.1}$$

The stability and convergence of a GS-Galerkin model is proven in Chapter 5.

Note that the difference between Equations 8.1 and 6.67 is that in the former there is no need for a term dealing with the essential boundary condition at Γ_u, because the PIM shape function possesses the Kronecker delta function property. The condition of Equation 6.2 will be satisfied on the entire essential boundary in the form of prescribed nodal displacements, which can be done by a simple procedure of row and column removal or some other methods as in FEM. In our formulation, we assume that the prescribed displacements in between two neighboring nodes are linear, which can be imposed precisely by performing linear interpolation on all the Γ_i^s segments that are on the essential boundary of the problem domain.

Using PIM shape functions, ϕ_I, for all the nodes, the assumed displacement at any point $\mathbf{x}$ in the problem domain can be expressed as

$$\mathbf{u}^h(\mathbf{x}) = \left\{ \begin{matrix} u(\mathbf{x}) \\ v(\mathbf{x}) \end{matrix} \right\}^h = \sum_{I \in S_n} \underbrace{\left[\begin{matrix} \phi_I(\mathbf{x}) & 0 \\ 0 & \phi_I(\mathbf{x}) \end{matrix} \right]}_{\boldsymbol{\Phi}_I(\mathbf{x})} \underbrace{\left\{ \begin{matrix} u_I \\ v_I \end{matrix} \right\}}_{\mathbf{u}_I} = \sum_{I \in S_n} \boldsymbol{\Phi}_I(\mathbf{x}) \mathbf{u}_I \tag{8.2}$$

where
  S_n is the set of local support nodes selected using a T-Scheme for a cell hosting $\mathbf{x}$
  $\boldsymbol{\Phi}_I$ is the matrix of PIM shape functions
  $\mathbf{u}_I$ is the vector of the nodal displacements for these local nodes

The smoothed strain $\bar{\boldsymbol{\varepsilon}}$ in Equation 8.1 is obtained using the generalized smoothing operation performed over the equally shared smoothing domains defined for each node in the problem domain. Figure 8.1 shows such a smoothing domain for node i. The strain at node i is obtained using

$$\bar{\boldsymbol{\varepsilon}}(\mathbf{x}_i) = \begin{cases} \dfrac{1}{A_i^s} \displaystyle\int_{\Omega_i^s} \tilde{\boldsymbol{\varepsilon}}(\mathbf{u}^h) \, \mathrm{d}\Omega = \dfrac{1}{A_i^s} \displaystyle\int_{\Omega_i^s} (\mathbf{L}_d \mathbf{u}^h) \, \mathrm{d}\Omega = \dfrac{1}{A_i^s} \displaystyle\int_{\Gamma_i^s} \mathbf{L}_n \mathbf{u}^h \, \mathrm{d}\Gamma, & \text{when } \mathbf{u}^h(\boldsymbol{\xi}) \in \mathbb{C}^0(\Omega_i^s) \\[4mm] \dfrac{1}{A_i^s} \displaystyle\int_{\Gamma_i^s} \mathbf{L}_n \mathbf{u}^h \, \mathrm{d}\Gamma, & \text{when } \mathbf{u}^h(\boldsymbol{\xi}) \in \mathbb{C}^{-1}(\Omega_i^s) \end{cases} \tag{8.3}$$

where
  $A_i^s = \int_{\Omega_i^s} \mathrm{d}\Omega$ is the area of smoothing domain for node i
  $\mathbf{L}_d$ is the differential operator defined in Equation 1.9
  $\mathbf{L}_n$ is the matrix of outward normal components
  $\mathbf{u}^h$ is the assumed displacement defined in Equation 8.2

In our formulation, we will always use the more general second equation in Equation 8.3. Substituting Equation 8.2 into Equation 8.3, the smoothed strain can be written in the following matrix form:

$$\bar{\boldsymbol{\varepsilon}}^h(\mathbf{x}_i) = \sum_{I \in S_s} \bar{\mathbf{B}}_I(\mathbf{x}_i)\mathbf{u}_I \tag{8.4}$$

where S_s is the support nodes for the smoothing domain that is a set of nodes involved in the interpolation for all points on Γ_i^s. The *smoothed* strain matrix has the form

$$\bar{\mathbf{B}}_I(\mathbf{x}_i) = \begin{bmatrix} \bar{\phi}_{I,x}(\mathbf{x}_i) & 0 \\ 0 & \bar{\phi}_{I,y}(\mathbf{x}_i) \\ \bar{\phi}_{I,y}(\mathbf{x}_i) & \bar{\phi}_{I,x}(\mathbf{x}_i) \end{bmatrix} \tag{8.5}$$

in which

$$\bar{\phi}_{I,l} = \frac{1}{A_i^s} \int_{\Gamma_i^s} \phi_I(\mathbf{x}) n_l(\mathbf{x}) d\Gamma, \quad (l = x, y) \tag{8.6}$$

The integration in Equation 8.6 is a curve integration that can be performed easily using the Gauss integration schemes, which can be written in a summation form of

$$\bar{\phi}_{I,l} = \frac{1}{A_i^s} \sum_{m=1}^{n_s} \left[\sum_{n=1}^{n_g} W_n(\phi_I(\mathbf{x}_{m,n}) n_l(\mathbf{x}_m)) \right] \tag{8.7}$$

where
$\quad n_l$ is the components of the unit outward normal on Γ_i^s
$\quad n_s$ is the number of line-segments of the boundary Γ_i^s
$\quad n_g$ is the number of Gauss points distributed in each segment
$\quad W_n$ is the corresponding Gauss weight

For example, for the node i shown in Figure 8.1, we have 10 line-segments on Γ_i^s: $n_s = 10$. If linear interpolation is used, the number of support nodes for each cell is 3. Each line-segment needs only one Gauss point ($n_g = 1$), and therefore, there are a total of 10 Gauss points ($gp1$–$gp10$) to be used for the entire smoothing domain. The number of support nodes for the smoothing domain is 6. In this simple linear interpolation case, the shape function values at all these 10 Gauss points can be tabulated in Table 8.2 by simple inspection. Of course, when higher order PIM shape functions are used, we need to compute the shape function values numerically, which can be programmed with ease. We show the simple linear interpolation case in the tabulated form to reveal explicitly how the line integration is performed in NS-PIM for computing the smoothed strains.

Note from Equation 8.6 that in computing the smoothed strain matrix and hence the stiffness matrix, we do not need to perform differentiation for the shape functions. This shows clearly that the consistence requirement of the shape functions is reduced: a weakened-weak formulation. Whether or not the shape functions are continuous within the smoothing domains is not a problem.

TABLE 8.2

Shape Function Values at Different Sites on the Smoothing Domain Boundary for Node i

Site	Node i	Node 1	Node 2	Node 3	Node 4	Node 5	Description
i	1.0	0	0	0	0	0	Field node
1	0	1.0	0	0	0	0	Field node
2	0	0	1.0	0	0	0	Field node
3	0	0	0	1.0	0	0	Field node
4	0	0	0	0	1.0	0	Field node
5	0	0	0	0	0	1.0	Field node
A	1/3	1/3	0	0	0	1/3	Centroid of cell
B	1/2	1/2	0	0	0	0	Mid-edge
C	1/3	1/3	1/3	0	0	0	Centroid of cell
D	1/2	0	1/2	0	0	0	Mid-edge
E	1/3	0	1/3	1/3	0	0	Centroid of cell
F	1/2	0	0	1/2	0	0	Mid-edge
G	1/3	0	0	1/3	1/3	0	Centroid of cell
H	1/2	0	0	0	$\dfrac{1}{2}$	0	Mid-edge
I	1/3	0	0	0	1/3	1/3	Centroid of cell
J	1/2	0	0	0	0	1/2	Mid-edge
$gp1$	5/12	5/12	0	0	0	1/6	**Mid-segment of Γ_i^s**
$gp2$	5/12	5/12	1/6	0	0	0	**Mid-segment of Γ_i^s**
$gp3$	5/12	1/6	5/12	0	0	0	**Mid-segment of Γ_i^s**
$gp4$	5/12	0	5/12	1/6	0	0	**Mid-segment of Γ_i^s**
$gp5$	5/12	0	1/6	5/12	0	0	**Mid-segment of Γ_i^s**
$gp6$	5/12	0	0	5/12	1/6	0	**Mid-segment of Γ_i^s**
$gp7$	5/12	0	0	1/6	5/12	0	**Mid-segment of Γ_i^s**
$gp8$	5/12	0	0	0	5/12	1/6	**Mid-segment of Γ_i^s**
$gp9$	5/12	0	0	0	1/6	5/12	**Mid-segment of Γ_i^s**
$gp10$	5/12	1/6	0	0	0	5/12	**Mid-segment of Γ_i^s**

We now further assume that the modified strain is the constant in the smoothing domain and the same as that given in Equation 8.3:

$$\bar{\boldsymbol{\varepsilon}}_i(\mathbf{x}) = \bar{\boldsymbol{\varepsilon}}(\mathbf{x}_i), \quad \forall \mathbf{x} \in \Omega_i^s \tag{8.8}$$

Now, substituting Equations 8.2 and 8.8 into Equation 8.1, following the similar procedure given in Section 6.1.1, the discretized system equation can be obtained in the following matrix form:

$$\bar{\mathbf{K}}\bar{\mathbf{U}} = \mathbf{F} \tag{8.9}$$

where $\bar{\mathbf{U}}$ is the vector of nodal displacements for all nodes in the problem domain, the stiffness matrix $\bar{\mathbf{K}}$ is assembled (in the same way discussed in Chapter 6) using the entries of the submatrix of stiffness

$$\bar{\mathbf{K}}_{IJ} = \sum_{i=1}^{N_s} A_i^s \bar{\mathbf{B}}_I^{\mathrm{T}} \mathbf{c} \bar{\mathbf{B}}_J \tag{8.10}$$

where $\bar{\mathbf{B}}_I$ is the *smoothed* strain matrix for node I defined in Equation 8.5. Note that $\bar{\mathbf{K}}_{IJ}$ needs to be computed only when nodes I and J share the same smoothing domain. Otherwise, it is zero. Hence, $\bar{\mathbf{K}}$ will be very sparse for an NS-PIM model, in addition to the obvious fact that it is symmetric. Based on the theorems presented in Chapter 5, $\bar{\mathbf{K}}$ is symmetric positive definite (SPD).

In Equation 8.9, the force vector $\mathbf{F}$ has the following entries of vectors:

$$\mathbf{f}_I = \int_{\Gamma_t} \phi_I \mathbf{t} d\Gamma + \int_{\Omega} \phi_I \mathbf{b} d\Omega \tag{8.11}$$

The force vectors are without bar-hat because no smoothing operation is applied to the linear functional in NS-PIM formulation. The force vectors are computed in exactly the same way as that in FEM or EFG.

Equation 8.9 can be solved using standard routines with ease because $\bar{\mathbf{K}}$ is SPD and sparse.

Note that $\bar{\mathbf{K}}$ will also be banded if the nodes are properly numbered, as in the FEM. For linear NS-PIM with T3-Scheme, the bandwidth of $\bar{\mathbf{K}}$ will be determined by the largest difference of node numbers of the nodes of all the triangular cells connected directly to the node. For 2D cases, a node-based smoothing domain is usually supported by four to eight nodes. Therefore, it is clear that the bandwidth of a NS-PIM model will be larger (about twice) than that of a linear FEM that is the smallest for the all possible numerical models (a triangular element involves only three nodes). In the EFG, an integration cell is supported by about 15–40 nodes (for ensuring the compatibility), therefore the bandwidth of an EFG model is roughly about two to four times that of an NS-PIM model. The efficiency of the NS-PIM is quite obvious from this rough analysis. Of course, we have not yet taken the solution accuracy and other properties into account.

8.1.4 Flowchart of the NS-PIM

A brief flowchart of NS-PIM is given below:

1. Loop over all the smoothing domains Ω_i^s.
2. Loop over surrounding cells directly connected to node i.
   a. Select the nodes for construction of PIM shape functions for the cell using T-Scheme.
   b. Compute $\bar{\phi}_{I,l}$ using Equation 8.7.
   c. Compute the nodal stiffness matrices and force vectors.
   d. Assemble the nodal contributions to the global matrices and vectors.
4. End cell loop.
5. End the smoothing domain loop.

Note that the integral in Equation 8.11 for nodal force vectors is the same as that in FEM and EFG, except for the way in which the shape function is computed. Note also that the NS-PIM formulation is variationally consistent, if the assumed displacement function is compatible (T3-Scheme) as proven in [13] and is stable and convergent if the assumed displacement function is incompatible (T6/3-Scheme) as proven in [28].

8.1.5 Comparison of NS-PIM, FEM, and NS-FEM

8.1.5.1 NS-PIM vs. FEM

When the same triangular mesh is used in the FEM and the NS-PIM, we have the following points:

1. The interpolation procedure in NS-PIM is based on a group of support nodes selected based on triangular cells. The support nodes can generally be selected from more than one cell, which may overlap with the support domains of other neighboring cells. The interpolation procedure in FEM is different from that in PIM, it is based strictly on elements: only the nodes of the element that contains the point of interest are used for the interpolation, and there is no overlapping in using nodes for computing shape functions. In addition, a proper mapping in FEM is a must (except for linear triangular elements) to ensure compatibility on the element interfaces.

2. In both FEM and PIM, the number of monomials used in the basis functions, m, is the same as the number of nodes, n. Therefore, the interpolation functions have the property of the Kronecker delta function. This feature of NS-PIM allows simple imposition of essential boundary conditions as in the standard FEM. No special treatments like the ones used in EFG and MLPG are needed.

3. The FEM uses the "compatible" strains in the element obtained using the strain–displacement relations, and hence it is a compatible mode. In the NS-PIM, however, the strains in the smoothing domains are constructed via the generalized smoothing operations, and hence the NS-PIM will not be compatible within the smoothing cells in terms of satisfying the displacement–strain relations.

4. Integration in the FEM is element based, while in the NS-PIM is, however, nodal smoothing domain based.

5. Both FEM and NS-PIM can reproduce linear displacement field exactly and hence pass the standard patch test (to machine accuracy).

6. The FEM solution does not in general satisfy the equilibrium conditions locally (either at any point in the elements or element-wise). The NS-PIM solution, on the other hand, satisfies the (homogeneous) equilibrium equations at any point within the smoothing domain, for each of the smoothing domains, and hence at any point in the entire problem domain, except on the interfaces of the smoothing domains. Therefore, the NS-PIM behavior is somewhat like an equilibrium model.

7. The standard FEM model is compatible everywhere (inside the elements and on the element interfaces, on the essential boundary, and is said to be fully compatible). The NS-PIM is compatible only on the interfaces of the smoothing domains and the essential boundary, but not in the smoothing domains. Because the equilibrium status is created inside the smoothing domains, a "complementary" situation is been achieved: where the compatibility condition is violated the equilibrium is ensured, and where the equilibrium is violated the compatibility is ensured. Such a complementary situation prevents any energy loss in the NS-PIM model, and hence can be said to be *energy consistent*.

8. The behavior of FEM model using triangular elements is very "stiff," and stress result is in general not very accurate; The behavior of the NS-PIM model using

exactly the same triangular mesh is much softer, and stress result is generally more accurate.

9. The linear FEM and linear NS-PIM models have the same set of nodes for nodal displacements and the same size in the discrete system equations and the number of unknowns. The stiffness matrices obtained using both FEM and NS-PIM are all SPD, if sufficient constraints are applied to eliminate the rigid body movement and the original problem is well posed. The proof of the SPD for FEM is based on the theory of weak formulation and that for the NS-PIM is based on the weakened-weak formulation [28].

10. The FEM provides the lower bound for the solution (in energy norm), and the NS-PIM can provide the upper bound of the solution, which will be discussed intensively in Section 8.3.

When quadratic polynomials are used in NS-PIM or RPIM shape functions are used, the sparsity of the stiff matrix of NS-PIM will be smaller than the FEM model using the same triangular mesh. We, however, still have the same number of unknowns.

8.1.5.2 NS-PIM vs. NS-FEM

The NS-PIM procedure can be extended for meshes of other types of elements including the general n-sided polygonal elements. If the same set of shape functions are used (the interpolation is confined using nodes only within the element containing the point of interest), the formulation leads to the node-based smoothed FEM (NS-FEM). A detailed formulation of NS-FEM was given in [14] for general n-sided polygonal elements, which was extended from the element-based smoothed FEM (SFEM) [15,16]. When only triangular elements and linear interpolations are used, the NS-PIM is identical to the NS-FEM and it gives the same results as the NIFEM proposed by Dohrmann et al. [17].

Note that the NS-PIM is more general than the NS-FEM and they are different in the following ways. The NS-PIM was basically conceived from the meshfree procedures: shape functions are constructed using nodes beyond the cells/elements, and they can be linear, quadratic, or even higher order depending on the number of nodes used in the support domain. In the case of NS-RPIM (see Section 8.2), the selection of nodes can practically be entirely free and the consistency of the shape functions can be arbitrarily high. Therefore, the NS-FEM can be considered as a special case of NS-PIM.

The NS-FEM is in turn more general formulation than the NIFEM. NS-FEM uses the strain smoothing technique, which transforms the node-based domain integration into the boundary integration. For 2D problems, the NS-FEM formulation can be applied for triangular, quadrilateral, and n-sided polygonal elements of any order because it needs only the shape function values and only on the boundaries of the smoothing domains. Such shape function values can be obtained by simple point interpolations. The NIFEM formulation is applicable only to uniform strain elements, and hence it can be only applied to linear triangular/tetrahedron elements. In addition, the numerical procedure of the NS-FEM and the NIFEM are also different. The NS-FEM uses only the value of shape function at points on the boundary of the smoothing domains associated with nodes, and no derivatives of the shape functions are needed. The NIFEM uses the derivative of shape functions, and the integration of the weak form is based on whole domain of the cell. Hence, the NIFEM can be viewed as a special case of the NS-FEM. Note also that, the NS-FEM is evolved from the SFEM and they all use the strain smoothing operation and the smoothed Galerkin weak form. The difference between NS-FEM and SFEM is on how the smoothing domains are created.

8.1.6 Numerical Examples for 2D Solids

Several numerical examples are studied in this section. The materials used are all linear elastic with Young's modulus $E = 3.0 \times 10^7$ and Poisson's ratio $\nu = 0.3$. The error indicators in displacement and energy are, respectively, defined as follows:

$$e_d = \sqrt{\frac{\sum_{i=1}^{N_n} \left(u_i^{\text{exact}} - u_i^{\text{num}} \right)^2}{\sum_{i=1}^{n} \left(u_i^{\text{exact}} \right)^2}} \tag{8.12}$$

$$e_e = \frac{1}{A} \sqrt{\frac{1}{2} \sum_{k=1}^{N_s} (\boldsymbol{\varepsilon}^{\text{exact}} - \bar{\boldsymbol{\varepsilon}}_k^{\text{num}})^{\text{T}} \mathbf{D} (\boldsymbol{\varepsilon}^{\text{exact}} - \bar{\boldsymbol{\varepsilon}}_k^{\text{num}}) A_k^s} \tag{8.13}$$

where the superscript "exact" denotes the exact or analytical solution, and "num" denotes the numerical solution obtained using a numerical method. For 2D problems, A is the area and for 3D problems, A becomes the volume of the problem domain. In Equation 8.12 the exact solution is sampled at the nodes, and in Equation 8.13 the exact solution is sampled at the center of the smoothing domains.

Example 8.1: 2D Patch Test

For a numerical method working for solid mechanics problems, the sufficient requirement for convergence is to pass the standard patch test [18]. Therefore, the first example is the standard patch test using the linear NS-PIM with T3-Scheme (Section 1.7.6). The patch is a square domain with the dimension of 10×10, and the patch is represented using regular and irregular nodes shown in Figure 8.3. The displacements are prescribed on all outside boundaries by the following linear function:

$$\begin{cases} u_x = 0.6x \\ u_y = 0.6y \end{cases} \tag{8.14}$$

The errors in displacement defined in Equation 8.12 are found to be 2.35×10^{-14} for the patch of regular nodes and 4.77×10^{-14} for the patch of irregular nodes. We also tested for quadratic NS-PIM with T6/3-Scheme and confirmed that it also passes the patch test, as predicated by the weakened-weak form theory [28]. This example demonstrates numerically, that the NS-PIM can reproduce linear fields exactly (to machine accuracy), which ensures second order convergence in displacement norm. Together with the proven stability, the NS-PIM solution will converge to the exact solution.

Example 8.2: Rectangular Cantilever

The NS-PIM is then benchmarked using Example 6.2. The analytical solution is available; it can be found in the textbook by Timoshenko and Goodier [19] and is listed in Equations 6.50 through 6.55. The beam is supported on the left edge by prescribing the displacements obtained using the analytical formula with $x = 0$. The beam is subjected to a parabolic traction at the free end given in Equation 6.56. The other properties are $D = 10$, $L = 50$, and $P = -1000$. To investigate the effect of the irregularity of nodal distribution, three models of 420 distributed nodes with different irregularity (shown in Figure 8.4) are used to examine the present method. The results of deflection along the neutral line of the beam are plotted in Figure 8.5 together with the analytical solutions.

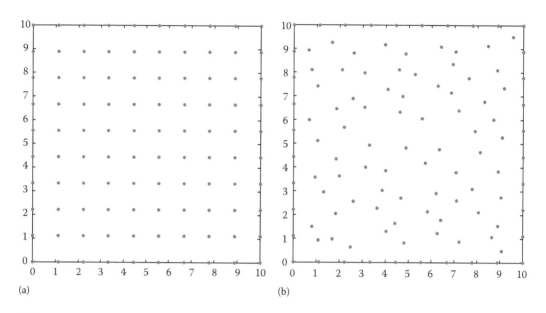

FIGURE 8.3
Nodal arrangement for patch tests: (a) patch with 100 nodes regularly distributed; and (b) patch with 109 nodes irregularly distributed.

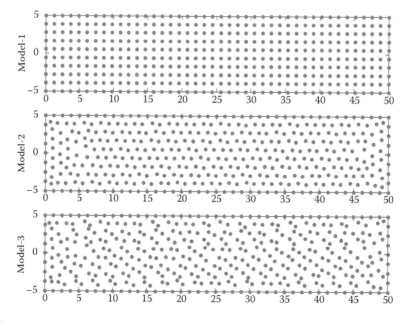

FIGURE 8.4
Three nodal distributions for the rectangular cantilever. Model-1: uniform distribution; Model-2: irregular distribution; and Model-3: highly irregular distribution.

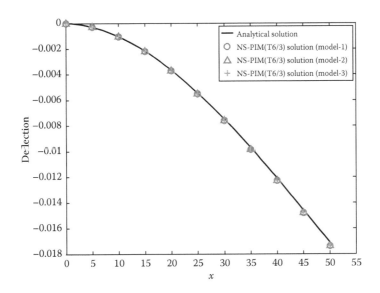

FIGURE 8.5
Deflection of the rectangular cantilever subject to a vertical force distributed at the free end. Computed using NS-PIM with three models of node distributions.

It can be found that the numerical results of these three models obtained using the quadratic NS-PIM are all in good agreement with the analytical ones, and the irregularity of the nodal distribution has little effect on the numerical results. Figure 8.6 plots the results of the shear stress along the line $(x = L/2)$ of the beam. It can be found again that the NS-PIM results are all in good agreement with the analytical ones, and the nodal irregularity has little effect on the numerical results.

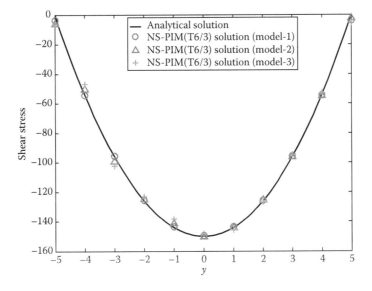

FIGURE 8.6
Shear stress distribution along the line $x = L/2$ on the rectangular cantilever subject to a vertical force distributed at the free end. Computed using NS-PIM with three models of node distributions.

To investigate the rate of convergence of the NS-PIM method to the exact solution, four models with different numbers of irregularly distributed nodes are used to compute the errors in both displacement and energy norms. For comparison, three methods—FEM-T3 linear, NS-PIM with T3-Scheme, and quadratic NS-PIM with T6/3-Scheme—are used for the same problem with the same sets of nodes. Figure 8.7a plots in the logarithm scale the solution error in displacement norm with different density of the mesh measured

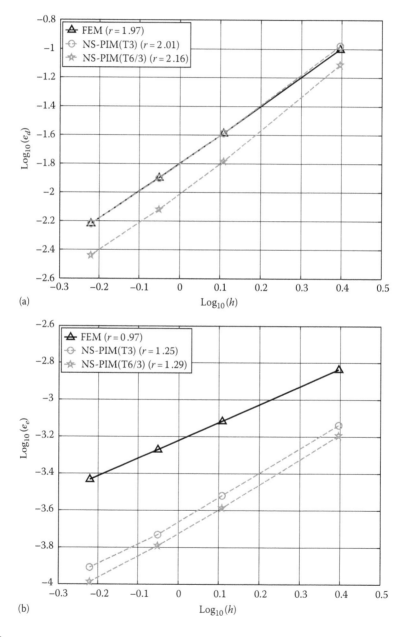

FIGURE 8.7

Comparison of convergence rates in displacement and energy norms for FEM-T3, linear NS-PIM, and quadratic NS-PIM for the rectangular cantilever subject to a vertical force distributed at the free end.

by the nodal spacing h. It is observed clearly for this case that the error decreases almost linearly with the decrease of h. We also estimated the convergence rates r from the slope of the plot in displacement norm for these three models. It is found the FEM achieved 1.97, which is very close to the theoretical rate of 2.0 for linear displacement weak form methods (see Section 3.5). The linear NS-PIM achieved a rate of 2.01 and the quadratic NS-PIM achieved 2.16. The rates for the linear NS-PIM are above the theoretical value of 2.0 showing a weak *superconvergence** even in displacement norm. In terms of accuracy, the linear NS-PIM has almost exactly the same accuracy as the FEM in displacement norm for this problem. The quadratic NS-PIM is about twice more accurate than the linear NS-PIM (and hence FEM).

Figure 8.7b also plots the solution error in energy norm with different nodal spacing h in the logarithm scale. It is observed again for this case that the error decreases almost linearly with the decrease of h. The convergence rates r are also estimated numerically from the slope of the plot for these three models. The FEM achieved a rate of 0.97 that is very close to the theoretical rate of 1.0 (see Section 3.5) for linear displacement weak form methods. The linear NS-PIM achieved a rate of 1.25 that is above the theoretical value of 1.0 showing a noticeable *superconvergence* in energy norm. It is smaller than the ideal theoretical convergence rate of 1.5 for a GS-Galerkin model (see Section 3.5). The rate of convergence for the quadratic NS-PIM is about 1.29, slightly higher than the linear NS-PIM. In terms of accuracy, the linear NS-PIM is about two to three times more accurate than the FEM in energy norm for this problem. The quadratic NS-PIM is only a little more accurate than the linear NS-PIM.

Remark 8.1: Superconvergence in Both Displacement and Energy Norms
Note that the so-called superconvergence refers to a phenomenon in FEM, in which solution in *energy norm* at some points in the elements converges faster than the predicted theoretical value [18]. In a weak formulation, such a superconvergence will not be observed in *displacement norm*. In a weakened-weak formulation (such as the NS-PIM), however, it is observed in both displacement norm and energy norm (see more below).

Example 8.3: Hole in an Infinite Plate

Example 7.4, examined using the MLPG method, is now reexamined here using the NS-PIM. The geometry of the plate is plotted in Figure 7.10. Because of the twofold symmetry, only a quarter of the plate is modeled with symmetric boundary conditions applied on $x = 0$ and $y = 0$. The parameters and the boundary conditions are exactly the same as those in Example 7.4. The exact solution for the displacement and the stress field within the plate are provided by Equations 7.41 through 7.46 in the polar coordinates (r, θ). The dimensionless[†] parameters used in this example are listed as follows:

Loading: $p = 10$
Young's modulus: $E = 3.0 \times 10^7$
Poisson's ratio: $v = 0.3$
The radius of the hole: $a = 10$
Width of the plate: $b = 50$

---

* The term "superconvergence" is defined as a convergence that is faster then the theoretical convergence rate of linear FEM model.
† This means that one may use any unit set as long as it is consistent.

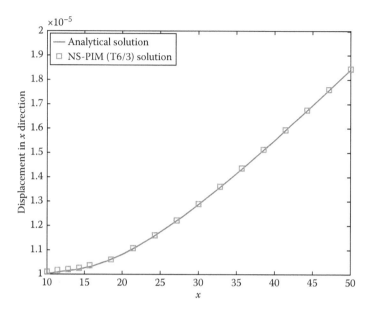

FIGURE 8.8

Distribution of u_x along the bottom edge of the one-quarter model of the plate with a circular central hole subjected to a unidirectional tensile load in the x-direction.

In this study, plane stress problem is considered and the domain is discretized using 411 irregularly distributed nodes. The computed displacements for nodes located along the bottom and left edges of the one-quarter model are calculated and plotted in Figures 8.8 and 8.9, respectively.

For precise error examination in the numerical solution, stresses-computed using Equations 7.44 through 7.46 on edges of $x = 50$ and $y = 50$ are used as the prescribed stress boundary condition in our numerical models. This is to make sure that the exact and numerical solutions are under the same conditions and hence compatible.

Figure 8.10 shows the distribution of normal stress σ_{xx} along the left edge of the quarter model. These figures show again that all the numerical results agree well with the analytical ones.

We study the volumetric locking issue using Example 8.3 with exactly the same settings except that $a = 1$, $b = 5$. Poisson's ratio varies from 0.4 to 0.4999999, and plane strain problem is considered. The NS-PIM is used to solve this problem together with the linear FEM for comparison.

Figure 8.11 plots the error in solution in displacement norm against Poisson's ratio changing from 0.4 to 0.4999999. Table 8.3 gives the detailed numbers. It is clearly seen that the FEM suffers from the volumetric locking: when Poisson's ratio approaches 0.5, the error in solution increases drastically starting from 0.4, as predicated in Remark 1.3. Special treatments are needed in FEM for this type of problems. However, the results show that the NS-PIM is naturally immune from the volumetric locking for nearly incompressible materials: we did not give any additional treatment for this problem, and accuracy has not been affected by the increasing incompressibility.

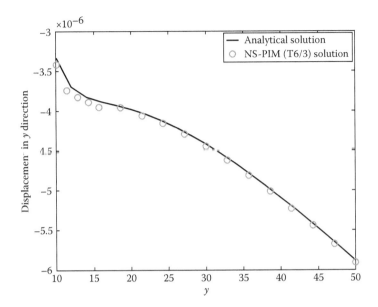

FIGURE 8.9
Distribution of u_y along the left edge of the one-quarter model of the plate with a circular central hole subjected to a unidirectional tensile load in the x-direction.

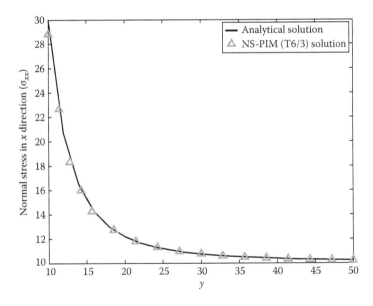

FIGURE 8.10
Distribution of stress along the left edge of the one-quarter model of the plate with a circular central hole subjected to a unidirectional tensile load in the x-direction.

Remark 8.2: Free from Volumetric Locking: A Property of NS-PIM
NS-PIM is naturally immune from the volumetric locking, and no special treatments are needed for solids of nearly incompressible materials.

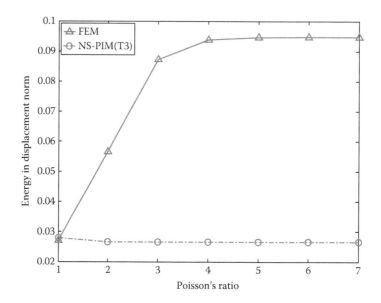

FIGURE 8.11
Error in solution in displacement norm against Poisson's ratio changing from 0.4 to 0.4999999. FEM solution is locked when Poisson's ratio approaches to 0.5, but the NS-PIM is naturally immune from the volumetric locking.

TABLE 8.3

Error in Solution in Displacement Norm for 2D Plane Strain Problem When Poisson's Ratio Changes from 0.4 to 0.4999999

Value of Poisson's Ratio ν	Displacement Error of FEM	Displacement Error of NS-PIM
0.4	0.27077147E−01	0.27917215E−01
0.49	0.56525782E−01	0.26611438E−01
0.499	0.87288403E−01	0.26621114E−01
0.4999	0.93841046E−01	0.26634752E−01
0.49999	0.94621628E−01	0.26636331E−01
0.499999	0.94701838E−01	0.26636492E−01
0.4999999	0.94709883E−01	0.26636508E−01

Example 8.4: Internal Pressured Thin Cylindrical Disk

As another benchmark problem, a thin cylindrical disk subjected to internal pressure is analyzed, as shown in Figure 8.12. The dimensionless* parameters used in this example are listed as follows:

Internal radius $a = 10$

Outer radius $b = 25$

Thickness $t = 1$

Internal pressure $p = 100$

---

* This means that one may use any unit set as long as it is consistent.

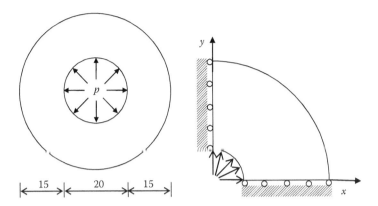

FIGURE 8.12
A thin disk subjected to internal pressure and its quarter model.

The analytical solution in polar coordinates is available for this plane stress problem [19] and is listed below:

$$u_r = \frac{pa^2}{E(b^2 - a^2)r}[(1 - v)r^2 + (1 + v)b^2] \tag{8.15}$$

$$\sigma_r = \frac{a^2 p}{b^2 - a^2}\left(1 - \frac{b^2}{r^2}\right) \tag{8.16}$$

$$\sigma_\theta = \frac{a^2 p}{b^2 - a^2}\left(1 + \frac{b^2}{r^2}\right) \tag{8.17}$$

In this study, the problem domain is represented with 441 irregularly distributed nodes and the numerical solutions using NS-PIM are plotted in Figures 8.13 and 8.14 together with the analytical solution. It can be observed that both the displacement and stress results are very accurate and stable, and in good agreement with the analytical ones.

Example 8.5: A Mechanical Part: 2D Rim

As an application of NS-PIM to practical mechanical component design, a typical rim of automotive component with a complicated shape is studied. As shown in Figure 8.15, the rim is fixed at all the nodes around the inner circle and a pressure of 100 units is applied along the lower arc edge of the rim. The rim is meshed triangular cells with 2608 nodes, as shown in Figure 8.15. As no analytical solutions are available for this problem, a reference solution obtained using the FEM with a very fine mesh of six-node triangular element (18,625 elements) is used. Displacement and stress results at the nodes along the lower half circle of the rim (dash line *m–n*) are computed using the quadratic NS-PIM and are plotted in Figures 8.16 through 8.20 together with the reference solutions. It is found that the NS-PIM solution in both displacements and stresses are in a good agreement with the reference solutions.

8.1.7 NS-PIM for 3D Solids

We now extend the NS-PIM for 3D solids by simply mentioning the major difference between the 2D and 3D. More detailed formulation can be found in [23].

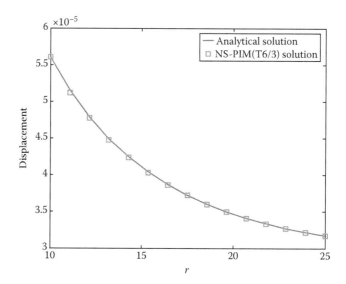

FIGURE 8.13
Displacement distribution along the left edge for the problem of internal pressurized thin disk.

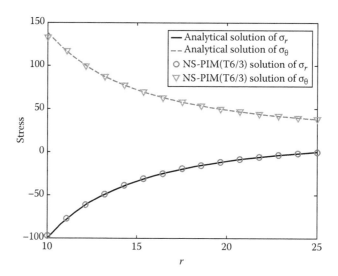

FIGURE 8.14
Stress distribution along the left edge for the problem of internal pressurized thin disk.

Theoretically, the NS-PIM for 3D is exactly the same as 2D, except that all the operations have to be extended to one more dimension. Most of the formulae presented in Section 8.1.3 have the same forms, but the following major changes are needed:

- The primary known variables of displacement components become three: u_x, u_y, and u_z, and the stress and strain components become six. For example, the smoothed strain matrix defined in Equation 8.4 should be changed to

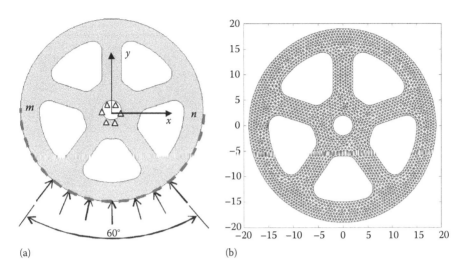

FIGURE 8.15
An automotive rim subjected to pressure along a portion of the lower arc edge. (a) problem setting; (b) nodes distribution.

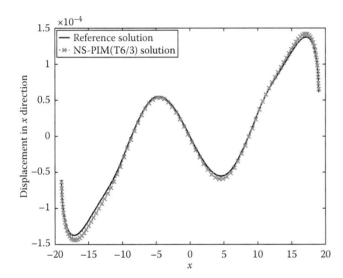

FIGURE 8.16
Distribution of displacement u_x along the edge m–n of the automotive rim.

$$
\bar{\mathbf{B}}_I(\mathbf{x}_i) =
\begin{bmatrix}
b_{Ix}(\mathbf{x}_i) & 0 & 0 \\
0 & b_{Iy}(\mathbf{x}_i) & 0 \\
0 & 0 & b_{Iz}(\mathbf{x}_i) \\
b_{Iy}(\mathbf{x}_i) & b_{Ix}(\mathbf{x}_i) & 0 \\
0 & b_{Iz}(\mathbf{x}_i) & b_{Iy}(\mathbf{x}_i) \\
b_{Iz}(\mathbf{x}_i) & 0 & b_{Ix}(\mathbf{x}_i)
\end{bmatrix}
\tag{8.18}
$$

All the other equations needed for the computation are practically the same in form as those given in Section 8.1.3.

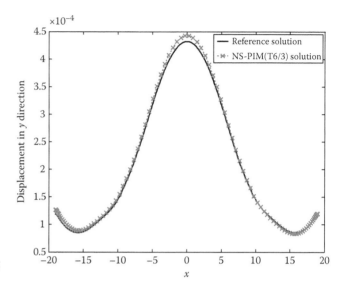

FIGURE 8.17
Distribution of displacement u_y along the edge m–n of the automotive rim.

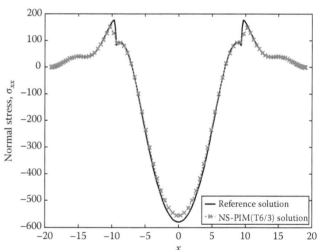

FIGURE 8.18
Distribution of stress σ_{xx} along the edge m–n of the automotive rim.

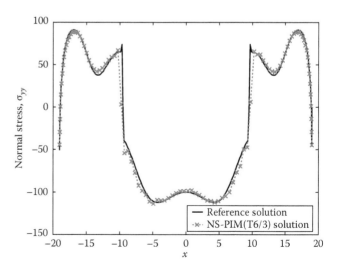

FIGURE 8.19
Distribution of stress σ_{yy} along the edge m–n of the automotive rim.

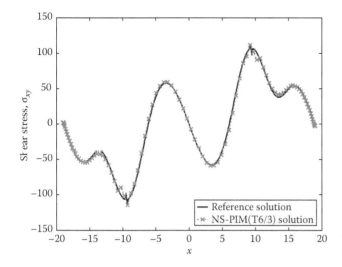

FIGURE 8.20
Distribution of σ_{xy} along the edge m–n of the automotive rim.

- The area in 2D becomes volume in 3D, the area integration in 2D now becomes volume integration in 3D, and the curve integration in 2D now becomes surface integration in 3D.

- The triangular cells for discretizing the domain now become tetrahedral cells. Figure 8.21 shows the formation of the nodal smoothing domain for node i. The smoothing domain consists of portions from all tetrahedral cells sharing the node. Figure 8.22 shows the four portions of a four-node tetrahedron cell J (with four nodes: i–k_2–k_3–k_6) that contributes, respectively, to four smoothing domains

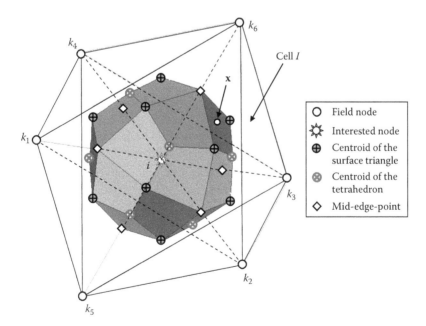

FIGURE 8.21
Illustration of the formation of nodal smoothing domain for node i. The smoothing domain consists of portions from all tetrahedral cells sharing the node. It looks quite complicated, but the formation can be programmed in a quite straightforward manner, as we use only tetrahedrons.

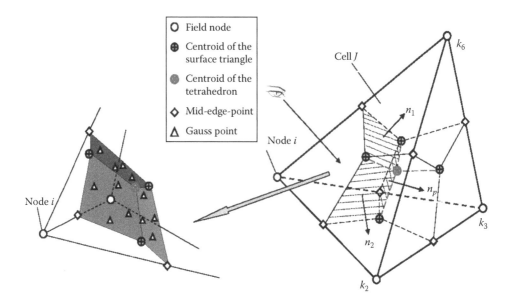

FIGURE 8.22
Four portions of a four-node tetrahedron cell J ($i-k_2-k_3-k_6$) that contributes, respectively, to four smoothing domains for nodes i, k_2, k_3 and k_6. The smoothing domains surface for the portion of the smoothing domain for node i contributed from cell J, is also shown. The smoothing domain is created by connecting the mid-edge-points, the centroids of the surface triangles and the centroid of the tetrahedron.

nodes i, k_2, k_3, and k_6. The smoothing domains surface for the portion of the smoothing domain for node i contributed from cell J, is also shown. The smoothing domain is created by connecting the mid-edge-points, the centroids of the surface triangles, and the centroid of the tetrahedron.

- In terms of node selection, four nodes of the tetrahedral cell containing the point of interest (usually the quadrature point) and basis for the interpolation becomes $1\ x\ y\ z$ and is used for linear NS-PIM. For higher order NS-PIM, more nodes should be selected, and it is a little tricky in node (or monomial basis term) selection, due to the possible singular moment matrix. Therefore, we suggest the use of the T2L-Scheme and RPIM shape functions (see Section 8.2), and that works for very irregularly distributed nodes without much special treatments.

- The macro flowchart for 3D NS-PIM code also follows that given in Section 8.1.4, with only the dimension changes in mind.

8.1.8 Numerical Examples for 3D Problems

A 3D code of linear NS-PIM has been developed, and it is examined using the following examples.

Example 8.6: A 3D Patch Test

The first example is the standard patch test using our in-house 3D NS-PIM code. A cubic patch with the dimension of $10 \times 10 \times 10$ is used, and the displacements are prescribed on all outside boundary surfaces by the following linear function:

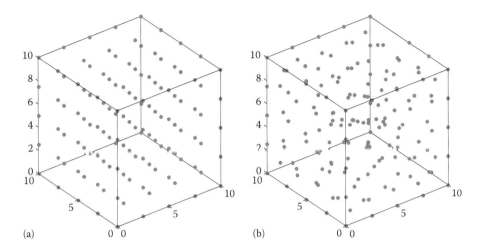

FIGURE 8.23
Node distribution of a cube for the 3D standard patch test: (a) regularly distributed nodes; and (b) irregularly distributed nodes.

$$\begin{cases} u_x = 0.6x \\ u_y = 0.6y \\ u_z = 0.6z \end{cases} \tag{8.19}$$

Two patches are represented using 125 regularly and 166 irregularly distributed nodes as shown in Figure 8.23. The errors in displacement defined in Equation 8.12 are found to be 1.2837×10^{-15} for the regular and 1.2036×10^{-15} for the irregular patch, which are almost the level of the machine precision. The results show that the displacements of all the interior nodes follow "exactly" the same function of the imposed displacement. This example demonstrates numerically that the 3D NS-PIM solution will have second order convergence due to its ability to reproduce linear fields.

Example 8.7: A 3D Cantilever

The NS-PIM is first benchmarked using the 3D cantilever problem under traction on the right surface, as shown in Figure 8.24. The parameters used in this example are listed as follows:

Loading: $P = -100.0$ N
Young's modulus: $E = 3.0 \times 10^7$ Pa

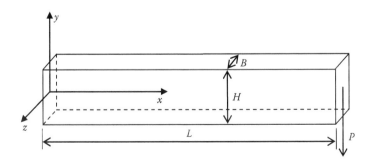

FIGURE 8.24
A 3D cantilever beam supported on the left surface ($x = 0$) and subjected to a traction on the right surface.

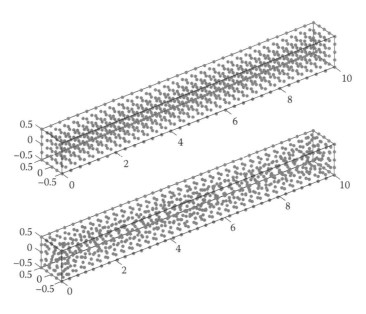

FIGURE 8.25
Nodal distributions in the 3D cantilever beam.

Poisson's ratio: $\nu = 0.3$
Height of the beam: $H = 1.0$ m
Length of the beam: $L = 10.0$ m
Width of the beam: $B = 1.0$ m

Since the beam is relatively thin, analytical solution listed in Equations 6.50 through 6.55 based on the plane stress theory can be used approximately as a reference solution for comparison purpose.

The beam is supported on the left surface by prescribing the displacements using the analytical solution. The traction applied on the right surface is also determined using the analytical formula given in Equation 6.56. The problem domain is presented using 874 irregularly distributed nodes as shown in Figure 8.25. The results of defection distribution along the neutral line ($y = 0$ and $z = 0$) and shear stress along the midline ($x = L/2$ and $z = 0$) are plotted together with the reference solutions in Figures 8.26 and 8.27, respectively. It can be found that the numerical results of the NS-PIM are in very good agreement with the reference ones.

Example 8.8: 3D Lame Problem

The 3D Lame problem is a hollow sphere with inner radius a and outer radius b and subjected to internal pressure P, as shown in Figure 8.28. For this benchmark problem, the analytical formulae for the solution are available in spherical coordinate system [19]:

$$u_r = \frac{pa^3 r}{E(b^3 - a^3)} \left[(1 - 2\nu) + (1 + \nu) \frac{b^3}{2r^3} \right] \tag{8.20}$$

$$\sigma_r = \frac{pa^3(b^3 - r^3)}{r^3(a^3 - b^3)} \tag{8.21}$$

$$\sigma_\theta = \frac{pa^3(b^3 + 2r^3)}{2r^3(b^3 - a^3)} \tag{8.22}$$

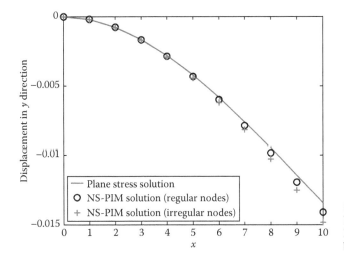

FIGURE 8.26
Deflection distribution along the neutral line ($y = 0$ and $z = 0$) in the 3D cantilever beam.

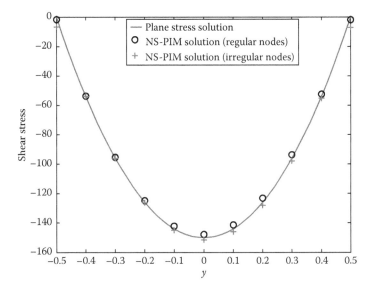

FIGURE 8.27
Shear stress distribution along the line ($x = L/2$ and $z = 0$) in the 3D cantilever beam.

where r is the radial distance from the centroid of the sphere to the point of interest in the sphere. As the problem is spherically symmetrical, only one-eighth of the sphere is modeled and symmetry conditions are imposed on the three planes of symmetry. The parameters used in this example are listed as follows:

Loading: $p = 1.0 \text{ N/m}^2$

Young's modulus: $E = 1.0$ Pa

Poisson's ratio: $\nu = 0.3$

Inner radius: $a = 1.0$ m

Outer radius: $b = 2.0$ m

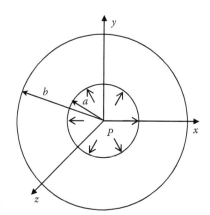

FIGURE 8.28
3D Lame problem of a hollow sphere under internal pressure.

The problem domain is presented using 1304 irregularly distributed nodes. The computed nodal displacements and stresses along the horizontal ($\theta = 0$) axis are plotted in Figures 8.29 and 8.30, respectively. It can be clearly seen that the numerical results agree well with the analytical ones.

To investigate the properties of convergence and efficiency of the NS-PIM, four models of 173, 317, 729, and 1304 irregularly distributed nodes are employed. For each of these four models, the error in energy norm of the numerical results is calculated according to the definition in Equation 8.13. For comparison, the FEM using linear four-node tetrahedron element is also employed to study the problem with the same nodes distributions. As shown in Figure 8.31, the results of error in energy norm against the average nodal spacing h are plotted for both the NS-PIM and the FEM. It is found that these two methods have similar rates of convergence,

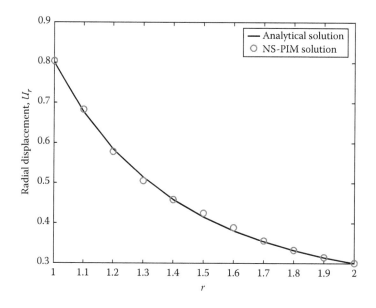

FIGURE 8.29
Distribution of the radial displacement along the horizontal axis ($\theta = 0$) in the 3D Lame problem.

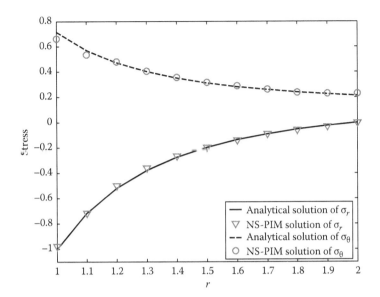

FIGURE 8.30
Distribution of radial and tangential stresses along the horizontal axis ($x = 0$) in the 3D Lame problem.

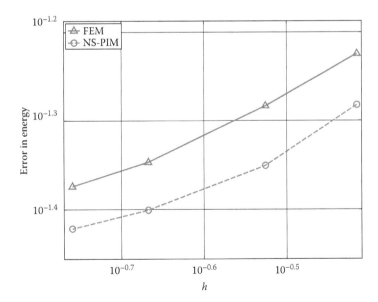

FIGURE 8.31
Comparison of convergence of NS-PIM with the FEM with the same nodes distribution for the Lame problem.

but the NS-PIM produces more accurate results compared with the linear FEM. Figure 8.32 plots the energy errors of the numerical results obtained using these two methods against the CPU time consumed (full matrix solver), which shows performance of numerical methods. It can be found that the NS-PIM is more efficient than the linear FEM.

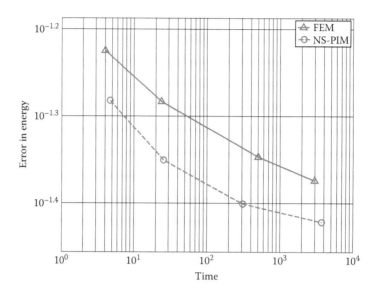

FIGURE 8.32
Comparison of efficiency of NS-PIM with the FEM with the same nodes distribution for the Lame problem.

Example 8.9: 3D Kirsch Problem

The 3D Kirsch problem of an infinite cube subjected to far field uniform tension is considered as shown in Figure 8.33. The NS-PIM code is used to compute the stress distribution in the vicinity of a small cavity in an infinite cube. The analytical solution for the normal stress σ_{zz} in the plane $z = 0$ is given as [19]

$$\sigma_{zz} = \sigma_0 \left[1 + \frac{4 - 5\nu}{2(7 - 5\nu)} \left(\frac{a}{r}\right)^3 + \frac{9}{2(7 - 5\nu)} \left(\frac{a}{r}\right)^5 \right] \tag{8.23}$$

where r is the radial distance from the centroid of the cube to the point of interest. The parameters used in this example are as follows:

Loading: $\sigma_0 = 1.0 \ \text{N/m}^2$
Young's modulus: $E = 3.0 \times 10^7 \text{Pa}$
Poisson's ratio: $\nu = 0.3$
Radius of the hole: $a = 1.0$ m
Width of the cube: $b = 10$ m

 The problem domain is presented as a set of tetrahedral cells with a total of 1256 nodes. Figure 8.34 shows the comparison between the analytical solution and the numerical solution of NS-PIM for the normal stress σ_{zz} along the *x*-axis. It can be clearly seen again that the NS-PIM solution is in an excellent agreement with the analytical ones.

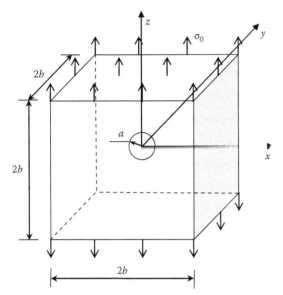

FIGURE 8.33
3D Kirsch problem: a cube with a spherical cavity subjected to a uniform tension.

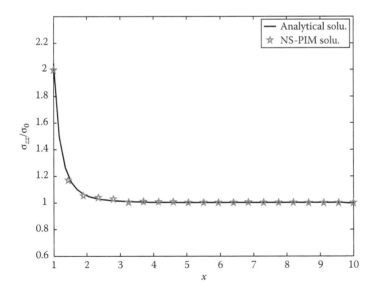

FIGURE 8.34
Distribution of normal stress in z-direction along the x-axis for the Kirsch problem.

Example 8.10: A Mechanical Part: 3D Rim Component

A typical rim component shown in Figure 8.35 used in mechanical system is modeled and studied using the NS-PIM method. The rim is constrained in three dimensions along the inner annulus and a uniform pressure of P is applied on the outer annulus over a range of 60°. Due to the complexity, we omit the details of the model here, but show directly the numerical results obtained using our 3D NS-PIM code in comparison with a reference solution obtained using FEM with very fine mesh with 29,835 nodes. In the NS-PIM, 7972 nodes are used. The numerical solutions of stress components at the nodes on the middle plane of $z = 0$ are plotted in the form of contour. Figures 8.36 through 8.38 show the comparison of the stress contours between the reference FEM solutions and the NS-PIM solutions for σ_{xx}, σ_{yy}, and τ_{xy}, respectively. It can be seen that the results obtained using the NS-PIM agree well with the reference ones.

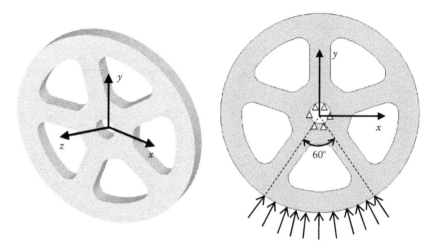

FIGURE 8.35
A simplified model of an automotive rim component.

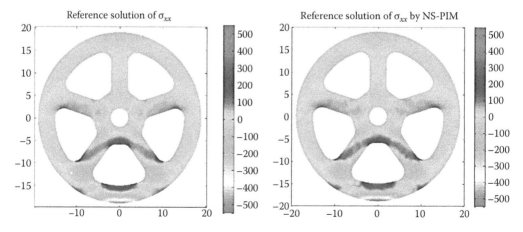

FIGURE 8.36
Contour of stress σ_{xx} on the plane $z = 0$ in the rim component.

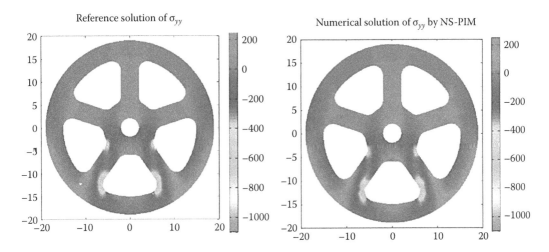

FIGURE 8.37
Contour of stress σ_{yy} on the plane $z = 0$ in the rim component.

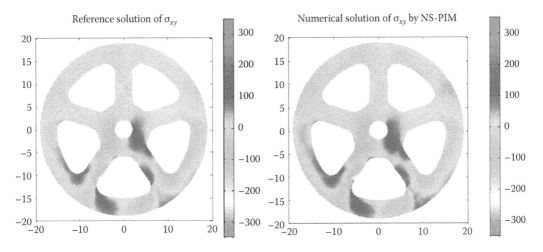

FIGURE 8.38
Contour of stress σ_{xy} on the plane $z = 0$ in the rim component.

Example 8.11: A 3D Riser Connector

The following example comes from a real offshore project of a Floating Production and Storage Unit (FPSO). Fluid of oil–gas–water mixture is transferred between the FPSO and subsea pipeline through a kind of flexible pipe called riser, which is attached to the FPSO shipside by a riser connector. The simplified model of a riser connector is shown in Figure 8.39. The load is applied on the top flange of the riser connector. The boundary conditions are defined at the end of I-beams where riser connector is supported by other structures. Due to the complexity of the actual structure, we omit the detailed specification of the model, but present the final results in comparison with the FEM solution.

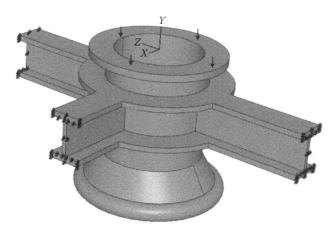

FIGURE 8.39
Simplified model of the three-dimensional riser connector in an offshore platform.

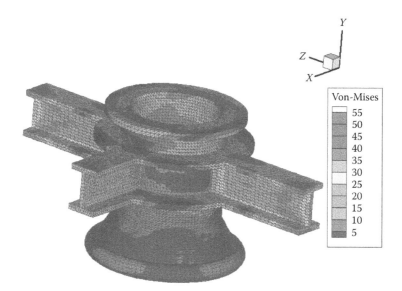

FIGURE 8.40
Reference solution of contour for elemental Von Mises stress obtained using FEM with fine mesh of 31,876 nodes.

Reference solution of this problem is obtained using the FEM model with very fine mesh (total 31,876 nodes), and the contour of elemental Von Mises stress is plotted in Figure 8.40 together with the deformed shape of the riser connector. For the purpose of comparison, this problem is studied using both the present NS-PIM and the linear FEM with the same nodes distribution (total 1718 nodes). The numerical results of the elemental Von Mises stress obtained using the fine FEM model is plotted in the form of contour in Figures 8.41 and 8.42. It can be found that, although the riser connector is presented with less than one-tenth of the number of nodes of the reference model, the NS-PIM solution matches with the reference solution very well. The NS-PIM solution is much closer to the reference solution compared to that of the linear FEM of the same coarse mesh.

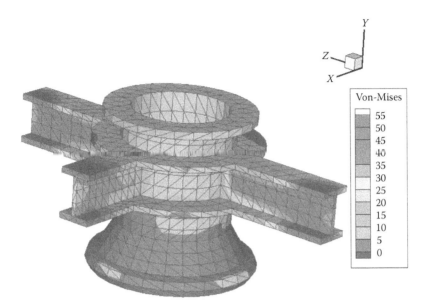

FIGURE 8.41
Contour of elemental Von Mises stress obtained using NS-PIM with a coarse mesh of 1718 nodes.

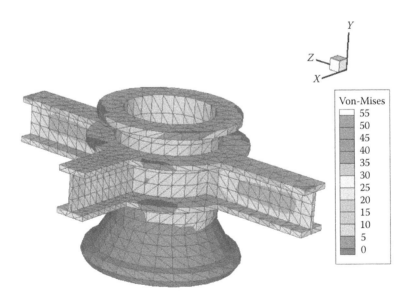

FIGURE 8.42
Contour of elemental Von Mises stress obtained using FEM with a coarse mesh of 1718 nodes.

8.2 NS-PIM Using Radial Basis Functions (NS-RPIM)

8.2.1 Considerations

Chapter 2 demonstrated that the use of radial functions as basis functions (with proper shape parameters) can guarantee a nonsingular moment matrix, and shape functions with delta function property can be created via simple point interpolation method for virtually randomly distributed nodes. This section introduces a node-based meshfree method that uses node-based smoothing operation and PIM shape functions constructed using radial basis functions instead of polynomial basis functions, which is called node-based smoothing radial point interpolation method or NS-RPIM in short. The material of this section is largely based on the work of Refs. [4,11,12]. The method was called originally the linearly conforming radial point interpolation method (LC-RPIM) because it is at least linearly conforming. We need now rename it as NS-RPIM because the other methods developed later, such as the ES-PIM, CS-PIM, etc. are all at least linear conforming but very much different in the use of smoothing domains and properties.

The formulation of NS-RPIM is largely the same as the NS-PIM, except the procedure of the node selection and shape function creation. In terms of smoothing domain, we can use exactly the same used for NS-PIM, but this section uses the alternative Voronoi cells detailed in Section 8.1.2.

8.2.2 Node Selection in NS-RPIM

NS-RPIM accommodates various schemes for node selection very flexibly, and the node distribution can be very irregular and virtually randomly distributed. In this section, we use RPIM shape function with linear polynomial basis for restoring the (polynomial) reproducibility. The following node selection schemes can be used:

1. Use T3- or T6-Schemes detailed in Section 1.7.6. When T3-Scheme is used, the NS-RPIM produces the same results as the linear NS-PIM, and hence we do not usually use the T3-Scheme for NS-RPIM, except for debugging purposes. When T6-Scheme is used, the NS-RPIM will be different from NS-PIM. T6-Scheme is quite well controlled, and hence is very robust and most efficient. The equally shared smoothing domain is preferred for this scheme, for efficiency and convenience in implementation.

2. Use T2L-Scheme. In this scheme, the three vertices of the home triangular cell of the point of interest (usually the quadrature point) and the nodes that are directly connected to the three vertices are used. T2L-Scheme is also well controlled, and can be used with a lot of freedom. This is a very robust scheme for node selection, which works well for virtually randomly distributed nodes, but less efficient than the T6-Scheme.

3. Use the support domain or influence domain method described in Chapter 1. This selection scheme is a general meshfree procedure used in EFG or SPH [24]. This scheme offers total freedom in node selection. It works well with both equally shared and Voronoi smoothing domains. It is used in this section. Theoretically, however, there is a chance for biased node selection leading to extrapolation. We do not recommend this scheme unless the nodal distribution can be well controlled.

Once the nodes are selected, techniques detailed in Section 2.7 are then used to compute the RPIM shape functions. Note again in the NS-PIM formulation, the derivatives of the shape functions are not required as in the NS-PIM. Once the RPIM shape functions are created, the rest of the procedure of NS-RPIM is the same as the NS-PIM, detailed in Section 8.1, except that NS-RPIM usually uses more local nodes for shape function construction leading to larger bandwidth in the discretized system equations.

8.2.3 Examples Solved Using NS-RPIM

A NS-RPIM code has been developed, and it is examined using the following examples. In these examples, the RPIM shape function is constructed using MQ-RBF with $\alpha_c = 4.0$ and $q = 1.03$ (see Chapter 2) and complete linear polynomial functions ($m = 3$) are included to ensure the polynomial linear consistency in the local displacement approximation. The numerical results are compared with those obtained using FEM with four-node isoparametric elements, and analytical solution whenever it is possible. For qualitative error analysis, in this section we use the relative error indicator defined in Equation 8.12 for displacement and the following relative error indicators for energy error:

$$e_e = \sqrt{\sum_{k=1}^{N_s} (\boldsymbol{\varepsilon}^{\text{exact}} - \bar{\boldsymbol{\varepsilon}}_k^{\text{num}})^{\text{T}} \mathbf{D}(\boldsymbol{\varepsilon}^{\text{exact}} - \bar{\boldsymbol{\varepsilon}}_k^{\text{num}}) A_k^s \Big/ \sum_{k=1}^{N_s} (\boldsymbol{\varepsilon}^{\text{exact}})^{\text{T}} \mathbf{D}(\boldsymbol{\varepsilon}^{\text{exact}}) A_k^s} \qquad (8.24)$$

Equation 8.24 is essentially the same as Equation 8.13 except the difference in scaling. Therefore, when looking at the convergence rates, they are exactly the same.

Example 8.12: Patch Test

This numerical example is the standard patch test for NS-RPIM using the patch of a unit square 1×1. The node distribution of both regularly and irregularly distributed 121 field nodes is shown in Figure 8.43. The irregular nodes are created by altering the coordinates of the regular nodes using the following equation:

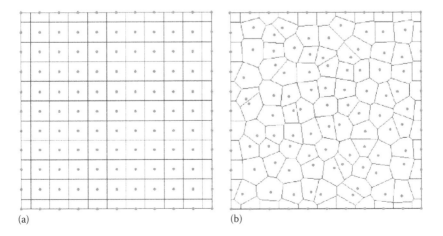

(a) (b)

FIGURE 8.43
A patch with 121 field nodes: (a) regular nodal distribution and the Voronoi smoothing domains and (b) irregular nodal distribution and the Voronoi diagrams ($\alpha_{ir} = 0.4$).

$$\begin{cases} x' = x + \Delta x \cdot r_c \cdot \alpha_{ir} \\ y' = y + \Delta y \cdot r_c \cdot \alpha_{ir} \end{cases} \qquad (8.25)$$

where

Δx and Δy are, respectively, the initial nodal spacings in the x- and y-directions
r_c is a computer-generated random number between -1.0 and 1.0
α_{ir} is the irregularity factor controlling the degree of the irregularity

In this standard patch test, the linear displacements are prescribed on all four outside boundaries by the linear functions of $u = x$ and $v = y$. The material parameters of the patch are $E = 1.0$ and $v = 0.25$, and the plane stress problem is considered. Satisfaction of the patch test requires that the displacements obtained by a numerical method at any interior point match those calculated using the same linear functions, the strains and stresses are constant in the entire patch.

8.2.3.1 Effect of Shape Parameters

As seen in Chapter 2, the radial basis functions have shape parameters that can affect the performance of a numerical method. We therefore first investigate the effects of the shape parameters α_c and q in the MQ-RBF. In this investigation, regularly distributed nodes and support domains of $\alpha_s = 2.5$ are used. First, we fix the value of the parameter q at 0.5, 1.03, and 1.3, and then have parameter α_c, which varies from -100 to 10. The relative errors of displacement and energy for the patch test example are computed and plotted in Figure 8.44. It is observed that the errors in both displacement and energy are in the order of 10^{-14} and is about the machine accuracy when the parameter $\alpha_c < 1.0$ for $q = 0.5$, 1.03, and 1.3. We now fix α_c at 0.1, 1.0, and 4.0, vary q from 0 to 2.0, and the relative errors of the displacement and energy are computed and plotted in Figure 8.45. It is observed again that the accuracy in displacement and energy is very stable except in the vicinity of the singular point (where q is an integer). The accuracy is much higher and is in the order of 10^{-14} for smaller parameter $\alpha_c = 0.1$ and 1.0. However, it is of the order of 10^{-11} when $\alpha_c = 4.0$ is used.

The results suggest that the shape parameter of the MQ-RBF should be: $\alpha_c \leq 0.25$ and $0.2 < q < 0.8$ and $1.2 < q < 1.8$. In the following computations, $\alpha_c = 0.1$ and $q = 0.5$ are thus used.

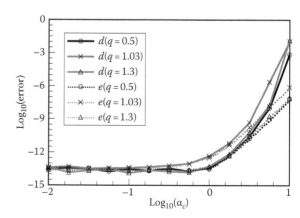

FIGURE 8.44
The effect of shape parameter α_c in MQ-RBF on the relative error in the standard patch tests using NS-RPIM with regular nodes. The dimension of the support domain is $\alpha_s = 2.5$.

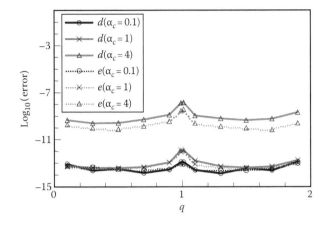

FIGURE 8.45
The effect of shape parameter q in MQ-RBF on the relative error in the standard patch tests using NS-RPIM with regular nodes. The dimension of the support domain is $\alpha_s = 2.5$.

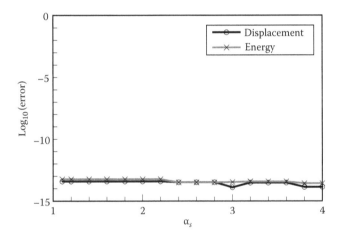

FIGURE 8.46
The effect of support domain size α_s on the relative error in the standard patch tests using NS-RPIM with regular nodes ($\alpha_c = 0.1$, $q = 0.5$, $\alpha_{ir} = 0$).

8.2.3.2 Effect of the Dimension of the Support Domain

In this study, we computed relative displacement and energy errors in the patch test and plotted in Figure 8.46 using regular nodes but different dimensions of the support domain α_s from 1.5 to 4.0. It is shown that the relative displacement and energy accuracy of the NS-RPIM is very stable with the order of 10^{-14}, regardless of the size of the support domains. This implies that passing the standard patch test does not depend on the size of the support domain, and hence more nodes for interpolation does not necessarily help to improve the accuracy because the results are already in the range of machine accuracy. This finding implies, in a way, that the NS-RPIM is very stable.

8.2.3.3 Effect of the Irregularity of a Nodal Distribution

We now fix the nodal support domain at $\alpha_s = 2.5$, and compute relative displacement and energy errors for varying irregularity factor α_{ir} for the interior nodes in the patch. The results are plotted in Figure 8.47. It is clearly shown that the accuracy of both displacement and energy of the NS-RPIM is very stable against the nodal irregularity. We in fact observe

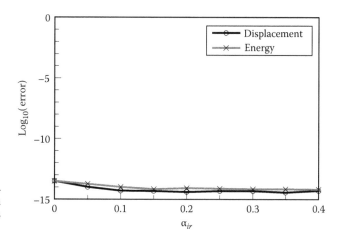

FIGURE 8.47
The effect of nodal irregularity factor α_{ir} on the relative error in the standard patch tests using NS-RPIM with regular nodes ($\alpha_c = 0.1$, $q = 0.5$, $\alpha_s = 2.5$).

a drop in error when the nodes get more irregular. For irregular nodal distribution, the displacement accuracy of the NS-RPIM is almost the same as its energy accuracy and of the order of 10^{-15}. This finding shows the robustness of the NS-RPIM against the nodal irregularity.

Example 8.13: Rectangular Cantilever

The NS-RPIM is next benchmarked using Example 6.2 again. The material properties and other parameters are taken as $E = 3.0 \times 10^4$ MPa, $\nu = 0.25$, $D = 1$ m, $L = 8$ m, and $P = -1000$ kN. Figure 8.48 shows the nodal distribution and its corresponding Voronoi diagrams for both regular and irregular node distributions.

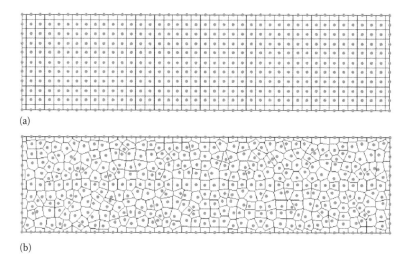

(a)

(b)

FIGURE 8.48
A rectangular cantilever with 451 field nodes and Voronoi diagrams: (a) regular; and (b) irregular ($\alpha_{ir} = 0.4$).

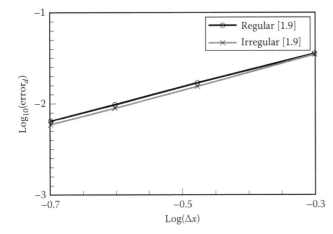

FIGURE 8.49
Convergence and rates in displacement norm for the cantilever beam solved using NS-RPIM with regular and irregular ($\alpha_{ir} = 0.2$) nodes ($\alpha_s = 2.5$).

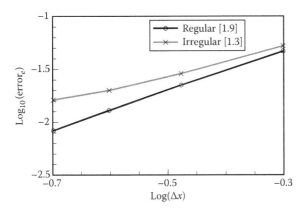

FIGURE 8.50
Convergence and rates in energy norm for the cantilever beam solved using NS-RPIM with regular and irregular ($\alpha_{ir} = 0.2$) nodes ($\alpha_s = 2.5$).

Figure 8.49 plots the solution error in displacement norm against the nodal spacing. Both regular and irregular ($\alpha_{ir} = 0.2$) nodes are used with $\alpha_s = 2.5$. It is found that the convergence and rates in the displacement norm are not affected by the nodal irregularity, and a numerical rate of 1.9 is achieved for both cases, which is very close to the theoretical value of 2.0 (see Section 3.5). Figure 8.50 plots the solution error in energy norm against the nodal spacing. A very high convergence rate of 1.9 is obtained for the case of regular node distribution, which is even more higher than the ideal theoretical rate of 1.5. For the case of irregular node distribution, the rate is 1.3 and still very close to the ideal theoretical rate. We observe clearly the superconvergence of NS-RPIM. It is observed that the convergence and rates in energy norm are affected by the nodal irregularity, which confirms the theoretical prediction given in Section 3.5.

Example 8.14: Hole in an Infinite Plate

We now revisit Example 8.3, with NS-RPIM. The parameters used in this example are listed as follows:

Loading: $p = 1$ MPa
Young's modulus: $E = 3.0 \times 10^4$ MPa
Poisson's ratio: $\nu = 0.25$
The radius of the hole: $a = 0.2$ m
Width of the plate: $b = 1$ m

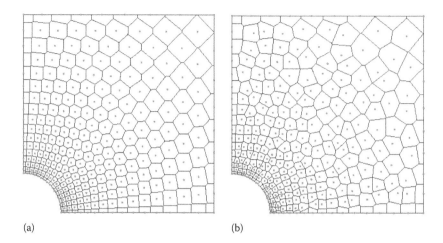

(a)   (b)

FIGURE 8.51
Node distribution and Voronoi smoothing domains for the quarter model of an infinite plate with a circular hole: (a) regular ($\alpha_{ir} = 0$); and (b) irregular ($\alpha_{ir} = 0.4$).

The background cells, nodal locations, and their Voronoi smoothing domains are plotted in Figure 8.51. To study the effects of the nodal irregularity, interior nodal coordinates for more irregular distribution (Figure 8.51a) are computed using the following equation:

$$\begin{cases} x' = x + L_{\min} r_R \cos(\pi r_\theta) \cdot \alpha_{ir} \\ y' = y + L_{\min} r_R \cos(\pi r_\theta) \cdot \alpha_{ir} \end{cases} \tag{8.26}$$

where
  $L_{\min}$ is the shortest distance of a node to its neighbor nodes
  r_R and r_θ are randomly generated numbers between -1.0 and 1.0 produced by computer
  α_{ir} is the irregularity factor that controls the irregularity

We consider for this problem, the node distributions given in Figure 8.51a (with $\alpha_{ir} = 0$) as "regular."

Figure 8.52 plots the solution error in displacement norm against the nodal spacing. Both regular and irregular ($\alpha_{ir} = 0.4$) nodes are used with $\alpha_s = 2.5$. It is found that the convergence and rates in the displacement norm are affected a little by the nodal irregularity, and a numerical rate of 1.9 is achieved for the regular node case, which is very close to the theoretical value of 2.0. For the irregular node case a rate of 1.7 is achieved and is still quite close to the theoretical value. Figure 8.53 plots the solution error in energy norm against the nodal spacing. A convergence rate of 1.4 is obtained for the case of regular node distribution, which is very close to the ideal theoretical rate of 1.5. For the case of irregular node distribution, the rate is 1.2, which still quite close to the ideal theoretical rate. It is observed once again that (1) strong superconvergence of NS-RPIM, and (2) the convergence and rates in energy norm are affected by the nodal irregularity, which confirms the theoretical prediction given in Section 3.5.

Figure 8.54 plots the displacement u along x-axis and the displacement v along y-axis, both obtained using NS-RPIM. In this investigation, two dimensions of the support domain of $\alpha_s = 2$ and $\alpha_s = 3$ are used. It is found that the effects of the size of the support domain are very small, and the numerical solution agrees well with the analytical one. Figure 8.55 shows the results for stress

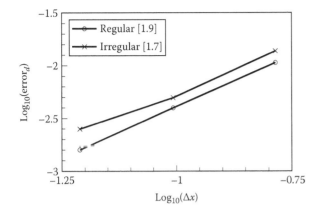

FIGURE 8.52
Convergence and rates in displacement norm for the cantilever beam solved using NS-RPIM with regular and irregular ($\alpha_{ir} = 0.2$) nodes ($\alpha_s = 2.5$).

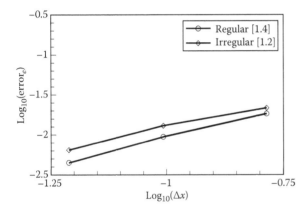

FIGURE 8.53
Convergence and rates in energy norm for the cantilever beam solved using NS-RPIM with regular and irregular ($\alpha_{ir} = 0.2$) nodes ($\alpha_s = 2.5$).

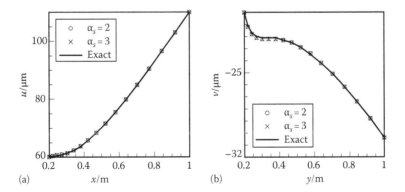

FIGURE 8.54
Displacement distribution in an infinite plate with a circular hole obtained using NS-RPIM ($\alpha_{ir} = 0$): (a) u along x-axis; and (b) v along y-axis.

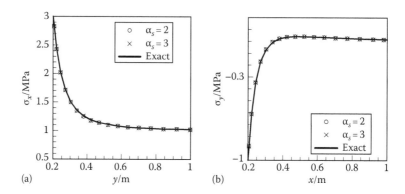

FIGURE 8.55

Stress distribution in an infinite plate with a circular hole obtained using NS-RPIM ($\alpha_{ir} = 0$): (a) σ_x along y-axis; and (b) σ_y along x-axis.

components σ_x along y-axis and σ_y along x-axis. It is found again that the effects of the size of the support domain on the stress solution are very small, and the numerical solution agrees well with the analytical one.

Example 8.15: Semi-Infinite Plane

We next use the NS-RPIM to study the problem of a semi-infinite plane subjected to a uniform pressure loading over $[-a, a]$, as shown in Figure 8.56a. Plane strain problem is considered, and the analytical solution of the stress is given by

$$\sigma_x = \frac{p}{2\pi}[2(\theta_1 - \theta_2) - \sin 2\theta_1 + \sin 2\theta_2]$$

$$\sigma_y = \frac{p}{2\pi}[2(\theta_1 - \theta_2) + \sin 2\theta_1 - \sin 2\theta_2] \tag{8.27}$$

$$\tau_{xy} = \frac{p}{2\pi}[\cos 2\theta_1 - \cos 2\theta_2]$$

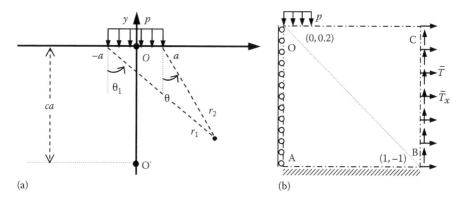

(a)   (b)

FIGURE 8.56

Semi-infinite plane subjected to a uniform pressure: (a) Coordinates and problem setting; and (b) half model ($p = 1$ MPa, $a = 0.2$ m).

The analytical solution of displacement for the plane stress is given by

$$u = \frac{p(1-\nu^2)}{\pi E}\left\{\frac{1-2\nu}{1-\nu}[(x+a)\theta_1 - (x-a)\theta_2] + 2y\ln\frac{r_1}{r_2}\right\}$$

$$v = \frac{p(1-\nu^2)}{\pi E}\left\{\frac{1-2\nu}{1-\nu}\left[y(\theta_1 - \theta_2) + 2ca\left(\arctan\frac{1}{c}\right)\right] + 2(x-a)\ln r_2\right.$$

$$\left. -2(x+a)\ln r_1 + 2a\ln[a^2(1+c^2)]\right\} \qquad (8.28)$$

where
$c = 100$
$a = 0.2$ m
ca is the distance from the origin to point O' and is fixed at $5a$

Due to the symmetry, we model only half of the domain with dimension of $5a \times 5a$. On the boundary of symmetry ($v = 0$), the displacement in the x-direction is fixed. At the bottom of the domain ($y = -1$), the displacements are fixed with prescribed values obtained using Equation 8.28, as shown in Figure 8.56b. On the right boundary ($x = 1$) the tractions are specified with values computed using Equation 8.27. Material parameters used in computation are $E = 100$ MPa and $\nu = 0.3$. The Voronoi smoothing domains in semi-infinite plane are shown in Figure 8.57. The dimensions of the support domain used are $\alpha_s = 2.0$ and $\alpha_s = 3.0$.

Figure 8.58a plots the solution error in displacement norm against the nodal spacing with $\alpha_s = 2.0$ and 3.0. In this investigation, the node distribution shown in Figure 8.57a is used. It is found that the convergence and rates in the displacement norm are affected a little by the nodal irregularity, and a numerical rates of 1.7 and 1.8 are achieved, respectively, for $\alpha_s = 2.0$ and 3.0, which are very close to the theoretical value of 2.0. Figure 8.58b plots the solution error in energy norm against the nodal spacing. Convergence rates of 0.9 is obtained for both the cases. It is observed that the convergence and rates in energy norm are affected by the size of the support domains.

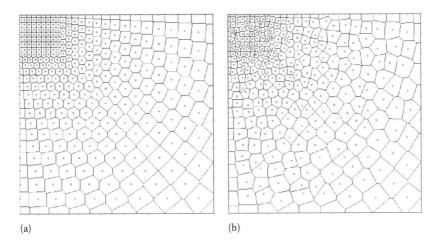

(a)   (b)

FIGURE 8.57
Voronoi smoothing domains for the semi-infinite plane: (a) $\alpha_{ir} = 0$; and (b) $\alpha_{ir} = 0.4$.

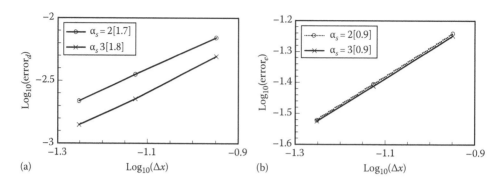

FIGURE 8.58
Convergence and rates in displacement norm for the cantilever beam solved using NS-RPIM with regular ($\alpha_{ir} = 0$) node distributions: (a) displacement norm; and (b) energy norm.

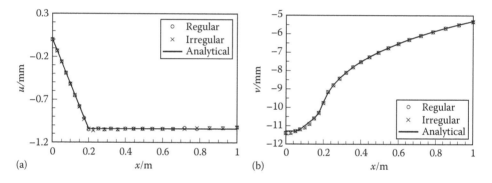

FIGURE 8.59
Displacement distribution along the surface of the semi-infinite plane obtained using NS-RPIM with $\alpha_s = 2.5$ and both regular ($\alpha_{ir} = 0$) and irregular ($\alpha_{ir} = 0.4$) node distributions: (a) displacement u; and (b) displacement v.

Figure 8.59 plots the displacement distribution along the surface of the semi-infinite plane obtained using NS-RPIM with node distribution shown in Figure 8.57a ($\alpha_{ir} = 0$) and Figure 8.57b ($\alpha_{ir} = 0.4$). The dimension of the support domain is $\alpha_s = 2.5$. It is found that the effects of the nodal regularity are very small and almost indistinguishable, and the numerical solution agrees well with the analytical one.

Figure 8.60 plots the results for stress components, and σ_{xx} and σ_{xy} are the diagonal line (O–B) of the semi-infinite plane. It is found that the effects of the nodal regularity are very small and almost negligible for this case, and the numerical solution agrees well with the analytical solutions.

Example 8.16: Triangular 2D Solid with a Heart-Shaped Hole

Finally, to examine the capability of NS-RPIM for problems with very complicated geometry, we artificially created a problem of a triangular 2D solid with a heart-shaped hole subjected to a uniform pressure of $p = 1.0$ MPa on the inclined right edge, as shown in Figure 8.61. Plane stress problem is considered, and the material constants of $E = 10^4$ MPa and $\nu = 0.25$ are used in

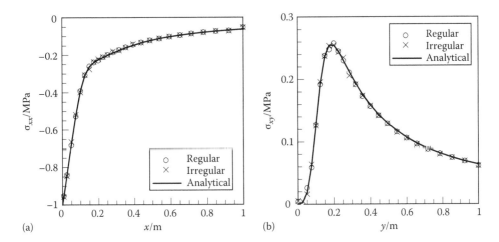

FIGURE 8.60
Stress distribution along the diagonal line (O–B) of the semi-infinite plane obtained using NS-RPIM with $\alpha_s = 2.5$ and both regular ($\alpha_{ir} = 0$)and irregular ($\alpha_{ir} = 0.4$) node distributions: (a) σ_{xx}; (b) σ_{xy}.

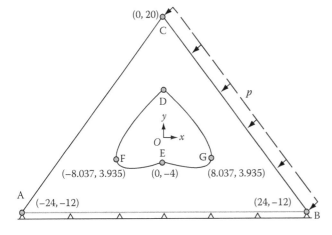

FIGURE 8.61
A triangular 2D solid with a heart-shaped hole formed using two splines through D–F–E and D–G–E. The solid is subjected to unilateral uniform pressure on the inclined right edge (all dimensions in meters).

computation. The nodal distribution (962 nodes) and Voronoi smoothing domains are shown in Figure 8.62. Since there is no analytical solution for this problem, a reference solution is obtained using FEM software ABAQUS® with a large number of (8140) nodes for comparison purpose.

In this study, three dimensions of the support domain are used: $\alpha_s = 1.5$, 2.5, and 3.5. Figure 8.63 plots the effect of the dimension of the support domains on the distribution of displacement u along free boundary (A–C) of the solid obtained using NS-RPIM together with the FEM of very fine mesh. It is found that the results agree very well and are almost indistinguishable from the figure. Figure 8.64 plots the same but for the displacement component v, which confirms the same finding.

Figure 8.65 plots the distribution of stress σ_{xx} along the vertical line (D–C) of the solid obtained using NS-RPIM together with the FEM of very fine mesh. The dimensions of the support domain are $\alpha_s = 1.5$, 2.5, and 3.5. Figure 8.66 plots the same but for stress σ_{yy}. It is

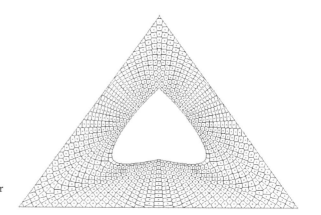

FIGURE 8.62
Voronoi smoothing domains for the triangular 2D solid with a heart-shaped hole.

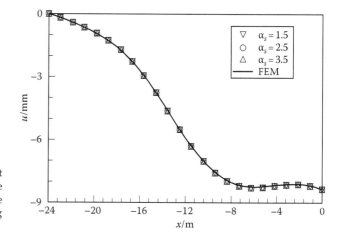

FIGURE 8.63
Effect of the dimension of the support domains on displacement u along free boundary (A–C) of the triangular plate with a heart-shaped hole obtained using NS-RPIM and FEM with very fine mesh.

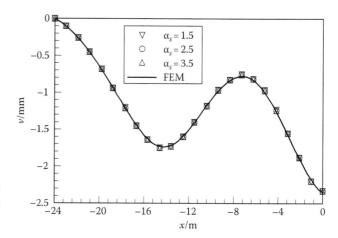

FIGURE 8.64
Effect of the nodal spacing on displacement v along free boundary (A–C) of the triangular plate with a heart-shaped hole obtained using NS-RPIM and FEM with very fine mesh.

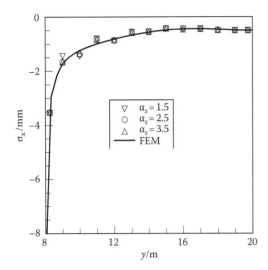

FIGURE 8.65
Distribution of stress component σ_{xx} along the vertical line (D–C) of the triangular plate with a heart-shaped hole obtained using NS-RPIM and FEM with very fine mesh.

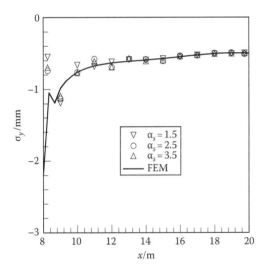

FIGURE 8.66
Distribution of stress component σ_{yy} along the vertical line (D–C) of the triangular plate with a heart-shaped hole obtained using NS-RPIM and FEM with very fine mesh.

found that the NS-RPIM results agree very well with that of FEM in areas away from the singularity point D. Near the singularity point, significant differences are observed. For NS-RPIM to obtain better results, more nodes are required. Note that the FEM model uses about eight times more nodes than the NS-RPIM. For problems with singularity a better approach is to use adaptive analysis, and even better together with shape functions with singularity terms.

8.2.4 Concluding Remarks

In this section, we studied NS-RPIM and found it very stable with lot of freedom in working with irregularly distributed nodes. The results are very accurate and superconvergence have also been observed. Owing to this excellent stability and accuracy, the NS-RPIM has also been applied to nonlinear problems such as contact problems. Interested readers may refer to [11].

8.3 Upper-Bound Properties of NS-PIM and NS-RPIM

8.3.1 Background

It is well known that the FEM [18,25] provides a lower bound in energy norm for the exact solution to elasticity (force driving) problems. It is, however, much more difficult to bound the solution from above for complicated problems, and it has been a dream of many decades to find a general systematical way to obtain an upper bound of the exact solution for complicated practical problems. It has been discovered recently that NS-PIM and NS-RPIM can provide an upper-bound solution in energy norm for elasticity problems, except a few trivial cases. The theorem on this issue has been presented in Chapters 4 and 5. This section discuss the upper-bound property of both NS-PIM and NS-RPIM, and demonstrates the upper-bound property of NS-PIM and NS-RPIM through a number of numerical examples. Using the NS-PIM or NS-RPIM together with the FEM, we now have a systematic way to numerically obtain both upper and lower bounds of the exact solution to elasticity problems, using the same mesh. The problem can be very complicated, as long as a FEM triangular mesh can be built. As the theoretical fundamentals for both NS-PIM and NS-RPIM are largely the same, our discussion will mainly focus on the NS-PIM.

8.3.2 Some Properties of FEM Model

Our discussion in this section is often conducted in comparison with the FEM model. For the convenience of our discussions, we first list some of the related properties of the fully compatible FEM models (see Remark 5.1).

Remark 8.3: Lower-Bound Property of FEM Model

The strain energy obtained from the FEM solution to a force driving problem based on assumed displacements that are fully compatible is a lower bound of the exact strain energy. This well-known property can be easily shown in the following.

The strain energy obtained from the FEM solution can be written as

$$\tilde{U}(\tilde{\mathbf{d}}) = \int_\Omega \frac{1}{2}\tilde{\boldsymbol{\varepsilon}}^{\mathrm{T}}\mathbf{D}\tilde{\boldsymbol{\varepsilon}}\mathrm{d}\Omega = \frac{1}{2}\tilde{\mathbf{d}}^{\mathrm{T}}\tilde{\mathbf{K}}\tilde{\mathbf{d}} \qquad (8.29)$$

Let the exact strain energy be defined as

$$U = \int_\Omega \frac{1}{2}\boldsymbol{\varepsilon}^{\mathrm{T}}\mathbf{D}\boldsymbol{\varepsilon}\mathrm{d}\Omega \qquad (8.30)$$

where $\boldsymbol{\varepsilon}$ is the exact solution of strains of the problem, which relates to the exact solution of displacement $\mathbf{u}$ in the form of $\boldsymbol{\varepsilon} = \mathbf{L}_d\mathbf{u}$. The exact solution satisfies the strong form equations given in Section 1.2 and traction and homogenous displacement boundary conditions.

For an FEM solution based on assumed displacement that is fully compatible, the total potential energy at the stationary point can be written as

$$\tilde{J}(\tilde{\mathbf{d}}) = \frac{1}{2}\tilde{\mathbf{d}}^{\mathrm{T}}\tilde{\mathbf{K}}\tilde{\mathbf{d}} - \tilde{\mathbf{d}}^{\mathrm{T}}\underbrace{\tilde{\mathbf{f}}}_{\tilde{\mathbf{K}}\tilde{\mathbf{d}}} = -\frac{1}{2}\tilde{\mathbf{d}}^{\mathrm{T}}\tilde{\mathbf{K}}\tilde{\mathbf{d}} = -\tilde{U}(\tilde{\mathbf{d}}) \qquad (8.31)$$

As the FEM solution of a compatible model is based on the minimum total potential energy principle, we have

$$\tilde{J} = -\tilde{U} \geq J_0 = -U_0 \tag{8.32}$$

or

$$\tilde{U}(\tilde{\boldsymbol{\varepsilon}}) \leq U(\boldsymbol{\varepsilon}) \tag{8.33}$$

which means that the strain energy obtained from an FEM solution is a lower bound of the exact solution of strain energy. ∎

Remark 8.3 implies that the strain energy obtained from the displacement-based fully compatible FEM solution is always an underestimate of the exact strain energy, and the displacement is always a lower bound of the exact solution in the "K norm" (or strain energy). The lower-bound property of FEM is valid for all types of elements as long as the model is fully compatible. In this section, however, we refer only linear triangular elements (that is fully compatible) when FEM is used.

8.3.3 Properties of NS-PIM Model

Remark 8.4: Upper Bound to FEM Model
When the same mesh and shape functions are used, the strain energy obtained from the NS-PIM solution is no less than that from the fully compatible FEM solution.

$$\underbrace{\frac{1}{2}\bar{\mathbf{U}}^{\mathrm{T}}\bar{\mathbf{K}}\bar{\mathbf{U}}}_{\bar{U}(\bar{\mathbf{U}})} \geq \underbrace{\frac{1}{2}\tilde{\mathbf{U}}^{\mathrm{T}}\tilde{\mathbf{K}}\tilde{\mathbf{U}}}_{\tilde{U}(\tilde{\mathbf{U}})} \tag{8.34}$$

This inequality is essentially the same as that given in Theorem 5.6, except that here it is expressed in terms of a solution of nodal displacements **U**. It was first presented and proven in [13]. An alternative proof based on variational formulation can be found in [27].

The equality is true when NS-PIM and FEM produce the exact solutions or the smoothing operation is performed independently for each individual part of the elements connecting to the node and hence the node-based smoothing has no effect.

Remark 8.5: Upper Bound to Exact Solution
Liu and coworkers have found that for force driving problems, not only $\bar{U}(\bar{\mathbf{U}}) \geq \tilde{U}(\tilde{\mathbf{U}})$ but also $\bar{U}(\bar{\mathbf{U}}) \geq U(\boldsymbol{\varepsilon}) \geq \tilde{U}(\tilde{\mathbf{U}})$ is true except for a few trivial cases. This means that the solution of an NS-PIM gives the upper bound of the exact solution in energy norm. This is discussed in Remark 5.13. Here we further discuss these exceptional cases for NS-PIM (or NS-RPIM) models, based on the argument of "the battle of softening and stiffening effects" [13,27].

Remark 8.6: The Battle of Softening and Stiffening Effects

The NS-PIM is a typical GS-Galerkin model based on $\mathbb{G}$ space theory. Remark 5.12 states that a GS-Galerkin model *can* produce an upper-bound solution by performing a proper smoothing operation. A NS-PIM model performs smoothing operation based on nodes and is often found capable of producing sufficient effects.

Remark 5.11 shows that an NS-PIM model can always provide an upper bound for the exact solution in energy norm, under the condition that the shape functions corresponding to the exact solution are used. For a general problem, however, finding the exact shape functions is not possible. Therefore, the NS-PIM can only use the usual PIM shape functions (or FEM shape functions). The use of any (compatible) shape functions in the place of the exact shape functions will cause, on the other hand, stiffening effects to the model. The battle between the softening and stiffening effects will determine whether a NS-PIM model can in fact provide an upper-bound solution to the problem.

Remark 8.7: Factors Affecting the Softening Effect

a. The number of elements that are connected to a node of a smoothing domain: the more the elements, the more the smoothing effects. In an extreme case, if the smoothing domain is defined for each (linear) element to perform the smoothing operation, there will be no softening effect at all. In this case the NS-PIM and FEM gives the same results, and the NS-PIM will not provide an upper-bound, but a lower-bound solution. Note that in an NS-PIM setting, the smoothing domains are tied together with the nodes.

b. The number of nodes being smoothed in the problem domain. In theory, one does not have to perform the smoothing operation for all the nodes. The softening effect will proportionally depend on the number of nodes that participated in the smoothing operation.

c. The dimension of the smoothing domain. In NS-PIM, the smoothing domains are usually "seamless," meaning that it is constructed in such a way that there is no gap and overlap in between the neighboring smoothing domains. If, however, one chooses to use a smaller or larger smoothing domain, the method will still work (as a mixed model) but the bound properties will change. In general, the softening effect will reduce, when the dimension of the smoothing domain is reduced. Example 8.17 will demonstrate such changes in the smoothing effect when the smoothing domain changes.

d. The number of nodes used in the problem domain. When a small number of nodes are used, the displacements approximated using PIM shape functions in a smoothing domain is far from the exact solution, resulting in a heavy smoothing to the strain field, and hence a strong softening effect. On the other hand, when a large number of nodes are used, the displacements approximated using the PIM shape functions in a smoothing domain is closer to the exact solution, resulting in less smoothing effects, and hence less softening effect. At the extreme, if infinitely small elements are used, the smoothing effects will diminish and the NS-PIM solution (also the FEM solution) will approach the exact solution.

Remark 8.8: Factors Affecting the Stiffening Effect

a. The stiffening effect depends on the order of the PIM shape functions used in the displacement approximation. When high-order PIM shape functions are used, the displacements approximated using the PIM shape functions in a smoothing domain is usually closer to the exact solution, which reduces the stiffening effect, and vice versa.

b. The stiffening effect depends on the number of nodes used in the problem domain. When a small number of nodes are used, the displacements approximated using the PIM shape functions in a smoothing domain is far from the exact solution, the stiffening effect is therefore small, and vice versa. At the extreme of infinitely small elements are used, the stiffening effects will diminish and the NS-PIM solution (also the FEM solution) will approach the exact solution.

Generally, the softening effect provided by the smoothing operation is found more significant than the stiffening effects in the NS-PIM setting. Therefore, the NS-PIM always produces an upper-bound solution for 1D, 2D, and 3D solids, except the following few special cases.

Remark 8.9: Upper Bound: A Few Exceptions

a. Too few cells are used. In an extreme case, when only one cell with linear interpolation is used, only one cell with constant strain participates in smoothing. In this case there should be no smoothing effects at all, and hence the solutions of NS-PIM and FEM are the same, and NS-PIM gives a lower-bound solution. In order to obtain sufficient smoothing effects to produce upper-bound solutions, the number of cells should not be too small.

b. Hanging elements are used in a model with small number of elements. As shown in Figure 8.67, there are three hanging triangular elements attached to the domain, and hence at the corner nodes of these three elements only one element for each node can participate in the smoothing. In such a case, there is no smoothing effect at all for these three smoothing domains. Note that such hanging elements are not

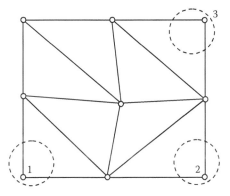

FIGURE 8.67
Hanging elements in 2D domains: smoothing operation on nodes 1, 2, and 3 has no effects.

supposed to be used even in the FEM model, as the stress there is zero when the nodes are free, which is equivalent to removing the entire corner elements! A similar situation can occur for elements on the two ends of a 1D domain.

In our numerical study, we found that NS-PIM can produce upper-bound solutions for all the problems we have studied, except the very special cases mentioned above.

Note that the discussions on NS-PIM are largely applicable to NS-RPIM, as they all share the same theoretical background, and the difference is only in the shape function used. Therefore, we will omit the detailed discussions on upper- bound properties of NS-RPIM, and refer the reader to paper [12] for more discussions.

In the following pages we will present a number of examples that confirm the properties of both the NS-PIM and NS-RPIM with a focus on the important upper- bound property.

8.3.4 Examples: Upper Bound and Convergence

Example 8.17: A 1D Bar Problem

Consider first a very simple problem of a bar with length l and of uniform cross-sectional area A. As shown in Figure 8.68, the bar is fixed at the left end and subjected to a uniform body force b. The parameters are taken as $l = 1$ m, $A = 1$ m^2, $B = 1$ N/m, and $E = 1$ Pa. Governing equation and boundary conditions are as follows:

$$E\frac{d^2 u}{dx^2} + 1 = 0 \tag{8.35}$$

$$u(x = 0) = 0$$
$$\sigma(x = 1) = 0 \tag{8.36}$$

The analytical solution that satisfies the above equations is obtained as

$$u(x) = -\frac{1}{2E}x^2 + \frac{1}{E}x \tag{8.37}$$

The exact strain energy of the problem can be calculated as follows:

$$U(\mathbf{u}) = \frac{1}{2}\int_l \boldsymbol{\varepsilon}^{\mathsf{T}} E \boldsymbol{\varepsilon}\, dl = \frac{1}{6E} \tag{8.38}$$

Although the problem is very simple, probably the simplest, it is very useful to show some of the important properties of NS-PIMs.

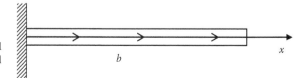

FIGURE 8.68
One-dimensional bar of uniform cross-sectional area A subjected to a uniformly distributed body force b along the x-axis.

To start, the effect of the dimension of smoothing domain is studied using this simple 1D problem. As shown in Figure 8.69, the problem domain of the bar is presented using three nodes: node 1 located at the left end, node 2 located at the midpoint of the problem domain, and node 3 located at the right end. Cells (1) and (2) are two background cells. In usual NS-PIM settings, the smoothing (domain) length for node 2 is obtained by connecting the midpoints of cell (1) and cell (2), which is $l/2$, and the lengths of the two smoothing domains for two end-nodes is $l/4$. Since we use linear interpolation based on each of these two cells, the strain is constant in each of the cells. Therefore, smoothing operations of nodes 1 and 3 have no effect and hence no need to perform. We now intentionally change the dimension of the smoothing length for node 2 by allowing the smoothing domain to shrink or stretch beyond the midpoints of the two cells, so that we can study the effect of the smoothing domain length on the strain energy of the NS-PIM solution.

Figure 8.69 plots the strain energy of the solution of the NS-PIM model for different smoothing domain length. It is found that the strain energy increases with the increase of the smoothing length that is measured as the ratio between the smoothing domain length of node 2 and the entire problem domain length l. When the smoothing length reduces to zero, the NS-PIM model becomes the FEM model of two elements with three nodes, which gives the lower-bound solution. When the ratio of the smoothing length increases to about 0.43, the NS-PIM solution of strain energy is larger than that to the exact solution: an upper-bound solution. Increase the smoothing domain further to 0.9, the NS-PIM model becomes very soft and the strain energy is very much higher than the exact one. This shows that one can in fact make the NS-PIM model as softer as we want by reducing the number but increasing the dimension of smoothing domain, which confirms Remark 5.12. The finding of this example also supports the discussions given in Remark 8.7. If we stretch the smoothing domain even further to the entire problem (equivalent to using

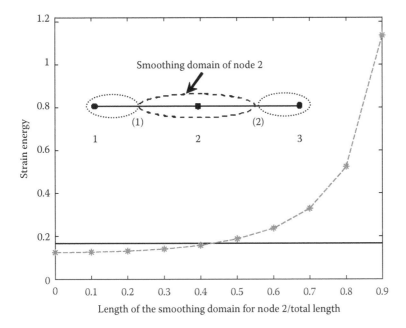

FIGURE 8.69
The effect of the length of the smoothing domain for 1D problem.

only one smoothing cell), the system matrix will become singular and the model becomes unstable, which confirms our theory on the minimum number of smoothing domains. From Table 3.1, we know that we need at least two smoothing domains for this 1D problem with two nodal degrees of freedoms (DOFs) for a stable GS-Galerkin model. In this book we will not discuss further about the smoothing domain size effects, and focus our discussion only on NS-PIM and NS-RPIM using seamless smoothing domains.

Next, for the same 1D problem we study the convergence issues, by increasing the number of nodes and examine the properties of the solutions of NS-PIM, NS-RPIM together with the FEM using exactly the same meshes. Six models of different numbers of uniformly distributed nodes are used with usual smoothing domains. The computed values of strain energy of the solution are plotted in Figure 8.70 against the number of nodes used together with the exact solution. It can be found that, NS-PIM and NS-RPIM produce the same results as FEM when only two nodes are used. In this case, two smoothing domains are used for, respectively, the two field nodes at the two ends of the bar, hence the smoothing has no effect at all to the problem, and the solution is the same as the FEM giving a lower bound (see, Remark 8.9a). When the number of nodes is bigger than 2, the smoothing takes effect, and the NS-PIM provides an upper-bound solution. With the increase of the number of nodes, the FEM solution approaches from below to the exact solution monotonically. NS-PIM and NS-RPIM solutions, however, approaches from above to the exact solution monotonically, due to the fact that the smoothing effects in the NS-PIM and NS-RPIM reduce as the displacement field approaches to the exact solution. These findings confirm Remark 8.5 on NS-PIMs models. This simple 1D example showed for the first time, clearly, the very important fact that we now can bound the exact solution from both sides. It is also noted that the upper bound provided by the NS-RPIM is much tighter than that provided by the NS-PIM. This is due to the higher order RPIM shape functions used in the NS-RPIM that reduces both the softening and stiffening effects pushing the solution closer to the exact one.

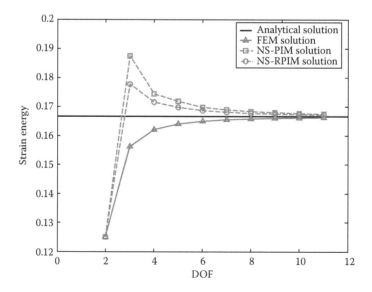

FIGURE 8.70
Upper-bound solution obtained using the linear NS-PIM and linear NS-RPIM for the 1D bar problem. The lower-bound solution is obtained using the FEM using linear elements of the same meshes.

Example 8.18: A 2D Problem: Rectangular Cantilever

In this example, we revisit the rectangular cantilever problem but with the focus on the examination of the upper-bound and convergence properties of the NS-PIM and NS-RPIM, in comparison with the linear FEM model using the same meshes. The settings, parameters and error norm definitions for this example problem are exactly the same as those given in Example 8.2. In this study, we use equally shared smoothing domains for both NS-PIMs. For NS-PIM we use T3-Scheme (linear interpolation), and for NS-RPIM we use support domain of $a_s = 3.0$ and MQ-RBF with shape parameters of $q = 1.03$ and $a_c = 4.0$.

Figure 8.71 shows the convergence status of displacement and energy norm errors against the average nodal spacing (h) for all the three methods used in this problem, i.e., NS-PIM, NS-RPIM

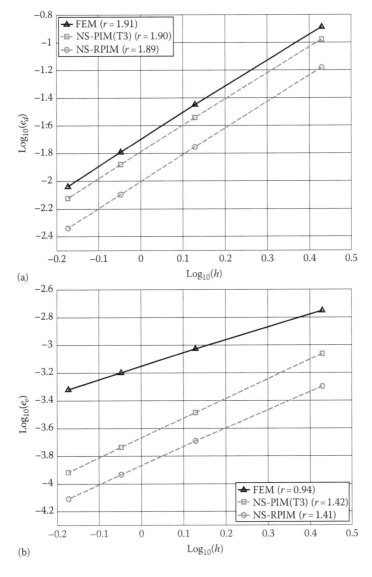

(a)

(b)

FIGURE 8.71
Upper-bound solution obtained using the linear NS-RPIM and linear NS-PIM for the rectangular cantilever. The lower-bound solution is obtained using the FEM with linear elements of the same meshes.

and FEM, using uniformly distributed nodes. The exact value of strain energy is calculated using the analytical solutions of stress components. We can observe the following: (1) both NS-PIM and NS-RPIM are more accurate than the FEM in both displacement and energy norms; (2) the rate of convergence in displacement norm is about 1.9 for all three models that is close to the theoretical value of 2.0; (3) the accuracy of the NS-RPIM is about two times in displacement norm and six times in energy norm of the FEM; (4) the accuracy of the NS-PIM is about the same in displacement norm and three times in energy norm of the FEM; (5) convergence rates of about 1.4 are achieved by both PIM models that is much higher than that of the FEM (0.94), even much higher than the theoretical value of FEM model (1.0): superconvergence, and very close to the ideal theoretical value (1.5) for $\mathbb{G}$ space theory (see, Section 3.5).

Comparing Figure 8.71 with Figures 8.49 and 8.50 where the Voronoi smoothing domains are used, we observed for the NS-RPIM that the effects of types of smoothing domains are not very significant, except very high (1.9) convergence rate for the NS-RPIM using regular Voronoi smoothing domains. Note that the difference in the shape parameters for the MQ-RBFs may also play a role.

Figure 8.72 plots the solution convergence process to the exact solution and the bound properties with the increase of the number of nodes used (DOFs) in all these three models using exactly the same mesh. It can be observed again that both the NS-RPIM and NS-PIM provide upper-bound solutions in energy norm, while the FEM gives a lower-bound solution for this 2D problem. Compared with the NS-PIM, the NS-RPIM provides again a tighter upper bound to the exact solution.

This benchmark 2D problem has again confirmed the properties of the NS-PIMs based on $\mathbb{G}$ space theory, which have been discussed in the previous sections.

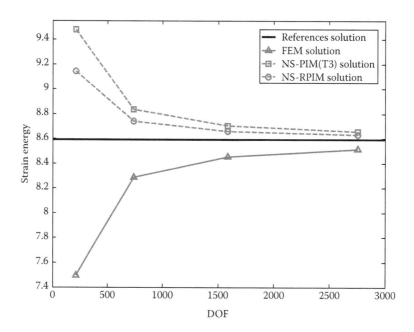

FIGURE 8.72
Upper-bound solution obtained using the linear NS-RPIM and linear NS-PIM for the rectangular cantilever. The lower-bound solution is obtained using the FEM with linear elements with the same set of nodes.

Example 8.19: Hole in an Infinite Plate

We now revisit the problem of a circular hole in an infinite plate but with the focus on the examination of the upper bound and convergence properties of the NS-PIM and NS-RPIM, in comparison with the linear FEM. The settings and the parameters for this example problem are the same as those given in Example 8.3, except the radius of the hole: $a = 1$ and the width of the plate: $b = 5$.

Four models of "regular" nodes distributions with 577, 1330, 2850, and 3578 nodes are used in this investigation. The convergence rates in both displacement norm and energy norm are showed in Figure 8.73. Similar conclusion can be drawn as the rectangular cantilever example: all the three models achieve almost-equal numerical convergence rate of about 1.9 in displacement norm, superconvergence is observed for both NS-PIMs, but this time NS-PIM outperforms in terms of energy norm.

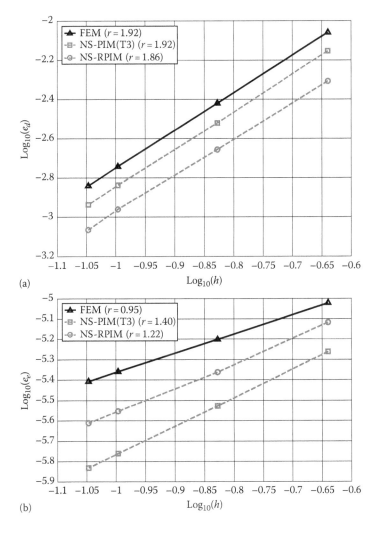

FIGURE 8.73
Comparison of convergence rates between the linear FEM, NS-PIM, and NS-RPIM for the problem of the 2D infinite plate with hole: (a) displacement error norm; and (b) energy error norm.

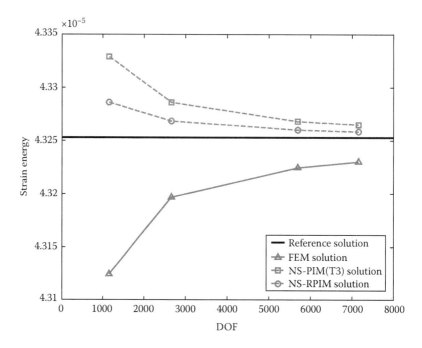

FIGURE 8.74
Upper-bound solutions obtained using NS-PIM and NS-RPIM for the 2D infinite plate with hole. The lower-bound solution is obtained using the FEM with linear elements with the same set of nodes.

Figure 8.74 plots the convergence process of strain energy solution obtained using these three models against the DOFs. It can be observed again that both NS-PIMs provide upper-bound solutions in energy norm. The NS-RPIM bound is tighter than that of the NS-PIM to the exact value of strain energy, while FEM gives a lower (and looser) solution bound.

Example 8.20: A 2D Square Solid Subjected to a Uniform Pressure and Body Force

A 2D square plate shown in Figure 8.75 subjected to uniform pressure and body force is now studied, with the focus on the examination of the upper-bound properties of the NS-PIM and NS-RPIM. This example is designed to show that the upper-bound properties of both NS-PIM and NS-RPIM hold even for the presence of the body force in the solid (still force driving). The plate is constrained on the left, the right, and the bottom edges, and subjected to uniform unit pressure and a uniformly distributed unit body force of $\mathbf{b}^T = \{0 \ -1\}$. We consider this problem as a plane stress problem with the following material constants: $E = 3.0 \times 10^7$ and $\nu = 0.3$.

The problem domain is discretized with four models of regular distributions of 41, 145, 545, and 2113 nodes. As the analytical solution for this problem is not available, a reference solution is obtained by using the FEM with a very fine mesh (8238 nodes). Figure 8.76 shows the computed strain energies using three models of NS-PIM, NS-RPIM, and FEM against the DOF, and it can be observed again that both the NS-PIM and the NS-RPIM give upper-bound solutions. The upper bound of the NS-RPIM is tighter than that of the NS-PIM. This example shows that the upper-bound properties hold for problems even when the body force is present.

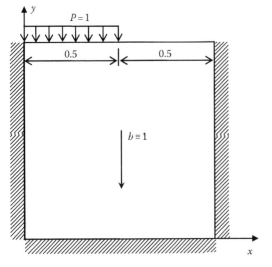

FIGURE 8.75
A 2D square solid subjected to uniform pressure on part of the top surface and a uniformly distributed body force over the entire solid. The solid is constrained on left, right, and bottom edges.

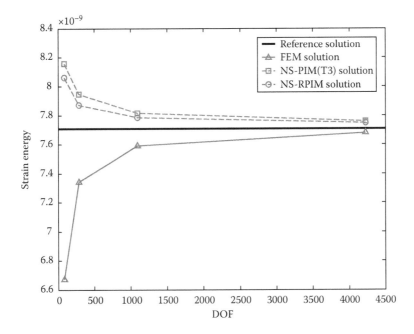

FIGURE 8.76
Upper-bound solutions obtained using NS-PIM and NS-RPIM for the 2D square solid problem; The lower-bound solution is obtained using the FEM with linear elements of the same mesh.

Example 8.21: A 3D Axletree Base

Finally, linear NS-PIM and FEM are used to solve a 3D practical problem of axletree base to examine the solution bounds. As shown in Figure 8.77, the axletree base is symmetric about the $y − z$ plane, fixed in the locations of four lower cylindrical holes and subjected to a uniform

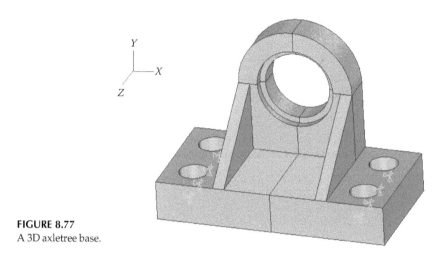

FIGURE 8.77
A 3D axletree base.

pressure of $P = 100$ N/m^2 applied on the concave annulus. The material constants used are $E = 3.0 \times 10^7$ Pa and $\nu = 0.3$.

Four models of 781, 1828, 2566, and 3675 nodes are used in the computation. Values of strain energy for both FEM and NS-PIM are plotted against the increase of DOFs. As no analytical solution is available for this problem, a reference solution is obtained using the FEM with a very fine mesh of 9963 nodes. Figure 8.78 shows the strain energy solution convergence process for this 3D practical problem with complicated shape. It is found that the NS-PIM has again produced an upper-bound solution that convergences to the exact solution from above with the increase of DOFs, while FEM solution converges from below.

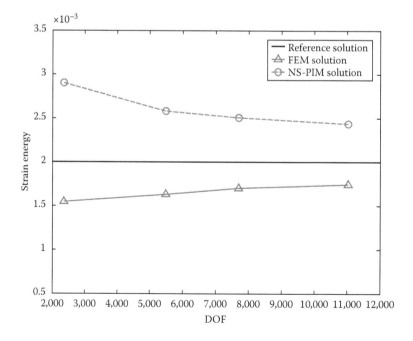

FIGURE 8.78
Upper-bound solution in strain energy obtained using the NS-PIM for the 3D axletree base problem; The lower-bound solution is obtained using the FEM using linear element.

Note in this example, the reference solution is obtained using FEM, and hence is itself in fact a lower-bound solution to the exact solution that we do not know. Therefore, the exact solution in Figure 8.78 should be higher than the reference solution, meaning that it should be closer to the NS-PIM solution. Similar issues exist for all the problems using FEM with very fine mesh as a reference solution.

8.3.5 Concluding Remarks

In this section, we studied the upper-bound property of the NS-PIMs for 1D, 2D, and 3D problems. In summary, some concluding remarks may be made as follows:

- The smoothing operation provides softening effects, and the assumption of displacement introduces stiffening effects to the numerical models.
- When node-based smoothing operations are performed on a numerical model of assumed displacement, the softening effects are found sufficiently strong leading to an upper-bound solution to the exact one.
- Both NS-PIM and NS-RPIM can provide upper-bound solutions in energy norm to force driving elasticity problems with homogenous displacement boundary conditions, except a few trivial cases.
- The upper bound of NS-RPIM solution is found tighter than that of NS-PIM.

Remark 8.10: NS-PIM: A Quasi-Equilibrium Model
The NS-PIM is not a fully compatible model, it is found free of volumetric locking, produces upper-bound solutions, and there exist spurious modes (see Section 8.4.8) at higher energy level. This behavior is a typical behavior of an equilibrium model. In fact, at any point in all these smoothing (open) domains, the equilibrium equations are satisfied in an NS-PIM model. It is however not an equilibrium model because the stresses right on these interfaces of the smoothing domains are not in equilibrium. Therefore, it is called a quasi-equilibrium model.

Remark 8.11: Solution Bounds for General Problems
In this section, we focus our discussion on force-driving problems. For *displacement driving* problems (zero external forces but nonzero prescribed displacement on the essential boundary), we expect the FEM and NS-PIM to swap their roles: NS-PIM gives the lower bound and FEM gives the upper bound. For general problems with mixed force and boundary conditions, we can still expect these two models to bound the exact solution from both sides, although which model is on which side will be problem dependent.

Remark 8.12: On Solution Bounds
Because we now have a practical means to obtain both upper and lower bounds, we do not really worry too much about where the exact solution is and use the approximated solution with confidence. In addition, how fine a mesh we should use can also be determined based

on the gap (error) of these two bounds. As soon as the error is acceptable for our design purpose, we stop further refining the model. This know-where-to-stop requires "certifying" the solution. It is very important because it give us confidence for the solution as well as prevents the use of unnecessarily large models in design and analysis, resulting in wastes of computational and manpower resources. The development of practical numerical methods for producing certified solutions will become more and more important to engineering design and analysis, and hence techniques that can provide upper-bound solutions like NS-PIMs are very much in demand. This is because of the simple argument that we can even use hundreds of millions of DOFs to solve a problem with extremely high accuracy, but if we cannot quantify the error, it is practically useless! On the other hand, when one uses a model of only 1000 DOFs to solve the problem with a *certified solution* of 10% error, it is in fact much more useful.

Remark 8.13: Extension to General *n*-sided Polygonal Cells/Elements
In the above discussions on NS-PIM, we focused on the use of triangular background cells/elements. However, the same idea can be applied for general *n*-sided cells/elements, as shown in Figure 8.79 for FEM settings [14].

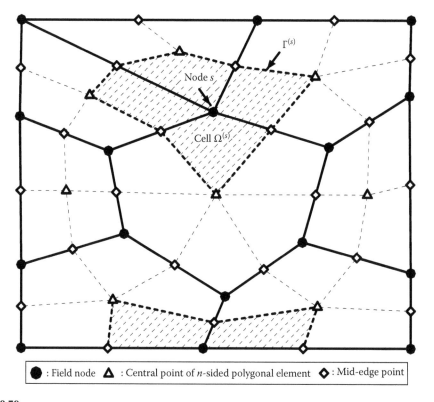

$\bullet$: Field node $\blacktriangle$: Central point of *n*-sided polygonal element $\diamondsuit$: Mid-edge point

FIGURE 8.79
An example of node-based smoothing domains for a mixed mesh of *n*-sided polygonal elements used in the NS-FEM [14].

8.4 Edge-Based Smoothed Point Interpolation Methods

8.4.1 Background

In the previous sections, we have presented node-based smoothed PIMs (NS-PIM and NS-RPIM), and examined their properties. These methods are very stable spatially (no zero-energy modes), work very well for static problems, and have very important upper-bound and superconvergence property in the strain energy. However, when these methods are applied to dynamic problems, *temporal* instability issues have been encountered. Such an instability issue is often observed at higher energy level as spurious (nonphysical) nonzero-energy modes in vibration analysis. The cause of such instability is the "overly softness" introduced by the node-based smoothing operations. The author believes that any method that has upper-bound property can have spurious modes at a higher energy level and hence instability for dynamic problems. Therefore, for dynamic analysis of solid and structures, we need alternatives or special treatments.

In this chapter, we introduce another important method called edge-based smoothed point interpolation method (ES-PIM) that does not have spurious nonzero energy modes, and is stable for even for dynamic problems. It is found also that the ES-PIM produce much more accurate results for static problems compared to NS-PIM and FEM using the same mesh.

8.4.2 The Idea of ES-PIM

In using the GS-Galerkin weak forms (see Chapter 5), we have additional instruments to develop a method of desired property: change of the smoothing domains. The node-based smoothing domain gives the NS-PIM a special upper bound property with a price of instability for dynamic problems [32]. The ES-PIM overcomes this *temporal* instability by simply changing a little in the construction of the smoothing domains. Instead of node-based smoothing domain, ES-PIM constructs the smoothing domains and performs the integration based on the edges of the elements/cells. The smoothing domain of an edge is created by connecting the nodes at the two ends of the edge to the two centroids of the two adjacent triangular cells. The edge-based smoothing domain provides some softening effect that improves the accuracy of the solution, but not so much softening effect as to avoid temporal instability. The detailed formulation is given in the following sections.

8.4.3 Approximation of Displacement Field

In an ES-PIM, the problem domain Ω is divided into a set of N_e background triangular cells with a set of N_{cg} edges and N_n nodes at the vertices of the triangular cells using triangulation described in Section 1.7.2, as in the NS-PIMs. The selection of nodes for displacement field interpolation and the construction of the shape function are performed exactly in the same way as in the NS-PIMs. We can also use either PIM or RPIM shape functions and hence ES-PIM and ES-RPIM formulations.

8.4.4 Construction for Edge-Based Smoothing Domains

Based on these triangular cells, the problem domain Ω is further divided into N_s smoothing domains associated with edges (sides) in a seamless and nonoverlapping manner, such

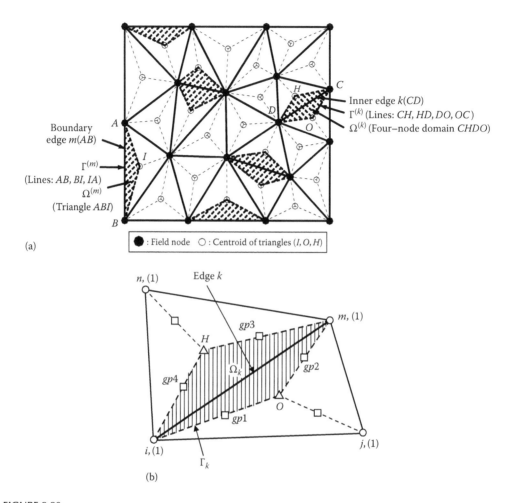

FIGURE 8.80
Triangular background triangular cells and the smoothing domains associated with the edges of the cells.

that $\overline{\Omega} = \cup_{i=1}^{N_s} \overline{\Omega_i^s}$ and $\Omega_i^s \cap \Omega_j^s = 0$, $\forall i \neq j$. The total number of smoothing domain is the total number of edges of triangular cells in the entire problem domain: $N_s = N_{cg}$. The smoothing domain Ω_i^s associated with the edge i is created by connecting the nodes at the ends of the edge to the two centroids of the two adjacent elements as shown in Figure 8.80a. Clearly such a division of smoothing domains is "legal" and satisfies the conditions given in Section 3.3.3. In an ES-PIMs, we assume also the strain is constant in each smoothing domain and hence the smoothing domain is stationary, and the $\mathbb{G}$ space theory applies.

8.4.5 Evaluation of Smoothed Strains

The strain in a smoothing domain is constant and is obtained via Gauss integration along these four segments (jO, Om, mH, and Hj) of the smoothing domain boundary Γ_k^s, as shown in Figure 8.80b. For linear interpolations, one gp sample at the middle of each segment suffices. The shape function values at all these Gauss points can be written out immediately using simple point interpolation, and are listed in Table 8.4. In this case there is no need to even create the shape functions. These shape function values listed in

TABLE 8.4

Shape Function Values at Different Sites for Two Cells Sharing the Edge

Site	Node i	Node j	Node m	Node n	Description
i	1.0	0	0	0	Field node
j	0	1.0	0	0	Field node
m	0	0	1.0	0	Field node
n	0	0	0	1.0	Field node
$\cap$	1/3	1/3	1/3	0	Centroid of cell
H	1/3	0	1/3	1/3	Centroid of cell
$gp1$	2/3	1/6	1/6	0	**Mid-segment of Γ_k^s**
$gp2$	1/6	1/6	2/3	0	**Mid-segment of Γ_k^s**
$gp3$	1/6	0	2/3	1/6	**Mid-segment of Γ_k^s**
$gp4$	2/3	0	1/6	1/6	**Mid-segment of Γ_k^s**

Table 8.4 are directly fed into Equation 8.7 to compute the entries for the smoothed strain matrix. Finally the smoothed strains are computed using Equations 8.4 and 8.8. The process is exactly the same as that in the NS-PIM. If quadratic PIM shape functions are used, the smoothed strain can be obtained in exactly the same way, except that we need to compute the shape function values numerically using Equation 8.3 with edge-based smoothing domains. When the smoothing domain is on the boundary of problem domain, we have only two segments to evaluate for the smoothed strains.

Once the smoothed strain is obtained, the GS-Galerkin can then be used to create the discretized system equations for our ES-PIM model, in exactly the same ways as in the NS-PIM. The ES-PIM model will always be spatially stable as long as the material is stable, based on the theories of space and the weakened-weak formulation.

Note that the ES-PIM works also for meshes of general polygonal elements as shown in Figure 8.81. The detailed procedure to evaluate the shape function values for such meshes can be found in Ref. [49].

Remark 8.14: Small Support Nodes for a Smoothing Domain: Efficiency for ES-PIM

Note in the ES-PIM formulation, the nodal DOFs for all the nodes "supporting" the smoothing domain of an edge will be related. For 2D cases, an edge-based smoothing domain is supported by three to four nodes. Because the bandwidth of **K** will be determined by the largest difference of node numbers of the nodes of all the triangular cells connected directly to the edge, it is clear that the bandwidth of a ES-PIM model will be smaller (about two times) than that of an NS-PIM model, but larger (about 1.5 times) than that of a linear FEM that is the smallest for the all possible numerical models (a triangular element involves only three nodes). In the EFG, an integration cell is supported by about 15–40 nodes (for ensuring the compatibility), therefore the bandwidth of an EFG model is roughly about 5–10 times that of an ES-PIM model. The efficiency of the ES-PIM is quite obvious from this rough analysis. Of course, we have not yet taken the solution accuracy into account.

We note that the idea of such an ES-PIM model can also be applied to general n-sided polygonal cells as shown in Figure 8.81. In this section however we focus on triangular background cells because (1) such cells can be generated much more easily in automatic ways for complicated geometry, and (2) ES-PIM models based on triangular cells work very well.

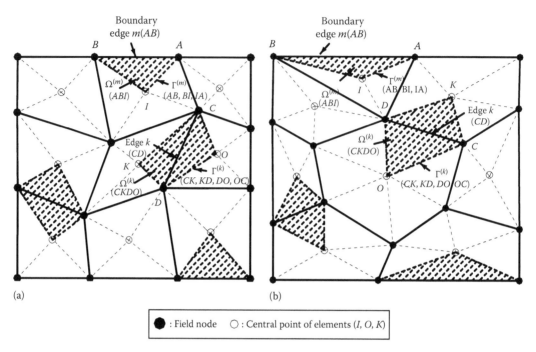

FIGURE 8.81
Possible edge-based smoothing domains for other meshes of n-sided polygonal cells [49].

8.4.6 ES-PIM Formulation for Dynamic Problems

After the smoothing domains are constructed, the rest of the formulation is exactly the same as the NS-PIM. Using the GS-Galerkin weak form Equation 8.1, and invoking the arbitrariness of virtual nodal displacements, yields the discretized algebraic system equation.

$$\bar{\mathbf{K}}\bar{\mathbf{U}} + \mathbf{C}\dot{\bar{\mathbf{U}}} + \mathbf{M}\ddot{\bar{\mathbf{U}}} = \mathbf{F} \tag{8.39}$$

where vector of nodal displacements $\bar{\mathbf{U}}$ and the stiffness matrix $\bar{\mathbf{K}}$ and force vector $\mathbf{F}$ have the same form as those in Equation 8.9, but $\bar{\mathbf{U}}$ and $\mathbf{F}$ can be now a function of time. Matrix $\mathbf{M}$ is mass matrix; $\mathbf{C}$ is the damping matrix that are assembled using the following entries:

$$\mathbf{m}_{IJ} = \int_{\Omega} \mathbf{\Phi}_I^{\mathrm{T}} \rho \mathbf{\Phi}_J \mathrm{d}\Omega \tag{8.40}$$

$$\mathbf{c}_{IJ} = \int_{\Omega} \mathbf{\Phi}_I^{\mathrm{T}} c \mathbf{\Phi}_J \mathrm{d}\Omega \tag{8.41}$$

where
   ρ is the mass density
   c is the damping parameter

Note that the computation of the $\mathbf{M}$, $\mathbf{C}$, and $\mathbf{F}$ are basically the same as we do in the FEM, and no smoothing operation is applied.

8.4.6.1 Static Analysis

For static problems, the equation can be obtained simply by removing the dynamic term in Equation 8.39:

$$\bar{\mathbf{K}}\mathbf{d} = \mathbf{F} \tag{8.42}$$

The solution procedure is exactly the same as in NS-PIM or FEM.

8.4.6.2 Free Vibration Analysis

For free vibration analysis, we do not consider the damping and the force terms, and hence Equation 8.39 reduces to

$$\bar{\mathbf{K}}\bar{\mathbf{U}} + \mathbf{M}\ddot{\bar{\mathbf{U}}} = 0 \tag{8.43}$$

A general solution of such an equation can be written as

$$\bar{\mathbf{U}} = \bar{\mathbf{U}}_A \exp(i\omega t) \tag{8.44}$$

where
  t indicates time
  $\bar{\mathbf{U}}_A$ is the amplitude of the nodal displacement or the eigenvector
  ω is the natural frequency that is found from

$$(-\omega^2 \mathbf{M} + \bar{\mathbf{K}})\bar{\mathbf{U}}_A = 0 \tag{8.45}$$

8.4.6.3 Forced Vibration Analysis

For forced vibration analysis, Equation 8.39 can be solved by direct integration methods in the same way as in the FEM. For simplicity, the Rayleigh damping is considered in this chapter, and the damping matrix $\bar{\mathbf{C}}$ is assumed to be a linear combination of $\mathbf{M}$ and $\bar{\mathbf{K}}$,

$$\bar{\mathbf{C}} = \alpha \mathbf{M} + \beta \bar{\mathbf{K}} \tag{8.46}$$

where α and β are the Rayleigh damping coefficients.

Many direct integration schemes can be used to solve the second-order time-dependent problems, Equation 8.39, such as the Newmark method, Crank–Nicholson method, etc. [18,25]. In this chapter, the Newmark method is used (see Section 7.2.6).

8.4.7 Examples of Static Problems

A code of ES-PIMs using triangular cells has been developed, and it is examined using the following examples. ES-PIM models include ES-PIM with T3-Scheme, ES-PIM with T6/3-Scheme, ES-RPIM with T6-Scheme, and ES-RPIM with T2L-Scheme. Note that the ES-PIM with T3-Scheme is essentially the same as the ES-FEM using triangular elements [46].

Example 8.22: 2D Patch Test

A square patch with dimension of 1×1 discretized using 214 irregularly distributed nodes, as shown in Figure 8.82, is studied using the ES-PIM with various point interpolation schemes: The displacements are prescribed on all the boundaries using Equation 8.14.

Table 8.5 lists the displacement norm errors defined in Equation 8.12 of numerical results of the standard patch tests obtained using the ES-PIM models. It is shown that all the four ES-PIM models can pass the patch tests to machine accuracy, including the compatible and incompatible ones. This implies all the models are capable of reproducing linear displacement fields "exactly." This is a numerical "proof" that the weakened-weak formulation based on the $\mathbb{G}$ space theory ensures the second order convergence of both compatible and incompatible methods, as long as the displacement function is from a $\mathbb{G}_h^1$ space. Note that to impose the linear essential boundary conditions exactly along the problem boundaries, linear interpolations should always be used for the points located on the patch boundaries.

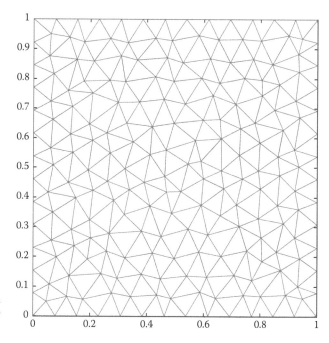

FIGURE 8.82
Patch of unique square discretized with 214 irregularly distributed nodes for the standard patch test.

TABLE 8.5

Error in Displacement Norm for the Standard Patch Using ES-PIM Models

ES-PIM Models	Error in Displacement Norm
ES-PIM (T3); compatible	1.7690582E−15
ES-PIM (T6/3); incompatible	2.6089673E−14
ES-RPIM (T6); incompatible	1.6020897E−15
ES-RPIM (T2L); incompatible	2.1726750E−15

Example 8.23: Rectangular Cantilever

We now revisit Example 8.2 with exactly the same settings with the ES-PIMs together with the linear FEM using the same triangular background cells/elements. The FEM can serve as a base model offering a "bottom line": any ES-PIM model that is established based on GS-Galerkin form is softer than the standard FEM model, which is built based on the standard Galerkin form. The NS-PIM is also included for comparison because we know it gives an upper-bound solution: a "skyline."

Figure 8.83 plots the convergence of the solutions in displacement norm for the rectangular cantilever solved using different methods. The ES-PIMs, together with FEM and NS-PIM, converge with reducing average nodal spacing (h) at rates ranging from 1.65 to 2.09 which are around the theoretical value of 2.0 for both the weak and weakened-weak formulations [28]. In terms of accuracy, except NS-PIM using T3-Scheme obtaining almost the same results as the FEM, all the other four ES-PIM models obtain about 4–10 times more accurate solutions than the FEM, with the ES-PIM-T3 giving the best performance. The big solution difference between NS-PIM-T3 and ES-PIM-T3 demonstrates clearly the significance in constructing different types of smoothing domains.

Figure 8.84 plots the convergence of the solutions in energy norm for the cantilever problem solved using these methods. It can be found that NS-PIM and ES-PIMs have better accuracy and converge faster compared to the FEM. We know that the theoretical convergence rate in energy norm of linear FEM is 1.0 and the W^2 formulation can have an ideal theoretical rate of 1.5. For the case of cantilever, the numerical convergence rate of FEM is 0.97, which is a little less than the theoretical one for weak formulation. The convergence rates of all these PIM methods based on W^2 formulation are between 1.02 and 1.5. In terms of both accuracy and convergence rate, ES-PIM-T6/3 performs better than ES-PIM-T3 and stands out together with ES-RPIM of T6-Scheme and T2L-Scheme. The solutions of ES-PIM-T6/3, ES-RPIM-T6, and ES-RPIM-T2L are about four times more accurate than that of the FEM.

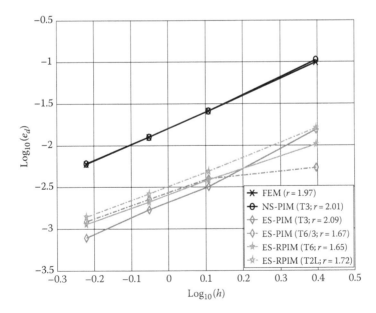

FIGURE 8.83
Convergence of the numerical results in displacement norm for the rectangular cantilever solved using different methods and same set of irregular triangular mesh.

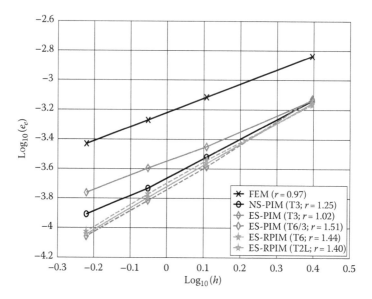

FIGURE 8.84
Convergence of the numerical results in energy norm for the rectangular cantilever solved using different methods and the same set of irregular triangular mesh.

A fair comparison should be in the form of computational efficiency. Figure 8.85 plots the errors in displacement norm against the computational cost (seconds, full matrix solver), which gives one, such as efficiency measure. Except NS-PIM-T3 that performs a little worse, all these ES-PIMs are more efficient than the FEM. ES-RPIM-T6 outperforms the ES-RPIM-T2L, and the ES-PIM-T3 performance is the best.

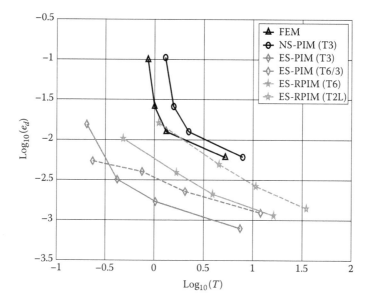

FIGURE 8.85
Comparison of computational efficiency (CPU time in seconds vs. error in displacement norm) between different methods solving the problem of rectangular cantilever.

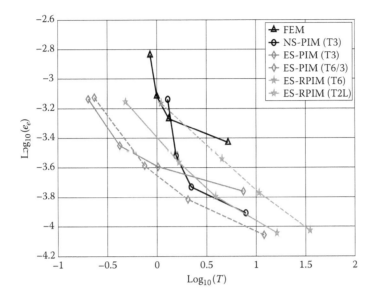

FIGURE 8.86
Comparison of computational efficiency (CPU time in seconds vs. error in energy norm) between different methods solving the problem of rectangular cantilever.

Figure 8.86 plots the computational efficiency in terms of energy norm for all these methods. It is found that all these ES-PIMs and the NS-PIM are more efficient than the FEM in energy, and ES-RPIM-T6 outperforms ES-RPIM-T2L. ES-PIM-T6/3 performed best in this energy error norm measure.

Figure 8.87 plots the process of the numerical solutions in strain energy converging to the exact solution for the cantilever problem solved using different methods. As expected, FEM and NS-PIM

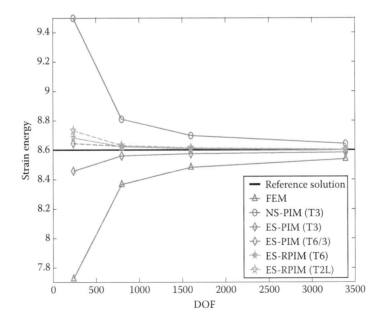

FIGURE 8.87
Solutions in strain energy converging to the exact solution for the rectangular cantilever obtained using different methods and same set of irregular triangular mesh.

offer, respectively, the bottom-line and skyline: giving lower and upper bounds. The solutions of these four ES-PIM models fall in between these two lines. ES-PIM of T3-Scheme performs softer than the FEM but stiffer than the NS-PIM, and gives lower-bound solution. As we discussed in Remark 8.7, one issue that affects the softness of the model is the order of shape functions used in the displacement approximation: when higher order shape functions are used, the displacements approximated in a smoothing domain are closer to the exact solution, which reduces the stiffening effect and vice versa. This argument is supported by the results shown in Figure 8.87 that ES-PIM-T6/3 behaves softer than ES-PIM-T3, and ES-RPIM-T2L behaves softer than ES-RPIM-T6. This confirms numerically our analysis on softening effects. We found also that three models, ES-PIMT6/3, ES-RPIM-T6, and ES-RPIM-T2L have produced upper-bound solutions for this problem. We cannot conclude that this will be true for other problems, as will be seen in the next example problem.

ES-PIM-T3 produces a lower-bound solution that is very close to the exact solution. This means that the ES-PIM-T3 is a weakly stiff model with very close-to-exact stiffness.

Example 8.24: Hole in an Infinite Plate

We now revisit Example 8.3 with exactly the same settings except that $a = 1$ and $b = 5$, using the ES-PIMs together with the linear FEM and NS-PIM using the same meshes of triangular background cells/elements. The FEM offers a "bottom line" and NS-PIM provides a "skyline."

Figure 8.88 plots the convergence of the solutions in displacement norm obtained using different methods. For this example, the ES-PIMs converge much faster than the FEM and NS-PIM, and the rates of convergence range from 2.18 to as high as 3.17 that is far above the theoretical value of 2.0 for both the weak and weakened-weak (W^2) formulations. We clearly observe superconvergence even in displacement norm (see Remark 8.1). In terms of accuracy, all the ES-PIMs are much more accurate than the FEM and NS-PIM. ES-RPIM-T2L solution is about 10 times more accurate than that of the FEM.

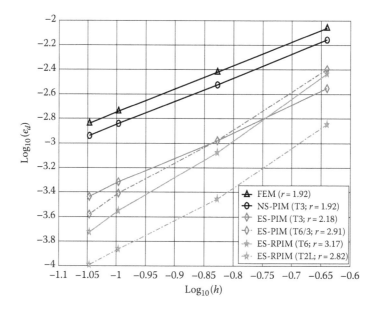

FIGURE 8.88
Convergence of the numerical results in displacement norm for the problem of an infinite plate with a hole solved using different methods with the same sets of triangular mesh.

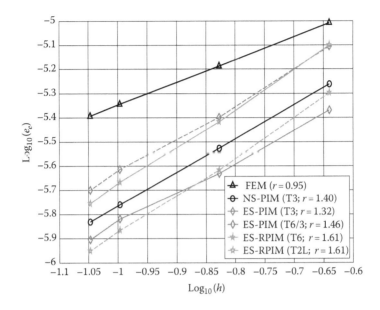

FIGURE 8.89

Convergence of the numerical results in energy norm for the problem of an infinite plate with a hole solved using different methods with the same sets of triangular mesh.

Figure 8.89 plots the convergence of the solutions in energy norm. It can be found that NS-PIM and all ES-PIMs have much better accuracy and converge much faster compared to the FEM. Their convergence rates range from 1.32 to 1.61, which are much higher than the theoretical rate of 1.0 of linear weak formulation, and around the ideal theoretical rate of 1.5 of W^2 formulation. We again observe superconvergence. In terms of accuracy, the ES-RPIM-T2L solution is about three times more accurate than that of the FEM.

Figure 8.90 plots the computational efficiency in terms of displacement norm. Except NS-PIM-T3 that performs a little worse, all the ES-PIMs are much more efficient than the FEM. ES-RPIM-T2L performed best followed by ES-RPIM-T6 and ES-PIM-T6/3.

Figure 8.91 plots the computational efficiency in terms of energy norm. It is found that all ES-PIMs and the NS-PIM are more efficient than the FEM. ES-RPIM-T2L and ES-PIM-T3 performed best.

Figure 8.92 plots the process of the numerical solutions in strain energy converging to the exact solution for different methods. As expected, FEM and NS-PIM offer, respectively, lower and upper bounds. The solutions of these four ES-PIM models fall in between the exact solution and the FEM solution. The ES-RPIM-T2L solution is most accurate for this problem.

Remark 8.15: ES-PIM-T3: An Ideal Candidate for Dynamic Problems

We found again from Example 8.24 that ES-PIM-T3 produces a lower-bound solution that is very close to the exact solution, as we have already observed from the previous example of rectangular beam. In fact, our group has not yet found a counter example for this property of ES-PIM-T3 for all different kinds of problems studied so far. Even if we found some counter examples in the future and possible existence of spurious modes, it is likely that these spurious modes can only occur at very high energy level and hence

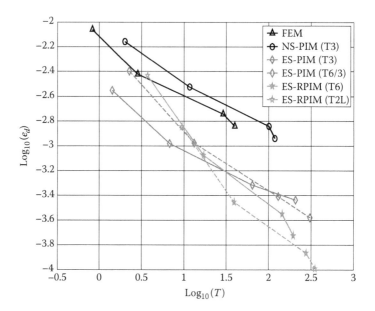

FIGURE 8.90
Comparison of computational efficiency (CPU time in seconds vs. error in displacement norm) between different methods solving the problem of an infinite plate with a hole.

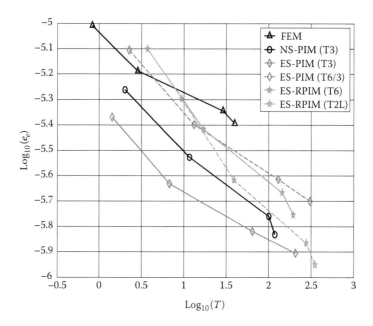

FIGURE 8.91
Comparison of computational efficiency (CPU time in seconds vs. error in energy norm) between different methods solving the problem of the infinite plate.

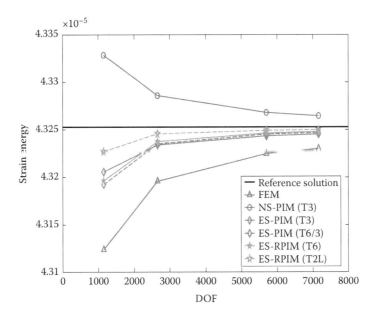

FIGURE 8.92
Solutions in strain energy converging to the exact solution for the problem of the infinite plate with a hole obtained using different methods with the same sets of triangular mesh.

unlikely to lead to significant temporal instability affecting the solution procedure for dynamic problems. These properties of weakly stiff and close-to-exact stiffness of ES-PIM-T3 make it an ideal candidate for solving dynamic problems.

Example 8.25: A Mechanical Part: 2D Rim

Finally, we study again Example 8.5: a typical 2D rim of automotive component using the ES-PIMs together with NS-PIM and FEM. As shown in Figure 8.15a, the rim is restricted along the inner circle and a uniform pressure of 100 units is applied along the outer arc edge of 60°. The rim is simulated as a plane stress problem using triangular mesh of background cells/elements.

Figure 8.93 plots the process of strain energy solution converging to the reference one. It can be found again that FEM provides lower bound; NS-PIM provides upper bound; and all these solutions of ES-PIMs fall in between. All these ES-PIMs and the NS-PIM produce lower solution bounds.

We found again from this example of quite complicated geometry that ES-PIM-T3 produces a lower-bound solution, which is very close to the exact solution: the properties of weakly stiff and close-to-exact stiffness. This reinforces Remark 8.15: ES-PIM-T3 is an ideal candidate for solving dynamic problems. In the following examples it will be used to solve dynamic problems.

8.4.8 Examples of Dynamic Problems

In this section, dynamic problems will be solved using ES-PIM-T3 together with FEM-T3, FEM-Q4 (four-node quadrilateral elements), and NS-PIM-T3 for comparison purposes.

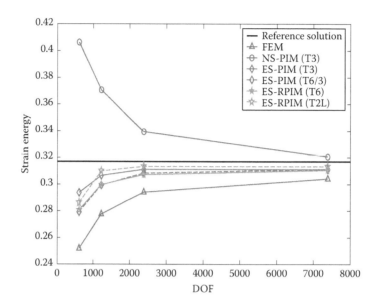

FIGURE 8.93
Solutions in strain energy converging to the exact solution for the 2D rim obtained using different methods with the same sets of triangular mesh.

Example 8.26: A Slender Rectangular Cantilever: Free Vibration

In this example, free vibration analysis is performed for a slender cantilever beam with $L = 100$ mm, $D = 10$ mm, thickness $t = 1.0$ mm, Young's modulus $E = 2.1 \times 10^4$ kgf/mm^4, Poisson's ratio $\nu = 0.3$, mass density $\rho = 8.0 \times 10^{-10}$ kgf s^2/mm^4. A plane stress problem is considered. Because the slenderness of the beam, we can use the Euler–Bernoulli beam theory to obtain the fundamental frequency $f_1 = 0.08276 \times 10^4$ Hz as a reference. Three regular meshes are used in the analysis. Numerical results using the FEM-Q4 with a very fine mesh (100×10) for the same problem are computed and used as reference solutions.

Eigenvalue Equation 8.45 is first established and solved using standard eigenvalue (symmetric) solvers for eigenvalues that give natural frequencies and eigenvectors that lead to vibration modes. Table 8.6 lists the first 12 natural frequencies of the cantilever. The first six corresponding vibration modes obtained using the NS-PIM are plotted in Figure 8.94a together with those using ES-PIM are

TABLE 8.6

First 12 Natural Frequencies (in 10 kHz) of the Slender Cantilever

Model	NS-PIM	ES-PIM	FEM-T3	FEM-Q4	Reference
Mesh: 10 × 1	0.0576	0.1048	0.1692	0.0992	0.0824
Nodes: 22	0.3243	0.6018	0.9163	0.5791	0.4944
Elements: 10 Q4 or 20 T3	0.7441	1.2833	1.2869	1.2834	1.2824
	0.9875	1.5177	2.1843	1.4830	1.3022
	1.0112	2.6362	3.5942	2.6183	2.3663
	1.1346	3.7724	3.8338	3.8140	3.6085
	1.2783	3.8559	5.0335	3.8824	3.8442
	1.5712	5.0349	6.2421	5.1924	4.9674

TABLE 8.6 (continued)

First 12 Natural Frequencies (in 10 kHz) of the Slender Cantilever

Model	NS-PIM	ES-PIM	FEM-T3	FEM-Q4	Reference
	2.3697	6.0827	6.4154	6.2345	6.3960
	3.2685	6.1520	7.5940	6.4846	6.4023
	3.7064	7.0519	8.4790	7.7039	7.8853
	3.8642	7.7212	8.7033	8.4632	8.9290
Mesh: 20 × 2	0.0675	0.0853	0.1117	0.0870	0.0821
Nodes: 63	0.4032	0.5078	0.6539	0.5199	0.4944
Elements: 40 Q4 or 80 T3	1.0518	1.2828	1.2843	1.2830	1.2824
	1.2810	1.3246	1.6748	1.3640	1.3022
	1.6467	2.3783	2.9554	2.4685	2.3663
	1.8786	3.5784	3.8424	3.7477	3.6085
	2.7823	3.8298	4.3866	3.8378	3.8442
	3.0926	4.8533	5.8836	5.1322	4.9674
	3.6783	6.1527	6.3751	6.3585	6.3960
	3.8089	6.3182	7.4046	6.5731	6.4023
	4.0543	7.4419	8.8210	8.0342	7.8853
	4.1605	8.6776	8.9411	8.8187	8.9290
Mesh: 40 × 4	0.0778	0.0827	0.0906	0.0835	0.0824
Nodes: 205	0.4654	0.4950	0.5409	0.5004	0.4944
Elements: 160 Q4 or 320 T3	1.2199	1.2826	1.2831	1.2827	1.2824
	1.2818	1.3006	1.4161	1.3174	1.3022
	1.6689	2.3554	2.5570	2.3926	2.3663
	2.2012	3.5778	3.8433	3.6462	3.6085
	3.2517	3.8408	3.8786	3.8431	3.8442
	3.3270	4.9029	5.3087	5.0150	4.9674
	3.8344	6.2867	6.3935	6.3883	6.3960
	4.5248	6.3774	6.8093	6.4561	6.4023
	4.6406	7.6987	8.3473	7.9398	7.8853
	5.3275	8.8751	8.9183	8.9057	8.9290

Note: Underlined are spurious nonzero energy modes.

plotted in Figure 8.94b. The 7th–12th modes are plotted in Figure 8.95a and b. It is observed that (1) four nonzero energy spurious modes have been found when coarse mesh is used for NS-PIM, due to the overly soft behavior; (2) the natural frequencies obtained using FEM-T3 are much larger than those using ES-PIM, due to the overly stiff behavior of FEM; (3) the ES-PIM does not have any spurious modes; and (4) the ES-PIM results are, in general, the closest to the reference solution, and they converge even faster than the FEM-Q4 using the same set of nodes. Because the natural frequencies can be used as a good indicator for assessing the stiffness of a model, the above findings confirm again that the ES-PIM model has a very close-to-exact stiffness.

Example 8.27: Free Vibration Analysis of a Shear Wall

We study again Example 7.8, but using ES-PIM, NS-PIM, and FEM models. Two types of meshes of triangular and quadrilateral elements are used as shown in Figure 8.96. Since there is no

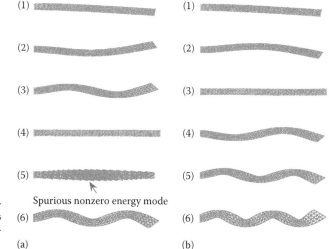

FIGURE 8.94

First six free vibration modes of the slender cantilever. (a) NS-PIM, sixth mode is spurious nonphysical mode; and (b) ES-PIM, no spurious modes.

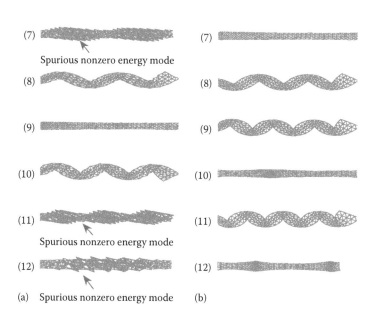

FIGURE 8.95

First 7th–12th free vibration modes of the slender cantilever. (a) NS-PIM, 7th, 11th, and 12th modes are spurious nonphysical modes; and (b) ES-PIM, no spurious modes.

exact solution for this problem, reference solutions are obtained using very fine and higher order FEM-Q8 with 6104 nodes and 1922 elements for comparison.

Table 8.7 lists the first 12 natural frequencies obtained using these methods. Figure 8.97 plots the first six corresponding vibration modes obtained using NS-PIM, together with those using ES-PIM. Figure 8.98 plots 7th–12th modes. From this table and figures, it is found that (1) five nonzero energy spurious modes and two like-hood spurious modes (modes 7 and 9) have been found in NS-PIM solution, due to the overly soft behavior; (2) the natural frequencies obtained using

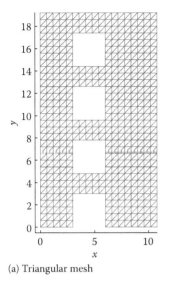

(a) Triangular mesh

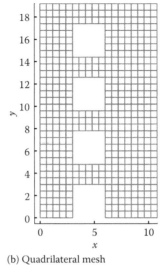

(b) Quadrilateral mesh

FIGURE 8.96
Meshes of three-node triangular and four-node quadrilateral elements for the shear wall with the same number of nodes.

TABLE 8.7

First 12 Natural Frequencies (in rad/s) of the Shear Wall

Model	NS-PIM	ES-PIM	FEM-T3	FEM-Q4	Reference (FEM-Q8)	Reference [43]
Nodes: 559	1.8271	2.0499	2.1444	2.0732	2.0107	2.079
Elements:	6.5113	7.0379	7.3185	7.0956	6.9517	7.181
476 Q4 or 952 T3	7.5146	7.6202	7.6506	7.6253	7.6001	7.644
	10.1828	11.7432	12.5535	11.9378	11.4706	11.833
	13.7334	15.1434	15.9434	15.3407	14.9716	15.947
	<u>14.7088</u>	18.2136	18.7625	18.3446	18.0660	18.644
	17.0319	19.7139	20.3822	19.8760	19.5809	20.268
	17.1035	21.9939	22.6764	22.2099	21.8718	22.765
	18.3597	22.7782	23.6401	23.0014	22.6358	
	18.8903	23.3486	24.1263	23.5515	23.2930	
	<u>19.4495</u>	25.0519	25.5337	25.1749	25.0178	
	<u>19.5377</u>	25.8374	26.8452	26.0713	25.8767	
Nodes: 2072	1.9351	2.0220	2.0632	2.0318	2.0107	2.079
Elements:	6.7760	6.9762	7.0865	6.9990	6.9517	7.181
1904 Q4 or 3808 T3	7.5658	7.6062	7.6200	7.6089	7.6001	7.644
	10.8946	11.5508	11.8795	11.6249	11.4706	11.833
	14.4676	15.0190	15.3218	15.0923	14.9716	15.947
	<u>15.3242</u>	18.1082	18.3171	18.1577	18.0660	18.644
	17.5528	19.6186	19.8617	19.6772	19.5809	20.268
	<u>17.8336</u>	21.9076	22.1972	21.9867	21.8718	22.765
	19.1518	22.6812	23.0004	22.7590	22.6358	
	<u>20.5110</u>	23.3083	23.5692	23.3800	23.2930	
	<u>21.0269</u>	25.0298	25.2055	25.0732	25.0178	
	<u>21.3027</u>	25.8752	26.2157	25.9562	25.8767	

Note: Underlined are spurious nonzero energy modes.

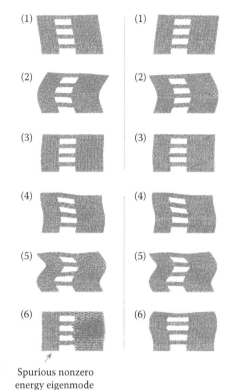

FIGURE 8.97
First six free vibration modes of the shear wall: (a) NS-PIM, sixth mode is spurious nonphysical mode; and (b) ES-PIM, no spurious modes.

Spurious nonzero
energy eigenmode

FEM-T3 are much larger than those using ES-PIM, due to the overly stiff behavior of FEM; (3) the ES-PIM does not have any spurious modes; (4) the ES-PIM results are the closest to the reference solution, and even more accurate than the FEM-Q4. Because the natural frequencies can be used as a good indicator on assessing the stiffness of a model, the above findings confirm again that the ES-PIM model has a very close-to-exact stiffness.

Example 8.28: Free Vibration Analysis of a Connecting Rod

A free vibration analysis of a connecting rod shown in Figure 8.99 is performed. Plane stress problem is considered with $E = 10$ GPa, $\nu = 0.3$, $\rho = 7.8 \times 10^3$ kg/m^3. The nodes on the left inner circumference are fixed in two directions. Two types of meshes using triangular (for NS- PIM and ES-PIM, FEM-T3) and quadrilateral elements (for FEM-Q4, FEM -Q8) are used as shown in Figure 8.100. Numerical results using the FEM-Q4 and FEM-Q8 for the same problem are computed and used as reference solutions.

From the results given in Table 8.8, it is observed that the ES-PIM gives the comparable results such as those of the FEM-Q4 using more nodes than the ES-PIM. The first six corresponding vibration modes obtained using NS-PIM are shown in Figure 8.101a, together with those using ES-PIM shown in Figure 8.101b. The 7th–12th modes are plotted in Figure 8.102. Four spurious modes and two likelihood instable modes (modes 7 and 11) are found. No spurious modes are found in the ES-PIM solution. This example of complicated geometry also confirms that the ES-PIM model has very close-to-exact stiffness, and is expected to perform well in dynamic analysis.

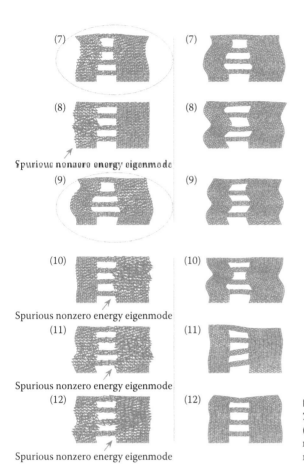

(7)   (7)

(8)   (8)

Spurious nonzero energy eigenmode

(9)   (9)

(10)   (10)

Spurious nonzero energy eigenmode

(11)   (11)

Spurious nonzero energy eigenmode

(12)   (12)

Spurious nonzero energy eigenmode

FIGURE 8.98
7th–12th free vibration modes of the shear wall:
(a) NS-PIM, 8th, 10th–12th modes are spurious
nonphysical modes; (b) ES-PIM, no spurious
modes.

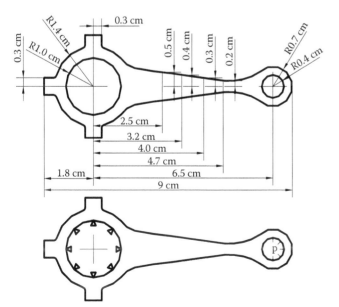

FIGURE 8.99
Geometric, loading, and boundary conditions of an automobile connecting bar.

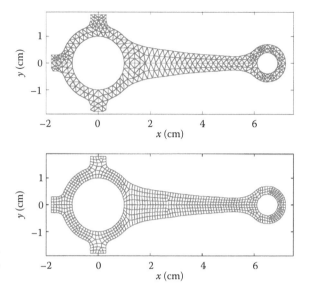

FIGURE 8.100

Meshes of three-node triangular and four-node quadrilateral elements of the automobile connecting bar.

TABLE 8.8

First 12 Natural Frequencies (Hz) of the Automobile Connecting Bar

Model	NS-PIM	ES-PIM	FEM-T3	Reference (FEM-Q4) 537 Nodes 429 Elements	Reference (FEM-Q4) 1455 Nodes 1256 Elements	Reference (FEM-Q8) 10,002 Nodes 3125 Elements
Nodes: 373	4.9420	5.1368	5.3174	5.1369		5.1222
Elements:	20.8051	22.0595	22.9448	22.050		21.840
574 T3	48.3890	49.3809	49.6982	49.299		49.115
	48.4864	52.0420	54.0642	52.232		51.395
	84.9250	92.7176	96.8632	93.609		91.787
	97.6804	109.5887	114.3134	108.59		106.15
	114.0340	132.6795	142.4456	134.64		130.14
	123.3202	158.2376	163.9687	159.45		156.14
	143.6428	158.9530	169.2762	160.59		157.70
	144.6607	201.3746	204.5709	203.52		200.06
	151.4276	204.8442	210.1202	208.68		204.41
	161.9533	209.2773	210.7405	209.02		204.99
Nodes: 1321	5.0481	5.1246	5.2084		5.1244	5.1222
Elements:	21.4886	21.8805	22.2661		21.909	21.840
2296 T3	48.8798	49.1726	49.3544		49.211	49.115
	50.4006	51.5181	52.4947		51.657	51.395
	89.6102	91.9305	93.8422		92.390	91.787
	92.6458	106.8473	109.2835		107.51	106.15

TABLE 8.8 (continued)

First 12 Natural Frequencies (Hz) of the Automobile Connecting Bar

Model	NS-PIM	ES-PIM	FEM-T3	Reference (FEM-Q4) 537 Nodes 429 Elements	Reference (FEM-Q4) 1455 Nodes 1256 Elements	Reference (FEM-Q8) 10,002 Nodes 3125 Elements
	103.4429	130.5546	134.5815		131.48	130.14
	125.6500	156.3497	159.7354		157.51	156.14
	151.6215	157.8186	159.9686		158.69	157.70
	152.0064	200.9013	203.3543		201.69	200.06
	155.5444	204.2601	207.5036		206.04	204.41
	188.5849	206.5273	209.1795		209.92	204.99

Note: Underlined are spurious nonzero energy modes.

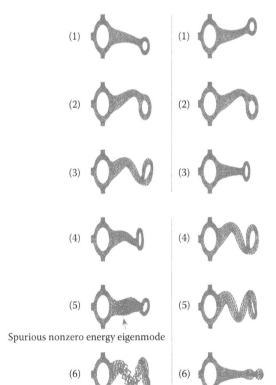

Spurious nonzero energy eigenmode

Spurious nonzero energy eigenmode

FIGURE 8.101

First six free vibration modes of the connecting bar. (a) NS-PIM, fifth and sixth modes are spurious nonphysical modes; and (b) ES-PIM, no spurious modes.

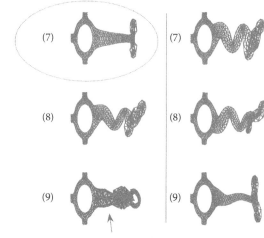

Spurious nonzero energy eigenmode

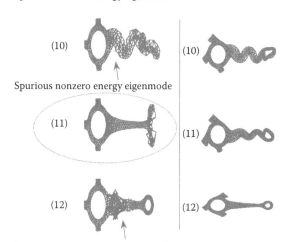

Spurious nonzero energy eigenmode

FIGURE 8.102
7th–12th free vibration modes of the connecting bar. (a) NS-PIM, 9th, 10th, and 12th modes are spurious nonphysical modes and (b) ES-PIM, no spurious modes.

Spurious nonzero energy eigenmode

Remark 8.16: ES-PIM: No Spurious Modes Found

In the above three examples, we have found no spurious modes for ES-PIM. This has been true also for all the examples studied in our group so far. Although, we still cannot rule it out completely because we could not yet prove this theoretically, it is unlikely that ES-PIM will produce spurious modes. Therefore, ES-PIM is a very good method for dynamic analysis.

Example 8.29: Forced Vibration Analysis of a Rectangular Cantilever

The Newmark method is implemented in these models to compute the transient response of the cantilever subjected to a harmonic loading $f(t) = P\cos\omega_f t$ in y-direction. Plane strain problem is considered with parameters of $L = 4.0$, $H = 1.0$, $t = 1.0$, $E = 1.0$, $\nu = 0.3$, $\rho = 1.0$, $\alpha = 0.005$, $\beta = 0.272$, $\omega_f = 0.05$, $\Delta t = 1.57$, and $\theta = 0.5$. Figure 8.103 plots the transient reposes in displace-

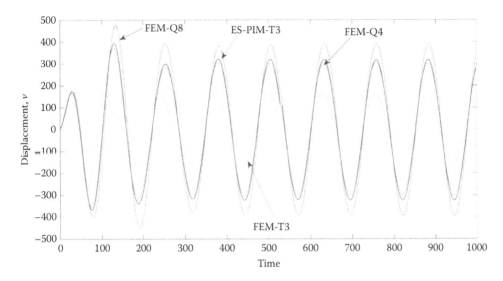

FIGURE 8.103
Transient displacement response of the cantilever subjected to a harmonic loading.

ment computed using the ES-PIM together with FEM-T3, FEM-Q4, and FEM-Q8 using the same mesh of 10×4 elements. It is seen that the results of the ES-PIM performs far better than the FEM-T3 and is closer to that of the FEM-Q8 even compared to the FEM-Q4.

Example 8.30: Forced Vibration Analysis of a Spherical Shell

A spherical shell shown in Figure 8.104 is now studied. The shell is subjected to a concentrated loading at its apex, and is modeled as 2D solid. One half of the shell is modeled using two types of meshes of triangular and quadrilateral elements as shown in Figure 8.105. The parameters used in the computation are $R = 12$, $t = 0.1$, $\phi = 10.9°$, $\theta = 0.5$, $E = 1.0$, $v = 0.3$, $\rho = 1.0$, and $\Delta t = 5$, and no damping effect is included. The Figure 8.106 plots the time history of the deflection at the apex

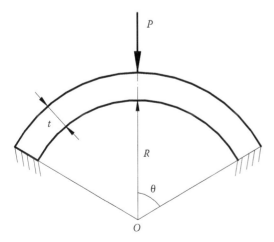

FIGURE 8.104
A schematic drawing of a spherical shell subjected to point load at the apex.

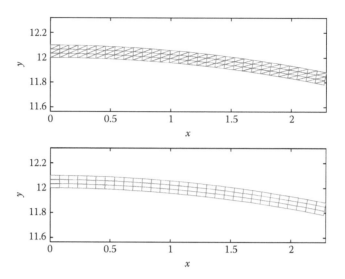

FIGURE 8.105
Meshes of three-node triangular and four-node quadrilateral elements for half a spherical shell.

of the shell excited by a harmonic loading of $f(t) = \cos \omega_f t$ with $\omega_f = 0.05$. It is found that the ES-PIM solution is much more accurate than that of FEM-T3, and comparable to that of the FEM-Q4. Figure 8.107 plots the same response of the shell excited by a heavy side step load of $f(t) = 1$ starting from $t = 0$. The red dashed line is for the results without damping, and it is seen that the amplitude of the deflection tends to a constant value with increasing time. When a Rayleigh damping of $\alpha = 0.005$ and $\beta = 0.272$ are considered, the response converges to constant.

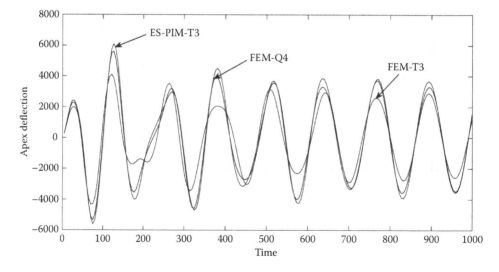

FIGURE 8.106
Solution of transient responses of the spherical shell subjected to a harmonic loading obtained using different methods.

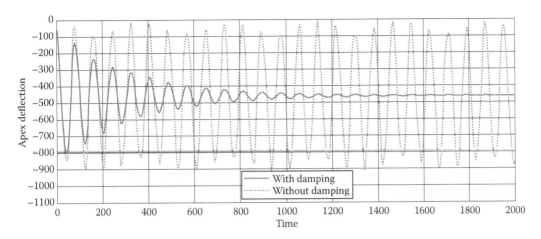

FIGURE 8.107
Solutions of transient responses of the spherical shell subjected to a step loading using the ES-FEM.

8.5 A Combined ES/NS Point Interpolation Method (ES/NS-PIM)

8.5.1 Background

From Remark 8.2 we know that the NS-PIM-T3 is naturally (without any special treatment) immune from volumetric locking for plane strain problems with Poisson's ratio approaches to 0.5. The ES-PIM formulated above is, however, subjected to volumetric locking. In this section we will present a combined ES/NS-PIM that is also immune from volumetric locking. The idea is originated from the selective formulations used in the conventional FEM [50], where the integration is performed selectively for two different material "parts" (μ-part and λ-part) to overcome such a locking. In our ES/NS-PIM, we use two different types of smoothing domains selectively for these two different material "parts." This is based on our finding that the node-based smoothing domains are effective in overcoming volumetric locking [13], and the λ-part is known as the culprit of the volumetric locking. Naturally, we use node-based domains for the λ-part and edge-based domains for the μ-part. The details are given below.

8.5.2 Formulation for ES/NS-PIM

In this section, we only use linear interpolation, PIMs with T3-Scheme. For 2D plane strain problems, we know from Remark 1.3 that the matrix constants $\mathbf{c}$ can be nearly singular when Poisson's ratio n approaches 0.5. This is the root of the volumetric locking behavior of a displacement method. Therefore, we decompose matrix $\mathbf{c}$ into two parts:

$$\mathbf{c} = \mathbf{c}_\lambda + \mathbf{c}_\mu \tag{8.47}$$

where
  $\mathbf{c}_\mu$ relates to the shearing modulus $\mu = E/[2(1+v)]$ and hence is termed as μ-part of $\mathbf{c}$
  $\mathbf{c}_\lambda$ relates to Lame's parameter $\lambda = \frac{2v\mu}{1-2v}$ and hence is termed as λ-part of $\mathbf{c}$

For plane strain cases, we have

$$\mathbf{c} = \begin{bmatrix} \lambda + 2\mu & \lambda & 0 \\ \lambda & \lambda + 2\mu & 0 \\ 0 & 0 & \mu \end{bmatrix} = \mu \underbrace{\begin{bmatrix} 2 & 0 & 0 \\ 0 & 2 & 0 \\ 0 & 0 & 1 \end{bmatrix}}_{\text{SPD}} + \lambda \underbrace{\begin{bmatrix} 1 & 1 & 0 \\ 1 & 1 & 0 \\ 0 & 0 & 0 \end{bmatrix}}_{\text{semi-SPD}} = \mathbf{c}_\lambda + \mathbf{c}_\mu \tag{8.48}$$

and for axis-symmetric problems:

$$\mathbf{c} = \mu \underbrace{\begin{bmatrix} 2 & 0 & 0 & 0 \\ 0 & 2 & 0 & 0 \\ 0 & 0 & 1 & 0 \\ 0 & 0 & 0 & 2 \end{bmatrix}}_{\text{SPD}} + \lambda \underbrace{\begin{bmatrix} 1 & 1 & 0 & 1 \\ 1 & 1 & 0 & 1 \\ 0 & 0 & 0 & 0 \\ 1 & 1 & 0 & 1 \end{bmatrix}}_{\text{semi-SPD}} = \mathbf{c}_\mu + \mathbf{c}_\lambda \tag{8.49}$$

In our ES/NS-PIM, we use the NS-PIM equations to calculate the stiffness matrix related to λ-part, following exactly the same procedure given in Section 8.1 but replacing $\mathbf{c}$ by $\mathbf{c}_\lambda$. For the μ-part, we use ES-PIM following exactly the same procedure given in Section 8.4 but replacing $\mathbf{c}$ by $\mathbf{c}_\mu$. The stiffness matrix of the combined ES/NS-PIM model becomes a superimposition of these two parts:

$$\bar{\mathbf{K}}^{\text{ES/NS-PIM}} = \underbrace{\bar{\mathbf{K}}^{\text{ES-PIM}}_\mu}_{\text{SPD}} + \underbrace{\bar{\mathbf{K}}^{\text{NS-PIM}}_\lambda}_{\text{semi-SPD}} \tag{8.50}$$

Since $\bar{\mathbf{K}}^{\text{ES-PIM}}_\mu$ will be SPD (after the essential boundary condition is imposed), and $\bar{\mathbf{K}}^{\text{NS-PIM}}_\lambda$ is semi-SPD, we can expect $\bar{\mathbf{K}}^{\text{ES/NS-PIM}}$ to be SPD, and therefore a unique stable solution.

Equation 8.50 suggests that we need double the effort in computing the stiffness matrix for an ES/NS-PIM model. This is the price we need to pay in dealing with 2D solids of incompressible materials. The ES/NS-PIM is a smoothing domain-based selective formulation.

8.5.3 Example for Volumetric Locking Problems

We now revisit Example 8.3 with exactly the same settings except that $a = 1, b = 5, E = 1000$, and $p = 1$. Poisson's ratio varies from 0.4 to 0.4999,999, and plane strain problem is considered. The ES/NS-PIM is used to solve this problem together with the linear NS-PIM and ES-PIM for comparison.

Figure 8.108 plots the error in solutions in displacement norm for nearly incompressible material when Poisson's ratio is changed from 0.4 to 0.4999,999. First, it is clearly seen that the ES-PIM suffers from the volumetric locking: when Poisson's ratio approaches 0.5, the error in solution increase drastically starting from 0.49. Compared to FEM (see Figure 8.11) it behaves better, due to the smoothing effect introduced in the ES-PIM, but still a quite strong locking. Second, the results show that the domain-based selective ES/NS-PIM model can overcome very successfully the volumetric locking for nearly incompressible materials. It gives even more accurate results than those of the NS-PIM, thanks to the superior accuracy of the ES-PIM used for the μ-part.

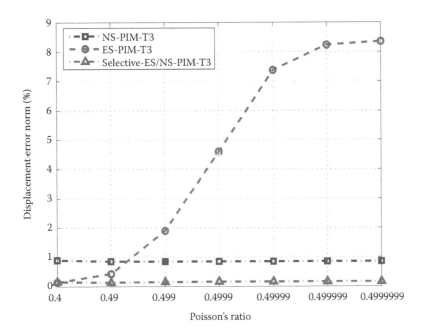

FIGURE 8.108
Displacement error for the infinite plate with a hole subjected to unidirectional tension. Poisson's ratio of the material varies near and up to 0.5.

Remark 8.17: Free from Volumetric Locking: A Property of ES/NS-PIM

The combined ES/NS-PIM can efficiently overcome the volumetric locking issue for 2D plane strain problems. The same trick still works well also for 3D problems. Readers are referred to [47], where an 3D FS/NS-FEM model that is a special case of 3D FS-/NS-PIM has been established.

Note that the ES/NS-PIM works, of course, also for compressible materials. The solution should be in between the ES-PIM and NS-PIM model, due to the simple superimposition nature. Therefore, the performance of the combined ES/NS-PIM model is expected to be between the ES-PIM and the NS-PIM. Because of the close-to-exact stiffness property, the ES-PIM usually performs very much better than the NS-PIM model. If Poisson's ratio is less than 0.4, using ES-PIM alone should give, in general, better accuracy. In addition, using ES-PIM alone is more efficient.

8.6 Strain-Constructed Point Interpolation Method

8.6.1 Background

In the previous sections, we have seen weakened-weak models based on GS-Galerkin formulation. It is in fact a special and the simplest strain-constructed model. In this section, we present more general SC-modes using PIM shape functions known as strain-constructed point interpolation methods (SC-PIMs), based on the general SC-Galerkin form. We will take much bolder approaches to construct the strain fields by simple means of point interpolation using smoothed strains at critical points over the node-based, edge-based, and/or cell-based smoothing domains. We propose and examine a total of six novel schemes for constructing strain fields, based on a set of triangular integration cells. All these are performed under conditions of (1) no additional DOFs are added in any way in the SC-model; (2) the formulation procedure (hence the computation of the stiffness matrix) should remain as simple as in the standard FEM; and (3) solutions converge to the exact solution. All these schemes proposed will be studied in great detail, via mainly numerical means, which leads to a number of SC-PIM formulations of excellent performance. The theoretical proof on the stability and convergence of the SC-Galerkin models has been given in Chapters 4 and 5.

An SC-PIM model, uses a triangular background cells, just as in the NS-PIM or ES-PIM. In this section, we present SC-models using only linear PIM shape functions.

8.6.2 Strain Field Construction

Strain field construction in SC-PIM models discussed in the section largely follows the steps discussed in Section 4.5.4

1. **Creation of quadrature/integration cells**

    Based on the background cells, a set of triangular integration cells are then constructed. The integration cells are usually created by subdividing the background cells into smaller cells in various ways. Such a nested division helps to make sure the strain norm equivalence and strain convergence conditions.

2. **Determine the strain $\breve{\varepsilon}$ at the points on the vertices of the integration cells**

    These strains are either the compatible strains $\tilde{\varepsilon}$ (from a continuous displacement field) or the smoothed strains $\bar{\varepsilon}$ obtained using a local smoothing domain. Here we need to make sure that the strain norm equivalence and strain convergence conditions are satisfied.

3. **Construct the strain field $\hat{\varepsilon}$ in the integration cell by interpolation**

In this section, we use linear interpolation (Equation 4.42) and area coordinates for the integration cell. This facilitates easy integration of energy over the cell: no numerical integration is needed.

Based on the above steps we designed the following schemes for strain field construction.

Scheme A (1 to 1 division)

For Scheme A, each triangular cell is exactly an integration cell, as shown in Figure 8.109. Thus the total number of the (quadrature) integration cells N_q equals the number of elements N_e (1 to 1 division). In each triangular integration cell, these three strains at the vertices are obtained using node-based smoothing domains that are the same as in the NS-PIM, except that the strain obtained are only for the node. The strain fields are linearly interpolated using

$$\hat{\varepsilon}(\mathbf{x}) = \sum_{i=1}^{3} \Phi_i(\mathbf{x}) \breve{\varepsilon}_i \qquad (8.51)$$

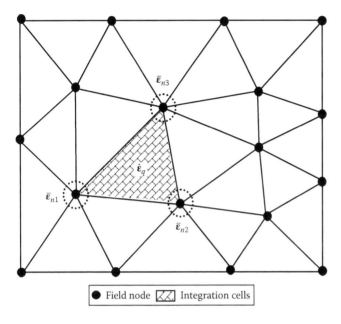

FIGURE 8.109
Integration cell q over which the strain field is constructed. Scheme A: entire background cell is used as a single integration cell.

with $\breve{\boldsymbol{\varepsilon}}_i = \bar{\boldsymbol{\varepsilon}}_{ni}, i = 1, 2, 3$. The subscript "$n$" stands for node-based smoothing strains. Then the strain energy potential can be obtained using Equation 4.47. Note that the strain field constructed in Scheme A is continuous in the entire problem domain.

Scheme B (1–3 division)

Figure 8.110 shows the construction of integration cells in Scheme B where a triangular background cell is divided into three integration cells (1–3 division). Each integration cell is formed by connecting two end-points of an edge and the centroid of the triangle, which contains the edge.

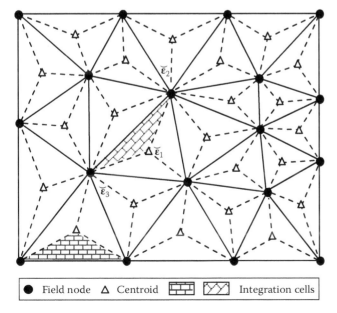

FIGURE 8.110
Integration cell q over which the strain field is constructed. Scheme B: a triangular background cell is further divided into three integration cells, and each integration cell is formed by connecting sequentially two vertices and the centroid of the background cell. Strain fields within an integration cell is linearly interpolated.

TABLE 8.9

Strains Used for Constructing the Modified Strain Fields:
Scheme B (Strain Continuous in the Entire Problem Domain)

Schemes	Strain at Vertex 1	Strain at Vertex 2	Strain at Vertex 3
Scheme B-1	$\breve{\boldsymbol{\varepsilon}}_1 = \tilde{\boldsymbol{\varepsilon}}$	$\breve{\boldsymbol{\varepsilon}}_2 = \bar{\boldsymbol{\varepsilon}}_e$	$\breve{\boldsymbol{\varepsilon}}_3 = \bar{\boldsymbol{\varepsilon}}_e$
Scheme B-2	$\breve{\boldsymbol{\varepsilon}}_1 = \tilde{\boldsymbol{\varepsilon}}$	$\breve{\boldsymbol{\varepsilon}}_2 = \bar{\boldsymbol{\varepsilon}}_{n2}$	$\breve{\boldsymbol{\varepsilon}}_3 = \bar{\boldsymbol{\varepsilon}}_{n3}$
Scheme B-3	$\breve{\boldsymbol{\varepsilon}}_1 = \bar{\boldsymbol{\varepsilon}}_e$	$\breve{\boldsymbol{\varepsilon}}_2 = \bar{\boldsymbol{\varepsilon}}_{n2}$	$\breve{\boldsymbol{\varepsilon}}_3 = \bar{\boldsymbol{\varepsilon}}_{n3}$

The strain fields within each integration cell are calculated using Equation 8.51. There are three editions for Scheme B, as listed in Table 8.9, where $\bar{\boldsymbol{\varepsilon}}_e$ is smoothed strain obtained for the mid-edge-point for the edge using the edge-based smoothing domains as in the ES-PIM. Note that the strain field constructed in Scheme B is continuous in the entire problem domain.

Scheme C (1 to 4 division)

For Scheme C, as shown in Figure 8.111, each triangular element is further divided into four integration cells, thus $N_Q = 4N_e$ (1–4 division). For each integration cell, the modified strain fields are constructed via linear interpolation using Equation 8.51 with node-based smoothed strains and edge-based smoothed strains. The edge-based smoothed strains are obtained using edge-based smoothing domains as in the ES-PIM, except that the strain obtained are only for the midpoint of the edge. The assignments of strains at the vertices points are listed in Table 8.10. Note that the strain field constructed in Scheme B is also continuous in the entire problem domain.

Scheme D (1 to 6 division)

The construction of integration cells of Scheme D is illustrated in Figure 8.112, where a triangular cell is further divided into six triangular integration cells (1–6 division). Each of them is formed by connecting the field node, the centroid of the triangle and the corresponding mid-edge-point. The strain field within each integration cell is linearly

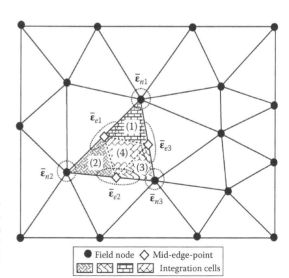

FIGURE 8.111
Integration cell q over which the strain field is constructed. Scheme B: each triangular element is further divided into four integration cells and the strain fields in each integration cell are linearly interpolated using the smoothed strains associated with the nodes and edges.

TABLE 8.10

Strains Used for Constructing Modified Strain Fields: Scheme C

Four Strain Fields within Each Integration Cell	Node-Based and Edge-Based Smoothed Strains Used for Strain Fields Approximation
Integration cell (1)	$\bar{\varepsilon}_1 = \bar{\varepsilon}_{n1}, \bar{\varepsilon}_2 = \bar{\varepsilon}_{e1}, \bar{\varepsilon}_3 = \bar{\varepsilon}_{e3}$
Integration cell (2)	$\bar{\varepsilon}_1 = \bar{\varepsilon}_{n2}, \bar{\varepsilon}_2 = \bar{\varepsilon}_{e2}, \bar{\varepsilon}_3 = \bar{\varepsilon}_{e1}$
Integration cell (3)	$\bar{\varepsilon}_1 = \bar{\varepsilon}_{n3}, \bar{\varepsilon}_2 = \bar{\varepsilon}_{e3}, \bar{\varepsilon}_3 = \bar{\varepsilon}_{e2}$
Integration cell (4)	$\bar{\varepsilon}_1 = \bar{\varepsilon}_{e1}, \bar{\varepsilon}_2 = \bar{\varepsilon}_{e2}, \bar{\varepsilon}_3 = \bar{\varepsilon}_{e3}$

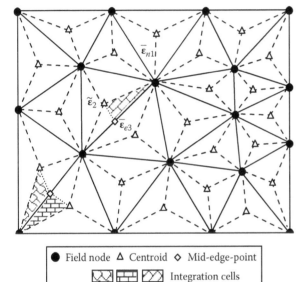

Field node ● Centroid △ Mid-edge-point ◇ Integration cells

FIGURE 8.112
Integration cell q over which the strain field is constructed. Scheme D: a triangular background cell is further divided into six integration cells and each one is formed by connecting the field node, the centroid of the triangular element, and the corresponding mid-end-point of the edge. Strain fields within each integration cell are linearly interpolated using the smoothed/compatible strains associated with the three vertices.

interpolated using Equation 8.51 with the three strains at the three vertices, which are the smoothed strain or compatible strains:

$$\breve{\varepsilon}_1 = \bar{\varepsilon}_{n1}, \quad \breve{\varepsilon}_2 = \tilde{\varepsilon}_2, \quad \breve{\varepsilon}_3 = \bar{\varepsilon}_{e3} \tag{8.52}$$

Note that the strain field constructed in Scheme D is continuous in the entire problem domain.

8.6.3 Example Problems

Example 8.31: 3D Patch Test

A square patch with dimension of 1×1 discretized using 109 irregularly distributed nodes is studied using the SC-PIM with various schemes for strain field construction. The displacements are prescribed on all the boundaries using Equation 8.14. Errors of numerical results in displacements norm are listed in Table 8.11. It is confirmed that all these SC-PIM schemes can pass the patch tests, and therefore should be able to produce second order convergent solutions.

TABLE 8.11

Displacement Errors of Results for the Linear Patch Test on Different Schemes of SC-PIM

Scheme	Error in Displacement Norm	Scheme	Error in Displacement Norm
Scheme A	3.1402142E−14	Scheme B-1	6.8463314E−16
Scheme C	9.0198460E−16	Scheme B-2	1.2696312E−15
Scheme D	9.6829514E−16	Scheme B-3	1.9473419E−15

Example 8.32: Rectangular Cantilever

We now revisit Example 8.2 with exactly the same settings as the SC-PIMs together with the linear FEM, NS-PIM, and ES-PIM using the same triangular background cells/elements. The fully compatible FEM can serve as a base model offering a "bottom line": any SC-PIM model that is established based on SC-Galerkin form should be softer than the FEM model. The NS-PIM is also included for comparison because we know it gives an upper-bound solution: a "skyline," and we have not yet found a more softer model than the NS-PIM. The ES-PIM is included because it is so far a "star" performer. Finding a model that beats ES-PIM has been a challenge.

Figure 8.113 plots the convergence of the solutions in displacement norm for the cantilever solved using different SC-PIM models presented in this section. The solutions of all these methods converge with the reducing nodal spacing at about the same convergence rate that is close to the theoretical value of 2.0 for both the weak and weakened-weak (W^2) formulations. In terms of accuracy, Scheme A is worse than the FEM measured in the displacement norm, Schemes B-2 and B-3 are about the same as FEM and all other schemes, i.e., Scheme C, Scheme D, and Scheme B-1 give more accurate solution compared to the FEM. As expected, the ES-PIM performs the best, and

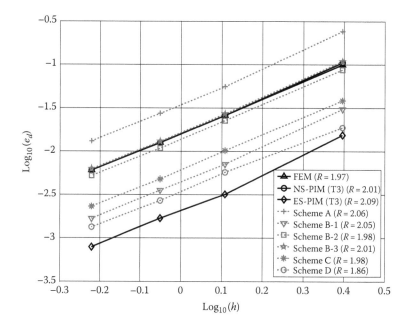

FIGURE 8.113

Convergence of the numerical results in displacement norm for the cantilever solved using different methods and same set of irregular triangular mesh.

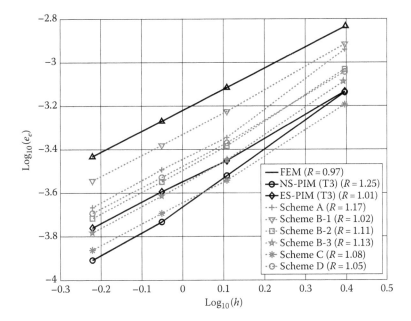

FIGURE 8.114

Convergence of the numerical results in energy norm for the cantilever solved using different methods and same set of irregular triangular mesh.

is about 10 times more accurate than the FEM. We note that Scheme D and B-1 are quite compatible to the ES-PIM, followed by Scheme C.

Figure 8.114 plots the convergence of the solutions in energy norm for the same cantilever solved using different methods. All these six SC-PIM models converge well and have much higher accuracy compared to the FEM, when the error is measured in energy norm. All the SC-PIM models achieve higher convergence rates, which are similar as linear NS-PIM and ES-PIM. In terms of both accuracy and convergence, the NS-PIM stands out. In terms of convergence rate, FEM achieved 0.97 that is a little less than the theoretical rate of 1.0. The rates of all the other schemes have convergence rates between 1.0 and 1.5 that is the ideal theoretical rate for the weakened-weak formulation based on $\mathbb{G}$ space theory. Again the NS-PIM stands out with a convergence rate of 1.25. Scheme C performed equally good as the NS-PIM. Note the ES-PIM is not the best performer in energy norm measure, but still among the best.

Figure 8.115 plots process of the solutions of strain energy converging to the exact solution for the cantilever beam obtained using different methods. We found that

1. All the six SC-PIM models, together with linear NS-PIM and ES-PIM, give upper-bound solution in energy norm with respect to the FEM solution: an SC-PIM model is always softer than FEM.

2. FEM, Scheme B-1, and linear ES-PIM give lower-bound solution, and all the other five schemes and NS-PIM provide upper-bound solution to the exact one.

3. Scheme D and ES-PIM give the best (tightest) pair of bounds to the exact solution.

4. The "softness-raking" of all these modes are Scheme A, Scheme B-3, NS-PIM, Scheme B-2, Scheme C, Scheme D, ES-PIM, Scheme B-1, and FEM.

We note that Scheme A is found even softer than NS-PIM that is known so far as the softest (spatially) stable model.

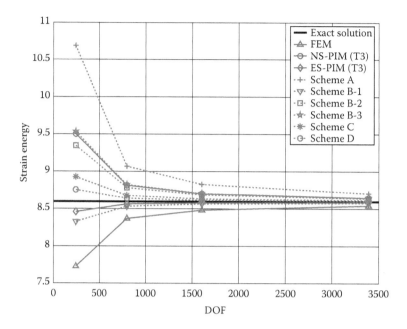

FIGURE 8.115

Solutions in strain energy converging to the exact solution for the cantilever obtained using different methods and same set of irregular triangular mesh.

Example 8.33: Hole in an Infinite Plate

We now revisit the Example 8.3 of a circular hole in an infinite plate with the same settings except $a = 1$, $b = 5$. Figure 8.116 plots the convergence of the solutions in displacement norm for this problem solved using different methods. Similar to the cantilever case, the solutions of all the methods converge well and have about the same convergence rate in terms of displacement norm. In terms of accuracy, Scheme A is again found to be worse than the FEM, and all other schemes are at least better than the FEM. Schemes B-2 and B-3 are about the same as FEM and linear NS-PIM, and Scheme C, Scheme D, and Scheme B-1 give much more accurate solution compared to FEM. For this problem, we found Scheme D standing out clearly, followed by Scheme C and then the ES-PIM. The accuracy of Scheme D is about 10 times more than the FEM.

Figure 8.117 plots the convergence of the solutions in energy norm for the problem of an infinite solid with a hole, solved using different methods. All the six schemes are found converging well and have much higher accuracy than the FEM. In terms of convergence rate, FEM gives a rate less than 1.0. All the SC-PIM modes achieve higher convergence rates similar as the linear NS-PIM and ES-PIM. Particularly, Scheme C and the linear ES-PIM outperform all the other schemes. The convergence rates for Schemes A, C, ES-PIM, and NS-PIM are around 1.3 that is very close to the theoretical rate of 1.5 for the W^2 formulations based on $\mathbb{G}$ space theory.

Figure 8.118 plots process of the solutions of strain energy converging to the exact solution for the problem of an infinite solid with a hole, obtained using different methods. We note the following:

1. All the six schemes, together with linear NS-PIM and ES-PIM, give upper-bound solution in energy norm with respect to the FEM solution.

2. FEM, Scheme B-1, linear ES-PIM, and Scheme D give lower-bound solution, and all the other four schemes and NS-PIM provide upper bound solution to the exact one.

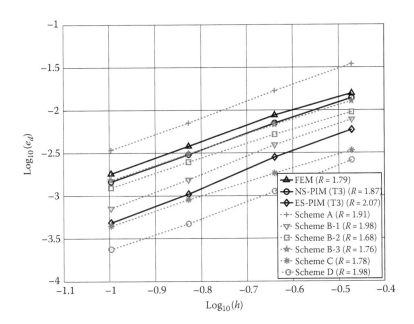

FIGURE 8.116

Convergence of the solution in displacement norm for the problem of an infinite solid with a hole solved using different methods and same set of irregular triangular mesh.

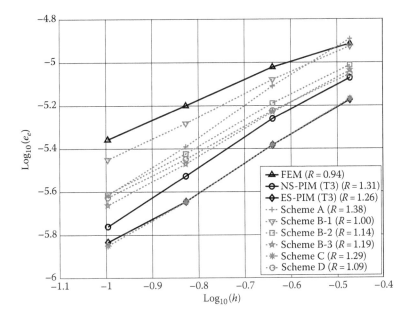

FIGURE 8.117

Convergence of the solution in energy norm for the problem of an infinite solid with a hole solved using different methods and same set of irregular triangular mesh.

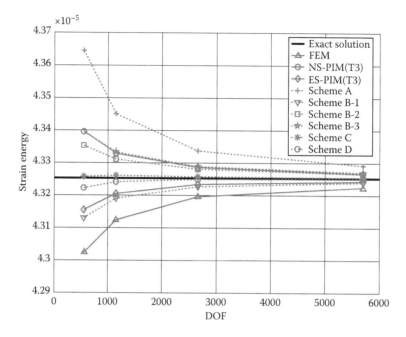

FIGURE 8.118
Solutions in strain energy converging to the exact solution for the problem of an infinite solid with a hole obtained using different methods and same set of irregular triangular mesh.

3. Scheme C and Scheme D give the best (tightest) pair of bounds to the exact solution.

4. The "softness-raking" of all these modes are unchanged: Scheme A, Scheme B-3, NS-PIM, Scheme B-2, Scheme C, Scheme D, ES-PIM, Scheme B-1, and FEM.

Scheme A is again found to be the softest (spatially) stable model that is even softer than NS-PIM.

Remark 8.18: SC-PIM: A Number of Good Properties
It is very clear now that we can find very good SC-PIM models that perform even better than ES-PIM. We can also use high-order PIM shape functions for SC-PIM, and much more models of good properties can be found. More discussions and examples, including using high-order PIM shape functions, can be found in a work very recently reported in [33].

8.7 A Comparison Study

8.7.1 Overhead and Solver Time

To conduct a meaningful and fair test on computational efficiency on these methods, we choose Example 8.3 of a square plate of 10×10 m with a central circular hole of 1 m subjected to a unidirectional tensile load of 10.0 N/m^2 in the x-direction. The material constants are $E = 1.0 \times 10^3$, $v = 0.3$. We choose this problem because the exact solution is not in a simple polynomial form and hence allow us to use many nodes without getting the exact solution too fast. In addition, we have the exact solution to quantify the error accurately.

The computational cost for a model consists of mainly two parts: "overhead" costs for all operations until the stiffness matrix is formed, and the "solver time" to solve the resultant system equations. For PIM models, the overhead time includes the smoothing operations. On one hand, they require some computational cost that the FEM does not. On the other hand it saves time for domain type Gauss integration and the computation of the derivatives of the shape functions needed in the FEM. Therefore, overall it should not have too much difference. To confirm this we conducted a detailed analysis for all linear models. The results for the overhead costs are listed in Table 8.12. The tests were conducted on a Dell PC of Intel® Pentium(R) CPU 2.80 GHz, 1.00 GB of RAM. In the test, FEM-T3, ES-PIM-T3, and NS-PIM-T3 all use exactly the same triangular mesh. It is clear that there is not much difference in the overhead computations for all these three models. Note that the linear FEM of triangular elements requires the least overhead time in all the FEM models of all types of elements, because the computation of the derivatives of linear shape functions are trivial (constant) and the integration is also very simple (elemental summation). The linear ES-PIM and NS-PIM is as efficient as the linear FEM in terms of overhead costs. Most meshfree methods, however, will surely lose out to FEM in this regard.

Note that the major cost in a computation of a not-too-small model is solving the system equations, and therefore we need to further examine the solver CPU time for these methods. Such a test will depend on the type of solver used. In the PIM modes, there is no increase in DOFs. If a full matrix solver is used the CPU time for PIMs should be the same as the FEM model using the same mesh. Because the solutions of PIMs are more accurate than that of the FEM, the computational efficiency of PIMs will be surely higher than the FEM, as discussed in Examples 8.23 and 8.24.

In this study, we conduct a test using a very efficient bandwidth solver coded in MFree2D©, and the results are also listed in Table 8.12. It is clear that the meshfree type methods consume more CPU time: ES-PIM is about three times and NS-PIM is about four times that of FEM. This is due to the larger bandwidth of the system matrix of meshfree methods. Note that compared to many other meshfree methods, the PIMs are among the ones with small bandwidth, and hence they are quite comparable to FEM methods, as shown in Table 8.12. Many other meshfree methods need about 20–50 times CPU time

TABLE 8.12

Comparison of the Estimated CPU Time (s) for Different Methods
(Three-Node Triangular Mesh with 22,930 Nodes)

Methods	FEM-T3	ES-PIM-T3	NS-PIM-T3
Overhead	2.19	2.33	2.20
Solver	24.64	77.82	103.6
Total	26.83	80.15	105.8
Ratio	1.0	2.99	3.94

compared to linear FEM of same mesh. We also note that the solver time is much more (10–50 times) than the overhead time: the overhead time is negligible and is a well-known fact in FEM analysis, if the model size is not too small.

8.7.2 Efficiency Comparison

A fair comparison should be the *computational efficiency*: CPU time needed for obtaining the results of the same accuracy. An efficiency test is therefore also conducted using MFree2D$^{©}$, using FEM-T3, EFG($\alpha_s = 2.5$), ES-PIM-T3, NS-PIM-T3, ES-PIM-T6/3, and ES-RPIM-T6. The results are plotted in Figure 8.119 in terms of displacement error norm and in Figure 8.120 in terms of energy error norm. The detailed data is listed in Tables 8.13 and 8.14. In this study, however, we neglected the insignificant overhead time and count only for the solver time. For this particular test, it is found that the ES-PIM is about eight times in displacement norm and 40 times in energy norm more efficient than the FEM model using the same mesh. The NS-PIM is about three times less efficient in displacement norm and 20 times more efficient in energy norm compared to the FEM model. This is in addition to many other important properties of PIMs discussed earlier. In this particular test, we also found that the ES-PIM is about 75 times in displacement norm and 25 times in energy norm more efficient than the EFG model using the same mesh. Note that EFG implemented in MFree2D using the penalty method for essential boundary conditions, because MFree2D uses bandwidth solver. It may not be as accurate as the Lagrange multiplayer approach, and hence EFG could be coded more efficiently.

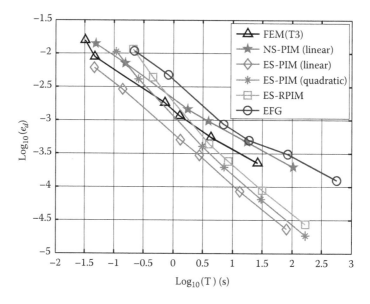

FIGURE 8.119

Comparison of the computational efficiency in terms of displacement error norm for the problem of an infinite plate with a circular hole. The ES-PIM (linear) is found to be the most efficient computationally.

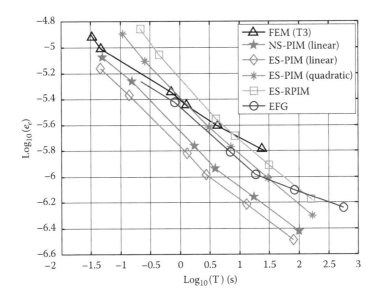

FIGURE 8.120

Comparison of the computational efficiency in terms of energy norm for the problem of an infinite plate with a circular hole. The ES-PIM (linear) is also found to be the most efficient computationally.

TABLE 8.13

Computational Efficiency: Estimated Solver CPU Time (s) Needed for Obtaining the Results of the Same Accuracy in Energy Norm (for Error in Solutions at $e_d = 1.0E{-}004$)

Methods	FEM-T3	EFG($\alpha_s = 2.5$)	ES-PIM-T3	NS-PIM-T3	ES-PIM-T6/3	ES-RPIM-T6
CPU time	92.358	873.57	11.09	270.59	19.021	28.923
Ratio/FEM	1	1/0.1057	1/8.328	1/0.3413	1/4.856	1/3.193
Ratio/EFG	0.1057	1	0.0127	0.3098	0.02177	0.03311
Ratio/ES-PIM	8.328	78.77	1	24.34	1.715	2.608

TABLE 8.14

Computational Efficiency: Estimated Solver CPU Time (s) Needed for Obtaining the Results of the Same Accuracy in Energy Norm (for Error in Solutions at $e_e = 6.3096E{-}007$)

Methods	FEM-T3	EFG($\alpha_s = 2.5$)	ES-PIM-T3	NS-PIM-T3	ES-PIM-T6/3	ES-RPIM-T6
CPU time	547.71	331.73	12.617	23.283	83.429	140.35
Ratio/FEM	1	1/1.651	1/43.41	1/23.53	1/6.565	1/3.902
Ratio/EFG	1.651	1	0.03803	0.07019	0.2515	0.4231
Ratio/ES-PIM	43.41	26.29	1	1.845	6.612	11.12

8.8 Summary

A research group led by G. R. Liu invented the PIM, and has been working very hard on its improvement over the past decade. We made lot of efforts, encountered lot of problems, and now made some progress with various versions of PIMs that work very well for various problems, as presented in this chapter. We are proud to say that our efforts have been paid off. Here we summarize a few points.

1. The PIM shape functions possess the Kronecker delta function property, and hence solve all the problems associated with the use of MLS approximation. The imposition of essential boundary conditions is as simple as in the standard FEM. They form, in general, an interpolant in a $\mathbb{G}$ space.

2. The weakened-weak (W^2) formulations are a recent key development in numerical analysis. In particular, the W^2 formulation based on the $\mathbb{G}$ space theory reveals very general and essential techniques on how to make a numerical model always stable (spatially). It provides a unified theoretical framework for both compatible and incompatible methods. The significance is that we can now always formulate a proven spatially stable numerical model, as long as a set of linearly independent shape functions with proper consistence (to ensure the positivity condition) can be found for a set of node distrusted in the problem domain: the compatibility is no more in question.

3. The SC-Galerkin (including GS-Galerkin) is used in the formulation of PIMs. Clearly, PIMs formulated using SC-Galerkin (with T-schemes for node selection), such as NS-PIMs, ES-PIMs, CS-PIMs, SC-PIMs, etc. are very well established under the general concept of W^2 formulation. In general, they are as stable as the FEM (weak form) models, all behave softer, much more accurate, higher convergence rate, capable of producing upper bound (NS-PIMs), work perfectly well with triangular types of meshes, and much more efficient. They are not entirely "meshfree," but need only triangular types of meshes that can be easily generated automatically, and hence achieve the ultimate real purpose of being meshfree perfectly well. They can all be easily made adaptive [28,52,53].

4. Using both the NS-PIM and FEM, we can now bound the exact solution from both sides to provide the so-called certified solutions, for all the well-posed linear elasticity problems as long as a triangular type of mesh can be built. This was not possible until the birth of the NS-PIM. In this solution bound regard, the NS-PIM is of unique importance for practical engineering applications.

5. The ES-PIM (with linear triangular mesh) is in general, to the best knowledge of the author, the most efficient method so far among all the FEM and meshfree models for 2D solid mechanics problems using linear triangular mesh. It is stable both spatially and temporally, and therefore is recommended for all solid mechanics problems. In the efficiency and adaptivity regards, the ES-PIM is of unique importance for practical engineering applications. The important role of edge-based smoothing is quite clear, and the part of the reasons could be the discontinuous (or even non-existence) of the derivatives of the (assumed) field variable along the edge of the cell for displacement assumed models. When the edge-based smoothing is performed, such as discontinuity is right in the middle of the smoothing domain and is being "smoothed" out, leading to good stability

and accuracy. The edge-based smoothing seems to be right on the root of the problem of a discretized model, and hence effective. It is a crucial piece of numerical trick that are important for W^2 formulations.

6. The recently developed CS-PIM [62] has been found quite competitive to the ES-PIM. Interested readers are referred to [62] for more details.

7. SC-PIMs have opened a window of opportunity for all sorts of methods with desired properties. It shows the power of modifying the strain field. SC-PIMs have been further developed for producing "exact" solution in a norm using a finite mesh. Examples of fine-designs of strain fields for methods of ultra-accuracy and bound properties can be found in [29–31]. The room for development in this direction is indeed huge.

8. When linear interpolation is used (using the T3-Schemes), the shape functions used in PIMs are the same as the linear FEM using triangular types of meshes. Therefore, such linear PIMs (NS-PIM and ES-PIM) have also been termed as NS-FEM and ES-FEM with triangular elements. Note that all the key numerical treatments in NS-FEM and ES-FEM are very much different from the standard FEM, and the only thing in common is the use of linear shape functions created using triangular types of meshes. NS-FEM and ES-FEM are special cases of PIMs, and belong to the general W^2 weak formulation (no gradient information of the shape functions are used).

9. The PIMs are applicable to nonlinear problems. Techniques developed in FEM for nonlinear problems can be implemented (with some modification) to PIMs with ease. Examples of these applications are given in [11] for contact problems, in [47] for 3D nonlinear problems, and in [48,63] for plates and shells.

10. The PIMs have also been applied to other types of problems, such as heat transfer [54,55], thermal-elasticity problems [56], and acoustic problem [57,58].

11. The PIMs have already been made adaptive, and implemented in the MFree2D. Chapter 15 will discuss issues related to adaptive analysis with a number of examples. More examples of works in this direction are reported in [28,52,53].

12. Using the upper bound property of NS-PIM, we can develop the so-called real-time computation models with error bounds using the reduced basis method [59,60].

13. For PDEs of asymmetric operators, the author prefers the GSM, which will be discussed in Chapter 9.

In the opinion of the author, we now have at least one clear and straightforward path of four steps for the development of more effective and practical numerical methods:

1. Create a set of triangular types of mesh of triangular/tetrahedral cells (Chapter 1).

2. Using a T-Scheme to select nodes, and then construct PIM (or RPIM) shape functions (Chapters 1 and 2).

3. Construct the strain field (Chapter 4).

4. Apply a SC-Galerkin weak form to establish a set of discrete system equations (Chapters 3, 5, and 8).

Note that we often found some good features using PIMs with higher order interpolation, such as better accuracy and tighter bounds, etc., as presented earlier. However, higher order PIMs do not always perform much better than those with the linear interpolation. We think there may be three reasons: (1) The linear PIMs are already very good. (2) The quadratic interpolation performed in PIMs are obtained by simply using more nodes, and the mesh density is not changed. Therefore, the improvement is not as significant as in the quadratic FEM models. (3) The constant (Heaviside) smoothing operations used have discounted some of the effects of higher order interpolation. More studies may be needed to explore the benefits of the higher order interpolations. One way could be the use of alternative smoothing functions with higher order of reproducibility, such as the ones presented in Chapter 2.

Note that when the local Petrov–Galerkin weak form is used to formulate a PIM model, there is no issue of compatibility, because the Petrov–Galerkin is basically a local weighted residual method that does not demand assumed displacement fields with global continuity. Therefore, both PIM and RPIM shape functions work well for the local Petrov–Galerkin weak form. The drawback of this formulation is that the resultant discrete system matrix is not symmetric even for PDEs of symmetric operators affecting the computational efficiency. In addition, the bandwidth of such a model is much larger compared to the W^2 models. Overall, it is about 50 times less efficient than the PIMs based on the SC-Galerkin. Our preference is obvious.

References

1. Liu, G. R. and Gu, Y. T., A point interpolation method, in *Proceedings of 4th Asia-Pacific Conference on Computational Mechanics*, Singapore, December 1999, pp. 1009–1014.
2. Liu, G. R. and Gu, Y. T., A point interpolation method for two-dimensional solids, *Int. J. Numerical Methods Eng.*, 50, 937–951, 2001.
3. Liu, G. R., Zhang, G. Y., Dai, K. Y, Wang, Y. Y., Zhong, Z. H., Li, G. Y., and Han, X., A linearly conforming point interpolation method (LC-PIM) for 2D solid mechanics problems, *Int. J. Comput. Methods*, 2(4), 645–665, 2005.
4. Liu, G. R., Li, Y., Dai, K. Y., Luan, M. T., and Xue, W., A Linearly conforming radial point interpolation method for solid mechanics problems, *Int. J. Comput. Methods*, 3, 401–428, 2006.
5. Liu, G. R. and Gu, Y. T., A local point interpolation method for stress analysis of two-dimensional solids, *Struct. Eng. Mech.*, 11(2), 221–236, 2001.
6. Liu, G. R. and Gu, Y. T., A local radial point interpolation method (LR-PIM) for free vibration analyses of 2-D solids, *J. Sound Vibration*, 246(1), 29–46, 2001.
7. Wang, J. G. and Liu, G. R., Radial point interpolation method for elastoplastic problems, in *Proceedings of 1st International Conference on Structural Stability and Dynamics*, Taipei, China, December 7–9, 2003, pp. 703–708.
8. Liu, G. R. and Gu, Y. T., A matrix triangularization algorithm for the polynomial point interpolation method, *Comput. Methods Appl. Mech. Eng.*, 192(19), 2269–2295, 2003.
9. Liu, G. R. and Gu, Y. T., A matrix triangularization algorithm for point interpolation method, in *Proceedings of Asia-Pacific Vibration Conference*, W. Bangchun, ed., Hangzhou, China, November 2001, pp. 1151–1154.
10. Chen, J. S., Wu, C. T., Yoon, S., and You, Y., A stabilized conforming nodal integration for Galerkin mesh-free methods, *Int. J. Numerical Methods Eng.*, 50, 435–466, 2001.
11. Li, Y., Liu, G. R., Luan, M. T., Dai, K. Y., Zhong, Z. H., Li, G. Y., and Han, X., Contact analysis for solids based on linearly conforming radial point interpolation method, *Comput. Mech.*, 39, 537–554, 2007.

12. Zhang, G. Y., Liu, G. R., Nguyen, T. T., Song, C. X., Han, X., Zhong, Z. H., and Li, G. Y., The upper-bound property for solid mechanics of the linearly conforming radial point interpolation method (LC-RPIM), *Int. J. Comput. Methods*, 4(3), 521–541, 2007.

13. Liu, G. R. and Zhang, G. Y., Upper-bound solution to elasticity problems: A unique property of the linearly conforming point interpolation method (LC-PIM), *Int. J. Numerical Methods Eng.*, 74, 1128–1161, 2008.

14. Liu, G. R., Nguyen-Thoi, T., Nguyen-Xuan, H., and Lam, K. Y., A node-based smoothed finite element method (NS-FEM) for upper-bound solutions to solid mechanics problems, *Comput. Struct.* 87, 14–26, 2009.

15. Liu, G. R., Dai, K. Y., and Nguyen, T. T., A smoothed finite element method for mechanics problems, *Comput. Mech.*, 39, 859–877, 2007.

16. Liu, G. R., Nguyen, T. T., Dai, K. Y., and Lam, K. Y., Theoretical aspects of the smoothed finite element method (SFEM), *Int. J. Numerical Methods Eng.*, 71, 902–930, 2007.

17. Dohrmann, C. R., Heinstein, M. W., Jung, J., Key, S. W., and Witkowski, W. R., Node-based uniform strain elements for three-node triangular and four-node tetrahedral meshes, *Int. J. Numerical Methods Eng.*, 47, 1549–1568, 2000.

18. Zienkiewicz, O. C. and Taylor R. L., *The Finite Element Method*, 5th ed., Butterworth Heimemann, Oxford, 2000.

19. Timoshenko, S. P. and Goodier, J. N., *Theory of Elasticity*, 3rd ed., McGraw-Hill, New York, 1970.

20. Wang, J. G. and Liu, G. R., On the optimal shape parameters of radial basis functions used for 2D meshless methods, *Comput. Method Appl. Mech. Eng.*, 191, 2611–2630, 2002.

21. Wu, Y. L. and Liu, G. R., A meshfree formulation of local radial point interpolation method (LRPIM) for impressible flow simulation, *Comput. Mech.*, 30(5–6), 355–365, 2003.

22. Liu, G. R. and Gu, Y. T., *An Introduction to Meshfree Methods and Their Programming*, Springer, Berlin, 2005.

23. Zhang, G. Y., Liu, G. R., Wang, Y. Y., Huang, H. T., Zhong, Z. H., Li, G. Y., and Han, X., A linearly conforming point interpolation method (LC-PIM) for three-dimensional elasticity problems, *Int. J. Numerical Methods Eng.*, 72, 1524–1543, 2007.

24. Liu, G. R. and Liu, M. B., *Smoothed Particle Hydrodynamics—A Meshfree Practical Method*, World Scientific, Singapore, 2003.

25. Liu, G. R. and Quek, S. S., *The Finite Element Method: A Practical Course*, Butterworth Heinemann, Oxford, 2003.

26. Oliveira Eduardo, R. and De Arantes, E., Theoretical foundations of the finite element method, *Int. J. Solids Struct.*, 4, 929–952, 1968.

27. Liu, G. R., A generalized gradient smoothing technique and the smoothed bilinear form for Galerkin formulation of a wide class of computational methods, *Int. J. Comput. Methods*, 5(2), 199–236, 2008.

28. Liu, G. R., A $\mathbb{G}$ space theory and weakened weak (W^2) form for a unified formulation of compatible and incompatible methods, Part I: Theory and Part II: Applications to solid mechanics problems, *Int. J. Numerical Methods Eng.*, 2009 (in press).

29. Liu, G. R., Xu, X., Zhang, G. Y., and Nguyen, T. T., A superconvergent point interpolation method (SC-PIM) with piecewise linear strain field using triangular mesh, *Int. J. Numerical Methods Eng.*, 77, 1439–1467, 2009.

30. Liu, G. R., Xu, X., Zhang, G. Y., and Gu, Y. T., An extended Galerkin weak form and a point interpolation method with continuous strain field and superconvergence (PIM-CS) using triangular mesh, *Comput. Mech.*, 43, 651–673, 2009.

31. Xu, X., Liu, G. R., and Zhang, G. Y., A point interpolation method with least square strain field (PIM-LSS) for solution bounds and ultra-accurate solutions using triangular mesh, *Computer Methods in Applied Mechanics and Engineering*, 198, 1486–1499, 2009.

32. Liu, G. R. and Zhang, G. Y., Edge-based smoothed point interpolation method (ES-PIM), *Int. J. Comput. Methods*, 5(4), 621–646, 2008.

33. Liu, G. R. and Zhang, G. Y., A Strain-constructed point interpolation method (SC-PIM) and strain field construction schemes for mechanics problems of solids and structures using triangular mesh, *Int. J. Solids Struct.* 5(4), 621–646, 2008.

34. Liu, G. R., Zhang, J., and Lam, K. Y., A gradient smoothing method (GSM) with directional correction for solid mechanics problems, *Comput. Mech.*, 41, 457–472, 2008.

35. Liu, G. R. and Xu, G. X., A gradient smoothing method (GSM) for fluid dynamics problems, *Int. J. Numerical Methods Fluids*, 56(10), 1101–1133, 2008.

36. Stroud, A. H. and Secrest, D., *Gaussian Quadrature Formulas*, Prentice-Hall, Englewood Cliffs, NJ, 1966.

37. Liu, G. R. and Gu, Y. T., A local point interpolation method for stress analysis of two-dimensional solids, *Struct. Eng. Mech.*, 11(2), 221–236, 2001.

38. Liu, G. R. and Kee, B. B. T., A stabilized least-squares radial point collocation method (LS-RPCM) for adaptive analysis, *Comput. Method Appl. Mech. Eng.*, 195, 4843–4861, 2006.

39. Kee, B. B. T., Liu, G. R., and Lu, C., A regularized least-squares radial point collocation method (RLS-RPCM) for adaptive analysis, *Comput. Mech.*, 40, 837–853, 2007.

40. Liu, G. R. and Gu, Y. T., A local radial point interpolation method (LR-PIM) for free vibration analyses of 2-D solids, *J. Sound Vibration*, 246(1), 29–46, 2001.

41. Xu, X. G. and Liu, G. R., A local-function approximation method for simulating two-dimensional incompressible flow, in *Proceedings of 4th International Asia-Pacific Conference on Computational Mechanics*, Singapore, December 1999, pp. 1021–1026.

42. Nagashima, T., Node-by-node meshless approach and its application to structural analyses, *Int. J. Numerical Methods Eng.*, 46, 341–385, 1999.

43. Brebbia, C. A. and Georgiou, P., Combination of boundary and finite elements in elastostatics, *Appl. Math. Model.*, 3, 212–219, 1979.

44. Liu, G. R., Yan, L., Wang, J. G., and Gu, Y. T., Point interpolation method based on local residual formulation using radial basis functions, *Struct. Eng. Mech.*, 14(6), 713–732, 2002.

45. Huang, W. X., *The Engineering Property of Soil*, Hydraulic & Electric Press, Beijing, 1983 (in Chinese).

46. Liu, G. R., Nguyen-Thoi, T., and Lam, K. Y., An edge-based smoothed finite element method (ES-FEM) for static, free and forced vibration analyses in solids, *J. Sound Vibration*, 320, 1100–1300, 2009.

47. Nguyen-Thoi, T., Liu, G. R., Lam, K. Y., and Zhang, G. Y., A Face-based Smoothed Finite Element Method (FS-FEM) for 3D linear and nonlinear solid mechanics problems using 4-node tetrahedral elements, *Int. J. Numerical Methods Eng.*, 78, 324–353, 2009.

48. Cui, X. Y., Liu, G. R., Li, G. Y., and Zhang, G., Analysis of plates and shells using an edge-based smoothed finite element method, *Computational Mechanics* (submitted). 2009.

49. Nguyen-Thoi, T., Liu, G. R., Nguyen-Xuan, H., and Lam, K. Y., An n-sided polygonal edge-based smoothed finite element method (nES-FEM) for solid mechanics, *Finite Elem. Anal. Des.* (submitted). 2008.

50. Hughes, T. J. R., *The Finite Element Method: Linear Static and Dynamic Finite Element Analysis*, Prentice-Hall, Englewood Cliffs, NJ, 1987.

51. Liu, G. R. and Quek, S. S. *The Finite Element Method: A Practical Course*, Butterworth Heinemann, Oxford, 2002.

52. Zhang, G. Y., Liu, G. R., and Li, Y., An efficient adaptive analysis procedure for certified solutions with exact bounds of strain energy for elasticity problems, *Finite Elem. Anal. Des.*, 44, 831–841, 2008.

53. Chen, L., Zhang, G. Y., and Zeng, K. Y., An adaptive analysis using the edge-based smoothed point interpolation methods (ES-PIM). *Engineering analysis with boundary elements*, 2009, (submitted).

54. Wu, S. C., Liu, G. R., Zhang, H. O., and Zhang, G. Y., A node-based smoothed point interpolation method (NS-PIM) for thermoelastic problems with solution bounds, *Int. J. Heat Mass Transfer*, 52(56), 1464–1471, 2009.

55. Wu, S. C., Liu, G. R., Zhang, H. O., and Zhang, G. Y., A node-based smoothed point interpolation method (NS-PIM) for three-dimensional thermoelastic problems, *Numerical Heat Transfer, Part A: Appl.*, 54(12), 1121–1147, 2008.

56. Wu, S. C., Liu, G. R., Zhang, H. O., Zhang, G. Y., Xu, X., and Li, Z. R., A node-based smoothed point interpolation method (NS-PIM) for three-dimensional heat transfer problems, *Int. J. Thermal Sci.*, 48(6), 1154–1165, 2009.
57. He, Z. C., Liu, G. R., Zhong, Z. H., Wu, S. C., and Zhang, G. Y., An edge-based smoothed finite element method (ES-FEM) for analyzing three-dimensional acoustic problems, *Comput. Methods Appl. Mech. Eng.* (revised), 2008.
58. He, Z. C., Liu, G. R., Zhong, Z. H., and Zhang, G. Y., Enclosure acoustic analysis using edge-based smoothed finite element method (ES-FEM), *Eng. Anal. Boundary Elem.* (submitted), 2008.
59. Liu, G. R., Khin Zaw, Wang, Y. Y., and Deng, B., A novel reduced-basis method with upper and lower bounds for real-time computation of solids mechanics problem, *Comput. Methods Appl. Mech. Eng.*, 198, 269–279, 2008.
60. Khin Zaw, Liu, G. R., Deng, B., and Tan, K. B. C., Rapid identification of elastic modulus of the interface tissue on dental implants surfaces using reduced-basis method and neural network, *J. Biomech.* 42, 634–641, 2009.
61. Liu, G. R. and Zhang, G. Y., A normed $\mathbb{G}$ space and weakened week (W^2) formulation of a cell-based smoothed point interpolation method, *Int. J. Comput. Methods*, 6(1), 147–149, 2009.
62. Liu, G. R., On $\mathbb{G}$ space theory, *Int. J. Comput. Methods*, 6(2), 1–33, 2009.
63. Cui, X. Y., Liu, G. R., Li, G. Y., and Zhang, G. Y., A cell-based smoothed radial point interpolation method (CS-RPIM) for static and free vibration of solids. *Engineering Analysis with Boundary Elements* (submitted). 2009.

9

Meshfree Methods for Fluid Dynamics Problems

9.1 Introduction

Previous chapters have discussed a number of meshfree methods to solve the problems of solid mechanics. Meshfree methods can also be applied to problems of fluid mechanics because they basically provide a means of discretizing partial differential equations (PDEs) in the spatial domain. This chapter discusses some of the meshfree methods that suit well and have been applied to solve computational fluid mechanics problems.

Simulation and analysis of problems of fluid dynamics have been generally performed using numerical methods such as the finite difference method (FDM), the finite volume method (FVM), and the finite element method (FEM). These numerical methods have been widely applied to practical problems and have dominated the subject of computational fluid dynamics (CFD). An important feature of these methods is that a corresponding Eulerian (for FDM and FVM) grid or a Lagrangian (for some FEM formulations) grid or both are required as a computational frame to solve the governing equations. When simulating some special problems with large distortions and moving material interfaces, deformable boundaries, and free surfaces, these methods can encounter many difficulties. Although many numerical schemes for the solution of fluid dynamics problems have emerged, difficulties still exist for problems with the above-mentioned features. Attempts have also been made to combine the good features of FDM, FVM, and FEM, and the two-grid systems of Lagrangian grid and Eulerian grid have been used [1]. Computational information is exchanged either by mapping or by special interface treatment between these two grids. This approach is rather complicated and can also cause problems related to stability and accuracy. The search for better methods and techniques is still going on.

Meshfree methods or techniques can offer some promising alternatives for solving problems of CFD. The most attractive feature of the meshfree methods is that there is no need for a mesh or that there is less reliance on the quality of the mesh. When mesh is used, a triangular mesh will often suffice. This opens a new opportunity to solve the above-mentioned problems and to conduct adaptive analyses, as it is demonstrated in this chapter. There are basically four types of methods that have been explored for CFD problems:

1. Integral representation methods such as the smoothed particle hydrodynamics (SPH) method [2–8] and the reproducing kernel particle methods [9]. The SPH method is a Lagrangian formulation and is one of the best choices for highly nonlinear, fast dynamic, multiphase, and momentum-driven types of problems. It uses particles and no mesh is needed during the computation.

2. Gradient smoothing method (GSM) is an Eulerian formulation for general CFD problems. It uses background triangular cells and hence is not entirely meshless. Its numerical operations are beyond the cells and carefully designed for excellent stability, efficiency, and accuracy. It works well with highly irregular triangular cells for both compressible and incompressible fluids [12,13]. Because of its excellent stability and efficiency, it has already been implemented for adaptive analysis [14].

3. Series representation methods including the meshless Petrov–Galerkin (MLPG) method [10,11] and the meshfree weak–strong (MWS) form method [15].

4. Differential representation methods including the finite point method [16] and the FDM with arbitrary irregular grids [17–20].

This chapter deals with the first two of these methods—SPH (Lagrangian) and GSM (Eulerian). For other methods, readers are referred to the references given above.

9.2 Navier–Stokes Equations

The strong form of governing equations for fluid dynamics problems are the well-known Navier–Stokes equations. They have two forms of expressions: Lagrangian and Eulerian forms. The SPH formulation deals with the Lagrangian form and the GSM deals with those of Eulerian form of Navier–Stokes equations. We now provide a brief description of these two forms of Navier–Stokes equation, which will be used in this chapter.

9.2.1 Lagrangian Form of Navier–Stokes Equations

In the following expressions, the Greek superscripts α and β are used to denote the coordinate directions. For fluid dynamics problems, the *continuity* equation for the fluid media can be given as

$$\frac{D\rho}{Dt} = -\rho \nabla \cdot \mathbf{v} \tag{9.1}$$

where
 $\mathbf{v}$ is the velocity vector
 ρ is the density of the fluid, where D/Dt is the *total time derivative* (or *material derivative*, or *global derivative*)
 ∇ is the gradient operator
 t is the time

The *momentum* equation is

$$\frac{Dv^{\alpha}}{Dt} = \frac{1}{\rho} \frac{\partial \sigma^{\alpha\beta}}{\partial x^{\beta}} + F^{\alpha} \tag{9.2}$$

where
 v is a component of velocity
 F is the external body force per unit mass
 σ is the total internal stress

The total internal stress is made up of two parts, one part comes from the isotropic pressure p and the other part is the viscous stress τ; i.e.,

$$\sigma^{\alpha\beta} = -p\delta^{\alpha\beta} + \tau^{\alpha\beta} \tag{9.3}$$

where the Kronecker operator is denoted by δ. The different definitions of the viscous stress lead to different SPH applications. For Newtonian fluids, the viscous stress is assumed to be proportional to the strain rate denoted by ε through the dynamic viscosity μ; i.e.,

$$\tau^{\alpha\beta} = \mu\varepsilon^{\alpha\beta} \tag{9.4}$$

where the strain rate, with the use of Stokes' hypothesis, takes the form

$$\varepsilon^{\alpha\beta} = \frac{\partial v^\beta}{\partial x^\alpha} + \frac{\partial v^\alpha}{\partial x^\beta} - \frac{2}{3}(\nabla \cdot \mathbf{v})\delta^{\alpha\beta} \tag{9.5}$$

The time rate of change of the specific total energy e comes from three parts: the work done by isotropic pressure multiplying the volumetric strain, the energy dissipation due to viscous forces, and the rate of heat lost by conduction. The first law of thermodynamics gives the energy equation as

$$\frac{De}{Dt} = -\frac{p}{\rho}\nabla \cdot \mathbf{v} + \frac{1}{\rho}\frac{\partial v^\beta}{\partial x^\alpha}\tau^{\alpha\beta} + \frac{1}{\rho}\frac{\partial}{\partial x^\alpha}\left(k\frac{\partial T}{\partial x^\alpha}\right) \tag{9.6}$$

where T and k denote, respectively, the temperature in the fluid and the thermal conductivity coefficient of the fluid media.

To close the loop of these equations, the equation of state (EOS) needs to be included for compressible flows. It relates the pressure and the temperature to the primitive variables of density, velocity, and energy in the form of

$$p = (\gamma - 1)\rho\left(e - \frac{1}{2}v^\alpha v^\alpha\right)$$
$$T = \gamma\left(e - \frac{1}{2}v^\alpha v^\alpha\right)\Big/ c_p \tag{9.7}$$

where c_p and γ denote, respectively, the specific heat at constant pressure and the specific heat coefficient of the fluid media.

9.2.2 Eulerian Form of Navier–Stokes Equations

The GSM uses the Eulerian form of the Navier–Stokes equations. They can be given using the relationship between the *total time derivative*, the *local derivative*, and the *convective derivative*:

$$\frac{D}{Dt} = \frac{\partial}{\partial t} + v^\alpha\frac{\partial}{\partial x^\alpha} \tag{9.8}$$

where
   $\partial/\partial t$ is the local derivative
   $v^\alpha \, \partial/\partial x^\alpha$ is the convective derivative

In the Eulerian form, the *continuity* equation for the fluid media can be given as

$$\frac{\partial \rho}{\partial t} + v^\alpha \frac{\partial \rho}{\partial x^\alpha} = -\rho \nabla \cdot \mathbf{v} \tag{9.9}$$

The *momentum* equation is

$$\frac{\partial v^\alpha}{\partial t} + v^\beta \frac{\partial v^\alpha}{\partial x^\beta} = \frac{1}{\rho} \frac{\partial \sigma^{\alpha\beta}}{\partial x^\beta} + F^\alpha \tag{9.10}$$

The energy equation becomes

$$\frac{\partial e}{\partial t} + v^\alpha \frac{\partial e}{\partial x^\alpha} = -\frac{p}{\rho} \nabla \cdot \mathbf{v} + \frac{1}{\rho} \frac{\partial v^\beta}{\partial x^\alpha} \tau^{\alpha\beta} + \frac{1}{\rho} \frac{\partial}{\partial x^\alpha} \left(k \frac{\partial T}{\partial x^\alpha} \right) \tag{9.11}$$

All the other equations are the same as those given in Section 9.2.1.

Navier–Stokes equations are quite general and hence complicated to solve. Proper numerical methods are therefore needed to obtain approximated solutions. On the other hand, simplification of these equations can be performed based on the types of physics problems. For example, for inviscid fluid flows, the so-called *Euler equations* are used. They have also both Lagrangian and Eulerian forms, but can be easily obtained by simply removing the terms related to the dynamic viscosity (thus need not to be repeated here). For problems that are fast dynamic in nature, the slow heat conduction may not be needed to be included in these equations (or solved separately), and these equations can be further simplified. This kind of simplification is applied across all engineering fields. It is important not only for better efficiency in obtaining the numerical solutions but also for the accuracy of the solution. We generally do not want physical phenomena that are too much different in scales to be mixed together in one set of coupled equations because it can often lead to very bad conditioning in the discretized system equations. On the other hand, too much simplification reduces the generality of the numerical methods. The numerical methods discussed here are for the general Navier–Stokes equations as well as the Euler equations. For fast dynamic problems, we may simply drop the heat conduction terms.

9.3 Smoothed Particle Hydrodynamics Method

SPH was developed and advanced by Lucy [2] and Gingold and Monaghan [3] to solve astrophysical problems in three-dimensional (3D) open space. Since its invention, SPH has been heavily studied and extended to dynamic response with material strength by Libersky and Petscheck [21,22] and Johnson et al. [23,24], fracture simulation [25], impact simulation [26,27], brittle solids [28], and metal-forming simulation [29]. SPH has also been explored for simulating dynamic fluid flows with large distortions [30], explosion processes [31], underwater shock [32–34], as well as other CFD problems [31,34–43]. Many issues related to SPH can be found in publications by Monaghan and coworkers [7,8,44–51]. A monograph detailing many of the key SPH techniques is also available [5]. As a meshfree, particle method of pure Lagrangian nature, SPH uses smoothed particles as interpolation points to represent materials at discrete locations, so it can easily trace material interfaces, free surfaces, and moving boundaries. The meshfree nature of SPH overcomes the difficulties due to large deformations because SPH uses particles or points rather than a mesh as a computational frame to perform approximation. These nice features of SPH make it fairly attractive, as can be seen from the large literature that has emerged during the last decade.

This section presents an implementation of the SPH to solve the Navier–Stokes equations for fluid dynamics applications. As the present SPH formulations are based on the Navier–Stokes equations, physical viscosity needs to be modeled. Some modifications and improvements in numerical techniques such as the smoothing kernel function, smoothing length, nearest neighboring particle searching, treatment of solid boundaries, and artificial compressibility are made to suit the needs of simulating dynamic fluid flows. The presented SPH implementation can simulate different flow scenarios such as inviscid or viscous flows, compressible or incompressible flows. Our SPH code is then applied to solve different fluid flow problems, including incompressible flows with solid boundaries, free surface flows, and complex compressible flow in explosion. Numerical examples show that the present SPH method can simulate these problems fairly well at reasonable accuracy with less computational effort. It is an effective addition and alternative to traditional numerical methods.

This chapter can only provide an abstracted version of SPH method. A much more comprehensive description of SPH is given in [5].

9.3.1 SPH Basics

Fundamental to SPH is the theory of integral representation of functions, which is discussed in the first two sections of Chapter 2. In SPH convention, the integral representation of function is often termed as *kernel approximation*.

In SPH implementation, the state of a system can be represented by a collection of arbitrarily distributed particles while forces are calculated through interparticle interactions in a smoothed fashion. These particles move freely in space, carry all the computational information, and thus can be regarded as interpolation points or field nodes, which form the computational frame for spatial discretization in solving PDEs. There are basically two steps in a SPH procedure:

1. *Kernel approximation.* Integration of the field variable functions multiplied by a given smoothing kernel function gives the kernel approximation of the function.
2. *Particle approximation.* Summation of the function values of the neighboring particles associated with a particle yields the particle approximation of the function at the particle.

As detailed in Chapter 2, a function f is approximated by multiplying f with a smoothing kernel function and then integrated over the smoothing domain, which is known as the integral representation or the kernel approximation of a function. Here we follow the conventional SPH notation to derive the system questions for CFD problems. The kernel approximation of f is denoted as $<f>$ and written in the form

$$\langle f(\mathbf{x}) \rangle = \int f(\boldsymbol{\xi}) \widehat{W}(\mathbf{x} - \boldsymbol{\xi}, h) d\boldsymbol{\xi} \qquad (9.12)$$

where
  $\mathbf{x}$ and $\boldsymbol{\xi}$ are the position vectors at different points
  $\widehat{W}(\mathbf{x} - \boldsymbol{\xi}, h)$ is termed as the (locally supported) smoothing function or smoothing kernel function in SPH or the weight function in general

The smoothing function is prescribed with the conditions listed in Equations 2.8 through 2.12. In Equation 9.12, h is the smoothing length representing the effective width of the smoothing kernel function, which is equivalent to the dimension of the support domain in the EFG (Chapter 6), MLPG (Chapter 7), and PIMs (Chapter 8). The determination and updating of h during the computation in SPH settings will be discussed in detail later.

In the discrete form, the smoothing kernel function is given by

$$\widehat{W}_{ij} = \widehat{W}(\mathbf{x}_i - \mathbf{x}_j, h) = \widehat{W}(|\mathbf{x}_i - \mathbf{x}_j|, h) \qquad (9.13)$$

$$\nabla_i \widehat{W}_{ij} = \frac{\mathbf{x}_i - \mathbf{x}_j}{r_{ij}} \frac{\partial \widehat{W}_{ij}}{\partial r_{ij}} = \frac{\mathbf{x}_{ij}}{r_{ij}} \frac{\partial \widehat{W}_{ij}}{\partial r_{ij}} \qquad (9.14)$$

where r_{ij} is the distance between particles i and j. Equation 9.12 is then discretized into a form of summation over all the nearest neighboring particles that are within the smoothing length h for a given particle i. In such a particle approximation, we shall have (locally)

$$\langle f_i \rangle = \sum_{j=1}^{N} \left(\frac{m_j}{\rho_j} \right) \times f_j \times \widehat{W}_{ij} \qquad (9.15)$$

where
  $f_j = f(\mathbf{x}_j)$
  m_j and ρ_j are the mass and density of particle j
  m_j/ρ_j is the volume associated with particle j
  N is the total number of the neighboring particles in the local support/smoothing domain of particle i

The approximation of spatial derivatives of the function of field variable can also be obtained in the same way in terms of the function values at particles, and is derived simply through the integration by parts to transform the differential operation on function f into the given smoothing kernel function. To this end, we have

$$\langle \nabla f_i \rangle = \sum_{j=1}^{N} \left(\frac{m_j}{\rho_j} \right) f_j \nabla_i \widehat{W}_{ij} \qquad (9.16)$$

From Equations 9.12, 9.15, and 9.16, the values of a function f and its spatial derivatives can be approximated over a collection of smoothed particles rather than over a mesh. This is the essence of the SPH method and the difference between the SPH method and the traditional numerical methods of FDM, FVM, and FEM.

The above equations have shown that the SPH is a very simple and straightforward numerical procedure to represent a function and its derivatives in a discrete form.

The following will introduce its applications to fluid dynamics problems governed by the Lagrangian form of Navier–Stokes equations.

9.3.2 SPH Formulation for Navier–Stokes Equations

The standard SPH method was generally used to solve the Euler equations for inviscid flows, since it is not easy to obtain the SPH expressions of the second derivatives for the physical viscous term in the general Navier–Stokes equations. Many techniques have been proposed in recent years to treat the physical viscosity. Monaghan [50] employed an SPH approximation of the viscous term to model heat conduction, which seems to work well for low velocity flows. Takeda et al. [52] directly used the second-order derivative of the smoothing kernel, which seems to work well for constant viscosity. Flebbe et al. [53] obtained an SPH expression for the physical viscosity, using a nested sum over concerned particles. Although the density evolution many be questionable [54], this approach in treating the physical viscosity is quite straightforward. In the present SPH implementation, we retain this simplicity merit in treating physical viscosity and try to improve the density evolution.

9.3.2.1 Density Evolution

There are two methods to evolve density in the standard SPH. The first one is the *density summation* method, which directly approximates the density of a given particle by simply substituting f in Equation 9.15 with ρ, and then summing over the neighboring particles within the effective width of the smoothing kernel; i.e.,

$$\langle \rho \rangle_i = \sum_{j=1}^{N} m_j \widehat{W}_{ij} \tag{9.17}$$

This implies that the density at particle i is approximated by a weighted average of those of the neighboring particles.

The second approach is to evolve the density from the continuity Equation 9.1, and after some simple transformation, we arrive at the following *continuity density* form:

$$\left\langle \frac{D\rho}{Dt} \right\rangle_i = \sum_{j=1}^{N} m_j (\mathbf{v}_i - \mathbf{v}_j) \cdot \nabla_i \widehat{W}_{ij} \tag{9.18}$$

There are advantages and disadvantages for both approaches. The density summation approach conserves the mass exactly, while the continuity density approach does not. However, the density summation approach has an edge effect when applied to particles at the domain-edge of the fluid, leading to spurious results. Another disadvantage of the density summation approach is that it requires more computational effort because the density must be evaluated before other parameters can be approximated.

In the present formulation, we implement both. For problems in which mass conservation plays a significant role, the density summation approach is used. For flows with free surfaces, a compromise is made to employ the continuity density approach for particles within the smoothing area of λh from the free surfaces (λ determines the actual dimension of the smoothing kernel and is described later), while other particles still use the density summation approach. This minimizes the edge effect of the density summation approach and the mass nonconservation of the continuity density approach.

9.3.2.2 Momentum Evolution

From Equation 9.2, the SPH approximation of the momentum equation for particle i can be written as

$$
\begin{aligned}
\left\langle \frac{Dv^\alpha}{Dt} \right\rangle_i &= \left\langle \frac{1}{\rho} \frac{\partial \sigma^{\alpha\beta}}{\partial x^\beta} \right\rangle_i + F_i^\alpha \\
&= \frac{1}{\rho_i} \left\langle \frac{\partial \sigma^{\alpha\beta}}{\partial x^\beta} \right\rangle_i + F_i^\alpha \\
&= \frac{1}{\rho_i} \sum_{j=1}^{N} m_j \frac{\sigma_j^{\alpha\beta}}{\rho_j} \frac{\partial \widehat{W}_{ij}}{\partial x_i^\beta} + F_i^\alpha
\end{aligned}
\tag{9.19}
$$

In order to symmetrize Equation 9.19, the following identity equation

$$
\sum_{j=1}^{N} m_j \frac{\sigma_i^{\alpha\beta}}{\rho_i \rho_j} \frac{\partial \widehat{W}_{ij}}{\partial x_i^\beta} = \frac{\sigma_i^{\alpha\beta}}{\rho_i} \sum_{j=1}^{N} \frac{m_j}{\rho_j} \frac{\partial \widehat{W}_{ij}}{\partial x_i^\beta} = \frac{\sigma_i^{\alpha\beta}}{\rho_i} \left\langle \frac{\partial 1}{\partial x^\beta} \right\rangle_i = 0
\tag{9.20}
$$

is used to obtain

$$
\left\langle \frac{Dv^\alpha}{Dt} \right\rangle_i = \sum_{j=1}^{N} m_j \frac{\sigma_i^{\alpha\beta} + \sigma_j^{\alpha\beta}}{\rho_i \rho_j} \frac{\partial \widehat{W}_{ij}}{\partial x_i^\beta} + F_i^\alpha
\tag{9.21}
$$

By substituting Equation 9.3 into Equation 9.21, the discretized moment equation can be written as

$$
\left\langle \frac{Dv^\alpha}{Dt} \right\rangle_i = -\sum_{j=1}^{N} m_j \frac{p_i + p_j}{\rho_i \rho_j} \frac{\partial \widehat{W}_{ij}}{\partial x_i^\alpha} + \sum_{j=1}^{N} m_j \frac{\mu_i \varepsilon_i^{\alpha\beta} + \mu_j \varepsilon_j^{\alpha\beta}}{\rho_i \rho_j} \frac{\partial \widehat{W}_{ij}}{\partial x_i^\beta} + F_i^\alpha
\tag{9.22}
$$

The first part of the right-hand side of Equation 9.22 is the standard SPH expression for pressure. It is the second part that concerns the physical viscosity. By using Equation 9.5, the SPH approximation of $\varepsilon^{\alpha\beta}$ for particle i can be approximated as

$$
\langle \varepsilon^{\alpha\beta} \rangle_i = \sum_{j=1}^{N} \frac{m_j}{\rho_j} v_j^\beta \frac{\partial \widehat{W}_{ij}}{\partial x_i^\alpha} + \sum_{j=1}^{N} \frac{m_j}{\rho_j} v_j^\alpha \frac{\partial \widehat{W}_{ij}}{\partial x_i^\beta} - \left(\frac{2}{3} \sum_{j=1}^{N} \frac{m_j}{\rho_j} v_j \cdot \nabla_i \widehat{W}_{ij} \right) \delta^{\alpha\beta}
\tag{9.23}
$$

The following identities are subtracted from Equation 9.23:

$$
\sum_{j=1}^{N} \frac{m_j}{\rho_j} v_i^\beta \frac{\partial \widehat{W}_{ij}}{\partial x_i^\alpha} = v_i^\beta \left\langle \frac{\partial 1}{\partial x^\alpha} \right\rangle_i = 0
\tag{9.24}
$$

$$
\sum_{j=1}^{N} \frac{m_j}{\rho_j} v_i^\alpha \frac{\partial \widehat{W}_{ij}}{\partial x_i^\beta} = v_i^\alpha \left\langle \frac{\partial 1}{\partial x^\beta} \right\rangle_i = 0
\tag{9.25}
$$

$$
\sum_{j=1}^{N} \frac{m_j}{\rho_j} \mathbf{v}_i \cdot \nabla_i \widehat{W}_{ij} = \mathbf{v}_i \cdot \langle \nabla 1 \rangle_i = 0
\tag{9.26}
$$

We obtain the final SPH approximation of $\varepsilon^{\alpha\beta}$ for particle i as

$$\langle\varepsilon^{\alpha\beta}\rangle_i = \sum_{j=1}^{N} \frac{m_j}{\rho_j} v_{ji}^{\beta} \frac{\partial \widehat{W}_{ij}}{\partial x_i^{\alpha}} + \sum_{j=1}^{N} \frac{m_j}{\rho_j} v_{ji}^{\alpha} \frac{\partial \widehat{W}_{ij}}{\partial x_i^{\beta}} - \left(\frac{2}{3} \sum_{j=1}^{N} \frac{m_j}{\rho_j} \mathbf{v}_{ji} \cdot \nabla_i \widehat{W}_{ij} \right) \delta^{\alpha\beta} \qquad (9.27)$$

which relates the velocity differences (v_{ji}) to the viscous strain rate and stress.

The SPH approximation of $\varepsilon^{\alpha\beta}$ for particle j can be obtained in a similar way by summing over the neighboring particles of j. After $\varepsilon^{\alpha\beta}$ for particles i and j have been calculated, the acceleration can by calculated be Equation 9.22. This approach is straightforward and can model variable viscosity and viscosities for different fluids.

9.3.2.3 Energy Equation

The SPH formulation for the discretized energy equation can be obtained by following a procedure similar to the momentum equation. The time rate of change of the total internal energy e for a particle i can be calculated once ε has been calculated:

$$\left\langle \frac{De}{Dt} \right\rangle_i = \frac{1}{2} \sum_{j=1}^{N} m_j \frac{p_i + p_j}{\rho_i \rho_j} \mathbf{v}_{ij} \cdot \nabla_i \widehat{W}_{ij} + \frac{\mu_i}{2\rho_i} \varepsilon_i^{\alpha\beta} \varepsilon_i^{\alpha\beta} \qquad (9.28)$$

Here we do not consider the slow phenomena of heat conduction. The procedure for solving this set of discretized system equations of SPH is rather standard and quite straightforward. However, the following implementation issues are very important to make SPH work. Readers interested in the coding of SPH are referred to Chapter 4 of [5] for more detailed implementation issues.

9.3.3 Major Numerical Implementation Issues

9.3.3.1 Smoothing Kernel

The smoothing kernel function is important in the SPH method because it determines the pattern of approximation for all the field variables, and the dimension of the influence area of a particle. The kernel function generally needs to satisfy conditions listed in Equations 2.3 through 2.7. In our SPH settings, however, Equation 2.4 is often written in the following form:

$$\widehat{W}(\mathbf{x} - \xi) = 0 \quad \text{for } |\mathbf{x} - \xi| > \lambda h \qquad (9.29)$$

where
  h is the smoothing length
  λ is a constant controlling the actual dimension of the smoothing domain

The most widely used smoothing kernel functions are the cubic and quartic spline functions listed in Section 2.2.2. For example, the cubic spline is given (in SPH convention) by

$$\widehat{W}_{ij}(S, h) = \alpha_D \times \begin{cases} \dfrac{2}{3} - S^2 + \dfrac{1}{2}S^3, & 0 \leq S \leq 1 \\ \dfrac{1}{6}(2 - S)^3, & 1 \leq S \leq 2 \\ 0, & S \geq 2 \end{cases} \tag{9.30}$$

where

$S = r_{ij}/h$, with r_{ij} the distance between two particles i and j

α_D is a factor depending on the dimension of the problem

For two-dimensional (2D) problems, $\alpha_D = 15/7\pi/h^2$. Note $\lambda = 2.0$ is used in the cubic spline kernel given in Equation 9.30.

The derivative of the cubic spline kernel can be easily obtained for $\lambda = 2.0$ as follows:

$$\widehat{W}'_{ij} = \alpha_D \times \begin{cases} \dfrac{1}{h}\left(-2S + \dfrac{3}{2}S^2\right), & 0 \leq S \leq 1 \\ -\dfrac{1}{2h}(2 - S)^2, & 1 \leq S \leq 2 \\ 0, & S \geq 2 \end{cases} \tag{9.31}$$

The shapes of this kernel function and its derivative are plotted in Figures 9.1 and 9.2. Figure 9.1 shows that the value of this cubic spline function increases as the two particles approach each other and it makes sense naturally: the closer the two neighboring particles, the greater mutual influence. However, its first derivative of the cubic spline function has its maximum value at the point $S = 2/3$, as shown in Figure 9.2. The first derivative decreases for $S < 2/3$ as the distance between the two particles decreases. This seems unnatural. This unnatural behavior of the cubic spline function sometimes can cause instability [55].

The SPH quadratic spline kernel function formulated for $\lambda = 2.0$ can be written in the form of [23]

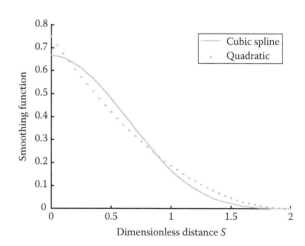

FIGURE 9.1

SPH smoothing functions $(\widetilde{W}_{ij}\alpha_D)$.

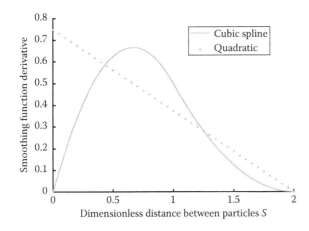

FIGURE 9.2
Derivatives of SPH smoothing functions $(W'_{ij}\bar{h}\alpha_D)$.

$$\widehat{W}_{ij} = \alpha_D \left[\frac{3}{16} S^2 - \frac{3}{4} S + \frac{3}{4} \right], \quad 0 \leq S \leq 2 \tag{9.32}$$

$$\widehat{W}'_{ij} = \alpha_D \left[\frac{3}{8} S - \frac{3}{4} \right], \quad 0 \leq S \leq 2 \tag{9.33}$$

where the dimension-dependent factor $\alpha_D = 2/\pi h^2$ for 2D cases. The quadratic smoothing function and its derivative are also shown in Figures 9.1 and 9.2. These figures show that a larger distance between two neighboring particles always leads to a smaller function value, as well as a smaller first derivative of the function value, and thus gives rise to less mutual influence. The problem with the quadratic smooth function is that it has a lower order of reproduction of functions.

The cubic spline smoothing functions and the quadratic smoothing functions in one or three dimensions behave in the same manner as their counterparts in two dimensions except for the difference in the dimension-dependent factor α_D, which can be determined by imposing the zero-order reproduction condition (or normalization condition) (Equation 2.10).

In our implementation of SPH, we coded with both cubic spline and quadratic smoothing functions. The cubic spline function is usually used, as it works well for "common" problems. For flows that may involve tensile instability, we switch to the quadratic smoothing function.

9.3.3.2 Artificial Viscosity

In the standard SPH expressions, the artificial viscosity is used to resolve shocks numerically and to prevent nonphysical particle penetration. The artificial viscosity term is added to the pressure term in the momentum and energy equation. The most commonly used artificial viscosity is

$$\Pi_{ij} = \begin{cases} \dfrac{-\alpha \bar{c}_{ij} \theta_{ij} + \beta \theta_{ij}^2}{\bar{\rho}_{ij}}, & \mathbf{v}_{ij} \cdot \mathbf{x}_{ij} < 0 \\ 0, & \mathbf{v}_{ij} \cdot \mathbf{x}_{ij} \geq 0 \end{cases} \tag{9.34}$$

where the parameters in the above equation are given by

$$\theta_{ij} = \frac{h_{ij}\mathbf{v}_{ij} \cdot \mathbf{x}_{ij}}{r_{ij}^2 + \eta^2} \tag{9.35}$$

$$\bar{c}_{ij} = \frac{1}{2}(c_i + c_j) \tag{9.36}$$

$$\bar{\rho}_{ij} = \frac{1}{2}(\rho_i + \rho_j) \tag{9.37}$$

$$h_{ij} = \frac{1}{2}(h_i + h_j) \tag{9.38}$$

$$\mathbf{v}_{ij} = \mathbf{v}_i - \mathbf{v}_j \tag{9.39}$$

In Equations 9.34 and 9.35, α, β, and η are constants that are typically set around 1, 1, and $0.1h_{ij}$, and c_i and c_j represent the speed of sound for particle i and j, respectively. The first term in Π_{ij} is similar to the Navier–Stokes shear and bulk viscosity, while the second term is similar to the Von Neumann–Richtmyer viscosity in FEM. The second term is very important in preventing nonphysical particle penetration, especially for particles that are approaching each other at high speed and almost head-on.

In our implementation, since the Navier–Stokes-based SPH formulation can resolve the general physical shear and bulk viscosity, it is not necessary to have the first term in Π_{ij}. However, the second term must be retained to prevent nonphysical particle penetration. This is different from the approach in [56], where the whole artificial viscosity is added to the pressure term in the corresponding SPH equations.

9.3.3.3 Artificial Compressibility

In standard SPH, for solving compressible flows, the particle motion is driven by the pressure gradient, while the particle pressure is calculated by the local particle density and internal energy through the equation of state. However, for incompressible flows, there is no equation of state for pressure. Moreover, the actual equation of state of the fluid will lead to prohibitive time steps. Although it is possible to include the constraint of the constant density in the SPH formulations, the resultant equations are too cumbersome to be solved.

In this implementation, the artificial compressibility technique is used. It is based on the fact that every incompressible fluid is in reality compressible and, therefore, it is feasible to use a quasi-incompressible equation of state to model the incompressible flow. The purpose of introducing the artificial compressibility is to produce a time derivative of pressure. Monaghan [47] applied the following equation of state for water to model free surface flow:

$$p = B\left[\left(\frac{\rho}{\rho_0}\right)^\gamma - 1\right] \tag{9.40}$$

where
   the constant $\gamma = 7$ is used in most circumstances
   ρ_0 is the reference density
   B is a problem-dependent parameter that exerts a limit for the maximum change

In the artificial compressibility technique, the sound speed should be much larger than the maximum speed of the bulk flow. The subtraction of 1 in Equation 9.40 can remove the boundary effect for free surface flows. It can be seen that small oscillation in density may result in large variation in pressure. Therefore, by employing the proper equation of state for the concerned fluid, this artificial compressibility technique models an incompressible flow as a slightly compressible fluid.

9.3.3.4 Smoothing Length

The smoothing length h is significant in SPH and has a direct influence on the efficiency and accuracy. If h is too small, there may not be enough particles in the designated smoothing range of λh to exert forces on the particle concerned. This lack of influence will result in low accuracy of the numerical solution. If the smoothing length is too large, all details of the particle or local properties may be smoothed out. This oversmoothing will also affect the accuracy. Various forms of smoothing length [57–59] have been suggested. They are generally problem dependent and usually not suitable for general CFD problems.

The smoothing length is directly related to fluid density. In different flow problems, the fluid density differs dramatically. To make SPH more adaptive to various flows, such as flows with large density inhomogeneity as well as those with uniformly distributed density or slightly changing density, a more robust smoothing length model is necessary. In this implementation, we use variable smoothing length [5].

9.3.3.5 Nearest Neighboring Particle Searching

For SPH implementations, the nearest neighboring particle search algorithm is another vital numerical aspect because the repeated calculation of interactions between the neighbors drastically affects the efficiency of the whole simulation, especially in simulations with a large number of particles. The complexity of comparing the interparticle distance with λh for all particles from the given particle is of order $O(N^2)$, thus it would incur an intolerable amount of computational time for large number of particles. In this implementation we use the hierarchy tree [60] method and the pairwise interaction technique [5] to improve the efficiency.

9.3.3.6 Solid Boundary Treatment

There is no specific formulation for the implementation of boundary conditions in SPH. For particles near the solid boundary, only those inside the boundary contribute to the summation of the particle interactions, and no contribution comes from outside since there are no particles. This one-sided contribution does not lead to a correct solution because on the solid surface, although the velocity is zero, other physical quantities such as density are not necessarily zero. In this implementation, we use two types of virtual (or ghost) particles to treat the solid boundary condition [5,47,61].

9.3.4 Applications

An in-house SPH code has been developed at the Centre for Advanced Computation in Engineering Science (ACES). The overall flowchart of the SPH code is schematically shown

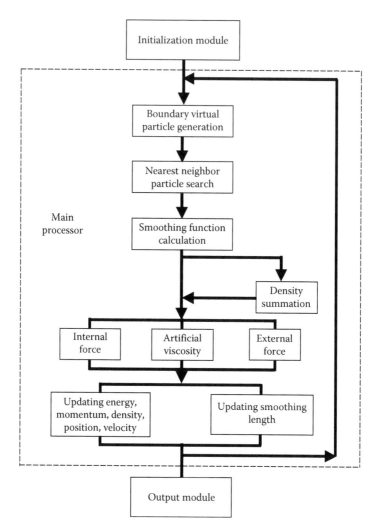

FIGURE 9.3
Flowchart of the present SPH code.

in Figure 9.3. A series of numerical tests have been carried out to test the ability and efficiency of the presented SPH method for simulating fluid dynamic problems, including incompressible flows, free surface flows, and explosion simulation.

9.3.4.1 Applications to Incompressible Flows

In FDM, incompressible flows have been widely studied because of their special features. In our SPH simulation of incompressible flows, the above-described artificial compressibility technique is employed to model the incompressible fluid as slightly compressible by selecting a proper equation of state. Three simple simulation cases, Poiseuille flow, Couette flow, and shear driven cavity flow, are simulated using our SPH code. The corresponding numerical results are presented below. In these numerical examples, an equation of state, $p = c^2\rho$ is used, where c is the sound speed.

Example 9.1: Poiseuille Flow

Poiseuille flow is an often-used benchmarking CFD problem. It is a steady flow between two stationary infinite plates placed at $y = 0$ and $y = l$. The originally stationary fluid driven by some body force F, gradually flows between the two plates and finally arrives at an equilibrium steady-flow state. In our simulation, the parameters are as follows:

Spacing of the plates: $l = 10^{-3}$ m

Kinetic viscosity of the fluid: $v = 10^{-6}$ m^2/s

Density of the fluid: $\rho = 10^3$ kg/m^3

Driven body force: $F = 2 \times 10^{-4}$ m/s^2

Peak fluid velocity: $v_0 = 2.5 \times 10^{-5}$ m/s, which corresponds to a Reynolds number of $Re = 2.5 \times 10^{-2}$

In our SPH simulation, a total of 101 particles are placed in the y direction. Figure 9.4 shows the comparison between the velocity profiles obtained by the SPH method and those by series solution [62,63] at $t = 0.01$ s, 0.05 s, 0.1 s, 0.2 s, and the final steady state at $t = \infty$. It is found that they are in good agreement and the difference is within 0.5%.

Example 9.2: Couette Flow

Couette flow is another often-used benchmarking CFD problem. It is a flow between two initially stationary infinite plates placed at $y = 0$ and $y = l$ when the upper plate moves at a certain constant velocity v_0. In our computation, we use $l = 10^{-3}$ m, $v = 10^{-6}$ m^2/s, $\rho = 10^3$ kg/m^3, $v_0 = 2.5 \times 10^{-5}$ m/s, and the corresponding Reynolds number is $Re = 2.5 \times 10^{-2}$. Again, 101 particles are placed in the span direction. Comparison between the velocity profiles obtained

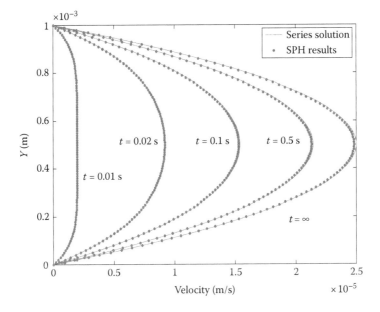

FIGURE 9.4
Velocity profiles for Poiseuille flow.

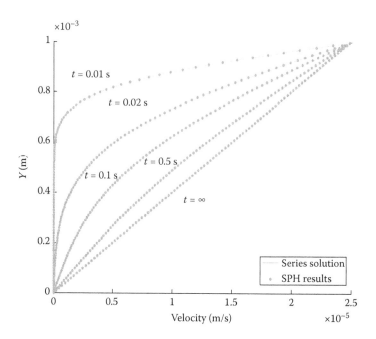

FIGURE 9.5
Velocity profiles for Couette flow.

by the SPH method and those by series solution [62,63] at $t=0.01$, 0.05, 0.1, 0.2 s, and the steady state at $t=\infty$ is shown in Figure 9.5. It is found that they are in good agreement and the difference is within 0.8%.

Example 9.3: Shear-Driven Cavity Problem

The classic shear-driven cavity problem is a flow within a closed square with the topside moving at a constant velocity V_{top} while the other three sides stay stationary. The flow reaches an equilibrium state, which behaves in a recirculation pattern. This is also a popular and critical benchmarking problem.

In our SPH simulation, a flow of the Reynolds number, 10, is considered. A total of $41 \times 41 = 1681$ field particles are initially placed in the square region. SPH results are compared with those by FDM with the same density of grids. Figure 9.6 shows the dimensionless vertical velocity profile along the horizontal centerline. Figure 9.7 shows the dimensionless horizontal velocity profile along the vertical centerline. It can be seen from Figures 9.6 and 9.7 that the results from the present implementation of SPH and those from FDM are comparable, while the SPH method slightly underpredicts the values compared to FDM.

Figure 9.8 shows the velocity distribution in the entire cavity computed using the present SPH code.

Example 9.4: Free Surface Flows

The study of free surface flows is very important in many industrial applications. Special treatment is necessary to deal with the arbitrary free surface. We have simulated a water discharge problem with the gate partly or fully opened. The viscous effect of the water is neglected here.

Figure 9.9 shows the particle distribution of water discharge at three representative instants after the gate is opened by 12% at the bottom of the gate. It can be seen that the particles distribute in an

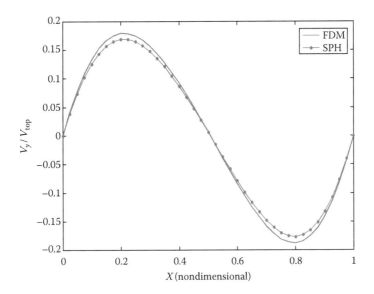

FIGURE 9.6
Dimensionless vertical velocity along the horizontal centerline.

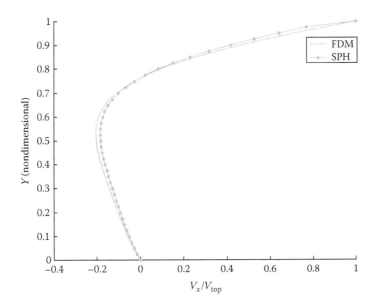

FIGURE 9.7
Dimensionless horizontal velocity along the vertical centerline.

orderly fashion with the flow of water before the gate. The streamline is obvious. Water particles eject off the gate bottom due to the pressure force, splash high outside due to the momentum, and finally fall to the ground due to gravity. Near the region of the gate bottom, the water flows rather evenly with potential energy transformed into kinetic energy. Because the water splashes high outside the gate and then falls to the ground, a cavity occurs during the course of the flow.

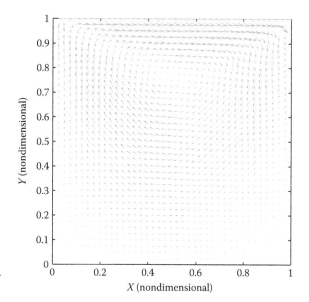

FIGURE 9.8
Velocity distributions for the shear-driven cavity problem.

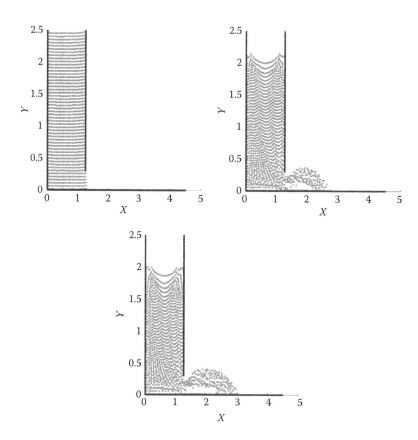

FIGURE 9.9
Particle distribution at three representative instants. Boundary virtual particle is not shown.

TABLE 9.1

Water Level H and Surge Front S Computed Using SPH Methods

Time	H_{exp}	H_m	H_p	S_{exp}	S_m	S_p
0.71	0.90	0.90	0.90	1.33	1.56	1.45
1.39	0.76	0.75	0.75	2.25	2.50	2.38
2.10	0.57	0.56	0.56	3.22	3.75	3.50
3.20	0.32	0.37	0.35	4.80	5.00	4.88

Note: Subscripts—exp, experimental results; p, present SPH code; m, results by [47].

The case of the water discharge with a fully opened gate is similar to the problem of a dam collapsing. In this case, the initial water level H_0 and the initial surge front S_0 are 25 m. The water particles flow forward in an orderly fashion with increasing surge front S and decreasing water level H. The numerical results from the present work (denoted by the subscript p), the experimental data (denoted by the subscript exp), and the results by Monaghan (denoted by the subscript m) are compared and shown in Table 9.1. The surge front S and the water level H at different instants are nondimensionalized by the initial water level H_0 while the time is nondimensionalized by $\sqrt{H_0/g}$, where g is the gravity constant. Our results, especially the surge fronts, are more accurate than the results Monaghan obtained. This is due to the use of two types of virtual particles together on the solid boundary [5].

9.3.4.2 Applications to Explosion Simulations

An explosion consists of a complicated sequence of energy-releasing processes and is difficult to simulate. In an explosion, especially an underwater explosion, there exist large deformations, large inhomogeneity, and moving material interfaces. In the light of such factors, some numerical simulations use two grids, one Eulerian grid for treating large deformations and inhomogeneity, the other Lagrangian grid for tracking different material interfaces. In this section, the SPH simulation of explosion is performed, and two numerical cases are presented. The first case is an explosion problem in a vacuum; the second is a water mitigation problem with moving gas/water/air interfaces.

Example 9.5: Explosion in Vacuum

In the first example, a clump of cylindrical explosive with high energy explodes in a vacuum. The initial total energy and density of explosive are 4.29×10^6 J/kg and 1630 kg/m^3, respectively, while the initial radius is 0.1 m. Figure 9.10 shows the density and pressure profiles in the radial direction at four different, randomly selected instances using the SPH method. The results are compared with those computed using the commercial software [64], which is based on FVM for fluid flow with explicit time marching. The good agreement between the results of the two approaches suggests that the presented SPH method can simulate the explosion process well.

We have also checked on the computational efficiency of our SPH code using this example in comparison with MSC/DYTRAN. We were not able to have our SPH code and MSC/DYTRAN run on a common computer for some trivial reasons, and therefore we cannot provide quantitative results for this comparison study. Nevertheless, we report here some indicative findings. The machine running MSC/DYTRAN was an SGI Origin 2000, and the machine running our SPH code was an HP workstation. The clock of these two machines is roughly the same. To run Example 9.5 to 0.1 ms, the MSC/DYTRAN code took about three times more wall-time compared with our SPH code.

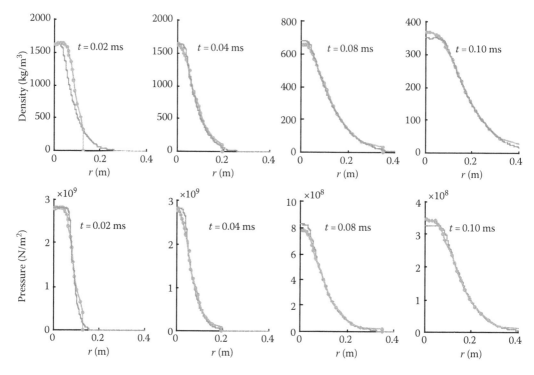

FIGURE 9.10

Density and pressure profiles for the 2D explosion problem at $t = 0.02$, 0.04, 0.08, and 0.10 ms. Lines with dots represent the results by SPH; other lines represent the results by DYTRAN.

Example 9.6: Simulation of Explosion Mitigated by Water

In this example, a more complicated case involving underwater shock is considered. To reduce potential damage resulting from an accidental explosion, high-energy explosives are sometimes stored with a layer of water cover, as shown in Figure 9.11. Outside the water is air. In the case

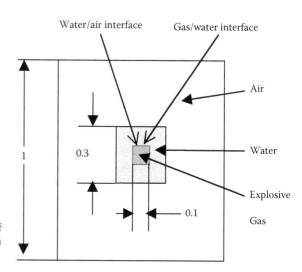

FIGURE 9.11

Initial explosive or water or air configuration of materials. The explosive is wrapped by water in a confined square space of air.

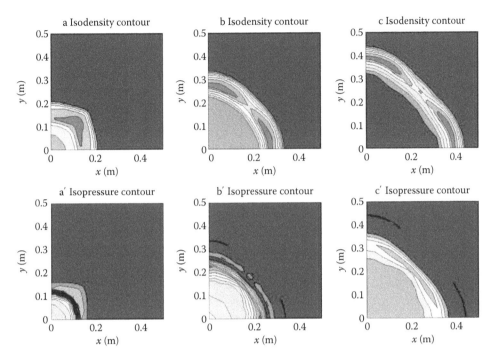

FIGURE 9.12
Density and pressure contour at $t = 0.09$, 0.21, and 0.30 ms in a quarter of the computational domain for the water mitigation problem, after the explosive detonated in water in a confined square space of air.

of an accidental explosion, the water layer covering the explosive may mitigate shock pressure greatly. At the beginning of the simulation, both the water and the air are at atmospheric condition. The simulation starts at the initial stage when the gas globe is surrounded by water. It is assumed that the energy contained in the gas globe is equal to that in the explosive charge. As the explosive charge detonates in water, it will drive a shock wave into the surrounding water. Figure 9.12 shows the density and pressure contours. With the progress of the explosion, the produced high-pressure gas pushes water to the outside and tends to occupy more space, while the water layer becomes thinner. The gas/water/air interfaces and the latter reflection waves can be seen clearly either from the density or pressure contours. The density around the four corners gradually becomes sparser. The final penetration of particles of different materials will first occur there, which clearly shows that particles of different materials should first mix near the place where the interactions between the same kinds of particles are the weakest.

Figure 9.13 gives the peak shock pressure curve vs. time obtained by the SPH method. It nearly coincides with the peak shock pressure curve obtained by MSC/DYTRAN. As the simulations by the SPH method and MSC/DYTRAN start at the same initial condition, the original pressures for the two simulations should be the same. Later, as time elapses, the shock wave moves farther from the center of the original explosive charge, and the peak shock pressure gradually decreases in an exponential fashion. The peak shock pressure obtained by MSC/DYTRAN seems to change more slowly than the peak shock pressure obtained by the SPH method, while as time goes on, the two curves gradually become closer. This simulation shows that the presented SPH method can provide a good prediction of the peak shock pressure for an underwater explosion both qualitatively and quantitatively.

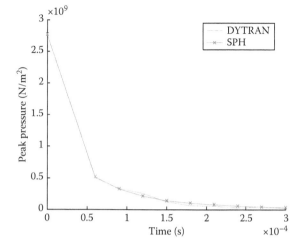

FIGURE 9.13
Time history of peak-shock pressure after the detonation of the explosive wrapped by water in a confined square space of air.

9.3.5 Some Remarks

An SPH formulation and implementation for solving Lagrangian form of Navier–Stokes equations has been introduced in this chapter. The SPH formulation is quite straightforward, and the approximation is quite "bold" leading to a Lagrangian type of meshfree particle approach. Although the consistence of function approximation in the standard SPH is even less than C°, the solution is found reasonably good. This is because the important role of error control of the smoothing operation used (see Remark 5.2).

Numerical tests have been carried out for different dynamic flow problems. For incompressible flows such as Poiseuille flow, Couette flow, and the shear driven cavity flow, the presented SPH method can yield satisfactory results. The advantages of the SPH method in treating free surface flows can be clearly seen in the simulation of the water discharge and the dam collapse problems. To test its ability to simulate problems with large deformations and large inhomogeneity, a 2D explosion problem in vacuum and a water mitigation problem are simulated. Compared with a grid-based method, the SPH method can successfully simulate such problems at reasonable accuracy with less computational effort. It can be concluded that SPH with proper modifications is an effective alternative and a valuable addition to traditional numerical methods for dynamic flow applications. More applications of SPH can be found in [5,6].

9.4 Gradient Smoothing Method

9.4.1 Background

In the previous chapters we have seen that meshfree methods based on weak and weakened-weak formulation using gradient smoothing techniques work very well for solid mechanics problems. In this section, we present the so-called GSM that is applied for strong form (differential form) governing equations. In GSM, all the unknowns of a field variable are stored at nodes and their derivatives at various locations are consistently and directly approximated with gradient smoothing operation using a set of properly defined gradient smoothing domains (GSDs). Once the derivatives are approximated, the implementation procedure of GSM is as simple as the traditional FDM [65]. The GSM, however, uses irregular mesh and hence can be applied very effectively with excellent stability for arbitrary geometries.

In the following sections, the theory of the GSM is first introduced. The GSM approximations to the gradients (first-order derivative) and Laplace operator (second-order derivative) of a field variable are presented in detail. Stencil analyses of coefficients of influence corresponding to various GSM schemes are then conducted. Important features of the stencils for the approximation of Laplace operator are discussed. Numerical solutions to Poisson equations are obtained using four favorable GSM schemes and investigated in detail to reveal the properties of convergence and stability. The computational efficiency, accuracy in results, and the robustness to the cell irregularity for GSM are also examined. Finally, the GSM results for some benchmarked compressible flow problems, e.g., inviscid flow over the NACA0012 airfoil, laminar flow over flat plate, and turbulent flow over the RAE2822 airfoil, are presented.

9.4.2 Derivative Approximation by Smoothing

In the GSM, derivatives at various locations, including nodes, centroids of cells, and midpoints of cell-edges, are approximated over relevant gradient smoothing domains using the gradient smoothing operations. The details about the theory, principle, and implementation procedure of the GSM are introduced in this section with the focus on the approximation of spatial derivatives.

For simplicity, a 2D problem is considered here to illustrate the gradient smoothing operation. Using techniques detailed in Section 3.3.5, or the Green's divergence theorem for continuous field variable (U), we shall have

$$\nabla U_i \approx \frac{1}{A_i} \oint_{\partial \Omega_i} U \vec{n} \, ds \tag{9.41}$$

which gives an approximation of gradients of a field variable U at a point $\mathbf{x}_i$ that is bounded by a local smoothing domain Ω_i, as shown in Figure 9.14. In Equation 9.41, A_i is the area of the smoothing domain, and $\vec{n}$ is the unit outward normal of the domain boundary, $\partial \Omega_i$. Equation (9.41) is widely used in many numerical operations, especially in the well-known finite volume methods. Analogously, by successively applying the gradient smoothing technique for second-order derivatives [71,72], the Laplace operator at $\mathbf{x}_i$ can be approximated as

$$\nabla \cdot (\nabla U_i) \approx \frac{1}{A_i} \oint_{\partial \Omega_i} \vec{n} \cdot \nabla U \, ds \tag{9.42}$$

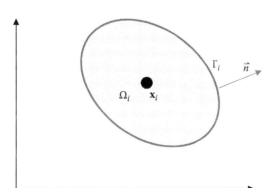

FIGURE 9.14
Generic gradient smoothing domain.

Hence, spatial derivatives at any point of interest can always be approximated using Equations 9.41 and 9.42 together with properly defined smoothing domains that will be discussed in Section 9.4.3.

9.4.3 Types of Smoothing Domains

In order to make GSM work properly and efficiently, the smoothing domains have to be carefully designed. In our GSM, the problem domain is first divided into a set of background cells via triangulation defined in Section 1.7.2. Based on the background cells, a smooth domain can be constructed for any point. Depending on the location of the point of interest, we use different types of smoothing domains based on the compact and conformal principle. As shown in Figure 9.15, three types of gradient smoothing domains are used for the approximation of spatial derivatives:

1. Node-associated GSD (nGSD): This type of smoothing domain is used for the approximation of derivatives at a node of interest. It is formed by connecting relevant centroids of triangles with midpoints of relevant cell-edges, similar to what we do in the NS-PIM.

2. Centroid-associated GSD (cGSD): This type of smoothing domains is used for approximating derivatives at the centroid of the cell. It is quite similar to what we do in the SM-PIM and in the cell-centered FVM [66].

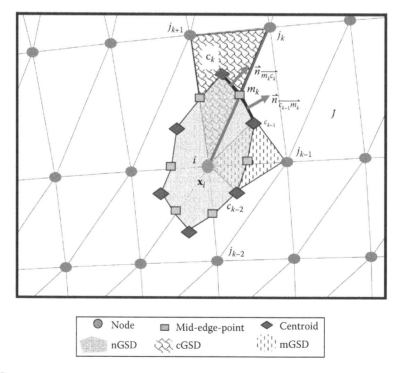

FIGURE 9.15
Types of gradient smoothing domains used in GSM constructed based on the background triangular cells.

3. Midpoint-associated GSD (mGSD): This is used for the calculation of the gradients at the midpoint of a cell-edge of interest, similar to the ES-PIM. The preferred mGSD is formed by connecting the end-nodes of the cell-edge with the centroids on the both sides of the cell-edge, as shown in Figure 9.15.

Using Equations 9.41 and 9.42, spatial derivatives at any point of interest can be approximated using the corresponding smoothing domain. The following different schemes can be devised for this purpose.

9.4.4 Schemes for Derivative Approximation

In using Equations 9.41 and 9.42, we need to accurately perform the integration along the boundaries of various types of GSDs. Both the one-point quadrature (rectangular rule) and two-point quadrature (trapezoidal rule) can be used for numerical integration. Two-point quadrature is always used to approximate derivatives at the midpoints and centroids in our study. As listed in Table 9.2, a total of eight schemes for spatial derivatives are developed, using combinations of different types of quadrature and methods of approximation.

In the schemes using one-point quadrature (I, II, and VII), the integrand along a smoothing domain-edge is simply evaluated by taking the value of a variable at the midpoint where the domain-edge intersects with a respective cell-edge. On the contrary, in the schemes using two-point quadrature (III, IV, V, VI, and VIII), values of the variable at the two end-points of the smoothing domain-edge of interest (the midpoint of the cell-edge and the centroid of the respective cGSD) are needed in numerical integration. In this work, both the first- and second-order derivatives at nodes are always approximated with gradient smoothing operation detailed in Section 2.1. The gradients at the midpoint of a cell-edge can be calculated in two ways: either by simple interpolation using the gradients at both the end-nodes of the cell-edge (I, II, III, IV, V, and VI) or by gradient smoothing operation over the respective mGSD (VII and VIII). Similarly, the gradients at a centroid can be obtained either by simple interpolation using the gradients at the three nodes of the cGSD (III and IV) or by gradient smoothing over the corresponding cGSD (V, VI, and VIII).

Note that when one-point quadrature schemes are used, there is no need to approximate the gradients at centroids, since the integrands in Equations 9.41 and 9.42 are evaluated only at the midpoints of cell-edges.

TABLE 9.2

Spatial Discretization Schemes for the Approximation of Derivatives

GSM Schemes	Quadrature Points Used	Type of GSD	Approximation of Derivatives at Midpoints	Approximation of Derivatives at Centroids	Use of Directional Correction
I	1-Point	nGSD	Interpolation	(Not required)	No
II	1-Point	nGSD	Interpolation	(Not required)	Yes
III	2-Point	nGSD	Interpolation	Interpolation	No
IV	2-Point	nGSD	Interpolation	Interpolation	Yes
V	2-Point	nGSD, cGSD	Interpolation	Gradient smoothing	No
VI	2-Point	nGSD, cGSD	Interpolation	Gradient smoothing	Yes
VII	1-Point	nGSD, mGSD	Gradient smoothing	(Not required)	No
VIII	2-Point	nGSD, mGSD, cGSD	Gradient smoothing	Gradient smoothing	No

9.4.5 Formulas for Spatial Derivatives

9.4.5.1 Two-Point Quadrature Schemes

First-order derivatives at nodes are obtained using Equation 9.41. For example, at node i, the first-order derivatives of the field variable U are given by

$$\frac{\partial U_i}{\partial x} \approx \frac{1}{A_i} \sum_{k=1}^{n_i} \left\{ \frac{1}{2} (\Delta S_x)_{ij_k}^{(L)} \left[(U_m)_{ij_k} + (U_c)_{\Delta ij_k j_{k+1}} \right] + \frac{1}{2} (\Delta S_x)_{ij_k}^{(R)} \left[(U_m)_{ij_k} + (U_c)_{\Delta ij_k j_{k-1}} \right] \right\} \quad (9.43)$$

$$\frac{\partial U_i}{\partial y} \approx \frac{1}{A_i} \sum_{k=1}^{n_i} \left\{ \frac{1}{2} (\Delta S_y)_{ij_k}^{(L)} \left[(U_m)_{ij_k} + (U_c)_{\Delta ij_k j_{k+1}} \right] + \frac{1}{2} (\Delta S_y)_{ij_k}^{(R)} \left[(U_m)_{ij_k} + (U_c)_{\Delta ij_k j_{k-1}} \right] \right\} \quad (9.44)$$

where A_i is the area of the smoothing domain, and

$$(\Delta S_x)_{ij_k}^{(L)} = \Delta S_{ij_k}^{(L)} (n_x)_{ij_k}^{(L)} \quad (9.45)$$

$$(\Delta S_y)_{ij_k}^{(L)} = \Delta S_{ij_k}^{(L)} (n_y)_{ij_k}^{(L)} \quad (9.46)$$

$$(\Delta S_x)_{ij_k}^{(R)} = \Delta S_{ij_k}^{(R)} (n_x)_{ij_k}^{(R)} \quad (9.47)$$

$$(\Delta S_y)_{ij_k}^{(R)} = \Delta S_{ij_k}^{(R)} (n_y)_{ij_k}^{(R)} \quad (9.48)$$

in the foregoing equations

U, U_m, and U_c denote values of the field variable U at nodes, midpoints of cell-edges, and centroids of triangular cells, respectively

ΔS_x and ΔS_y are the two components of a domain-edge vector

n_x and n_y represent the two components of the unit outward normal vector of the domain-edge

i denotes the node of interest

j_k is the other end-node of the cell-edge linked to node i (see Figure 9.15)

superscripts (L) and (R) stand for the two domain-edges on two sides of the cell-edge ij_k

n_i is the total number of supporting nodes within the stencil of the node i

subscripts $\Delta ij_k j_{k+1}$ and $\Delta ij_k j_{k-1}$ stand for the left-side and right-side triangles connected with the cell-edge ij_k

In the computation, these geometrical parameters are calculated and stored before the intensive calculation starts. The values of the field variable U at nonstorage locations, i.e., at midpoints and centroids, are computed by simple interpolation of function values at the nodes, respectively.

First-order derivatives at midpoints and centroids can be obtained in a similar manner. The gradients at midpoints $((\nabla U_m)_{ij_k})$ of cell-edges and centroids $((\nabla U_c)_{\Delta ij_k j_{k+1}}$ and $(\nabla U_c)_{\Delta ij_k j_{k-1}})$ of cells can also be approximated using Equation 9.41, but based on the related mGSDs and cGSDs, respectively. Such a treatment is adopted for the approximation of gradients at midpoints in Schemes VII and VIII, and gradients at centroids in Schemes V, VI, VII, and VIII. Similarly, the geometrical parameters including the areas, domain-edge vectors, and normal vectors of domain-edges related to mGSDs and cGSDs should be predetermined and stored for the late use in iterative process of solving the algebraic equations.

Alternatively, the gradients at these nonstorage locations can be approximated by simple interpolation of the gradients at relevant nodes. This manner is used to approximate the gradients at the midpoints in Schemes I, II, III, IV, V, and VI, and the gradients at centroids in Schemes III and IV.

Second-order derivatives at nodes can then be obtained using Equation 9.42 and given in the following form:

$$
\begin{aligned}
\frac{\partial^2 U_i}{\partial x^2} + \frac{\partial^2 U_i}{\partial y^2} \approx \frac{1}{\Omega_i} \sum_{k=1}^{n_i} \frac{1}{2} &\left\{ \left[\frac{\partial}{\partial x}(U_m)_{ij_k} + \frac{\partial}{\partial x}(U_c)_{\Delta ij_k j_{k+1}} \right] (\Delta S_x)_{ij_k}^{(L)} \right. \\
&+ \left[\frac{\partial}{\partial y}(U_m)_{ij_k} + \frac{\partial}{\partial y}(U_c)_{\Delta ij_k j_{k+1}} \right] (\Delta S_y)_{ij_k}^{(L)} \\
&+ \left[\frac{\partial}{\partial x}(U_m)_{ij_k} + \frac{\partial}{\partial x}(U_c)_{\Delta ij_k j_{k-1}} \right] (\Delta S_x)_{ij_k}^{(R)} \\
&+ \left. \left[\frac{\partial}{\partial y}(U_m)_{ij_k} + \frac{\partial}{\partial y}(U_c)_{\Delta ij_k j_{k-1}} \right] (\Delta S_y)_{ij_k}^{(R)} \right\}
\end{aligned}
\tag{9.49}
$$

9.4.5.2 One-Point Quadrature Schemes

In one-point quadrature schemes (I, II, and VII), it is assumed that

$$
(U_c)_{\Delta ij_k j_{k+1}} = (U_c)_{\Delta ij_k j_{k-1}} = (U_m)_{ij_k}
\tag{9.50}
$$

and

$$
(\nabla U_c)_{\Delta ij_k j_{k+1}} = (\nabla U_c)_{\Delta ij_k j_{k-1}} = (\nabla U_m)_{ij_k}
\tag{9.51}
$$

First-order derivatives at nodes are then approximated in a simplified form as

$$
\frac{\partial U_i}{\partial x} \approx \frac{1}{A_i} \sum_{k=1}^{n_i} (\Delta S_x)_{ij_k} (U_m)_{ij_k}
\tag{9.52}
$$

and

$$
\frac{\partial U_i}{\partial y} \approx \frac{1}{A_i} \sum_{k=1}^{n_i} (\Delta S_y)_{ij_k} (U_m)_{ij_k}
\tag{9.53}
$$

Second-order derivatives at nodes are simply estimated as

$$
\nabla \cdot (\nabla U) \approx \frac{1}{A_i} \sum_{k=1}^{n_i} \left[\frac{\partial}{\partial x}(U_m)_{ij_k}(\Delta S_x)_{ij_k} + \frac{\partial}{\partial y}(U_m)_{ij_k}(\Delta S_y)_{ij_k} \right]
\tag{9.54}
$$

where

$$
(\Delta S_x)_{ij_k} = (\Delta S_x)_{ij_k}^{(L)} + (\Delta S_x)_{ij_k}^{(R)}
\tag{9.55}
$$

and

$$
(\Delta S_y)_{ij_k} = (\Delta S_y)_{ij_k}^{(L)} + (\Delta S_y)_{ij_k}^{(R)}
\tag{9.56}
$$

As shown in Equations 9.52 through 9.54 in one-point quadrature schemes, only field variable values and the gradients at the midpoints of cell-edges are needed in the approximations. Thus, the vectors for a pair of domain-edges connected with the cell-edge ij_k can be lumped together, reducing storage space.

It should be noted that the gradients at midpoints in Equation 9.54 are approximated using gradient smoothing operation based on mGSDs in Scheme VII, while they are approximated using the simple interpolation approach in Schemes I and II.

The one-point quadrature schemes are clearly simpler and much more cost effective in terms of both computation flops and storage. Schemes based on two-point quadrature demand more in computation and storage for values of variables at centroids and domain-edge vectors for cGSDs. When mGSDs are used for prediction of gradients at midpoints of cell-edges, such demands become even higher. However, theoretically, schemes based on two-point quadrature can give more accurate results, which will be discussed later in the numerical examples.

9.4.5.3 Directional Correction

As described in the preceding section, when the second-order derivatives are needed, it is essential to approximate the gradients at midpoints of cell-edges in GSM procedure. One option is to simply take the arithmetic average of gradients at the two end-nodes of a cell-edge of interest. We find that this can lead to decoupling solutions (known as *checkerboard* problem) which will be illustrated in detail in Section 9.4.9. To overcome such a problem, we use the directional correction technique as proposed by Crumpton et al. [73]:

$$(\nabla \tilde{U}_m)_{ij_k} = (\nabla U_m)_{ij_k} - \left[(\nabla U_m)_{ij_k} \cdot \vec{t}_{ij_k} - \left(\frac{\partial U}{\partial l} \right)_{ij_k} \right] \vec{t}_{ij_k} \qquad (9.57)$$

where

$$\left(\frac{\partial U}{\partial l} \right)_{ij_k} \approx \frac{U_{j_k} - U_i}{\Delta l_{ij_k}}$$

$$\vec{t}_{ij_k} = \frac{\vec{r}_{ij_k}}{\Delta l_{ij_k}}$$

$$\vec{r}_{ij_k} = \mathbf{x}_{j_k} - \mathbf{x}_i$$

$$\Delta l_{ij_k} = |\mathbf{x}_{j_k} - \mathbf{x}_i|$$

Here $\mathbf{x}_i$ and $\mathbf{x}_{j_k}$ denotes the spatial locations of node i and j_k, respectively. This technique is adopted in Schemes II, IV, and VI, where the gradients at the midpoints of cell-edges are obtained by simple interpolation. More details about the role of directional correction will be addressed in Section 9.4.7.

9.4.6 Time Marching Approach

A time marching approach that is based on explicit multistage Runge–Kutta method [74] is adopted in our study for solutions to attain algebraic equations. For a transient or pseudotransient problem, the system of governing equations can be simply rewritten in the form of

$$\frac{\partial \mathbf{U}}{\partial t} = -\mathbf{R} \qquad (9.58)$$

where
    $\mathbf{U}$ denotes the array of primitive variables
    $\mathbf{R}$ represents the relevant residuals

With the explicit five-stage Runge–Kutta (*RK5*) method, the solutions at the time step $(n+1)$ can be sequentially and explicitly updated in the following fashion:

$$
\begin{aligned}
\mathbf{U}_i^{(0)} &= \mathbf{U}_i^n \\
\mathbf{U}_i^{(1)} &= \mathbf{U}_i^{(0)} - \alpha_1 \Delta t \mathbf{R}_i^{(0)} \\
\mathbf{U}_i^{(2)} &= \mathbf{U}_i^{(0)} - \alpha_2 \Delta t \mathbf{R}_i^{(1)} \\
\mathbf{U}_i^{(3)} &= \mathbf{U}_i^{(0)} - \alpha_3 \Delta t \mathbf{R}_i^{(2)} \\
\mathbf{U}_i^{(4)} &= \mathbf{U}_i^{(0)} - \alpha_4 \Delta t \mathbf{R}_i^{(3)} \\
\mathbf{U}_i^{(5)} &= \mathbf{U}_i^{(0)} - \alpha_5 \Delta t \mathbf{R}_i^{(4)}
\end{aligned}
\qquad (9.59)
$$

in this equation
    $\mathbf{U}_i^n$ denotes the solutions at node i at the time step (n)
    $\mathbf{R}_i^{(k)}$ represents the residuals that is evaluated with the kth-stage solutions and their derivatives
    Δt stands for the time-step
    the coefficients adopted in current study are $\alpha_1 = 0.0695$, $\alpha_2 = 0.1602$, $\alpha_3 = 0.2898$, $\alpha_4 = 0.5060$, and $\alpha_5 = 1.000$

In using the *RK5* method, only the zeroth- and fifth-stage solutions at nodes should be stored in memory. The *RK5* method has been widely used in the simulations of many transient fluid flow problems, because of its attractive efficiency and stability.

The edge-based data structure coupled with the gather–scatter procedure [69,70] is also adopted in our GSM solver, which essentially eliminates certain computational redundancies often encountered with node-based or cell-based data structures [75]. As a result, our GSM solver exhibits outstanding efficiency.

9.4.7 Analyses of Approximation Stencil

In the weak and weakened-weak formulations, we had systematical ways to ensure the stability and convergence of the solution. In a strong formulation, we need proper measures to deal with these issues. Here we conduct a careful examination of the stencils of supporting nodes for various schemes proposed for the GSM. The coefficients of influence of a node where derivatives are approximated are derived using the GSM schemes for regular node settings. The objective for stencil analyses is to select the most suitable schemes that satisfy the basic principles of numerical discretization. For simplicity, the stencils for approximating the Laplace operator based on cells in both square and equilateral triangle shapes are studied.

9.4.7.1 Basic Consideration for Stencil Assessment

The following five basic considerations or rules are used to assess the quality of a stencil:

(a) Consistency at each interface of the two adjacent gradient smoothing domains
(b) Positivity of coefficients of influence
(c) Zero-sum of the coefficients of influence
(d) Negative-slope linearization of the source term
(e) The compactness of the stencil

The first four rules are for solutions with physically realistic behavior and overall balance [76]. To satisfy Rule (a), it requires that the same expression of approximation must be used on the interface of two adjacent GSDs. Rule (b) requires that the coefficient for the node of interest and the coefficients of influence must be positive, when the discretization equation is written in the form of

$$a_{ii}U_i + \sum_{k=1}^{n_i} a_{ij_k} U_{j_k} = b_i \tag{9.60}$$

Rule (c) requires that

$$a_{ii} = -\sum_{k=1}^{n_i} a_{ij_k} \tag{9.61}$$

which ensures the constant field reproducibility. Rule (d) relates to the treatment of the source terms. As addressed by Patankar [76], it is essential to keep the slope of linearization negative, since a positive slope can lead to computational instabilities and physically unrealistic solutions. A good discretization stencil needs also to satisfy Rule (e) for the concerns about numerical accuracy and efficiency, as commented by Barth [70]. The very first layer of nodes surrounding the node of interest should be included in the discretization stencil. Moreover, as the stencil becomes wider, not only the computational cost increases, but eventually the accuracy decreases as less valid data from further away is brought into approximation. We do not want our numerical scheme to be too "dispersive."

Barth [70] has proposed a few lemmas to address the necessity of positivity of coefficients to satisfy a discrete maximum principle that is a key tool in the design and analysis of numerical schemes suitable for nonoscillatory discontinuity (e.g., shock). At steady state, nonnegativity of the coefficients becomes sufficient to satisfy a discrete maximum principle that can be applied successively to obtain global maximum principle and stable results. His statements reiterate the importance of Rule (b) as mentioned by Patankar [76].

In our GSM, when the gradient smoothing operation is applied to the GSDs, Rule (a) is automatically satisfied, meaning that the local conservation of quantities is guaranteed as much as for the global conservation once proper boundary conditions are used. We, therefore, focus our discussion on Rules (b), (c), and (e) for the GSM schemes.

9.4.7.2 Stencils for Gradient Approximations

The stencils for gradient approximation using the eight schemes are derived using regular background cells.

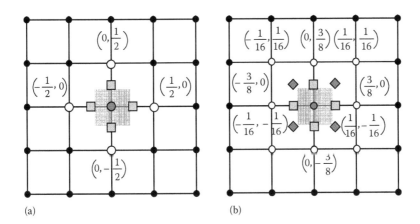

FIGURE 9.16
Stencils for derivative approximation based on square cells. (a) Schemes I, II, and VII; and (b) Schemes III, IV, V, VI, and VIII.

Square cells: Using the GSM schemes listed in Table 9.2, the coefficients of influence based on square cells are shown in Figure 9.16. We find that the three one-point quadrature schemes (I, II, and VII) correspond to the same stencil, as shown in Figure 9.16a. This stencil is also identical to that of 2-point based central-difference scheme in the FDM. We also observe that the stencil for all two-point quadrature schemes is the same, as shown in Figure 9.16b. This stencil is identical to that of 6-point based central-differencing scheme in the FDM [91]. These findings confirm that when cells in square shape are used, the GSM is identical to the FDM. The GSM, however, works also for cells in irregular shapes.

Equilateral triangular cells: Using the GSM schemes listed in Table 9.2, the coefficients of influence based on equilateral triangular cells are shown in Figure 9.17. It is interesting to find out that all the GSM schemes yield an identical stencil. This stencil is the same as that of interpolation method using six surrounding nodes [91]. Note that for irregular triangular cells, the interpolation method can in general fail [7], but our GSM still performs well,

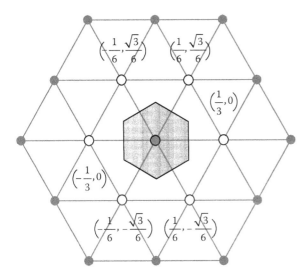

FIGURE 9.17
Stencils for derivative approximation based on equilateral triangular cells for all eight schemes (I–VIII).

as will be demonstrated in the numerical examples. This is because of the crucial stability provided by the smoothing operation.

9.4.7.3 Stencils for Approximated Laplace Operators

Square cells: The stencils for the approximated Laplace operator with GSM schemes based on uniform square cells are derived and listed in Figures 9.18 and 9.19. It is found that

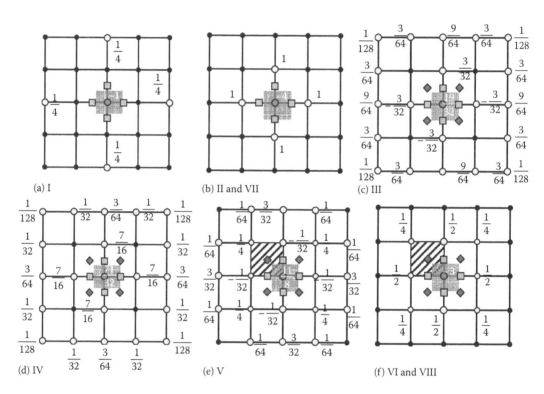

FIGURE 9.18
Stencils for the approximated Laplace operator on uniform square cells.

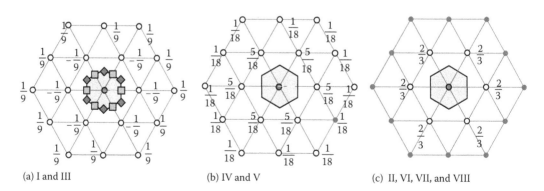

FIGURE 9.19
Stencils for the approximated Laplace operator on uniform equilateral triangular cells.

Schemes I, III, and V, as shown in Figure 9.18a, c, and e, result in wide stencils with unfavorable weighting coefficients (zero and negative values at the first-layer nodes). With such kinds of stencils, unexpected decoupling solutions may be produced [67]. This is also confirmed in the numerical examples. As mentioned by Monier [74], such kinds of stencils cannot damp out high-frequency numerical errors. The situation can get worse near the boundary where the viscous effect becomes dominant. Therefore, Schemes I, III, and V are regarded unfavorable.

Such unfavorable stencils are resulted from the simple interpolation used to approximate the gradients at the midpoints of cell-edges in the three schemes. As the directional correction for gradients approximation at midpoints is included, Schemes II and VI become relatively compact with favorable coefficients, as depicted in Figure 9.18b and f. Scheme II is a 5-point stencil and Scheme VI corresponds to 9-node compact stencil. They are the same as those for central-difference scheme in the FDM. However, even with directional correction, an unfavorable stencil in Scheme IV is noticed, where four zero-valued coefficients in the first layer of nodes and fully populated nonzero coefficients for all nodes on the second layer are found, as shown in Figure 3.3d. Therefore, Schemes II and VI are found to be favorable, while Scheme IV is considered as unfavorable.

The compact and favorable stencils are also obtained using Schemes VII and VIII where the gradients at midpoints of cell-edges are approximated by gradient smoothing operation over respective mGSDs, as seen in Figure 9.18b and f. This implies that the gradient smoothing operation over the corresponding mGSD is a good alternative to the simple interpolation modified with directional correction technique. Following the consistency rule on the approximation of derivatives at different locations and the stability (against the node irregularity) feature offered by the gradient smoothing technique, Schemes VII and VIII are more preferable than Schemes II and VI.

Equilateral triangular cells: The analyses are also conducted for all the eight schemes on uniform equilateral triangular cells and the resultant stencils are shown in Figure 9.19. It is found again that Schemes I and III result in an unfavorable stencil, because the coefficients at the very first layer of neighboring nodes are negative as seen in Figure 9.19a, which violates the basic Rule (b). Schemes IV and V, and Schemes II, VI, VII, and VIII produce two sets of stencils with favorable coefficients, as shown in Figure 9.19b and c. Schemes IV and V consist of two layers of neighboring nodes, while Schemes II, VI, VII, and VIII use only the first layer of neighboring nodes. According to Rule (e), Schemes II, VI, VII, and VIII are more favorable than Schemes IV and V, because of their relatively compact stencils.

We note that Rule (c) is satisfied in stencils for all schemes studied here. As a summary, based on the stencil analyses, four GSM schemes, i.e., II, VI, VII, and VIII, are regarded as favorable schemes, because they consistently produce compact stencils with favorable coefficients on the both regular cells of squares and equilateral triangles. The four favorable schemes are examined further in numerical examples to be addressed later.

9.4.8 Truncation Errors

Using the Taylor series expansion as in the standard FDM, we can easily find the truncation errors for the four selected GSM schemes based on square and equilateral triangular cells. Table 9.3 summarizes the results for the first-order derivative approximation, and Table 9.4 shows those for the second-order derivatives. It is clear that all these schemes are of second-order accuracy. The truncation errors for Scheme VII are identical to those for Scheme II. Schemes VI and VIII have the same truncation errors. All these theoretical findings will be further conformed numerically in later sections.

TABLE 9.3

Truncation Errors in the Approximation of First-Derivatives in GSM

Schemes	Truncation Error
II and VII (square cells)	$O_x(h^2) = -\dfrac{h^2}{6}\dfrac{\partial^3 U_{ij}}{\partial x^3} + O(h^3)$
	$O_y(h^2) = -\dfrac{h^2}{6}\dfrac{\partial^3 U_{ij}}{\partial y^3} + O(h^3)$
VI and VIII (square cells)	$O_x(h^2) = -h^2\left(\dfrac{5}{24}\dfrac{\partial^3 U_{ij}}{\partial x^3} + \dfrac{1}{2}\dfrac{\partial^3 U_{ij}}{\partial x\partial y^2}\right) + O(h^3)$
	$O_y(h^2) = -h^2\left(\dfrac{5}{24}\dfrac{\partial^3 U_{ij}}{\partial y^3} + \dfrac{1}{2}\dfrac{\partial^3 U_{ij}}{\partial x^2\partial y}\right) + O(h^3)$
II, VI, VII, and VIII (equilateral triangular cells)	$O_x(h^2) = -h^2\left(\dfrac{1}{24}\dfrac{\partial^3 U_i}{\partial x^3} + \dfrac{1}{8}\dfrac{\partial^3 U_i}{\partial x\partial y^2}\right) + O(h^3)$
	$O_y(h^2) = -h^2\left(\dfrac{1}{24}\dfrac{\partial^3 U_i}{\partial y^3} + \dfrac{1}{8}\dfrac{\partial^3 U_i}{\partial x^2\partial y}\right) + O(h^3)$

TABLE 9.4

Truncation Errors in the Approximation of the Laplace Operator in GSM

Schemes	Truncation Error
II and VII (square cells)	$O(h^2) = -\dfrac{h^2}{12}\left(\dfrac{\partial^4 U_{ij}}{\partial x^4} + \dfrac{\partial^4 U_{ij}}{\partial y^4}\right) + O(h^3)$
VI and VIII (square cells)	$O(h^2) = -\dfrac{h^2}{12}\left(\dfrac{\partial^4 U_{ij}}{\partial x^4} + 3\dfrac{\partial^4 U_{ij}}{\partial x^2\partial y^2} + \dfrac{\partial^4 U_{ij}}{\partial y^4}\right) + O(h^3)$
II, VI, VII, and VIII (equilateral triangular cells)	$O(h^2) = -\dfrac{h^2}{16}\left(\dfrac{\partial^4 U_i}{\partial x^4} + 2\dfrac{\partial^4 U_i}{\partial x^2\partial y^2} + \dfrac{\partial^4 U_i}{\partial y^4}\right) + O(h^3)$

9.4.9 Numerical Examples: Benchmarking with Poisson Problems

Numerical examples are used to examine the four selected GSM schemes (II, VI, VII, and VIII). A 2D Poisson equation is first conducted using our GSM code. Different spatial discretization schemes are tested and compared with one another in terms of numerical accuracy and computational efficiency. The roles of directional correction and gradient smoothing operation used for the approximation of the gradients at midpoints of cell-edges are numerically verified. In addition, the effects of the shape, the density, and the irregularity of cells upon the accuracy and stability are intensively investigated.

Example 9.7: Poisson Problems

The Poisson equation is first solved with our GSM code. Poisson equation governs many physical problems, such as the heat conduction problems with sources, where the temperature is the unknown field function. Dirichlet conditions are applied on the boundaries, i.e., the values of the field function at the boundaries are prescribed. The pseudotransient time marching approach based on *RK5* is applied to obtain steady-state solutions. The maximum allowable time step is obtained subjected to numerical stability. A typical convergence history is shown in Figure 9.20.

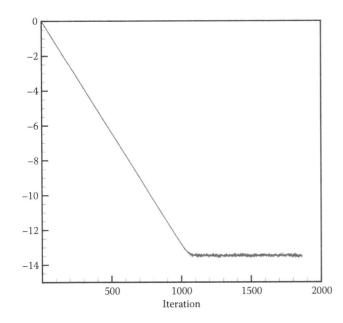

FIGURE 9.20
A typical convergence history in a pseudotransient time matching approach.

The Poisson equation takes the following form:

$$\frac{\partial U}{\partial t} = \frac{\partial^2 U}{\partial x^2} + \frac{\partial^2 U}{\partial y^2} - f(x,y), \quad (0 \le x \le 1, 0 \le y \le 1) \tag{9.62}$$

We consider two Poisson's problem with different boundary and initial conditions.
The first problem has a source and initial conditions prescribed as

$$\left. \begin{aligned} f(x,y,t) &= 13\exp(-2x+3y) \\ U(x,y,0) &= 0 \end{aligned} \right\}, \quad (0 \le x \le 1, 0 \le y \le 1) \tag{9.63}$$

For this problem, the analytical solution is

$$\hat{U}(x,y) = e^{(-2x+3y)}, \quad (0 \le x \le 1, 0 \le y \le 1) \tag{9.64}$$

The second problem has a source, initial conditions given as

$$\left. \begin{aligned} f(x,y,t) &= \sin(\pi x)\sin(\pi y) \\ U(x,y,0) &= 0 \end{aligned} \right\}, \quad (0 \le x \le 1, 0 \le y \le 1) \tag{9.65}$$

The analytical solution to the second problem is expressed as

$$\hat{U}(x,y) = -\frac{1}{2\pi^2}\sin(\pi x)\sin(\pi y) \quad (0 \le x \le 1, 0 \le y \le 1) \tag{9.66}$$

These analytical solutions are used for the assessment of numerical errors in the GSM solutions. Three indicators of numerical errors are used in this study. The convergence error index, ε_{con}, is defined as

$$\varepsilon_{con} = \sqrt{\sum_{i=1}^{N} \left(U_i^{(n+1)} - U_i^{(n)} \right)^2} \Bigg/ \sqrt{\sum_{i=1}^{N} \left(U_i^{(1)} - U_i^{(0)} \right)^2} \qquad (9.67)$$

where
$U_i^{(n)}$ denotes the predicted value of the field variable at node i at the nth iteration
N is the total number of nodes in the domain

The value of ε_{con} is monitored during iterations and used to stop the iterative process. In most simulations, in order to exclude the effect due to the temporal discretization, computations are stopped till ε_{con} becomes stabilized.

The numerical error in a GSM solution for the overall field is defined using L^2-norm of error:

$$error = \sqrt{\sum_{i=1}^{N_n} \left(U_i^{exact} - U_i^{num} \right)^2} \Bigg/ \sqrt{\sum_{i=1}^{N_n} \left(U_i^{exact} \right)^2} \qquad (9.68)$$

where U_i^{num} and U_i^{exact} are numerical and analytical solutions at node i, respectively. This type of error is used to compare the accuracy among different schemes. The third type of error is the node-wise relative error defined as

$$rerror_i = |U_i^{num} - U_i^{exact}| / |U_i^{exact}| \qquad (9.69)$$

The node-wise relative errors distributed over the computational domain are used to identify problematic regions in simulations.

Four types of background cells, square, right triangle, "regular" triangle, and irregular triangle shown in Figure 9.21, are used in the study of Poisson's problems. The irregular triangles are designed for the study of robustness of the GSM to the irregularity of triangular cells.

9.4.9.1 The Role of Directional Correction

As expected, the so-called decoupled solution will occur when Scheme I is applied onto square cells in solving the first Poisson problem. Saw-toothed numerical errors (*checkerboard* problem) are generated and cannot be dampened out, as shown in Figure 9.22a. The only difference of Scheme II from Scheme I is the inclusion of direction correction. As shown in Figure 9.22b, the *checkerboard* problem encountered in Scheme I is successfully overcome in Scheme II. This supports numerically the findings in theoretical stencil analyses. To show the results quantitatively, Table 9.5 lists the numerical errors for Schemes I and II. It is evident that when the directional correction is used, the overall numerical error is significantly reduced: the magnitude of overall errors with Scheme II reduced by about five times, compared to Scheme I. However, such benefit in Scheme II is achieved at the cost of relatively longer computational time, due to smaller allowable time-step restricted in numerical stability, as compared to Scheme I.

9.4.9.2 Comparison of Four Selected Schemes

We now conduct a quite thorough test on the four selected favorable schemes, using different types of cells with various node densities for the Poisson problems. Profiles of numerical error vs. averaged node spacing for the first Poisson problem are plotted in

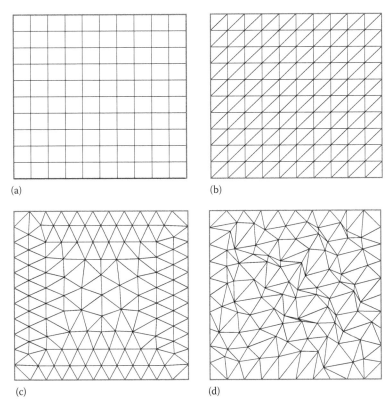

FIGURE 9.21
Four types of background cells used in the study of Poisson's problems; (a) square, (b) right triangle, (c) "regular" triangle, and (d) irregular triangle.

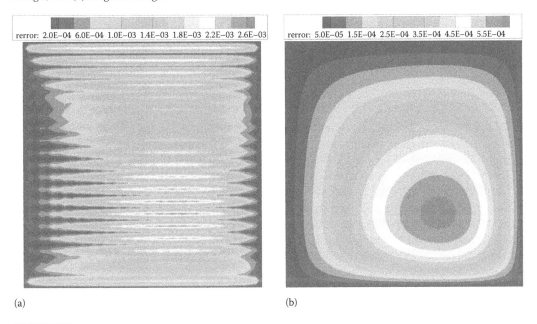

FIGURE 9.22
Contour plots of relative errors in the solution for the first Poisson problem using square cells: (a) Scheme I and (b) Scheme II.

TABLE 9.5

Comparison of Numerical Errors for the First Poisson Problem Using Uniform Square Cells

	Scheme I		Scheme II	
No. of Nodes	Error	Iteration	Error	Iteration
36	1.96e−2	20	5.07e−3	24
121	8.58e−3	89	1.66e−3	82
441	2.63e−3	202	4.87e−4	303
1681	7.16e−4	728	1.33e−4	1379
6561	1.86e−4	2679	3.38e−5	4299

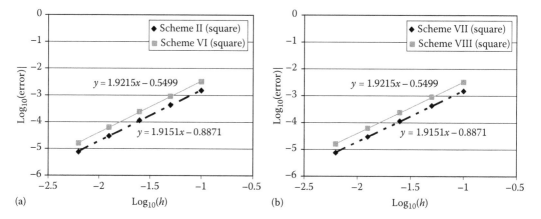

FIGURE 9.23
Convergence of numerical error with reducing averaged node spacing based on square cells.

Figures 9.23 through 9.25, together with fitted straight lines. Note that the errors in these plots are evaluated using Equation 9.68, and the averaged node spacing, h, is evaluated using the equation $h = \dfrac{V}{\sqrt{n_{node}} - 1}$. The slopes of the fitted straight-lines give the numerical convergence rates.

9.4.9.2.1 On Cells in Square Shape

As shown in Figure 9.23 when square cells are used, Scheme VII is as accurate as Scheme II, because the two schemes result in the same approximation stencil. This is also true for the Schemes VI and VIII. We note that the two-point quadrature schemes (VI and VIII) give relatively lower accuracy than the one-point quadrature schemes (II and VII). They also result in higher computational costs than the one-point quadrature schemes. Therefore, when square cells are used, Schemes II and VII are equivalent and they are superior to Schemes VI and VIII.

From the slopes of the fitted straight-lines, it is found from Figure 9.23 that the numerical convergence rates are all quite close to 2.0, which confirm that these four schemes are of second-order accuracy, as discussed in the analyses of truncation errors.

9.4.9.2.2 On Cells in Right Triangle Shape

Figure 9.24 plots the convergence of the numerical error against averaged node spacing, when right triangular cells are used. These results reveal that Scheme VI gives a slightly more accurate solution than Scheme II. Scheme VII is as accurate as Scheme VIII.

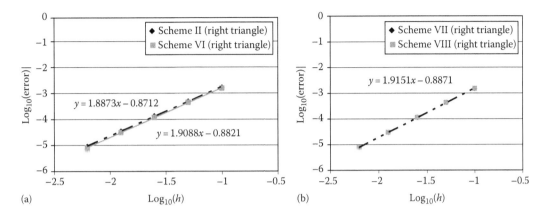

FIGURE 9.24
Convergence of numerical error with reducing averaged node spacing based on right triangular cells.

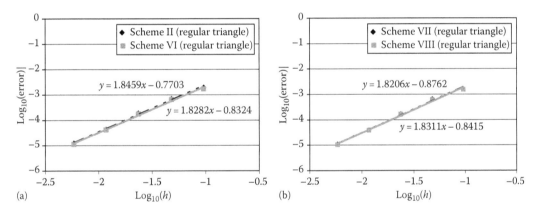

FIGURE 9.25
Convergence of numerical error with reducing averaged node spacing based on "regular" triangular cells.

9.4.9.2.3 On Cells in "Regular" Triangle Shape

Figure 9.25 plots the results of the use of "regular" triangular cells. In this case, both two-point quadrature schemes (VI and VIII) produce slightly more accurate results than one-point quadrature schemes. This is consistent with the findings for right triangular cells. Table 9.6 summarizes the numerical errors for the four selected schemes when "regular" triangular cells are used. It is clear that Scheme VII is slightly more accurate than Scheme II, and Scheme VIII is more accurate than Scheme VI. Such discrepancies in accuracy are related to the approximation of gradient at boundary nodes. In Schemes II and VI, the gradients at boundary nodes need to be approximated because they are necessarily required by the simple interpolation approach used for the approximation of gradients at midpoints of internal cell-edges linked to boundary nodes. Thus, additional errors may be introduced due to the approximation of gradients at boundary nodes. In contrast, in Schemes VII and VIII, subjected to Dirichlet boundary conditions, approximations of gradients at boundary nodes are absolutely avoided, since the gradient smoothing operation over the respective mGSDs only use the field values at the boundary nodes.

TABLE 9.6

Comparison of Numerical Errors in Solutions Obtained Using "Regular" Triangular Cells
for the First Poisson Problem with Four Selected Schemes

		Numerical Errors in Solutions			
No. of Nodes	Time Step Δt	Scheme II	Scheme VII	Scheme VI	Scheme VIII
131	0.008	1.95e−3	1.65e−3	1.68e−3	1.55e−3
478	0.001	7.02e−4	6.53e−4	6.46e−4	6.20e−4
1887	0.0005	1.89e−4	1.70e−4	1.74e−4	1.65e−4
7457	0.0001	4.38e−5	3.98e−5	4.12e−5	3.94e−5
29629	0.00003	1.22e−5	1.10e−5	1.11e−5	1.05e−5

We now state without showing detailed data that this analysis applies to the right
triangular cells also.

In terms of computational cost, Scheme VII and Scheme II are almost the same, and
Scheme VIII and Scheme VI are very close to each other. Schemes VI and VIII require about
twice as much computational time as Schemes II and VII. Although Schemes VI and VIII
are more accurate than Schemes II and VII, the improvement in accuracy is not significant.
Therefore, overall the two one-point quadrature schemes (II and VII) are more preferable
for further exploration, especially for realistic fluid flow problems with a large number
of cells.

The same tests have been conducted for second Poisson problem and consistent findings
described above are also observed.

9.4.9.3 Robustness to Irregularity of Cells

We now know that the GSM works as good as the FDM for regular cells. Next, we use the
second Poisson problem to conduct a study on effects of irregularity of triangular cells. It is
well-known that triangular cells have best adaptivity to complex geometries of problem
domains and can be generated automatically in very efficient ways. Because the GSM is
purposely designed for problem domains of complicated geometry for engineering problems, we prefer to use triangular cells. In addition, the GSM has to be very robust against
the irregularity of the cell shape. Our objective of this study is thus to further examine the
sensitivity of the GSM to cell irregularity. Based on our findings so far, we now can zoom-
in on the two one-point quadrature schemes (II and VII).

To study this in a systematic manner, we first define the irregularity factor γ for all
triangular cells used in the background mesh:

$$\gamma = \frac{\sum_{i=i}^{n_e} \frac{(a_i - b_i)^2 + (b_i - c_i)^2 + (c_i - a_i)^2}{a_i^2 + b_i^2 + c_i^2}}{n_e} \tag{9.70}$$

where

a_i, b_i, and c_i, denote, respectively, the lengths of the three constitutive cell-edges of a
triangular cell

n_e stands for the total number of triangular cells throughout the problem domain

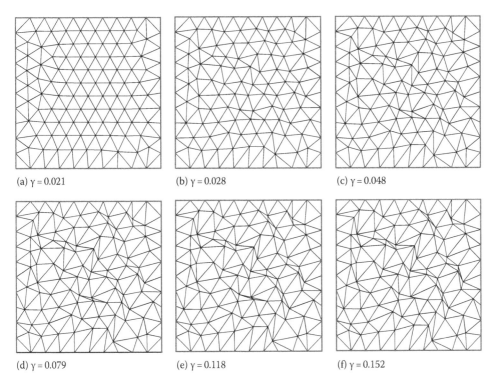

(a) γ = 0.021   (b) γ = 0.028   (c) γ = 0.048

(d) γ = 0.079   (e) γ = 0.118   (f) γ = 0.152

FIGURE 9.26
Triangular cells with various irregularity factors.

Equation 9.70 is derived from the formula proposed by Stillinger et al. [77] for a single triangle. Using Equation 9.70, we see that the irregularity γ vanishes for equilateral triangles and is always positive for all other shapes including isosceles triangles.

Figure 9.26 shows six meshes of triangular cells generated by triangulation with same number of nodes but various irregularities factor γ. It is obvious that as the irregularity factor γ increases, the mesh becomes more and more distorted. We notice that when the irregularity factor γ becomes larger than 0.152, overlapped (or negative volume) cells are generated in the domain, as shown in Figure 9.27 for $\gamma = 0.16$. In actual practice, one can always prevent overlapping in a triangular mesh using standard triangulation algorithms such as the Voronoi algorithm. Therefore, $\gamma = 0.152$ is considered as a very extreme case of irregularity. Nevertheless, in this study, all the six meshes of triangular cells shown in Figure 9.26 and other cases with irregularity factor up to 0.17 are used to test the robustness of the GSM solver. Note that the irregularity of cells can also change the maximum allowable time step, in addition to the accuracy of the solution.

Figure 9.28 plots the contour of the numerical solutions using Scheme VII and two meshes with irregular triangular cells. The effects of the irregularity are almost unnoticeable from these figures.

Table 9.7 lists in detail both the allowable time step and the errors in the convergent solutions obtained using the GSM with Schemes II and VII with these irregular meshes. We first notice from Table 9.7 that as cell irregularity increases, the time-step (Δt) has to be reduced for stable and convergent results. Meanwhile, the accuracy of the SGM solutions also reduces accordingly, but the solutions are still reasonably accurate on all sets of

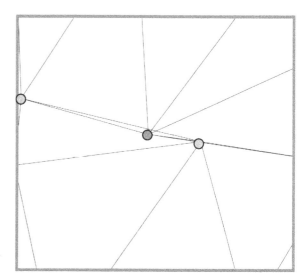

FIGURE 9.27
Overlapped cells generated in a problem
domain when the irregularity factor $\gamma = 0.16$.

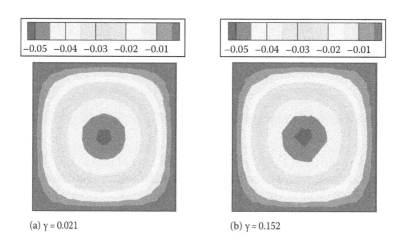

(a) $\gamma = 0.021$ (b) $\gamma = 0.152$

FIGURE 9.28
Contour plots of solutions to the second Poisson problem obtained using GSM with Scheme VII using two meshes
of irregular triangular cells.

irregular cells including the extreme case of $\gamma = 0.152$. We also notice that Scheme VII performs better than Scheme II in terms of accuracy.

Figure 9.29 plots the numerical errors in GSM solutions with Schemes II and VII to the second Poisson problem vs. the irregularity factor γ. These results show that for cases without nonoverlapped cells, the numerical errors in the GSM solution are very stable and do not vary so much, as the irregularity of cells increases. Only when the overlapped cells are generated in the domain that noticeable sudden jumps in numerical errors are observed. Even in such extreme cases, stable solutions are still attainable for both Schemes II and VII. In addition, Scheme VII shows much better stability and accuracy than Scheme II among all irregular cells examined here. This suggests that the GSM with Scheme VII is remarkably robust and insensitive to cell irregularity. Such an attractive

TABLE 9.7

Comparison of Allowable Maximum Time Step and Numerical Error for Irregular Triangular Cells Used in the Two Preferable GSM Schemes

Triangular Mesh	Irregularity Factor γ	Maximum Allowable Time Step (Δt)	Error in the Convergent Solution	
			Scheme VII	Scheme II
(a)	0.021	0.01	0.0163	0.0172
(b)	0.028	0.01	0.0169	0.0177
(c)	0.048	0.009	0.0179	0.0188
(d)	0.079	0.0075	0.0194	0.0206
(e)	0.118	0.004	0.0214	0.0231
(f)	0.152	0.0005	0.0234	0.0259

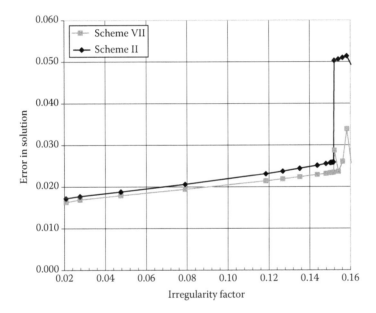

FIGURE 9.29
Numerical errors in GSM solutions to the second Poisson problem obtained using Schemes II and VII with respect to irregularity of cells.

feature is attributed to the consistent use of smoothing operations in Scheme VII, leading to the crucial stability and robustness to the system of algebraic equations.

9.4.10 Numerical Examples: Application to Navier–Stokes Equations

The GSM with Scheme VII is now applied for solutions to some benchmarked compressible flows. The full Navier–Stokes equations with respect to conservative variables are solved using the GSM solver with a number of important numerical tricks:

1. The well-known second-order Roe flux-difference splitting scheme [78] is used to evaluate the convective fluxes, because of its high accuracy for boundary layers and good resolution of shocks [67].

2. The left and right states are predicted with Barth and Jespersen method [68].

3. To avoid oscillation and nonphysical solutions, Venkatakrishnan's limiter [79] is used.

4. The flow turbulence is simulated with the Spalart–Allmaras one-equation turbulence model [80] implemented in our GSM code.

In this study, the results for inviscid flow over the NACA0012 airfoil, laminar flow over a flat plate, and turbulent flow over the RAE2822 airfoil, are presented.

Example 9.8: Inviscid Flow over the NACA0012 Airfoil

An Euler solver with the proposed GSM with Scheme VII has been developed and it is applied for the solutions to an inviscid flow over the NACA0012 airfoil. The parameters used in this study are for the freestream, $T_\infty = 288$ K, $p_\infty = 1.0 \times 10^5$ Pa, Ma $= 0.8$, and $\alpha = 1.25°$, where T_∞, p_∞, Ma, and α denote, respectively, the temperature, the static pressure, the Mach number, and the angle of attack of the freestream. More details on governing equations, treatments on boundary conditions for the GSM model can be found in [84].

Figure 9.30 shows the unstructured rectangular cells used in GSM. Figures 9.31 through 9.33, respectively, plot the contours of the predicted density, static pressure, and Mach number in the flow field. It is evident that the strong shock occurring on the upper surface of the airfoil is well resolved. The weak shock on the lower surface is also captured by our GSM solver. The results agree well with published results by Barth [69].

Example 9.9: Laminar Flow over a Flat Plate

Next we apply our GSM solver to simulate the laminar flow over a flat plate. The freestream conditions are: Re $= 5000$, Ma $= 0.5$, and $\alpha = 0°$. A rectangular computational domain is generated and gridded by right triangles. Nonslip conditions are imposed onto the surface of the flat plate. Symmetry conditions are applied to the face ahead of the plate along the x-axis. Farfield conditions [67] are set at the other external boundaries.

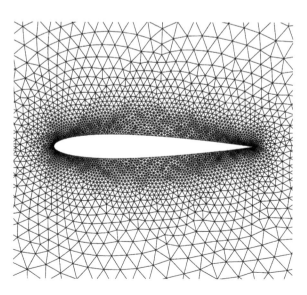

FIGURE 9.30
Irregular triangular cells near the NACA0012 airfoil used in the computation using GSM with Scheme VII.

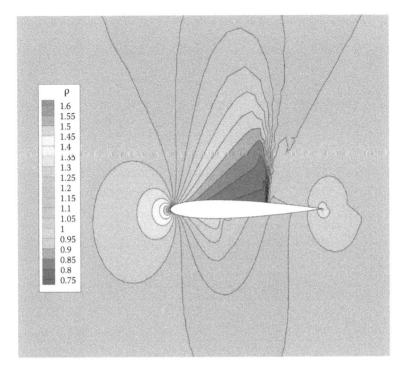

FIGURE 9.31
GSM solution of spatial distribution of the density over the NACA0012 airfoil.

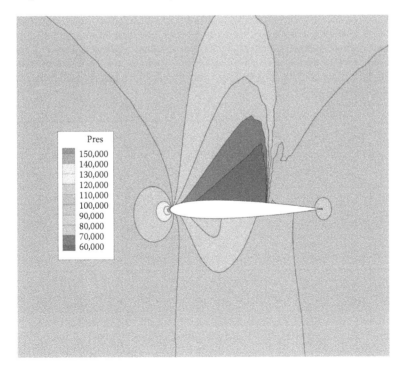

FIGURE 9.32
GSM solution of spatial distribution of the static pressure over the NACA0012 airfoil.

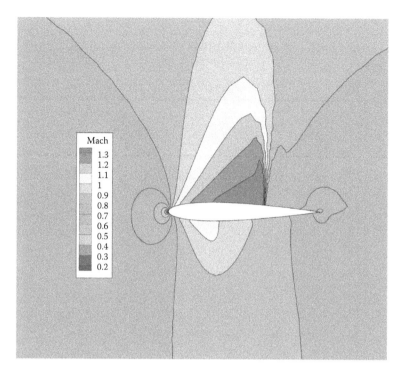

FIGURE 9.33
GSM solution of spatial distribution of Mach number over the NACA0012 airfoil.

Figure 9.34 shows the convergence history of the error in solution with respect to iteration number. Figure 9.35 plots the numerical solution of contour of the Mach number together with velocity vector field, where the boundary layer effect is clearly observed. Figure 9.36 plots the profile of the skin friction coefficient C_f varied with the distance x in comparison to the Blasius analytical solution. It is clear that our GSM numerical results agree remarkably well with the analytical solution.

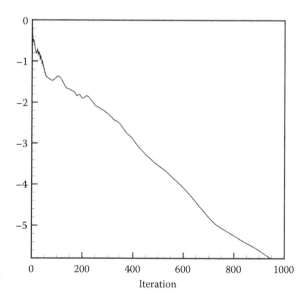

FIGURE 9.34
Convergence history of the error in solution with respect to iteration number.

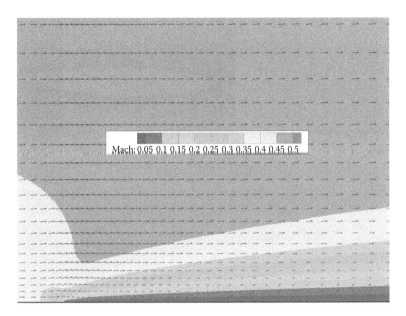

FIGURE 9.35
GSM solution of laminar flow over a flat plate: contour of Mach number and the velocity vector field.

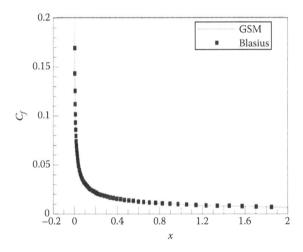

FIGURE 9.36
Comparison of wall friction coefficient predicated by GSM with the Blasius analytical solution.

Example 9.10: Turbulence Flow over the RAE2822 Airfoil

Our GSM solver is also used to simulate the turbulence flow over the RAE2822 airfoil. The parameters for the freestream are

$$T_\infty = 255.556 \text{ K}, \ p_\infty = 1.0756256 \times 10^5 \text{ Pa}, \ Re = 6.5 \times 10^6, \ Ma = 0.729, \text{ and } \alpha = 2.31°.$$

Figure 9.37 plots the unstructured triangular cells used in this simulation. Figure 9.38 plots the GSM solution of Mach number contours in this case. The shock occurring on the upper side of the airfoil is well captured by the GSM. The predicted pressure coefficient (*Cp*) distributed on

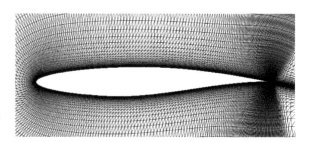

FIGURE 9.37
Unstructured triangular cells around the RAE2822 airfoil used in the GSM simulation.

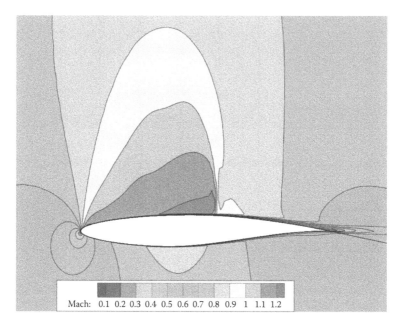

FIGURE 9.38
GSM solution of contour of Mach number for the flow over the RAE2822 airfoil.

the airfoil surface is shown in Figure 9.39. It is seen that the GSM solution agrees well with that from the commercial CFD package—FLUENT (FVM-based) [81], and experimental data [82]. We also notice that the GSM result is more accurate than FLUENT solution, especially in regions near the leading-edge. Some noticeable differences between our GSM solution and experimental data occur around the shock region. This may be due to the insufficient resolution of cells used near the shock, suggesting that a solution-based adaptive analysis is needed to improve the resolution.

9.4.11 Some Remarks

In this section a GSM is introduced to solve the Navier–Strokes equations in Eulerian form. The solution procedure of the GSM is similar to that of the FDM, and hence it belongs to a strong form method. However, the technique for gradient approximation is done in integral from, which is more like a weak form formulation. Therefore, the GSM may be regarded as a weak-form-like method. The GSM is conservative and efficient and works well with heavily distorted triangular cells, and hence applicable to fluid flow problems

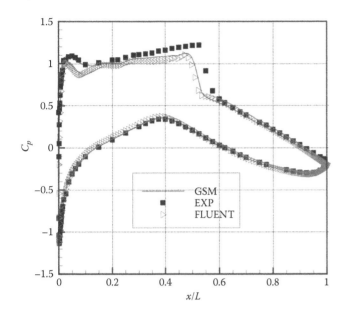

FIGURE 9.39
Comparison of profiles of pressure coefficients on the surface of the RAE2822 airfoil.

with arbitrary geometry. We have formulated and examined a number of GSM schemes and found that

- Schemes II, VI, VII, and VIII are favorable because of their compact stencils with positive coefficients of influence.

- Schemes VII and VIII that use gradient smoothing operation over mGSDs outperform Schemes II and VI in terms of robustness, stability, and accuracy. This is because schemes VII and VIII use only gradient smoothing to approximate all the derivatives providing the best error control (see Remark 5.2).

- The one-point quadrature based schemes (Schemes II and VII) have well balanced performance in terms of both efficiency and accuracy.

- Scheme VII is superior to Scheme II from the point of view of consistency in derivative approximation at various locations and robustness against irregularity of cells. Therefore, the Scheme VII is most preferable in practice, especially for large scale problems.

- We believe that the excellent performance of Schemes VII and VIII is largely due to the use of mGSD (that is essentially the same as the edge-based smoothing domain). This finding is inline with our earlier finding for W^2 models that the ES-PIM performed best (see, Chapter 8). The important role of edge-based smoothing is quite clear now, and the reason behind could be the discontinuous (or even non-existence) of the derivatives of the (assumed) field variable along the edge of the cell. When the mGSD (or ES-) is used, such a discontinuity is right in the middle of the smoothing domain and is being "smoothed" out, leading to good stability and accuracy. Therefore, additional techniques such as the "directional correction" are not required in Schemes VII and VIII. The edge-based smoothing seems to be right on the root of the problem of a discretized model, and is a crucial piece of numerical trick that are important for both strong and W^2 formulations.

Comparing the GSM with the standard FDM we note the following:

- The spatial derivatives are approximated using Taylor-series expansion in the FDM. In the GSM, we use the gradient smoothing operation, which is in fact a boundary flux approximation (see Remark 4.2). After the derivatives are approximated, the follow up treatment in GSM and FEM are essentially the same.

- The standard FDM works well usually for structured meshes. The GSM, however, works well also for unstructured mesh with irregular triangular cells and hence applicable to domains of arbitrarily complicated geometry.

- The FDM can be made for less regular domains via *domain transformation*. However, there exist problems and limitations with such a transformation, and it is far from efficient for complicated problem domains. On the contrary, in the GSM, the governing equations are always discretized directly on the physical space, no transformation is needed, and naturally no related complications.

- Most FDM schemes are usually not conservative, while all the proposed GSM schemes are locally conservative and can be made globally conservative with proper treatment on boundary conditions.

- When regular mesh is used, and a proper set of GSDs are used, the GSM becomes an FDM. Therefore, an FDM can be viewed as a special case of GSM in this regard.

The GSM has many similarities to the FVM [25] and thus many techniques implemented in the FVM can be utilized in the GSM procedure. However, they are distinct from each other in the following ways:

- The FVM was derived from physical conservation laws with respect to physical quantities such as mass, momentum, and energy using control volumes. FVM works without knowing the strong form PDEs. The GSM was originated from the gradient smoothing operation to approximate the directives of *any function* regardless of its physical background. It is a purely mathematical treatment applied directly to the strong form PDEs. GSM works only when the strong form equations are available. The procedure of GSM is in fact more like the FDM in this regard.

- The FVM is a typical weak form method based on physical laws, while the GSM is more like a strong form method with a weak formulation flavor and is called weak-form-like method.

- The traditional FVM uses the original elements/cells directly as control volumes, to which the governing equations are discretized. In the GSM, the original cells formed by triangulation are used as background cells. All sorts of different smoothing domains (nGSD, mGSD, and cGSD) are then formed based on these background cells, and the strong form PDEs are discretized using all these GSDs in various ways.

- In the GSM formulation, we are not confined to use the Heaviside (piecewise constant) smoothing function. When different smoothing functions (like the ones used in the SPH) are used, we could have many more alternatives. For example, the piecewise linear smoothing function has already been tried in the GSM [84]. Of course, when the smoothing function becomes more sophisticated, the numerical treatments can be more complicated, which may require special techniques.

- When GSM with Heaviside type smoothing function and proper selected smoothing domains are used, the GSM becomes the FVM. In this regard, the FVM may be viewed as a special case of GSM.

Finally, we note that the GSM and SPH have a same root at the integral representation of functions, but they depart when the SPH uses the particle approximation. Because GSM carries out the integral representation accurately (no particle approximation) in a well-controlled manner using properly defined GSDs, it is much more stable than the SPH for general CFD problems.

9.5 Adaptive Gradient Smoothing Method (A-GSM)

Because of the excellent stability and accuracy features of GSM in using irregular triangular background mesh, it is an ideal candidate for adaptive analysis. This section presents such a GSM.

9.5.1 Adaptive Remeshing Technique

The adaptive process aims at yielding a set of "optimal" mesh on which solutions with desired accuracy can be achieved. Ideally, such a mesh can produce a solution with equally distributed errors across the field [85], so that the number of nodes used can be a minimum. In this study, an adaptive process based on the error equidistribution strategy is explored to enhance the GSM solver in resolving abrupt changes occurring in the fluid field. In the proposed adaptive process, a directional error indicator at each node in a current mesh is evaluated first, followed by the determination of meshing parameters based on the error equidistribution strategy. Once the meshing parameters are obtained, the whole field will be remeshed using the well-known advancing front technique. The whole process is carried out in an iterative way till the desired accuracy is achieved or the allowable maximum number of adaptive iterations is reached.

9.5.1.1 Directional Error Indicator

In an adaptive process, we need an error indicator to identify the regions either for further refinement or coarsening. The direction-oriented error indicator [86] is applied for such a purpose. It can also assist to determine the coordinates of nodes for remeshing the field. Consider a one-dimensional problem, using FEM procedure, the local interpolation error e^I within a primitive or background cell can be expressed as

$$e^I = \frac{1}{2}\xi(h - \xi)\left|\frac{d^2 U}{dx^2}\right| \tag{9.71}$$

where ξ and h, respectively, denote a local coordinate defined for the cell and the cell length. Equation 9.71 relates the error with the cell length and the second derivative of the field variable at point x in the cell. The L^2-norm of the error over the cell can then be computed as

$$\|e^I\|_{L^2} = \sqrt{\int_0^h \frac{E_e^2}{h} dU} = \frac{1}{\sqrt{120}}h^2\left|\frac{d^2 U}{dx^2}\right| \tag{9.72}$$

According to error equidistribution strategy, a set of "optimal" mesh corresponding to the case where the errors are equally distributed across the field, requires

$$h^2 \left| \frac{d^2 U}{dx^2} \right| = C \qquad (9.73)$$

where C is a positive constant that can be interpreted as the anticipated small error for the adaptive process. If δ is used to denote the "optimal" spacing, for the set of "optimal" mesh, we shall have

$$\delta^2 \left| \frac{d^2 U}{dx^2} \right| = C \qquad (9.74)$$

This equation gives a simple way to determine an "optimal" spacing around a node, so long as the second derivatives on a current mesh are known or can be approximated. Equation 9.74 can be directly extended to 2D or 3D problems. For 2D problems of our concerns, it can be written in the quadratic form as

$$\delta_\alpha^2 \left(\sum_{i,j=1}^{2} m_{ij} \alpha_i \alpha_j \right) = C \qquad (9.75)$$

where
   α is an arbitrary unit vector
   δ_α is the spacing along the direction of α
   m_{ij} are the entries of a 2×2 symmetric matrix of second derivatives, which is

$$m_{ij} = \frac{\partial^2 U}{\partial x_i \partial x_j} \qquad (9.76)$$

Figure 9.40 shows how the values of the spacing δ_α in the α direction can be obtained: it is the distance from the origin to the intersection point of the vector α with the boundary of an ellipse. The ellipse has lengths of semimajor axis (δ_1) and semiminor axis (δ_2), respectively, in the α_1 and α_2 directions. These two axes can be calculated once the eigenvalues of the matrix **m** are obtained.

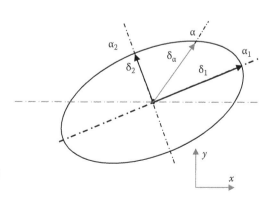

FIGURE 9.40
Illustration of the determination of spacing δ along direction α.

For fluid dynamics problems, it makes much more sense to use the first derivatives instead of the second to estimate the numerical error because the first derivatives have their physical meanings [87]. The use of the product of first derivatives is also cheaper especially for inviscid flows, since the computed field gradients can be directly used. The entries in matrix **m** can then be replaced with

$$m_{ij} = \frac{\partial U}{\partial x_i} \frac{\partial U}{\partial x_j} \tag{9.77}$$

In this study, both the options for approximating matrix **m** are examined later via numerical examples.

The approach described above is an asymptotic method based on the assumption that the mesh to be attained will be "optimal" when the local errors equal at all nodes. Though these error indicators have no rigorous mathematical proof, considerable success has been achieved in practical applications [86,88]. In addition, the error indicator is direction dependent, and hence anisotropic mesh can be generated by taking the direction-dependent feature fluid flows into account.

9.5.1.2 Meshing Parameters

Once the first or second derivatives of the interested field variable are approximated based on the current mesh and a value for constant C is specified, Equation 9.75 can then be solved for "optimal" spacing. In our adaptive process, three meshing parameters for each node are used:

$$\delta = \delta_2 = \sqrt{\frac{C}{\lambda_2}} \tag{9.78}$$

is the node spacing that equals to the length of the semiminor axis of the ellipse.

$$s = \frac{\delta_1}{\delta_2} = \sqrt{\frac{\lambda_1}{\lambda_2}} \tag{9.79}$$

is the stretching ratio in two directions, and

$$\alpha = \alpha_1 \tag{9.80}$$

is the stretching direction along the major axis. In Equations 9.78 and 9.79, λ_1 and λ_2 denote the two eigenvalues of the matrix **m** at a node. Since major and minor axes are orthogonal with each other, only the major axis direction α_1 needs to be computed as the eigenvectors of the matrix **m**.

In addition, two threshold values of δ_{min} and δ_{max} are specified as the bounds and used to control the nodal spacing:

$$\delta_{min} \leq \delta \leq \delta_{max} \tag{9.81}$$

where the upper bound δ_{max} is mainly used to prevent meaninglessly large spacing due to a vanishing eigenvalue **m**, and δ_{max} is usually chosen as the cell spacing in regions with uniform flow behaviors. The minimum value δ_{min} is used for avoiding extremely small nodal spacing in regions with too large gradients such as in shock region. Since we are interested only in the isotropic mesh here, the stretching ratio is fixed at 1.0 throughout the numerical examples presented in this section. In this study, C is defined as

$$C = \lambda_{max}\delta_{min} \tag{9.82}$$

where λ_{max} is the maximum eigenvalue for all the nodes in the entire problem domain.

With the attained meshing parameters as described above, the adaptive mesh can be regenerated with the advanced front technique.

9.5.1.3 Advancing Front Technique

Once these meshing parameters are determined, the advancing front technique [89] can be used for remeshing, and the cells and nodes are generated simultaneously. The advancing front technique can generate cells of variable size and stretching, and it differs from the Delaunay algorithms [90], which usually connect the nodes that are already distributed in space. In general, the advancing front technique can result in high-quality meshes, and also offers the flexibility in generating anisotropic mesh and the liability in handling moving components. Therefore, this technique is adopted in the development of adaptive GSM.

The advancing front technique is a bottom-up approach for mesh generation with the following steps:

1. First, each boundary curve is discretized. Nodes are placed on the boundary curve components and then contiguous nodes are joined with straight line segments. These segments form the edges of triangular cells in the later stage. The length of these segments must, therefore, be consistent with the desired local distribution of node spacings.
2. Next, the triangular cells are generated. For a 2D domain, all the sides produced in the first step are assembled as initial front. At any given time, the front contains the set of all the sides, which are currently available to form a triangular cell. Thus, the front is a dynamic data structure which is updated continuously during the generation process. A side is selected from the front and a triangular element is generated according to the computed meshing parameters. The triangulation process may involve creating a new node or simply connecting the side to an existing node.
3. Once the triangle is formed, the front is updated by removing the old side out of the front list and adding the new sides into the front list. Then the triangulation proceeds until the contents in the front become empty.

Figure 9.41 depicts the triangulation process using the advancing front technique for a square planar domain. The initial front and the form of the mesh at various stages are illustrated.

A more detailed description about the advancing front technique can be found in [90].

9.5.1.4 The Adaptive GSM Procedure

The overall adaptive procedure used in A-GSM is carried out in the following consequent fashion:

Step 1: GSM solutions are obtained using an initial mesh, which gives the approximated first or second derivatives on the initial mesh.

Step 2: Three meshing parameters δ, s, and α are then calculated based on the current grid.

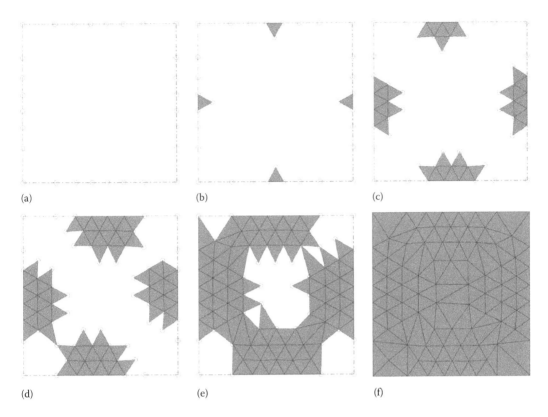

FIGURE 9.41
A triangulation process in the adaptive GSM based on the advancing front technique.

Step 3: Discretization of boundary curves is next performed using meshing parameters obtained in Step 2. The discretized boundaries are used as the initial fronts for the advancing front method.

Step 4: The triangulation process is carried out with the advancing front technique resulting in a new mesh.

Step 5: "Cosmetic" techniques including diagonal swapping, removal of worse cells, and grid smoothing are applied to the new mesh to improve the quality.

Step 6: Interpolate the GSM solution for the current mesh onto the new mesh with the help of quad-tree searching technique [75], and the weak Lagrange–Galerkin procedure [87].

Step 7: Compute the solutions with GSM solver based on the new mesh and interpolated solutions obtained in Step 6 as initial solutions for iteration.

Step 8: Repeat Steps 2 through 7, till the expected accuracy or the maximum number of adaptive cycles is achieved.

Figure 9.42 shows the workflow of the overall adaptive GSM procedure. Numerical tests for a 2D Poisson problem and Euler equation are conducted using our adaptive GSM code.

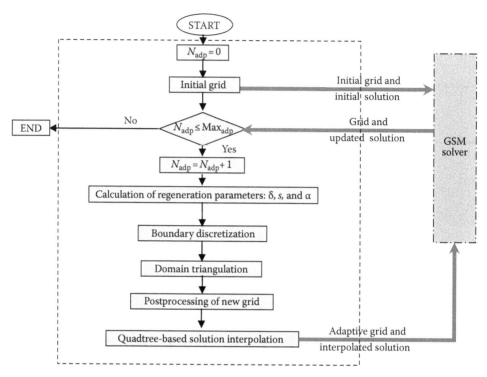

FIGURE 9.42
Workflow of the overall adaptive GSM procedure.

9.5.2 Numerical Examples: Adaptive Analysis

Example 9.11: Adaptive Analysis of Poisson Problems

We revisit the second Poisson problem defined by Equations 9.62, 9.65, and 9.66 with our adaptive GSM solver. Figure 9.43 shows the contours of the analytical solution for U and the magnitude of the gradient of U. From Figure 9.43b we shall expect to see more nodes deployed in areas near the four edges of the domain where higher gradient are observed, when the adaptive process is applied to analyze this problem.

An adaptive analysis is then conducted for the Poisson problem with an initial relatively uniform mesh shown in Figure 9.44a, and the results at three adaptive analysis stages are plotted in Figure 9.44b through d. At each adaptive remeshing stage, the threshold values of δ_{min} and δ_{max} are usually reduced by half so that finer grids can be regenerated. As shown in Figure 9.44b through d, the nodes in the regions with larger gradients are refined correspondingly, as anticipated.

In the adaptive analysis process, "cosmetic" treatment, including removal of redundant cells, diagonal swapping of cell-edges, and mesh smoothing, and shifting of the interior nodes have been used to improve the mesh quality. Details on these techniques implemented in GSM can be found in [14].

Figure 9.45 plots the decreasing process of solution errors during the adaptive analysis of the second Poisson problem, together with that of globally uniform refinement. It is shown that the adaptive GSM can also produce results at desired accuracy with less number of nodes than the nonadaptive GSM. In this particular case, as the problem is quite "mild" (the field variable U does not change very drastically), the differences in efficiency of the adaptive GSM and the nonadaptive GSM is small.

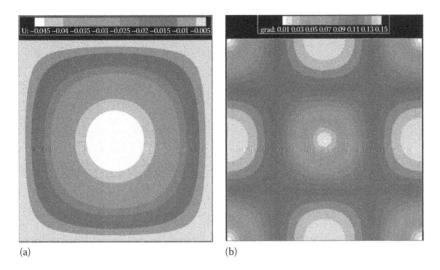

FIGURE 9.43
Contour of the analytical solutions to the second Poisson problem: (a) field function and (b) magnitude of resultant gradient.

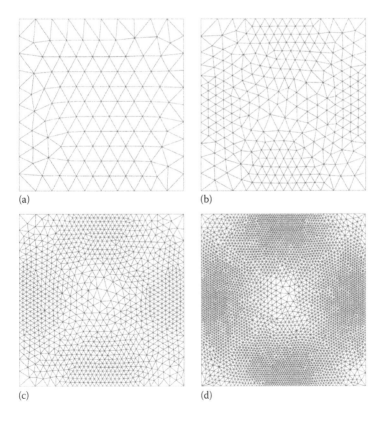

FIGURE 9.44
Meshes at different adaptive analysis steps. (a) Initial grid, (b) first adaptive grid, (c) second adaptive grid, and (d) third adaptive grid.

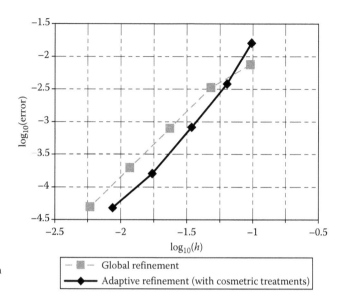

FIGURE 9.45
Reduction of solution errors when mesh is refined.

Example 9.12: Inviscid Flow over the NACA0012 Airfoil

We next revisit the Example 9.8 with the adaptive GSM solver. All the parameters are kept the same. Figure 9.46a plots the initial solution using an initial mesh. It is seen that a strong shock is observed on the upper surface and a weak shock on the lower surface of the airfoil. However, the strong shock is pretty wide because the resolution of grid near the shock region is not very high.

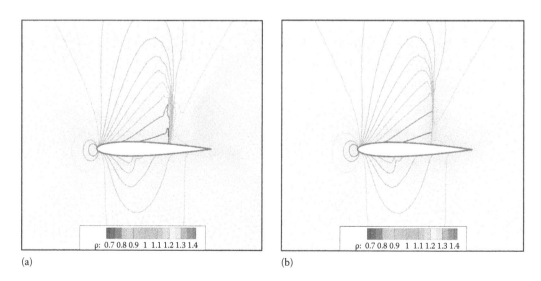

(a)   (b)

FIGURE 9.46
Initial and final meshes and contours of density obtained with the adaptive GSM solver based on errors in the first derivatives of density.

With the density chosen as the field variable and the first derivatives in Equation 9.77 used as error indicator, an adaptive GSM is applied for this problem and the resultant solutions at the final adaptive mesh are shown in Figure 9.46b. It is seen this time that the cells around the strong shock are sufficiently refined and the predicted shock front becomes thinner as expected. However, the mesh resolution around the weak shock is not improved. The similar observation has also been reported in [87].

We now use the error indicators based on the second derivatives of density, as shown in Equation 9.76. The adaptive analysis starts with quite coarse mesh as shown in Figure 9.47a. Broader shock zones are observed at the first step, as shown in Figure 9.47b, where shock front is not sufficiently refined, but its adjacent regions are refined instead. As the adaption progresses, the narrower regions across the shocks are successively refined and thus the shock front is better captured, as shown in Figure 9.47c through e. All key features for the transonic flow over the NACA0012 airfoil are successfully simulated, i.e., the two shock regions, the leading and trailing edges are all well resolved in the adaptive analyses. In summary, this example shows that the error indicator based on second derivatives is more robust in capturing the key features of the flows than that based on first derivatives.

Figure 9.48 shows the predicted pressure coefficients on the surface of the airfoil obtained using the adaptive GSM with two different error indicators. Using the second derivatives of density as error indicator, the weak shock region is precisely resolved, the strong shock region, leading edge, and trailing edge are also correctly captured. Our adaptive GSM results agree very well with the FVM solution using unstructured mesh [69].

9.5.3 Some Remarks

Because of the excellent stability of the GSM method working with irregular cells, we further developed an adaptive GSM, using the Scheme VII. Our study has indicated the following:

- The adaptive GSM scheme is very robust and stable so that it consistently results in accurate results even for a set of mesh with highly distorted cells.
- The adaptive GSM provides more accurate solutions with remarkably less number of nodes.
- The adaptive GSM is very stable during the overall adaptive process.

9.6 A Discussion on GSM for Incompressible Flows

The GSM has already been further extended to solve incompressible flows using a number of existing techniques [13]. In our implementation, the governing equations are enhanced with artificial compressibility and full pseudotime temporal terms. Thus, the time marching approach that is adopted in original GSM procedure can be continuously used. In addition, the dual time stepping approach has also been employed for improving the computational efficiency for solutions to time-dependent incompressible flow problems.

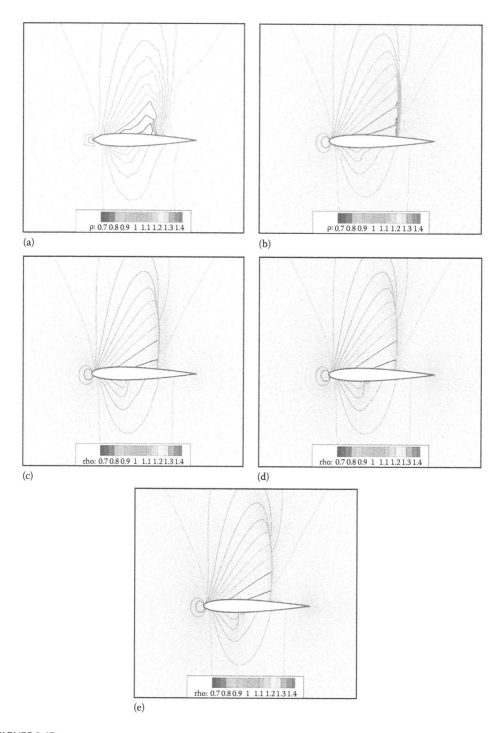

FIGURE 9.47
Meshes generated at the steps of the adaptive analysis and the solution of contours of density obtained using the adaptive GSM solver using error indicator based on the second derivatives of density: (a) initial mesh; (b) first step; (c) second step; (d) third step, and (e) final step.

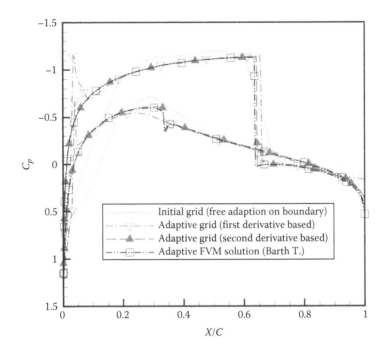

FIGURE 9.48
Profiles of pressure coefficients on the NACA0012 airfoil surface obtained using GSM with difference error indicators.

Benchmarking tests repeatedly reveal that the incompressible GSM is as accurate, robust, and efficient as the compressible GSM. Hence, the GSM is a versatile method that is valid for a wide range of fluid flow problems, such as inviscid and viscous flows, and compressible and incompressible flows.

9.7 Other Improvements on GSM

In order to further boost the efficiency of the GSM, we have recently also integrated some convergence acceleration techniques into the GSM procedure, including the local time stepping, the implicit residual smoothing, and the implicit matrix-free lower upper symmetric Gauss Seidel (LUSGS) method. With these additional techniques implemented, much faster convergence to final solutions has been achieved and thus the turnaround time in result delivery is remarkably shortened. In particular, with the help of the matrix-free implicit LUSGS solver, the demands in computational resources have also been greatly reduced. This implies that the GSM solver can now be very efficiently applied for solutions to realistic fluid problems with a considerably large number of nodes. Details about these recent progress and advances on GSM can be found in [83,84].

Finally, we mention that for problems with special features such as the convection-dominated problems, corresponding special techniques such as the well-known up-wind schemes need to be used. This type of schemes can be implemented without much difficulty in general meshfree settings as shown in [91,92].

References

1. Shin, Y. S. and Chisum, J. E., Modeling and simulation of underwater shock problems using a coupled Lagrangian–Eulerian analysis approach, *Shock Vibration*, 4, 1–10, 1997.
2. Lucy, L., A numerical approach to testing the fission hypothesis, *Astronomical J.*, 82, 1013–1024, 1977.
3. Gingold, R. A. and Monaghan, J. J., Smooth particle hydrodynamics: Theory and applications to non-spherical stars, *Mon. Notices R. Astronomical Soc.*, 181, 375–389, 1977.
4. Gingold, R. A. and Monaghan, J. J., Kernel estimates as a basis for general particle methods in hydrodynamics, *J. Comput. Phys.*, 46, 429–453, 1982.
5. Liu, G. R. and Liu, M. B., *Smoothed Particle Hydrodynamics: A Meshfree Particle Method*, World Scientific, New Jersey, 2003.
6. Liu, M. B., Liu, G. R., and Zong, Z., An overview on smoothed particle hydrodynamics, *Int. J. Comput. Methods*, 5(1), 135–188, 2008.
7. Monaghan, J. J., SPH meets the shocks of Noh, Monash University Paper, Melbourne, Australia, 1987.
8. Monaghan, J. J., An introduction to SPH, *Comput. Phys. Commun.*, 48, 89–96, 1988.
9. Liu, W. K., Jun, S., and Zhang, Y., Reproducing kernel particle methods, *Int. J. Numerical Methods Fluids*, 20, 1081–1106, 1995.
10. Lin, H. and Atluri, S. N., The meshless local Petrov–Galerkin (MLPG) method for solving incompressible Navier-Stokers equations, *Comput. Model. Eng. Sci.*, 2, 117–142, 2001.
11. Liu, G. R., Wu, Y. L., and Gu, Y. T., Application of meshless local Petrov–Galerkin (MLPG) approach to fluid flow problem, in *Computational Mechanics—New Frontiers for New Millennium*, S. Valliappan, and N. Khalili, eds., Elsevier, Amsterdam, 2001.
12. Liu, G. R. and Xu, G. X., A gradient smoothing method (GSM) for fluid dynamics problems, *Int. J. Numerical Methods Fluids*, 56(10), 1101–1133, 2008.
13. Xu, G. X., Liu, G. R., and Lee, K. H., Application of gradient smoothing method (GSM) for steady and unsteady incompressible flow problems using irregular triangles, Submitted to *Int. J. Numerical Methods Fluids*, 2008.
14. Xu, G. X. and Liu G. R., An adaptive gradient smoothing method (GSM) for fluid dynamics problems, *Int. J. Numerical Methods Fluids* (accepted), 2008.
15. Liu, G. R., Wu Y. L., and Ding, H., Meshfree weak-strong (MWS) form method and its application to imcompressible flow problems, *Int. J. Numerical Methods Fluids*, 46, 1025–1047, 2004.
16. Onate, E., Idelsohn, S., Zienkiewicz, O. C., and Taylor, R. L., A finite point method in computational mechanics applications to convective transport and fluid flow, *Int. J. Numerical Methods Eng.*, 39, 3839–3866, 1996.
17. Liszka, T. and Orkisz, J., The finite difference method at arbitrary irregular grids and its application in applied mechanics, *Comput. Struct.*, 11, 83–95, 1980.
18. Jensen, P. S., Finite difference techniques for variable grids, *Comput. Struct.*, 2, 17–29, 1980.
19. Orkisz, J., Finite difference method, in *Handbook of Computational Solid Mechanics*, M. Kleiber, ed., Springer-Verlag, Berlin, 1998.
20. Liu, G. R., Zhang, J., Li, H., Lam, K. Y., and Kee, B. B. T., Radial point interpolation based finite difference method for mechanics problems, *Int. J. Numerical Methods Eng.*, 68, 728–754, 2006.
21. Libersky, L. D. and Petscheck, A. G., Smoothed particle hydrodynamics with strength of materials, in *Proceedings of the Next Free Lagrange Conference*, Vol. 395, H. Trease, J. Fritts, and W. Crowley, eds., Springer-Verlag, New York, 1991, pp. 248–257.
22. Libersky, L. D. and Petscheck, A. G., High strain Lagrangian hydrodynamics—A three-dimensional SPH code for dynamic material response, *J. Comput. Phys.*, 109, 67–75, 1993.
23. Johnson, G. R. and Beissel, S. R., Normalized smoothing functions for SPH impact computations, *Int. J. Numerical Methods Eng.*, 39(16), 2725–2741, 1996.
24. Johnson, G. R., Stryk, R. A., and Beissel, S. R., SPH for high velocity impact computations, *Comput. Methods Appl. Mech. Eng.*, 139, 347–373, 1996.

25. Benz, W. and Asphaug, E., Explicit 3D continuum fracture modeling with smoothed particle hydrodynamics, in *Proceedings of 24th Lunar and Planetary Science Conference*, Lunar and Planetary Institute, Houston, TX, 1993, pp. 99–100.

26. Benz, W. and Asphaug, E., Impact simulations with fracture. I. Methods and tests, *Icarus*, 107, 98–116, 1994.

27. Randles, P. W., Carney, T. C., Libersky, L. D., Renick, J. D., and Petschek, A. G., Calculation of oblique impact and fracture of tungsten cubes using smoothed particle hydrodynamics, *Int. J. Impact Eng.*, 17, 1995.

28. Benz, W. and Asphaug, E., Simulations of brittle solids using smoothed particle hydrodynamics, *Comput. Phys. Commun.*, 87, 253–265, 1995.

29. Bonet, J. and Kulasegaram, S., Correction and stabilization of smoothed particle hydrodynamics methods with applications in metal forming simulations, *Int. J. Numerical Methods Eng.*, 47, 1189–1214, 2000.

30. Swegle, J. W. and Attaway, S. W., On the feasibility of using smoothed particle hydrodynamics for underwater explosion calculations, *Comput. Mech.*, 17, 151–168, 1995.

31. Liu, M. B., Liu, G. R., and Lam, K. Y., Investigations on water mitigations by using meshless particle method, *Shock Waves*, 12(3), 181–195, 2002.

32. Lam, K. Y., Liu, G. R., Liu, M. B., and Zong Z., Smoothed particle hydrodynamics for fluid dynamic problems, presented at International Symposium on Supercomputing and Fluid Science, Institute of Fluid Science, Tohoku University, Sendai, Japan (Plenary Lecture), August 2000.

33. Liu, M. B., Liu, G. R., Zong, Z., and Lam, K. Y., Numerical simulation of underwater explosion by SPH, in *Advances in Computational Engineering & Science*, A. N. Atluri and F. W. Brust, eds., Tech Science Press, Palmdale, CA, 2000, pp. 1475–1480.

34. Liu, M. B., Liu, G. R., Zong, Z., and Lam, K. Y., Numerical simulation of underwater shock using SPH methodology, presented at Computational Mechanics and Simulation of Underwater Explosion Effects—US/Singapore Workshop, Washington, DC, November 2000.

35. Campbell, P.M., Some new algorithms for boundary value problems in smoothed particle hydrodynamics, DNA Report, DNA-88-286, 1989.

36. Liu, W. K., Jun, S., Sihling, D. T., Chen, Y. J., and Hao, W., Multiresolution reproducing kernel particle method for computational fluid dynamics, *Int. J. Numerical Methods Fluids*, 24, 1–25, 1997.

37. Liu, W. K., Jun, S., Sihling, D. T., Chen, Y. J., and Hao, W., Multiresolution reproducing kernel particle method for computational fluid dynamics, *Int. J. Numerical Methods Fluids*, 24, 1391–1415, 1997.

38. Cleary, P. W., Modeling confined multi-material heat and mass flows using SPH, *Appl. Math. Modeling*, 22, 981–993, 1998.

39. Chen, J. K., Beraun, J. E., and Carney, T. C., A corrective smoothed particle method for boundary value problems in heat conduction, *Int. J. Numerical Methods Eng.*, 46, 231–252, 1999.

40. Chen, J. K., Beraun, J. E., and Jih, C. J., An improvement for tensile instability in smoothed particle hydrodynamics, *Comput. Mech.*, 23, 279–287, 1999.

41. Liu, M. B., Liu, G. R., and Lam, K. Y., A new technique to treat material interfaces for smoothed particle hydrodynamics, in *Proceedings of 1st Asia-Pacific Congress on Computational Mechanics*, Sydney, Australia, November 20–23, 2001, pp. 997–982.

42. Liu, M. B., Liu, G. R., Zong, Z., and Lam, K. Y., Numerical simulation of incompressible flows by SPH, Presented at International Conference on Scientific & Engineering Computing, Beijing, China, March 2001.

43. Campbell, J., Vignjevic, R., and Libersky, L., A contact algorithm for smoothed particle hydrodynamics, *Comput. Methods Appl. Mech. Eng.*, 184, 49–65, 2000.

44. Monaghan, J. J., Why particle methods work, *SIAM J. Sci. Stat. Comput.*, 3(4), 423–433, 1982.

45. Monaghan, J. J., On the problem of penetration in particle methods, *J. Comput. Phys.*, 82, 1–15, 1989.

46. Monaghan, J. J., Smoothed particle hydrodynamics, *Annu. Rev. Astronomy Astrophy.*, 30, 543–574, 1992.

47. Monaghan, J. J., Simulating free surface flows with SPH, *J. Comput. Phys.*, 110, 399–406, 1994.
48. Monaghan, J. J., An introduction to SPH, *Comput. Phys. Commun.*, 48, 89–96, 1998.
49. Monaghan, J. J. and Gingold, R. A., Shock simulation by the particle method SPH, *J. Comput. Phys.*, 52, 374–389, 1993.
50. Monaghan, J. J. and Kocharyan, A., SPH simulation of multi-phase flow, *Comput. Phys. Commun.*, 87, 225–235, 1995.
51. Monaghan, J. J. and Poinracic, J., Artificial viscosity for particle methods, *Appl. Numerical Math.*, 1, 187–194, 1985.
52. Takeda, H., Miyama, S. M., and Sekiya, M., Numerical simulation of viscous flow by smoothed particle hydrodynamics, *Prog. Theor. Phys.*, 92(5), 939–959, 1994.
53. Flebbe, O., Muenzel, S., Herold, H., Riffert, H., and Ruder, H., Smoothed particle hydrodynamics—physical viscosity and the simulation of accretion disks, *Astrophysical J.*, 431, 754–760, 1994.
54. Whitworth, A. P., Bhattal, A. S., Turner, J. A., and Watkins, S.J., Estimating density in smoothed particle hydrodynamics, *Astronomy Astrophys.*, 301, 929–932, 1995.
55. Swegle, J. W. and Attaway, S. W., On the feasibility of using smoothed particle hydrodynamics for underwater explosion calculations, *Comput. Mech.*, 17, 151–168, 1995.
56. Libersky, L. D. and Petscheck, A. G., High strain Lagrangian hydrodynamics—a three-dimensional SPH code for dynamic material response, *J. Comput. Phys.*, 109, 67–75, 1993.
57. Hernquist, L. and Katz, N., TreeSPH—a unification of SPH with the hierarchical tree method, *Astrophys. J. Suppl. Ser.*, 70, 419–446, 1989.
58. Steinmetz, M. and Muller, E., On the capabilities and limits of smoothed particle hydrodynamics, *Astronomy Astrophys.*, 268, 391–410, 1993.
59. Nelson, R. P. and Papaloizou, J. C. B., Variable smoothing lengths and energy conservation in smoothed particle hydrodynamics, *Mon. Notices R. Astronomical Soc.*, 270, 1–20, 1994.
61. Randles, P. W. and Libersky, L. D., Smoothed particle hydrodynamics—some recent improvements and applications, *Comput. Methods Appl. Mech. Eng.*, 138, 375–408, 1996.
62. Morris, J. P. and Monaghan, J. J., A switch to reduce SPH viscosity, *J. Comput. Phys.*, 136, 41–50, 1997.
63. Morris, J. P., Patrick, J. F., and Zhu, Y., Modeling lower Reynolds number incompressible flows using SPH, *J. Comput. Phys.*, 136, 214–226, 1997.
64. *MSC/Dytran User's Manual*, version 4, The MacNeal-Schwendler Corporation, 1997.
65. Richardson, L. F., The approximate arithmetical solution by finite differences of physical problems involving differential equations, with an application to the stresses in a masonry dam, *Philos. Trans. R Soc. A*, 210, 307–357, 1910.
66. Versteeg, H. K. and Malalasekera, W., *An Introduction to Computational Fluid Dynamics—The Finite Volume Method*, Longman Scientific & Technical, Harlow, England, 1995.
67. Blazek, J., *Computational Fluid Dynamics: Principles and Application*, 1st ed., Elsevier Press, Oxford, 2001.
68. Barth, T. J. and Jespersen, D. C., The design and application of upwind schemes on unstructured grids, *AIAA paper*, 0366, 1989.
69. Barth., T. J., Aspects of unstructured grids and finite-volume solvers for the Euler and Navier–Stokes Equations, VKI Lecture Series in CFD Course 1994, Von Karman Institute, Belgium, 1994.
70. Barth., T. J., Numerical Methods for Conservation Laws on Structured and Unstructured Grids, VKI Lecture Series in CFD course 2003, Von Karman Institute, Belgium, 2003.
71. Liu, G. R., Dai, K. Y., and Nguyen, T. T., A smoothed finite element method for mechanics problems, *Comput. Mech.*, 39, 859–877, 2007.
72. Liu, G. R., Nguyen, T. T., Dai, K. Y., and Lam, K. Y., Theoretical aspects of the smoothed finite element method (SFEM), *Int. J. Numerical Methods Eng.*, 71, 902–930, 2007.
73. Crumpton, P. I., Moinier, P., and Giles, M. B., An unstructured algorithm for high Reynolds number flows on highly-stretched grids, in *Proceedings of the 10th International Conference on Numerical Methods in Laminar and Turbulent Flows*, Swansea, England, 1997.

74. Moinier, P., Algorithm developments for an unstructured viscous flow solver, PhD thesis, University of Oxford, Oxford, 1999.

75. Lohner, R., *Applied Computational Fluid Dynamics Techniques: An Introduction Based on Finite Element Methods*, John Wiley & Sons, London, 2001.

76. Patankar, S. V., *Numerical Heat Transfer and Fluid Flow*, McGraw-Hill, New York, 1980.

77. Stillinger, D. K., Stillinger, F. H., Torquato, S., Truskett, T. M., and Debenedetti, P. G., Triangle distribution and equation of state for classical rigid disks, *J. Stat. Phys.*, 100(1–2), 49–71, 2000.

78. Roe, P. L., Approximate Riemann solvers, parameter vectors, and difference schemes, *J. Comput. Phys.*, 43, 357–382, 1981.

79. Venkatakrishnan, V., Convergence to steady-state solutions of the Euler equations on unstructured grids with limiters, *J. Comput. Phys.*, 118, 120–130, 1995.

80. Spalart, P. R. and Allmaras, S. A., One-equation turbulence model for aerodynamic flows, *La Recherche Aerospatiale*, 1, 5–21, 1994.

81. *FLUENT 6.3 User's Guide*, ANSYS, Inc, 2006.

82. Cook, P. H., McDonald, M. A. and Firmin, M. C. P., Aerofoil RAE 2822—Pressure Distributions, and Boundary Layer and Wake Measurements, Experimental Data Base for Computer Program Assessment, AGARD Report AR 138, 1979.

83. Xu, G. X. and Liu, G. R., A matrix-free implicit gradient smoothing method (GSM) for compressible flow problems, submitted to *Int. J. Numerical Methods Fluids*, 2008.

84. Xu, G. X., Development of gradient smoothing method for fluid flow problems, PhD thesis, National University of Singapore, Singapore, 2008.

85. Babuska, I. and Rheinboldt, W., Error estimates for adaptive finite element computations, *SIAM J. Numerical Anal*, 15, 736–754, 1978.

86. Peiro, J., Peraire, J., and Morgan, K., *FELISA SYSTEM Version 1.1—Reference Manual*, 1994.

87. Giraldo, F. X., A space marching adaptive remeshing technique applied to the 3D Euler equations for supersonic flow, PhD thesis, University of Virginia, Charlottesville, VA, 1995.

88. Peraire, J., Vahdati, M., Morgan, K., and Zienkiewics, O. C., Adaptive remeshing for compressible flow computations, *J. Comput. Phys.*, 72, 449–466, 1987.

89. Knabner, P. and Angermann, L., *Numerical Methods for Elliptic and Parabolic Partial Differential Equations*, Springer, New York, 2003.

90. Thompson, J. F., Soni, B. K., and Weatherill, N. P., *Handbook of Grid Generation*, CRC Press, Boca Raton, FL, 1999.

91. Liu, G. R. and Gu, Y. T., *An Introduction to MFree Methods and Their Programming*, Springer, Berlin, 2005.

92. Gu, Y. T. and Liu, G. R., Meshless techniques for convection dominated problems, *Comput. Mech.*, 38(2), 171–182, 2006.

10

Meshfree Methods for Beams

A beam is a simple but very common and important structural component. A huge volume of earlier research works have focused on analysis of beams, which is still one of the most essential topics in mechanical engineering training. The finite element method (FEM) is now the mainstream method for analysis of all kinds of problems involving beams [1,2]. In recent years, meshfree methods have also been applied to analyze beams, such as the EFG method for modal analyses of Euler–Bernoulli beams and Kirchhoff plates [3], point interpolation methods (PIMs) [4], MLPG for thin beams [6], and local PIMs (LPIM and LRPIM) for both thin and thick beams [7].

As a beam is one-dimensional (1D) spatially, PIM works perfectly well, and there is no singular issue in the moment matrix, as long as there are no duplicate nodes. Using PIM shape functions and the Galerkin weak form, discrete system equations can be established easily. In fact the procedure is almost the same as FEM for beams, if a Hermite interpolation is used. The primary difference is that FEM uses only the nodes of the element to create shape functions, but PIM may use nodes beyond the integration cells. When the GS-Galerkin weakened-weak (W^2) form is used, we need only the first-order consistence for the assumed functions, as predicted in Chapter 3. We will materialize this by formulating the NS-PIM for the fourth-order differential equations that is capable to produce upper-bound solutions.

This chapter deals only with straight beams governed by the Euler–Bernoulli beam theory. We consider only the bending deformation, and it is assumed that the beam is planar, meaning it deforms only within the x–y plane.

10.1 PIM Shape Function for Thin Beams

10.1.1 Formulation

The procedure for constructing the PIM shape function is largely the same as that detailed in Chapter 2. The major point to note is that one generally needs to use higher-order terms of polynomial basis, because the governing equations for thin beams are of the fourth order in contrast to those of the second order for 2D solids.

A thin beam is represented by a line of its neutral axis that is defined in a 1D domain Ω, as shown in Figure 10.1. This neutral axis line is discretized by a set of nodes properly distributed on the axis line. This set of nodes is often termed field nodes, as it is used to register the values of the field variables. Two neighboring nodes form a cell. For such a 1D interpolation, we have two choices: Lagrange interpolation and Hermite interpolation. When the Lagrange interpolation is used, we have only one variable (the deflection of the beam) for each field node, and the procedure for obtaining PIM shape functions is exactly the same as discussed in Chapter 2 for 1D cases. When the Hermite interpolation is used, we shall have two variables (the deflection, and the rotation that is the derivative or slope of the deflection) for each field node, and the procedure for obtaining PIM shape

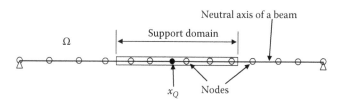

FIGURE 10.1
A straight beam represented by its neutral axis, which is represented by a set of field nodes. PIM shape functions are constructed using a subset of the nodes included in the support domain of a point of interest x_Q or the host cell of x_Q that can be a quadrature point or the center of an integration cell.

functions is a little different, but still quite straightforward. In the following, we brief this process of creating PIM shape functions using the Hermite interpolation.

Consider a Euler–Bernoulli beam governed by a fourth-order differential equation. The deflection of the beam is denoted by $v(x)$, and the rotation of the cross section of the beam (or slope of the deflection) is denoted by $\theta(x)$. The deflection $v(x)$ is approximated by using polynomial PIM shape functions, which are to be constructed using a set of nodes included in the 1D support domain of a point of interest x_Q (or the host cell of x_Q), as shown in Figure 10.1:

$$v(x, x_Q) = \sum_{i=1}^{2n} p_i(x) a_i(x_Q) = \mathbf{p}^T(x) \mathbf{a}(x_Q) \tag{10.1}$$

where
$p_i(x)$ are the monomials of x
n is the number of nodes in the support domain of x_Q
$a_i(x_Q)$ is the coefficient of $p_i(x)$, corresponding to the given point x_Q

Matrix $\mathbf{p}^T(x)$ in Equation 10.1 consists of monomials in 1D space in the form

$$\mathbf{p}^T(x) = \{1, x, x^2, x^3, x^4, \ldots, x^{2n-1}\} \tag{10.2}$$

The number of terms in the matrix $\mathbf{p}^T(x)$ depends on the number of nodes included in the support domain. In the Hermite interpolation, we choose to have two variables, deflection and rotation for each node. Therefore, if there are n nodes in the support domain, we should have $2n$ terms in the matrix $\mathbf{p}^T(x)$. In the practical implementation, one can also choose nodes based on the requirement of the number of terms in $\mathbf{p}^T(x)$.

In our thin beam formulation, the first derivative of the deflection is required. We assume, in general, that the derivatives up to an order l of the field variable are required. By using Equation 10.1, these derivatives of deflection can be obtained simply, as follows:

$$v^{(l)}(x, x_Q) = \sum_{i=1}^{2n} p_i^{(l)}(x) a_i(x_Q) = \{\mathbf{p}^{(l)}(x)\}^T \mathbf{a}(x_Q) \tag{10.3}$$

where $v^{(l)}(x, x_Q)$ and $\{p^{(l)}(x)\}^T$ are the *l*th-order derivatives of the field variable $v(x, x_Q)$ and the basis function $\mathbf{p}^T(x)$.

According to the Euler–Bernoulli beam theory, the rotation of the cross section θ of the beam can be obtained from the first derivative of the deflection, i.e.,

$$\theta(x) = -\frac{dv}{dx} = -\sum_{i=1}^{2n} p_i^{(1)}(x)a_i(x_Q) = -\{\mathbf{p}^{(1)}(x)\}^T \mathbf{a}(x_Q) \tag{10.4}$$

where

$$\mathbf{p}^{(1)} = \left\{ \frac{dp_1(x)}{dx} \quad \frac{dp_2(x)}{dx} \quad \frac{dp_2(x)}{dx} \quad \cdots \quad \frac{dp_{2n}(x)}{dx} \right\}^T$$

$$= \left\{ 0 \quad 1 \quad 2x \quad \cdots \quad (2n-1)x^{2n-2} \right\}^T \tag{10.5}$$

The coefficients a_i in Equations 10.1 and 10.4 can be determined by enforcing these equations to be satisfied at the n nodes surrounding point x_Q. At node i we have equation

$$\mathbf{v}_i = \left\{ \begin{array}{c} v_i \\ \theta_i \end{array} \right\} = \left\{ \begin{array}{c} \sum_{i=1}^{2n} p_i(x_i)a_i \\ -\sum_{i=1}^{2n} \frac{dp_i(x_i)}{dx} a_i \end{array} \right\} = \left\{ \begin{array}{c} \mathbf{p}^T(x_i)\mathbf{a} \\ -[\mathbf{p}^{(1)}(x_i)]^T \mathbf{a} \end{array} \right\} \tag{10.6}$$

where v_i and θ_i are the nodal values of v and θ at $x = x_i$. Equation 10.6 can be written in the following matrix form:

$$\mathbf{d}_s = \mathbf{P}_Q \mathbf{a} \tag{10.7}$$

where $\mathbf{d}_s$ is the (generalized) displacement vector for the nodes in the support domain arranged in the form

$$\mathbf{d}_s = [v_1, \theta_1, v_2, \theta_2, \dots, v_n, \theta_n]^T \tag{10.8}$$

$\mathbf{P}_Q$ is the (generalized) moment matrix formed using alternately vectors p and $p^{(1)}$ evaluated at n nodes at x_i ($i = 1, \dots, n$) in the support domain:

$$\mathbf{P}_Q^T = [\mathbf{p}(x_1), \mathbf{p}^{(1)}(x_1), \mathbf{p}(x_2), \mathbf{p}^{(1)}(x_2), \dots, \mathbf{p}(x_n), \mathbf{p}^{(1)}(x_n)] \tag{10.9}$$

Solving Equation 10.7 for $\mathbf{a}$, we have

$$\mathbf{a} = \mathbf{P}_Q^{-1} \mathbf{d}_s \tag{10.10}$$

Note that for 1D point interpolations, $\mathbf{P}_Q^{-1}$ always exists unless there are duplicated nodes in the support domain, which will not happen in practical implementation, because there is no reason to create a mode that has duplicated nodes. Even where there are duplicated nodes, they should be detected and eliminated.

Substituting Equation 10.10 into Equation 10.1, we have

$$v(x) = \mathbf{\Phi}^T(x)\mathbf{d}_s \tag{10.11}$$

where $\mathbf{\Phi}(x)$ is the matrix of shape functions arranged in the form

$$\mathbf{\Phi}^T(x) = [\phi_1(x) \quad \phi_2(x) \quad \phi_3(x) \quad \phi_4(x) \quad \cdots \quad \phi_{2n-1}(x) \quad \phi_{2n}(x)]$$

$$= [\phi_{v1}(x) \quad \phi_{\theta1}(x) \quad \phi_{v2}(x) \quad \phi_{\theta2}(x) \quad \cdots \quad \phi_{vn}(x) \quad \phi_{\theta n}(x)] \tag{10.12}$$

The above equation can be rewritten in the following form similar to that used in FEM [1]:

$$v(x) = \mathbf{\Phi}_v^T(x)\mathbf{v}_s + \mathbf{\Phi}_\theta^T(x)\mathbf{\theta}_s \tag{10.13}$$

where
  $\mathbf{v}_s$ collects only the deflections at the nodes in the support domain
  $\mathbf{\theta}_s$ collects only the slopes at the nodes in the support domain

The shape function matrix corresponding to the deflection is then given by

$$\mathbf{\Phi}_v^T(x) = \mathbf{p}^T(x)\mathbf{P}_Q^{-1} = [\phi_{w1}(x), \phi_{w2}(x), \dots, \phi_{wn}(x)] \tag{10.14}$$

The shape function matrix corresponding to the slope is given by

$$\mathbf{\Phi}_\theta^T(x) = \mathbf{p}_x^T(x)\mathbf{P}_Q^{-1} = [\phi_{\theta1}(x), \phi_{\theta2}(x), \dots, \phi_{\theta n}(x)] \tag{10.15}$$

10.1.2 Example

Example 10.1: Hamilton PIM Shape Functions for Thin Beams

The computation for shape functions in $\Phi(x)$, or in $\Phi_v(x)$, and $\Phi_\theta(x)$ is straightforward. It requires a numerical inversion of the moment matrix $\mathbf{P}_Q$. Typical shape functions of $\Phi_v(x)$, $\Phi_\theta(x)$, and their derivatives obtained using evenly distributed nodes and $n=3$ are shown in Figure 10.2. The first and second derivatives of the shape functions are shown in Figures 10.3 and 10.4, respectively. It can be found that the shape functions $\Phi_w(x)$ and $\Phi_\theta(x)$ obtained through the above procedure satisfy

$$\phi_i(x_j) = \delta_{ij} \tag{10.16}$$

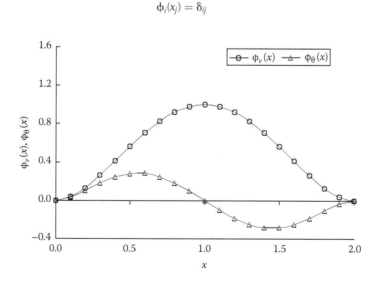

FIGURE 10.2
PIM shape functions for field variables of displacement and rotation (slope). Computed using three nodes at $x = 0.0$, 1.0, 2.0, based on the deflection at node at $x = 1.0$. (From Gu, Y.T. and Liu, G.R., *Comput. Methods Appl. Mech. Eng.*, 190, 5515, 2001. With permission.)

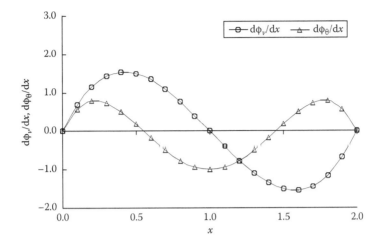

FIGURE 10.3
The first derivatives of PIM shape functions for field variables of displacement and rotation (slope). Obtained using evenly distributed nodes and $n = 3$ for the deflection at node $x = 1.0$. (From Gu, Y.T. and Liu, G.R., *Comput. Methods Appl. Mech. Eng.*, 190, 5515, 2001. With permission.)

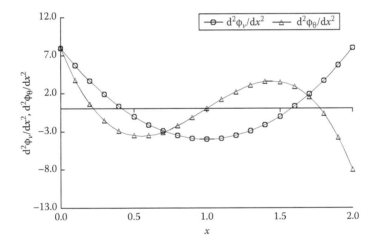

FIGURE 10.4
The second derivatives of PIM shape functions for field variables of displacement and rotation (slope). Obtained using evenly distributed nodes and $n = 3$ for the deflection at node at $x = 1.0$. (From Gu, Y.T. and Liu, G.R., *Comput. Methods Appl. Mech. Eng.*, 190, 5515, 2001. With permission.)

and

$$\sum_{i=1}^{2n} \phi_i = 1 \tag{10.17}$$

where δ_{ij} is the Kronecker delta function given by

$$\delta_{ij} = \begin{cases} 1 & i = j \\ 0 & i \neq j \end{cases} \tag{10.18}$$

Therefore, the shape functions constructed possess the delta function property, and the essential boundary conditions can be easily imposed to a numerical model using PIM shape functions.

10.2 Strong Form Equations

Consider a beam occupying domain Ω bounded by Γ, which consists of four portions: Γ_v, Γ_θ, Γ_Q, and Γ_M. The strong form of the governing equation for thin beams based on Euler–Bernoulli thin beam theory is a fourth-order differential equation. For beams of the constant bending stiffness, EI, it can be written as

$$EI\frac{\mathrm{d}^4 v}{\mathrm{d}x^4} = b_y \quad \text{in domain } \Omega \tag{10.19}$$

where
 v is the transverse deflection of the beam
 b_y is the distributed load over the beam

A single-span Euler–Bernoulli beam has four boundary conditions, two at each end. The boundary conditions can be a combination of the following:

$$v(x_0) = v_\Gamma, \quad \text{on } \Gamma_v \tag{10.20}$$

$$-\frac{\mathrm{d}v(x_0)}{\mathrm{d}x} = \theta_\Gamma, \quad \text{on } \Gamma_\theta \tag{10.21}$$

$$M(x_0) = EI\frac{\mathrm{d}^2 v}{\mathrm{d}x^2} = M_\Gamma, \quad \text{on } \Gamma_M \tag{10.22}$$

$$V(x_0) = -EI\frac{\mathrm{d}^3 v}{\mathrm{d}x^3} = V_\Gamma, \quad \text{on } \Gamma_v \tag{10.23}$$

where
 M and V denote the moment and the shear force, respectively
 Γ_v, Γ_θ, Γ_M, and Γ_V denote, respectively, the portions of the boundary where deflection, slope, moment, and shear force are specified

In this chapter, we use PIM shape functions, and therefore, the handling of the essential boundary conditions is the same as in the standard FEM.

10.3 Weak Formulation: Galerkin Formulation

Similar to Section 5.3, by multiplying with a test function δv, the weak form associated with Equation 10.19 can be obtained as

$$\int_{\Omega} \delta v \left(EI \frac{d^4 v}{dx^4} - b_y \right) dx = 0 \tag{10.24}$$

Applying Green divergence theorem, Equation 10.24 can be written as by

$$\int_{\Omega} EI \frac{d^2(\delta v)}{dx^2} \frac{d^2 v}{dx^2} dx - \int_{\Omega} \delta v b_y dx + nEI \delta v \frac{d^3 v}{dx^3}\bigg|_{\Gamma} - nEI \frac{d(\delta v)}{dx} \frac{d^2 v}{dx^2}\bigg|_{\Gamma} = 0 \tag{10.25}$$

where n is the unit outward normal on the boundary Γ. For our 1D problem, it is either 1 or -1.

As the test function δv vanishes on the prescribed essential boundary, only the natural boundary conditions are effective, and Equation 10.25 can be rewritten as

$$\int_{\Omega} EI \frac{d^2(\delta v)}{dx^2} \frac{d^2 v}{dx^2} dx = \int_{\Omega} \delta v b_y dx + V_{\Gamma} \delta v|_{\Gamma_V} - M_{\Gamma} \frac{d(\delta v)}{dx}\bigg|_{\Gamma_M} \tag{10.26}$$

This is the standard Galerkin weak formulation for beams. Using PIM shape functions, a Galerkin PIM can be formulated, and a set discretized system equation can be obtained for this 1D problem, as we do in the standard FEM. When we use only the nodes of a cell, the PIM becomes the FEM, and the cells are called elements.

10.4 Weakened-Weak Formulation: GS-Galerkin

As discussed in Chapters 3 and 4, a GS-Galerkin formulation needs a set of smoothing domains constructed on top of the cells. For our 1D domain, the division of the problem domain Ω is shown in Figure 10.5. The problem domain is first divided into N_c cells with a total of N_n nodes. For each node, a smoothing domain is bounded by the two centers of the two neighboring cells, and hence we shall have N_s smoothing domains, and $N_s = N_n$ in this 1D setting. The formation is also seamless: $\overline{\Omega} = \cup_{i=1}^{N_s} \overline{\Omega}_i^s$ and $\Omega_i^s \cap \Omega_j^s = 0$, $\forall i \neq j$. The smoothing domains for the two boundary nodes are formed only with half a cell on the boundary. This is exactly the analogy of NS-PIM in the 1D case.

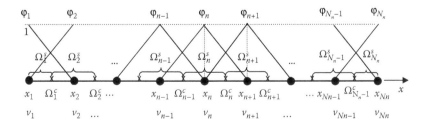

FIGURE 10.5
Discretization of the problem domain into cells and smoothing domains associated with nodes.

In this 1D NS-PIM for a beam, the displacement interpolation is cell based, but the integration is based on the smoothing domains associated with the nodes, and each (interior) smoothing domain is supported by two cells (and hence three nodes).

Using the generalized smoothing operations (see Chapter 3), the smoothed "curvature" or the second derivative of the deflection function for the smoothing domain Ω_i^s can be given by (see Equation 3.53)

$$\overline{\frac{d^2 v}{dx^2}} = \underbrace{\frac{1}{l_i^s} \int_{\Gamma_i^s} \left(n(x) \frac{dv}{dx} \right) dx}_{\text{constant in } \Omega_i^s} \tag{10.27}$$

where
l_i^s is the length of the ith smoothing domain
$n(x)$ is the unit outward normal on Γ_i^s that is the boundary of Ω_i^s

Similarly, we have the same equation for the test function:

$$\overline{\frac{d^2 (\delta v)}{dx^2}} = \underbrace{\frac{1}{l_i^s} \int_{\Gamma_i^s} \left(n(x) \frac{d(\delta v)}{dx} \right) dx}_{\text{constant in } \Omega_i^s} \tag{10.28}$$

The GS-Galerkin weakened-weak form for beams becomes

$$\sum_{i=1}^{N_n} \frac{1}{l_i^s} \left[\int_{\Gamma_i^s} \left(n(x) \frac{d(\delta v)}{dx} \right) dx \right] \left[EI \int_{\Gamma_i^s} \left(n(x) \frac{dv}{dx} \right) dx \right]$$

$$= \int_\Omega \delta v b_y dx + V_\Gamma \delta v|_{\Gamma_V} - M_\Gamma \frac{d(\delta v)}{dx} \bigg|_{\Gamma_M} \tag{10.29}$$

With the weak form and weakened-weak form being established, we can now obtain discrete models using the PIM shape functions. Since the PIM model based on the Galerkin formulation is very much the same as the standard FEM, which is familiar to many, we will keep the Galerkin formulation. Instead, we focus on the PIM model based on the GS-Galerkin formulation with node-based smoothing domains: 1D NS-PIM for beams that is an analogy of the NS-PIM discussed in Chapter 8 for 2D and 3D solids.

Remark 10.1: Order of Consistence: Strong, Weak, and Weakened-Weak Formulations
Comparison of Equations 10.19, 10.26, and 10.29 for our 1D problem of the fourth-order differential equation (DE) reveals clearly that the consistence requirement on the assumed field variables (deflection) is different for strong, weak, and weakened-weak formulations. We now list these differences in Table 10.1. It is clear that for a fourth-order differential equation, the GS-Galerkin needs only first $(= 4/2 - 1)^{\text{th}}$-order consistence! It is also clear

TABLE 10.1

Consistence Requirement on the Assumed Deflection Function for Fourth-Order Differential Equations

Items	Formulations	Consistence Requirement for the Function	Coverage of an Equation
1.	Strong form Fourth-order DE	Fourth order at any point in the problem domain	At each point in the problem domain
2.	Weak form Galerkin	Second order in cells/elements, first order on cell/element interfaces	Over the cell/element
3.	W^2 form GS-Galerkin	First order on the interfaces of the smoothing domains	Over support cells of the smoothing domain

that when we relax the requirement on continuity, we need to establish our equations over a bigger area/region. For the strong form method (collocation, for example), each equation is established for a point or node in the problem domain. For the Galerkin formulation, the basic unit of integration for the energy equation is the element or cell. For the GS-Galerkin, the basic unit of integration is the smoothing domain, which consists of a group of cells that support the smoothing domain.

10.5 Three Models

Based on Table 10.1, we can have a number of ways to solve the fourth-order differential equation for thin beams, as listed in Table 10.2.

TABLE 10.2

Possible Models for Thin Beams

Models	Approaches	Description
Model (−1)	Analytical	This can be done for many types of external forces, has been well studied in the course of mechanics of materials, and will be skipped here.
Model (0) FDM, Collocation	Strong formulation	Well studied in early years. Issues on boundary conditions are quite tricky [5]. We will not discuss this here.
Model-1 FEM, 2-DOF PIM	Galerkin weak form	Using Hermite interpolation and standard formulation in FEM, using 2-node elements and 2 DOFs at one node [1,2].
Model-2 1-DOF PIM	Galerkin weak form	Integration based on cell. Using Lagrangian PIM shape function constructed using four nodes (two nodes from the home cells, one from the neighboring cell on the left, and one on the right). At each node, only 1-DOF (deflection) is used. (No rotational DOF).
Model-3 NS-PIM	GS-Galerkin W^2 form	Integration based on smoothing domains associated with the nodes. Using Lagrangian PIM shape function constructed using two nodes from the home cells (linear interpolation). At each node, only 1-DOF (deflection) is used. (No rotational DOF).

In this chapter, we study only Model-1, Model-2, and Model-3. As the formulation for Model-1 is quite standard and can be found in any FEM book, we will not discuss it in detail. We will simply use it in our example problems.

Model-2 is similar to Model-1, except for the treatment on the boundary conditions. Because the rotational DOF is not used in the formulation, fixing the rotation on the boundary needs some caution. This situation is similar to that encountered in the collocation methods [5]. In this work, we simply use fictitious nodes on the boundary nodes [5].

We now focus on the NS-PIM formulation for beams [4] based on the weakened-weak form using GS-Galerkin.

10.6 Formulation for NS-PIM for Thin Beams

It is clear from Equation 10.29 that our W^2 formulation needs only first-order differentiation for the assumed functions of deflection of the beam, whose strong form equation is a fourth-order DE. To demonstrate this theory, we use only linear interpolation for our NS-PIM. The deflection v in each cell (say, cell Ω_n^c in Figure 10.5) can be expressed as

$$v(x) = \{\, \phi_n(x) \quad \phi_{n+1}(x) \,\} \left\{ \begin{matrix} v_n \\ v_{n+1} \end{matrix} \right\}, \quad x \in \Omega_n^c \tag{10.30}$$

where the linear shape function can be simply given as

$$\begin{aligned} \phi_n(x) &= 1 - (x - x_n)/l_n^c \\ \phi_{n+1}(x) &= (x - x_n)/l_n^c \end{aligned}, \quad x \in \Omega_n^c \tag{10.31}$$

where
$l_n^c = x_{n+1} - x_n$ is the length of the cell Ω_n^c
v_n and v_{n+1} denote the nodal deflections at nodes n and $n+1$:

$$\frac{dv(x)}{dx} = \{\, \phi_{n,x}(x) \quad \phi_{(n+1),x}(x) \,\} \left\{ \begin{matrix} v_n \\ v_{n+1} \end{matrix} \right\}, \quad x \in \Omega_n^c \tag{10.32}$$

On the left boundary $(n = 1)$, we shall have

$$\frac{dv(x)}{dx} = \{\, \phi_{1,x}(x) \quad \phi_{2,x}(x) \,\} \left\{ \begin{matrix} v_1 \\ v_2 \end{matrix} \right\}, \quad x \in \Omega_1^c \tag{10.33}$$

where we use "," to represent differentiation. Substituting Equation 10.30 into Equation 10.27 (or Equation 10.28), the smoothed curvature for the smoothing domain Ω_n^s can be given as

$$\underbrace{\frac{1}{l_n^s}\int_{\Gamma_n^s}\left(n(x)\frac{dv}{dx}\right)dx}_{\text{Constant in }\Omega_n^s} = \frac{1}{l_n^s}\left(-\left\{\phi_{(n-1),x}\left(x_{n-\frac{1}{2}}\right)\quad \phi_{(n),x}\left(x_{n-\frac{1}{2}}\right)\right\}\left\{\begin{array}{c}v_{n-1}\\v_n\end{array}\right\}\right.$$

$$\left.+\left\{\phi_{(n),x}\left(x_{n+\frac{1}{2}}\right)\quad \phi_{(n+1),x}\left(x_{n+\frac{1}{2}}\right)\right\}\left\{\begin{array}{c}v_n\\v_{n+1}\end{array}\right\}\right)$$

$$=\frac{1}{l_n^s}\underbrace{\left\{-\phi_{(n-1),x}\left(x_{n-\frac{1}{2}}\right)\quad \left(\phi_{(n),x}\left(x_{n+\frac{1}{2}}\right)-\phi_{(n),x}\left(x_{n-\frac{1}{2}}\right)\right)\quad \phi_{(n+1),x}\left(x_{n+\frac{1}{2}}\right)\right\}}_{\bar{\mathbf{B}}_n}$$

$$\times\underbrace{\left\{\begin{array}{c}v_{n-1}\\v_n\\v_{n+1}\end{array}\right\}}_{\mathbf{d}_n}=\bar{\mathbf{B}}_n\mathbf{d}_n \tag{10.34}$$

where

$$n=2,\ldots,(N_n-1)$$

$$x_{n+\frac{1}{2}}=x_n+\frac{l_n^c}{2}$$

On the left boundary ($n=1$), we shall have

$$\underbrace{\frac{1}{l_1^s}\int_{\Gamma_1^s}\left(n(x)\frac{dv}{dx}\right)dx}_{\text{Constant in }\Omega_1^s}=\frac{1}{l_1^s}\left(-\left\{\phi_{1,x}(x_1)\quad\phi_{2,x}(x_1)\right\}\left\{\begin{array}{c}v_1\\v_2\end{array}\right\}+\left\{\phi_{1,x}\left(x_{1+\frac{1}{2}}\right)\quad\phi_{2,x}\left(x_{1+\frac{1}{2}}\right)\right\}\left\{\begin{array}{c}v_1\\v_2\end{array}\right\}\right)$$

$$=\frac{1}{l_1^s}\left\{\left(\phi_{1,x}\left(x_{1+\frac{1}{2}}\right)-\phi_{1,x}(x_1)\right)\quad\left(\phi_{2,x}\left(x_{1+\frac{1}{2}}\right)-\phi_{2,x}(x_1)\right)\right\}\left\{\begin{array}{c}v_1\\v_2\end{array}\right\}=\bar{\mathbf{B}}_1\mathbf{d}_1 \tag{10.35}$$

Similarly, on the right boundary ($n=N_n$), we shall have

$$\underbrace{\frac{1}{l_{N_n}^s}\int_{\Gamma_{N_n}^s}\left(n(x)\frac{dv}{dx}\right)dx}_{\text{Constant in }\Omega_{N_n}^s}=\frac{1}{l_{N_n}^s}\left(-\left\{\phi_{(N_n-1),x}\left(x_{N_n-\frac{1}{2}}\right)\quad\phi_{(N_n),x}\left(x_{N_n-\frac{1}{2}}\right)\right\}\left\{\begin{array}{c}v_{N_n-1}\\v_{N_n}\end{array}\right\}\right.$$

$$\left.+\left\{\phi_{(N_n-1),x}(x_{N_n})\quad\phi_{(N_n-1),x}(x_{N_n})\right\}\left\{\begin{array}{c}v_{N_n-1}\\v_{N_n}\end{array}\right\}\right)$$

$$=\frac{1}{l_{N_n}^s}\underbrace{\left\{\left(\phi_{(N_n-1),x}(x_{N_n})-\phi_{(N_n-1),x}\left(x_{N_n-\frac{1}{2}}\right)\right)\quad\left(\phi_{(N_n-1),x}(x_{N_n})-\phi_{(N_n),x}\left(x_{N_n-\frac{1}{2}}\right)\right)\right\}}_{\bar{\mathbf{B}}_{N_n}}$$

$$\times\underbrace{\left\{\begin{array}{c}v_{N_n-1}\\v_{N_n}\end{array}\right\}}_{\mathbf{d}_{N_n}}=\bar{\mathbf{B}}_{N_n}\mathbf{d}_{N_n} \tag{10.36}$$

For simply supported boundary conditions, we simply set zero deflection at the corresponding boundary point, as we do in the FEM. For clamped boundary conditions, however, the situation is a little tricky because our formulation does not have rotation as a DOF. Hence, the following treatment is needed.

Let us consider a clamped boundary on the left ($n = 1$). We know that the deflection can be obtained using Equation 10.33, and hence the smoothed curvature there can be given by

$$
\frac{1}{l_1^s} \int_{\Gamma_1^s} \left(n(x) \frac{dv}{dx} \right) dx = \frac{1}{l_1^s} \left(\underbrace{(-1) \frac{dv}{dx}\bigg|_{x_1} + (1) \frac{dv}{dx}\bigg|_{x_{1+\frac{1}{2}}}}_{=0} \right) = \frac{1}{l_1^s} \frac{dv}{dx}\bigg|_{x_{1+\frac{1}{2}}}
$$

$$
= \frac{1}{l_1^s} \underbrace{\left\{ \phi_{1,x}(x_{1+\frac{1}{2}}) \quad \phi_{2,x}(x_{1+\frac{1}{2}}) \right\}}_{\bar{\mathbf{B}}_1^C} \underbrace{\begin{Bmatrix} v_1 \\ v_2 \end{Bmatrix}}_{\mathbf{d}_1} = \bar{\mathbf{B}}_1^R \mathbf{d}_1 \tag{10.37}
$$

where we have imposed the zero rotation condition. Similarly, if the clamped boundary is on the right end ($n = N_n$), we have

$$
\frac{1}{l_{N_n}^s} \int_{\Gamma_{N_n}^s} \left(n(x) \frac{dv}{dx} \right) dx = \frac{1}{l_{N_n}^s} \underbrace{\left\{ -\phi_{(N_n-1),x}(x_{N_n-\frac{1}{2}}) \quad -\phi_{(N_n),x}(x_{N_n-\frac{1}{2}}) \right\}}_{\bar{\mathbf{B}}_{N_n}^C} \underbrace{\begin{Bmatrix} v_{N_n-1} \\ v_{N_n} \end{Bmatrix}}_{\mathbf{d}_{N_n}}
$$

$$
= \bar{\mathbf{B}}_{N_n}^R \mathbf{d}_{N_n} \tag{10.38}
$$

With the use of Equations 10.37 and 10.38 in place of Equations 10.35 and 10.36, the zero rotation boundary conditions can be imposed naturally. For imposing clamped boundary conditions, all we need to do is to further impose zero deflection at the corresponding boundary nodes, as we do in the FEM.

Substituting Equations 10.30, 10.35, and 10.36 (or Equations 10.37 and 10.38 for zero rotation boundary), into Equation 10.29, a set of discretized algebraic system equations can now be easily obtained in the following matrix form:

$$
\bar{\mathbf{K}} \bar{\mathbf{U}} = \mathbf{F} \tag{10.39}
$$

where $\bar{\mathbf{U}} = \{v_1, v_2, \ldots, v_{N_n}\}^{\mathrm{T}}$ is the vector of the nodal deflections at all the nodes in the model, and $\mathbf{F}$ is the force vector defined as

$$
\mathbf{F} = \int_{\Omega} \mathbf{\Phi}^{\mathrm{T}}(\mathbf{x}) b_y \, dx + \int_{\Gamma_V} \mathbf{\Phi}^{\mathrm{T}}(\mathbf{x}) V_\Gamma \, d\Gamma - \int_{\Gamma_M} \mathbf{\Phi}_{,x}^{\mathrm{T}}(\mathbf{x}) M_\Gamma \, d\Gamma \tag{10.40}
$$

where $\mathbf{\Phi}(\mathbf{x})$ is the global matrix of nodal shape functions as arranged in Equation 10.31. The global stiffness matrix $\bar{\mathbf{K}}$ is assembled in the form

$$
\bar{\mathbf{K}} = \sum_{n=1}^{N_n} \bar{\mathbf{K}}_n \tag{10.41}
$$

where
   the summation means an assembly process, which is the same as that practiced in the
      NS-PIM or the FEM
$\bar{\mathbf{K}}_n$ is the stiffness matrix associated with Ω_n^s that is computed using

$$\bar{\mathbf{K}}_n = EI\bar{\mathbf{B}}_n^T\bar{\mathbf{B}}_n l_n^s \tag{10.42}$$

10.7 Formulation for Dynamic Problems

We now brief the formulations for dynamic problems. The strong form governing equation
for free vibration of thin beams is given by

$$EI\frac{d^4v(x,t)}{dx^4} + \rho A_0\frac{d^2v(x,t)}{dt^2} = 0 \quad \text{in domain } \Omega \tag{10.43}$$

where
   $v(x,t)$ is the deflection of the beam
   ρ is the mass density
   A_0 is the cross-section area of the beam

The boundary conditions remain the same as those for the static problems.
   The Galerkin weak form can be rewritten as

$$\int_\Omega EI\frac{d^2(\delta v)}{dx^2}\frac{d^2v}{dx^2}dx + \int_\Omega \rho A_0\delta v\frac{d^2v(x,t)}{dt^2}dx = \int_\Omega \delta v b_y dx + V_\Gamma\delta v|_{\Gamma_V} - M_\Gamma\frac{d(\delta v)}{dx}\Big|_{\Gamma_M} \tag{10.44}$$

The GS-Galerkin weak form can be given as

$$\sum_{i=1}^{N_n}\frac{1}{l_i^s}\left[\int_{\Gamma_i^s}\left(n(x)\frac{d(\delta v)}{dx}\right)dx\right]\left[EI\int_{\Gamma_i^s}\left(n(x)\frac{dv}{dx}\right)dx\right] - \int_\Omega \rho A_0\delta v\frac{d^2v(x,t)}{dt^2}dx$$

$$= \int_\Omega \delta v b_y dx + V_\Gamma\delta v|_{\Gamma_V} - M_\Gamma\frac{d(\delta v)}{dx}\Big|_{\Gamma_M} \tag{10.45}$$

For NS-PIM, substituting Equations 10.30, 10.35, and 10.36 (or Equations 10.37 and 10.38
for zero rotation boundary), into Equation 10.45, a set of discretized algebraic system
equations can be obtained in the following matrix form:

$$\bar{\mathbf{K}}\bar{\mathbf{U}} + \mathbf{M}\ddot{\bar{\mathbf{U}}} = \mathbf{F} \tag{10.46}$$

where $\bar{\mathbf{K}}$ and $\mathbf{F}$ are the same as those in Equation 10.39. The mass matrix $\mathbf{M}$ can be obtained
in exactly the same way as in the FEM [1,2]. Here, we simply use the so-called lumped
mass scheme:

$$\mathbf{M} = \text{diag}\{m_1, m_2, \ldots, m_n, \ldots, m_{N_n}\} \qquad (10.47)$$

where m_n is the mass of the smoothing domain corresponding to the node n and given by

$$m_n = \rho A_0 l_n^s \qquad (10.48)$$

For free vibration analysis, all external forces will be zero, and a general solution to Equation 10.46 can be written as

$$\bar{\mathbf{U}} = \bar{\mathbf{U}}_A e^{i\omega t} \qquad (10.49)$$

where
  $\bar{\mathbf{U}}_A$ is the vector of the amplitude of nodal deflections
  ω is the angular frequency

Substituting Equation 10.49 into Equation 10.46 yields the eigenvalue equation

$$(\bar{\mathbf{K}} - \omega^2 \mathbf{M})\bar{\mathbf{U}}_A = 0 \qquad (10.50)$$

which gives eigenvalues related to the natural frequencies and eigenvectors related to vibration modes.

10.8 Numerical Examples of Static Analysis

Example 10.2: Static Deformation and Moments in Thin Beams

A thin beam of length $L = 1.0$ subjected to different boundary conditions is considered in this subsection. The parameters are taken as $EI = 1.0$ and $q_0 = 1.0$. In this study, we use three models: 2-DOF PIM that is exactly same as the 2-DOF FEM, 1-DOF PIM, and NS-PIM. The descriptions for these three models are given in Table 10.2.

Figure 10.6 shows computed results for the deflection, rotation, and moment obtained using these models, together with the exact solution. The beam is simply-simply supported and subjected to a uniform load over the entire span. Overall, these numerical results agree very well, and the differences are not distinguishable on these figures, especially for deflections and rotations. When we look at the detail on the two pinned boundary points, we observe that NS-PIM gives the best prediction on the moments.

Figure 10.7 shows computed results for a beam subjected to a concentrated load at the mid-span. Again, these results agree very well and the differences are not distinguishable on these figures, especially for deflections and rotations. When we look at the detail on the two pinned boundary points and the loading point, we observe that NS-PIM gives the best prediction on the moments. This shows that smoothing operations can improve solutions at nonsmoothing points for field variables.

Figure 10.8 shows computed results for a clamped-clamped beam subjected to a uniform load over the beam span. In this example, the results for deflection agree very well for all

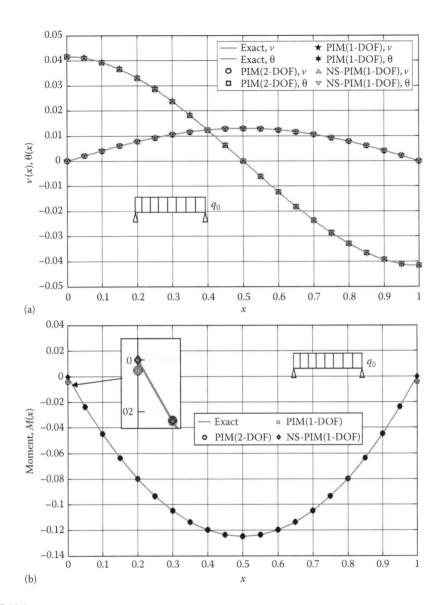

FIGURE 10.6
Results obtained using three numerical models and the exact solution for the simply-simply supported beam subjected to a uniform load: (a) deflection and rotation; and (b) moment.

the methods. The differences for the rotation at boundary points are visible. The 1-DOF PIM clearly has a problem in directly enforcing the rotation at the clamped points, and gives less accurate moment solutions. We observe again that NS-PIM gives the best prediction on all the results: deflection, rotation, and moment, at these nonsmoothing points.

Figure 10.9 shows computed results for a clamped-clamped beam subjected to a concentrated load at the mid-span. In addition to what we have observed from Figure 10.8, we observe that NS-PIM gives the best prediction at the loading point.

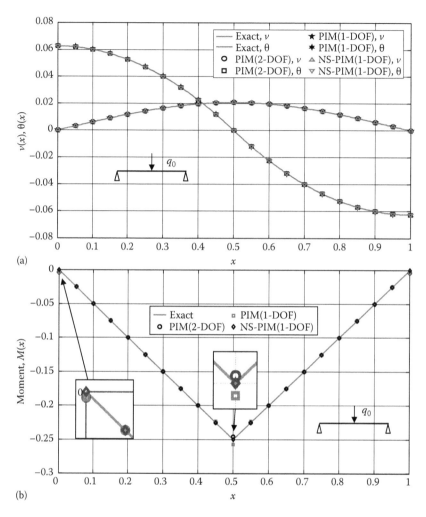

FIGURE 10.7
Results obtained using three numerical models and the exact solution for the simply-simply supported beam subjected to a concentrated load at the mid-span: (a) deflection and rotation; and (b) moment.

Figure 10.10 shows computed results for a cantilever beam subjected to a uniform load over the span of the beam. The results reconfirm that NS-PIM gives the best prediction for all the results for the moment at the clamped point.

10.9 Numerical Examples: Upper-Bound Solution

It is known that a fully compatible FEM model produces a lower-bound solution to the exact solution for force-driving problems. We have also seen NS-PIMs are capable of producing upper-bound solutions, in Chapter 8. We confirm this again for thin beams.

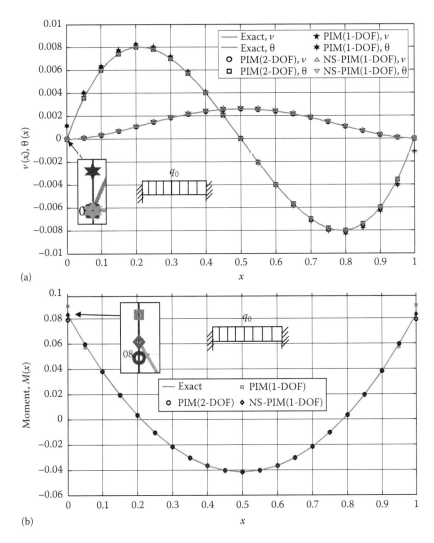

FIGURE 10.8
Results obtained using three numerical models and the exact solution for the clamped-clamped beam subjected to a uniform load: (a) deflection and rotation; and (b) moment.

Figure 10.11 plots the numerical solution of strain energy for a simply-simply supported thin beam subjected to a concentrated load at the mid-span, together with the exact solution. The solutions are obtained using 2-DOF PIM (FEM) and NS-PIM with increasing total DOFs. It is seen that the FEM gives very accurate solutions in an energy norm, and it does not change much when the total DOFs increase. In fact, it is very close to the exact solution due to the very high (third)-order interpolation used. In checking the numbers we confirmed that it gives lower-bound solutions. On the other hand, the NS-PIM gives less accurate solutions in an energy norm due to the lower (linear)-order interpolation used, but it produces the important upper-bound solutions.

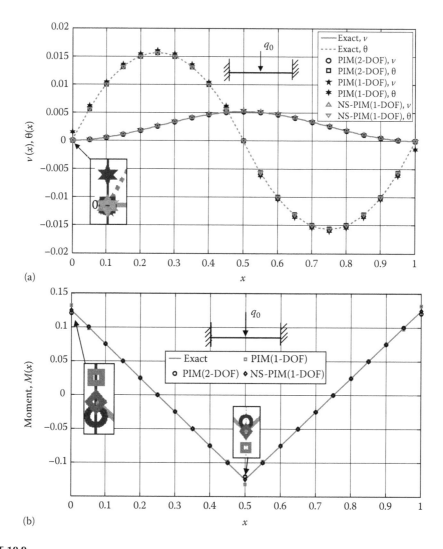

FIGURE 10.9

Results obtained using three numerical models and the exact solution for the clamped-clamped beam subjected to a concentrated load at the mid-span: (a) deflection and rotation; and (b) moment.

It approaches very quickly to the exact solution as shown in Figure 10.12. The same upper-bound solution is also observed for a clamped-clamped beam subjected to a concentrated load at the mid-span (see Figure 10.12). When the load becomes uniformly distributed, FEM needs more nodes, but still converges very fast to the exact solution as shown in Figure 10.13. The NS-PIM solutions are less accurate in an energy norm, but are upper bounds to the exact solutions.

Figure 10.14 plots the numerical solution of strain energy for a simply-simply supported thin beam subjected to a uniformly distributed load over the beam span. It is hard to tell

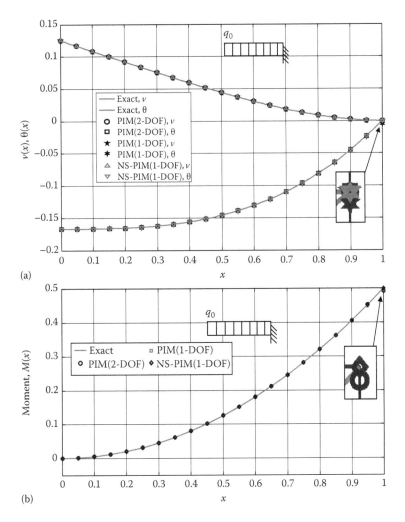

FIGURE 10.10
Results obtained using three numerical models and the exact solution for the cantilever beam subjected to a uniform load. (a) deflection and rotation; and (b) moment.

from this figure as to whether it still gives an upper-bound solution, because these curves are too close. In checking the numbers we found that the NS-PIM is a lower bound, but with four digits being the same as the exact values. Therefore, it seems to be a bottom-line case. As discussed in Chapter 8, we know that two nodes on the boundary do not receive any smoothing effects, which could be the cause of the problem. To confirm this we deliberately reduce the length of these two cells on the boundary to one-fourth of that of the other cells. This should reduce the boundary effects, and we should obtain upper-bound solutions. The test results are plotted in Figure 10.15: the NS-PIM, indeed, produces upper-bound solutions, after reducing the boundary effects.

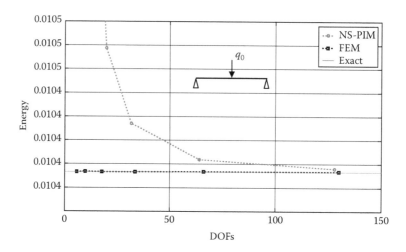

FIGURE 10.11
Solution bounds in strain energy for a simply-simply supported beam subjected to a concentrated load. FEM (2-DOF PIM) gives lower bounds and NS-PIM gives upper bounds.

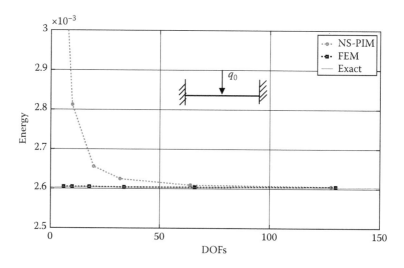

FIGURE 10.12
Solution bounds in strain energy for a clamped-clamped beam subjected to a concentrated load. FEM (2-DOF PIM) gives lower bounds and NS-PIM gives upper bounds.

10.10 Numerical Examples of Free Vibration Analysis

Example 10.3: Free Vibration of Thin Beams

For free vibration analysis, thin beams with different boundary conditions are considered here. The geometrical and material parameters are the same as those given in Section 10.8. For providing a fair comparison, we use meshes with the same number of total DOFs for all the methods used,

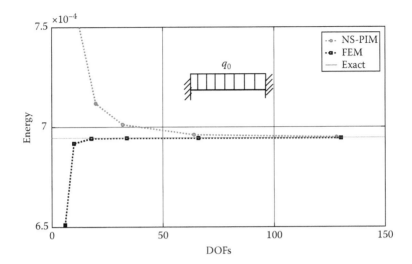

FIGURE 10.13
Solution bounds in strain energy for a clamped-clamped beam subjected to a uniform load. FEM (2-DOF PIM) gives lower bounds and NS-PIM gives upper bounds.

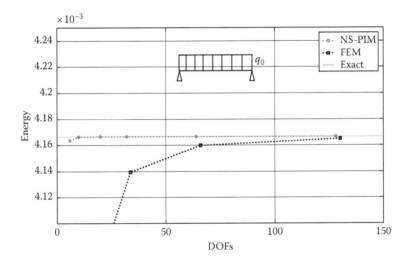

FIGURE 10.14
Solution bounds in strain energy for a simply-simply supported beam subjected to a uniform load. FEM (2-DOF PIM) gives lower bounds, but it gives an extremely tight lower bound, due to the boundary effects.

meaning that for 2-DOF PIM we use 41 nodes, but for 1-DOF PIM and NS-PIM we use 81 nodes.

All the results are given in a dimensionless parameter of frequencies: $\beta_i = \sqrt{\omega_i \sqrt{\rho/EI}}$. The results of thin beams with simply-simply, clamped-clamped, cantilever, and simply-clamped supports are computed and listed in Tables 10.3 through 10.6. Examination of these data leads to the following remarks.

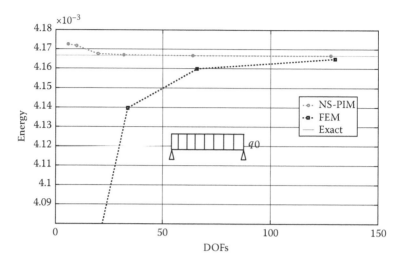

FIGURE 10.15

Solution bounds in strain energy for a simply-simply supported beam subjected to a uniform load. FEM (2-DOF PIM) gives lower bounds and NS-PIM gives upper bounds, after reducing the boundary effects by reducing the length of the two boundary cells to one-fourth.

TABLE 10.3

Natural Frequencies for Simply-Simply Supported Thin Beam $\left(\beta_i = \sqrt{\omega_i \sqrt{\rho/EI}}\right)$

Modes	2-DOF PIM (FEM)	Analytical	NS-PIM	1-DOF PIM
1	3.14159	3.14159	3.14139	3.14119
2	6.28319	6.28318	6.28157	6.27996
3	9.42479	9.42477	9.41933	9.41388
4	12.56641	12.56636	12.55346	12.54056
5	15.70809	15.70795	15.68274	15.65757
6	18.84988	18.84954	18.80598	18.76253
7	21.99184	21.99113	21.92197	21.85306
8	25.13409	25.13272	25.02951	24.92680

TABLE 10.4

Natural Frequencies for Clamped-Clamped Thin Beam $\left(\beta_i = \sqrt{\omega_i \sqrt{\rho/EI}}\right)$

Modes	2-DOF PIM (FEM)	NS-PIM	1-DOF PIM
1	4.73004	4.72852	4.72696
2	7.85321	7.84764	7.84191
3	10.99563	10.98226	10.96846
4	14.13724	14.11103	14.08399
5	17.27897	17.23366	17.18699
6	20.42083	20.34892	20.27501
7	23.56292	23.45562	23.34571
8	26.70536	26.55258	26.39675

TABLE 10.5

Natural Frequencies for Cantilever Thin Beam $\left(\beta_i = \sqrt{\omega_i \sqrt{\rho/EI}} \right)$

Modes	2-DOF PIM (FEM)	Analytical	NS-PIM	1-DOF PIM
1	1.87510	1.87510	1.87498	1.87492
2	4.69409	4.69406	4.69252	4.69151
3	7.85476	7.85398	7.84921	7.84517
4	10.99556	10.99557	10.98219	10.97175
5	14.13724	14.13717	14.11103	14.08964
6	17.27897	17.27876	17.23366	17.19553
7	20.42083	20.42035	20.34892	20.28710
8	23.56292	23.56194	23.45562	23.36197

TABLE 10.6

Natural Frequencies for Simply-Clamped Thin Beam $\left(\beta_i = \sqrt{\omega_i \sqrt{\rho/EI}} \right)$

Modes	2-DOF PIM (FEM)	Analytical	NS-PIM	1-DOF PIM
1	3.926<u>60</u>	3.926<u>99</u>	3.92591	3.92520
2	7.06859	7.06858	7.06531	7.06197
3	10.21019	10.21018	10.20122	10.19209
4	13.35183	13.35177	13.33282	13.31349
5	16.49353	16.49336	16.45890	16.42378
6	19.63535	19.63495	19.57828	19.52055
7	22.77737	22.77655	22.68974	22.60143
8	25.91971	25.91814	25.79211	25.66408

Remark 10.2: Softness of the Models

1. The 2-DOF PIM or FEM is found "stiffer" (larger natural frequencies) than the exact model. The only exception is the mode 1 underlined in Table 10.6.

2. The NS-PIM is found "softer" (smaller natural frequencies) than the exact model. This confirms our softness theory, which explains also why NS-PIM produces upper-bound solutions.

3. The 1-DOF PIM is found even softer than the NS-PIM model.

10.11 Concluding Remarks

In this chapter, point interpolation methods for analyzing Bernoulli–Euler beams governed by fourth-order differential equations are presented using the Galerkin weak form and the GS-Galerkin weakened-weak form. Three models, 2-DOF PIM (FEM), 1-DOF PIM, and

NS-PIM have been established for static and dynamic analysis. We mention the following key points.

1. NS-PIM based on the weakened-weak form solves well fourth-order beam equations with assumed functions of only first-order consistence. The C^1 continuity requirement in weak formulation is now removed.

2. The NS-PIM is found "softer" (smaller natural frequencies) than the exact model and capable in producing upper-bound solutions in an energy norm for force-driving problems. This confirms our softness theory.

3. The NS-PIM produces good results, and stands clearly out for solutions at non-smoothing points.

4. The 1-DOF PIM is found even softer than the NS-PIM.

The NS-PIM formulation can be used for higher-order shape functions for thin beams. It can also be extended to higher-order beam theories without much technical difficulty, using only linear or higher-order interpolations. We demonstrate this for more complicated cases of plates in Chapter 11 and shells in Chapter 12.

PIMs based on the local Petrov–Galerkin weak form or LPIM works well also for beams, but is less efficient for asymmetric system equations [1]. We prefer to use the FEM (2-DOF PIM) for lower bound and NS-PIM for upper bound for beams, because of their efficiency and accuracy.

References

1. Petyt, M., *Introduction to Finite Element Vibration Analysis*, Cambridge University Press, Cambridge, U.K., 1990.
2. Liu, G. R. and Quek, S. S., *The Finite Element Method: A Practical Course*, Butterworth Heinemann, Oxford, 2002.
3. Krysl, P. and Belytschko, T., Analysis of thin plates by the element-free Galerkin method, *Comput. Mech.*, 17, 26–35, 1996.
4. Cui, X. Y., Liu, G. R., Li, G. Y., and Zhang, G., A rotation free formulation for static and free vibration analysis of this beams using gradient smoothing technique, *CMES-Compat. Model. Eng. Sci.*, 38(3), 217–229, 2008.
5. Liu, G. R. and Gu, Y. T., *An Introduction to Meshfree Methods and Their Programming* (Chapter 6), Springer, Dordrecht, the Netherlands, 2005.
6. Atluri, S. N., Cho, J. Y., and Kim, H. G., Analysis of thin beams, using the meshless local Petrov–Galerkin method, with generalized moving least squares interpolations, *Comput. Mech.*, 24, 334–347, 1999.
7. Gu, Y. T. and Liu, G. R., A local point interpolation method for static and dynamic analysis of thin beams, *Comput. Methods Appl. Mech. Eng.*, 190, 5515–5528, 2001.

11

Meshfree Methods for Plates

With the wide engineering applications of plate structures with complex geometry, static and dynamic analyses of plates with complicated shapes become very important. However, exact analyses of such plates are usually very difficult. Therefore, numerical techniques with different discretization schemes such as the finite element method (FEM) have been developed. FEM has achieved remarkable success in static and dynamic analyses of plates.

Meshfree methods have also been developed for analysis of plates. Krysl and Belytschko [1,2] have extended the element-free Galerkin (EFG) method to static analysis of thin plates and shells. In their work, the essential boundary conditions are enforced by a method of Lagrange multipliers. An EFG method has also been formulated for modal analyses of Euler–Bernoulli beams and Kirchhoff plates [3]. EFG has also been formulated for dynamic problems [4], buckling problems of thin plates [5,6], as well as composite laminates [5–7]. In these works, the essential boundary conditions are imposed using orthogonal transform techniques.

EFG formulations for thick plates were presented for both static and dynamic problems based on the first- and third-order shear deformation theories [4]. When using higher-order plate theories, the well-known issue of shear locking arises. Therefore, different methods for eliminating shear locking have also been discussed.

Point interpolation methods (PIMs) based on weakened-weak (W^2) formulations and PIM shape functions with the Kronecker delta function property have also been formulated for plates. The work is still in progress, and some of the current works on edge-based smoothed point interpolation method (ES-PIM) that are very stable and efficient will be presented in this chapter.

This chapter is dedicated to meshfree methods for analysis of plates. Both formulations and applications of EFG and PIMs will be presented in detail. Problems considered include static deformation, buckling, and dynamic response of thin and thick plates, including composite laminated plates.

11.1 Mechanics for Plates

A plate is geometrically similar to a two-dimensional (2D) solid occupying a 2D domain Ω bounded by Γ, as shown in Figure 11.1. The difference is that the forces/loads applied on a plate are in a direction perpendicular to the plane of the plate. A plate can also be viewed as a 2D analog of a beam. A plate experiences a bending resulting in a deflection w in the z-direction as a function of x and y. The stress σ_{zz} in a plate is assumed to be zero. In this chapter, we consider plates undergoing only bending deformation, by which we mean that there exists a *neutral plane* in the plate where no in-plane deformation occurs.

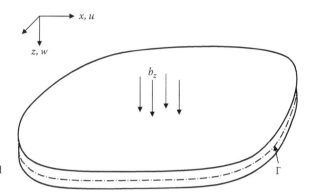

FIGURE 11.1
Plate of thickness h subjected to transverse loads.

Similar to beams, there are several theories for analyzing the deflection in plates. These theories can also be basically divided into two major categories: theory for *thin* plates and theory for *thick* plates. This chapter addresses the following:

- Thin plate theory, often called *classic* plate theory (CPT) and also known as *Kirchhoff* plate theory
- The *first-order shear deformation theory* (FSDT) known as *Mindlin* plate theory
- The *third-order shear deformation theory* (TSDT)

11.1.1 Thin Plates: Classic Plate Theory

11.1.1.1 Deformation Theory

The CPT assumes that the normal to the neutral surface of the undeformed plate remains straight and normal to the neutral surface during deformation or bending. This assumption is often called the *Kirchhoff* assumption. The Kirchhoff assumption results in

$$\gamma_{xz} = 0, \quad \gamma_{yz} = 0 \tag{11.1}$$

Thus, we have only three strains—ε_{xx}, ε_{yy}, and γ_{xy}—to deal with and they are all in the plane of the plate. Therefore, it can share the same constitutive equation of 2D solids of plane stress, Equation 1.17. In addition, the displacement components in the x- and y-directions, u and v, at a distance z from the neutral surface can be expressed by

$$u = -z\frac{\partial w}{\partial x} \tag{11.2}$$

$$v = -z\frac{\partial w}{\partial y} \tag{11.3}$$

where w is the deflection of the neutral plane of the plate in the z-direction.

Using Equations 11.2 and 11.3, we obtain a simple relationship:

$$\mathbf{u} = \left\{ \begin{array}{c} u \\ v \\ w \end{array} \right\} = \underbrace{\left\{ \begin{array}{c} -z\frac{\partial}{\partial x} \\ -z\frac{\partial}{\partial y} \\ 1 \end{array} \right\}}_{\mathbf{L}_u} w = \mathbf{L}_u w \tag{11.4}$$

11.1.1.2 Stress and Strain

It is clearly shown here that all three displacement components are expressed in terms of the deflection w, due to the Kirchhoff assumption. The relationship between the three components of strain and the deflection can be given by

$$\varepsilon_{xx} = \frac{\partial u}{\partial x} = -z\frac{\partial^2 w}{\partial x^2} \tag{11.5}$$

$$\varepsilon_{yy} = \frac{\partial v}{\partial y} = -z\frac{\partial^2 w}{\partial y^2} \tag{11.6}$$

$$\gamma_{xy} = \frac{\partial u}{\partial y} + \frac{\partial v}{\partial x} = -2z\frac{\partial^2 w}{\partial x \partial y} \tag{11.7}$$

or in matrix form

$$\boldsymbol{\varepsilon} = z\mathbf{L}_d w \tag{11.8}$$

where $\boldsymbol{\varepsilon}$ is the vector of in-plane strains defined by

$$\boldsymbol{\varepsilon} = \left\{ \begin{array}{c} \varepsilon_{xx} \\ \varepsilon_{yy} \\ \gamma_{xy} \end{array} \right\} \tag{11.9}$$

where

$\gamma_{xy} = 2\varepsilon_{xy}$ is "engineering" shear strain
$\mathbf{L}_d$ is the differential operator matrix for CPT plates given by

$$\mathbf{L}_d = \begin{bmatrix} -\dfrac{\partial^2}{\partial x^2} \\[2mm] -\dfrac{\partial^2}{\partial y^2} \\[2mm] -2\dfrac{\partial^2}{\partial x \partial y} \end{bmatrix} \quad \text{for CPT} \tag{11.10}$$

We now define the *pseudostrain* as

$$\boldsymbol{\varepsilon}_p = \mathbf{L}_d w \tag{11.11}$$

that is, the strain evaluated at the neutral plane of the plate and hence is independent of the coordinate z. The in-plane (normal and shear) stresses σ_{xx}, σ_{yy}, and σ_{xy} can be obtained by substituting Equation 11.8 into the constitutive equation:

$$\boldsymbol{\sigma} = z\mathbf{c}\mathbf{L}_d w \tag{11.12}$$

It is seen from the above equation that the in-plane stresses vary linearly in the vertical direction on the cross sections of the plate, due to the Kirchhoff assumption.

11.1.1.3 Moments and Shear Forces

Figure 11.2a shows the stresses on the cross sections of a small representative cell of $dx \times dy$ from a plate of thickness h. The plate cell is subjected to external force b_z, and inertial force $\rho h \ddot{w}$, where ρ is the density of the material. Figure 11.2b shows the moments M_{xx}, M_{yy}, M_{zz}, and M_{xy}, and shear forces V_{xz} and V_{yz}. The moments are resulted from the distributed in-plane stresses σ_{xx}, σ_{yy}, and σ_{xy} on the cross section, and can be calculated by the following integration:

$$\left\{ \begin{array}{c} M_{xx} \\ M_{yy} \\ M_{xy} \end{array} \right\} = \int_{-h/2}^{h/2} \boldsymbol{\sigma} z \, dz = \left(\int_{-h/2}^{h/2} \mathbf{c} z^2 dz \right) \mathbf{L}_d w = \left(\int_{-h/2}^{h/2} \mathbf{c} z^2 dz \right) \boldsymbol{\varepsilon}_p = \mathbf{D} \boldsymbol{\varepsilon}_p \qquad (11.13)$$

where $\mathbf{D}$ is a matrix of the constants related to the material property and the thickness of the plate. For inhomogeneous plates such as laminated plates, $\mathbf{c}$ can be a function of z, $\mathbf{D}$ will then depend on the configuration (stacking sequence of the layers) of the plate, and it will have the following general form:

$$\mathbf{D} = \int_{-h/2}^{h/2} \mathbf{c} z^2 dz = \begin{bmatrix} D_{11} & D_{12} & D_{16} \\ D_{12} & D_{22} & D_{26} \\ D_{16} & D_{26} & D_{66} \end{bmatrix} \qquad (11.14)$$

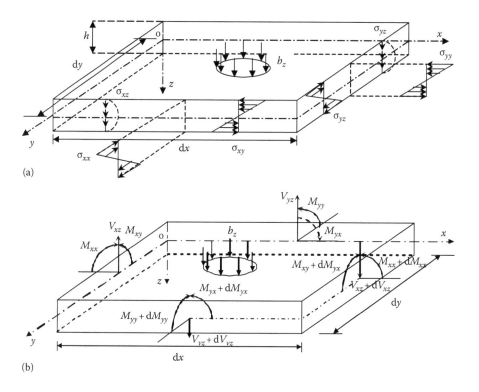

(a)

(b)

FIGURE 11.2
An isolated representative cell of $dx \times dy$ in a plate. (a) Stresses on the cross sections; and (b) shear forces and moments on the cross sections.

where D_{ij} ($i, j = 1, 2, 6$) are the constants that can be determined once the material and the layer configuration of the plate are defined. For homogeneous plates, we simply have

$$\mathbf{D} = \int_{-h/2}^{h/2} \mathbf{c}z^2 dz = \frac{h^3}{12}\mathbf{c} \tag{11.15}$$

Using Equation 1.17, we obtain

$$\mathbf{D} = \begin{bmatrix} D_{11} & D_{12} & D_{16} \\ D_{12} & D_{22} & D_{26} \\ D_{16} & D_{26} & D_{66} \end{bmatrix} = \underbrace{\frac{Eh^3}{12(1-\nu^2)}}_{D} \begin{bmatrix} 1 & \nu & 0 \\ \nu & 1 & 0 \\ 0 & 0 & (1-\nu)/2 \end{bmatrix} \tag{11.16}$$

where D is the flexural rigidity of the plate:

$$D = \frac{Eh^3}{12(1-\nu^2)} \tag{11.17}$$

We define now the *pseudostress* as

$$\boldsymbol{\sigma}_p = \begin{Bmatrix} M_{xx} \\ M_{yy} \\ M_{xy} \end{Bmatrix} \tag{11.18}$$

which is also independent of coordinate z.

11.1.1.4 Constitutive Equation for a Thin Plate

By using Equations 11.13 and 11.18, the generalized Hooke's law for a thin plate becomes

$$\boldsymbol{\sigma}_p = \mathbf{D}\boldsymbol{\varepsilon}_p \tag{11.19}$$

11.1.1.5 Dynamic Equilibrium Equations

Using Equations 11.13, 11.11, and 11.16, we can have explicit expressions for the moments in homogeneous plates of isotropic materials:

$$M_{xx} = -\left(D_{11}\frac{\partial^2 w}{\partial x^2} + D_{12}\frac{\partial^2 w}{\partial y^2}\right) \tag{11.20}$$

$$M_{yy} = -\left(D_{12}\frac{\partial^2 w}{\partial x^2} + D_{22}\frac{\partial^2 w}{\partial y^2}\right) \tag{11.21}$$

$$M_{xy} = -2D_{66}\frac{\partial^2 w}{\partial x \partial y} \tag{11.22}$$

In deriving the system equilibrium equations, first we consider the equilibrium of the small plate cell in the z-direction (see Figure 11.2b). Using $dV_{xz} = \frac{\partial V_{xz}}{\partial x}dx$ and $dV_{yz} = \frac{\partial V_{yz}}{\partial y}dy$, we have

$$\left(\frac{\partial V_{xz}}{\partial x} dx\right) dy + \left(\frac{\partial V_{yz}}{\partial y} dy\right) dx + (b_z - \rho h \ddot{w}) dx\, dy = 0 \tag{11.23}$$

or

$$\frac{\partial V_{xz}}{\partial x} + \frac{\partial V_{yz}}{\partial y} + b_z = \rho h \ddot{w} \tag{11.24}$$

Considering then the moment equilibrium of the plate cell with respect to the x-axis, and neglecting the second-order small terms leads to a formula for shear force V_{xz} given in terms of moments.

$$V_{xz} = \frac{\partial M_{xx}}{\partial x} + \frac{\partial M_{xy}}{\partial y} \tag{11.25}$$

Finally, considering the moment equilibrium of the plate cell with respect to the y-axis and neglecting the second-order small terms gives

$$V_{yz} = \frac{\partial M_{xy}}{\partial x} + \frac{\partial M_{yy}}{\partial y} \tag{11.26}$$

The dynamic equilibrium equation for plates can be obtained by substituting Equation 11.13 into Equations 11.25 and 11.26, and then into Equation 11.24. For homogeneous and isotropic plates we have

$$D\left(\frac{\partial^4 w}{\partial x^4} + 2\frac{\partial^4 w}{\partial x^2 \partial y^2} + \frac{\partial^4 w}{\partial y^4}\right) + \rho h \ddot{w} = b_z \tag{11.27}$$

where D is the flexural rigidity of the plate. A more general form of partial differential equation (PDE) for governing symmetric laminates of anisotropic materials can be derived in a similar manner, and is given in [11].

$$\frac{\partial^2}{\partial x^2}\left(D_{11}\frac{\partial^2 w}{\partial x^2} + D_{12}\frac{\partial^2 w}{\partial y^2}\right) + \frac{\partial^2}{\partial y^2}\left(D_{12}\frac{\partial^2 w}{\partial x^2} + D_{22}\frac{\partial^2 w}{\partial y^2}\right)$$
$$+ 2\frac{\partial^2}{\partial x \partial y}\left(2D_{66}\frac{\partial^2 w}{\partial x \partial y}\right) - b_z(x,y) + I_0\frac{\partial^2 w}{\partial t^2} - I_2\frac{\partial^2}{\partial t^2}\left(\frac{\partial w}{\partial x} + \frac{\partial w}{\partial y}\right) = 0 \tag{11.28}$$

where D_{ij} can be obtained using Equation 11.14. In Equation 11.28, I_0 is the mass per unit area of the plate defined by

$$I_0 = \int\limits_{-h/2}^{h/2} \rho\, dz \tag{11.29}$$

For plates of homogeneous materials, we have

$$I_0 = \int\limits_{-h/2}^{h/2} \rho\, dz = \rho h \tag{11.30}$$

In Equation 11.28, I_2 is the mass moments of inertia given by

$$I_2 = \int_{-h/2}^{h/2} \rho z^2 dz \tag{11.31}$$

For plates of homogeneous materials, we have

$$I_2 = \int_{-h/2}^{h/2} \rho z^2 dz = \rho h^3/12 \tag{11.32}$$

The static equilibrium equation for plates of isotropic material can be obtained by dropping the dynamic terms in Equation 11.27:

$$D\left(\frac{\partial^4 w}{\partial x^4} + 2\frac{\partial^4 w}{\partial x^2 \partial y^2} + \frac{\partial^4 w}{\partial y^4}\right) = b_z \tag{11.33}$$

which is a fourth-order PDE. The static equilibrium equation for symmetric laminates of anisotropic materials can be obtained by dropping the dynamic terms in Equation 11.28:

$$\frac{\partial^2}{\partial x^2}\left(D_{11}\frac{\partial^2 w}{\partial x^2} + D_{12}\frac{\partial^2 w}{\partial y^2}\right) + \frac{\partial^2}{\partial y^2}\left(D_{12}\frac{\partial^2 w}{\partial x^2} + D_{22}\frac{\partial^2 w}{\partial y^2}\right)$$
$$+ 2\frac{\partial^2}{\partial x \partial y}\left(2D_{66}\frac{\partial^2 w}{\partial x \partial y}\right) - b_z(x,y) = 0 \tag{11.34}$$

which is also a fourth-order PDE.

11.1.1.6 Boundary Conditions

The boundary conditions for plates are given at boundary Γ as follows. The conditions on the essential boundaries are

$$w = w_\Gamma, \qquad \text{on essential boundary } \Gamma_w \tag{11.35}$$

$$\varphi_n = \frac{\partial w}{\partial n} = \varphi_\Gamma, \quad \text{on essential boundary } \Gamma_\theta \tag{11.36}$$

where
φ_n is the rotation on the boundary about the boundary line
w_Γ and φ_Γ are the prescribed deflection and rotation on the essential boundaries
n denotes the normal of the boundary of the problem domain Ω

On the natural boundaries, we have

$$M_n = M_{\Gamma n}, \quad \text{on } \Gamma_M \tag{11.37}$$

$$M_{nt} = M_{\Gamma nt}, \quad \text{on } \Gamma_M \tag{11.38}$$

$$V_n = V_{\Gamma_n}, \quad \text{on } \Gamma_V \tag{11.39}$$

where M_n, M_{nt}, and V_n denote the moment, torsional moment, and shear force on the boundary of the plate, which are defined by

$$M_n = n_x^2 M_{xx} + 2n_x n_y M_{xy} + n_y^2 M_{yy} \tag{11.40}$$

$$M_{nt} = -n_y n_x M_{xx} + n_x n_y M_{yy} \tag{11.41}$$

$$V_n = n_x V_{xz} + n_y V_{yz} \tag{11.42}$$

in which $\{n_x, n_y\}$ is the unit outward normal vector on the boundary. In Equations 11.37 through 11.39, $M_{\Gamma n}$, $M_{\Gamma nt}$, and $V_{\Gamma n}$ are, respectively, the prescribed moment, torsional moment, and shear force on the plate edges of natural boundary.

The essential boundary condition can be written in a concise form of

$$\mathbf{u}_b = \mathbf{u}_\Gamma \quad \text{on essential boundary} \quad \Gamma_u = \Gamma_w \cup \Gamma_\theta \tag{11.43}$$

where $\mathbf{u}_b$ is a vector consisting of the prescribed deflection and rotations at the essential boundary of the plate given by

$$\mathbf{u}_b = \mathbf{L}_b w \tag{11.44}$$

where $\mathbf{L}_b$ is a vector of differential operators.

For CPT plates, $\mathbf{L}_b$ is given by

$$\mathbf{L}_b = \left\{ \begin{matrix} 1 \\ \dfrac{\partial}{\partial n} \end{matrix} \right\} \quad \text{for clamped boundary} \tag{11.45}$$

and

$$\mathbf{L}_b = \left\{ \begin{matrix} 1 \\ 0 \end{matrix} \right\} \quad \text{for simply supported boundary} \tag{11.46}$$

Note in Equation 11.46 that the zero entry in $\mathbf{L}_b$ ensures that the deflection is constrained for simply supported boundaries.

11.1.2 Thicker Plates: A First-Order Shear Deformation Theory

The Mindlin plate theory is applied for thicker plates, as the shear deformation and rotary inertia effects can be included. The Mindlin plate theory is also known as the FSDT. The Mindlin theory does not demand that the cross section be perpendicular to the neutral plane after deformation. The situation is very similar to that of the Timoshenko beam, but is extended in one more dimension. Thus, we usually have $\gamma_{xz} \neq 0$, $\gamma_{yz} \neq 0$. Therefore, we will have five components of strains and stresses to deal with.

The displacements u and v, which are parallel to the undeformed neutral surface, at a distance z from the neutral plan, can be expressed by

$$u = z\varphi_x \tag{11.47}$$

$$v = z\varphi_y \tag{11.48}$$

where φ_x and $-\varphi_y$ denote the rotations of the cross section of the plate about the y- and x-axes, respectively. The deflection of the plate is still represented by the deflection at the neutral plane of the plate, and is denoted by w. The vector of the displacements can be expressed as

$$\left\{ \begin{array}{c} u \\ v \\ w \end{array} \right\} = \underbrace{\left[\begin{array}{ccc} 0 & z & 0 \\ 0 & 0 & z \\ 1 & 0 & 0 \end{array} \right]}_{\mathbf{L}_u} \underbrace{\left\{ \begin{array}{c} w \\ \varphi_x \\ \varphi_y \end{array} \right\}}_{\mathbf{u}} = \mathbf{L}_u \mathbf{u} \tag{11.49}$$

where

$$\mathbf{u} = \left\{ \begin{array}{c} w \\ \varphi_x \\ \varphi_y \end{array} \right\} \tag{11.50}$$

is the vector of three independent field variables for Mindlin plates.

By using Equation 1.21 for general solids, and removing the components ε_{zz}, the strains in the Mindlin plate are expressed as follows:

$$\underbrace{\left\{ \begin{array}{c} \varepsilon_{xx} \\ \varepsilon_{yy} \\ \gamma_{xy} \\ \gamma_{xz} \\ \gamma_{yz} \end{array} \right\}}_{\boldsymbol{\varepsilon}} = \underbrace{\left[\begin{array}{ccc} 0 & z\dfrac{\partial}{\partial x} & 0 \\[2mm] 0 & 0 & z\dfrac{\partial}{\partial y} \\[2mm] 0 & z\dfrac{\partial}{\partial y} & z\dfrac{\partial}{\partial x} \\[2mm] \dfrac{\partial}{\partial x} & 1 & 0 \\[2mm] \dfrac{\partial}{\partial y} & 0 & 1 \end{array} \right]}_{\mathbf{L}_d} \left\{ \begin{array}{c} w \\ \varphi_x \\ \varphi_y \end{array} \right\} = \mathbf{L}_d \mathbf{u} \tag{11.51}$$

The strain (or stress) components given in Equation 11.51 can be divided into in-plane strains and off-plane (transverse) strains. Such a division is useful in some of the energy formulations when we need to account for in-plane and off-plane strain energies in a separate manner (see Section 11.5). The in-plane strains can be extracted from Equation 11.51, and given by

$$\boldsymbol{\varepsilon}_z = z\mathbf{L}_B\boldsymbol{\varphi} \tag{11.52}$$

where

$$\mathbf{L}_B = \left[\begin{array}{cc} \dfrac{\partial}{\partial x} & 0 \\[2mm] 0 & \dfrac{\partial}{\partial y} \\[2mm] \dfrac{\partial}{\partial x} & \dfrac{\partial}{\partial y} \end{array} \right] \tag{11.53}$$

$$\boldsymbol{\varphi} = \left\{ \begin{array}{c} \varphi_x \\ \varphi_y \end{array} \right\} \tag{11.54}$$

The off-plane transverse shear strains γ_{xz} and γ_{yz} are also extracted from Equation 11.51:

$$\boldsymbol{\gamma} = \left\{ \begin{matrix} \gamma_{xz} \\ \gamma_{yz} \end{matrix} \right\} = \left\{ \begin{matrix} \varphi_x + \dfrac{\partial w}{\partial x} \\ \varphi_y + \dfrac{\partial w}{\partial y} \end{matrix} \right\} \tag{11.55}$$

Note that if the transverse shear strains are negligible, the above equation will lead to

$$\varphi_y = -\frac{\partial w}{\partial y} \tag{11.56}$$

$$\varphi_x = -\frac{\partial w}{\partial x} \tag{11.57}$$

The FSDT becomes the CPT. This implies that when we construct the functions for the rotations in a similar manner as in Equations 11.56 and 11.57, the model based on FSDT can also work for thin plates. This technique will be used in Example 11.16.

11.1.2.1 Constitutive Equation for Thick Plates

For isotropic linear elastic materials, the stresses can be obtained using Equations 1.10 and 1.14 and by simply removing the components σ_{zz}:

$$\underbrace{\left\{ \begin{matrix} \sigma_{xx} \\ \sigma_{yy} \\ \sigma_{xy} \\ \sigma_{xz} \\ \sigma_{yz} \end{matrix} \right\}}_{\boldsymbol{\sigma}} = \underbrace{\frac{E}{1-\nu^2} \begin{bmatrix} 1 & \nu & 0 & 0 & 0 \\ \nu & 1 & 0 & 0 & 0 \\ 0 & 0 & (1-\nu)/2 & 0 & 0 \\ 0 & 0 & 0 & \kappa(1-\nu)/2 & 0 \\ 0 & 0 & 0 & 0 & \kappa(1-\nu)/2 \end{bmatrix}}_{\mathbf{D}} \underbrace{\left\{ \begin{matrix} \varepsilon_{xx} \\ \varepsilon_{yy} \\ \gamma_{xy} \\ \gamma_{xz} \\ \gamma_{yz} \end{matrix} \right\}}_{\boldsymbol{\varepsilon}} = \mathbf{D}\boldsymbol{\varepsilon} \tag{11.58}$$

where
   ν is Poisson's ratio
   κ is the shear effectiveness factor
   $\kappa = 5/6$ that is often used for Mindlin plates

For the in-plane portion we have

$$\underbrace{\left\{ \begin{matrix} \sigma_{xx} \\ \sigma_{yy} \\ \sigma_{xy} \end{matrix} \right\}}_{\boldsymbol{\sigma}_M} = \underbrace{\frac{E}{1-\nu^2} \begin{bmatrix} 1 & \nu & 0 \\ \nu & 1 & 0 \\ 0 & 0 & (1-\nu)/2 \end{bmatrix}}_{\mathbf{c}} \underbrace{\left\{ \begin{matrix} \varepsilon_{xx} \\ \varepsilon_{yy} \\ \gamma_{xy} \end{matrix} \right\}}_{\boldsymbol{\varepsilon}_M} = \mathbf{c}\boldsymbol{\varepsilon}_M \tag{11.59}$$

where subscript "*M*" stands for membrane.

Similar to the CPT, the in-plane *bending-pseudostress* can be defined by using Equation 11.18:

$$\boldsymbol{\sigma}_B = \left\{ \begin{array}{c} M_{xx} \\ M_{yy} \\ M_{xy} \end{array} \right\} \tag{11.60}$$

which is independent of the coordinate z. We now define *bending-pseudostrain* as

$$\boldsymbol{\varepsilon}_B = \mathbf{L}_B \boldsymbol{\varphi} \tag{11.61}$$

that is strains evaluated at the neutral plane of the plate and hence is independent of the coordinate z. Using Equations 11.60 and 11.61, we have the generalized Hooke's law for bending stresses and strains of the FSDT plates:

$$\boldsymbol{\sigma}_B = \mathbf{D} \boldsymbol{\varepsilon}_B \tag{11.62}$$

where $\mathbf{D}$ is given in Equation 11.14 for general cases (laminated plates) and Equation 11.16 for homogenous plates.

The off-plane transverse average shear stress relates to the transverse shear strain in the form:

$$\left\{ \begin{array}{c} \sigma_{xz} \\ \sigma_{yz} \end{array} \right\} = \kappa \underbrace{\left[\begin{array}{cc} G & 0 \\ 0 & G \end{array} \right]}_{\mathbf{G}} \underbrace{\left\{ \begin{array}{c} \gamma_{xz} \\ \gamma_{yz} \end{array} \right\}}_{\boldsymbol{\gamma}_S} = \kappa \mathbf{G} \boldsymbol{\gamma}_S \tag{11.63}$$

where subscript "S" stands for shear, and G is the shear modulus. Both off-plane stresses and strains are all assumed to be constant over the thickness. Therefore, the off-plane *shearing-pseudostress* can be defined

$$\boldsymbol{\sigma}_S = h \left\{ \begin{array}{c} \sigma_{xz} \\ \sigma_{yz} \end{array} \right\} \tag{11.64}$$

We have the generalized Hooke's law for shear stresses and strains of FSDT plates as

$$\boldsymbol{\sigma}_S = \kappa h \mathbf{G} \boldsymbol{\gamma}_S \tag{11.65}$$

where

$$\mathbf{G} = \left[\begin{array}{cc} G & 0 \\ 0 & G \end{array} \right] \tag{11.66}$$

11.1.2.2 Essential Boundary Conditions

At simply supported edges, for plates based on FSDT, we have

$$w\big|_{\text{at edges}} = 0 \tag{11.67}$$

where w is the deflection on the neutral plane of the plate, and

$$\varphi_t|_{\text{at edges}} = 0 \tag{11.68}$$

where φ_t is the rotation with respect to the axis perpendicular to the normal direction of the boundary.

At clamped edges, we have

$$w|_{\text{at edges}} = 0 \tag{11.69}$$

$$\varphi_n|_{\text{at edges}} = 0 \tag{11.70}$$

$$\varphi_t|_{\text{at edges}} = 0 \tag{11.71}$$

where φ_n is the rotation with respect to the axis parallel to the normal direction of the boundary.

For shear deformable plates, there are two kinds of essential boundary conditions: soft type and hard type [18]. The hard-type conditions constrain both deflection and rotations, and the soft-type conditions constrain only the deflection.

11.1.3 Thick Plates: A TSDT

11.1.3.1 Deformation Theory

The Mindlin plate theory or the FSDT has some problems in the solution, such that the transverse shear forces obtained are not zero on the plate surfaces, which contradicts the actual physical situation. A more accurate theory for thick plates, called TSDT [11], has been proposed. Here, we briefly describe TSDT following the formulation in [20]. Based on TSDT, the displacement field of the plate can be expressed as

$$\left\{ \begin{array}{c} u \\ v \\ w \end{array} \right\} = \underbrace{\left[\begin{array}{ccc} -\alpha z^3 \dfrac{\partial}{\partial x} & z - \alpha z^3 & 0 \\ -\alpha z^3 \dfrac{\partial}{\partial y} & 0 & z - \alpha z^3 \\ 1 & 0 & 0 \end{array} \right]}_{\mathbf{L}_u} \underbrace{\left\{ \begin{array}{c} w \\ \varphi_x \\ \varphi_y \end{array} \right\}}_{\mathbf{u}} = \mathbf{L}_u \mathbf{u} \tag{11.72}$$

where
$\alpha = 4/(3h^2)$
$\mathbf{u}$ is the same as in Equation 11.50
w is the transverse deflection of the neutral plane of the plate
φ_x and $-\varphi_y$ denote the rotations of the cross section of the plate about the y- and x-axes, respectively

It can be easily seen that by setting $\alpha = 0$, Equation 11.72 becomes Equation 11.49, and the displacement field based on FSDT can be obtained. Furthermore, if we let $\alpha = 0$, $\varphi_x = -\dfrac{\partial w}{\partial x}$, and $\varphi_y = -\dfrac{\partial w_0}{\partial y}$, the displacement field based on CPT can be obtained. The independent variables are w, φ_x, and φ_y for both plates based on FSDT and TSDT.

The strains for plates of TSDT are as follows:

$$
\underbrace{\left\{\begin{array}{c} \varepsilon_{xx} \\ \varepsilon_{yy} \\ \gamma_{xy} \\ \gamma_{xz} \\ \gamma_{yz} \end{array}\right\}}_{\boldsymbol{\varepsilon}} = \underbrace{\begin{bmatrix} -\alpha z^3 \dfrac{\partial^2}{\partial x^2} & (z - \alpha z^3)\dfrac{\partial}{\partial x} & 0 \\[2ex] -\alpha z^3 \dfrac{\partial^2}{\partial y^2} & 0 & (z - \alpha z^3)\dfrac{\partial}{\partial y} \\[2ex] -2\alpha z^3 \dfrac{\partial^2}{\partial x \partial y} & (z - \alpha z^3)\dfrac{\partial}{\partial y} & (z - \alpha z^3)\dfrac{\partial}{\partial x} \\[2ex] (1 - \beta z^2)\dfrac{\partial}{\partial x} & (1 - \beta z^2) & 0 \\[2ex] (1 - \beta z^2)\dfrac{\partial}{\partial y} & 0 & (1 - \beta z^2) \end{bmatrix}}_{\mathbf{L}_d} \left\{\begin{array}{c} w \\ \varphi_x \\ \varphi_y \end{array}\right\} = \mathbf{L}_d \mathbf{u} \tag{11.73}
$$

where $\beta = 3\alpha$. Again, if we set $\alpha = 0$, the foregoing equation becomes Equation 11.51, that is, for plates of FSDT. The stress–strain relationship is the same as that for the FSDT plates given in Equation 11.58 with $\kappa = 1$.

11.1.3.2 Essential Boundary Conditions

For plates based on TSDT, the essential boundary conditions are given as follows.

At simply supported edges the conditions are same as those given in Equations 11.67 and 11.68. At clamped edges, we shall have

$$
w\big|_{\text{at edges}} = 0 \tag{11.74}
$$

$$
\varphi_n\big|_{\text{at edges}} = 0 \tag{11.75}
$$

$$
\varphi_t\big|_{\text{at edges}} = 0 \tag{11.76}
$$

$$
\frac{\partial w}{\partial n}\bigg|_{\text{at edges}} = 0 \tag{11.77}
$$

Again, there are two kinds of essential boundary conditions: soft type and hard type.

11.2 EFG Method for Thin Plates

This section presents an EFG method for static and free-vibration analyses of plates following the work presented in [4,5]. The discretized system equations based on Kirchhoff's thin plate theory (or CPT) are provided. Methods are introduced to impose essential boundary conditions. For static deflection of thin plates, a penalty method is formulated. For analysis of free vibration of thin plates, the essential boundary conditions are formulated via a weak form, separated from the weak form of the system equation. The boundary conditions are then imposed using orthogonal transform techniques. The eigenvalue equation derived using this approach has a smaller dimension than that in FEM.

A number of numerical examples are presented. Static deflections of thin rectangular plates with fully clamped and simply supported boundaries are computed using the present EFG. Natural frequencies of thin square, elliptical, hexagonal, and complicated plates with various boundaries such as free, simply supported, and fully clamped are also calculated. Both regularly and irregularly distributed nodes are used in the computation to examine the sensitivity of the results to the irregularity of the nodes. Examples are presented to demonstrate the convergence and validation of the present EFG formulation compared with analytical solutions.

It is shown that the present EFG formulation has a clear advantage over element-based formulations, as it is a rotation-free formulation: only the deflection is a nodal variable compared to three in the element-based formulation (one deflection and two rotations). The dimension of the discretized system equations generated using meshfree methods is therefore one-third of that generated using FEM. In addition, in EFG there is no "conformability" issue, which exists on the interface between the finite elements, as long as sufficient nodes are used in the local moving least squares (MLS) approximation. Higher-order consistency can also be achieved in meshfree methods by including higher-order terms in the polynomial basis and the use of more local nodes.

11.2.1 Approximation of Deflection

11.2.1.1 Shape Function

Consider a plate of Ω shown in Figure 11.1. A Cartesian coordinate system is used to establish the system equations. The plate is represented by its neutral plane. The deflections of the plate in the x-, y-, and z-directions are denoted as u, v, w, respectively.

Based on Kirchhoff's thin plate assumption, the deflection $w(\mathbf{x})$ of its neutral plane at $\mathbf{x} = \{x, y\}^{\mathrm{T}}$ can be taken as the independent variable, and the other two displacement components $u(\mathbf{x})$ and $v(\mathbf{x})$ can be expressed in terms of $w(\mathbf{x})$.

The plate is represented by a set of field nodes scattered in the domain of the plate. MLS approximation is used to approximate $w(\mathbf{x})$ using nodes included in the support domain of $\mathbf{x}$, and hence the two rotations are also interpreted by the approximated deflection. The process of approximating the deflection is the same as that discussed in Section 2.4. The differences are as follows:

1. For thin plates, we should use higher-order polynomial basis functions, because the governing equation can be fourth-order PDE. Therefore, a higher order of consistency is required. In the examples given in this section, quadratic ($m = 6$) polynomial basis functions are used.

2. Higher-order derivatives of shape functions may be required in deriving system equations.

Using the MLS shape functions, the deflection of the plates can be approximated by the parameters of nodal deflection, w_I, in the following form:

$$w^h(\mathbf{x}) = \sum_{I \in S_n} \phi_I(\mathbf{x}) w_I \tag{11.78}$$

where
   S_n is the set of nodes in the support domain of a point of interest $\mathbf{x}$
   ϕ_I is the MLS shape function for the Ith node obtained using quadratic polynomial basis

11.2.2 Variational Forms

Because the Kronecker delta condition is not satisfied by the MLS shape function, the essential boundary conditions (Equation 11.43) need to be imposed in a proper manner. For static deflection analysis of thin plates, we use the penalty method to enforce essential boundary conditions. The constrained Galerkin weak form for the thin plates based on the CPT can be written as

$$\int_A \delta\boldsymbol{\varepsilon}_p^T \boldsymbol{\sigma}_p dA - \int_\Omega \delta\mathbf{u}^T \mathbf{b} d\Omega - \int_{\Gamma_t} \delta\mathbf{u}^T \mathbf{t}_\Gamma dS - \delta\frac{1}{2}\int_{\Gamma_u}(\mathbf{u}_b - \mathbf{u}_\Gamma)^T\boldsymbol{\alpha}(\mathbf{u}_b - \mathbf{u}_\Gamma)d\Gamma = 0 \qquad (11.79)$$

where
   A is the area of the plate
   Γ_t is the natural boundary
   $\boldsymbol{\varepsilon}_p$ is the pseudostrain defined by Equation 11.9
   $\mathbf{b}$ is a body force vector
   $\boldsymbol{\alpha}$ is a diagonal matrix of penalty factors, which are usually very large numbers

The relationship between the pseudostrain and stress is given by Equation 11.19.

In Equation 11.79, the first term relates to the virtual work done by the internal stress in the thin plate. The second term relates to the virtual work done by the body force that may be distributed over the entire volume of the plates. The third term relates to the virtual work done by the forces applied on the natural boundary. The last term counts for that on the essential boundaries.

For free-vibration analyses of thin plates, the Galerkin weak form associated with the elastodynamic undamped equilibrium equations can be written as follows:

$$\int_A \delta\boldsymbol{\varepsilon}_p^T \boldsymbol{\sigma}_p dA + \int_\Omega \delta\mathbf{u}^T \rho\ddot{\mathbf{u}} \, d\Omega = 0 \qquad (11.80)$$

where ρ is the mass density. Because the plate is free of external forces, it is easy to understand that the second and third terms in Equation 11.79 vanish. However, we need to justify why we left out the last term in Equation 11.79.

For free-vibration analysis, one method is to formulate the boundary condition equation separately from the system equation. The weak form of the essential boundary conditions with Lagrange multipliers is used to produce the discretized essential boundary conditions. It is given below:

$$\int_{\Gamma_u} \delta\boldsymbol{\lambda}^T(\mathbf{u}_b - \mathbf{u}_\Gamma)d\Gamma = 0 \qquad (11.81)$$

where $\boldsymbol{\lambda}$ is a vector of Lagrange multipliers each of which can be interpolated as follows:

$$\lambda(\mathbf{x}) = N_I(s)\lambda_I, \quad \mathbf{x} \in \Gamma_u \qquad (11.82)$$

where s and $N_I(s)$ are, respectively, the curvilinear coordinate along the boundary and the Lagrange interpolant, which we have detailed in Chapter 6. The variation of the Lagrange multiplier can be written as

$$\delta\lambda(\mathbf{x}) = N_I(s)\delta\lambda_I, \quad \mathbf{x} \in \Gamma_u \qquad (11.83)$$

Using Equations 11.4 and 11.11, the constrained Galerkin weak form for static problems of thin plates can be rewritten in terms of deflection as

$$\int_A \delta(\mathbf{L}_d w)^{\mathrm{T}} \mathbf{D}(\mathbf{L}_d w) \mathrm{d}A - \int_\Omega \delta(\mathbf{L}_u w)^{\mathrm{T}} \mathbf{b} \mathrm{d}\Omega - \int_{\Gamma_t} \delta(\mathbf{L}_u w)^{\mathrm{T}} \mathbf{t}_\Gamma \cdot \mathrm{d}S$$

$$- \delta \frac{1}{2} \int_{\Gamma_u} (\mathbf{L}_b w - \mathbf{u}_\Gamma)^{\mathrm{T}} \boldsymbol{\alpha} (\mathbf{L}_b w - \mathbf{u}_\Gamma) \mathrm{d}\Gamma = 0 \qquad (11.84)$$

The Galerkin weak form for free vibration of Equation 11.80 can be rewritten as

$$\int_A \delta(\mathbf{L}_d w)^{\mathrm{T}} \mathbf{D}(\mathbf{L}_d w) \mathrm{d}A + \int_\Omega \rho \delta (\mathbf{L}_u w)^{\mathrm{T}} \mathbf{L}_u \ddot{w} \mathrm{d}\Omega = 0 \qquad (11.85)$$

with the weak form for essential boundary condition of Equation 11.81 rewritten as

$$\int_{\Gamma_u} \delta \boldsymbol{\lambda}^{\mathrm{T}} (\mathbf{L}_b w - \mathbf{u}_\Gamma) \mathrm{d}\Gamma = 0 \qquad (11.86)$$

11.2.3 Discrete Equations

Substituting the displacement field, Equation 11.78, into Equation 11.84 leads to the following discrete system equation:

$$(\mathbf{K} + \mathbf{K}^\alpha)\mathbf{U} = \mathbf{F} \qquad (11.87)$$

where the global stiffness matrix $\mathbf{K}$ is assembled using the nodal stiffness (a scalar due to the 1 degree of freedom (DOF) rotation-free formulation) defined by

$$K_{IJ} = \int_A \mathbf{B}_I^{\mathrm{T}} \mathbf{D} \mathbf{B}_J \mathrm{d}A \qquad (11.88)$$

in which

$$\mathbf{B}_I = \mathbf{L}_d \boldsymbol{\phi}_I = \begin{Bmatrix} -\phi_{I,xx} \\ -\phi_{I,yy} \\ -2\phi_{I,xy} \end{Bmatrix} \qquad (11.89)$$

Matrix $\mathbf{K}^\alpha$ in Equation 11.87 is obtained using the scalar entries of

$$K^\alpha_{IJ} = \int_{\Gamma_u} \mathbf{\Psi}^{\mathrm{T}}_I \boldsymbol{\alpha} \mathbf{\Psi}_J \, \mathrm{d}\Gamma \tag{11.90}$$

where

$$\mathbf{\Psi}_I = \left\{ \begin{array}{c} \phi_I \\ \phi_{I,n} \end{array} \right\} \quad \text{for clamped boundary} \tag{11.91}$$

and

$$\mathbf{\Psi}_I = \left\{ \begin{array}{c} \phi_I \\ 0 \end{array} \right\} \quad \text{for simply supported boundary} \tag{11.92}$$

in which n is the unit normal on the essential boundary surface Γ_u.

The force vector $\mathbf{F}$ is assembled using the nodal force given by

$$f_I = \int_\Omega \underbrace{(\mathbf{L}_u \phi_I)^{\mathrm{T}}}_{\mathbf{B}_u} \mathbf{b} \, \mathrm{d}\Omega + \int_{\Gamma_t} \underbrace{(\mathbf{L}_u \phi_I)^{\mathrm{T}}}_{\mathbf{B}_u} \mathbf{t}_\Gamma \, \mathrm{d}S = \int_\Omega \mathbf{B}^{\mathrm{T}}_u \mathbf{b} \, \mathrm{d}\Omega + \int_{\Gamma_t} \mathbf{B}^{\mathrm{T}}_u \mathbf{t}_\Gamma \, \mathrm{d}S \tag{11.93}$$

where

$$\mathbf{B}_u = \left\{ \begin{array}{c} -z\phi_{I,x} \\ -z\phi_{I,y} \\ \phi_I \end{array} \right\} \tag{11.94}$$

Because the plate considered here is only subject to a transverse (in the z-direction) load, we have

$$\mathbf{b} = \left\{ \begin{array}{c} 0 \\ 0 \\ b_z \end{array} \right\} \tag{11.95}$$

The external traction on the edge of the plate $\mathbf{t}_\Gamma$ can be given by the stresses on the surface of the edge:

$$\mathbf{t}_\Gamma = \left[\begin{array}{ccccc} n_x & 0 & 0 & n_y & 0 \\ 0 & n_y & 0 & 0 & n_x \\ 0 & 0 & n_y & n_x & 0 \end{array} \right] \left\{ \begin{array}{c} \sigma_{xx} \\ \sigma_{yy} \\ \sigma_{yz} \\ \sigma_{xz} \\ \sigma_{xy} \end{array} \right\} = \left\{ \begin{array}{c} \sigma_{xx} n_x + \sigma_{xz} n_y \\ \sigma_{xy} n_x + \sigma_{yy} n_y \\ \sigma_{xz} n_x + \sigma_{yz} n_y \end{array} \right\} \tag{11.96}$$

The nodal force can now be written in detail as

$$
f_I = \int_\Omega \phi_I b_z \, d\Omega + \int_{\Gamma_t} \{-z\phi_{I,x} \quad -z\phi_{I,y} \quad \phi_I\} \begin{Bmatrix} \sigma_{xx} n_x + \sigma_{xy} n_y \\ \sigma_{xy} n_x + \sigma_{yy} n_y \\ \sigma_{xz} n_x + \sigma_{yz} n_y \end{Bmatrix} dS
$$

$$
= \int_\Omega \phi_I b_z \, d\Omega + \int_{\Gamma_t} \left[-\phi_{I,x} \left(n_x \underbrace{\int_{-h/2}^{h/2} z\sigma_{xx} dz}_{M_{xx}} + n_y \underbrace{\int_{-h/2}^{h/2} z\sigma_{xy} dz}_{M_{xy}} \right) - \phi_{I,y} \left(n_x \underbrace{\int_{-h/2}^{h/2} z\sigma_{xy} dz}_{M_{xy}} + n_y \underbrace{\int_{-h/2}^{h/2} z\sigma_{yy} dz}_{M_{yy}} \right) \right.
$$

$$
\left. + \phi_I \left(n_x \underbrace{\int_{-h/2}^{h/2} \sigma_{xz} dz}_{V_{xz}} + n_y \underbrace{\int_{-h/2}^{h/2} \sigma_{xz} dz}_{V_{yz}} \right) \right] d\Gamma
$$

$$
= h \int_A \phi_I p_z \, dA + \int_{\Gamma_t} \left[-\phi_{I,x} \underbrace{(n_x M_{xx} + n_y M_{xy})}_{M_{\Gamma xn}} - \phi_{I,y} \underbrace{(n_x M_{xy} + n_y M_{yy})}_{M_{\Gamma yn}} + \phi_I \underbrace{(n_x V_{xz} + n_y V_{yz})}_{V_{\Gamma zn}} \right] d\Gamma
$$

$$(11.97)$$

where p_z is the pressure (force per unit area) applied on the neutral plane of the plate. Note in Equation 11.97 that the integration over the surface of the plate edge has been changed to a curve integration on the natural boundary (on the neutral plane). We now examine the first two terms in the integrand of the last integral in Equation 11.97. Note that on the boundary of the plate, we should have

$$
\begin{aligned}
\phi_{I,x} &= \phi_{I,n} n_x - \phi_{I,s} n_y \\
\phi_{I,y} &= \phi_{I,n} n_y + \phi_{I,s} n_x
\end{aligned}
\tag{11.98}
$$

Therefore,

$$
\begin{aligned}
&\phi_{I,x}(n_x M_{xx} + n_y M_{xy}) + \phi_{I,y}(n_x M_{xy} + n_y M_{yy}) \\
&= (\phi_{I,n} n_x - \phi_{I,s} n_y)(n_x M_{xx} + n_y M_{xy}) + (\phi_{I,n} n_y + \phi_{I,s} n_x)(n_x M_{xy} + n_y M_{yy}) \\
&= \phi_{I,n} \underbrace{[n_x n_x M_{xx} + 2n_x n_y M_{xy} + n_y n_y M_{yy}]}_{M_{\Gamma n}} + \phi_{I,s} \underbrace{[-n_y n_x M_{xx} + n_x n_y M_{yy}]}_{M_{\Gamma s}} \\
&= \phi_{I,n} M_{\Gamma n} + \phi_{I,s} M_{\Gamma s}
\end{aligned}
\tag{11.99}
$$

where $M_{\Gamma n}$ and $M_{\Gamma s}$ are, respectively, the specified moment and torsional moment on the natural boundary (see Equations 11.37 and 11.38).

Substituting Equation 11.99 into Equation 11.97 leads to

$$
f_I = h \int_A \phi_I p_z \, dA + \int_{\Gamma_t} [-\phi_{I,n} M_{\Gamma n} - \phi_{I,s} M_{\Gamma s} + \phi_I V_{\Gamma z}] d\Gamma
\tag{11.100}
$$

The vector $\mathbf{U}$ in Equation 11.87 has the form:

$$\mathbf{U} = \{w_1, \ldots, w_{N_n}\}^{\mathrm{T}} \tag{11.101}$$

where N_n is the total number of nodes in the entire domain of the plate. Note that the total degrees of freedom of the thin plates are the number of nodes, and the rotations are not involved as unknown variables in the global system equation.

Similarly, by substituting the displacement field (Equation 11.78) into the variational form (Equation 11.85), the dynamic discrete system equation for free vibration of thin plates is obtained as follows:

$$\mathbf{M\ddot{U}} + \mathbf{KU} = 0 \tag{11.102}$$

where $\mathbf{M}$ is the global mass matrix that is assembled using the nodal mass given by the following integral form over the entire area of the plate:

$$M_{IJ} = \int_A (\rho \phi_I \phi_J h + \phi_{I,x} \phi_{J,x} I + \phi_{I,y} \phi_{J,y} I) \mathrm{d}A \tag{11.103}$$

where I is the mass moment of inertial for the plate computed by

$$I = \int_{-h/2}^{h/2} \rho z^2 \mathrm{d}z \tag{11.104}$$

For homogeneous plates, $I = \rho(h^3/12)$.

The first integral in Equation 11.103 is the mass inertial corresponding to the vertical translational vibration of the plate, and the second and third terms are rotational inertia corresponding to the rotational vibrations of the cross section of the thin plate. For *very* thin plates, these two terms for rotational vibrations can be neglected (see, e.g., [8]).

11.2.3.1 Equations for Essential Boundary Conditions

Substituting the displacement expression of Equation 11.78 into the boundary condition weak form of Equation 11.86 yields a set of linear algebraic constraint equations:

$$\mathbf{H}_{(2n_b \times N_n)} \mathbf{U}_{(N_n \times 1)} = \mathbf{q}_{(2n_b \times 1)} \tag{11.105}$$

where n_b is the number of the nodes that are included in the union of all support domains of all quadrature points on the essential boundary. Nonzero entries in the matrix $\mathbf{H}$ are computed by

$$\mathbf{H}_{KI} = \int_{\Gamma_u} N_K \Psi_I \mathrm{d}\Gamma \tag{11.106}$$

in which $\boldsymbol{\Psi}_I$ is the matrix of MLS shape functions given by Equation 11.91 or Equation 11.92, depending on the type of boundary condition, and $\mathbf{N}_K$ has the form

$$\mathbf{N}_K = \begin{bmatrix} N_K & 0 \\ 0 & N_K \end{bmatrix} \tag{11.107}$$

where N_K is the Lagrange interpolation function for node K on the essential boundary. A Lagrange interpolant formed by Equation 6.18 can be used. Vector $\mathbf{q}$ in Equation 11.105 is defined by

$$\mathbf{q}_K = \int_{\Gamma_u} N_K \mathbf{u}_\Gamma \mathrm{d}s \tag{11.108}$$

Note that matrix in Equation 11.105 has n columns and $2n_b$ rows, where n_b is usually a very small number compared with N_n. All entries in $\mathbf{H}$ without the contribution from Equation 11.106 will be zero. Thus, $\mathbf{H}$ is in general a very "short-fat" and sparse matrix.

For free-vibration analysis, the essential boundary conditions have to be homogeneous, that is, $\mathbf{u}_\Gamma = 0$ on the essential boundary. Using Equation 11.108, we have $\mathbf{q} = 0$ and, hence,

$$\mathbf{HU} = 0 \tag{11.109}$$

This is the discretized essential boundary condition for free-vibration analysis.

11.2.4 Eigenvalue Problem

11.2.4.1 Eigenvalue Equation

Consider now that the plate is undergoing a harmonic vibration. The deflection $\mathbf{U}$ can be expressed in the form

$$\mathbf{U} = \mathbf{U}_A \exp(i\omega t) \tag{11.110}$$

where
    i is the imaginary unit
    ω is the angular frequency
    $\mathbf{U}_A$ is the amplitude of the vibration

Substitution of the foregoing equation into Equation 11.102 leads to the following eigenvalue equation:

$$(\mathbf{K} - \omega^2 \mathbf{M})\mathbf{U}_A = 0 \tag{11.111}$$

where $\mathbf{U}_A$ is an eigenvector. For a system with N_n nodes, the dimension of matrices $\mathbf{K}$ and $\mathbf{M}$ should be $N_n \times N_n$, and there should be N_n eigenvalues ω_i ($i = 1, 2, \ldots, N_n$), which correspond to N_n natural frequencies, and N_n eigenvectors. We shall have

$$\mathbf{U}_A = \{U_{A1} \quad U_{A2} \quad \cdots \quad U_{AN_n}\}^\mathrm{T} \tag{11.112}$$

Note, however, that Equation 11.111 has to be solved subject to the constraints of Equation 11.109 that can be restated for free vibration as

$$\mathbf{HU}_A = \mathbf{0} \tag{11.113}$$

Using singular value decomposition, $\mathbf{H}$ can be decomposed as

$$\mathbf{H}_{2n_b \times N_n} = \mathbf{R}_{2n_b \times 2n_b} \begin{bmatrix} \Sigma_{r \times r} & \mathbf{0} \\ \mathbf{0} & \mathbf{0} \end{bmatrix}_{2n_b \times N_n} \mathbf{V}^{\mathrm{T}}_{N_n \times N_n} \tag{11.114}$$

where
  $\mathbf{R}$ and $\mathbf{V}$ are orthogonal matrices
  $\Sigma_{r \times r}$ has diagonal form, in which diagonal elements are equal to singular values of $\mathbf{H}$
  r is the rank of $\mathbf{H}$, which is the same as the number of independent constraints

The orthogonal matrix $\mathbf{V}$ can be partitioned as follows:

$$\mathbf{V}^{\mathrm{T}} = \{\mathbf{V}_{N_n \times r},\ \mathbf{V}_{N_n \times (N_n - r)}\}^{\mathrm{T}} \tag{11.115}$$

where $\mathbf{V}_{N_n \times r}$ corresponds to the portion of $\Sigma_{r \times r}$. We now examine

$$\mathbf{HV} = \mathbf{H}\begin{bmatrix} \mathbf{V}_{N_n \times r} & \mathbf{V}_{N_n \times (N_n - r)} \end{bmatrix} = \begin{bmatrix} \mathbf{HV}_{N_n \times r} & \mathbf{HV}_{N_n \times (N_n - r)} \end{bmatrix}$$
$$= \mathbf{R}_{2n_b \times 2n_b} \begin{bmatrix} \Sigma_{r \times r} & \mathbf{0} \\ \mathbf{0} & \mathbf{0} \end{bmatrix}_{2n_b \times N_n} \underbrace{\mathbf{V}^{\mathrm{T}}_{N_n \times N_n} \mathbf{V}_{N_n \times N_n}}_{\mathbf{I}} \tag{11.116}$$

Invoking the orthogonal condition of $\mathbf{V}$, we have $\mathbf{V}^{\mathrm{T}}_{N_n \times N_n} \mathbf{V}_{N_n \times N_n} = \mathbf{I}$. The foregoing equation becomes

$$\mathbf{HV} = \mathbf{H}\begin{bmatrix} \mathbf{V}_{N_n \times r} & \mathbf{V}_{N_n \times (N_n - r)} \end{bmatrix} = \begin{bmatrix} \mathbf{HV}_{N_n \times r} & \mathbf{HV}_{N_n \times (N_n - r)} \end{bmatrix}$$
$$= \mathbf{R}_{2n_b \times 2n_b} \begin{bmatrix} \Sigma_{r \times r} & 0 \\ 0 & \underbrace{0}_{\mathbf{HV}_{N_n \times (N_n - r)}} \end{bmatrix} \tag{11.117}$$

This implies

$$\mathbf{HV}_{N_n \times (N_n - r)} = \mathbf{0}_{N_n \times (N_n - r)} \tag{11.118}$$

or a null transformation, or $\mathbf{V}_{N_n \times (N_n - r)}$ is a basis of the null space of the linear transformation $\mathbf{H}$. Therefore, the following orthogonal matrix transformation:

$$\mathbf{U}_A = \mathbf{V}_{N_n \times (N_n - r)}\tilde{\mathbf{U}}_A \tag{11.119}$$

satisfies Equation 11.113. Substituting Equation 11.119 into Equation 11.111 and premultiplying $\mathbf{V}^{\mathrm{T}}_{N_n \times (N_n - r)}$ to the resultant equation lead to

$$[\tilde{\mathbf{K}} - \omega^2 \tilde{\mathbf{M}}]\tilde{\mathbf{U}}_A = 0 \tag{11.120}$$

which is the condensed eigenvalue equations, where

$$\tilde{\mathbf{K}} = \mathbf{V}^{\mathrm{T}}_{N_n \times (N_n - r)} \mathbf{K} \mathbf{V}_{N_n \times (N_n - r)} \tag{11.121}$$

is the condensed stiffness matrix and

$$\tilde{\mathbf{M}} = \mathbf{V}^{\mathrm{T}}_{N_n \times (N_n - r)} \mathbf{M} \mathbf{V} \mathbf{V}_{N_n \times (N_n - r)} \tag{11.122}$$

is the condensed mass matrix.

Note that these condensed matrices are, in general, nonnegative definite. Solving Equation 11.120 using standard routines of eigenvalue equation solvers gives natural frequencies of the free vibration of thin plates. This orthogonal transformation technique was used by Ouatouati and Johnson [3] for imposing constraints for eigenvalue equations formulated using the EFG method.

Note that matrix **H** in Equation 11.113 is formed by the weak form of the constraint equation, which requires integration on the boundary and ensures the satisfaction of the essential boundary conditions. We can obtain the discrete constraint equations directly using MLS approximation and obtain the **H** matrix, as shown in [3].

11.2.5 Numerical Examples

In the examples given in this section, the complete quadratic polynomial basis function is used ($m = 6$) for constructing MLS shape functions. Cells of background mesh are used for integrating the system matrices and the Gauss quadrature scheme is utilized. The dimension of support domain is chosen as 3.5–3.9 times the nodal distance.

Example 11.1: Static Deflection of Rectangular Thin Plates

Consider now a benchmark problem of a rectangular plate, as shown in Figure 11.3. Analytical solutions are available for this problem [9]. A concentrated force of $P = 100.0$ N is applied at the center of the plate, and the following parameters are used:

Length in x-direction: $a = 0.6$ m for rectangular plates with various widths
Length: $a = b = 0.6$ m for square plate

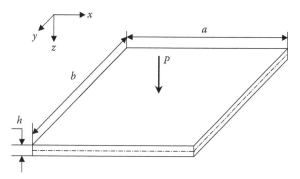

FIGURE 11.3
Thin rectangular plate with uniform thickness h.
(From Liu, G.R. and Chen, X.L., *J. Sound Vib.*,
241(5), 839, 2001. With permission.)

Thickness: $h = 0.001$ m

Young's modulus: $E = 1.0 \times 10^9$ N/m^2

Poisson's ratio: $\nu = 0.3$

A dimensionless deflection coefficient β of the center of a thin rectangular plate is defined as $\beta = w_{max}D/pa^2$, where w_{max} is the deflection at center of the plates and $D = Eh^3/[12(1-\nu^2)]$ is the flexural rigidity of the plate.

To analyze the convergence of the present EFG method, we calculate deflections of a square plate using different densities of nodes. Two kinds of boundary conditions are imposed: simply supported and fully clamped. In the notation of the boundary conditions, S denotes simply supported and C means clamped. The edges of the plate are denoted clockwise using S or C depending on the type of boundary on the edge. For example, notation SCSC means that the left and right edges of the plate are simply supported while the upper and lower edges are clamped, and notation SCCS means that the left and lower edges are simply supported while the upper and right edges are clamped.

The numerical and analytical results of deflection of the square plate are shown in Table 11.1. Good convergence has been achieved.

Further examinations are performed for plates with different width:length ratios. The deflections are calculated using 16×16 nodes in the present EFG method. The results are shown in Tables 11.2 and 11.3. Compared with the analytical results [9], good agreement has been achieved for all tested cases.

TABLE 11.1

Deflection of a Square Plate $\beta = W_{max}D/pa^2$

	EFG					Ref. [9]
Nodes	6×6	9×9	12×12	15×15	18×18	
SSSS	0.01032	0.01141	0.01145	0.01155	0.01157	0.01160
CCCC	0.00452	0.00538	0.00546	0.00552	0.00554	0.00560

Source: Liu, G.R. and Chen, X.L., *J. Sound Vib.*, 241(5), 839, 2001. With permission.

TABLE 11.2

Deflection of Simply Supported Rectangular Plates $\beta = W_{max}D/pa^2$

b/a	1.0	1.2	1.4	1.6	1.8	2.0
EFG	0.01157	0.01344	0.01476	0.01556	0.01603	0.01632
Ref. [9]	0.01160	0.01353	0.01484	0.01570	0.01620	0.01651

Source: Liu, G.R. and Chen, X.L., *J. Sound Vib.*, 241(5), 839, 2001. With permission.

TABLE 11.3

Deflection of Fully Clamped Rectangular Plates

b/a	1.0	1.2	1.4	1.6	1.8	2.0
β (EFG)	0.00552	0.00637	0.00680	0.00698	0.00703	0.00704
β [9]	0.00560	0.00647	0.00691	0.00712	0.00720	0.00722

Source: Liu, G.R. and Chen, X.L., *J. Sound Vib.*, 241(5), 839, 2001. With permission.

Example 11.2: Natural Frequency Analysis of Thin Square Plates

Consider now a square plate with the following parameters:

Length: $a = b = 10.0$ m
Thickness: $h = 0.05$ m
Young's modulus: $E = 200 \times 10^9$ N/m^2
Poisson's ratio: $\nu = 0.3$
Mass density: $\rho = 8000$ kg/m^3

Case 1: Free Thin Square Plate

We calculate the natural frequencies of free vibration of a free thin square plate. The frequency coefficients are computed using regular nodes of different density, and the results are shown in Table 11.4 together with the FEM results. In the FEM results, HOE denotes an eight-noded semiloof thin shell element (4×4 mesh); LOE denotes a four-noded isoparametric shell element (8×8 mesh). The first three frequencies corresponding to the rigid displacements are zero, and therefore are not listed in the table. The results obtained using the present EFG method are between those of FEMs using HOE and LOE. The present results show good convergence and good agreement with the analytical solution. When a 9×9 nodal density is used, the present results are more accurate than either FEM result.

Case 2: Simply Supported and Fully Clamped Thin Square Plate

Natural frequencies of lateral free vibration of a simply supported and fully clamped thin square plate are computed using the present EFG method. To analyze the effectiveness of the present EFG method using irregular nodes, we calculate frequencies using 13×13 regular nodes shown in Figure 11.4a and 169 irregular nodes shown in Figure 11.4b. The results are shown in Tables 11.5 and 11.6. It is found that the results of using both regular and irregular nodes show good agreement with each other and with the analytical solutions.

TABLE 11.4

Natural Frequency Coefficients $\bar{\omega} = (\omega^2 \rho h a^4 / D)^{1/4}$ of Lateral Free Vibration of a Free Square Plate

Mode	Analytical Solution [40]	EFG				FEM [40]	
		5×5	9×9	13×13	17×17	HOE	LOE
4	3.670	3.700	3.670	3.670	3.670	3.567	3.682
5	4.427	4.468	4.434	4.430	4.429	4.423	4.466
6	4.926	5.000	4.939	4.933	4.930	4.875	4.997
7	5.929	6.010	5.907	5.903	5.901	5.851	5.942
8	5.929	6.010	5.907	5.903	5.901	5.851	5.942
9	7.848	8.189	7.855	7.840	7.832	7.820	8.079

Source: Liu, G.R. and Chen, X.L., *J. Sound Vib.*, 241(5), 839, 2001. With permission.
Note: HOE denotes an eight-noded semiloof thin shell element (4×4 mesh); LOE denotes a four-noded isoparametric shell element (8×8 mesh).

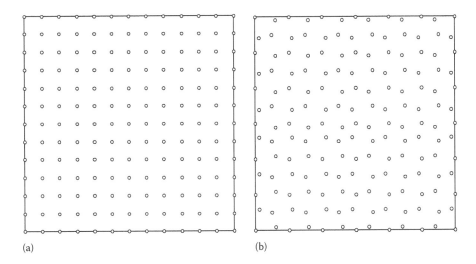

(a)   (b)

FIGURE 11.4
Node distribution in a square plate. (a) $13 \times 13 = 169$ regular nodes; and (b) 169 irregular nodes. (From Liu, G.R. and Chen, X.L., *J. Sound Vib.*, 241(5), 839, 2001. With permission.)

TABLE 11.5

Natural Frequency Coefficients $\bar{\omega} = (\omega^2 \rho h a^4 / D)^{1/4}$ of Lateral Free Vibration of a Simply Supported Square Plate

Mode	Analytical Solution [40]	EFG Regular Nodes 13×13	EFG Irregular Nodes 169
1	4.443	4.443	4.453
2	7.025	7.031	7.033
3	7.025	7.036	7.120
4	8.886	8.892	8.912
5	9.935	9.959	9.966
6	9.935	9.966	10.010
7	11.327	11.341	11.345
8	11.327	11.347	11.540
9	—	13.032	12.994
10	—	13.036	13.064

Source: Liu, G.R. and Chen, X.L., *J. Sound Vib.*, 241(5), 839, 2001. With permission.

Example 11.3: Natural Frequency Analysis of Elliptical Plates

An elliptical plate with radii $a = 5.0$ m and $b = 2.5$ m is investigated. Other parameters are the same as the square plate examined in Example 11.2.

Frequency coefficients of free vibration are computed. Table 11.7 shows the frequencies of a free thin elliptical plate using regular nodes. The first three frequencies corresponding to rigid displacements are zero and are not listed in the table. Good convergence has been achieved.

TABLE 11.6

Natural Frequency Coefficients $\bar{\omega} = (\omega^2 \rho h a^4/D)^{1/4}$ of Lateral Free Vibration of a Fully Clamped Square Plate

		EFG	
Mode	Analytical Solution [42]	Regular Nodes 13 × 13	Irregular Nodes 169
1	5.999	6.017	5.999
2	8.568	8.606	8.596
3	8.568	8.606	8.602
4	10.407	10.439	10.421
5	11.472	11.533	11.507
6	11.498	11.562	11.528
7	—	12.893	12.925
8	—	12.896	12.986
9	—	14.605	14.570
10	—	14.606	14.604

Source: Liu, G.R. and Chen, X.L., *J. Sound Vib.*, 241(5), 839, 2001. With permission.

TABLE 11.7

Natural Frequency Coefficients $\bar{\omega} = (\omega^2 \rho h (2a)^4/D)^{1/4}$ of Lateral Free Vibration of a Free Elliptical Plate

	EFG		
Mode	97 Nodes	241 Nodes	289 Nodes
4	5.197	5.176	5.173
5	6.533	6.509	6.505
6	8.288	8.244	8.234
7	9.451	9.405	9.397
8	10.602	10.559	10.547
9	11.333	11.256	11.249
10	12.223	12.168	12.160

Source: Liu, G.R. and Chen, X.L., *J. Sound Vib.*, 241(5), 839, 2001. With permission.

Table 11.8 shows the frequencies of a fully clamped thin elliptical plate. The frequencies are calculated using regular nodes (Figure 11.5a) and irregular nodes (Figure 11.5b). Good agreement between the results using regular and irregular nodes has been observed.

Example 11.4: Natural Frequency Analysis of Polygonal Plates

Free-vibration analysis of a square and hexagonal plate is performed. The length of each side is $a = 10.0$ m. Other parameters are the same as the square plate examined in Example 11.2.

The natural frequencies of a square plate with fully clamped boundaries are first calculated. A total of 524 irregular nodes, as shown in Figure 11.6, are used. The frequency coefficients are defined as

$$\bar{\omega} = \left(\frac{\omega^2 \rho h a_p^4}{D}\right)^{1/4}$$

TABLE 11.8

Natural Frequency Coefficients $\bar{\omega} = (\omega^2 \rho h (2a)^4 / D)^{1/4}$ of Lateral Free Vibration of a Fully Clamped Elliptical Plate

	EFG	
Mode	Regular Nodes 201	Irregular Nodes 201
1	10.467	10.454
2	12.619	12.621
3	15.009	14.992
4	16.726	16.716
5	17.629	17.658
6	18.838	18.840
7	20.604	20.508
8	21.081	21.060
9	22.913	22.890
10	23.610	23.591

Source: Liu, G.R. and Chen, X.L., *J. Sound Vib.*, 241(5), 839, 2001. With permission.

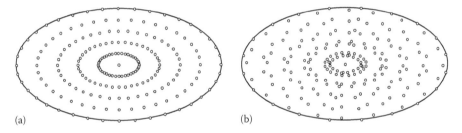

(a)   (b)

FIGURE 11.5
An elliptical plate with 201 nodes: (a) "regular"; and (b) irregular. (From Liu, G.R. and Chen, X.L., *J. Sound Vib.*, 241(5), 839, 2001. With permission.)

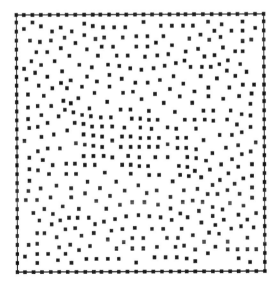

FIGURE 11.6
A square plate with 524 irregular nodes. (From Liu, G.R. and Chen, X.L., *J. Sound Vib.*, 241(5), 839, 2001. With permission.)

where a_p is the radius of the inscribing circle for regular polygonal plates. The natural frequencies of a hexagonal plate with fully clamped boundaries are also calculated, where 380 irregular nodes, as shown in Figure 11.7, are used. Table 11.9 lists the natural frequency coefficients of the lowest 10 modes for these two plates. Table 11.10 shows a comparison between the

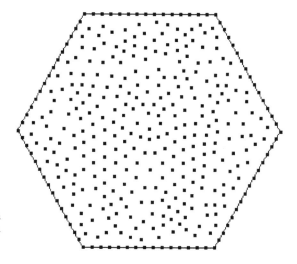

FIGURE 11.7
A hexagonal plate with 380 irregular nodes. (From Liu, G.R. and Chen, X.L., *J. Sound Vib.*, 241(5), 839, 2001. With permission.)

TABLE 11.9

Natural Frequency Coefficients $\bar{\omega} = (\omega^2 \rho h a_p^4 / D)^{1/4}$ of Lateral Free Vibration of Fully Clamped Regular Polygonal Plate

	EFG	
Mode	Square	Hexagon
1	9.089	9.042
2	18.700	17.805
3	18.829	20.586
4	28.121	29.802
5	33.515	34.740
6	33.649	37.368
7	42.550	46.622
8	43.529	51.846
9	53.896	55.799
10	54.156	59.138

Source: Liu, G.R. and Chen, X.L., *J. Sound Vib.*, 241(5), 839, 2001. With permission.

TABLE 11.10

Comparison of Natural Frequency Coefficients $\bar{\omega} = (\omega^2 \rho h a_p^4 / D)^{1/4}$ of First Mode of Fully Clamped Regular Polygonal Plates

	EFG	Exact [10]	Numerical [10]
Square	9.089	8.997	9.122
Hexagon	9.042	—	9.638

Source: Liu, G.R. and Chen, X.L., *J. Sound Vib.*, 241(5), 839, 2001. With permission.

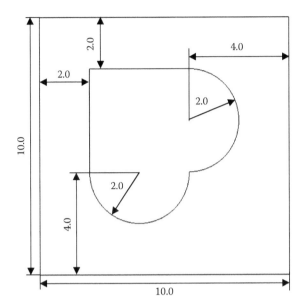

FIGURE 11.8
Plate with a hole of complicated shape. The unit is meters. (From Liu, G.R. and Chen, X.L., *J. Sound Vib.*, 241(5), 839, 2001. With permission.)

frequency coefficients using the EFG method and those given in [10]. For the square plate, the natural frequency coefficient of the first mode using the EFG method agrees very well with the exact solution and the numerical result in the textbook [10]. For the hexagonal plate, the natural frequency coefficient of the first mode using the EFG method is slightly smaller than the numerical result in the textbook [10].

Example 11.5: Natural Frequency Analysis of a Plate of Complex Shape

A plate with a very complicated shape is also studied. The geometric parameters are shown in Figure 11.8. The unit is meters. Other parameters are the same as the square plate examined in Example 11.2. The plate is chosen hypothetically, but it serves the purpose of demonstrating the applicability of the present EFG method to the plates with complicated shape.

Different boundary conditions are considered to examine the present EFG method in imposing boundary conditions. The nodal distribution is plotted in Figure 11.9. Table 11.11 lists the frequencies obtained for the plate with different boundary conditions. As expected, the natural frequencies of the plate with clamped boundaries are generally higher than those with simply supported boundaries.

11.3 EFG Method for Thin Composite Laminates

Composite laminates are widely used in modern structures due to their advantages of high strength, high stiffness, and low weight. Therefore, buckling analysis of laminates becomes very important in the process of designing such composite structures. Exact buckling solutions of laminates for arbitrary geometries and lamination schemes are usually very difficult. Therefore, numerical methods such as FEM have been used for analyzing laminated plate problems. However, it is not easy to construct conventionally

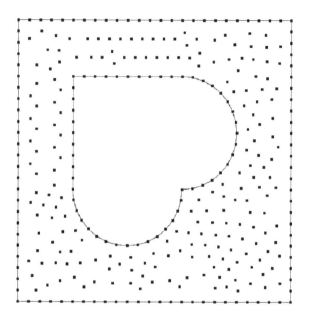

FIGURE 11.9
Nodal distribution in a plate with a hole with complicated shape. (From Liu, G.R. and Chen, X.L., *J. Sound Vib.*, 241(5), 839, 2001. With permission.)

TABLE 11.11

Natural Frequency Coefficients $\bar{\omega} = (\omega^2 \rho h a^4 / D)^{1/4}$ of Lateral Free Vibration of a Plate with a Hole of Complicated Shape

	EFG			
Mode	SSSS	CCCC	SCSC	SCCS
1	5.453	7.548	7.170	6.079
2	8.069	10.764	10.343	9.204
3	9.554	11.113	11.415	10.837
4	10.099	11.328	12.572	11.273
5	11.328	12.862	12.811	12.278
6	12.765	13.300	13.272	13.322
7	13.685	14.168	13.997	14.308
8	14.305	15.369	14.627	14.900
9	15.721	16.205	15.743	15.170
10	17.079	17.137	16.391	16.302

Source: Liu, G.R. and Chen, X.L., *J. Sound Vib.*, 241(5), 839, 2001. With permission.

conformable high-order plate elements as required for thin plates (C^1 consistency), and it requires element connectivity to form the finite element equations, whose generation is often a time-consuming procedure.

In Section 12.2, an EFG method has been formulated for analyzing static deflection and free vibration of thin plates. This section introduces an EFG formulation to solve the static buckling problems of thin plates and symmetrically laminated composite plates. The eigenvalue equations of the static buckling of the plates are established by applying energy principles and Kirchhoff plate theory. Similar to the formulations in Section 12.2, the deflection of plates is the only unknown variable at a node; therefore, the dimension of the discrete eigenvalue equations obtained by the present formulation is only one-third of that in FEM. Thus, solving the eigenvalue equation is more computationally efficient compared with FEM. To demonstrate the efficiency of the present EFG method, static buckling loads for thin plates have been calculated and examined in detail by comparing with other available solutions. The convergence of static buckling loads of thin plates is analyzed. The static buckling loads for thin plates of complicated shape and symmetrically laminated composite plates with different boundaries have also been calculated using the present EFG method.

11.3.1 Governing Equation for Buckling

A symmetrically laminated composite plate with thickness h in the z-direction is shown schematically in Figure 11.10. The laminated plate may consist of n_L layers of plies. The reference plane $z = 0$ is located at the undeformed neutral plane of the laminated plate. The fiber orientation of a layer is indicated by α, as shown in Figure 11.11. The laminate plate is subjected to in-plane forces within the plane of symmetry of the plate on its edges. It is assumed that the applied edge forces are independent of each other. We can then write [11]

$$N_x = -N_0, \quad N_y = -\mu_1 N_0, \quad N_{xy} = -\mu_2 N_0 \tag{11.123}$$

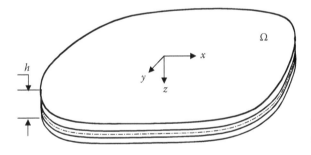

FIGURE 11.10
Thin laminate with symmetrically stacked layers of composites and its coordinate system.

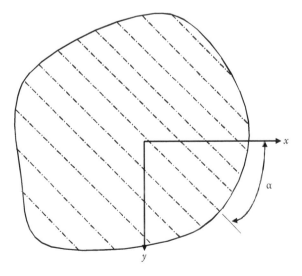

FIGURE 11.11
Ply with a fiber orientation of α in a laminated plate.

where
    N_0 is a constant
    μ_1 and μ_2 are possibly the functions of the coordinates

The strain energy potential related to the pseudostrain and stress fields can be obtained using

$$U_{PE} - \frac{1}{2} \int_A \boldsymbol{\varepsilon}_p^{\mathrm{T}} \boldsymbol{\sigma}_p dA \tag{11.124}$$

where
    A stands for the area of the plate
    $\boldsymbol{\varepsilon}_p$ is the pseudostrain defined by Equation 11.11
    $\boldsymbol{\sigma}_p$ is the pseudostress defined by Equation 11.18

Substituting Equations 11.11 and 11.18 into Equation 11.124, we obtain

$$\begin{aligned} U_{PE} = \frac{1}{2} \int_A \Bigg[& D_{11}\left(\frac{\partial^2 w}{\partial x^2}\right)^2 + 2D_{12}\frac{\partial^2 w}{\partial x^2}\frac{\partial^2 w}{\partial y^2} + D_{22}\left(\frac{\partial^2 w}{\partial y^2}\right)^2 + 4D_{66}\left(\frac{\partial^2 w}{\partial x \partial y}\right)^2 \\ & + 4\left(D_{16}\frac{\partial^2 w}{\partial x^2} + D_{26}\frac{\partial^2 w}{\partial y^2}\right)s\frac{\partial^2 w}{\partial x \partial y} \Bigg] dA \end{aligned} \tag{11.125}$$

The energy potential created by the in-plane forces can be expressed by

$$U_N = \frac{1}{2} \int_A \left[N_x\left(\frac{\partial w}{\partial x}\right)^2 + N_y\left(\frac{\partial w}{\partial y}\right)^2 + 2N_{xy}\frac{\partial w}{\partial x}\frac{\partial w}{\partial y} \right] dA \tag{11.126}$$

The matrices of the elastic constants of the laminates **D** can be obtained using Equation 11.14, and are given as follows:

$$D_{IJ} = \frac{1}{3}\sum_{K=1}^{n_L} (\bar{Q}_{IJ})_K(z_K^3 - z_{K-1}^3), \quad I,J = 1,2,6 \tag{11.127}$$

$$\bar{Q}_{11} = Q_{11}\cos^4\alpha + 2(Q_{12} + 2Q_{66})\sin^2\alpha\cos^2\alpha + Q_{22}\sin^4\alpha \tag{11.128}$$

$$\bar{Q}_{12} = (Q_{11} + Q_{22} - 4Q_{66})\sin^2\alpha\cos^2\alpha + Q_{12}(\sin^4\alpha + \cos^4\alpha) \tag{11.129}$$

$$\bar{Q}_{16} = (Q_{11} - Q_{12} - 2Q_{66})\sin\alpha\cos^3\alpha + (Q_{12} - Q_{22} + 2Q_{66})\sin^3\alpha\cos\alpha \tag{11.130}$$

$$\bar{Q}_{22} = Q_{11}\sin^4\alpha + 2(Q_{12} + 2Q_{66})\sin^2\alpha\cos^2\alpha + Q_{22}\cos^4\alpha \tag{11.131}$$

$$\bar{Q}_{26} = (Q_{11} - Q_{12} - 2Q_{66})\sin^3\alpha\cos\alpha + (Q_{12} - Q_{22} + 2Q_{66})\sin\alpha\cos^3\alpha \tag{11.132}$$

$$\bar{Q}_{66} = (Q_{11} + Q_{22} - 2Q_{12} - 2Q_{66})\sin^2\alpha\cos^2\alpha + Q_{66}(\sin^4\alpha + \cos^4\alpha) \tag{11.133}$$

$$Q_{11} = \frac{E_1}{1 - \nu_{12}\nu_{21}} \tag{11.134}$$

$$Q_{12} = \frac{\nu_{12}E_2}{1 - \nu_{12}\nu_{21}} \tag{11.135}$$

$$Q_{22} = \frac{E_2}{1 - \nu_{12}\nu_{21}} \tag{11.136}$$

$$Q_{66} = G_{12} \tag{11.137}$$

$$\nu_{21}E_1 = \nu_{12}E_2 \tag{11.138}$$

where

E_1 and E_2 are Young's moduli parallel to and perpendicular to the orientation of fibers
ν_{12} and ν_{21} are the corresponding Poisson's ratios
α is the angle of fiber orientation of the ply

The total potential energy of the laminated composite becomes

$$\Pi_{PE} = U_{PE} + U_N \tag{11.139}$$

11.3.2 Discrete Equation for Buckling Analysis

The laminated plate is represented using a set of nodes on the symmetric plane ($z = 0$). The deflection at any point in the plate can be approximated using Equation 11.78. The expression of the total energy can be obtained by substituting the deflection w of Equation 11.78 into Equations 11.125 and 11.126, and then Equation 11.139. The stationary condition for the total potential energy gives a set of discrete eigenvalue equations for the laminated plate:

$$[\mathbf{K} - N_0\mathbf{B}]\mathbf{W} = 0 \tag{11.140}$$

where

$$\mathbf{W} = \{w_1, w_2, \dots, w_{N_n}\}^{\mathrm{T}} \tag{11.141}$$

is the deflection vector consisting of deflections at all the N_n nodes and $\mathbf{K}$ is the global stiffness matrix which is assembled using the nodal stiffness (a scalar) defined by

$$
\begin{aligned}
K_{ij} = \frac{1}{2} \int_A &\left[2D_{11}\frac{\partial^2\phi_i}{\partial x^2}\frac{\partial^2\phi_j}{\partial x^2} + 2D_{12}\left(\frac{\partial^2\phi_i}{\partial x^2}\frac{\partial^2\phi_j}{\partial y^2} + \frac{\partial^2\phi_j}{\partial x^2}\frac{\partial^2\phi_i}{\partial y^2}\right) + 2D_{22}\frac{\partial^2\phi_i}{\partial y^2}\frac{\partial^2\phi_j}{\partial y^2} \right. \\
&+ 8D_{66}\frac{\partial^2\phi_i}{\partial x\partial y}\frac{\partial^2\phi_j}{\partial x\partial y} + 4D_{16}\left(\frac{\partial^2\phi_i}{\partial x^2}\frac{\partial^2\phi_j}{\partial x\partial y} + \frac{\partial^2\phi_j}{\partial x^2}\frac{\partial^2\phi_i}{\partial x\partial y}\right) \\
&\left. + 4D_{26}\left(\frac{\partial^2\phi_i}{\partial y^2}\frac{\partial^2\phi_j}{\partial x\partial y} + \frac{\partial^2\phi_j}{\partial y^2}\frac{\partial^2\phi_i}{\partial x\partial y}\right) \right] dA
\end{aligned} \tag{11.142}
$$

$$B_{ij} = \frac{1}{2}\int_A \left[2\frac{\partial\phi_i}{\partial x}\frac{\partial\phi_j}{\partial x} + 2\mu_1\frac{\partial\phi_i}{\partial y}\frac{\partial\phi_j}{\partial y} + 2\mu_2\left(\frac{\partial\phi_i}{\partial x}\frac{\partial\phi_j}{\partial y} + \frac{\partial\phi_j}{\partial x}\frac{\partial\phi_i}{\partial y}\right) \right] dA \tag{11.143}$$

For isotropic thin plates, the expression of potential energy of bending can be reduced as

$$U_{PE} = \frac{D}{2} \int_A \left[\left(\frac{\partial^2 w}{\partial x^2} + \frac{\partial^2 w}{\partial y^2} \right)^2 - 2(1-v) \left\{ \frac{\partial^2 w}{\partial x^2} \frac{\partial^2 w}{\partial y^2} - \left(\frac{\partial^2 w}{\partial x \partial y} \right)^2 \right\} \right] dA \qquad (11.144)$$

The strain energy caused by in-plane forces is the same as Equation 11.126. Minimizing the total potential energy yields the same form of discrete eigenvalue equations as Equation 11.140, where matrix **B** is defined by Equation 11.143 and the elements in matrix **K** have a much simpler form:

$$K_{ij} = \frac{D}{2} \int_A \left[2\left(\frac{\partial^2 \phi_i}{\partial x^2} + \frac{\partial^2 \phi_i}{\partial y^2} \right)\left(\frac{\partial^2 \phi_j}{\partial x^2} + \frac{\partial^2 \phi_j}{\partial y^2} \right) \right.$$
$$\left. - 2(1-v) \left\{ \left(\frac{\partial^2 \phi_i}{\partial x^2} \frac{\partial^2 \phi_j}{\partial y^2} + \frac{\partial^2 \phi_j}{\partial x^2} \frac{\partial^2 \phi_i}{\partial y^2} \right) - 2\frac{\partial^2 \phi_i}{\partial x \partial y} \frac{\partial^2 \phi_j}{\partial x \partial y} \right\} \right] dA \qquad (11.145)$$

where D is the flexural rigidity of the plate defined by Equation 11.17. It can be easily confirmed that the foregoing equation is an alternative form of Equation 11.88.

The treatment on essential boundary conditions can be performed exactly as in Section 11.2.4.

11.3.2.1 Eigenvalue Equations for Buckling

By performing orthogonal matrix transformation

$$\mathbf{W} = \mathbf{V}_{N_n \times (N_n - r)} \tilde{\mathbf{W}} \qquad (11.146)$$

to Equation 11.140, the condensed eigenvalue equation of the static buckling can be rewritten as

$$[\tilde{\mathbf{K}} - N_0 \tilde{\mathbf{B}}]\tilde{\mathbf{W}} = 0 \qquad (11.147)$$

where $\tilde{\mathbf{W}}$ is an eigenvector,

$$\tilde{\mathbf{K}} = \mathbf{V}_{(N_n - r) \times N_n}^{T} \mathbf{K} \mathbf{V}_{N_n \times (N_n - r)} \qquad (11.148)$$

is the dimension stiffness matrix, and

$$\tilde{\mathbf{B}} = \mathbf{V}_{(N_n - r) \times N_n}^{T} \mathbf{B} \mathbf{V}_{N_n \times (N_n - r)} \qquad (11.149)$$

Solving Equation 11.147 with standard routines of eigenvalue solvers gives the static buckling values of thin laminated plates including thin plates of isotropic homogeneous materials.

11.3.3 Discrete Equation for Free-Vibration Analysis

The discrete dynamic equation has the same form as Equation 11.87. The difference is that the stiffness matrix **K** needs to be replaced with that in Equation 11.140 for composite laminates. The treatment of essential boundary conditions is also the same as in Section 11.1.4, and the final eigenvalue equation is given by Equation 11.120, where the stiffness matrix **K** needs to be replaced with that in Equation 11.140 for composite laminated plates. The mass matrix formulated using Equation 11.103 is still valid, but the mass moment of inertial **I** needs to be computed using Equation 11.104 with the consideration that ρ could be a function of z for composite laminated plates.

11.3.4 Numerical Examples of Buckling Analysis

In all the examples given in this section, complete second-order polynomial basis functions are used ($m = 6$) for constructing MLS shape functions. The dimension of the support domain is chosen as 3.5–3.9 times the averaged nodal distance.

Example 11.6: Static Buckling of Rectangular Plates (Validation)

The convergence of static buckling loads of square plates obtained by the present EFG method is studied. The dimensions of the rectangular plates are noted by a and b, respectively, for the dimensions in the x- and y-directions. An in-plane compressive load is applied in the x-direction. The static buckling loads of thin plates with different aspect ratios and boundaries are calculated and compared with analytical results [12]. The effects of different boundary conditions are also investigated. In the notation of the boundary conditions, the same convention used in the previous section is adopted. The buckling loads are represented via the factor of buckling load defined as $k = N_0 b^2 \pi^2 D$, where $D = Eh^3/[12(1-\nu^2)]$.

The geometry of the plates is shown in Figure 11.12. The geometric parameters and material properties of the thin rectangular plates are as follows: length $b = 10.0$ m, thickness $h = 0.06$ m, and the width of the plate can be determined by the aspect ratio, which changes from case to case. The material constants are Young's modulus $E = 2.45 \times 10^6$ N/m^2, Poisson's ratio $\nu = 0.23$.

The buckling factors for square plates with three kinds of aspect ratios and simply supported boundaries are calculated using different densities of field nodes. The results are listed in Table 11.12. It is seen that highly accurate results can be obtained using a small number ($6 \times 6 = 36$) of field nodes. The convergence is very fast.

Table 11.13 shows the buckling factors of rectangular plates of different aspect ratios with simply supported boundaries. Field nodes (13×13) regularly distributed in the plates

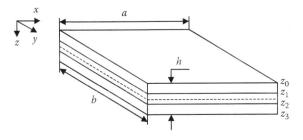

FIGURE 11.12
Symmetrically laminated composite plate and its coordinate system.

TABLE 11.12

Convergence of Buckling Factor $k = N_0 b^2 / \pi^2 D$ for Rectangular Plates ($b = 10.0$ m, $h = 0.06$ m; BC: SSSS)

	Nodes							
	5×5	6×6	7×7	8×8	9×9	10×10	11×11	Ref. [12]
$a/b = 0.8$	4.60	4.20	4.17	4.20	4.19	4.20	4.20	4.20
$a/b = 1.0$	4.25	3.99	3.97	4.01	4.03	4.02	4.00	4.00
$a/b = 1.2$	4.03	4.13	4.14	4.15	4.12	4.14	4.14	4.13

TABLE 11.13

Buckling Factor $k = N_0 b^2 / \pi^2 D$ for Rectangular Plates ($b = 10.0$ m, $h = 0.06$ m; Nodes: 13×13, BC: SSSS)

	a/b													
	0.2	0.3	0.4	0.5	0.6	0.7	0.8	0.9	1.0	1.1	1.2	1.3	1.4	1.41
EFG	27.09	13.22	8.42	6.25	5.13	4.53	4.20	4.04	4.04	4.05	4.13	4.28	4.48	4.48
Ref. [12]	27.0	13.2	8.41	6.25	5.14	4.53	4.20	4.04	4.00	4.04	4.13	4.28	4.47	4.49
Error (%)	0.33	0.15	0.12	0.00	−0.19	0.00	0.00	0.00	1.00	0.25	0.00	0.00	0.22	−0.22

are used. The buckling factors obtained using the present EFG method agrees very well with Timoshenko's analytical solutions. The buckling factor decreases as the aspect ratio increases from 0.2 to 1.0, while the buckling factor increases as the aspect ratio increases from 1.0 to 1.41. The minimum buckling factor occurs when the aspect ratio is 1.0, namely, when the plate is squared.

Table 11.14 shows the buckling factors of plates with clamped boundaries. A total of 169 ($= 13 \times 13$) regularly distributed nodes are used in the computation. Very good agreement between the buckling factors in the EFG method and Timoshenko's analytical solutions has also been achieved. The buckling factor obtained using the EFG method is slightly larger than the analytical solution and decreases as the aspect ratio increases from 0.75 to 4.0. The minimum buckling factor occurs when the aspect ratio is 4.0. The error between the EFG result and Timoshenko's solution for plates with clamped boundaries is larger than that for plates with simply supported boundaries.

TABLE 11.14

Buckling Factor $k = N_0 b^2 / \pi^2 D$ for Rectangular Plates ($b = 10.0$ m, $h = 0.06$ m; Nodes: 13×13, BC: CCCC)

	a/b													
	0.75	1.00	1.25	1.50	1.75	2.00	2.25	2.50	2.75	3.00	3.25	3.50	3.75	4.00
EFG	11.79	10.18	9.42	8.47	8.22	8.00	7.75	7.70	7.57	7.49	7.48	7.41	7.39	7.38
Ref. [12]	11.69	10.07	9.25	8.33	8.11	7.88	7.63	7.57	7.44	7.37	7.35	7.27	7.24	7.23
Error (%)	0.86	1.09	1.84	1.68	1.36	1.52	1.57	1.72	1.75	1.63	1.77	1.93	2.07	2.07

TABLE 11.15

Buckling Factor $k = N_0 b^2 / \pi^2 D$ for Rectangular Plates ($b = 10.0$ m, $h = 0.06$ m; Nodes: 13×13, BC: SCSC)

	a/b						
	0.4	0.5	0.6	0.7	0.8	0.9	1.0
EFG	9.48	7.72	7.11	7.05	7.38	7.96	7.77
Ref. [12]	9.44	7.69	7.05	7.00	7.29	7.83	7.69
Error (%)	0.42	0.39	0.85	0.71	1.23	1.66	1.04

TABLE 11.16

Buckling Factor $k = N_0 b^2 / \pi^2 D$ for Rectangular Plates (BC: SCCS)

	a/b					
	1.0	1.2	1.4	1.6	1.8	2.0
EFG	6.28	6.26	6.07	5.98	5.77	5.92

Table 11.15 shows the buckling factors for SCSC boundaries. The nodal distribution is also 13×13 (regular). The errors between the EFG results and analytic solutions are very small. The minimum buckling factor is observed at the aspect ratio of 0.7.

Table 11.16 shows the buckling factors for SCCS boundaries. Nodes of 13×13 that are regularly distributed in the plates are used. The buckling factors from aspect ratio 1.0 to 2.0 are calculated using the EFG method. The minimum buckling factor occurs at the aspect ratio of 1.8.

Example 11.7: Static Buckling of a Square Plate (Efficiency)

To study the efficiency of the EFG method, the buckling of a thin square plate with clamped boundaries is studied using both EFG and an FEM. In FEM, four-node isoparametric elements are used. The dimensions of the plate are length $a = b = 10.0$ m, thickness $h = 0.7$ m. The material properties are elastic constant $E = 1.0 \times 10^9$ N/m^2, Poisson's ratio $v = 0.3$. The results obtained using EFG and FEM for different densities of nodes are shown in Table 11.17.

Note that the dimension of the eigenvalue equation produced by EFG is only one-third of that produced by FEM. Because of the smaller dimension of the eigenvalue equation in the EFG formulation, the central processing unit (CPU) time for solving the eigenvalue equation in EFG is much less than that in FEM, as shown in Table 11.17, especially when the node number is large. Note that the total CPU time for solving the problem is affected by many factors, such as the construction of shape functions, imposition of boundary conditions, and the number of eigenvalues required. In EFG, selecting the nodes in the support domain of a quadrature point and constructing the shape functions for the point are much more expensive procedures as compared with FEM. The total CPU time of EFG is therefore greater than that of FEM for the same density of nodes. This is the price EFG has to pay for not using an element mesh. However, the results obtained by EFG are much more accurate than those of FEM. For this square plate the analytical result of buckling factor is $k = 10.07$. EFG produces a result of $k = 10.41$ using only 81 nodes, but FEM requires 361 nodes to obtain a result with the same accuracy. Therefore, to obtain a result with the same accuracy, the EFG method is still much more computationally efficient than FEM for solving static buckling problems.

TABLE 11.17

Comparison of Results of $k = N_0 b^2 / \pi^2 D$ for a Square Plate Obtained by EFG
and FEM ($a = b = 10.0$ m, $h = 0.7$ m, BC: CCCC)

		Nodes					
		9×9	11×11	13×13	15×15	17×17	19×19
Dimension of equation	EFG	81	121	169	225	289	361
	FEM	243	363	507	675	867	1083
CPU time (s) for eigenvalue solver	EFG	0.02	0.04	0.1	0.3	0.6	1.2
	FEM	0.0	0.7	1.4	5.2	16.6	63.8
Total CPU time (s)	EFG	**14.3**	47.2	155.7	531.5	1302.3	3789.8
	FEM	8.5	33.2	76.0	217.5	554.6	**1530.1**
Buckling factor k	EFG	**10.41**	10.23	10.18	10.16	10.14	10.13
	FEM	16.23	13.60	12.20	11.37	10.84	**10.47**

Example 11.8: Buckling of a Plate with Complicated Shape

To demonstrate the applicability of the present EFG method to the plates with complicated shape, the static buckling loads of a plate with a hole with complicated shape are also calculated. The geometries of the plates shown in Figure 11.13 are calculated. The field nodes irregularly distributed in the plate are shown in Figure 11.14. The number of the nodes is 339. Triangular background meshes with 493 nodes are used for integration and all the field nodes are in the vertex of the integration cells. It can be seen from Table 11.18 that the plate with SSSS boundaries has the lowest buckling factor, and the plate with CCCC boundaries has the highest buckling factor. The static buckling factors of the plates with a hole are smaller compared with those of the corresponding plates without a hole.

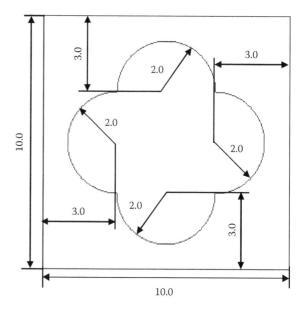

FIGURE 11.13
Plate with a hole of complicated shape.

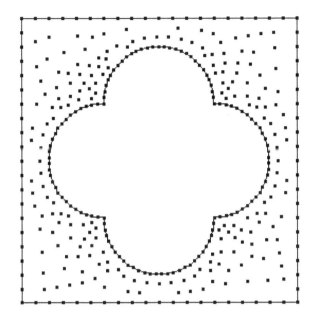

FIGURE 11.14
Nodal distribution in a plate with a hole of complicated shape.

TABLE 11.18

Buckling Factor $k = N_0 b^2 / \pi^2 D$ for the Plate with a Hole of Complicated Shape

	Boundaries			
	SSSS	CCCC	SCSC	SCCS
k	2.64	16.66	5.43	3.97

Example 11.9: Static Buckling of a Laminated Plate

The static buckling loads of three layers of symmetrically laminated composite plates of E-glass/epoxy materials for four cases of boundaries are calculated, as shown in Table 11.19. The in-plane compressive loads are applied in the x-direction. The factor of the buckling load is defined by $k = N_0 b^2 \pi^2 D_0$, where $D_0 = Eh^3 / [12(1 - \nu_{12}\nu_{21})]$. The geometric parameters and material properties of the laminates are length $a = b = 10.0$ m, thickness $h = 0.06$ m, ratio of elastic constant $E_1/E_2 = 2.45$, ratio of elastic constant $G_{12}/E_2 = 0.48$, Poisson's ratio $\nu_{12} = 0.23$. The buckling load factors of the laminates in the EFG method are calculated using 13×13 nodes regularly distributed in the plate domain. For simply supported boundaries, the buckling value increases as the ply angle increases. In contrast, for clamped boundaries, the buckling value decreases as the ply angle increases. For SSSS, CCCC, and SCCS boundaries, the buckling value for the plies of $(0°, 0°, 0°)$ is the same as that for the plies of $(0°, 90°, 0°)$. For SCCS boundaries, the buckling factor is little affected by the angle ply. As expected, the buckling values of the laminates with clamped boundaries are generally larger than those with simply supported boundaries.

TABLE 11.19

Buckling Factor $k = N_0 b^2 / \pi^2 D_0$ for Symmetrically Laminated Composite Square Plates with Different Angle Plies and Boundaries ($a = b = 10.0$ m, $h = 0.06$ m, Nodes: 13×13)

Angle Ply	BC	SSSS	CCCC	SCSC	SCCS
(0°, 0°, 0°)		2.39	6.78	4.34	3.97
(10°, −10°, 10°)		2.42	6.72	4.39	3.97
(15°, −15°, 15°)		2.45	6.64	4.46	3.96
(20°, −20°, 20°)		2.49	6.55	4.56	3.96
(30°, −30°, 30°)		2.57	6.36	4.84	3.96
(40°, −40°, 40°)		2.63	6.21	4.91	3.94
(45°, −45°, 45°)		2.64	6.16	4.79	3.93
(0°, 90°, 0°)		2.39	6.78	4.43	3.97

11.3.5 Numerical Examples for Free-Vibration Analysis

Example 11.10: Free-Vibration Analysis of Orthotropic Square Plates

Consider a square plate of a single layer of orthotropic material. The following parameters are used in the computation:

Length: $a = b = 10.0$ m

Thickness: $h = 0.05$ m

Mass density: $\rho = 8000$ kg/m^3

To compare the results obtained by the present EFG method with those given by Chen [13], we use the same elastic constants as in [13] for the following two cases: (1) If $E_1 \geq E_2$, $\nu_{12} = 0.3$; (2) If $E_1 < E_2$, $\nu_{21} = 0.3$.

The dimensionless fundamental frequency is defined as $\xi = (\omega^2 a^4 \rho h / D_3)^{1/4}$, where $D_3 = D_{12} + 2D_{66}$. Regular rectangular background meshes (12×12) and regularly distributed field nodes (13×13) are used. The vertices of the background mesh are used as the field nodes. The dimensionless fundamental frequencies are listed in Tables 11.20 through 11.23 for four cases of different boundary conditions. In these tables, the subscript of ξ indicates the number of mode. In each table, frequencies for different ratios of elastic constants are listed. Table 11.20 shows the results for a square plate simply supported at all edges. Good agreement between the frequencies of the first mode obtained using the present EFG method and those given in [13] has been observed. Table 11.21 shows the results for a square plate clamped at all edges. The frequencies of the first mode obtained by the present EFG method fall in between those of [13] and [14]. The results shown in Tables 11.22 and 11.23 also indicate that the frequencies of the first mode obtained by the present EFG method agree well with those of [13]. In all four cases, the frequencies of the first mode obtained by the present EFG method are slightly larger than those given in [13].

Example 11.11: Natural Frequency Analysis of Composite Laminated Plates

Symmetric laminated composite plates with three layers of E-glass/epoxy materials are considered. Natural frequencies of square, elliptical, and complicated shaped plates are calculated using the present EFG method. The dimensionless frequency parameters are $\beta = (\rho h \omega^2 a^4 / D_0)^{1/2}$, where $D_0 = Eh^3 / [12(1 - \nu_{12}\nu_{21})]$.

TABLE 11.20

The Dimensionless Fundamental Natural Frequencies ξ for Orthotropic Square Plates (BC: SSSS)

| D_{22}/D_3 | | 0.5 | | | 1.0 | | | 2.0 | | |
D_{11}/D_3		0.5	1.0	2.0	0.5	1.0	2.0	0.5	1.0	2.0
EFG	ξ_1	4.130	4.295	4.576	4.295	4.443	4.700	4.576	4.700	4.921
	ξ_2	6.333	6.387	6.479	6.387	7.031	7.104	6.479	7.104	8.008
	ξ_3	6.341	6.996	7.936	6.996	7.036	7.961	7.936	7.961	8.011
	ξ_4	8.273	8.600	8.781	8.600	8.892	9.404	8.781	9.404	9.043
	ξ_5	8.714	8.743	9.159	8.743	9.959	9.988	9.159	9.988	11.575
	ξ_6	8.732	9.949	10.903	9.949	9.966	11.560	10.903	11.560	11.578
	ξ_7	10.411	10.587	11.277	10.587	11.341	11.604	11.277	11.604	12.713
	ξ_8	10.422	11.205	11.552	11.205	11.347	12.518	11.552	12.518	12.714
	ξ_9	11.242	11.259	12.416	11.259	13.032	13.048	12.416	13.048	14.775
	ξ_{10}	11.249	12.891	13.515	12.891	13.036	14.110	13.515	14.110	15.285
ξ_1 [13]		4.118	4.279	4.557	4.279	4.425	4.678	4.557	4.678	4.897

TABLE 11.21

The Dimensionless Fundamental Natural Frequencies ξ for Orthotropic Square Plates (BC: CCCC)

| D_{22}/D_3 | | 0.5 | | | 1.0 | | | 2.0 | | |
D_{11}/D_3		0.5	1.0	2.0	0.5	1.0	2.0	0.5	1.0	2.0
EFG	ξ_1	5.312	5.697	6.288	5.697	6.017	6.532	6.288	6.532	6.948
	ξ_2	7.554	7.710	7.982	7.710	8.605	8.805	7.982	8.805	9.976
	ξ_3	7.554	8.495	9.767	8.495	8.606	9.841	9.767	9.841	9.976
	ξ_4	9.430	9.973	10.203	9.973	10.439	11.209	10.203	11.209	11.847
	ξ_5	9.978	10.069	10.840	10.069	11.533	11.637	10.840	11.637	13.500
	ξ_6	10.006	11.497	12.459	11.497	11.562	13.334	12.459	13.334	13.525
	ξ_7	11.592	11.911	12.644	11.911	12.893	13.455	12.644	13.455	14.690
	ξ_8	11.597	12.565	13.423	12.565	12.896	14.367	13.423	14.367	14.692
	ξ_9	12.524	12.651	14.192	12.651	14.605	14.655	14.192	14.655	16.761
	ξ_{10}	12.525	14.202	14.584	14.202	14.606	15.902	14.584	15.902	17.185
ξ_1 [14]		5.324	5.712	6.307	5.712	6.034	6.553	6.307	6.553	6.972
ξ_1 [15]		5.171	5.551	6.131	5.551	5.866	6.371	6.131	6.371	6.779

The geometric parameters and material properties of the square plates are as follows:

Length: $a = b = 10.0$ m

Thickness: $h = 0.06$ m

Mass density: $\rho = 8000$ kg/m^3

Ratio of elastic constant: $E_1/E_2 = 2.45$

Ratio of elastic constant: $G_{12}/E_2 = 0.48$

Poisson's ratio: $\nu_{12} = 0.23$

The radii of the elliptical plates are $a = 5.0$ m and $b = 2.5$ m, respectively. Other geometric parameters and material properties are the same as the square plates.

TABLE 11.22

The Dimensionless Fundamental Natural Frequencies ξ for Orthotropic Square Plates (BC: SCSC)

D_{22}/D_3		0.5			1.0			2.0		
D_{11}/D_3		0.5	1.0	2.0	0.5	1.0	2.0	0.5	1.0	2.0
EFG	ξ_1	4.829	4.936	5.130	5.319	5.400	5.551	6.017	6.073	6.181
	ξ_2	6.655	7.239	7.468	6.876	7.413	8.231	7.246	7.715	8.456
	ξ_3	7.360	7.397	8.106	8.362	8.388	8.439	9.206	9.701	9.735
	ξ_4	8.901	9.215	9.690	9.020	9.799	10.205	9.685	10.287	11.032
	ξ_5	8.954	9.909	9.952	9.577	10.156	11.469	10.545	10.717	11.776
	ξ_6	9.891	10.081	11.628	11.262	11.445	11.689	11.505	12.474	13.376
	ξ_7	10.881	11.415	11.703	11.405	11.919	12.720	11.913	13.197	13.444
	ξ_8	11.285	11.609	12.466	11.439	12.525	12.982	13.387	13.396	14.259
	ξ_9	11.347	12.453	12.727	12.414	13.131	14.541	13.810	14.109	15.380
	ξ_{10}	12.446	13.093	13.908	13.384	14.214	14.879	13.993	14.915	15.921
ξ_1 [13]		4.729	4.838	5.037	5.205	5.288	5.443	5.883	5.941	6.052

TABLE 11.23

The Dimensionless Fundamental Natural Frequencies ξ for Orthotropic Square Plates (BC: SCCS)

D_{22}/D_3		0.5			1.0			2.0		
D_{11}/D_3		0.5	1.0	2.0	0.5	1.0	2.0	0.5	1.0	2.0
EFG	ξ_1	4.726	4.997	5.428	4.997	5.228	5.611	5.428	5.611	5.927
	ξ_2	6.935	7.037	7.213	7.037	7.807	7.941	7.213	7.941	8.980
	ξ_3	6.953	7.750	8.855	7.750	7.827	8.905	8.855	8.905	8.997
	ξ_4	8.844	9.277	9.489	9.277	9.656	10.296	9.489	10.296	10.836
	ξ_5	9.356	9.407	9.988	9.407	10.752	10.811	9.988	10.811	12.542
	ξ_6	9.376	10.735	11.679	10.735	10.770	12.469	11.679	12.469	12.556
	ξ_7	11.004	11.250	11.957	11.250	12.120	12.518	11.957	12.518	13.703
	ξ_8	11.011	11.909	12.498	11.909	12.121	13.445	12.498	13.445	13.705
	ξ_9	11.885	11.930	13.305	11.930	13.821	13.850	13.305	13.850	15.771
	ξ_{10}	11.895	13.458	13.725	13.458	13.828	15.011	13.725	15.011	16.237
ξ_1 [13]		4.637	4.905	5.332	4.905	5.136	5.515	5.332	5.515	5.829

11.3.5.1 Convergence of Natural Frequency

We use different densities of regularly distributed field nodes in the present EFG method to calculate natural frequencies of the laminated composite square plates with simply supported boundaries. Regular rectangular meshes are used and all the field nodes are in the vertexes of the meshes. The angle ply is arranged as (30°, 30°, 30°). Table 11.24 shows the frequency parameters. Good convergence can be observed from Table 11.24 for the present EFG method.

11.3.5.2 Effectiveness of Irregularly Distributed Nodes

To analyze the effectiveness of the present EFG method using irregularly distributed nodes, we calculate the frequencies of the square laminates using $17 \times 17 = 289$ regularly

TABLE 11.24

Convergence of Frequency of Laminated Composite Square Plates (BC: SSSS)

	Field Nodes				
	9×9	11×11	13×13	15×15	17×17
β	20.02	16.05	15.88	15.89	15.86

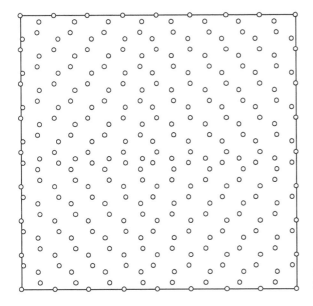

FIGURE 11.15
Square plate with 289 irregularly distributed nodes.

distributed nodes and 289 irregularly distributed nodes (Figure 11.15). A uniform mesh of 16×16 mesh is used for the integration for the square laminates. A regular mesh of 180 cells is used for integration for the elliptical laminates. The angle ply of both laminates is arranged as $(30°, -30°, 30°)$. The results are shown in Tables 11.25 and 11.26. It is found from the Tables 11.25 and 11.26 that the frequency parameters of using both regularly and irregularly distributed nodes show very good agreement with each other. This confirms that node irregularity does not affect the results significantly, provided there is no significant variation on nodal density across the domain.

11.3.5.3 Square Plate

Four cases of different boundary conditions together with five cases of different angle plies are considered. Regular rectangular background meshes (12×12) and regularly distributed field nodes (13×13) are used. Table 11.27 shows the results for laminates simply supported at all edges. The present frequencies agree very well with those given in [15,16]. Table 11.28 shows the results for laminates clamped at all edges. The present frequencies are very close to and slightly larger than those given in [15]. Table 11.29 shows the effect of fiber orientation on the frequency parameters of the laminated plates with SCSC boundary conditions. Maximum natural frequency parameters of the fifth, seventh, and ninth modes are observed at the plate with $\alpha = 30°$; the first, third, fourth, sixth, and eighth modes occur

TABLE 11.25

Natural Frequency Parameters β of Laminated Composite
Square Plates (BC: SSSS)

Modes	EFG	
	Regular Nodes 289	Irregular Nodes 289
1	15.86	15.92
2	35.86	35.99
3	42.58	42.66
4	61.47	61.66
5	71.83	72.35
6	85.97	86.45
7	93.95	94.19
8	109.00	109.06
9	119.71	120.45
10	133.43	133.58

TABLE 11.26

Natural Frequency Parameters β of Laminated Composite
Elliptical Plates (BC: CCCC)

Modes	EFG	
	Regular Nodes 201	Irregular Nodes 201
1	19.89	19.87
2	30.44	30.44
3	44.34	44.25
4	49.95	49.90
5	61.94	61.98
6	65.77	65.76
7	84.63	84.19
8	84.81	84.75
9	93.36	93.12
10	107.33	107.06

at the plate with $\alpha = 45°$; and the second and tenth modes occur at the plate of $(0°, 90°, 0°)$. Table 11.30 shows the effect of fiber orientation on the natural frequency parameters of laminated plates with SCCS boundary conditions. Maximum natural frequency parameters of the third and seventh modes are observed at $\alpha = 0°$; the sixth mode at $\alpha = 15°$; the first, second, fifth, and ninth modes occur at the plate with $\alpha = 45°$; and the fourth, eighth, and tenth modes are registered at the plate with $(0°, 90°, 0°)$.

11.3.5.4 Elliptical Plate

The natural frequencies are computed for the elliptical laminates using the present EFG method. Regularly distributed nodes (201) and regular meshes (180) are used. Table 11.31 shows the effect of fiber orientation on the natural frequency parameters of laminated

TABLE 11.27

Natural Frequency Parameters β of Laminated Composite Square Plates (BC: SSSS)

Angle Ply	Source of Result	Modes									
		1	2	3	4	5	6	7	8	9	10
(0°, 0°, 0°)	EFG	15.18	33.34	44.51	60.79	64.80	90.39	94.23	109.4	109.9	136.8
	[15]	15.19	33.31	44.52	60.78	64.55	90.31	93.69	108.7	—	—
	[16]	15.19	33.30	44.42	60.77	64.53	90.29	93.66	108.6	—	—
(15°, −15°, 15°)	EFG	15.41	34.15	43.93	60.91	66.94	91.74	92.01	109.3	112.4	135.0
	[15]	15.37	34.03	43.80	60.80	66.56	91.40	91.51	108.9	—	—
	[16]	15.43	34.09	43.87	60.85	66.67	91.40	91.56	108.9	—	—
(30°, −30°, 30°)	EFG	15.88	35.95	42.63	61.54	72.12	86.32	94.08	109.2	120.6	134.3
	[15]	15.86	35.77	42.48	61.27	71.41	85.67	93.60	108.9	—	—
	[16]	15.90	35.86	42.62	61.45	71.71	85.72	93.74	108.9	—	—
(45°, −45°, 45°)	EFG	16.11	37.04	41.80	61.94	78.03	80.11	95.07	109.3	132.3	134.1
	[15]	16.08	36.83	41.67	61.65	76.76	79.74	94.40	109.0	—	—
	[16]	16.14	36.93	41.81	61.85	77.04	80.00	94.68	109.0	—	—
(0°, 90°, 0°)	EFG	15.18	33.82	44.14	60.79	66.12	91.16	93.31	108.8	112.4	136.8

TABLE 11.28

Natural Frequency Parameters β of Laminated Composite Square Plates (BC: CCCC)

Angle Ply	Source of Result	Modes									
		1	2	3	4	5	6	7	8	9	10
(0°, 0°, 0°)	EFG	29.27	51.21	67.94	86.25	87.97	119.3	127.6	138.5	144.0	169.8
	[15]	29.13	50.82	67.29	85.67	87.14	118.6	126.2	137.5	—	—
(15°, −15°, 15°)	EFG	29.07	51.82	66.54	85.17	90.56	120.0	124.1	140.8	143.2	167.6
	[15]	28.92	51.43	65.92	84.55	89.76	119.3	122.7	139.9	—	—
(30°, −30°, 30°)	EFG	28.69	53.57	63.24	84.43	96.13	115.4	121.5	139.6	150.8	166.3
	[15]	28.55	53.15	62.71	83.83	95.21	114.1	120.7	138.6	—	—
(45°, −45°, 45°)	EFG	28.50	55.11	60.91	84.25	103.2	106.7	122.3	138.3	165.2	166.1
	[15]	28.38	54.65	60.45	83.65	102.0	105.6	121.4	137.3	—	—
(0°, 90°, 0°)	EFG	29.27	51.93	67.40	86.25	89.76	120.3	126.4	141.6	143.2	172.2

TABLE 11.29

Natural Frequency Parameters β of Laminated Composite Square Plates (BC: SCSC)

Angle Ply	Modes									
	1	2	3	4	5	6	7	8	9	10
(0°, 0°, 0°)	20.55	46.41	47.18	70.52	85.02	96.06	107.5	116.7	136.0	150.6
(15°, −15°, 15°)	20.89	45.70	48.28	71.23	86.87	94.03	108.7	117.1	138.7	149.7
(30°, −30°, 30°)	21.85	44.45	51.19	72.63	88.65	93.23	109.0	121.1	147.7	150.0
(45°, −45°, 45°)	23.16	43.08	55.56	73.61	83.50	102.6	108.4	126.3	136.4	151.0
(0°, 90°, 0°)	20.77	46.92	47.32	70.90	86.92	95.22	108.8	116.3	139.3	151.2

TABLE 11.30

Natural Frequency Parameters β of Laminated Composite Square Plates (BC: SCCS)

Angle Ply	Modes									
	1	2	3	4	5	6	7	8	9	10
(0°, 0°, 0°)	21.62	41.64	55.66	72.81	75.92	104.3	110.5	123.7	126.2	150.3
(15°, −15°, 15°)	21.69	42.40	54.72	72.48	78.18	105.5	107.7	125.3	126.9	149.5
(30°, −30°, 30°)	21.81	44.27	52.49	72.48	83.53	100.6	107.4	124.4	135.3	149.7
(45°, −45°, 45°)	21.88	45.70	50.89	72.62	89.90	93.18	108.3	123.9	148.4	150.0
(0°, 90°, 0°)	21.62	42.24	55.21	72.82	77.47	105.2	109.4	125.4	126.6	152.5

TABLE 11.31

Natural Frequency Parameters β of Laminated Composite Elliptical Plates (BC: CCCC)

Angle Ply	Modes									
	1	2	3	4	5	6	7	8	9	10
(0°, 0°, 0°)	18.48	29.38	44.97	45.72	60.44	65.33	79.24	85.31	91.50	102.8
(15°, −15°, 15°)	18.83	29.70	44.73	46.72	62.06	64.07	81.09	87.14	88.90	104.5
(30°, −30°, 30°)	19.89	30.44	44.34	49.95	61.94	65.77	84.63	84.81	93.36	107.3
(45°, −45°, 45°)	21.60	31.38	44.11	55.17	60.19	70.21	81.64	88.25	103.7	110.2
(0°, 90°, 0°)	18.81	29.58	44.99	46.72	61.34	65.14	79.99	87.23	91.16	103.4

plates clamped at all edges. Maximum frequency parameters of the fifth mode are registered at the plate with $\alpha = 15°$; the seventh mode at the plate with $\alpha = 30°$; the first, second, fourth, sixth, eighth, ninth, and tenth modes at the plate with $\alpha = 45°$; and the third mode at the plate with (0°, 90°, 0°).

11.3.5.5 Plate of Complicated Shape

The natural frequencies are calculated for the laminates with a hole of complicated shape, as shown in Figure 11.16. The 310 nodes and 462 triangular background meshes are plotted in Figure 11.17. Four cases of different boundary conditions together with five cases of different angle ply were considered. Table 11.32 shows the effect of fiber orientation on the frequency parameters of laminates simply supported at all edges. Maximum frequency parameters of the fifth mode occur at $\alpha = 0°$; the third mode occurs at $\alpha = 15°$; the fourth and seventh modes occur at $\alpha = 30°$; the first and tenth modes occur at $\alpha = 45°$; and the second, sixth, eighth, and ninth modes occur at (0°, 90°, 0°). Table 11.33 shows the effect of fiber orientation on the frequency parameters of laminates clamped at all edges. Maximum frequency parameters of the first and second modes occur at the plate with $\alpha = 15°$; the third and sixth modes occur at the plate with $\alpha = 45°$; and the fourth, fifth, seventh, eighth, ninth, and tenth modes occur at the plate of (0°, 90°, 0°). Table 11.34 shows the effect of fiber orientation on the frequency parameters of laminates for SCSC boundary conditions. Maximum frequency parameters of the fifth mode occur at the plate with $\alpha = 15°$; the first, second, third, sixth, and tenth modes occur at the plate with $\alpha = 45°$; and the fourth,

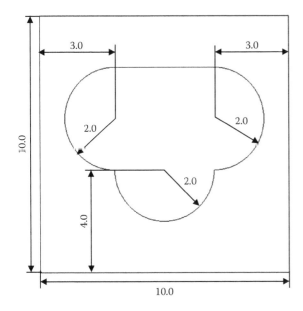

FIGURE 11.16
Square plate with a hole of complicated shape.

seventh, eighth, and ninth modes occur at the plate of (0°, 90°, 0°). Table 11.35 shows the effect of fiber orientation on the frequency parameters of laminates for SCCS boundary conditions. Maximum frequency parameters of the second, seventh, and tenth modes occur at the plate with $\alpha = 15°$; the third, fourth, and ninth modes occur at the plate with $\alpha = 45°$; and the first, fifth, sixth, and eighth modes occur at the plate with (0°, 90°, 0°). It can also be seen that the natural frequencies of the laminates with clamped boundaries are generally higher than those with simply supported boundaries.

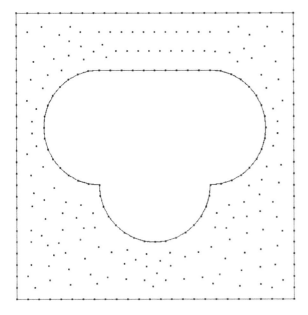

FIGURE 11.17
Nodal distribution of the square plate with a hole of complicated shape.

TABLE 11.32

Frequency Parameters β of Laminated Composite Plates of Complicated Shape (BC: SSSS)

Angle Ply	Modes									
	1	2	3	4	5	6	7	8	9	10
(0°, 0°, 0°)	25.64	79.94	88.32	118.2	169.2	179.3	205.0	241.0	259.1	288.1
(15°, −15°, 15°)	26.51	79.31	88.37	118.3	168.0	175.4	210.8	240.1	250.6	291.0
(30°, −30°, 30°)	28.33	79.48	88.18	118.7	165.7	172.1	213.0	238.4	247.0	294.1
(45°, −45°, 45°)	29.53	79.28	86.85	118.6	161.4	171.5	210.9	237.5	245.2	295.1
(0°, 90°, 0°)	25.70	79.95	88.30	118.4	169.0	180.1	206.1	241.7	259.9	288.7

TABLE 11.33

Frequency Parameters β of Laminated Composite Plates of Complicated Shape (BC: CCCC)

Angle Ply	Modes									
	1	2	3	4	5	6	7	8	9	10
(0°, 0°, 0°)	88.48	95.16	107.1	157.0	166.9	185.2	250.1	291.0	299.3	325.3
(15°, −15°, 15°)	88.86	95.35	107.8	156.4	166.5	186.8	244.8	290.4	297.8	323.5
(30°, −30°, 30°)	88.62	94.62	110.0	155.8	166.8	191.1	240.0	284.8	294.2	320.1
(45°, −45°, 45°)	87.74	93.19	113.3	156.1	167.4	194.7	235.4	281.4	290.4	316.9
(0°, 90°, 0°)	88.67	95.15	107.9	157.8	167.5	186.4	250.3	291.3	299.8	325.4

TABLE 11.34

Frequency Parameters β of Laminated Composite Plates of Complicated Shape (BC: SCSC)

Angle Ply	Modes									
	1	2	3	4	5	6	7	8	9	10
(0°, 0°, 0°)	64.07	72.44	98.25	139.3	160.0	170.8	230.7	254.6	266.0	273.0
(15°, −15°, 15°)	64.73	72.00	100.0	138.6	163.3	168.7	229.3	250.4	264.0	274.1
(30°, −30°, 30°)	66.36	72.99	103.5	138.4	161.5	171.1	228.0	247.4	259.9	276.5
(45°, −45°, 45°)	67.69	74.39	108.5	139.7	158.3	172.8	230.6	244.8	254.7	277.3
(0°, 90°, 0°)	64.37	72.86	99.32	140.1	160.5	171.2	231.9	255.0	266.3	273.4

TABLE 11.35

Frequency Parameters β of Laminated Composite Plates of Complicated Shape (BC: SCCS)

Angle Ply	Modes									
	1	2	3	4	5	6	7	8	9	10
(0°, 0°, 0°)	68.95	93.87	114.2	140.6	170.2	191.5	221.9	251.5	264.3	301.7
(15°, −15°, 15°)	68.43	94.46	115.2	141.0	166.8	190.6	223.2	244.8	265.9	304.1
(30°, −30°, 30°)	68.64	94.08	117.1	141.9	164.8	189.8	223.1	238.9	270.8	301.6
(45°, −45°, 45°)	68.95	93.10	119.8	142.6	164.4	188.4	221.1	236.6	273.6	292.3
(0°, 90°, 0°)	69.17	94.02	114.9	141.1	170.9	191.8	222.5	251.9	265.2	302.5

11.4 EFG Method for Thick Plates

In Sections 12.2 and 12.3, we discussed EFG formulations for static, buckling, and the vibration of thin plates that are governed by the simplest displacement-based theory: CPT. In CPT, the transverse shear strain is neglected. For analyses of thin plates with a thickness-to-width ratio of up to about 1:10, CPT can often give good results with sufficient accuracy for engineering application, because the effect of the transverse shear deformation is very small. For thick plates, however, transverse shear strains cannot be neglected. The CPT underpredicts deflections and overpredicts natural frequencies and buckling loads for thick plates.

Thick plates are very important structural elements and have wide applications. There are a few displacement-based theories for the analyses of plates. A slightly more complex displacement-based theory, often termed the FSDT or the Mindlin plate theory, must be used to take into account the transverse shear strain. In FSDT, in-plane displacements through the thickness are assumed to vary linearly. FSDT gives a state of constant shear strain through the thickness, which cannot satisfy the natural condition that the shear strains in the top and bottom surfaces of plates should vanish. In FSDT, a shear correction factor that depends on material property, geometry, and boundary conditions is introduced to correct for the discrepancy in the shear forces. However, FSDT still does not reflect the high-order variation of shear strain through the thickness. To ensure the vanishing of the transverse shear strains on the top and bottom surfaces of the plates and to more accurately reflect the high-order variation of shear strain through the thickness, the TSDT was developed. In the formulations of finite plate element based on TSDT, C^1 displacement interpolants must be constructed, which can be difficult. Moreover, the application of high-order displacement-based elements makes mesh generation algorithms much more complicated, leading to an increment in computational costs.

Meshfree methods have been applied for shear deformable beams and plates [17]. This section formulates the EFG method to solve free-vibration and static buckling problems for shear deformable plates. The material presented in this section is based on the work done in [4–7]. Based on both FSDT and TSDT of plates, weak forms of governing equation for free vibration and static buckling are established. The essential boundary conditions are formulated through a weak form separate from the system equation and are imposed using the orthogonal transform techniques used in Section 11.1. To examine the validity of the EFG method in the application of eigenvalue problems for the shear deformable plates, natural frequencies of the square thick plates with different hard-type boundary conditions are calculated. Buckling loads of the square thick plates with different boundary conditions and a square thick plate with a circular hole are also investigated.

11.4.1 Approximation of Field Variables

Using the EFG method, the plate is represented by a set of nodes scattered in the domain of the plate. The field variables should be the deflections and the rotations at all the nodes. MLS approximation is used to approximate all these field variables w, φ_x, and φ_y, i.e.,

$$w = \sum_{I=1}^{n} \phi_I w_I \qquad (11.150)$$

$$\varphi_x = \sum_{I=1}^{n} \phi_{xI}\varphi_{xI} \tag{11.151}$$

$$\varphi_y = \sum_{I=1}^{n} \phi_{yI}\varphi_{yI} \tag{11.152}$$

where ϕ_I, ϕ_{xI}, and ϕ_{yI} are the shape functions for the three field variables w, φ_x, and φ_y, respectively. These shape functions do not have to be the same. Formulations that use

$$\phi_{xI}(x,y) = \frac{\partial \phi_I(x,y)}{\partial x} \tag{11.153}$$

$$\phi_{yI}(x,y) = \frac{\partial \phi_I(x,y)}{\partial y} \tag{11.154}$$

can avoid shear locking [19]. In our formulation, we will assume that ϕ_I, ϕ_{xI}, and ϕ_{yI} are the different shape functions and are independent of each other. Equations 11.150 through 11.152 can be written in the following matrix form:

$$\mathbf{u} = \begin{Bmatrix} w \\ \varphi_x \\ \varphi_y \end{Bmatrix} = \sum_{I=1}^{n} \underbrace{\begin{bmatrix} \phi_I & 0 & 0 \\ 0 & \phi_{xI} & 0 \\ 0 & 0 & \phi_{yI} \end{bmatrix}}_{\mathbf{\Phi}_I} \underbrace{\begin{Bmatrix} w_I \\ \varphi_{xI} \\ \varphi_{yI} \end{Bmatrix}}_{\mathbf{u}_I} \tag{11.155}$$

where
  w is the transverse deflection of the neutral plane of the plate
  φ_x and $-\varphi_y$ denote the rotations of the cross section of the plate about the y- and x-axes, respectively

Matrix $\mathbf{\Phi}_I$ contains the shape functions arranged as

$$\mathbf{\Phi}_I = \begin{bmatrix} \phi_I & 0 & 0 \\ 0 & \phi_{xI} & 0 \\ 0 & 0 & \phi_{yI} \end{bmatrix} \tag{11.156}$$

and $\mathbf{u}_I$ is the vector of nodal variables for node I given by

$$\mathbf{u}_I = \begin{Bmatrix} w_I \\ \varphi_{xI} \\ \varphi_{yI} \end{Bmatrix} \tag{11.157}$$

11.4.2 Variational Forms of System Equations

11.4.2.1 Free-Vibration Analysis

For free-vibration analysis of shear deformable plates, we can start with the Galerkin weak form of elastic solids:

$$\int_{\Omega} \delta\boldsymbol{\varepsilon}^{\mathrm{T}}\boldsymbol{\sigma}\mathrm{d}\Omega + \int_{\Omega} \delta\mathbf{u}^{\mathrm{T}}\rho\ddot{\mathbf{u}}\mathrm{d}\Omega = 0 \qquad (11.158)$$

where ρ is the mass density. Note that the terms related to the external forces have been removed, because we are interested only in free vibration, and hence the external forces are not considered. The formulation can then be based on the TSDT, as the FSDT is a special case of the TSDT.

Using the expressions for the displacement (Equation 11.72), strains (Equation 11.73), and stresses (Equation 11.58), we can easily arrive at the following equations for thick plate structures:

$$\int_{\Omega} \delta(\mathbf{L}_d\mathbf{u})^{\mathrm{T}}\mathbf{D}\mathbf{L}_d\mathbf{u}\mathrm{d}\Omega + \int_{\Omega} \rho\delta(\mathbf{L}_u\mathbf{u})^{\mathrm{T}}\mathbf{L}_u\ddot{\mathbf{u}}\mathrm{d}\Omega = 0 \qquad (11.159)$$

11.4.2.2 Static Buckling Analysis

For static buckling analysis of shear deformable plates, the total strain energy potential is [20]

$$\Pi_{\mathrm{PE}} = \underbrace{\frac{1}{2}\int_{\Omega} \boldsymbol{\varepsilon}^{\mathrm{T}}\mathbf{D}\boldsymbol{\varepsilon}\mathrm{d}\Omega}_{U_{\mathrm{PE}}} + \underbrace{\int_{\Omega} \boldsymbol{\varepsilon}_N^{\mathrm{T}}\boldsymbol{\tau}_N\mathrm{d}\Omega}_{U_N} \qquad (11.160)$$

The first term is the same as that in Equation 11.158 representing the strain energy potential resulted from the linear strain field, and $\boldsymbol{\varepsilon}_N$ denotes the nonlinear strains that are needed for modeling the buckling. The higher-order terms associated with the in-plane displacements in $\boldsymbol{\varepsilon}_N$ are neglected, and $\boldsymbol{\varepsilon}_N$ has the following form:

$$\boldsymbol{\varepsilon}_N^{\mathrm{T}} = \left\{ \frac{1}{2}\left(\frac{\partial w}{\partial x}\right)^2 \quad \frac{1}{2}\left(\frac{\partial w}{\partial y}\right)^2 \quad \frac{\partial w}{\partial x}\frac{\partial w}{\partial y} \quad 0 \quad 0 \right\} \qquad (11.161)$$

The associated stress $\boldsymbol{\tau}_N$ is given by

$$\boldsymbol{\tau}_N = \left\{ \begin{array}{c} -\dfrac{N_x}{h} \\[2mm] -\dfrac{N_y}{h} \\[2mm] -\dfrac{N_{xy}}{h} \\[2mm] 0 \\[1mm] 0 \end{array} \right\} = -\frac{1}{h}\mathbf{P} \qquad (11.162)$$

The variational form of the total potential energy is

$$\delta\Pi_{\mathrm{PE}} = \delta(U_{\mathrm{PE}} + U_N) = \int_{\Omega} \delta\boldsymbol{\varepsilon}^{\mathrm{T}}\boldsymbol{\sigma}\mathrm{d}\Omega + \int_{\Omega} \delta\boldsymbol{\varepsilon}_N^{\mathrm{T}}\boldsymbol{\tau}_N\mathrm{d}\Omega = 0 \qquad (11.163)$$

Using Equation 11.163 as well as Equations 11.161 and 11.162, we can easily arrive at the following weak form for the static buckling problem of thick plates.

$$\int_\Omega \delta(\mathbf{L}_d\mathbf{u})^\mathrm{T}\mathbf{D}\mathbf{L}_d\mathbf{u}d\Omega - \int_A \delta(\mathbf{L}_{N1}w)^\mathrm{T}(\mathbf{L}_{N2}^\mathrm{T}w)\mathbf{P}dA = 0 \tag{11.164}$$

where

$$\mathbf{L}_{N1}^\mathrm{T} = \left\{ \frac{\partial}{\partial x}, \ \frac{\partial}{\partial y} \right\} \tag{11.165}$$

$$\mathbf{L}_{N2}^\mathrm{T} = \begin{bmatrix} \dfrac{\partial}{\partial x} & 0 & \dfrac{\partial}{\partial y} & 0 & 0 \\ 0 & \dfrac{\partial}{\partial y} & \dfrac{\partial}{\partial x} & 0 & 0 \end{bmatrix} \tag{11.166}$$

$$\mathbf{P} = \left\{ \begin{array}{c} N_x \\ N_y \\ N_{xy} \\ 0 \\ 0 \end{array} \right\} \tag{11.167}$$

In deriving Equation 11.164, the integration in the z-direction has been carried out for the second term.

11.4.3 Discrete System Equations

Substituting the displacement interpolants Equations 11.150 through 11.152 into Equation 11.159 or 11.164, we obtain the eigenvalue equation for both free vibration and static buckling of plate structures as follows:

$$[\mathbf{K} - \eta\mathbf{G}]\mathbf{q} = 0 \tag{11.168}$$

where η is the eigenvalues. For free-vibration problems, $\eta = \omega^2$, where ω represents the angular frequencies, and for static buckling problems, $\eta = N_0$, where N_0 is the buckling load. Vector $\mathbf{q}$ is the eigenvector.

In Equation 11.168, $\mathbf{K}$ is the stiffness matrix for both free-vibration and static buckling problems, which is assembled using a nodal stiffness matrix of 3×3 given by

$$\mathbf{K}_{IJ} = \int_\Omega \mathbf{B}_I^\mathrm{T}\mathbf{D}\mathbf{B}_J d\Omega \tag{11.169}$$

where the matrix **B** is given by

$$\mathbf{B}_I = \mathbf{L}_d \mathbf{\Phi}_I = \begin{bmatrix} -\alpha z^3 \dfrac{\partial^2}{\partial x^2} & (z-\alpha z^3)\dfrac{\partial}{\partial x} & 0 \\[2mm] -\alpha z^3 \dfrac{\partial^2}{\partial y^2} & 0 & (z-\alpha z^3)\dfrac{\partial}{\partial y} \\[2mm] -2\alpha z^3 \dfrac{\partial^2}{\partial x \partial y} & (z-\alpha z^3)\dfrac{\partial}{\partial y} & (z-\alpha z^3)\dfrac{\partial}{\partial x} \\[2mm] (1-\beta z^2)\dfrac{\partial}{\partial x} & (1-\beta z^2) & 0 \\[2mm] (1-\beta z^2)\dfrac{\partial}{\partial y} & 0 & (1-\beta z^2) \end{bmatrix} \begin{bmatrix} \phi_I & 0 & 0 \\ 0 & \phi_{xI} & 0 \\ 0 & 0 & \phi_{yI} \end{bmatrix} \tag{11.170}$$

In Equation 11.168, **G** is the mass matrix for the free-vibration problem,

$$\mathbf{G}_{IJ} = \int_{\Omega} \rho \mathbf{B}_{uI}^{\mathrm{T}} \mathbf{B}_{uJ} \, \mathrm{d}\Omega \tag{11.171}$$

where

$$\mathbf{B}_{uI} = \mathbf{L}_u \mathbf{\Phi}_I = \begin{bmatrix} -\alpha z^3 \dfrac{\partial}{\partial x} & z-\alpha z^3 & 0 \\[2mm] -\alpha z^3 \dfrac{\partial}{\partial y} & 0 & z-\alpha z^3 \\[2mm] 1 & 0 & 0 \end{bmatrix} \begin{bmatrix} \phi_I & 0 & 0 \\ 0 & \phi_{xI} & 0 \\ 0 & 0 & \phi_{yI} \end{bmatrix} \tag{11.172}$$

For the static buckling problem, **G** is assembled using a nodal matrix of

$$\mathbf{G}_{IJ} = \begin{bmatrix} G_{IJ}^w & 0 & 0 \\ 0 & 0 & 0 \\ 0 & 0 & 0 \end{bmatrix} \tag{11.173}$$

where

$$G_{IJ}^w = \frac{1}{2}\int_A \left[2\frac{\partial \phi_I}{\partial x}\frac{\partial \phi_J}{\partial x} + 2\mu_1 \frac{\partial \phi_I}{\partial y}\frac{\partial \phi_J}{\partial y} + 2\mu_2 \left(\frac{\partial \phi_I}{\partial x}\frac{\partial \phi_J}{\partial y} + \frac{\partial \phi_I}{\partial x}\frac{\partial \phi_J}{\partial y} \right) \right] \mathrm{d}A \tag{11.174}$$

where μ_1 and μ_2 are defined in Equation 11.123.

Equation 11.168 is the eigenvalue equation for thick plates without essential boundary conditions.

11.4.4 Discrete Form of Essential Boundary Conditions

Because the Kronecker delta condition $\phi_I(x_J) = \delta_{IJ}$ at each node is not satisfied by the MLS shape function, the essential boundary conditions are imposed in a manner similar to that described in Section 11.1. The essential boundary conditions are represented by a weak form with Lagrange multipliers to produce the discretized essential boundary conditions as given below:

$$\int_{\Gamma_u} \delta\lambda^T(\mathbf{u}_b - \mathbf{u}_\Gamma)d\Gamma = 0 \tag{11.175}$$

where
  λ is a vector of Lagrange multipliers
  $\mathbf{u}_\Gamma$ is the prescribed essential boundary conditions

For free-vibration and static buckling analyses, we should have $\mathbf{u}_\Gamma = 0$, because the problem is homogeneous. Vector $\mathbf{u}$ is the displacement approximated on the essential boundary conditions. For the essential boundary of plates based on FSDT,

$$\mathbf{u}_b = \left\{ \begin{array}{c} w \\ \varphi_n \\ \varphi_s \end{array} \right\} = \underbrace{\begin{bmatrix} 1 & 0 & 0 \\ 0 & n_x & n_y \\ 0 & -n_y & n_x \end{bmatrix}}_{\mathbf{L}_b} \underbrace{\left\{ \begin{array}{c} w \\ \varphi_x \\ \varphi_y \end{array} \right\}}_{\mathbf{u}} = \mathbf{L}_b\mathbf{u} \tag{11.176}$$

where n_x and n_y are the direction cosines of the outward normal on the boundary.
  For the essential boundary of plates based on TSDT,

$$\mathbf{u}_b = \left\{ \begin{array}{c} w \\ \varphi_n \\ \varphi_s \\ \dfrac{\partial w}{\partial n} \end{array} \right\} = \underbrace{\begin{bmatrix} 1 & 0 & 0 \\ 0 & n_x & n_y \\ 0 & -n_y & n_x \\ \dfrac{\partial}{\partial n} & 0 & 0 \end{bmatrix}}_{\mathbf{L}_b} \underbrace{\left\{ \begin{array}{c} w \\ \varphi_x \\ \varphi_y \end{array} \right\}}_{\mathbf{u}} = \mathbf{L}_b\mathbf{u} \tag{11.177}$$

The discrete essential boundary conditions derived from Equation 11.175 can be written in the form of (for cases of TSDT)

$$\mathbf{H}_{4n_b\times 3N_n}\mathbf{Q}_{3N_n\times 1} = 0 \tag{11.178}$$

where n_b is the number of constraint points on the supported boundaries. In computing $\mathbf{H}$, one-point Gauss quadrature is used along each span between the constraint points. For example, for a clamped edge of plates based on TSDT and any one span, $\mathbf{H}$ is assembled from the nodal contributions defined as

$$\mathbf{H}_{KI} = \int_{\Gamma_u} \begin{bmatrix} N_K\phi_I & 0 & 0 \\ 0 & n_x N_K\phi_{xI} & n_y N_K\phi_{yI} \\ 0 & -n_y N_K\phi_{xI} & n_x N_K\phi_{yI} \\ \phi_{I,n} & 0 & 0 \end{bmatrix} d\Gamma \tag{11.179}$$

where N_K ($K = 1, 2$) are Lagrange linear interpolations given in Equation 6.15 for the span between two constraint points on the essential boundary.

Generally, the matrix $\mathbf{H}$ is sparse and singular. Using singular-value decomposition technique, it can be decomposed as

$$\mathbf{H}_{4n_b \times 3N_n} = \mathbf{U}_{4n_b \times 4n_b} \begin{bmatrix} \boldsymbol{\Sigma}_{r \times r} & 0 \\ 0 & 0 \end{bmatrix}_{4n_b \times 3N_n} \mathbf{V}^{\mathrm{T}}_{3N_n \times 3N_n} \tag{11.180}$$

where
$\mathbf{R}$ and $\mathbf{V}$ are the orthogonal matrices
$\boldsymbol{\Sigma}$ is the singular value of $\mathbf{H}$
r is the rank of $\mathbf{H}$, which represents the number of independent constraints

The matrix $\mathbf{V}$ can be written as

$$\mathbf{V}^{\mathrm{T}}_{3N_n \times 3N_n} = \left\{ \mathbf{V}_{3N_n \times r}, \ \mathbf{V}_{3N_n \times (3N_n - r)} \right\}^{\mathrm{T}} \tag{11.181}$$

By performing orthogonal transformation in Equation 11.168,

$$\mathbf{q} = \mathbf{V}_{3N_n \times (3N_n - r)} \tilde{\mathbf{q}} \tag{11.182}$$

The condensed eigenvalue equation of both free vibration and static buckling can be expressed as

$$(\tilde{\mathbf{K}} - \eta \tilde{\mathbf{G}}) \tilde{\mathbf{q}} = 0 \tag{11.183}$$

where

$$\tilde{\mathbf{K}} = \mathbf{V}^{\mathrm{T}}_{3N_n \times (3N_n - r)} \mathbf{K}_{3N_n \times 3N_n} \mathbf{V}_{3N_n \times (3N_n - r)} \tag{11.184}$$

and

$$\tilde{\mathbf{G}} = \mathbf{V}^{\mathrm{T}}_{3N_n \times (3N_n - r)} \mathbf{G}_{3N_n \times 3N_n} \mathbf{V}_{3N_n \times (3N_n - r)} \tag{11.185}$$

For the plates of FSDT, the above procedure is valid, except $4n_b$ should be changed to $3n_b$, and $\mathbf{H}_{KI}$ is given by

$$\mathbf{H}_{KI} = \int_{\Gamma_u} \begin{bmatrix} N_K \phi_I & 0 & 0 \\ 0 & n_x N_K \phi_{xI} & n_y N_K \phi_{yI} \\ 0 & -n_y N_K \phi_{xI} & n_x N_K \phi_{yI} \end{bmatrix} d\Gamma \tag{11.186}$$

Solving Equation 11.183 for the eigenvalues yields the square of the circle frequencies for the free-vibration problem and the buckling loads for the static buckling problem. The eigenvectors, after being transformed back using Equation 11.182, give the vibration modes or buckling modes.

11.4.5 Equations for Static Deformation Analyses

For static deflection analysis of thick plates, the penalty method can be used to enforce the essential boundary conditions by adding an additional essential boundary condition term in the Galerkin weak form of the static elastic equilibrium. The constrained Galerkin weak form for thick plates can be written as

$$\int_{\Omega} \delta(\mathbf{L}_d \mathbf{u})^T \mathbf{c} \mathbf{L}_d \mathbf{u} d\Omega - \int_{\Omega} \delta(\mathbf{L}_u \mathbf{u})^T \mathbf{b} d\Omega - \int_{\Gamma_t} \delta(\mathbf{L}_u \mathbf{u})^T \mathbf{t}_\Gamma dS + \delta \int_{\Gamma_u} \frac{1}{2}(\mathbf{u}_b - \mathbf{u}_\Gamma)^T \boldsymbol{\alpha}(\mathbf{u}_b - \mathbf{u}_\Gamma)d\Gamma = 0$$

(11.187)

where
  Γ_t is the edge surface of the plate where the natural boundary condition is specified
  $\boldsymbol{\alpha}$ is a diagonal matrix of the penalty factors
  vector $\mathbf{u}_b$ is given by either Equation 11.176 or 11.177, depending on the plate theory used

The dimension of $\boldsymbol{\alpha}$ is 3×3 or 4×4 depending also on the plate theory used.
  The discrete system equation can be expressed as

$$[\mathbf{K} + \mathbf{K}^\alpha]\mathbf{U} = \mathbf{F}$$

(11.188)

where global stiffness matrix $\mathbf{K}$ is the same as that in Equation 11.168. The additional stiffness matrix $\mathbf{K}^\alpha$ is formed using

$$\mathbf{K}_{IJ}^\alpha = \int_{\Gamma_u} \boldsymbol{\Phi}_{bI}^T \boldsymbol{\alpha} \boldsymbol{\Phi}_{bJ} d\Gamma$$

(11.189)

and $\boldsymbol{\Phi}_{bI}$ is the matrix of the shape functions for node I. For FSDT, it can be written as

$$\boldsymbol{\Phi}_{bI} = \begin{bmatrix} \phi_I & 0 & 0 \\ 0 & n_x \phi_{xI} & n_y \phi_{yI} \\ 0 & -n_y \phi_{xI} & n_x \phi_{yI} \end{bmatrix}$$

(11.190)

For TSDT,

$$\boldsymbol{\Phi}_{bI} = \begin{bmatrix} \phi_I & 0 & 0 \\ 0 & n_x \phi_{xI} & n_y \phi_{yI} \\ 0 & -n_y \phi_{xI} & n_x \phi_{yI} \\ \dfrac{\partial \phi_I}{\partial n} & 0 & 0 \end{bmatrix}$$

(11.191)

The force vector $\mathbf{F}$ in Equation 11.188 is the global force vector assembled using the nodal force vector of

$$\mathbf{f}_I = \int_{\Omega} (\mathbf{L}_u \boldsymbol{\Phi}_I)^T \mathbf{b} d\Omega + \int_{\Gamma_t} (\mathbf{L}_u \boldsymbol{\Phi}_I)^T \mathbf{t}_\Gamma dS + \int_{\Gamma_u} \boldsymbol{\Phi}_{bI}^T \boldsymbol{\alpha} \mathbf{u}_\Gamma d\Gamma$$

(11.192)

Further derivation on the first two terms in Equation 11.192 can be performed by following the procedure given in Section 11.1.3.

11.4.6 Numerical Examples of Static Deflection Analyses

To examine the efficiency of the present formulation of the EFG method for the static deflection analyses of shear deformable thick plates, Examples 11.12 through 11.16 are studied. The common geometric and material property parameters used in these five examples are as follows:

Length: $a = b = 10.0$ m

Young's modulus: $E = 1.0 \times 10^9$ N/m²

Poisson's ratio: $\nu = 0.3$

Mass density: $\rho = 8000$ kg/m³

The deflection coefficient $\xi = w_{max} D / P b^2$ is defined for the concentrated load P, and $\xi = w_{max} D / q b^4$ for the uniform load q, where w_{max} is the maximum deflection at the center of the plates. Elastic rigidity of the plate is $D = E h^3 / [12(1 - \nu^2)]$.

In all the following examples, the size of the support domain is chosen to be 3.9 times the average nodal distance, except for special illustration. The nodes used in the following three examples are regularly distributed nodes. Uniform rectangular cells of background mesh are used for the integration, and the vertices of the background cells coincide with the field nodes.

In the following examples, the three shape functions used are the same, i.e.,

$$\phi_I(x, y) = \phi_{xI}(x, y) = \phi_{yI}(x, y) \tag{11.193}$$

The issue of shear locking is addressed in Example 11.16.

Example 11.12: Thin and Thick Square Plates: A Comparison Study

Using 21 × 21 nodes in the EFG method, two types of simply supported boundary conditions—soft type and hard type—are considered, and the maximum deflections of thin plate ($h/a = 0.01$) and thick plate ($h/a = 0.1$) are computed. The results for the thin plate are shown in Table 11.36. The deflections calculated based on both FSDT and TSDT are very close to Timoshenko's solution [9]. It is also seen that the deflections of thin plate calculated using the soft-type simply supported boundary condition are very close to those obtained using the hard-type simply supported boundary condition.

The maximum deflections of a thick plate are shown in Table 11.37. The deflections calculated using FSDT are very close to those obtained using TSDT. The deflections of thick plate calculated

TABLE 11.36

Maximum Deflection Coefficients ξ of Simply Supported Thin Plate ($h/a = 0.01$)

Load	Boundary Condition	FSDT	TSDT	Ref. [9]
Concentrated	$w = 0$ (soft)	0.01161	0.01162	0.01160
	$w = 0$, $\varphi_t = 0$ (hard)	0.01156	0.01157	
Uniform	$w = 0$ (soft)	0.004084	0.004085	0.004062
	$w = 0$, $\varphi_t = 0$ (hard)	0.004062	0.004063	

TABLE 11.37

Maximum Deflection Coefficients ξ of Simply Supported Thick Plate ($h/a = 0.1$)

Load	Boundary Condition	FSDT	TSDT
Concentrated	$w = 0$ (soft)	0.01398	0.01391
	$w = 0$, $\varphi_s = 0$ (hard)	0.01323	0.01318
Uniform	$w = 0$ (soft)	0.004619	0.004615
	$w = 0$, $\varphi_s = 0$ (hard)	0.004273	0.004275

using the soft-type simply supported boundary condition are significantly larger than those obtained using the hard-type simply supported boundary condition, especially for the uniform load.

Example 11.13: Thin Square Plates: Convergence Study

The convergence of the maximum deflections at the center of a square thin plate applied with a concentrated load at the center of the plate is analyzed. The plate is simply supported with soft-type. The thickness of $h = 0.1$ m and the aspect ratio of $h/a = 0.01$ are considered. Various nodal densities in the plate domain are used to calculate the deflections based on FSDT, TSDT, and CPT in the present EFG method. The maximum deflections are shown in Table 11.38. The present EFG deflections based on both FSDT and TSDT rapidly converge to Timoshenko's solution. However, the convergence rate of the deflection solutions based on CPT is higher than that based on FSDT and TSDT. All the deflections have a monotonous convergence.

Example 11.14: Thick Square Plates: Convergence Study

The convergence on the maximum deflections at the center of a hard-type simply supported square thick plate subjected to a uniform load is analyzed. A thickness $h = 1.0$ m and an aspect ratio of $h/a = 0.1$ are used. Based on FSDT and TSDT in the present EFG method, the various nodal densities are used to calculate the deflections, as shown in Table 11.39. The present EFG deflections based on both FSDT and TSDT rapidly converge to the analytical solution [21].

Example 11.15: Thick Plates: Effects of Boundary Conditions

Thick plates with thickness $h = 1.0$ m and aspect ratio $h/a = 0.1$ are considered. Based on CPT, FSDT, and TSDT in the present EFG method, 21×21 nodes are used to calculate the

TABLE 11.38

Maximum Deflection Coefficients ξ of Soft-Type Simply Supported Thin Plate ($h/a = 0.01$) Subjected to a Concentrated Load at the Center of the Plate

	Nodes						Ref. [9]
	6×6	9×9	12×12	15×15	18×18	21×21	
FSDT	0.009835	0.01085	0.01139	0.01151	0.01158	0.01161	0.01160
TSDT	0.01001	0.01102	0.01144	0.01154	0.01159	0.01162	
CPT	0.01032	0.01141	0.01145	0.01155	0.01157	0.01157	

TABLE 11.39

Maximum Deflection Coefficients ξ of Hard-Type Simply Supported Thick Plate ($h/a = 0.1$) Subjected to a Uniform Load

| | \multicolumn{7}{c}{Nodes} | | | | | | |
	4 × 4	6 × 6	8 × 8	10 × 10	12 × 12	14 × 14	Ref. [21]
FSDT	0.003948	0.004306	0.004281	0.004272	0.004275	0.004273	0.004249
TSDT	0.003945	0.004311	0.004303	0.004278	0.004279	0.004276	

TABLE 11.40

Maximum Deflection Coefficients ξ of Thick Plate ($h/a = 0.1$)

Load	Boundaries	CPT	FSDT	TSDT	Ref. [21]
Concentrated	SSSS	0.01159	0.01323	0.01318	
	CCCC	0.005566	0.007322	0.007248	
	SCSC	0.006990	0.008845	0.008768	
	SCCS	0.007379	0.009190	0.009115	
Uniform	SSSS	0.004066	0.004273	0.004275	0.004249
	CCCC	0.001259	0.001505	0.001494	0.001496
	SCSC	0.001910	0.002209	0.002198	0.002191
	SCCS	0.002098	0.002375	0.002366	

maximum deflections of the thick plates for four kinds of hard-type boundary conditions and two types of loads. The maximum deflections are listed in Table 11.40. The present EFG results of deflections calculated based on FSDT and TSDT are very close to each other and larger than those obtained based on CPT. The present deflections based on FSDT and TSDT agree well with the available analytical solutions [21]. This demonstrates that the present EFG formulation based on the shear deformation plate theory is accurate in calculating deflections of thick plates.

Example 11.16: Elimination of Shear Locking: A Few Techniques

It is well known that models based on the thick plate theory is used to simulate thin plates, erroneous solutions will be obtained known as the shear-locking phenomenon. Techniques to avoid shear locking have been well developed in FEM. An excellent discussion of this issue can be found in the textbook [11]. All these techniques, listed as follows, can be adopted here for our meshfree methods:

1. Use sufficiently high-order elements.
2. Construct shape functions for the rotations φ_x and φ_y from the first-order derivatives of the shape function used for the transverse displacement w.
3. Evaluate the shear energy in a proper manner.

The effectiveness of the second method has also recently been proved for EFG method [19]. Here we study the first two methods in detail.

11.4.6.1 Use of High-Order Basis

A soft-type simply supported Mindlin plate with the uniform load of $p = 1.0$ N/m^2 is considered to analyze the shear-locking problem. To visualize the shear-locking phenomenon clearly, an extremely small aspect ratio of $h/a = 1.0 \times 10^{-4}$ is used. The other parameters are as follows:

Lengths of two sides: $a = b = 10.0$ m
Young's modulus: $E = 1.0 \times 10^9$ N/m^2
Poisson's ratio: $\nu = 0.3$

Regularly distributed 21×21 nodes are used. The dimensionless support domain is $\alpha_s = 3.9$–4.1. Three polynomial terms $(1, x, y)$ are used for both methods. The deflection coefficient is defined as $\xi = w_{\text{max}}D/b_z b^4$, where w_{max} is the maximum deflection of the center of the plates. The analytical solution for the maximum deflection at the center of the plates is $\xi = 0.004062$ (Timoshenko's solution).

The same shape functions are used for the approximations of the deflection w, and the rotations φ_x, and φ_y. Different orders of polynomial basis functions listed in Table 11.41 are used to construct MLS shape functions. The deflections of the plates computed by EFG based on FSDT are shown in Figure 11.18. The shear-locking phenomenon is observed very clearly when only three polynomial bases are used in MLS approximation. When the polynomial terms up to 18 are used, shear locking is eliminated, and the deflections of the thin plate calculated based on FSDT are very close to and slightly larger than those obtained using EFG based on CPT. As the polynomial terms increase, more nodes need to be included in the influence domain. In this study, the dimension of the support domain is chosen to be 4.1 times the average nodal distance for 18 polynomial terms. Note that a *plate* of an aspect ratio of $h/a = 10 \times 10^{-4}$ is quite an extreme.

11.4.6.2 Comparison Study

Using the present EFG method for Mindlin plates (FSDT), we perform a comparison study to further analyze the shear-locking issue. The following two schemes are used:

Scheme 1
Use high-order basis functions but the same set of shape functions for three field variables, as we have done above.

Scheme 2
The shape functions for rotations are the first-order derivatives of that for deflection.

TABLE 11.41

Basis Functions for MLS Displacement Approximation

m	$\mathbf{P}^T$
3	$1, x, y$
6	$1, x, y, x^2, xy, y^2$
10	$1, x, y, x^2, xy, y^2, x^3, x^2y, xy^2, y^3$
15	$1, x, y, x^2, xy, y^2, x^3, x^2y, xy^2, y^3, x^3y, x^2y^2, xy^3, x^3y^2, x^2y^3$
18	$1, x, y, x^2, xy, y^2, x^3, x^2y, xy^2, y^3, x^3y, x^2y^2, xy^3, x^3y^2, x^2y^3, x^3y^3, x^4, y^4$

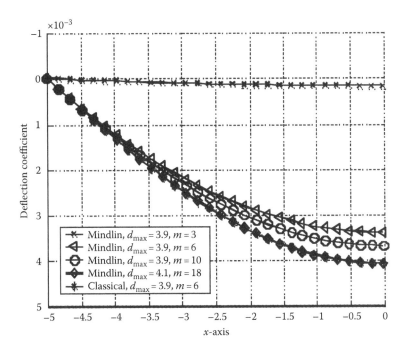

FIGURE 11.18
Shear locking in simply supported thin plate of a thickness-to-width ratio of $h/a = 1.0 \times 10^{-4}$. The plate is subjected to a uniform load. Results are computed using EFG based on FSDT.

$$w = \sum_{I=1}^{n} \phi_I w_I, \quad \varphi_x = \sum_{I=1}^{n} \phi_{I,x} \varphi_{xI}, \quad \varphi_y = \sum_{I=1}^{n} \phi_{I,y} \varphi_{yI} \qquad (11.194)$$

This approach has been used in [19] in the EFG method. A simply supported square plate under uniform load is considered for this analysis of the shear-locking problem, and all the parameters are exactly the same as the previous case.

The results are listed in Table 11.42 and plotted in Figure 11.19. We make the following points:

1. There is no shear locking, if $h/a > 0.01$, even if $m = 3$ is used. Usually, we use at least $m = 6$ for plate problems.

2. The higher the order of polynomial used, the less the shear locking. When $m = 18$, no shear locking is observed until $h/a = 1.0 \ 10^{-5}$, which is not actually a plate.

3. When Scheme 2 is used, no shear locking is observed.

Although Scheme 1 can practically solve the shear-locking problem, Scheme 2 is a better way to eliminate this issue completely.

11.4.7 Numerical Examples of Vibration Analyses

To examine the efficiency of the EFG method for eigenvalue analyses of the shear deformable thick plates, some numerical examples on free vibration and static buckling

TABLE 11.42

Maximum Deflections $\xi = w_{maxD}/b_z b^4$ of a Square Plate with Different Aspect Ratio

		1.0×10^{-1}	1.0×10^{-2}	h/a 1.0×10^{-3}	1.0×10^{-4}	1.0×10^{-5}
Scheme 1	$m = 3$	0.4618×10^{-2}	0.4068×10^{-2}	$\mathbf{0.3307 \times 10^{-2}}$	$\mathbf{0.1715 \times 10^{-3}}$	$\mathbf{0.1791 \times 10^{-5}}$
	$m = 6$	0.4618×10^{-2}	0.4084×10^{-2}	$\mathbf{0.3928 \times 10^{-2}}$	$\mathbf{0.3401 \times 10^{-2}}$	$\mathbf{0.5461 \times 10^{-3}}$
	$m = 10$	0.4620×10^{-2}	0.4088×10^{-2}	0.4065×10^{-2}	$\mathbf{0.3691 \times 10^{-2}}$	$\mathbf{0.2639 \times 10^{-2}}$
	$m = 15$	0.4621×10^{-2}	0.4090×10^{-2}	0.4069×10^{-2}	$\mathbf{0.3810 \times 10^{-2}}$	$\mathbf{0.3556 \times 10^{-2}}$
	$m = 18$	0.4621×10^{-2}	0.4088×10^{-2}	0.4064×10^{-2}	$\mathbf{0.4089 \times 10^{-2}}$	$\mathbf{0.4009 \times 10^{-2}}$
Scheme 2	$m = 3$	0.004612	0.004073	0.004059	0.004059	0.004025
	$m = 6$	0.004617	0.004085	0.004066	0.004066	0.004065
	$m = 10$	0.004620	0.004088	0.004066	0.004066	0.004066
	$m = 15$	0.004621	0.004090	0.004068	0.004067	0.004067
	$m = 18$	0.004621	0.004089	0.004068	0.004068	0.004068
Timoshenko				0.004062		

of plates are presented. The common geometric and material property parameters are as follows:

Length: $a = b = 10.0$ m

Thickness: $h = 1.0$ m

Young's modulus: $E = 200 \times 10^9$ N/m^2

Poisson's ratio: $\nu = 0.3$

Mass density: $\rho = 8000$ kg/m^3

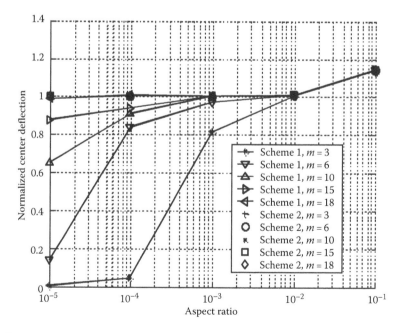

FIGURE 11.19

Normalized maximum deflections of a square plate with different aspect ratios.

In all the tables of numerical examples, the dimensionless frequency coefficient is $\bar{\omega} = (\omega^2 \rho h a^4/D)^{1/4}$ and the factor of buckling load is $k = N_0 b^2 \pi^2 D$ where the flexural rigidity is $D = Eh^3/[12(1-\nu^2)]$.

Example 11.17: Frequency Analysis of Thick Plates (FSDT)

Based on FSDT, the natural dimensionless frequencies of a simply supported square thick plate are calculated using different densities of nodes regularly distributed in the plate. Quadrilateral background meshes are applied for the Gauss integration. In each background mesh, 4×4 Gauss points are used. All the nodes are in the vertices of the background meshes. The natural dimensionless frequencies are shown in Table 11.43 together with the analytical and FEM solutions. In the FEM results, HOE denotes an eight-node isoparametric thick shell element (4×4 elements, 65 nodes); LOE denotes a four-node isoparametric shell element (8×8 elements, 9×9 nodes). The frequencies obtained using the present EFG method rapidly converge and are in very good agreement with the analytical solutions for all eight modes. Given the same density of nodes 9×9 in the plate, the present EFG results are better than the FEM results using LOE when compared with the analytical solution. The present EFG method results using 7×7 nodes are also closer to the analytical solutions than the FEM results using HOE with 65 nodes.

Example 11.18: Frequency Analysis of Thick Plates (FSDT and TSDT)

Based on FSDT and TSDT, the natural dimensionless frequencies of a square thick plate with different boundaries are calculated using 10×10 nodes regularly distributed in the plate. Quadrilateral background meshes are also applied for the Gauss integration and 4×4 Gauss points are used in each background mesh. All the nodes are at the vertices of the background meshes. The frequencies are shown in Table 11.44. In the notation of boundary conditions, FFFF denotes fully free at all edges. For all the five cases of boundaries, the frequencies of the square plate based on FSDT agree well with those based on TSDT for all modes. The frequencies for FFFF boundaries are lowest. The frequencies for the plate with SSSS boundaries are lower than those with CCCC boundaries. Both frequencies for the plate with SCSC and SCCS boundaries are between those with SSSS and CCCC boundaries.

TABLE 11.43

Natural Dimensionless Frequencies $\bar{\omega} = (\omega^2 \rho h a^4/D)^{1/4}$ of Free Vibration of a Simply Supported Square Thick Plate Based on FSDT

Modes	Analytical Solution [40]	EFG				FEM [40]	
		5×5	7×7	9×9	11×11	HOE	LOE
1	4.37	4.43	4.37	4.37	4.37	4.37	4.40
2	6.74	7.33	6.75	6.75	6.75	6.77	6.94
3	6.74	7.53	6.77	6.75	6.75	6.77	6.94
4	8.35	9.79	8.38	8.36	8.36	8.41	8.59
5	9.22	10.91	9.31	9.24	9.23	9.40	9.84
6	9.22	11.14	9.33	9.24	9.23	9.40	9.84
7	10.32	15.07	10.41	10.35	10.33	10.59	10.85
8	10.32	15.99	10.48	10.35	10.33	10.59	10.85

Note: HOE denotes an eight-noded isoparametric thick shell element (16 elements, 65 nodes); LOE denotes a four-noded isoparametric shell element (64 elements, 81 nodes).

TABLE 11.44

Natural Dimensionless Frequencies $\bar{\omega} = (\omega^2 \rho h a^4 / D)^{1/4}$ of Free Vibration of a Square Thick Plate with Different Boundaries Based on FSDT and TSDT

Boundary	Theory	Modes							
		1	2	3	4	5	6	7	8
FFFF	FSDT	0	0	0	3.57	4.35	4.83	5.65	5.65
	TSDT	0	0	0	3.57	4.35	4.83	5.66	5.66
SSSS	FSDT	4.37	6.75	6.75	8.36	9.23	9.23	10.34	10.34
	TSDT	4.37	6.75	6.75	8.36	9.23	9.23	10.34	10.34
CCCC	FSDT	5.71	7.88	7.88	9.33	10.13	10.18	11.14	11.14
	TSDT	5.72	7.92	7.92	9.39	10.19	10.23	11.21	11.22
SCSC	FSDT	5.17	7.01	7.70	8.88	9.33	10.08	10.60	10.92
	TSDT	5.17	7.02	7.73	8.92	9.33	10.13	10.63	10.96
SCCS	FSDT	5.03	7.31	7.33	8.85	9.70	9.71	10.74	10.76
	TSDT	5.04	7.33	7.35	8.87	9.72	9.74	10.77	10.79

11.4.8 Numerical Examples of Buckling Analyses

Example 11.19: Buckling Analysis of Thick Plates (FSDT and TSDT)

Based on FSDT and TSDT, the factors $\bar{\omega} = (\omega^2 \rho h a^4 / D)^{1/4}$ of axial buckling loads along the *x*-axis of a simply supported square plate are calculated using different densities of nodes regularly distributed in the plate. Quadrilateral background meshes and 4×4 Gauss points in each background mesh are chosen for the Gauss integration as for the frequency analyses. All the nodes are located at the vertices of the background meshes. The results of buckling factors are shown in Table 11.45. The factors based on both FSDT and TSDT have good convergences, agree very well with each other and with the analytical results given in [21] and are a little larger than the analytical results given in [24].

Example 11.20: Buckling Analysis: Effects of Boundary Conditions

Based on FSDT and TSDT, the factors of buckling loads of a square plate with different loads and boundaries are calculated using 9×9 nodes regularly distributed in the plate. The choices of background meshes, Gauss points, and nodal position are the same as Example 11.17. The factors of axial buckling loads along the *x*-axis and shear buckling loads and biaxial buckling loads along both the *x*- and *y*-axes for different boundaries are listed in Tables 11.46 through 11.48. The factors for SSSS boundaries are smaller than those for CCCC boundaries. The factors for both SCSC and

TABLE 11.45

Convergence of Axial Buckling Factor $k = N_0 b^2 / \pi^2 D$ along the *x*-Axis for a Square Thick Plate Based on FSDT and TSDT (BC: SSSS)

	Nodes						Ref. [24]	Ref. [21]
	4×4	5×5	6×6	7×7	8×8	9×9		
FSDT	4.060	3.803	3.783	3.790	3.787	3.788	3.741	3.787
TSDT	4.095	3.797	3.780	3.785	3.785	3.785		

TABLE 11.46

Axial Buckling Factor $k = N_0 b^2 / \pi^2 D$ along the x-Axis for a Square Thick Plate Based on FSDT and TSDT with Different Boundaries

Theory	Boundaries			
	SSSS	CCCC	SCSC	SCCS
FSDT	3.788	8.327	6.398	5.546
TSDT	3.785	8.471	6.451	5.569

TABLE 11.47

Shear Buckling Factor $k = N_0 b^2 / \pi^2 D$ for a Square Thick Plate Based on FSDT and TSDT with Different Boundaries

Theory	Boundaries			
	SSSS	CCCC	SCSC	SCCS
FSDT	7.859	10.757	9.693	9.226
TSDT	7.853	11.669	10.542	9.349

TABLE 11.48

Biaxial Buckling Factor $k = N_0 b^2 / \pi^2 D$ along the x- and y-Axes for a Square Thick Plate Based on FSDT and TSDT with Different Boundaries ($N_x = N_y = N$)

Theory	Boundaries			
	SSSS	CCCC	SCSC	SCCS
FSDT	1.894	4.563	3.384	2.941
TSDT	1.892	4.657	3.396	2.944

SCCS boundaries are between those for simply supported and clamped boundaries. Except for CCCC and SCSC boundaries, the factors of shear buckling loads based on TSDT are slightly larger than those based on FSDT; the factors based on TSDT are very close to those based on FSDT in Tables 11.46 through 11.48.

Example 11.21: Buckling Loads of a Square Plate with a Circular Hole

The factors of buckling loads are computed for a square plate with a circular hole, as shown in Figure 11.20. The plate is chosen to demonstrate the applicability of the present EFG method to plates with complicated shape. Triangular background meshes (196) for Gauss integration are applied as shown in Figure 11.21. Three Gauss points are used for each background mesh. At the vertices of the background meshes, 144 nodes are located. The influence domain is chosen

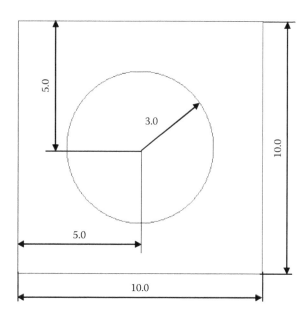

FIGURE 11.20
Square plate with a circular hole.

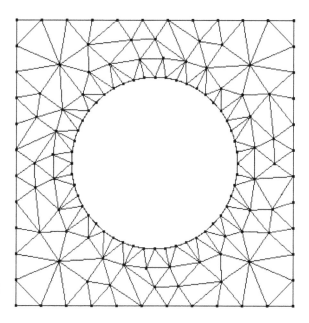

FIGURE 11.21
Distributions of 144 nodes and 196 background
meshes in a plate with a circular hole.

to be 3.5 times the average nodal distance. The factors of axial buckling loads along the *x*-axis and shear buckling loads and biaxial buckling loads along both the *x*- and *y*-axes for the different boundaries are listed in Tables 11.49 through 11.51. Except for CCCC boundaries, the factor of the shear buckling load based on TSDT is slightly larger than that based on FSDT; the factors based on TSDT are very close to those based on FSDT in Tables 11.49 through 11.51. Factors of shear buckling loads are larger than those of biaxial buckling loads along both the *x*- and *y*-axes. Factors of axial buckling loads along the *x*-axis are between them.

TABLE 11.49

Axial Buckling Factor $k = N_0 b^2 / \pi^2 D$ along the x-Axis for a Square Thick Plate with a Circular Hole Based on FSDT and TSDT with Different Boundaries

Theory	Boundaries			
	SSSS	CCCC	SCSC	SCCS
FSDT	1.986	7.995	4.096	3.221
TSDT	1.969	8.097	4.130	3.226

TABLE 11.50

Shear Buckling Factor $k = N_0 b^2 / \pi^2 D$ for a Square Thick Plate with a Circular Hole Based on FSDT and TSDT with Different Boundaries

Theory	Boundaries			
	SSSS	CCCC	SCSC	SCCS
FSDT	7.867	12.669	10.596	8.968
TSDT	7.873	13.357	10.861	9.043

TABLE 11.51

Biaxial Buckling Factor $k = N_0 b^2 / \pi^2 D$ along the x- and y-Axes for a Square Thick Plate with a Circular Hole Based on FSDT and TSDT with Different Boundaries ($N_x = N_y = N$)

Theory	Boundaries			
	SSSS	CCCC	SCSC	SCCS
FSDT	1.032	4.781	2.559	1.763
TSDT	1.021	4.860	2.556	1.774

11.5 ES-PIM for Plates

In Chapter 8, we have presented a family of PIMs based on triangular mesh. It was found that, due to the use of the weakened-weak (W^2) formulation based on the $\mathbb{G}$ space theory, the consistence on the assumed displacement functions has been further reduced. Therefore, for second order PDEs, we can allow discontinuous functions.

We also found that the edge-based smoothed PIM or ES-PIM with linear shape function is one of the best performers in all the meshfree methods examined in terms of accuracy, efficiency, simplicity, robustness, and applicability to problems with complicated geometries. An intensive comparison study (see Section 8.4) has shown that ES-PIM is much more efficient than the linear FEM for a benchmark problem of 2D solid mechanics. In addition, the formulation and implementation of ES-PIM is very simple, and works well for triangular mesh. Therefore, the linear ES-PIM is so far the most promising method for complicated engineering problems.

This section formulates an ES-PIM for plates based on the FSDT, and demonstrates that the ES-PIM works well for both thin and thick plates, because it is free from shear locking. Again, we only use triangular background cells in our ES-PIM formulation.

11.5.1 Creation of an ES-PIM Model

An ES-PIM model for a plate can be established using the following key techniques.

1. The problem domain of a plate is divided into N_c background triangular cells with N_n nodes and N_{cg} edges defined in Section 1.7.2. PIM shape functions are then constructed using the PIM with a T-scheme.

2. Based on these triangular cells, a set of N_s smoothing domains Ω_k^s $(k = 1, \ldots, N_s)$ bounded by Γ_k^s are created, which are associated with the edges of the triangles, as shown in Figure 8.80. In this edge-based case, we have $N_s = N_{cg}$.

3. The smoothing domains serve also as a base for integration. Therefore, the integration of the strain energy in the weakened-weak form (GS-Galerkin weak form) becomes a simple summation of strain energy of all the N_s smoothing domains.

4. The strains in a smoothing domain are assumed constant, and thus the strain energy in a smoothing cell will also be constant. Because FSDT is used, there will be possible shear locking when the plate gets very thin. To overcome this, we divide the strain energy into two terms: *bending* energy and *shear* energy, by grouping the strains into in-plane and off-plane components.

5. The constant in-plane strains for each smoothing domain are obtained using the strain smoothing operation as discussed in Chapter 4, by line integrations along the boundary of the smoothing domain Γ_k^s. Such integration requires only shape function values on Γ_k^s, which can be obtained using, in general, PIM with a T-scheme. When linear interpolation (T3-scheme) is used, these values can be obtained by simple inspection and no need to create the PIM shape function explicitly.

6. The constant strains for the shear deformation can be obtained using the so-called discrete shear gap (DSG) method [23].

The rest is routine.

11.5.2 Formulation

11.5.2.1 Weakened-Weak Form

In this formulation, we consider plates only undergoing bending deformation, that is, there exists a *neutral-plane* in the plate where no in-plane deformation occurs. Following the discussion in Chapter 5 and using all pseudostresses and pseudostrains defined in Section 11.1.2, the smoothed Galerkin weak form for plates based on the FSDT can be written as

$$\underbrace{\frac{h^3}{12}\int_A \delta\bar{\boldsymbol{\varepsilon}}_B^{\mathrm{T}}(\mathbf{u})\mathbf{c}\bar{\boldsymbol{\varepsilon}}_B(\mathbf{u})\mathrm{d}A}_{\text{Bending}} + \underbrace{\kappa h\int_A \delta\bar{\boldsymbol{\gamma}}_S^{\mathrm{T}}(\mathbf{u})\mathbf{G}\bar{\boldsymbol{\gamma}}_S(\mathbf{u})\mathrm{d}A}_{\text{Shearing}}$$

$$-\underbrace{\int_A \delta\mathbf{u}^{\mathrm{T}}\mathbf{b}_z\mathrm{d}A}_{\text{Body force}} - \underbrace{\int_{\Gamma_t} \delta\mathbf{u}^{\mathrm{T}}\mathbf{b}_{z\Gamma}\mathrm{d}\Gamma}_{\text{Boundary force}} = 0 \qquad (11.195)$$

where **u** is the *generalized* displacement vector consisting of the deflection and two rotations defined on the mid-plane of the plate. It is arranged in form of

$$\mathbf{u} = \left\{ \begin{array}{c} u_1 \\ u_2 \\ u_3 \end{array} \right\} = \left\{ \begin{array}{c} w \\ \varphi_x \\ \varphi_y \end{array} \right\} \tag{11.196}$$

and the generalized body force vector $\mathbf{q}_z$ can be written as

$$\mathbf{q}_z = \left\{ \begin{array}{c} q_1 \\ q_2 \\ q_3 \end{array} \right\} = \int_{-h/2}^{h/2} \left\{ \begin{array}{c} b_z \\ -z b_x \\ -z b_y \end{array} \right\} dz = \left\{ \begin{array}{c} h b_z \\ m_{xx} \\ m_{yy} \end{array} \right\} \tag{11.197}$$

where
q_1 is the total vertical force applied over the plate (force/area)
q_2 and q_3 are the distributed moments applied over the plate

Because all the strains are smoothed over the smoothing domain, the first two terms in Equation 11.195 can be written in the following simple summation form (see Chapter 4):

$$\underbrace{\frac{h^3}{12} \sum_{k=1}^{N_s} \delta\bar{\boldsymbol{\varepsilon}}_{Bk}^T \mathbf{c} \bar{\boldsymbol{\varepsilon}}_{Bk}}_{\text{Bending}} + \underbrace{\kappa h \sum_{k=1}^{N_s} \delta\bar{\boldsymbol{\gamma}}_{Sk}^T \mathbf{G} \bar{\boldsymbol{\gamma}}_{Sk}}_{\text{Shearing}} - \underbrace{\int_A \delta\mathbf{u}^T \mathbf{b}_z dA}_{\text{Body force}} - \underbrace{\int_{\Gamma_t} \delta\mathbf{u}^T \mathbf{b}_{z\Gamma} d\Gamma}_{\text{Boundary force}} = 0 \tag{11.198}$$

Remark 11.1: Shear Locking: An Obvious Observation

Shear locking is not a so-easy-to-understand phenomenon to many engineers. However, we can actually observe this very clearly from Equation 11.198. It shows that the bending related energy term has a third-order dependence on the thickness of the plate; while shear related energy term has only a first-order dependence on the thickness of the plate. When $h \to 0$, the shear related energy will strongly dominate and the bending stiffness vanishes: the solution of this model will be *locked* on the shear deformation. This implies that our weakened-weak form model defined by Equation 11.198 will converge to a solution that is different from the strong form Equation 11.33 for thin plates. This analysis applies also to all the standard Galerkin formulations used in the FEM and EFG models: the h-dependence nature in these formulations is exactly the same! The root of the problem is the use of FSDT (or TSDT) when the rotational degrees of freedom are introduced with off-plane shear strains for *thick* plates (not meant for thin plates where these shear strains vanish when $h \to 0$). When a numerical model is built based on these theories, it has no choice but locked at the shear deformation when $h \to 0$. Therefore, special techniques are needed to prevent this from happening, if we want our numerical model based on thick plate theories works also for thin plates.

Writing the weak formulation in the form of Equation 11.198 or 11.95 helps us to predict some of the behaviors of a numerical model.

11.5.2.2 Strain Field Construction

In the first term of Equation 11.198, $\bar{\boldsymbol{\varepsilon}}_B$ is the smoothed in-plane strains resulting from the bending deformation. Using Equation 11.61, we have for the kth smoothing domain

$$
\bar{\boldsymbol{\varepsilon}}_B =
\begin{cases}
\dfrac{1}{A_s} \displaystyle\int_{\Omega_k^s} \boldsymbol{\varepsilon}_B d\Omega = \dfrac{1}{A_s} \displaystyle\int_{\Omega_k^s} \mathbf{L}_B \boldsymbol{\varphi}^h d\Omega = \dfrac{1}{A_k^s} \displaystyle\int_{\Gamma_k^s} \mathbf{L}_n \boldsymbol{\varphi}^h d\Gamma, & \text{when } \boldsymbol{\varphi}^h(\xi) \in \mathbb{C}^0(\Omega_k^s) \\[6pt]
\dfrac{1}{A_k^s} \displaystyle\int_{\Gamma_k^s} \mathbf{L}_B \boldsymbol{\varphi}^h d\Gamma, & \text{when } \boldsymbol{\varphi}^h(\xi) \in \mathbb{C}^{-1}(\Omega_k^s)
\end{cases}
\tag{11.199}
$$

Because $\boldsymbol{\varepsilon}_B$ is the pseudo-bending-strain (independent of z) and is the same as that for the 2D plane stress solids (Chapter 8), and $\mathbf{L}_B$ defined in Equation 11.53 is also exactly the same as $\mathbf{L}_d$ defined in Equation 1.9 for 2D solids, the computation of the first term in Equation 11.198 should be the same as we have done for the ES-PIM for 2D solids (see, Section 8.4). Therefore, there is no need for any further elaborations.

In the second term in Equation 11.198, $\bar{\boldsymbol{\gamma}}_S$ is the smoothed off-plane strains resulted from the shear deformation. To make our ES-PIM "locking-free," we simply adopt the DSG method that is used in the FEM [23]. The procedure is very simple and is given as follows.

In a triangular cell with area of A_e, the coordinates of these three nodes are first denoted (counter-clockwise) as (x_i, y_i), $i = 1, 2, 3$, and then the shear strain is defined as [23]

$$
\gamma_{xz} = \sum_{i=1}^{3} b_i (\Delta w_{xi} + \Delta w_{yi})
$$
$$
\gamma_{yz} = \sum_{i=1}^{3} c_i (\Delta w_{xi} + \Delta w_{yi})
\tag{11.200}
$$

where

$$
b_1 = \frac{1}{2A_e}(y_2 - y_3), \quad b_2 = \frac{1}{2A_e}(y_3 - y_1), \quad b_3 = \frac{1}{2A_e}(y_1 - y_2)
$$
$$
c_1 = \frac{1}{2A_e}(x_2 - x_3), \quad c_2 = \frac{1}{2A_e}(x_3 - x_1), \quad c_3 = \frac{1}{2A_e}(x_1 - x_2)
\tag{11.201}
$$

In Equation 11.200, Δw_{xi} and Δw_{yi} are the DSGs at the node i ($=1, 2, 3$) of the cell, and are given by

$$
\Delta w_{x1} = \Delta w_{x3} = \Delta w_{y1} = \Delta w_{y2} = 0
$$
$$
\Delta w_{x2} = (w_2 - w_1) + \frac{1}{2}a(\varphi_{x1} + \varphi_{x2}) + \frac{1}{2}b(\varphi_{y1} + \varphi_{y2})
$$
$$
\Delta w_{y3} = (w_3 - w_1) + \frac{1}{2}c(\varphi_{x1} + \varphi_{x3}) + \frac{1}{2}d(\varphi_{y1} + \varphi_{y3})
\tag{11.202}
$$

where w_i, φ_{xi}, and φ_{yi} are the nodal values of the deflection and rotations for node i, and

$$
a = x_2 - x_1, \quad b = y_2 - y_1
$$
$$
c = x_3 - x_1, \quad d = y_3 - y_1
\tag{11.203}
$$

Equation 11.200 can now be written in the matrix form of

$$\gamma_s = \begin{Bmatrix} \gamma_{xz} \\ \gamma_{yz} \end{Bmatrix} = \mathbf{B}_s \mathbf{d}_e \tag{11.204}$$

where $\mathbf{d}_e$ is the vector contains all the nodal "displacements" at the three nodes of the triangular cell:

$$\mathbf{d}_e = \left\{ \underbrace{w_1, \varphi_{x1}, \varphi_{y1}}_{\text{node 1}} \quad \underbrace{w_2, \varphi_{x2}, \varphi_{y2}}_{\text{node 2}} \quad \underbrace{w_3, \varphi_{x3}, \varphi_{y3}}_{\text{node 3}} \right\}^{\mathrm{T}} \tag{11.205}$$

and $\mathbf{B}_s$ is the shear strain matrix given by

$$\mathbf{B}_s = \begin{bmatrix} \mathbf{B}_{s1} & \mathbf{B}_{s2} & \mathbf{B}_{s3} \end{bmatrix} \tag{11.206}$$

with

$$
\mathbf{B}_{s1} = \frac{1}{2A_e} \begin{bmatrix} b-d & A_e & 0 \\ c-a & 0 & A_e \end{bmatrix}
$$

$$
\mathbf{B}_{s2} = \frac{1}{2A_e} \begin{bmatrix} d & \dfrac{ad}{2} & \dfrac{bd}{2} \\ -c & -\dfrac{ac}{2} & -\dfrac{bc}{2} \end{bmatrix} \tag{11.207}
$$

$$
\mathbf{B}_{s3} = \frac{1}{2A_e} \begin{bmatrix} -b & -\dfrac{bc}{2} & -\dfrac{bd}{2} \\ a & \dfrac{ac}{2} & \dfrac{ad}{2} \end{bmatrix}
$$

In our ES-PIM formulation, for the smoothing domain Ω_k^s associated with the k-th edge, the smoothed shear strain $\bar{\gamma}_{Sk}$ is obtained by the simple average of the shear strains for these two neighboring cells sharing the edge k:

$$\bar{\gamma}_{Sk} = \frac{1}{A_k}\left(A_{k_+}\gamma_{Sk_+} + A_{k_-}\gamma_{Sk_-}\right) = \frac{A_{k^+}}{A_k}\mathbf{B}_{Sk_+}\mathbf{d}_{k_+} \oplus \frac{A_{k^-}}{A_k}\mathbf{B}_{Sk_-}\mathbf{d}_{k_-} = \bar{\mathbf{B}}_{sk}\mathbf{d}_k \tag{11.208}$$

where the subscripts "+" and "−"stand for the two neighboring cells sharing the edge k. Matrix $\bar{\mathbf{B}}_{sk}$ is assembled using the area-weighted $\mathbf{B}_{Sk^+}$ and $\mathbf{B}_{Sk^-}$, and A_k, A_{k+} and A_k, are the areas of the smoothing domain of the edge k and those of the two neighboring cells sharing the edge k. We used $\oplus$ to represent an assembly (location matched summation). The nodal "displacement" vector $\mathbf{d}_k$ contains the nodal values of four nodes "supporting" the smoothing domain of edge k, and it can be written as

$$\mathbf{d}_k = \left\{ \underbrace{w_1, \varphi_{x1}, \varphi_{y1}}_{\text{node 1}} \quad \underbrace{w_2, \varphi_{x2}, \varphi_{y2}}_{\text{node 2}} \quad \underbrace{w_3, \varphi_{x3}, \varphi_{y3}}_{\text{node 3}} \quad \underbrace{w_4, \varphi_{x4}, \varphi_{y4}}_{\text{node 4}} \right\}^{\mathrm{T}} \tag{11.209}$$

Naturally, such averaging operation is not needed for edges on the problem boundary.

With these strains given in Equations 11.199 and 11.208, the strain potentials can be easily evaluated and hence the stiffness matrix of our ES-PIM model can be derived.

11.5.2.3 Force Terms

As a GS-Galerkin does not change force potentials (see Chapter 5), these terms can be dealt with in the same way as in the FEM. The procedure is the same as those given in Chapter 8. Finally, the discrete system equation can be obtained as

$$\bar{\mathbf{K}}\bar{\mathbf{U}} = \mathbf{F} \tag{11.210}$$

which can be solved using standard routines for $\bar{\mathbf{U}}$ that contains all the nodal values of the generalized displacements.

Note in this ES-PIM formulation for plates, we require slight changes to the ES-PIM code for 2D solids discussed in Chapter 8.

11.5.3 Numerical Examples

We now present a number of numerical examples solved using our ES-PIM code for plates. We use only linear interpolation for the field variables of deflection w and rotations φ_x and φ_y. Therefore, only one Gauss point for each of the line segment of Γ_k^s is needed and the interpolation values are given directly in Table 8.4, and thus there is no need to obtain the PIM shape functions explicitly. This simple point interpolation trick was originally used in the smoothed finite element method (SFEM) settings for different shapes of elements [25–28], and applicable to all PIMs using linear interpolation (see Chapter 8).

When such a linear interpolation is used, the ES-PIM is the same as edge-based smoothed finite element method (ES-FEM) using the same triangular mesh. The numerical results presented in this section are the same as those given in [22].

Example 11.22: Patch Test for Plates

The first numerical example is the standard patch test for plates undergoing bending deformation. A square plate patch with a dimension of 10×10 and a thickness $h = 0.001$, shown in Figure 11.22, is triangulated with 6 triangular cells, 6 nodes, and 11 edges, and is used for the standard pure bending patch test. The plate patch is subjected to the prescribed values of deflection and rotations at the four nodes on boundary of the patch. These values are computed using

$$w = 1.0 \times 10^{-3}(1 + x + y + x^2 + xy + y^2)$$

$$\varphi_x = -\frac{\partial w}{\partial x} = -1.0 \times 10^{-3}(1 + y + 2x) \tag{11.211}$$

$$\varphi_y = -\frac{\partial w}{\partial y} = -1.0 \times 10^{-3}(1 + x + 2y)$$

The plate patch is homogenous and made of a material with $E = 1.0 \times 10^6$ and $\nu = 0.25$. To satisfy the patch test, the deflection and rotations at any interior nodes computed by numerical method should be exactly the same as the analytic ones given in Equation 11.211. To examine the numerical error quantitatively, an error norm is defined as

$$e_u = \sqrt{\sum_{i=1}^{6} \left(\frac{\left(\mathbf{u}_i^{num} - \mathbf{u}_i^{exact}\right)^{\mathrm{T}}\left(\mathbf{u}_i^{num} - \mathbf{u}_i^{exact}\right)}{\left(\mathbf{u}_i^{exact}\right)^{\mathrm{T}}\left(\mathbf{u}_i^{exact}\right)} \right)} \tag{11.212}$$

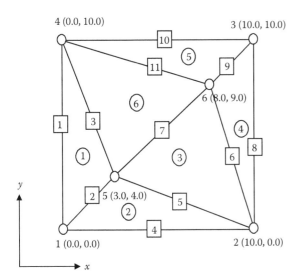

FIGURE 11.22
A square plate patch with 6 triangular cells, 6 nodes, and 11 edges for the standard patch test.

where the superscript "num" stands for numerical solution and "exact" stands for the exact solution given by Equation 11.211. Our numerical test has found $e_u = 4.474065656539186 \times 10^{-15}$ that is of the order of the machine accuracy. This shows that our ES-PIM has successfully passed the standard pure bending patch test.

Example 11.23: Thick and Thin Square Plates: ES-PIM is Locking-free

A square plate with the length $L = 10.0$ and different thickness and with different boundary conditions is considered in this example to examine the overall performance and to demonstrate the locking-free feature of our ES-PIM. The material properties are $E = 3.0 \times 10^7$ and $v = 0.3$. Owing to the twofold symmetry of the plate, only a quarter of the plate in the first quadrant is modeled. Triangular meshes of different densities are shown in Figure 11.23. The results of the center deflection w_c are normalized using $\hat{w} = w_c D/qL^4$ [29], where q is a vertical uniformly distributed load. For the plate subjected to a concentrated load at the plate center, the deflection is normalized as $\hat{w} = w_c D/PL^2$ where P is the concentrated load. Thin plates with an aspect ratio of $L/h = 100$ and a thick plate of $L/h = 5$ with different boundary conditions are investigated.

11.5.3.1 Overall Performance Examination

For the purpose of examining the overall performance of our ES-PIM, the numerical results obtained using our ES-PIM are compared with the analytic solutions from [32], as well as the solutions of FEM models using several specially designed triangular plate elements including HCT [35], HSM [30], BCIZ [34], DKT [33], RDKTM [31], and DSG [23].

Numerical results of the normalized deflection at the plate center for different cases are listed in Tables 11.52 through 11.56. The relative errors of these results are also plotted in Figures 11.24 through 11.26 for easy comparisons. From these tables and figures, it can be seen that comparing with all these FEM models, the overall performance of ES-PIM is outstanding in terms of accuracy and convergence for both the thick and thin plates with different boundary conditions. For all these studied cases, the ES-PIM is not the best for each of these cases, but its performance is very consistent and it is always among the best for all these cases: showing clearly the stable and well-balanced feature of the ES-PIM.

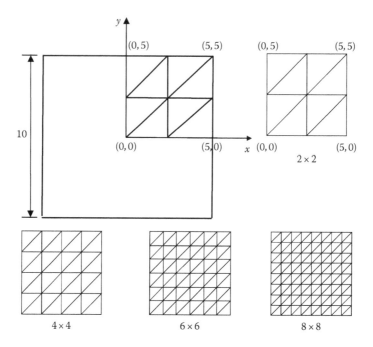

FIGURE 11.23
A square plate, quarter model and triangular meshes of cells/elements used.

TABLE 11.52

Numerical Results of the Normalized Deflection at the Center
of a Simply Supported Square Thin Plate Subjected Uniform
Load ($L/h = 100$, Unit: 10^{-3}, Analytical Solution: 4.064)

Mesh	DKT	RDKTM	DSG	ES-PIM
2×2	4.056	4.058	3.705	4.078
4×4	4.065	4.069	3.975	4.068
6×6	4.064	4.066	4.024	4.066
8×8	4.064	4.065	4.042	4.065

TABLE 11.53

Numerical Results of the Normalized Deflection at the Center
of a Simply Supported Square Thick Plate Subjected Uniform
Load ($L/h = 5$, Unit: 10^{-3}, Analytical Solution: 4.907)

Mesh	DKT	RDKTM	DSG	ES-PIM
2×2	4.056	4.902	4.499	4.960
4×4	4.065	4.904	4.804	4.920
6×6	4.064	4.906	4.860	4.912
8×8	4.064	4.906	4.879	4.909

TABLE 11.54

Numerical Results of the Normalized Deflection at the Center of Clamped Square Thin Plate Subjected Uniform Load ($L/h = 100$, Unit: 10^{-3}, Analytical Solution: 1.265)

Mesh	DKT	RDKTM	DSG	ES-PIM
2×2	1.547	1.550	1.070	1.350
4×4	1.347	1.350	1.213	1.299
6×6	1.303	1.305	1.243	1.285
8×8	1.287	1.289	1.254	1.279

TABLE 11.55

Numerical Results of the Normalized Deflection at the Center of a Clamped Square Thick Plate Subjected Uniform Load ($L/h = 5$, Unit: 10^{-3}, Analytical Solution: 2.17)

Mesh	DKT	RDKTM	DSG	ES-PIM
2×2	1.547	2.423	2.226	1.862
4×4	1.347	2.243	2.205	2.093
6×6	1.303	2.205	2.190	2.136
8×8	1.287	2.191	2.183	2.152

TABLE 11.56

Numerical Results of the Normalized Deflection at the Center of Clamped Square Thin Plate Subjected Concentrated Central Load ($L/h = 100$, Unit: 10^{-3})

Mesh	Simply Supported		Clamped	
	DSG	ES-PIM	DSG	ES-PIM
2×2	1.0624	1.1986	4.252	5.393
4×4	1.1285	1.1750	5.205	5.658
6×6	1.1445	1.1685	5.416	5.657
8×8	1.1507	1.1656	5.496	5.647
Analytic solutions	1.1601		5.612	

Note that the linear ES-PIM is very simple in the formulation and no extra sampling points are introduced, comparing with FEM models of special elements.

11.5.3.2 ES-PIM: Free of Shear Locking

From Equation 11.198, Remark 11.1 has discussed an observation of the shear-locking phenomenon when the higher order plate theories applied to simulate thin plates. We now use the ES-PIM to examine this phenomenon numerically. A simply supported or clamped square plate subjected to a uniform loading is used in this study. The geometry

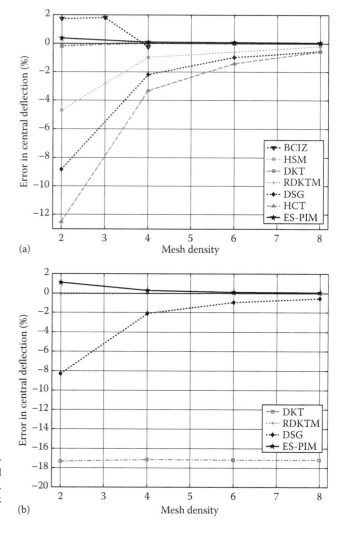

FIGURE 11.24
Relative error in the solution for the deflection at center of a simply supported square plate subjected to a uniform load. (a) Thin plate ($L/h = 100$); and (b) thick plate ($L/h = 5$).

and material properties are the same as those used in the previous study, and the computed results using a 6×6 mesh are plotted in Figure 11.27. The results show clearly that the ES-PIM produces almost identical solutions to the analytical ones, regardless of the aspect ratios of the plate in a wide range of $L/h \in [5, 10^7]$. No shear locking has been observed even when the aspect ratio of the plate becomes 10^7. At such a huge aspect ratio, the plate is not exactly a "plate" rather a very thin membrane. On the other hand, when $L/h = 5$, the plate is in fact no longer a reasonable plate, but more like a bulky 2D solid. This confirms that the locking-free formulation in our ES-PIM using DSG method is a success, and hence the ES-PIM can be applied to plates of all thickness, as long as it can be called a plate within reason.

11.5.3.3 ES-PIM: Robustness in Mesh Distortion

To study the sensitivity of ES-PIM to mesh distortion, the same square plate with simply supported boundary conditions and subjected to a uniform load is studied using a

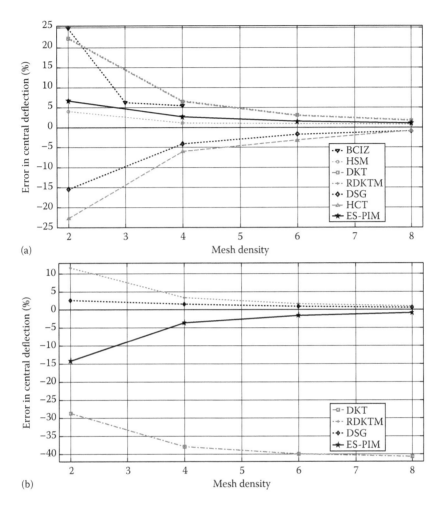

FIGURE 11.25
Relative error in the solution for the deflection at center of a clamped square plate subjected to a uniform load.
(a) Thin plate ($L/h = 100$); (b) thick plate ($L/h = 5$).

16×16 mesh. Distorted meshes are created by altering the coordinates of the regular nodes using

$$x_{ir} = x + \Delta x \cdot r_c \cdot \alpha_{ir}$$
$$y_{ir} = y + \Delta y \cdot r_c \cdot \alpha_{ir}$$
(11.213)

where
    Δx and Δy are the initial regular nodal spacing in x- and y-directions, respectively
    r_c is a computer-generated random number between -1.0 and 1.0
    α_{ir} is a prescribed irregularity factor ranging from 0.0 to 0.5

Regular and irregular cells with three different α_{ir} values are plotted in Figure 11.28.
It is seen that when $\alpha_{ir} = 0.5$, the mesh is extremely distorted. Four cases with different

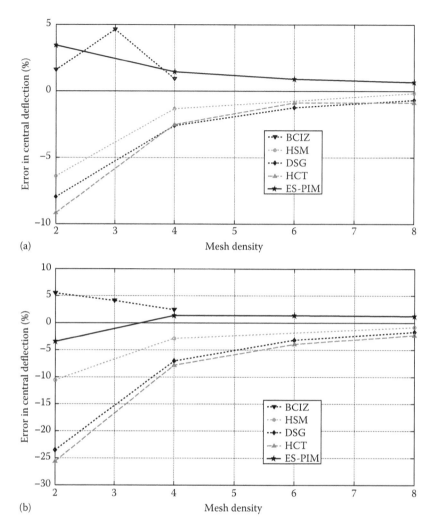

FIGURE 11.26
Relative error in the solution for the deflection at center for a thin square plate ($L/t = 100$) subjected to a concentrated load at the center of the plate. (a) Simply supported square plate; and (b) clamped square plate.

aspect ratios, $L/h = 5$, 20, 102, and 105, are studied and the relative errors in the ES-PIM solution of the central deflection are computed and plotted in Figure 11.29. It is clearly shown that the ES-PIM can obtain satisfactory results for all cases with distorted meshes. The relative errors of the central deflection are all less than 1% even when the cells are severely distorted with $\alpha_{ir} = 0.5$. This study demonstrates that the ES-PIM is not sensitive to the mesh distortion: a typical feature of a GS-Galerkin model.

Example 11.24: Thick and Thin Circular Plates: Further Examination on ES-PIM

A simply supported and clamped circular plate subjected to a uniformly distributed vertical loading is now analyzed. The plate has a radius of $R = 5$ and two thicknesses of $h = 0.1$ and

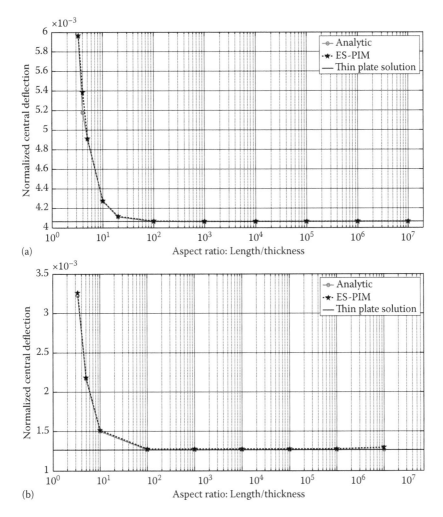

FIGURE 11.27
Shear locking test on a square plate subjected to a uniformly distributed load. (a) Simply supported; and (b) clamped.

$h = 1$, and the material properties are $E = 3 \times 10^5$ and $v = 0.3$. The setting of the problem and the meshes are shown in Figure 11.30. Because of the symmetry, only a quarter of the plate is modeled using three meshes with 6, 24, and 96 triangular cells. The boundary conditions for this problem are given by

$$\begin{array}{ll} \text{Simply supported: } w = 0 & \text{at all nodes on the boundary} \\ \text{Clamped: } w = \varphi_x = \varphi_y = 0 & \text{at all nodes on the boundary} \end{array} \qquad (11.214)$$

For a simply supported or clamped thin circular plate subjected to a uniform load, the problem is in fact axial symmetry, and the analytical solutions of the deflection and moments at the center of the plate can be found in [36]. For simply supported plates, we have

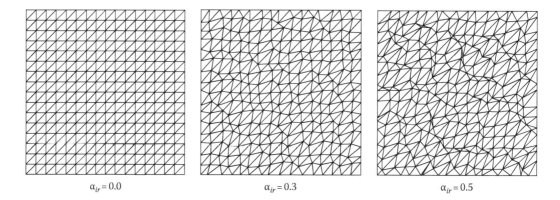

$\alpha_{ir} = 0.0$ $\alpha_{ir} = 0.3$ $\alpha_{ir} = 0.5$

FIGURE 11.28
Distorted meshes used in one quarter of the square plate.

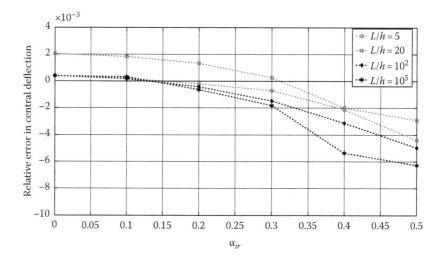

FIGURE 11.29
Error in the solution of deflection at the center of the simply supported square plate subjected uniform load: effects of mesh distortion.

$$w = \frac{qR^4}{64D}\left(\frac{5+v}{1+v} + \frac{8}{3k(1-v)}\left(\frac{h}{R}\right)^2\right)$$

$$M_r = \frac{qR^2}{16}(3+v)$$

(11.215)

For clamped plate:

$$w = \frac{qR^4}{64D}\left(1 + \frac{8}{3k(1-v)}\left(\frac{h}{R}\right)^2\right)$$

$$M_r = \frac{qR^2}{16}(1+v)$$

(11.216)

where D is the flexural rigidity of the plate defined in Equation 11.17.

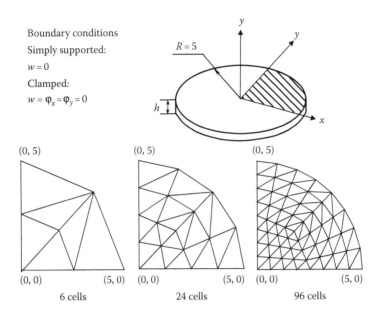

Boundary conditions
Simply supported:
$w = 0$
Clamped:
$w = \varphi_x = \varphi_y = 0$

$R = 5$

(0, 5) (0, 5) (0, 5)

(0, 0) (5, 0) (0, 0) (5, 0) (0, 0) (5, 0)

6 cells   24 cells   96 cells

FIGURE 11.30
Circular plate subjected to a uniform load. Problem setting and mesh of cells/elements.

Numerical results obtained using ES-PIM are compared with those using several FEM models of triangular elements including DKT [33], RDKTM [31], DSG [23], MiSP3 [37], BST-BK [38], and BST-BL [39]. All the numerical results for the deflection at the plate center (0,0) are normalized as w_c/w_{ref} where w_{ref} is the analytical solutions. Figure 11.31 plots the solution of the deflection at the center of a thin circular plate ($R/h = 50$) with the uniformly distributed load, against the cell/element numbers. It is observed that the ES-PIM performs best in this case. The results for the moment at the center of the plate are plotted in Figure 11.32. It is seen in the case that the ES-PIM is among the best. We also observed that the ES-PIM needs a reasonably number of cells (24 and 96) for good results. When too few cells are used (6, for example), the results of the present EFG method are found only better than the plain FEM elements but generally worse than those carefully designed elements. This is because the ES-PIM relies on a sufficient number of edges for the smoothing operations to take effect. When the model has too few edges, the ES-PIM is not much different than the plain elements. For the case of 6 cells, there are a total of 12 edges and only half of them (6 interior edges) produce smoothing effects. Therefore, we cannot expect any outstanding performance from such a coarse ES-PIM model. This finding agrees with those for the methods using generalized smoothed Galerkin (GS-Galerkin) formulations. With the increase of the number of cells (e.g., 24 and 96), the ES-PIM can outperform other FEM models. Note that in any practical problems, we use a lot more than tens of cells, and hence ES-PIM are expected to perform well in solving practical engineering problems.

Clamped thin ($R/h = 50$) and thick ($R/h = 5$) circular plates subjected to a uniform load are analyzed to obtain the distribution of the bending moments (M_r, M_θ) and shear force (V_z), using 96 cells/elements. The numerical results obtained using ES-PIM are plotted

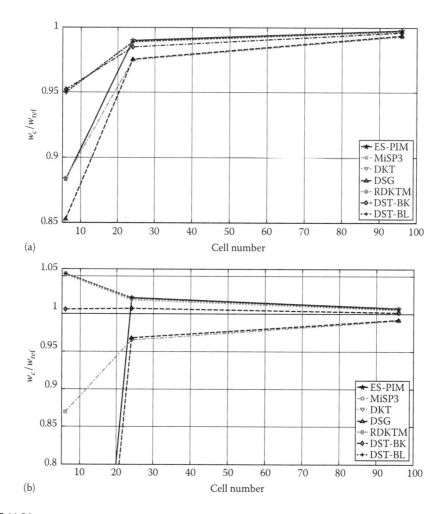

FIGURE 11.31
Solution of the deflection at center of a thin circular plate ($R/h = 50$) subjected to a uniformly distributed load. (a) Simply supported plate; and (b) clamped plate.

in Figures 11.33 through 11.35, together with the results obtained using FEM models of BST-BK, BST-BL, and MiSP3 with the same mesh. The analytical solutions are given as follows [36]:

$$M_r(r) = \frac{q}{16}[(1 + v)R^2 - (3 + v)r^2]$$

$$M_\theta(r) = \frac{q}{16}[(1 + v)R^2 - (1 + 3v)r^2] \qquad (11.217)$$

$$V_Z(r) = -\frac{qr}{2}$$

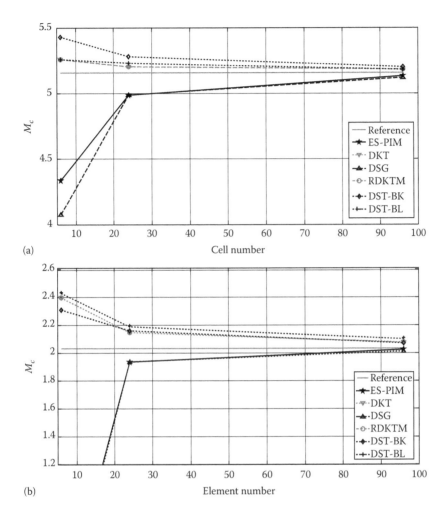

FIGURE 11.32
Solution of the moment at center of a thin circular plate $(R/h = 50)$ subjected to a uniform load. (a) Simply supported plate; and (b) clamped plate.

where
  r is the distance measured from the plate center
  D is the flexural rigidity of the plate

Values of M_r, M_θ, and V_z at a node are obtained in the ES-PIM by averaging these values of all the smoothing domains connected to the node. These forces in the MiSP3 element are computed at the nodes, and in the BST-BL and BST-BK elements are computed at the centroid of the elements. It is found that the ES-PIM performs among the bests for all the cases of thin and thick plates. For bending moments, numerical results of all the methods studied here are all in good agreements with the reference solutions. For shear forces, the ES-PIM and MiSP3 outperform others in terms of solution accuracy.

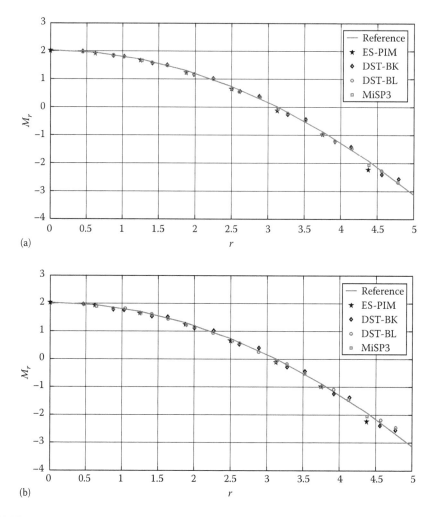

FIGURE 11.33
Solution of the moment M_r at center of a thin circular plate ($R/h = 50$) subjected to a uniform load. (a) Simply supported plate; and (b) clamped plate.

11.5.4 Some Remarks

In this section, an ES-PIM is formulated to analyze the deformation of plates based on the FSDT using simple three-node triangular cells.

- Our formulation has shown again that the ES-PIM based on the W^2 formulation can solve second-order PDEs using discontinuous functions. This is, however, yet to be confirmed numerically. We are still working on the coding of quadratic PIMs (T6/3-scheme) and RPIMs (T6- or T2L-schemes) for plates.

- In the ES-PIM formulation, no extra sampling points are introduced to evaluate the stiffness matrix, and no increase of total degrees of freedom.

- The ES-PIM is found to be very stable, accurate, and the performance is very consistent.

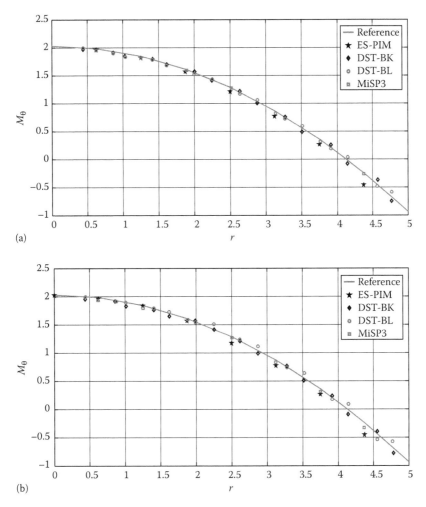

FIGURE 11.34
Solution of the moment M_θ at center of a thin circular plate ($R/h = 50$) subjected to a uniform load. (a) Simply supported plate; and (b) clamped plate.

- The linear ES-PIM can pass the pure bending patch test, which ensures the convergence of the method numerically.
- Using the DSG method, the ES-PIM is free from shear locking, and hence it works well for both thin and thick plates.

Moving forward, ES-PIM can be further improved and extended for other applications.

- The linear ES-PIM has been further coded for geometrically nonlinear analysis of thin and thick plates [22] and extensions to other types of nonlinear problems should be straightforward, because these techniques developed in FEM can be utilized with minor changes.

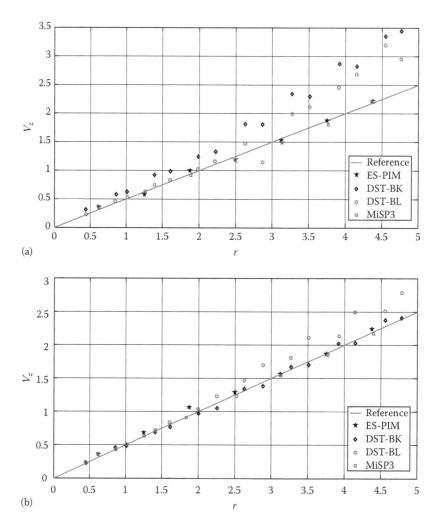

FIGURE 11.35
Solution of the shear force V_z on the edge of the clamped circular plate with uniform load. (a) Thin plate ($t/R = 0.02$); and (b) thick plate ($t/R = 0.2$).

- The linear ES-PIM can be easily extended to meshes of general n-sided polygonal cells [43].
- Other GS-Galerkin models can also be developed for plates in similar ways presented in this section.

References

1. Krysl, P. and Belytschko, T., Analysis of thin plates by the element-free Galerkin method, *Comput. Mech.*, 17, 26–35, 1996.
2. Krysl, P. and Belytschko, T., Analysis of thin shells by the element-free Galerkin method, *Int. J. Solids Struct.*, 33, 3057–3080, 1996.

3. Ouatouati, A. E. and Johnson, D. A., A new approach for numerical modal analysis using the element free method, *Int. J. Numerical Methods Eng.*, 46, 1–27, 1999.

4. Liu, G. R. and Chen, X. L., A mesh-free method for static and free vibration analyses of thin plates of complicated shape, *J. Sound Vib.*, 241(5), 839–855, 2001.

5. Liu, G. R. and Chen, X. L., Static buckling of composite laminates using EFG method, in *Proceedings of the 1st International Conference on Structural Stability and Dynamics*, Taipei, China, December 7–9, 2000, pp. 321–326.

6. Liu, G. R. and Chen, X. L., Bucking of symmetrically laminated composite plates using the element-free Galerkin method, *Int. J. Struct. Stability Dyn.*, 2(3), 2002, 281–294.

7. Chen, X. L., Liu, G. R., and Lim, S. P., An element free Galerkin method for the free vibration analysis of composite laminates of complicated shape, *Compos. Struct.*, 59(2), 279–289, 2003.

8. Liu, G. R. and Xi, Z. C., *Elastic Waves in Anisotropic Laminates*, CRC Press, Boca Raton, FL, 2001.

9. Timoshenko, S. P. and Woinowsky-Krieger, S., *Theory of Plates and Shells*, 2nd ed., McGraw-Hill, New York, 1995.

10. Schinzinger, R. and Laura, P. A. A., *Conformal Mapping Methods and Applications*, Elsevier Science, New York, 1991.

11. Reddy, J. N., *Mechanics of Laminated Composite Plates: Theory and Analysis*, CRC Press, Boca Raton, FL, 1997.

12. Timoshenko, S. P. and James, M. G., *Theory of Elastic Stability*, McGraw-Hill, Singapore, 1985.

13. Chen, Y. Z., Evaluation of fundamental vibration frequency of an orthotropic bending plate by using an iterative approach, *Comput. Methods Appl. Mech. Eng.*, 161, 289–296, 1998.

14. Gontkewitz, V. C., Natural vibration of plates and shells, *Nayka Dymka* (Kiev), 1964 (in Russian).

15. Chow, S. T., Liew, K. M., and Lam, K. Y., Transverse vibration of symmetrically laminated rectangular composite plates, *Compos. Struct.*, 20, 213–226, 1992.

16. Leissa, A. W. and Narita, Y., Vibration studies for simply supported symmetrically laminated regular plates, *Compos. Struct.*, 12, 113–132, 1989.

17. Donning, B. M. and Liu, W. K., Meshless methods for shear-deformable beams and plates, *Comput. Methods Appl. Mech. Eng.*, 152, 47–71, 1998.

18. Haggblad, B. and Bathe, K. J., Specifications of boundary conditions for Reissner/Mindlin plate bending finite elements, *Int. J. Numerical Methods Eng.*, 30, 981–1011, 1990.

19. Kanok-Nukulchai, W., Barry, W., Saran-Yasoontorn, K., and Bouillard, P. H., On elimination of shear locking in the element-free Galerkin method, *Int. J. Numerical Methods Eng.*, 52, 705–725, 2001.

20. Wang, C. M. and Kitipornchai, S., Buckling of rectangular Mindlin plates with internal line supports, *Int. J. Solids Struct.*, 30, 1–17, 1993.

21. Senthilnathan, N. R., A simple higher-order shear deformation plate theory, PhD thesis, National University of Singapore, Singapore, 1989.

22. Cui, X. Y., Liu, G. R., Li, G. Y., Zhang, G. Y., and Zheng, G., Analysis of plates and shells using an edge-based smoothed finite element method. *Comput. Mech.*, 2009 (submitted).

23. Bletzinger, K. U., Bischoff, M., and Ramm, E., A unified approach for shear-locking-free triangular and rectangular shell finite elements, *Comput. Struct.*, 75, 321–334, 2000.

24. Srinivas, S., Rao, A. K., and Rao, C. V. J., Flexure of simply-supported thick homogeneous and laminated rectangular plates, *ZAMM*, 49, 449–458, 1969.

25. Liu, G. R., Dai, K. Y., and Nguyen, T. T., A smoothed finite element method for mechanics problems, *Comput. Mech.* 39, 859–877, 2007.

26. Dai, K. Y. and Liu, G. R., Free and forced vibration analysis using the smoothed finite element method (SFEM), *J. Sound Vib.*, 301, 803–820, 2007.

27. Dai, K. Y., Liu, G. R., and Nguyen, T. T., An n-sided polygonal smoothed finite element method (nSFEM) for solid mechanics, *Finite Elem. Anal. Des.*, 43, 847–860, 2007.

28. Liu, G. R., Nguyen, T. T., Dai, K. Y., and Lam, K. Y., Theoretical aspects of the smoothed finite element method (SFEM), *Int. J. Numerical Methods Eng.*, 71, 902–930, 2007.

29. Zienkiewicz, O. C. and Taylor, R. L., *The Finite Element Method: The fifth edition*, Vol. 1: *The Basis*, Butterworth-Heinemann, Oxford, 2000.

30. Alwood, R. J. and Cornes, G. M., A polygonal finite element for plate bending problems using the assumed stress approach, *Int. J. Numerical Methods Eng.*, 1, 135–149, 1969.
31. Chen, W. J. and Cheung, Y. K., Refined 9-DOF triangular Mindin plate elements, *Int. J. Numerical Methods Eng.*, 51, 1259–1281, 2001.
32. Timoshenko, S. and Woinowsky-Krieger, S., *Theory of Plates and Shells*, McGraw-Hill, New York, 1940.
33. Batoz, J. L., An explicit formulation for an efficient triangular plate-bending element, *Int. J. Numerical Methods Eng.*, 18, 1077–1089, 1982.
34. Bazeley, G. P., Cheung, Y. K., Irons, B. M., and Zienkiewicz, O. C., Triangular elements in plate bending conforming and non-conforming solutions, in *Proceedings of the Conference on Matrix Methods in Structural Mechanics*, Air Force Institute of Technology, Wright-Patterson A. F. Base, OH, 1965, pp. 547–577.
35. Clough, R. W. and Tocher, J. L., Finite element stiffness matrices for analysis of plates in bending, in *Proceedings of the Conference on Matrix Methods in Structural Mechanics*, Air Force Institute of Technology, Wright-Patterson A. F. Base, OH, 1965, pp. 515–545.
36. Ugural, A. C., *Stresses in Plates and Shells*, McGraw-Hill, New York, 1981.
37. Ayad, R., Dhatt, G., and Batoz, J. L., A new hybrid-mixed variational approach for Reissner-Mindlin plates, The MiSP model, *Int. J. Numerical Methods Eng.*, 42, 1149–1479, 1998.
38. Batoz, J. L. and Katili, I., On a simple triangular Reissner/Mindlin plate element based on incompatible modes and discrete constraints, *Int. J. Numerical Methods Eng.*, 35, 1603–1632, 1992.
39. Batoz, J. L. and Lardeur, P., A discrete shear triangular nine DOF element for the analysis of thick to very thin plates, *Int. J. Numerical Methods Eng.*, 28, 533–560, 1989.
40. Abbassian, F., Dawswell, D. J., and Knowles, N. C., *Free Vibration Benchmarks*, Atkins Engineering Sciences, Glasgow, U.K., 1987.
41. Lu, Y. Y., Belytschko, T., and Gu, L., A new implementation of the element free Galerkin method, *Comput. Methods Appl. Mech. Eng.*, 113, 397–414, 1994.
42. Robert, D. B., *Formulas for Natural Frequency and Mode Shape*, Van Nostrand Reinhold, New York, 1979.
43. Nguyen-Thoi, T., Liu, G. R., and Lam, K. Y., An n-sided polygonal edge-based smoothed finite element method (nESFEM) for solid mechanics, 2008 (submitted).

12

Meshfree Methods for Shells

Previous chapters have introduced a number of meshfree methods for solids, fluids, beams, and plates. This chapter formulates two meshfree methods for shell structures: element-free Galerkin (EFG) and edge-based smoothed point interpolation method (ES-PIM).

Spatial thin shell structures are used very extensively in many engineering structures, including aircraft, pressure vessels, storage tanks, and so on, due to their outstanding efficiency in utilizing materials. Because of the complex nature, both structurally and in mechanics, numerical means have to be utilized for analyses of shells during the design process. The finite element method (FEM) remains the most popular numerical technique for such analyses [1,20]. However, FEM often requires quality meshes, creating which is a tedious, costly, and time-consuming process.

Meshfree methods present a promising alternative to FEM, as they offer opportunities to relieve the manual meshing process in modeling a structure. This is particularly important for shells, as shell structures are very complex both in field variable variation and in geometric configuration. Meshfree methods can offer a very important capability in representing the complex curved geometry of shell structures. The meshfree approximations both in field variables and in the structure itself can provide more accurate results compared to the standard FEM.

Very few works have been reported in the development of meshfree methods for shell structures. The first contribution in this regard was made by Krysl and Belytschko [2] based on the thin shell theory using moving least squares (MLS) approximation with Lagrange multipliers for essential boundary conditions. Noguchi et al. [3] developed a formulation for thick shell using MLS approximation with penalty method for handling essential boundary conditions. Li et al. [4] formulated a meshfree method based on the reproducing kernel particle method for thin shells with large deformation. In this work, the essential boundary condition is imposed by modifying shape functions for nodes near and on the essential boundaries. Other works are reported in [5,6] on EFG, and recently on ES-PIM [19]. This chapter covers the following two topics:

- Formulation of the EFG method for shell structures. The materials on EFG presented here are based on the works in [5,6], where both the field variables and geometry of the shell are all approximated using the MLS approximation.
- Formulation of the ES-PIM for shells based on our recent work [19]. The first-order shear deformation theory or FSDT is used together with the discrete shear gap (DSG) method for eliminating shear locking. The ES-PIM is formulated based on triangular mesh that can be generated automatically, and has features of simplicity, free of shear locking, robustness, accuracy, and efficiency.

12.1 EFG Method for Spatial Thin Shells

This section formulates an EFG method for thin shells governed by Kirchhoff–Love shell theory. In the EFG method, the generalized displacement (deflections and rotations) at an arbitrary point is approximated from nodal displacements using MLS approximation. A compact support domain is used to determine the field nodes to be used for constructing MLS shape functions. As discussed in Chapter 6, use of MLS approximation requires special treatment for essential boundary conditions for the lack the Kronecker delta property. These techniques include the penalty method [3], Lagrange multipliers [2], and a method that modifies shape functions for nodes near and on the essential boundaries [4]. This chapter discusses both the penalty method and Lagrange multipliers method for analyzing static problems. For dynamic analysis, the Lagrange multipliers method is used for transient analyses and the orthogonal transform method is used for free-vibration analyses.

The formulation presented in this section is based on [5,6]. It begins with a brief discussion of MLS approximation. The governing equations for the analysis of general shells and membrane structures are then introduced. Numerical formulations based on a geometrically exact theory accounting for the Kirchhoff hypothesis are presented. This is followed by the definition of curved surfaces, kinematics of shells, stress and strain measures, and the constitutive relations adopted in the formulation. The final discrete equations for static, free vibration, and transient vibration are then obtained. For free vibration, the essential boundary conditions are imposed using orthogonal transform techniques to solve the eigenvalue equation [7,8]. For static problems, essential boundary conditions are imposed through the Lagrange multipliers method and the penalty method. Finally, the method is applied to several numerical examples of shells with different geometries to illustrate the efficiency and accuracy of the present EFG method.

12.1.1 Moving Least Squares Approximation

The derivation of shape functions from the MLS approximation method is the same as that provided in Chapter 2, except that a higher order polynomial basis needs to be included. A two-dimensional (2D) field approximation is needed for modeling thin shells. A component of the displacement vector is approximated by a polynomial function as follows:

$$u^h(\mathbf{x}) = \sum_{I \in S_n} \phi_I(\mathbf{x}) u_I \tag{12.1}$$

where
S_n is the set of nodes in the support domain of $\mathbf{x}$, and is the shape function
$\phi(\mathbf{x})$ is the MLS shape function

Because the partial differential equations (PDEs) for shells are of fourth order, the order of the polynomial basis should be higher than that for 2D solids, and at least quadratic in EFG. In this chapter, the two different orders of polynomial basis are primarily used. The following six terms of basis functions up to quadratic terms are used for shells where the shear effect is significant.

$$\mathbf{P}^T(\mathbf{x}) = \{1, x, y, x^2, xy, y^2\} \tag{12.2}$$

The following 15 terms of basis functions up to quartic terms are used to ameliorate membrane locking in bending-dominated cases:

$$\mathbf{P}^{\mathrm{T}}(\mathbf{x}) = \{1, x, y, x^2, xy, y^2, x^3, x^2y, xy^2, y^3, x^4, x^3y, x^2y^2, xy^3, y^4\} \tag{12.3}$$

In computing the MLS shape functions, the quartic spline weight function given in Equation 2.17 is used in this section because of the requirements on the continuity of the MLS shape functions and their derivatives.

12.1.2 Governing Equation for a Thin Shell

The shells considered in this section are assumed to be thin with arbitrary depth and Gaussian curvature governed by Kirchhoff–Love theory. The governing equations used in this section are based on geometrically exact theory of shells formulated by Simo and Fox [9] with some modifications to account for the Kirchhoff hypothesis. Here, we outline only the basic concepts of the formulation. Details can be found in [9].

12.1.2.1 Kinematic Description of a Shell

The reference frame coordinates are illustrated in Figure 12.1. The shell in three-dimensional (3D) space is described in a global Cartesian coordinate system, $\mathbf{x}$. The Gauss intrinsic coordinates defined locally are used to describe the configuration of the shell. $\boldsymbol{\varphi}(\xi^1, \xi^2)$ gives the position of the point on the shell neutral surface, and $\mathbf{t}(\xi^1, \xi^2)$ is a direction unit vector normal to the shell neutral surface both in the unformed reference and deformed states according to the Kirchhoff–Love hypothesis. The pair $(\boldsymbol{\varphi}, \mathbf{t})$ defines the position of an arbitrary point in the shell. The configuration of the shell can be expressed mathematically as

$$\psi = \{\mathbf{x} \in \mathbb{R}^3 | \mathbf{x} = \boldsymbol{\varphi}(\xi^1, \xi^2) + \xi \mathbf{t}(\xi^1, \xi^2) \quad \text{with} \quad \xi^1, \xi^2 \in \mathbb{A} \quad \text{and} \quad \xi \in \langle h^-, h^+ \rangle\} \tag{12.4}$$

Here $\mathbb{A}$ denotes the parametric space for the shell; (h^-, h^+) are the distances of the "lower" and "upper" surfaces of the shell measured from the shell neutral surface.

The convective basis vectors $\mathbf{g}_I$ are defined as

$$\nabla \mathbf{x} = \frac{\partial \mathbf{x}}{\partial \xi^I} \otimes \mathbf{E}^I \equiv \mathbf{g}_I \otimes \mathbf{E}^I \tag{12.5}$$

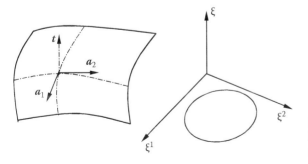

FIGURE 12.1
Reference frames of coordinates on the neutral surface of a thin shell.

where
  ∇ denotes the gradient operator
  $\otimes$ denotes the tensor product
  $\mathbf{E}$ is the unit basis vector in the global Cartesian basis

A contravariant basis $\mathbf{g}^J$ can be obtained from the standard relation of

$$\mathbf{g}^J \cdot \mathbf{g}_I = \delta_I^J \tag{12.6}$$

The unit area in the neutral surface is defined by the differential two-form:

$$d\mathbb{A} = \boldsymbol{\varphi}_{,1} \times \boldsymbol{\varphi}_{,2} \, d\xi^1 \, d\xi^2 \tag{12.7}$$

where "$\times$" denotes vector cross product. The determinants of the tangent maps in the deformed and reference configuration will be denoted subsequently as j and j^0, respectively, with $\bar{j}$ and $\bar{j}^0$ denoting the Jacobians on the reference surface

$$j = \det[\nabla \mathbf{x}], \quad j^0 = \det[\nabla \mathbf{x}^0], \quad \bar{j} = j|_{\xi=0}, \quad \bar{j}^0 = j^0|_{\xi=0} \tag{12.8}$$

where the superscript "0" is used to denote quantities in the reference configuration. The surface-convected frame, which spans the tangent space to the neutral surface, is defined as $\alpha_\alpha = \boldsymbol{\varphi}_{,\alpha}$ ($\alpha = 1, 2$). Hence, the first fundamental form on the reference surface is

$$a = a_{\alpha\beta} a^\alpha \otimes a^\beta, \quad a_{\alpha\beta} = \boldsymbol{\varphi}_{,\alpha} \cdot \boldsymbol{\varphi}_{,\beta} \tag{12.9}$$

where
  α^I ($I = 1, 2$) denotes the dual surface-convected basis through the standard relation
  $\mathbf{a}^I \cdot \mathbf{a}_J = \delta_J^I$
  "$\cdot$" denotes dot product. The second fundamental form is defined as

$$\kappa_{\alpha\beta} = \boldsymbol{\varphi}_{,\alpha} \cdot \mathbf{t}_{,\beta} \tag{12.10}$$

12.1.2.2 Strain Measures

The linear membrane and bending strain measures can be derived from the kinematic variables in Equations 12.9 and 12.10 as

$$\varepsilon_{\alpha\beta} = \frac{1}{2}\left(\boldsymbol{\varphi}_{,\alpha}^0 \cdot \mathbf{u}_{,\beta} + \boldsymbol{\varphi}_{,\beta}^0 \cdot \mathbf{u}_{,\alpha} \right) \tag{12.11}$$

and

$$\rho_{\alpha\beta} = \frac{1}{2}\left(\boldsymbol{\varphi}_{,\alpha}^0 \cdot \Delta \mathbf{t}_{,\beta} + \boldsymbol{\varphi}_{,\beta}^0 \cdot \Delta \mathbf{t}_{,\alpha} + \mathbf{u}_{,\alpha} \cdot \mathbf{t}_{,\beta}^0 + \mathbf{u}_{,\beta} \cdot \mathbf{t}_{,\alpha}^0 \right) \tag{12.12}$$

where only the symmetric part of the bending strain measure is considered.

The Kirchhoff–Love hypothesis needs to be introduced explicitly to obtain the definite forms for the strain measures. The mathematical form of this hypothesis is expressed as

$$\mathbf{t} = (\bar{j})^{-1}(\boldsymbol{\varphi}_{,1} \times \boldsymbol{\varphi}_{,2}), \quad \|\mathbf{t}\| = 1 \tag{12.13}$$

where $\| \ \|$ denotes the norm of a vector. Hence, the derivatives of the normal vector in the reference configuration $\mathbf{t}^0$ and the partial derivatives of the increment $\Delta \mathbf{t}$ can be

$$\mathbf{t}^0_{,\alpha} = (\bar{j}^0)^{-1}\left(\boldsymbol{\varphi}^0_{,1\alpha} \times \boldsymbol{\varphi}^0_{,2} + \boldsymbol{\varphi}^0_{,1} \times \boldsymbol{\varphi}^0_{,2\alpha}\right) \tag{12.14}$$

$$\Delta \mathbf{t}_{,\alpha} = (\bar{j}^0)^{-1}\left(\mathbf{u}_{,1\alpha} \times \boldsymbol{\varphi}^0_{,2} + \mathbf{u}^0_{,1} \times \boldsymbol{\varphi}^0_{,2\alpha} + \boldsymbol{\varphi}^0_{,1\alpha} \times \mathbf{u}_{,2} + \boldsymbol{\varphi}^0_{,1} \times \mathbf{u}_{,2\alpha}\right) \tag{12.15}$$

Note that the membrane strain measures of Equation 12.11 are not affected by the introduction of the Kirchhoff–Love hypothesis. Considering the symmetry with respect to partial differentiation $\boldsymbol{\varphi}^0_{,12} = \boldsymbol{\varphi}^0_{,21}$ and $\mathbf{u}_{,12} = \mathbf{u}_{,21}$, the bending strain measures can be rewritten as

$$\rho_{11} = -\mathbf{u}_{,11} \cdot \mathbf{t}^0 + (\bar{j}^0)^{-1}\left[\mathbf{u}_{,1} \cdot \left(\boldsymbol{\varphi}^0_{,11} \times \boldsymbol{\varphi}^0_{,2}\right) + \mathbf{u}_{,2} \cdot \left(\boldsymbol{\varphi}^0_{,1} \times \boldsymbol{\varphi}^0_{,11}\right)\right] \tag{12.16}$$

$$\rho_{22} = -\mathbf{u}_{,22} \cdot \mathbf{t}^0 + (\bar{j}^0)^{-1}\left[\mathbf{u}_{,1} \cdot \left(\boldsymbol{\varphi}^0_{,22} \times \boldsymbol{\varphi}^0_{,2}\right) + \mathbf{u}_{,2} \cdot \left(\boldsymbol{\varphi}^0_{,1} \times \boldsymbol{\varphi}^0_{,22}\right)\right] \tag{12.17}$$

$$
\begin{aligned}
\rho_{12}s &= -\frac{1}{2}(\mathbf{u}_{,12} + \mathbf{u}_{,21}) \cdot \mathbf{t}^0 + \frac{1}{2}(\bar{j}^0)^{-1}\left[\mathbf{u}_{,1} \cdot \left(\left(\boldsymbol{\varphi}^0_{,12} + \boldsymbol{\varphi}^0_{,21}\right) \times \boldsymbol{\varphi}^0_{,2}\right) + \mathbf{u}_{,2} \cdot \left(\boldsymbol{\varphi}^0_{,1} \times \left(\boldsymbol{\varphi}^0_{,12} + \boldsymbol{\varphi}^0_{,21}\right)\right)\right] \\
&= -\mathbf{u}_{,12} \cdot \mathbf{t}^0 + (\bar{j}^0)^{-1}\left[\mathbf{u}_{,1} \cdot \left(\boldsymbol{\varphi}^0_{,12} \times \boldsymbol{\varphi}^0_{,2}\right) + \mathbf{u}_{,2} \cdot \left(\boldsymbol{\varphi}^0_{,1} \times \boldsymbol{\varphi}^0_{,12}\right)\right]
\end{aligned} \tag{12.18}
$$

12.1.2.3 Stress Resultants and Stress Couples

A section in the current configuration is described by

$$\psi^\alpha = \{\mathbf{x} \in \mathbb{R}^3 | \mathbf{x} = \mathbf{x}|_{\xi^\alpha = \text{const}}\}, \quad \alpha = 1, 2 \tag{12.19}$$

The stress resultants and resultant couples are defined by normalizing the force and torque with the surface Jacobian $j = \|\boldsymbol{\varphi}_{,1} \times \boldsymbol{\varphi}_{,2}\|$ as follows:

$$\mathbf{n}^\alpha = (\bar{j})^{-1} \int_{h^-}^{h^+} \boldsymbol{\sigma} \mathbf{g}^\alpha \mathrm{d}\xi, \quad \alpha = 1, 2 \tag{12.20}$$

$$\mathbf{m}^\alpha = (\bar{j})^{-1} \int_{h^-}^{h^+} (\mathbf{x} - \boldsymbol{\varphi}) \times \boldsymbol{\sigma} \mathbf{g}^\alpha j \mathrm{d}\xi, \quad \alpha = 1, 2 \tag{12.21}$$

The director stress couple $\tilde{\mathbf{m}}^\alpha$ can also be defined through the expression

$$\mathbf{m}^\alpha = \mathbf{t} \times \tilde{\mathbf{m}}^\alpha \quad \Rightarrow \quad \tilde{\mathbf{m}}^\alpha = (\bar{j})^{-1} \int_h^{h^+} \xi \boldsymbol{\sigma} \mathbf{g}^\alpha j \, \mathrm{d}\xi \tag{12.22}$$

The through-thickness stress resultant has been omitted because it does not play a role in Kirchhoff–Love theory.

12.1.2.4 Constitutive Equations

For the isotropic elastic shell structures, the effective membrane and stress couple resultant for the isotropic hyperelastic material can be written as

$$\left\{ \begin{array}{c} \tilde{n}^{11} \\ \tilde{n}^{22} \\ \tilde{n}^{12} \end{array} \right\} = \frac{Eh}{1 - v^2} \mathbf{C} \left\{ \begin{array}{c} \varepsilon_{11} \\ \varepsilon_{22} \\ 2\varepsilon_{12} \end{array} \right\} \tag{12.23}$$

$$\left\{ \begin{array}{c} \tilde{m}^{11} \\ \tilde{m}^{22} \\ \tilde{m}^{12} \end{array} \right\} = \frac{Eh^3}{12(1 - v^2)} \mathbf{C} \left\{ \begin{array}{c} \rho_{11} \\ \rho_{22} \\ 2\rho_{12} \end{array} \right\} \tag{12.24}$$

where the matrix $\mathbf{C}$ is given by

$$\mathbf{C} = \begin{bmatrix} (a^{011})^2 & \left(va^{011}a^{022} + (1-v)(a^{012})^2\right) & a^{011}a^{012} \\ & (a^{022})^2 & a^{022}a^{012} \\ \text{symm} & & \frac{1}{2}\left((1+v)(a^{012})^2 + (1-v)a^{011}a^{022}\right) \end{bmatrix} \tag{12.25}$$

Here $a^{0\alpha\beta}$ are the components of the first fundamental form in the dual basis.

12.1.3 Strain–Displacement Relations

The displacement vector can be expressed in the global Cartesian basis $\mathbf{E}_K$ as

$$\mathbf{u}(\zeta) = \sum_{I \in S_n} \phi_I(\zeta)[u_I \mathbf{E}_1 + v_I \mathbf{E}_2 + w_I \mathbf{E}_3] \tag{12.26}$$

where
  S_n is the set of nodes in the support domain of ζ
  u_I, v_I, and w_I are the components of the displacement vector at the Ith node in $\mathbf{E}_1$, $\mathbf{E}_2$, and $\mathbf{E}_3$ directions, respectively

The membrane strain–displacement relation for the Ith node is obtained by substituting the displacement approximation Equation 12.26 into Equation 12.11 to give

$$\left\{ \begin{array}{c} \varepsilon_{11} \\ \varepsilon_{22} \\ 2\varepsilon_{12} \end{array} \right\} = \begin{bmatrix} \phi_{I,1}\varphi^0_{,1} \cdot \mathbf{E}_1 & \phi_{I,1}\varphi^0_{,1} \cdot \mathbf{E}_2 & \phi_{I,1}\varphi^0_{,1} \cdot \mathbf{E}_3 \\ \phi_{I,2}\varphi^0_{,2} \cdot \mathbf{E}_1 & \phi_{I,2}\varphi^0_{,2} \cdot \mathbf{E}_2 & \phi_{I,2}\varphi^0_{,2} \cdot \mathbf{E}_2 \\ \left(\phi_{I,1}\varphi^0_{,2} + \phi_{I,2}\varphi^0_{,1}\right) \cdot \mathbf{E}_1 & \left(\phi_{I,1}\varphi^0_{,2} + \phi_{I,2}\varphi^0_{,1}\right) \cdot \mathbf{E}_2 & \left(\phi_{I,1}\varphi^0_{,2} + \phi_{I,2}\varphi^0_{,1}\right) \cdot \mathbf{E}_2 \end{bmatrix}$$
$$\times \left\{ \begin{array}{c} u_I \\ v_I \\ w_I \end{array} \right\} \tag{12.27}$$

The bending strain–displacement matrix $[\mathbf{B}_{(b)I}]$ for the Ith node is obtained by substituting the displacement approximation Equation 12.26 into Equation 12.12

$$\left\{ \begin{array}{c} \rho_{11} \\ \rho_{22} \\ 2\rho_{12} \end{array} \right\} = [\mathbf{B}_{(b)I}] \left\{ \begin{array}{c} u_I \\ v_I \\ w_I \end{array} \right\} \tag{12.28}$$

where the elements of the strain–displacement matrix $[\mathbf{B}_{(b)I}]$ are given by

$$[\mathbf{B}_{(b)I}]_{1m} = (\bar{j}^0)^{-1} \left[-\bar{j}^0 \phi_{I,11} \mathbf{t}^0 + \phi_{I,1} \left(\boldsymbol{\varphi}^0_{,11} \times \boldsymbol{\varphi}^0_{,2} \right) + \phi_{I,2} \left(\boldsymbol{\varphi}^0_{,1} \times \boldsymbol{\varphi}^0_{,11} \right) \right] \cdot \mathbf{E}_m$$

$$[\mathbf{B}_{(b)I}]_{2m} = (\bar{j}^0)^{-1} \left[-\bar{j}^0 \phi_{I,22} \mathbf{t}^0 + \phi_{I,1} \left(\boldsymbol{\varphi}^0_{,22} \times \boldsymbol{\varphi}^0_{,2} \right) + \phi_{I,2} \left(\boldsymbol{\varphi}^0_{,1} \times \boldsymbol{\varphi}^0_{,22} \right) \right] \cdot \mathbf{F}_m \tag{12.29}$$

$$[\mathbf{B}_{(b)I}]_{3m} = 2(\bar{j}^0)^{-1} \left[-\bar{j}^0 \phi_{I,12} \mathbf{t}^0 + \phi_{I,1} \left(\boldsymbol{\varphi}^0_{,12} \times \boldsymbol{\varphi}^0_{,2} \right) + \phi_{I,2} \left(\boldsymbol{\varphi}^0_{,1} \times \boldsymbol{\varphi}^0_{,12} \right) \right] \cdot \mathbf{E}_m$$

12.1.4 Principle of Virtual Work

The effective membrane and bending forces are defined to describe the weak formulation of shells

$$\tilde{\mathbf{n}} = \tilde{n}^{\beta\alpha} \mathbf{a}_\beta \otimes \mathbf{a}_\alpha \tag{12.30}$$

$$\tilde{\mathbf{m}} = \tilde{m}^{\beta\alpha} \mathbf{a}_\beta \otimes \mathbf{a}_\alpha \tag{12.31}$$

By making use of the basic kinematic assumption (Equation 12.4), the weak form of the governing equation for thin shells under static load can be written as

$$W_{\text{Sta}}(\delta \mathbf{x}) = \int_{\mathbb{A}} \left[\tilde{n}^{\beta\alpha} \cdot \delta\varepsilon_{\beta\alpha} + \tilde{m}^{\beta\alpha} \cdot \delta\rho_{\beta\alpha} \right] d\mathbb{A} - W_{\text{ext}}(\delta \mathbf{x}) \tag{12.32}$$

Here W_{ext} is the virtual work of the external loading given by

$$W_{\text{ext}} = \int_{\mathbb{A}} [\bar{\mathbf{n}} \cdot \delta\boldsymbol{\varphi} + \bar{\mathbf{m}} \cdot \delta\bar{\mathbf{t}}] d\mathbb{A} + \int_{\Gamma_n} \bar{\bar{\mathbf{n}}} \cdot \delta\boldsymbol{\varphi}\bar{j} \, d\Gamma + \int_{\Gamma_m} \bar{\bar{\mathbf{m}}}\delta\bar{\mathbf{t}}\bar{j} d\Gamma \tag{12.33}$$

where
  $\bar{\mathbf{n}}$ is the applied resultant force per unit length
  $\bar{\mathbf{m}}$ is the applied direct couple per unit length
  $\bar{\bar{\mathbf{n}}}$ and $\bar{\bar{\mathbf{m}}}$ are the prescribed resultant force and the prescribed director couple on the boundaries Γ_n and Γ_m, respectively

For static analysis of thin shells, the penalty method is used to enforce essential boundary conditions by adding an additional boundary condition term to Equation 12.33 to obtain

$$W_{\text{Sta}}(\delta \mathbf{x}) - \int_{\Gamma_u} \boldsymbol{\alpha} \cdot (\mathbf{u} - \mathbf{u}_\Gamma)\delta\mathbf{u} d\Gamma = 0 \tag{12.34}$$

Here $\mathbf{u}$ and $\mathbf{u}_\Gamma$ are the nodal displacement vector and the prescribed displacement vector on the essential boundary Γ_u, and $\boldsymbol{\alpha}$ is a diagonal matrix of penalty factors, which are usually very large numbers.

For free-vibration analyses of the thin shells, the discrete system equation and the boundary conditions are formulated separately. The variational form of the elastic dynamic undamped equilibrium equation can be written as follows:

$$W_{\mathrm{Dya}}(\delta \mathbf{x}) = \int_{\mathbb{A}} \left[\tilde{\mathbf{n}}^{\beta\alpha} \cdot \delta\varepsilon_{\beta\alpha} + \tilde{\mathbf{m}}^{\beta\alpha} \cdot \delta\rho_{\beta\alpha} \right] d\mathbb{A} + \int_{\mathbb{R}} \delta\mathbf{u} \cdot \rho\ddot{\mathbf{u}} d\mathbb{R} - W_{\mathrm{ext}}(\delta\mathbf{x}) \tag{12.35}$$

The weak form of the essential boundary conditions with Lagrange multipliers is employed to obtain the discretized essential boundary conditions

$$\int_{\Gamma_u} \delta\boldsymbol{\lambda}^{\mathrm{T}} \cdot (\mathbf{u} - \bar{\mathbf{u}}) d\Gamma = 0 \tag{12.36}$$

where $\boldsymbol{\lambda}$ is a matrix of Lagrange multipliers, each of them can be interpolated on the essential boundary using its nodal values.

$$\boldsymbol{\lambda}(\mathbf{x}) = \sum_I N_I(s)\boldsymbol{\lambda}_I, \quad \mathbf{x} \in \Gamma_u \tag{12.37}$$

and

$$\delta\boldsymbol{\lambda}(\mathbf{x}) = \sum_I N_I(s)\delta\boldsymbol{\lambda}_I, \quad \mathbf{x} \in \Gamma_u \tag{12.38}$$

where
   s is the arc length along the essential boundaries
   $N_I(s)$ are the Lagrange interpolates discussed in Equation 6.15

12.1.5 Surface Approximation

The (neutral) surface of the shell is also approximated using MLS shape functions. The procedure is exactly the same as for approximating the displacement field variables. The approximated surface can be described by

$$\boldsymbol{\varphi}(\zeta) = \phi_I(\zeta)\mathbf{x}_I \tag{12.39}$$

where ζ is the coordinate in the parameter space for the neutral surface of the shell. A deficiency of this approximation is that the constructed surface does not pass through the prescribed points, unlike that in finite element meshes. This is due to the use of MLS shape functions.

12.1.6 Discretized Equations

For static analyses of shells, the penalty term to impose essential boundary conditions in Equation 12.34 can be discretized as follows:

$$\alpha \int_{\Gamma_u} \mathbf{u} \cdot \delta\mathbf{u} d\Gamma = \sum_{I \in S_\lambda} \delta\tilde{\mathbf{u}}^{\mathrm{T}} \alpha \boldsymbol{\Phi}_I \boldsymbol{\Phi}_I^{\mathrm{T}} \tilde{\mathbf{u}} \tag{12.40}$$

and

$$\alpha \int_{\Gamma_u} \mathbf{u}_\Gamma \cdot \delta \mathbf{u} d\Gamma = \sum_{I \in S_\lambda} \delta \tilde{\mathbf{u}}^T \mathbf{u}_{\Gamma I} \alpha \Phi_I \qquad (12.41)$$

where
  S_λ is the set of sampling points for integration on surface Γ_u
  $\tilde{\mathbf{u}}$ is the vector corresponding to a degree of freedom in the translation and rotation fields

The penalty matrix in Equation 12.40 and the penalty vector in Equation 12.41 are assembled into the global stiffness matrix and the global external force vector, respectively.

For dynamic analysis of shells, Equations 12.1, 12.23, 12.24, 12.27, and 12.28 are substituted into the variational weak form (Equation 12.35). This gives the dynamic discrete equation

$$\mathbf{M\ddot{U} + C\dot{U} + KU = F} \qquad (12.42)$$

where $\mathbf{K}$, $\mathbf{M}$, and $\mathbf{F}$ are, respectively, the global stiffness, global mass matrices, and global force vector, which are assembled using the corresponding nodal matrices and vectors formed in the similar manner as those for plates (see Chapter 11).

12.1.7 Static Analysis

For static analysis of shells, all the terms in Equation 12.42 related to dynamic effects should vanish, and the equation system is simplified as

$$\mathbf{KU = F} \qquad (12.43)$$

which is a set of linear algebraic equations that can be solved for the deflection using standard routines of equation solvers.

12.1.8 Free Vibration

For free-vibration analysis, the external force vector $\mathbf{F}$ should vanish, and the damping should not be considered; we then have

$$\mathbf{M\ddot{U} + KU = 0} \qquad (12.44)$$

Considering harmonic vibration, the eigenvalue equations for shells derived from Equation 12.44 is of the form

$$(\mathbf{K} - \omega^2 \mathbf{M})\mathbf{Q} = 0 \qquad (12.45)$$

where
  ω is the natural frequency
  $\mathbf{Q}$ is a vector that collects the nodal values corresponding the amplitudes of the displacements given by

$$\mathbf{Q} = \{Q_1, \ldots, Q_{N_u}\}^{\mathrm{T}} \tag{12.46}$$

where N_u is the total number of nodal degrees of freedom (DOFs) (unconstrained). We now derive the discretized form of the essential boundary conditions. Substituting the displacement field $\mathbf{u}$ of Equation 12.1 into the weak form (Equation 12.36) yields a set of linear algebraic constraint equations:

$$\mathbf{Bu} = \bar{\mathbf{B}} \tag{12.47}$$

where

$$\mathbf{B}_{IJ} = \phi_J(x_I), \quad \bar{\mathbf{B}}_I = \mathbf{u}_\Gamma(x_I), \quad I \in \Gamma_u, \quad J \in \mathbb{A} \tag{12.48}$$

In general, $\mathbf{B}$ is a very sparse and singular matrix with a dimension of $n_c \times N_u$, where n_c is the total number of DOFs for all the nodes involved in the essential boundary (participated in the construction of Equation 12.47). For eigenvalue analysis, the essential boundary conditions are homogeneous, i.e., $\bar{\mathbf{B}} = 0$.

In treating the essential boundary conditions, we follow the procedure described in Section 11.2.4. An orthogonal transformation is performed to produce a positive-definite stiffness matrix for the eigenvalue equation in computing natural frequencies. Using singular-value decomposition, $\mathbf{B}$ can be decomposed as

$$\mathbf{B}_{n_c \times N_u} = \mathbf{W}_{n_c \times r} \mathbf{\Sigma}_{r \times r} (\mathbf{V}_{N_u \times r})^{\mathrm{T}} \tag{12.49}$$

where
   $\mathbf{W}$ and $\mathbf{V}$ are orthogonal matrices
   r is the number of nonzero singular values of $\mathbf{B}$
   $\mathbf{\Sigma}$ is a diagonal collecting these nonzero singular values

The matrix $\mathbf{V}$ can be written in two submatrices:

$$(\mathbf{V}_{N_u \times r})^{\mathrm{T}} = (\mathbf{V}_{N_u \times r}, \mathbf{V}_{N_u \times (N_u - r)})^{\mathrm{T}} \tag{12.50}$$

where the rank r of $\mathbf{B}$ is equal to the number of independent constraints, and the others are redundant. The following change of coordinates satisfied the constraint Equation 12.47:

$$\mathbf{Q}_{N_u \times N_u} = \mathbf{V}_{N_u \times (N_u - r)} \tilde{\mathbf{Q}}_{(N_u - r) \times N_u} \tag{12.51}$$

By substituting Equation 12.51 into Equation 12.45 and left-multiplying the result by the transpose of $\mathbf{V}_{N_u \times (N_u - r)}$, the reduced order eigenvalue problem for the structure is obtained

$$(\tilde{\mathbf{K}} - \omega^2 \tilde{\mathbf{M}}) \tilde{\mathbf{Q}} = 0 \tag{12.52}$$

where $\tilde{\mathbf{K}} = (\mathbf{V}_{N_u \times (N_u - r)})^{\mathrm{T}} \mathbf{K} \mathbf{V}_{N_u \times (N_u - r)}$ and $\tilde{\mathbf{M}} = (\mathbf{V}_{N_u \times (N_u - r)})^{\mathrm{T}} \mathbf{M} \mathbf{V}_{N_u \times (N_u - r)}$ are the dimension reduced (condensed) stiffness and mass matrices. Equation 12.52 is now nonnegative definite and can be solved using standard eigenvalue solvers for eigenvalues that relate to the natural frequencies and eigenvectors that relate to the vibration modes of the thin shells.

12.1.9 Forced (Transient) Vibration

For transient response analysis, Equation 12.42 can be solved using conventional direct integration techniques based on finite difference approaches, where the time space is divided into small time steps. The Newmark method is applied in the section for time integration to obtain the time history of the displacement response of the shell. The procedure is the same as that described in Chapter 7 for 2D solids.

12.1.10 Numerical Example of Static Problems

Example 12.1: Deflection of a Barrel Vault Roof under Gravity Force

The present EFG method is used to investigate the response of a barrel vault roof under self-weight. The problem has been analyzed by several researchers using the EFG method [2,3] and FEM [9]. The barrel vault roof is a standard benchmark test because it undergoes complex membrane and inextensional bending states of stress. The example is used to evaluate the convergence and accuracy of the present EFG method for the static analysis of shells. Figure 12.2 shows the barrel roof and defines the parameters used in the description of its geometry. The two curved edges of the roof are diaphragm supported, which allows displacement in the axial direction and rotation about the tangent to shell boundary. The following parameters are used in the computation:

Length	Radius	Thickness	Semispan Angle	Young's Modulus	Poisson's Ratio	Mass Density
$L=600$	$R=300$	$H=3.0$	$\theta=40°$	$E=3.0 \times 10^6$	$\nu=0$	$\rho=0.20833$

Due to the symmetry, only a quarter of the vault roof is modeled, and symmetric boundary conditions are introduced along the planes of symmetry. To evaluate the effectiveness of the present EFG method, both regular and irregular nodal arrangements in the parametric space shown in Figure 12.3 are used in the analyses. Both quadratic ($m=6$) and quartic ($m=15$) basis functions are used in the analysis. Figure 12.4a and b show, respectively, the distributions of the

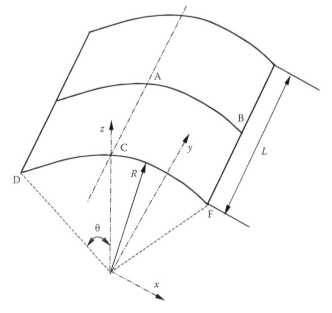

FIGURE 12.2
Barrel vault roof and the coordinate system.

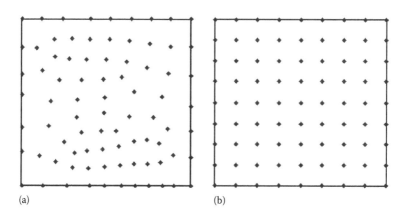

FIGURE 12.3
Nodal arrangement in parameter space.

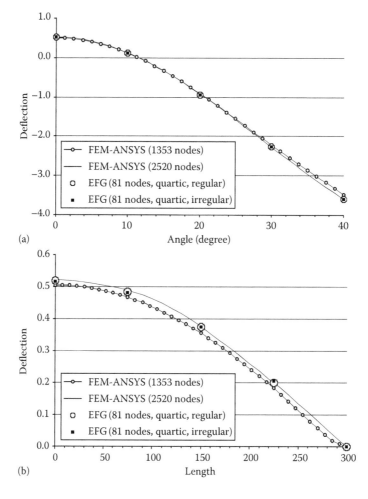

FIGURE 12.4
(a) Vertical deflection of the barrel vault roof along section AB (see Figure 12.2); and (b) vertical deflection along section AC.

vertical deflection along edges AB and AC, where the FEM solution is obtained using a general-purpose program, i.e., ANSYS[C]. The EFG results using 81 nodes agree well with the finite element results of ANSYS using 2520 nodes. The reason for the demand by FEM for a very fine element mesh may be the simplification of the geometry, because the field variable (deflection) does not vary drastically and, as shown in Figure 12.4, a coarse mesh should produce a good approximation. However, in ANSYS, the shell is modeled using flat shell elements. This simplification cannot model the mechanics coupling effects of bending forces and membrane forces, which is very significant for shell structures, unless a very fine mesh is used. In our EFG code, however, the geometry of the shell is modeled using MLS approximation, which very accurately represents the curvature of the shell surface, and hence produces very accurate results using very coarse nodes. This example clearly demonstrates the advantage of the meshfree method in modeling shell structures. Note also that under the gravity force, which is downward, the central point on the roof moves upward against the direction of body force. This fact provides very clear evidence of the importance of the coupling of bending force and membrane force. The existence of the membrane force causes the rooftop to rise.

The convergence of the EFG results is also investigated in detail, and the results are summarized in Figure 12.5, where the curves are normalized using the converged value of the vertical deflection at B, that is, −3.618. It can be seen that the performance of the EFG method compares well with the finite element solution of [9,10]. The results given by quartic polynomial basis are usually better than that by quadratic basis when nodal points are sufficiently dense. However, worse results are obtained when low nodal densities are employed. The reason could be again the approximation of the geometry of the shell. When the nodes are too coarse, even EFG will fail to approximate the geometry well.

12.1.10.1 *Lagrange Multiplier Method vs. Penalty Method*

The penalty method and Lagrange multipliers can both be used to enforce the essential boundary conditions. Here we discuss the advantages and disadvantages of these two

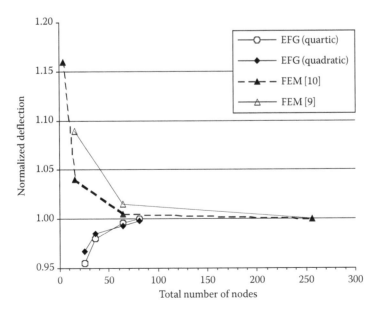

FIGURE 12.5
Convergence of results of vertical deflection at B.

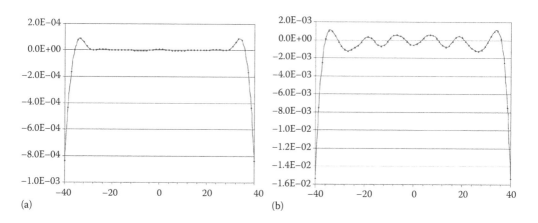

FIGURE 12.6
EFG solution of vertical deflection of the barrel vault roof along section DF. The vertical deflections are supposed to be zero. The numerical solution gives very small values. The largest errors were observed at corners. The Lagrange multiplier method is more accurate than the penalty method. (a) Lagrange multipliers method; and (b) penalty method.

methods using the example of barrel vault roof. Figure 12.6 shows the distribution of the vertical deflection along edge DF, where the deflection is supposed to be zero because a zero essential boundary condition was imposed in the EFG analysis. It can be seen from Figure 12.6 that both methods could not give exact results for the zero displacement. Very small errors are observed especially at the two corners. The Lagrange multipliers method is more accurate than that of the penalty method. In the penalty method, a penalty factor of 5000 times the value of the elastic modulus is used. Note again that the penalty method is much cheaper to implement.

12.1.11 Numerical Examples of Free Vibration of Thin Shells

The performance of the EFG method for free-vibration analysis is also evaluated. The results of several examples are presented in comparison with analytical solutions and other results.

Example 12.2: Free Vibration of a Clamped Cylindrical Shell Panel

A panel of cylindrical thin shallow shell, which has been investigated by Petyt [11], is also examined here to benchmark our EFG code for free-vibration analysis. The thin shell panel is clamped at all edges, and the natural frequencies are computed using the present EFG. The geometry and boundary conditions of the shell panel are shown in Figure 12.7 and the following parameters are used in the analysis:

Length, L (mm)	Radius, R (mm)	Thickness, h (mm)	Span Angle, θ	Young's Modulus, E (N/m^2)	Poisson's Ratio, ν	Mass Density, ρ (kg/m^3)
76.2	762	0.33	7.64°	6.8948×10^{10}	0.33	2657.3

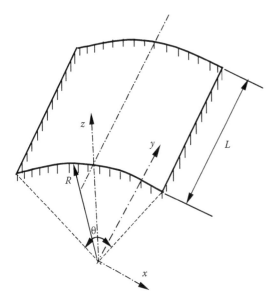

FIGURE 12.7
Clamped panel of cylindrical thin shell.

Regular nodes of different densities are used to investigate the convergence characteristics of natural frequencies. Results are given in terms of a frequency parameter defined as $\lambda = \sqrt{Q} = (\omega^2 \rho h R^4 / D)^{1/8}$, where $D = Eh^3/(12(1 - \nu^2))$, and are shown in Figure 12.8. As can be seen from Figure 12.8, very high convergence rates for the various vibration modes are obtained. The values of λ are also tabulated in Table 12.1 together with experimental results and results obtained by other different numerical methods [11]. In this table, ERR denotes the extended Raleigh–Ritz method, FET stands for the triangular FEM, and FER denotes rectangular FEM. It is found that the EFG results show good convergence and good agreement with other methods.

Example 12.3: Free Vibration of a Hyperbolical Shell

Free-vibration analysis of the clamp-free hyperbolical shell shown in Figure 12.9 is performed. The shell geometry is defined by the following equation of its meridian:

$$\left(\frac{R}{a}\right)^2 - \left(\frac{L-d}{b}\right)^2 = 1 \tag{12.53}$$

where b is a characteristic dimension of the shell, which can be calculated as $b = ad/\sqrt{R_2^2 - a^2}$. The following geometric and material properties are used in the analysis:

Length, L (m)	Height, H (m)	Height, d (m)	Radius, a (m)	Radius, b (m)	Young's Modulus, E (N/m^2)	Poisson's Ratio, ν	Mass Density, ρ (kg/m^3)
100.787	0.127	82.194	25.603	63.906	2.069×10^{10}	0.15	2405

A regular nodal arrangement in the axial and circumferential direction is used in the analysis. The results of the present study using different numbers of nodes are given in Table 12.2. Results obtained in [12] using a numerical integration technique and in [13] using an FEM are also listed in the same table for an easy comparison. Close agreements are again observed.

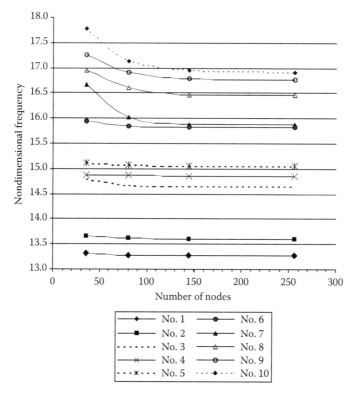

FIGURE 12.8
Convergence of results for natural frequency parameter λ of a clamped panel of a cylindrical shallow thin shell.

TABLE 12.1

Natural Frequency Parameter λ of a Clamped Panel of Cylindrical Thin Shell

Mode	Experimental Results	ERR	FET	FER	EFG			
					6×6	9×9	12×12	16×16
1	13.06	13.28	13.28	13.35	13.32	13.28	13.28	13.28
2	13.54	13.60	13.60	13.65	13.67	13.62	13.60	13.60
3	14.57	14.65	14.65	14.71	14.79	14.68	14.65	14.65
4	14.70	14.86	14.85	14.88	14.88	14.87	14.86	14.85
5	15.09	15.06	15.06	17.21	15.11	15.08	15.06	15.06
6	15.93	15.82	15.83	15.87	15.95	15.85	15.83	15.82
7	15.78/15.86	15.91	15.88	15.96	16.67	16.03	15.89	15.88
8	16.55	16.46	16.46	16.49	16.95	16.61	16.47	16.46
9	16.79	16.78	16.78	16.81	17.25	16.91	16.80	16.78
10	16.89	16.93	16.92	16.96	17.77	17.14	16.95	16.92

Note: ERR, extended Raleigh–Ritz method; FET, triangular finite element method; FER, rectangular finite element method.

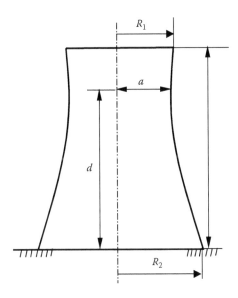

FIGURE 12.9
Geometry of a hyperbolical shell that is clamped at the bottom circular edge and free at the top circular edge.

TABLE 12.2

Natural Frequencies (Hz) of a Clamp-Free Hyperbolical Shell

			EFG	
Mode	Ref. [13]	Ref. [12]	12 × 16	18 × 24
1	1.0354	1.0348	1.0351	1.0325
2	1.1508	1.1467	1.1486	1.1450
3	1.1826	1.1808	1.1809	1.1780
4	1.3061	1.3015	1.3043	1.2998
5	1.3293	1.3231	1.3254	1.3223
6	1.3758	1.3749	1.3799	1.3753
7	1.4329	1.4293	1.4284	1.4259
8	1.4488	1.4475	1.4497	1.4470

Example 12.4: Free Vibration of a Cylindrical Shell

The natural frequencies of a cylindrical shell are examined using the present EFG code. The geometry of the shell is schematically drawn in Figure 12.10. The shell is clamped at one edge and free at the other. The following geometry and material properties are used:

Length, L (mm)	Thickness, h (mm)	Radius, R (mm)	Young's Modulus, E (N/m^2)	Poisson's Ratio, ν	Mass Density, ρ (kg/m^3)
226.8	1.021	106.1	2.069×10^{11}	0.3	7868

The nodes used in our EFG are regularly arranged in the axial and circumferential directions. The EFG results are tabulated in Table 12.3. The results obtained in [14] and [15] using high-precision finite element method (HPFEM) and standard FEM, respectively, are also listed in the same table for comparison. Again, good agreement is evident.

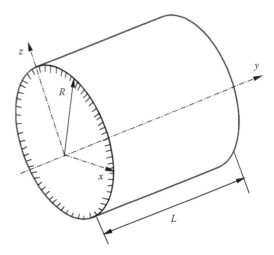

FIGURE 12.10
Geometry of a cylindrical shell.

TABLE 12.3

Natural Frequencies (Hz) of the Clamp-Free Cylinder

			FEG	
Mode	EFG [14]	HPFEM [15]	8×16	12×24
1	487	482	490	483
2	565	561	564	562
3	621	616	629	624
4	NR	NR	875	869
5	982	981	979	980

Note: NR, no results were given.

12.1.12 Numerical Examples of Forced Vibration of Thin Shells

To investigate the accuracy as well as the capability of the EFG method for forced vibration problems of thin shells, numerical examples for thin shells subjected to different transient excitations are presented, and the results obtained are compared with those of ordinary FEM and other numerical methods.

Example 12.5: Clamped Circular Plate Subject to an Impulsive Load

A code developed for shells should also be able to work for plates. The first test case is a circular plate subjected to a rectangular impulsive force. The parameters used in this calculation are given as follows:

Thickness, h (mm)	Radius, R (mm)	Young's Modulus, E (N/m²)	Poisson's Ratio, ν	Mass Density, ρ (kg/m³)
0.05	1.1	1000.0	0.3	0.229

The natural frequencies are first computed by the EFG method, and the results are compared with those obtained using the boundary element method (BEM) and FEM. We state without showing the results that they are in very good agreement.

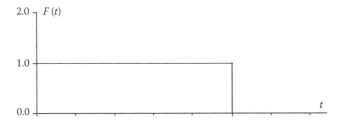

FIGURE 12.11
Transient force of a rectangular pulse applied to the circular plate.

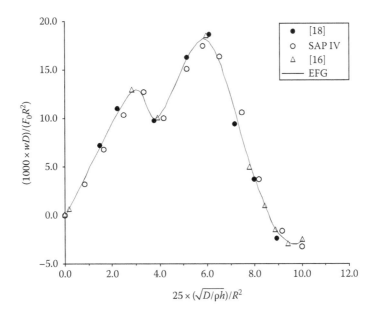

FIGURE 12.12
The time history of the deflection at the center of a clamped circular plate subjected to a rectangular impulsive excitation.

The circular plate is loaded by a concentrated vertical impulsive force $F(t)$ at its center. The magnitude of the force is 1 N, and the duration of pulse is $t_0 = 0.121$ s as shown in Figure 12.11. In time stepping, the time increment used is $\Delta t = 0.001$ s. Figure 12.12 shows the history of the vertical deflection at the center of the plate obtained by the presented method, the analytical solution [16], SAP IV [17], and the domain-BEM [18]. It is observed that the results obtained using the present EFG code agree well with all these methods including the analytical results in [16].

Example 12.6: Clamped Cylindrical Shell Subject to a Sine Load

The transient response of a clamped cylindrical shell as shown in Figure 12.10 is investigated. The following geometry and material properties are used:

Length, L (mm)	Thickness, h (mm)	Radius, R (mm)	Young's Modulus, E (N/m²)	Poisson's Ratio, ν	Mass Density, ρ (kg/m³)
600	3.0	300.0	2.1×10^{11}	0.3	7868

Figure 12.13 shows the history of the transient excitation of half a sine function of time. The excitation is at the center of the meridian ($y = L/2$, $z = R$). The force is expressed as

$$F = F_0 \sin(1000t) \tag{12.54}$$

where $F_0 = 1000.0$ N. Good agreement is observed. In performing the time marching, the time step $\Delta t = 2.5e^{-5}$ s is used, which is almost $1/35$ of the fundamental period of the cylinder. Regular nodes (12×16) are arranged in the axial and circumferential directions. Figure 12.14 shows the transient response of a clamped cylinder subjected to a transient excitation of half a sine function of time. The results are obtained using the present EFG code and FEM.

Figure 12.15 shows the transient response for the cylindrical shell, but the external excitation force becomes $F = F_0 \sin(2000t)$. Again, very good agreement between the results of EFG and FEM is obtained.

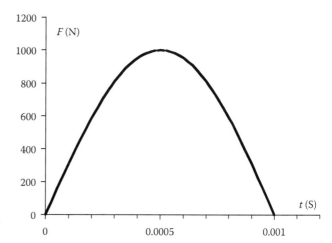

FIGURE 12.13
Time history of the external force excitation of half a sine function of time.

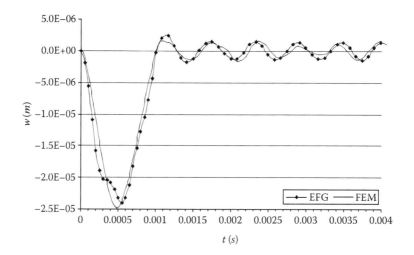

FIGURE 12.14
Transient response of the vertical displacement at the central point in the meridian of the cylindrical shell ($y = L/2$, $z = R$).

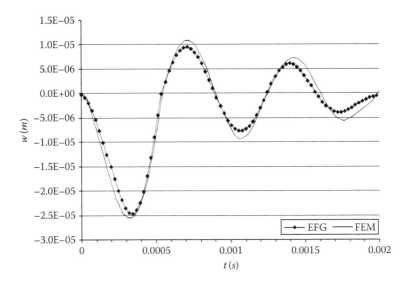

FIGURE 12.15
Same as Figure 12.14, but $F = F_0 \sin(2000t)$.

Example 12.7: Clamped Spherical Shell Subject to a Sine Curve Load

A spherical cap of thin shell with a central angle of $\theta = 120°$ shown in Figure 12.16 is investigated. The shell is subjected to an excitation of a force of half-a-circle sine function of time at the apex. The time history of the loading is given by Equation 12.54, but $F_0 = 200.0$ N. This spherical cap is clamped on the circular boundary at the bottom. The following geometry and material properties are used in the computation:

Thickness, h (mm)	Radius, R (mm)	Young's Modulus, E (N/m²)	Poisson's Ratio, v	Mass Density, ρ (kg/m³)
10	900.0	$2.0e^{11}$	0.3	7800

The time step used is $\Delta t = 5.0e^{-5}$ s, which is almost 1/25 of the fundamental period of the spherical cap; 185 nodes are used in the calculation. Figure 12.17 shows the history of the vertical displacement response at the apex of the spherical cap obtained by both the present EFG method and FEM. Very good agreement is obtained.

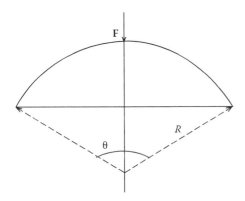

FIGURE 12.16
Cross section of the spherical cap of a thin shell.

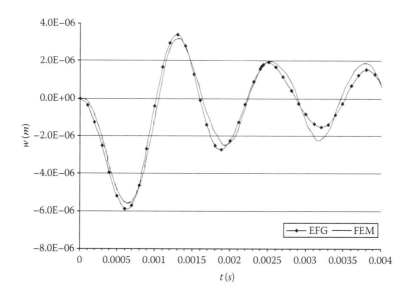

FIGURE 12.17
Transient response of the vertical displacement at the apex of the spherical cap.

12.1.13 Some Remarks

The EFG method has been formulated for static, free-vibration, and forced-vibration analysis of spatial thin shell structures. In the EFG method, the MLS technique is used for approximations of both the surface geometry of the shell and the field variables. The present EFG results are benchmarked with a number of examples by comparison with the results obtained by other methods. It is found that the EFG method offers (1) accurate geometry representation and (2) fast convergence.

12.2 EFG Method for Thick Shells

The Kirchhoff–Love shell theory works only for very thin shells. The reason is similar to that for the theories for plates (see Chapter 11). For thick shells, formulations need to be based on thick shell theories. The formulation for thick shells in this section is based on the geometrically exact theory of flexible shells proposed by Simo et al. [9]. We collect the necessary equations in the following sections, but refer the reader to [9] for details.

12.2.1 Fundamental Relations

The Gauss intrinsic coordinates are used to describe the configuration of the shell as described in the previous sections. Making use of the definition of spatial tensors, the corresponding linearized strain measures are defined relative to the dual spatial surface basis as

$$\varepsilon_{\alpha\beta} = \frac{1}{2} \left(\boldsymbol{\varphi}^0_{,\alpha} \cdot \mathbf{u}_{,\beta} + \boldsymbol{\varphi}^0_{,\beta} \cdot \mathbf{u}_{,\alpha} \right) \tag{12.55}$$

$$\gamma_\alpha = \left(\boldsymbol{\varphi}^0_{,\alpha} \cdot \Delta \mathbf{t} + \mathbf{u}_{,\alpha} \cdot \mathbf{t}^0\right) \tag{12.56}$$

$$\rho_{\alpha\beta} = \frac{1}{2}\left(\boldsymbol{\varphi}^0_{,\alpha} \cdot \Delta \mathbf{t}_{,\beta} + \boldsymbol{\varphi}^0_{,\beta} \cdot \Delta \mathbf{t}_{,\alpha} + \mathbf{u}_{,\alpha} \cdot \mathbf{t}^0_{,\beta} + \mathbf{u}_{,\beta} \cdot \mathbf{t}^0_{,\alpha}\right) \tag{12.57}$$

Here, $\Delta \mathbf{t}$ is the incremental spatial rotation defined by

$$\Delta \mathbf{t} = \bar{\boldsymbol{\Lambda}} \cdot \Delta \mathbf{T} \tag{12.58}$$

where

$$\Delta \mathbf{T} = \Delta T^1 \mathbf{E}_1 + \Delta T^2 \mathbf{E}_2 \tag{12.59}$$

and $\bar{\boldsymbol{\Lambda}}$ is the (3×2) matrix given by

$$\bar{\boldsymbol{\Lambda}} = [t_1 \ t_2] = \begin{bmatrix} \Lambda_{11} & \Lambda_{12} \\ \Lambda_{21} & \Lambda_{22} \\ \Lambda_{31} & \Lambda_{32} \end{bmatrix} \tag{12.60}$$

which can be obtained by deleting the third column of $\bar{\boldsymbol{\Lambda}}$, where $\boldsymbol{\Lambda}: \mathbb{A} \subset \mathbb{R}^2 \to S_E^2$ is the orthogonal transformation, and can be expressed as

$$\boldsymbol{\Lambda} = (\mathbf{t} \cdot \mathbf{E})\mathbf{1} + [\widehat{\mathbf{E} \times \mathbf{t}}] + \frac{1}{1 + \mathbf{t} \cdot \mathbf{E}}(\mathbf{E} \times \mathbf{t}) \otimes (\mathbf{E} \times \mathbf{t}) \tag{12.61}$$

The constitutive relations for the effective membrane $\tilde{\mathbf{n}}$ and for the stress couple resultant $\tilde{\mathbf{m}}$ have the form of Equations 12.23 and 12.24. The constitutive relations for the effective shear stress resultants $\tilde{\mathbf{q}}$ can be written as

$$\begin{Bmatrix} \tilde{q}^1 \\ \tilde{q}^2 \end{Bmatrix} = \kappa G h \begin{Bmatrix} \gamma^1 \\ \gamma^2 \end{Bmatrix} \tag{12.62}$$

Here k is the shear reduction coefficient and G is the shear modulus.

In calculating the effective stress couple resultant $\tilde{\mathbf{m}}$, Equations 12.24 and 12.31 should be used. In calculating the effective membrane $\tilde{\mathbf{n}}$, Equation 12.30 and the following equation should be used.

$$\tilde{n}^{\beta\alpha} = n^{\beta\alpha} - \lambda^\beta_\mu \tilde{m}^{\alpha\mu} \tag{12.63}$$

In calculating the effective shear resultant forces, we use

$$\tilde{\mathbf{q}} = \tilde{q}^\alpha \boldsymbol{\varphi}_{,\alpha} \tag{12.64}$$

where $\tilde{q}^\alpha$ is defined as

$$\tilde{q}^\alpha = q^\alpha - \lambda^3_\mu \tilde{m}^{\alpha\mu} \tag{12.65}$$

In calculating $t_{,\alpha}$, we use

$$\mathbf{t}_{,\alpha} = \lambda_\alpha^\beta \boldsymbol{\varphi}_{,\beta} + \lambda_\alpha^3 \mathbf{t} \tag{12.66}$$

The above equation shows that the difference between thick shell theory and thin shell theory is that the shear effects have been taken into account in the thick shell theory.

12.2.2 Principle of Virtual Work

The weak form of the governing equation for the thick shells under static load can be written as

$$W_{\text{Sta}}(\delta \mathbf{x}) = \int_{\mathbb{A}} \left[\tilde{n}^{\beta\alpha} \cdot \delta \varepsilon_{\beta\alpha} + \tilde{m}^{\beta\alpha} \cdot \delta \kappa_{\beta\alpha} + \tilde{q}^\alpha \delta \gamma_\alpha \right] d\mathbb{A} - W_{\text{ext}}(\delta \mathbf{x}) \tag{12.67}$$

Here $d\mathbb{A} = \bar{j} d\xi^1 d\xi^2$ is the current surface measure. W_{ext} is the virtual work of the external loading given by

$$W_{\text{ext}} = \int_{\mathbb{A}} [\bar{\mathbf{n}} \cdot \delta \boldsymbol{\varphi} + \bar{\bar{\mathbf{m}}} \cdot \delta \mathbf{t}] d\mathbb{A} + \int_{\Gamma_n} \bar{\bar{\mathbf{n}}} \cdot \delta \boldsymbol{\varphi} \bar{j} d\Gamma + \int_{\Gamma_m} \bar{\bar{\mathbf{m}}} \delta t_j d\Gamma \tag{12.68}$$

where
  $\bar{\mathbf{n}}$ is the applied resultant force per unit length
  $\bar{\bar{\mathbf{m}}}$ is the applied direct couple per unit length
  $\bar{\bar{\mathbf{n}}}$ and $\bar{\bar{\mathbf{m}}}$ are the prescribed resultant force and the prescribed director couple on their
    corresponding boundaries Γ_n and Γ_m, respectively

The MLS shape functions are then used to approximate the displacement fields. Substituting Equation 12.1 into the weak forms leads to a set of global system equations. The procedure and the treatment of essential boundaries are the same as the previous section.

We need different ways to handle the essential boundary condition for different problems, whenever MLS approximation is employed.

12.2.3 Numerical Examples

Example 12.8: Static Deflection of a Barrel Vault Roof under Gravity Force

Example 12.1 is reexamined using the EFG formulation for thick shells. The barrel vault roof is shown in Figure 12.2. All the parameters and conditions are exactly the same as those in Example 12.1. In constructing the MLS shape function, both quadratic and quartic basis functions are used. Figure 12.18 shows the results of convergence of vertical displacement at point B in the barrel vault roof. Figure 12.18 shows that the convergence by the present analysis is excellent in comparison with the results of using high-performance FEM [9] and EFG for thin shells [2]. It can also be seen that the convergent rate for quartic basis always exceeds that for quadratic basis. This was not very clear from the results using thin shell theory (see Figure 12.5). The result for quartic basis functions approaches the exact result from below with the increase of EFG nodes, while that for quadratic basis functions oscillates along the exact result with the increase of EFG nodes.

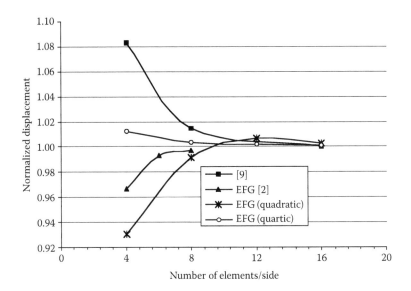

FIGURE 12.18
Convergence of vertical displacement at point B in the barrel vault roof.

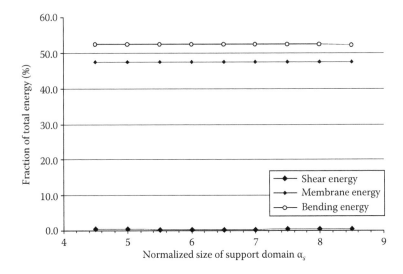

FIGURE 12.19
Variation of membrane, shear, and bending energy with respect to size of the support domain.

Figure 12.19 shows the variation of the fractions of shear, membrane, and bend energies in total energy, with respect to the size of the support domain. The computation is performed using quartic basis function. It can be seen that the membrane and bending energies are stable with the increase of the support domain size; the difference between the membrane and bending energies are no more than 4.0%. The shear energy is very small and approaches zero with the increase of EFG nodes, which means that the shear stress plays a very minor role in deformation of the vault roof.

Figures 12.20 and 12.21 show the distribution of the vertical deflection in the roof along the edges AB and AC, respectively. The central point A moves upward, that is, in the reverse direction of the body force, for the effects of the membrane forces. It can be seen that thousands of nodes are needed to obtain the converged solution using a general-purpose program, for example, ANSYS, for the reason mentioned in Example 12.1. Fewer nodes are needed for the present EFG method.

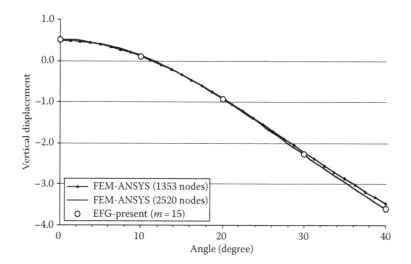

FIGURE 12.20
Vertical displacement in the barrel vault roof along AB.

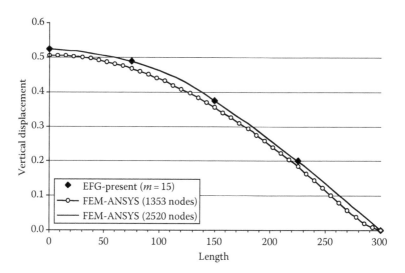

FIGURE 12.21
Vertical displacement in the barrel vault roof along AC.

Example 12.9: Pinched Cylindrical Shell

The deformation of the cylindrical shell shown in Figure 12.22 is analyzed using the present EFG method. The shell is loaded by a pair of centrally located and diametrically opposed concentrated forces. The cylindrical boundaries are supported by a rigid diaphragm that allows displacement in the axial direction and rotation about the tangent to the shell boundary. The parameters for this problem are given in the following table:

Length, L (mm)	Thickness, h (mm)	Radius, R (mm)	Young's Modulus, E (N/m^2)	Poisson's Ratio, ν
600	3.0	300.0	3.0×10^6	0.3

This problem is one of the most critical tests for numerical methods for both inextensional bending and complex membrane states of stress. Because of its double symmetry, only 1/8 of the cylinder has been modeled. The results are shown in Figure 12.23, which are normalized by the value of 1.8248e^{-5}, which is the convergent numerical solution of the radial displacement magnitude at the loaded points. The convergence of the present EFG method is excellent in comparison with the results by FEM [9] and by EFG for thin shells [2]. It is seen that the EFG results reported in [2] converge faster than the present EFG code.

The effects of the dimension of the support domain on the radial displacement at the loading point are also investigated, and the results are shown in Figure 12.24. The shortest nodal distance between nodes is used to normalize the size of the support domain. The fluctuation of the displacement is visible but it is less than 3% when the support domain is larger than 3.8. Effects of the dimension of the support domain on the energy components (bending, membrane, and shear energies) are also investigated, and the results are summarized in Figure 12.25. The energies show a trend of fluctuation similar to that of the displacement. Figure 12.26 shows the convergence of membrane (M), shear (S), and bending (B) energy fractions in the total energy. The results are calculated using both quartic and quadratic polynomial basis functions. As can be seen, for quartic polynomial basis, the shear energy converges to a value of less than 4.0% as the field nodes increase, which means shear stress plays but a small role in the deformation. The membrane energy

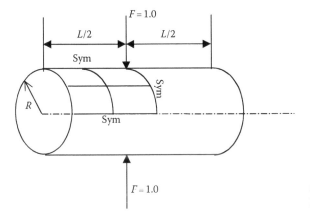

FIGURE 12.22
Pinched cylindrical shell.

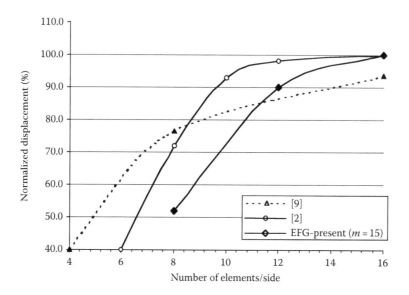

FIGURE 12.23
Convergence of the results of the radial displacement at the center of the cylindrical shell.

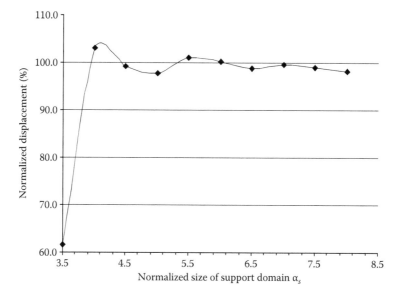

FIGURE 12.24
Effects of the dimension of the support domain on the radial displacement at the loading point.

fraction approaches approximately 40.0% which gives very satisfactory results in comparison with the results using nine-noded gamma FEM by Belytschko et al. [10]. For the quadratic polynomial basis, the membrane fraction approaches 40.0%, the same as that for the quartic polynomial. However, the shear energy increases with the increase of field nodes. It can also be observed that for coarse field nodes, membrane locking is dominant while for the fine field nodes, shear locking becomes more important.

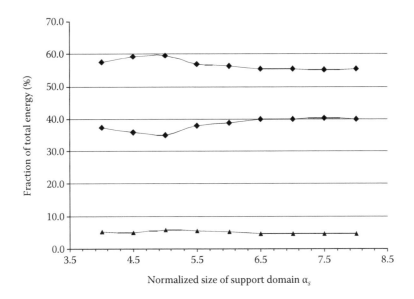

FIGURE 12.25
Effects of the dimension of the support domain on the energy components (from top to bottom: bending, membrane, and shear energy).

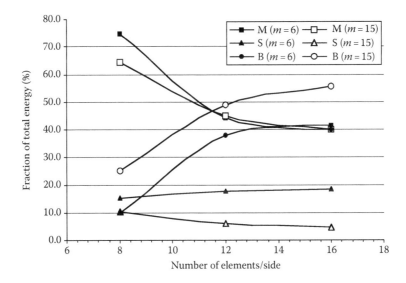

FIGURE 12.26
Convergence of membrane (M), shear (S), and bending (B) energies.

Example 12.10: Pinched Hemispherical Shell

In this example, a pinched hemispherical shell shown in Figure 12.27 is analyzed using the EFG method. The shell is pinched by two pairs of opposed radial point loads of magnitude $P = 2.0$. The bottom circumferential edge of the hemisphere is free. The material parameters are $E = 6.825 \times 10^7$ N/m^2 and $\nu = 0.3$. The sphere radius is $R = 10.0$ and the thickness $t = 0.04$.

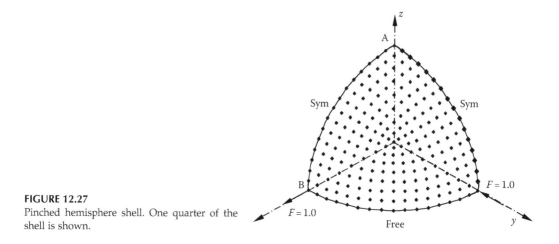

FIGURE 12.27
Pinched hemisphere shell. One quarter of the shell is shown.

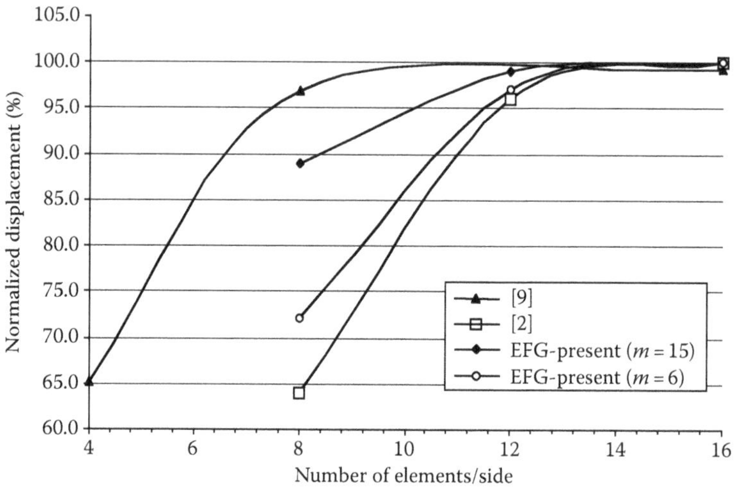

FIGURE 12.28
Convergence of the radial displacement at the loading point on the hemisphere shell.

This problem is also a challenging benchmark problem to check whether the formulation of the shell can represent inextensional modes of deformation, as it exhibits almost no membrane strains. This problem is a less critical test with regard to inextensional bending, compared to the pinched cylinder problem. However, it is a very useful problem for checking the ability of the present EFG method to handle rigid body rotations normal to the shell surface. Large sections of this shell rotate almost as rigid bodies in response to this load, and the ability to accurately model rigid body motion is essential for any numerical method.

Due to symmetry, only a quarter of the hemisphere is modeled. The results are shown in Figure 12.28. The radial displacements in the direction of loads at loaded points are the same and are found analytically to be 0.0924. This value is used to normalize the numerical results presented in Figure 12.28. The convergence is even better than that of the high-performance results using mixed formulation [9]. Among the EFG solutions, the present

EFG method also gives better accuracy than that of the thin shell formulation [2]. The results obtained using the quartic polynomial basis functions are better than those using quadratic polynomial basis functions. However, poor accuracy is obtained when very coarse nodes are used in EFG compared with FEM using the same density of nodes. Figure 12.29 shows the effects of the dimension of the support domain on the radial displacement at the loading point on the hemispherical shell. The present EFG results using quartic basis functions do not depend on the size of the support domain and gives an almost exact solution, in the range of our investigation. The results obtained using quadratic basis functions are stable and accurate only in a narrow range ($\alpha_s = 4.5$–6.5). Figure 12.30 shows the effects of the dimension of the support domain on membrane (M)

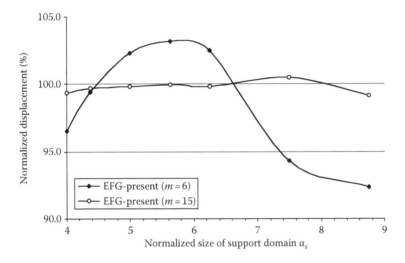

FIGURE 12.29
Effects of the dimension of the support domain on the radial displacement at the loading point on the hemisphere shell.

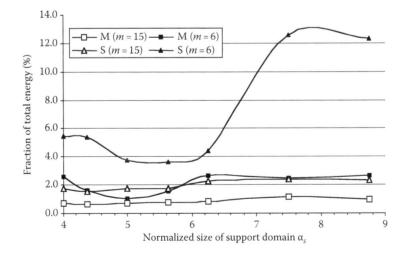

FIGURE 12.30
Effects of the dimension of the support domain on membrane (M) and shear (S) energies in the hemisphere shell.

and shear (S) energies in the hemisphere shell. Figure 12.31 shows the same effects but for the bending (B) energy in the hemisphere shell. It can be seen that both the shear and the membrane energy are very small when the quartic basis functions are used. Although the shear energy obtained using the quadratic basis functions is very small, the membrane energy increases with the increase of the support domain. This means that the membrane energy is over predicted with the increase of field nodes, which is an indication of membrane locking.

Figures 12.32 and 12.33 show the fractions of shear, membrane, and bending energies in the total strain energy calculated using both quartic and quadratic polynomial basis

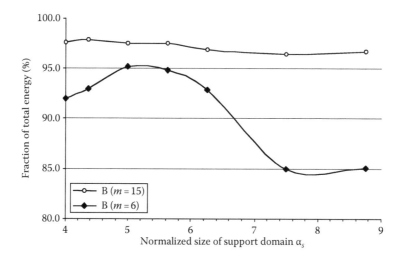

FIGURE 12.31
Effects of the dimension of the support domain on bending (B) energy in the hemisphere shell.

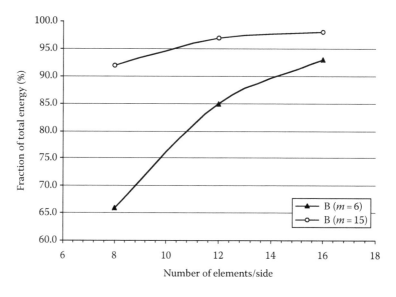

FIGURE 12.32
Convergence of bending energy.

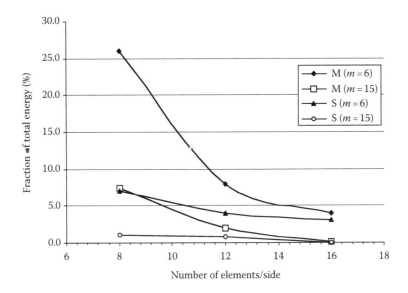

FIGURE 12.33
Convergence of membrane and shear energies.

functions. It is found that when quartic polynomial basis functions are used, both membrane and shear energy tend to zero as the field nodes increase. This implies that the shear stress and membrane stress play trivial roles in the deformation of the shell, and the bending energy is dominant. It is also found that membrane locking is dominant for coarse field nodes, but is quickly eliminated with the increase of field nodes.

12.2.4 Remarks

The EFG method has been extended to analyze thick spatial shells based on the stress-resultant shell theory, which has 5 DOF assigned to every point of the shell.

For thick shells, the formulation allows transverse shear strain and results in the Mindlin plate when the surface becomes flat and the membrane state is negligible. To avoid shear locking and membrane locking, quartic basis function is recommended for MLS approximation. Using high-order basis functions is necessary to avoid shear locking. A good alternative is to use different orders of shape function for the deflection and rotation, as discussed in Chapter 11 for plates (see Example 11.16).

12.3 ES-PIM for Thick and Thin Shells

In Chapters 8, 10, and 11 we have presented PIMs for solids, beams, and plates based on triangular mesh. It was found that due to the use of the weakened-weak (W^2) formulation based on $\mathbb{G}$ space theory, the consistency on the assumed displacement functions has been further reduced. Therefore, for second-order PDEs, we can allow discontinuous functions, as shown in Chapter 8 for solids and Chapter 11 for plates. For fourth-order PDEs, we can use function of only first-order consistency, as shown in Chapter 10 for beams.

We also found that for 2D solids, the ES-PIM with linear shape function is one of the best performers in all the meshfree methods examined in terms of accuracy, efficiency, simplicity, robustness, and applicability to problems of complicated geometries. An intensive comparison study has shown that its efficiency is about eight times in displacement error norm and 40 times in energy error norm that of the linear FEM using the same mesh (see Chapter 8).

This section formulates an ES-PIM for shells based on the FSDT and the DSG scheme, and demonstrates that the ES-PIM works well for both thin and thick shells, because it is free from shear locking. Again, we use only triangular background cells aiming for simplicity, robustness, adaptive to complicated geometry.

12.3.1 Overall Strategy

In this section, we consider shells undergoing all three possible deformations in 3D space: in-plane stretching (membrane), off-plane bending and shearing deformation. We are aware of that a 3D shells structure can be very complicated in geometry, and geometry modeling can be a challenge. The most practical approach to deal with this domain complexity is, in the opinion of the author, triangulation with fine mesh. Our formulation is, therefore, based on a mesh of flat triangular cells. The curvature of the shell is achieved via the orientation changes of these flat cells. For each cell, some of the basic techniques and formulation in the FEM can be directly utilized. Edge-based smoothing is then performed to these normals of the two neighboring cells shearing the edge, in addition to these three deformations of in-plane stretching, off-plane bending, and shearing. This approach is straightforward, well-managed for the necessary issues, and hence ensuring stable and convergent solutions. Because our ES-PIM will be very efficient, we can take effort to use very fine mesh to accurately represent the original geometry of the shell for accurate solutions. We can even perform adaptive analysis for solutions of desired accuracy.

For each cell, the FSDT for plate will be used. To overcome shear locking when the shell gets very thin, we divide the strain energy into three terms: *Membrane* energy, *bending* energy, and *shear* energy, and the DSG scheme [21] is used to evaluate the shear strains.

In this section, we discuss only the linear ES-PIM, which is also termed as ES-FEM and presented in [19].

12.3.2 ES-PIM Formulation

12.3.2.1 Edge-Based Smoothing Model

In this section, an edge-based strain smoothing technique is now introduced for shells. An ES-PIM model for the shell is established in the following steps.

First, the problem domain for a shell is divided into N_c flat triangular cells with N_n nodes and N_{cg} edges using triangulation defined in Section 1.7.2. Figure 12.34 shows a representative flat triangular cell. We consider homogenous shells, and hence all the field variables can be defined on the mid-surface (that is also the neutral surface) of the shell. The three nodes of the cell are located at the three vertices of the mid-surface of the shell element. A local coordinate system for the element can be defined as

- The origin at node 1
- x'-Axis along the side 1–2
- z'-Axis parallel to the normal of the mid-surface
- y'-Axis perpendicular to the x'–z' plane

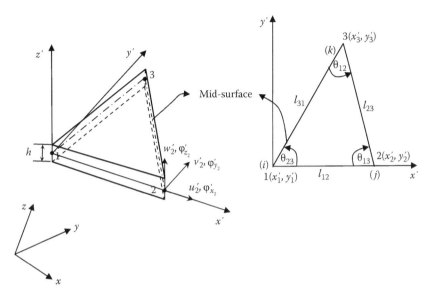

FIGURE 12.34
A typical three-node flat triangular cell on shell with 6 DOFs per node.

The 6 DOFs at each node are defined as

- u': the displacement in the x'-direction
- v': the displacement in the y'-direction
- w': the displacement in the z'-direction
- φ'_x: the rotation about the x'-axis
- φ'_y: the rotation about the y'-axis
- φ'_z: the rotation about the z'-axis

Second, based on these triangular cells, a set of N_s smoothing domains Ω_k^s bounded by Γ_k^s are created associated with the edges of the triangular cells, as shown in Figure 8.80. In this edge-based case, we have $N_s = N_{cg}$. For each edge, the smoothing domain is formed by sequentially connecting two end points of the edge and centroids of the neighboring triangles shearing the edge. For an interior edge, it is shared by two cells, and for an edge on the problem boundary, it is shared only by one cell, as shown in Figure 12.35.

12.3.2.2 Coordinate Transformation

The DOFs with respect to the global coordination system are denoted as: $u, v, w, \varphi_x, \varphi_y,$ and φ_z. The local DOFs of nodal displacements and rotations relates to these global DOFs via a coordinate transform matrix $\mathbf{T}$:

$$\mathbf{d}' = \mathbf{T}\mathbf{d} \tag{12.69}$$

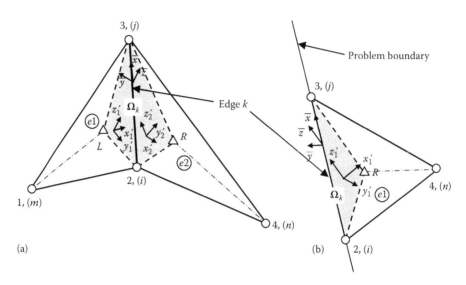

FIGURE 12.35
Edge-based smoothing domain for edge k. (a) Inside the problem domain; and (b) on the problem boundary.

where

$$\mathbf{d}' = \begin{bmatrix} \mathbf{d}'_1 \\ \mathbf{d}'_2 \\ \mathbf{d}'_3 \end{bmatrix} \begin{matrix} \text{for node 1} \\ \text{for node 2,} \\ \text{for node 3} \end{matrix} \quad \mathbf{d} = \begin{bmatrix} \mathbf{d}_1 \\ \mathbf{d}_2 \\ \mathbf{d}_3 \end{bmatrix} \begin{matrix} \text{for node 1} \\ \text{for node 2} \\ \text{for node 3} \end{matrix} \quad (12.70)$$

Each of the vector for nodal DOFs can be written in details as

$$\begin{aligned} \mathbf{d}'_i &= \begin{bmatrix} u'_i & v'_i & w'_i & \varphi'_{xi} & \varphi'_{yi} & \varphi'_{zi} \end{bmatrix}^{\mathrm{T}} \\ \mathbf{d}_i &= \begin{bmatrix} u_i & v_i & w_i & \varphi_{xi} & \varphi_{yi} & \varphi_{zi} \end{bmatrix}^{\mathrm{T}} \end{aligned} \quad i = 1, 2, 3 \quad (12.71)$$

The transformation matrix $\mathbf{T}$ is defined as (see, e.g., [20])

$$\mathbf{T} = \mathrm{diag} \begin{bmatrix} \underbrace{\mathbf{T}_3}_{\substack{\text{translation} \\ \text{node 1}}} , \underbrace{\mathbf{T}_3}_{\substack{\text{rotation} \\ \text{node 1}}} , \underbrace{\mathbf{T}_3}_{\substack{\text{translation} \\ \text{node 2}}} , \underbrace{\mathbf{T}_3}_{\substack{\text{rotation} \\ \text{node 2}}} , \underbrace{\mathbf{T}_3}_{\substack{\text{translation} \\ \text{node 3}}} , \underbrace{\mathbf{T}_3}_{\substack{\text{rotation} \\ \text{node 3}}} \end{bmatrix} \quad (12.72)$$

where $\mathbf{T}_3$ is a 3 by 3 submatrix given as

$$\mathbf{T}_3 = \begin{bmatrix} c_{x'x} & c_{x'y} & c_{x'z} \\ c_{y'x} & c_{y'y} & c_{y'z} \\ c_{z'x} & c_{z'y} & c_{z'z} \end{bmatrix} \quad (12.73)$$

where $c_{x'x}$ is the direction cosines of the angle between the x'- and x-axes, and the similar is applied to the other entries in $\mathbf{T}_3$.

In the local coordinate system, the coordinates for these three nodes 1, 2, and 3 are denoted as (x'_1, y'_1), (x'_2, y'_2), and (x'_3, y'_3), respectively, and these coordinate values can be found as

$$x_1' = y_1' = 0$$
$$x_2' = l_{12}, \quad y_2' = 0$$
$$x_3' = l_{31} \cos \theta_{23}, \quad y_3' = l_{31} \sin \theta_{23}$$

(12.74)

The lengths of each edge of the triangle can be computed using

$$l_{ij} = |\mathbf{X}_j - \mathbf{X}_i|$$

(12.75)

where

$$\mathbf{X}_i = (x_i, y_i), \quad i = 1, 2, 3$$

(12.76)

is the coordinate in the global coordinate system. In Equation 12.74, θ_{ij} is the inner angle of the triangle shown in Figure 12.34, and is computed using

$$\cos \theta_{ij} = \frac{(\mathbf{X}_i - \mathbf{X}_k) \cdot (\mathbf{X}_j - \mathbf{X}_k)}{|\mathbf{X}_i - \mathbf{X}_k||\mathbf{X}_j - \mathbf{X}_k|},$$

(12.77)

The subscript i varies from one to three, j and k are determined by the cyclic permutation in the order of i, j, and k.

12.3.2.3 Interpolation of Field Variables

Using interpolation in the local coordinate system, we have

$$\underbrace{\begin{Bmatrix} u'(\mathbf{x}') \\ v'(\mathbf{x}') \\ w'(\mathbf{x}') \\ \varphi_x'(\mathbf{x}') \\ \varphi_y'(\mathbf{x}') \\ \varphi_z'(\mathbf{x}') \end{Bmatrix}}_{\mathbf{u}'(\mathbf{x}')} = \begin{bmatrix} \mathbf{\Psi}_{1i}^b(\mathbf{x}') & \mathbf{\Psi}_2^b(\mathbf{x}') & \mathbf{\Psi}_3^b(\mathbf{x}') \end{bmatrix} \begin{bmatrix} \mathbf{d}_1' \\ \mathbf{d}_2' \\ \mathbf{d}_3' \end{bmatrix}$$

(12.78)

where $\mathbf{\Psi}_i^b(\mathbf{x}')$ is a diagonal matrix of shape functions. In this section, we use linear shape functions created by T3-scheme, which gives the same linear FEM shape function using three-node element (the base model), and $\mathbf{\Psi}_i^b(\mathbf{x}')$ can be written as

$$\mathbf{\Psi}_i^b(\mathbf{x}') = \mathrm{diag}[\phi_i^b(\mathbf{x}'), \phi_i^b(\mathbf{x}'), \phi_i^b(\mathbf{x}'), \phi_i^b(\mathbf{x}'), \phi_i^b(\mathbf{x}'), \phi_i^b(\mathbf{x}')]$$

(12.79)

in which $\phi_i^b(\mathbf{x}')$ is the shape function for node i is

$$\phi_i^b(\mathbf{x}') = a_i + b_i x' + c_i y'$$
$$a_i = \frac{1}{2A_e}(x_j' y_k' - x_k' y_j'), \quad b_i = \frac{1}{2A_e}(y_j' - y_k'), \quad c_i = \frac{1}{2A_e}(x_k' - x_j')$$

(12.80)

where A_e is the area of the triangular cell. Here the cyclic permutation also applies to i, j, and k.

In the local coordinate, the shell can be treated as a combination of a plate and a 2D solid. Using the FSDT for plates (see Chapter 11), the strains in the local coordinate system can be given as

$$\boldsymbol{\varepsilon}' = \begin{Bmatrix} \varepsilon'_{xx} \\ \varepsilon'_{yy} \\ \gamma'_{xy} \\ \gamma'_{xz} \\ \gamma'_{yz} \end{Bmatrix}_{5\times 1} = \underbrace{\begin{Bmatrix} \boldsymbol{\varepsilon}'_m \\ \mathbf{0}_{2\times 1} \end{Bmatrix}}_{\substack{\text{2D solid} \\ \text{membrane}}} + z \underbrace{\begin{Bmatrix} \boldsymbol{\varepsilon}'_B \\ \mathbf{0}_{2\times 1} \end{Bmatrix}}_{\substack{\text{plate} \\ \text{bending}}} + \underbrace{\begin{Bmatrix} \mathbf{0}_{3\times 1} \\ \boldsymbol{\gamma}_S \end{Bmatrix}}_{\substack{\text{plate} \\ \text{shear}}} \tag{12.81}$$

in which

$$\boldsymbol{\varepsilon}'_m = \begin{Bmatrix} \dfrac{\partial u'}{\partial x'} \\[2mm] \dfrac{\partial v'}{\partial y'} \\[2mm] \dfrac{\partial u'}{\partial y'} + \dfrac{\partial v'}{\partial x'} \end{Bmatrix}, \quad \boldsymbol{\varepsilon}'_b = \begin{Bmatrix} -\dfrac{\partial \varphi'_y}{\partial x'} \\[2mm] \dfrac{\partial \varphi'_x}{\partial y'} \\[2mm] \dfrac{\partial \varphi'_x}{\partial y'} - \dfrac{\partial \varphi'_y}{\partial x'} \end{Bmatrix}, \quad \boldsymbol{\varepsilon}'_s = \begin{Bmatrix} \gamma'_{xz} \\ \gamma'_{yz} \end{Bmatrix} \tag{12.82}$$

Substituting Equation 12.78 into Equation 12.82, we obtain

$$\boldsymbol{\varepsilon}'_m = \mathbf{B}'_m \mathbf{d}' = [\mathbf{B}'_{m1}, \mathbf{B}'_{m2}, \mathbf{B}'_{m3}] \begin{bmatrix} \mathbf{d}'_1 \\ \mathbf{d}'_2 \\ \mathbf{d}'_3 \end{bmatrix} \tag{12.83}$$

$$\boldsymbol{\varepsilon}'_B = \mathbf{B}'_B \mathbf{d}' = [\mathbf{B}'_{B1}, \mathbf{B}'_{B2}, \mathbf{B}'_{B3}] \begin{bmatrix} \mathbf{d}'_1 \\ \mathbf{d}'_2 \\ \mathbf{d}'_3 \end{bmatrix} \tag{12.84}$$

in which

$$\mathbf{B}'_{mi} = \begin{bmatrix} \phi^b_{i,x'} & 0 & 0 & 0 & 0 & 0 \\ 0 & \phi^b_{i,y'} & 0 & 0 & 0 & 0 \\ \phi^b_{i,y'} & \phi^b_{i,x'} & 0 & 0 & 0 & 0 \end{bmatrix}, \quad i = 1, 2, 3 \tag{12.85}$$

$$\mathbf{B}'_{Bi} = \begin{bmatrix} 0 & 0 & 0 & 0 & \phi^b_{i,x'} & 0 \\ 0 & 0 & 0 & -\phi^b_{i,y'} & 0 & 0 \\ 0 & 0 & 0 & -\phi^b_{i,x'} & \phi^b_{i,y'} & 0 \end{bmatrix}, \quad i = 1, 2, 3 \tag{12.86}$$

where subscript "," indicates differentiation.

In order to eliminate the shear locking, the DSG scheme [21] is used. Then in each triangular cell, the shear strain can be given as

$$\gamma'_{xz} = \sum_{i=1}^{3} b_i(\Delta w_{x'i} + \Delta w_{y'i})$$

$$\gamma'_{yz} = \sum_{i=1}^{3} c_i(\Delta w_{x'i} + \Delta w_{y'i}) \tag{12.87}$$

where Δw_{xi} and Δw_{yi} are the DSGs at the node i ($=1,2,3$) of the cell, and they are defined as

$$\Delta w_{x'1} = \Delta w_{x'3} = \Delta w_{y'1} = \Delta w_{y'2} = 0$$

$$\Delta w_{x'2} = (w'_2 - w'_1) + \frac{1}{2}a(\varphi_{x'1} + \varphi_{x'2}) + \frac{1}{2}b(\varphi_{y'1} + \varphi_{y'2}) \tag{12.88}$$

$$\Delta w_{y'3} = (w'_3 - w'_1) + \frac{1}{2}c(\varphi_{x'1} + \varphi_{x'3}) + \frac{1}{2}d(\varphi_{y'1} + \varphi_{y'3})$$

where

$$\begin{array}{ll}
a = x'_2 - x'_1, & b = y'_2 - y'_1 \\
c = x'_3 - x'_1, & d = y'_3 - y'_1
\end{array} \tag{12.89}$$

The shear strain $\boldsymbol{\gamma}'_s$ in each cell can be written as

$$\boldsymbol{\gamma}'_s = \left\{ \begin{array}{c} \gamma'_{xz} \\ \gamma'_{yz} \end{array} \right\} = \mathbf{B}'_s \mathbf{d}' \tag{12.90}$$

where the shear strain matrix is given by

$$\mathbf{B}'_s = \begin{bmatrix} \mathbf{B}'_{s1} & \mathbf{B}'_{s2} & \mathbf{B}'_{s3} \end{bmatrix} \tag{12.91}$$

in which

$$\mathbf{B}'_{s1} = \frac{1}{2A_e} \begin{bmatrix} 0 & 0 & b-d & 0 & A_e & 0 \\ 0 & 0 & c-a & -A_e & 0 & 0 \end{bmatrix}$$

$$\mathbf{B}'_{s2} = \frac{1}{2A_e} \begin{bmatrix} 0 & 0 & d & -\dfrac{bd}{2} & \dfrac{ad}{2} & 0 \\ 0 & 0 & -c & \dfrac{bc}{2} & -\dfrac{ac}{2} & 0 \end{bmatrix} \tag{12.92}$$

$$\mathbf{B}'_{s3} = \frac{1}{2A_e} \begin{bmatrix} 0 & 0 & -b & \dfrac{bd}{2} & -\dfrac{bc}{2} & 0 \\ 0 & 0 & a & -\dfrac{ad}{2} & \dfrac{ac}{2} & 0 \end{bmatrix}$$

12.3.2.4 Edge-Based Smoothing Operation

In our ES-PIM formulation, we perform the smoothing operations based on one common coordinate system defined for each interior edge that is shared by two cells. For an interior edge, these two neighboring cells have their own local coordinates that are, in general, different from each other. A common coordinate system $(\bar{x}, \bar{y}, \bar{z})$ associated with the edge (called edge coordinate system) is, therefore, created by letting

- $\bar{x}$ be coinciding with the edge k
- $\bar{z}$ be the averaged normal direction of the two neighboring cells sharing edge k
- $\bar{y}$ be the cross product of the two unit vectors in the $\bar{z}$- and $\bar{x}$-directions

Once the edge coordinate system is created, we then perform the coordinates transformation for any of the two neighboring elements sharing the edge. The strains in the edge coordinate system of a cell sharing the edge can be written as

$$
\begin{aligned}
\bar{\boldsymbol{\varepsilon}}_m &= \mathbf{R}_{m1}\mathbf{R}_{m2}\boldsymbol{\varepsilon}'_m \\
\bar{\boldsymbol{\varepsilon}}_B &= \mathbf{R}_{B1}\mathbf{R}_{B2}\boldsymbol{\varepsilon}'_B \\
\bar{\boldsymbol{\varepsilon}}_S &= \mathbf{R}_{S1}\mathbf{R}_{S2}\boldsymbol{\gamma}'_S
\end{aligned}
\tag{12.93}
$$

in which

$$
\mathbf{R}_{m1} = \mathbf{R}_{b1}
$$

$$
= \begin{bmatrix}
c_{\bar{x}x}^2 & c_{\bar{x}y}^2 & c_{\bar{x}z}^2 & c_{\bar{x}x}c_{\bar{x}y} & c_{\bar{x}y}c_{\bar{x}z} & c_{\bar{x}x}c_{\bar{x}z} \\
c_{\bar{y}x}^2 & c_{\bar{y}y}^2 & c_{\bar{y}z}^2 & c_{\bar{y}x}c_{\bar{y}y} & c_{\bar{y}y}c_{\bar{y}z} & c_{\bar{y}x}c_{\bar{y}z} \\
2c_{\bar{x}x}c_{\bar{y}x} & 2c_{\bar{x}y}c_{\bar{y}y} & 2c_{\bar{x}z}c_{\bar{y}z} & c_{\bar{x}x}c_{\bar{y}y}+c_{\bar{y}x}c_{\bar{x}y} & c_{\bar{x}z}c_{\bar{y}y}+c_{\bar{y}z}c_{\bar{x}y} & c_{\bar{x}x}c_{\bar{y}z}+c_{\bar{y}x}c_{\bar{x}z}
\end{bmatrix}
\tag{12.94}
$$

$$
\mathbf{R}_{S1} = \begin{bmatrix}
2c_{\bar{x}x}c_{\bar{z}x} & 2c_{\bar{x}y}c_{\bar{z}y} & 2c_{\bar{x}z}c_{\bar{z}z} & c_{\bar{x}x}c_{\bar{z}y}+c_{\bar{z}x}c_{\bar{x}y} & c_{\bar{x}z}c_{\bar{z}y}+c_{\bar{z}z}c_{\bar{x}y} & c_{\bar{x}x}c_{\bar{z}z}+c_{\bar{z}x}c_{\bar{x}z} \\
2c_{\bar{y}x}c_{\bar{z}x} & 2c_{\bar{y}y}c_{\bar{z}y} & 2c_{\bar{y}z}c_{\bar{z}z} & c_{\bar{y}x}c_{\bar{z}y}+c_{\bar{z}x}c_{\bar{y}y} & c_{\bar{y}z}c_{\bar{z}y}+c_{\bar{z}z}c_{\bar{y}y} & c_{\bar{y}x}c_{\bar{z}z}+c_{\bar{z}x}c_{\bar{y}z}
\end{bmatrix}
\tag{12.95}
$$

$$
\mathbf{R}_{m2} = \mathbf{R}_{b2} = \begin{bmatrix}
c_{x'x}^2 & c_{y'x}^2 & c_{x'x}c_{y'x} \\
c_{x'y}^2 & c_{y'y}^2 & c_{x'y}c_{y'y} \\
c_{x'z}^2 & c_{y'z}^2 & c_{x'z}c_{y'z} \\
2c_{x'x}c_{x'y} & 2c_{y'x}c_{y'y} & c_{x'x}c_{y'y}+c_{x'y}c_{y'x} \\
2c_{x'y}c_{x'z} & 2c_{y'y}c_{y'z} & c_{x'y}c_{y'z}+c_{x'z}c_{y'y} \\
2c_{x'x}c_{x'z} & 2c_{y'x}c_{y'z} & c_{x'x}c_{y'z}+c_{x'z}c_{y'x}
\end{bmatrix}
\tag{12.96}
$$

$$
\mathbf{R}_{S2} = \begin{bmatrix}
c_{x'x}c_{z'x} & c_{y'x}c_{z'x} \\
c_{x'y}c_{z'y} & c_{y'y}c_{z'y} \\
c_{x'z}c_{z'z} & c_{y'z}c_{z'z} \\
c_{x'x}c_{z'y}+c_{x'y}c_{z'x} & c_{y'x}c_{z'y}+c_{y'y}c_{z'x} \\
c_{x'z}c_{z'y}+c_{x'y}c_{z'z} & c_{y'z}c_{z'y}+c_{y'y}c_{z'z} \\
c_{x'x}c_{z'z}+c_{x'z}c_{z'x} & c_{y'x}c_{z'z}+c_{y'z}c_{z'x}
\end{bmatrix}
\tag{12.97}
$$

where $c_{\bar{x}x}$ is the direction cosine of the angle between the $\bar{x}$- and x-axes, etc.

Substituting Equations 12.69, 12.83, 12.84, and 12.90 into Equation 12.93, the strain in the edge coordinate system of a cell sharing the edge can be rewritten as

$$
\begin{aligned}
\bar{\boldsymbol{\varepsilon}}_m &= \mathbf{R}_{m1}\mathbf{R}_{m2}\mathbf{B}'_m\mathbf{T}\mathbf{d} \\
\bar{\boldsymbol{\varepsilon}}_B &= \mathbf{R}_{B1}\mathbf{R}_{B2}\mathbf{B}'_B\mathbf{T}\mathbf{d} \\
\bar{\boldsymbol{\varepsilon}}_S &= \mathbf{R}_{S1}\mathbf{R}_{S2}\mathbf{B}'_S\mathbf{T}\mathbf{d}
\end{aligned}
\tag{12.98}
$$

As shown in Figure 12.35, the smoothing domain Ω_k corresponding to an interior edge k is shared by two cells. The smoothed shear strain for the entire smoothing domain is then obtained by the area weighted average of these strains of the two neighboring cells sharing the edge k:

$$\bar{\boldsymbol{\varepsilon}}_{mk} = \frac{1}{A_k}\left(A_{k^+}\bar{\boldsymbol{\varepsilon}}_{mk^+} + A_{k^-}\bar{\boldsymbol{\varepsilon}}_{mk^-}\right)$$

$$\bar{\boldsymbol{\varepsilon}}_{Bk} = \frac{1}{A_k}\left(A_{k^+}\bar{\boldsymbol{\varepsilon}}_{Bk^+} + A_{k^-}\bar{\boldsymbol{\varepsilon}}_{Bk^-}\right) \tag{12.99}$$

$$\bar{\boldsymbol{\gamma}}_{Sk} = \frac{1}{A_k}\left(A_{k^+}\boldsymbol{\gamma}_{Sk^+} + A_{k^-}\boldsymbol{\gamma}_{Sk^-}\right)$$

where
the subscripts "+" and "−" stand for the two neighboring cells sharing edge k
A_k, A_{k^+}, and A_{k^-} are the areas of the smoothing domain of edge k and those of the two neighboring cells sharing the edge k

Using Equations 12.98 and 12.99, the strains for smoothing domain associated with the kth edge can be written in the matrix form:

$$\bar{\boldsymbol{\varepsilon}}_{mk} = \bar{\mathbf{B}}_{mk}\mathbf{d}_k$$

$$\bar{\boldsymbol{\varepsilon}}_{Bk} = \bar{\mathbf{B}}_{Bk}\mathbf{d}_k \tag{12.100}$$

$$\bar{\boldsymbol{\gamma}}_{Sk} = \bar{\mathbf{B}}_{Sk}\mathbf{d}_k$$

in which

$$\bar{\mathbf{B}}_{mk} = \frac{A_{k^+}}{A_k}\mathbf{B}_{mk^+} \oplus \frac{A_{k^-}}{A_k}\mathbf{B}_{mk^-}$$

$$\bar{\mathbf{B}}_{Bk} = \frac{A_{k^+}}{A_k}\mathbf{B}_{Bk^+} \oplus \frac{A_{k^-}}{A_k}\mathbf{B}_{Bk^-} \tag{12.101}$$

$$\bar{\mathbf{B}}_{Sk} = \frac{A_{k^+}}{A_k}\mathbf{B}_{Sk^+} \oplus \frac{A_{k^-}}{A_k}\mathbf{B}_{Sk^-}$$

where we use $\oplus$ to represent an assembly (a location matched summation). The generalized nodal displacement vector $\mathbf{d}_k$ for the smoothing domain k contains the nodal values of four nodes "supporting" the smoothing domain, and it can be written as (see Figure 12.35a)

$$\mathbf{d}_k = \left\{ \underbrace{\mathbf{d}_m}_{\text{node 1}} \quad \underbrace{\mathbf{d}_i}_{\text{node 2}} \quad \underbrace{\mathbf{d}_n}_{\text{node 3}} \quad \underbrace{\mathbf{d}_j}_{\text{node 4}} \right\}^{\mathrm{T}} \tag{12.102}$$

Naturally, such averaging operation is not needed for any edge that is on the problem boundary, as there is only one cell associated with the edge (see Figure 12.35b).

12.3.2.5 Stress–Strain Relations for Shells

We consider general shells undergoing all three possible deformations in 3D space: in-plane stretching (membrane), off-plane bending, and shearing deformation. For the membrane deformation, we simply use the stress–strain relation for 2D solids (plane stress, Chapter 1):

$$\boldsymbol{\sigma}_m = h\mathbf{c}\boldsymbol{\varepsilon}_m \tag{12.103}$$

where $\mathbf{c}$ is given in Equation 1.17. For the bending deformation, we simply use the pseudo-stress–strain relation for bending deformation of plates based on FSDT (Chapter 11):

$$\boldsymbol{\sigma}_B = \frac{h^3}{12}\mathbf{c}\boldsymbol{\varepsilon}_B \tag{12.104}$$

For the off-plane shear deformation, we also use the pseudo-stress–strain relation for shear deformation for plates based on FSDT (Chapter 11):

$$\boldsymbol{\sigma}_S = \kappa h\mathbf{G}\boldsymbol{\gamma}_S \tag{12.105}$$

where
   $\mathbf{G}$ is given in Equation 11.66
   $\kappa = 5/6$ is the shear correction factor

All these stress and strains are now independent of z, and they have the form of

$$\boldsymbol{\sigma}_m = \left\{ \begin{array}{c} h\sigma_{xx} \\ h\sigma_{yy} \\ h\sigma_{xy} \end{array} \right\} = \left\{ \begin{array}{c} N_x \\ N_y \\ N_{xy} \end{array} \right\} \tag{12.106}$$

$$\boldsymbol{\sigma}_B = \left\{ \begin{array}{c} \int\limits_{-h/2}^{h/2} z\sigma_{xx}dz \\ \int\limits_{-h/2}^{h/2} z\sigma_{yy}dz \\ \int\limits_{-h/2}^{h/2} z\sigma_{xy}dz \end{array} \right\} = \left\{ \begin{array}{c} M_{xx} \\ M_{yy} \\ M_{xy} \end{array} \right\} \tag{12.107}$$

$$\boldsymbol{\sigma}_s = \left\{ \begin{array}{c} h\sigma_{xz} \\ h\sigma_{yz} \end{array} \right\} = \left\{ \begin{array}{c} V_{xz} \\ V_{yz} \end{array} \right\} \tag{12.108}$$

Note that all these stress–strain relations given in Equations 12.103 through 12.105 are applicable also to the smoothed strains and stresses: The smoothing operation has no effects on the constitutive relations.

12.3.2.6 Weakened-Weak Form

Following the discussions in Chapter 5 and using all these pseudo-stresses and strains defined above, the smoothed Galerkin weak form for shells based on the FSDT can be written as

$$h \int_A \delta\bar{\boldsymbol{\varepsilon}}_m^T(\mathbf{u})\mathbf{c}\bar{\boldsymbol{\varepsilon}}_m(\mathbf{u})\mathrm{d}A + \underbrace{\frac{h^3}{12} \int_A \delta\bar{\boldsymbol{\varepsilon}}_B^T(\mathbf{u})\mathbf{c}\bar{\boldsymbol{\varepsilon}}_B(\mathbf{u})\mathrm{d}A}_{\text{Bending}} + \underbrace{\kappa h \int_A \delta\bar{\boldsymbol{\gamma}}_S^T(\mathbf{u})\mathbf{G}\bar{\boldsymbol{\gamma}}_S(\mathbf{u})\mathrm{d}A}_{\text{Shearing}}$$

$$\underbrace{- \int_A \delta\mathbf{u}^T\mathbf{b}\mathrm{d}A}_{\text{Body force}} - \underbrace{\int_{\Gamma_t} \delta\mathbf{u}^T\mathbf{t}_\Gamma\mathrm{d}\Gamma}_{\text{Boundary force}} = 0 \qquad (12.109)$$

where **u** is the *generalized* displacement vector consisting of three translational displacements and three rotations defined on the mid-plane of the plate. It is arranged in form of

$$\mathbf{u} = \begin{Bmatrix} u_1 \\ u_2 \\ u_3 \\ u_4 \\ u_5 \\ u_6 \end{Bmatrix} = \begin{Bmatrix} u \\ v \\ w \\ \varphi_x \\ \varphi_y \\ \varphi_z \end{Bmatrix} \qquad (12.110)$$

and the generalized body force vector **q** can be written as

$$\mathbf{q}_z = \begin{Bmatrix} q_1 \\ q_2 \\ q_3 \\ q_4 \\ q_5 \\ q_6 \end{Bmatrix} = \begin{Bmatrix} hb_x \\ hb_y \\ hb_z \\ m_{xx} \\ m_{yy} \\ m_{xy} \end{Bmatrix} \qquad (12.111)$$

Because all the strains are smoothed over the smoothing domain, the first three terms in the foregoing equation can be written in the simple summation form (see Chapter 4):

$$\underbrace{h \sum_{k=1}^{N_s} \delta\bar{\boldsymbol{\varepsilon}}_{mk}^T(\mathbf{u})\mathbf{c}\bar{\boldsymbol{\varepsilon}}_{mk}(\mathbf{u})}_{\text{Membrane}} + \underbrace{\frac{h^3}{12} \sum_{k=1}^{N_s} \delta\bar{\boldsymbol{\varepsilon}}_{Bk}^T(\mathbf{u})\mathbf{c}\bar{\boldsymbol{\varepsilon}}_{Bk}(\mathbf{u})}_{\text{Bending}} + \underbrace{\kappa h \sum_{k=1}^{N_s} \delta\bar{\boldsymbol{\gamma}}_{Sk}^T(\mathbf{u})\mathbf{G}\bar{\boldsymbol{\gamma}}_{Sk}(\mathbf{u})}_{\text{Shearing}}$$

$$\underbrace{- \int_A \delta\mathbf{u}^T\mathbf{b}\mathrm{d}A}_{\text{Body force}} - \underbrace{\int_{\Gamma_t} \delta\mathbf{u}^T\mathbf{t}_\Gamma\mathrm{d}\Gamma}_{\text{Boundary force}} = 0 \qquad (12.112)$$

With these strains given in Equation 12.100, these strain potentials can be easily evaluated and hence the stiffness matrix of our ES-PIM model can be evaluated.

12.3.2.7 Force Terms

As a generalized smoothed Galerkin (GS-Galerkin) do not change force potentials (see Chapter 5), these terms can be dealt with in exactly the same way as in the FEM.

The procedure is the same as those given in Chapter 8. Finally, the discrete system equation can be obtained as

$$\bar{\mathbf{K}}\bar{\mathbf{U}} = \mathbf{F} \tag{12.113}$$

which can be solved using standard routines for $\bar{\mathbf{U}}$ that contains all the nodal values of the generalized displacements.

12.3.3 Numerical Examples

We now present a number of numerical examples solved using our ES-PIM code for shells. We will start with several benchmark problems to examine the performance of the linear ES-PIM. Numerical results of the ES-PIM are compared with those of several FEM shell elements, including

- QPH (One point quadrature quadrilateral shell element with physical hourglass control) [22]
- Four-node SRI (A standard four-node Mindlin element with selective reduced integration) [23]
- MITC4 (A four-node fully integrated shell element using mixed interpolated tensorial components) [24]
- DSG3 (DSG triangular element) [21]
- DKT-CST (triangular flat element superimposing the DKT plate bending element with the CST plane stress element)
- DKT-CST* (triangular curved element superimposing the DKT plate bending element with the membrane element of Carpenter et al.) [25]

Example 12.11: Scordelis-Lo Roof

The Scordelis-Lo roof shown in Figure 12.36 is a benchmark problem to test numerical methods for shell analysis. This problem is very useful for determining the ability of a numerical model in representing the complicated states of the membrane strain. The length of the shell $L = 25$ ft, the radius $R = 25$ ft, the thickness $h = 0.25$ ft, and the span angle $\theta_0 = 40°$. The material properties are: Poisson's ratio $\nu = 0.0$ and Young's modulus $E = 4.32 \times 10^8$ N/ft^2. The boundary conditions at each end of the roof are supported by a rigid diaphragm. The loading is a uniform vertical gravity load of 90 N/ft^2. Owing to the symmetry, only a quarter of the roof is modeled. Three meshes, 4×4, 8×8, and 16×16 are used in this problem and the typical 4×4 mesh is shown in Figure 12.36.

For this problem a reference solution for the vertical deflection at the center of the free edge is found to be 0.3024 ft. All the numerical solutions obtained are normalized with this reference value. The results of the linear ES-PIM are listed in Table 12.4, together with those of other FEM models. It is observed that among the methods using triangular mesh, the ES-PIM performs best. The ES-PIM solutions are quite compatible with those using quadrilateral mesh. We also found that the ES-PIM converges very fast and the results are very accurate when 16×16 mesh is used. The vertical displacement along the centre line of the roof is plotted in Figure 12.37, and the longitudinal displacement along the support end of the roof is plotted in Figure 12.38. It is clearly shown that the ES-PIM solution agrees well with the analytic solutions [26].

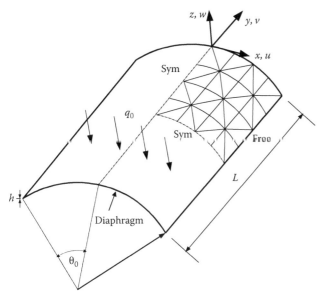

FIGURE 12.36
Problem setting for the Scordelis-Lo roof
with a 4×4 mesh for the quarter model.

TABLE 12.4

Normalized Vertical Displacement at Midpoint of the Free Edge
on the Scordelis-Lo Roof (Normalized with 0.3024 ft)

	Quadrilateral Mesh			Triangular Mesh			
Mesh	QPH	MIT4	4-SRI	DSG	DKT-CST	DKT-CST	ES-PIM
4×4	0.940	0.940	0.960	0.725	0.707	0.646	0.884
8×8	0.980	0.970	0.980	0.886	0.866	0.846	0.954
16×16	1.010	1.000	1.000	0.979	0.955	0.950	1.000

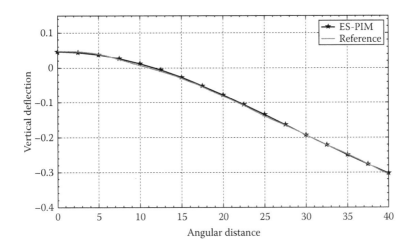

FIGURE 12.37
Vertical deflection along the central line on the Scordelis-Lo roof.

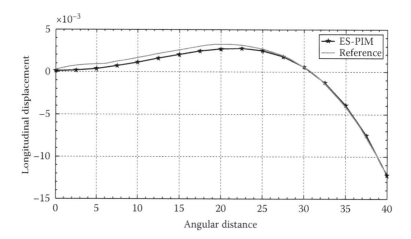

FIGURE 12.38
Longitudinal displacement along the support line on the Scordelis-Lo roof edge.

Example 12.12: Pinched Cylinder with End Diaphragms

A pinched cylinder supported at each end by rigid diaphragm shown in Figure 12.39 is considered in this section. It is a widely used benchmark problem for determining the ability of a numerical model to represent inextensional bending and complex membrane states. The length of the pinched cylinder $L = 600$ in., the radius $R = 300$ in., the thickness $h = 3$ in. The material properties are: Poisson's ratio $\nu = 0.3$, and Young's modulus $E = 3.0 \times 10^6$ N/in². The loading is a pair of pinching loads $F = 1.0$ N. Owing to the symmetry, only 1/8 of the problem is modeled. Three meshes, 4×4, 8×8, and 16×16, are used in the investigation. The 4×4 mesh is shown in Figure 12.39. In this case, the exact solution of the radial displacement under the point load is found to be 1.8248×10^{-5} in. The numerical solutions are normalized with this value. The results are listed in Table 12.5. It can be seen that the ES-PIM is performed third for this problem, behind the two DKT-CST models.

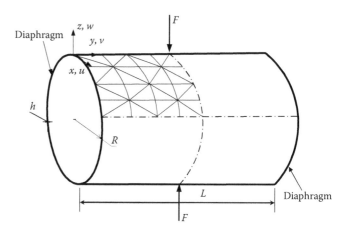

FIGURE 12.39
Problem setting for the Scordelis-Lo roof with a 4×4 mesh for the quarter model.

TABLE 12.5

Normalized Displacement at Load Point on the Pinched Cylinder with End Diaphragms

	Quadrilateral Mesh			Triangular Mesh			
Mesh	QPH	MIT4	4-SRI	DSG	DKT-CST	DKT-CST	ES-PIM
4×4	0.370	0.370	0.370	0.316	0.462	0.773	0.405
8×8	0.740	0.740	0.750	0.695	0.860	0.947	0.813
16×16	0.930	0.930	0.940	0.904	0.994	1.015	0.983

Note: The value used for normalization is 1.8248×10^{-5} in.

Example 12.13: Hemispherical Shell

A hemispherical shell loaded antisymmetrically by point loads, shown in Figure 12.40 is now examined. This problem has almost no membrane strains but it is a challenging test on the ability of numerical model to handle rigid body rotations about the normal to the shell's surface. The geometric parameters are radius $R = 10$ m and thickness $h = 0.04$ m. The material properties are $v = 0.3$ and $E = 6.825 \times 10^7$ Pa. The point loading is $F = 1$ N. The solutions are obtained using three meshes each with 5, 9, and 17 nodes per side. A typical five nodes per side mesh is shown in Figure 12.40. The exact solution of radial deflection at point load is 0.0924 m. The numerical solutions listed in Table 12.6 are normalized with this exact value. It can be observed that the ES-PIM solutions agree well with analytical solutions. Its overall performance is about the third, but quite close to the solutions of two DKT-CST models.

Example 12.14: Hood of an Automobile

An actual structure component of a car hood shown in Figure 12.41 is studied using the ES-PIM. The dent resistance of the car hood is one of the important considerations in the process of car design, and need to be analyzed in detail. In this example, all of the boundary nodes of the hood are fixed, and a concentrated load $F = 150$ N is applied in the x-direction. The thickness of the shell is 0.8 mm, and is very thin considering the large dimensions in other two directions.

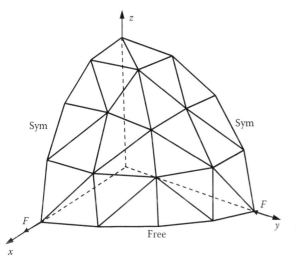

FIGURE 12.40
Pinched cylindrical with end diaphragms with a mesh of five nodes per side.

TABLE 12.6

Normalized Displacement at Load Point on the Hemispherical Shell

Node/Side	Quadrilateral Mesh			Triangular Mesh			
	QPH	MIT4	4-SRI	DSG	DKT-CST	DKT-CST	ES-PIM
5	0.280	0.390	0.410	0.912	1.028	1.064	0.996
9	0.860	0.910	0.930	0.843	1.016	1.018	0.961
17	0.990	0.990	0.980	0.959	0.996	0.995	0.992

Note: The value used for normalization is 0.0924 m.

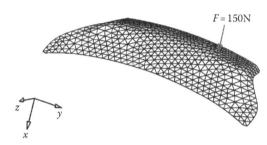

FIGURE 12.41
Triangular mesh for the hood of an automobile.

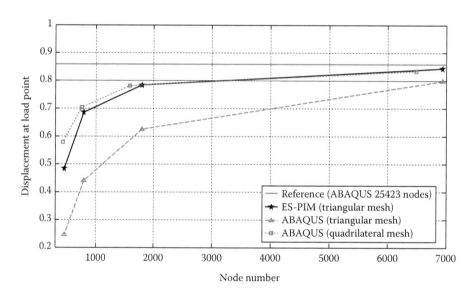

FIGURE 12.42
Displacements in *x*-direction at the loading point on the hood of an automobile.

Poisson's ratio of the material is 0.3, and the Young's modulus is 2.1e5 N/mm^2. In order to quantify the accuracy of the ES-PIM, the solutions obtained using ABAQUS$^©$ with triangular elements and quadrilateral elements are used for comparison. The reference solution is also obtained using ABAQUS with quadrilateral shell elements of large number of (25423) nodes. All the solutions are plotted together in Figure 12.42. It is shown that the ES-PIM using triangular mesh achieves the same level of accuracy as the ABAQUS solution using quadrilateral shell elements, and a much higher accuracy than that using triangular shell elements.

12.3.4 Concluding Remarks

In this section, a linear ES-PIM is formulated to analyze the deformation of shells based on the FSDT using simple three-node triangular mesh with 6 DOF per node:

- In the ES-PIM formulation, no extra sampling points are introduced to evaluate the stiffness matrix, and no increase of total DOF.
- The ES-PIM is found very stable, the results are very accurate, and the performance is very good and consistent for all the tests conducted. It often achieves the same level of accuracy as the FEM quadrilateral shell elements.
- Using the DSG method, the ES-PIM is free from shear locking, and hence it works well for both thin and thick shells.

Moving forward, ES-PIM can be further improved and extended for other applications:

- Extensions to nonlinear problems should be straightforward, because these techniques developed in FEM can be utilized with minor changes.
- The linear ES-PIM can be easily extended to meshes of general n-sided polygonal cells [27].
- Extension to quadratic PIMs (T6/3-scheme) and radial point interpolation methods (RPIMs) (T6- or T2L-schemes) for shells can offer a variety of ways to model the shells. The cells do not have to be flat, the shell curvature can be modeled, and ways to deal with the membrane and shear locking can also be devised. The work in this direction is still ongoing.
- Other SC-Galerkin models can also be developed for shells in similar ways presented in this section.

12.4 Summary

In this chapter, we formulated the EFG method for both thin and thick shells and ES-PIM for thick shells but applicable to thin shells. We have shown through a number of benchmarking problems that they work very well in their ways, and have their own features.

- The EFG can give accurate approximation to both the field variables and geometry of the shell using the MLS approximation. It, however, uses a very large number of nodes in the local approximation, which reduces the efficiency.
- The ES-PIM uses flat triangular cells that can be generated automatically. It uses only three local nodes for displacement and four nodes for strain approximations. It has features of simplicity, free of shear locking, robustness, accuracy, adaptivity, and efficiency. However, the shell is modeled as flat triangles, and hence fine mesh may be needed.

References

1. Yang, H. T. Y., Saigal, S., and Liaw, D. G., Advances of thin shell finite elements and some applications—Version I, *Comput. Struct.*, 35, 481–504, 1990.
2. Krysl, P. and Belytschko, T., Analysis of thin shells by the element-free Galerkin method, *Int. J. Solids Struct.*, 33, 3057–3080, 1996.
3. Noguchi, H., Kawashima, T., and Miyamura, T., Element free analyses of shell and spatial structures, *Int. J. Numerical Methods Eng.*, 47, 1215–1240, 2000.
4. Li, S., Hao, W., and Liu, W. K., Numerical simulations of large deformation of thin shell structures using meshfree methods, *Comput. Mech.*, 25, 102–116, 2000.
5. Liu, L., Liu, G.R., and Tan, V. B. C., Element free analyses for static and free vibration of thin shells, in *Proceedings of the Asia-Pacific Vibration Conference*, W. Bangchun, ed., Hangzhou, China, November 2001.
6. Liu, L., Liu, G. R., and Tan, V. B. C., Element free method for static and free vibration analysis of spatial thin shell structures, *Comput. Methods Appl. Mech. Eng.*, 191, 5923–5942, 2002.
7. Ouatouati, A. E. and Johnson, D. A., A new approach for numerical modal analysis using the element free method, *Int. J. Numerical Methods Eng.*, 46, 1–27, 1999.
8. Liu, G. R. and Chen, X. L., A mesh-free method for static and free vibration analyses of thin plates of complicated shape, *J. Sound Vib.*, 241(5), 839–855, 2001.
9. Simo, J. and Fox, D. D., On a stress resultant geometrically exact shell model, Part I: Formulation and optimal parameterization, *Comput. Methods Appl. Mech. Eng.*, 72, 267–304, 1989.
10. Belytschko, T., Stolarski, H., Liu, W. K., Carpenter, N., and Ong, J. S.-J., Stress projection for membrane and shear locking in shell finite elements, *Comput. Methods Appl. Mech. Eng.*, 51, 221–258, 1985.
11. Petyt, M., Vibration of curved plates, *J. Sound Vib.*, 15, 381–395, 1971.
12. Carter, R. L., Robinson, A. R., and Schnobrich, W. C., 1. Free vibrations of hyperboloidal shells of revolution, *J. Eng. Mech.*, 93, 1033–1053, 1969.
13. Ozakca, M. and Hinton, E., Free vibration analysis and optimization of axisymmetrical plates and shells—I. Finite element formulation, *Comput. Struct.*, 52, 1181–1197, 1994.
14. Tou, S. K. and Wong, K. K., High-precision finite element analysis of cylindrical shells, *Comput. Struct.*, 26, 847–854, 1987.
15. Subir, K. S. and Gould, P. L., Free vibration of shells of revolution using FEM, *J. Eng. Mech. Div. ASCE EM2*, 100, 283–303, 1974.
16. Sneddon, I. N., The symmetrical vibrations of a thin elastic plate, *Proc. Cambridge Philos. Soc.*, 41, 27–43, 1945.
17. Bathe, K. J., Wilson, E. L., Paterson, F. E., and Sap, I. V., A Structural Analysis Program for Static and Dynamic Response of Linear Systems, Report EERC 73–11, University of California, Berkeley, CA, 1973.
18. Beskos, D. E., Dynamic analysis of plates and shallow shells by the D/BEM, in *Advances in the Theory of Plates and Shells*, Elsevier, Oxford, 1990, pp. 177–196.
19. Cui, X. Y., Liu, G. R., Li, G. Y., Zhang, G. Y., and Zheng, G., Analysis of plates and shells using an edge-based smoothed finite element method. *Computational Mechanics*, (submitted), 2009.
20. Liu, G. R. and Quek, S. S., *The Finite Element Method: A Practical Course*, Butterworth Heinemann, Oxford, 2002.
21. Bletzinger, K. U., Bischoff, M., and Ramm, E., A unified approach for shear-locking-free triangular and rectangular shell finite elements, *Comput. Struct.* 75, 321–334, 2000.
22. Belytschko, T., Leviathan I. Physical stabilization of the 4-node shell element with one point quadrature. *Comput. Methods Appl. Mech. Eng.*, 113, 321–350, 1994.
23. Hughes, T. J. R. and Liu, W. K., Nonlinear finite element analysis shells: Part II, Two-dimensional shells, *Comput. Methods Appl. Mech. Eng.*, 26, 331–362, 1981.
24. Bathe, K. J. and Dvorkin, E. N., A formulation of general shell elements—The use of mixed interpolation of tensorial components. *Int. J. Numerical Methods Eng.*, 22, 697–722, 1986.

25. Carpenter, N., Stolarski, H., and Belytschko, T., Improvements in 3-node triangular shell elements. *Int. J. Numerical Methods Eng.*, 23, 1643–1667, 1986.
26. Scordelis, A. C. and Lo, K. S., Computer analysis of cylindrical shells. *J. Am. Concrete Inst.* 61, 539–561, 1969.
27. Dai, K. Y., Liu, G. R., and Nguyen, T. T., An n-sided polygonal smoothed finite element method (nSFEM) for solid mechanics. *Finite Elem. Anal. Des.*, 43, 847–860, 2007.

13

Boundary Meshfree Methods

The boundary element method (BEM) is a numerical technique based on the boundary integral equation (BIE), which has been developed in the 1960s. For many (especially linear) problems, BEM is undoubtedly superior to the "domain discretization" type of methods, such as the finite element method (FEM) and the finite difference method (FDM). In BEM, for example, only the two-dimensional (2D) bounding surface of a three-dimensional (3D) body needs to be discretized. BEM is applicable to all those problems for which the fundamental solution (Green's functions) is known in a reasonably simple form, preferably in a closed form.

The meshfree idea has also been used in BIE, such as the boundary node method (BNM) by Mukherjee et al. [1–4] and the local boundary integral equation (LBIE) method by Zhu and Atluri [5]. These boundary-type meshfree methods are formulated using the moving least squares (MLS) approximations and techniques of BIE. In BNM, the surface of the problem domain is discretized by properly scattered nodes. The BNM has been applied to 3D problems of both potential theory and the theory of elastostatics [3,4]. Very good results have been obtained in these problems. However, because the shape functions based on the MLS approximation lack the delta function property, extra efforts are needed to satisfy accurately the boundary conditions in BNM. This problem becomes even more serious in BNM because of the large number of boundary conditions, compared with the total number of nodes for the problem. The method used in [2] to impose boundary conditions resulted in a set of system equations that were doubled in number. The advantage of the boundary-type method is therefore eroded and discounted to a certain degree, making BNM computationally much more expensive than the conventional BEM.

A boundary-type meshfree method called the boundary point interpolation method (BPIM) has been formulated [6,7], where the point interpolation method (PIM) [8,9] was used to construct shape functions. It is confirmed that there is no need at all to use MLS in boundary-type meshfree methods, at least for 2D problems. PIM works much more efficiently in constructing shape functions, and all the PIM shape functions possess the Kronecker delta function property. This removes the issue of treatment of boundary conditions, which is especially beneficial to methods based on BIE. The dimension of the equation system of BPIM is equivalent to that of BEM. For 2D problems, BPIM works perfectly well without any special trick and is superior to BNM in simplicity, accuracy, and computational efficiency. For 3D problems, for which 2D shape functions need to be constructed, there could be an issue of singular moment matrices. In such cases, the special techniques discussed in Section 2.5.4 should be applied. A robust way of constructing PIM shape functions is to use radial functions as the basis. The advantage of using a radial function basis is that it guarantees the existence of the inverse moment matrix. The methods formulated are termed as boundary radial PIM (BRPIM) [10]. A good alternative could be the use of T-schemes (Section 1.7.6).

This chapter focuses on the introduction of boundary meshfree methods formulated using both MLS shape functions (BNM) and PIM shape functions (BPIM and BRPIM). These boundary meshfree methods can be formulated in the same manner, except that in formulating BNM using MLS shape functions proper treatments of essential boundary conditions [12,2] are required.

Only 2D problems are discussed in this chapter. In all these boundary meshfree methods, only the boundary of the problem domain is represented by properly scattered nodes. BIE for 2D elastostatics is then discretized using meshfree shape functions based only on a group of arbitrarily distributed boundary nodes that are included in the support domain of a point of interest. For 3D BNM, readers are referred to the work by Chati et al. [3].

13.1 BPIM Using Polynomial Basis

A BPIM using polynomial basis in the construction of shape functions is first presented in this chapter for solving boundary value problems of solid mechanics. The method was presented in [11]. The PIM shape functions are constructed in a curvilinear coordinate system, and possess the delta function property. The boundary conditions can be implemented with ease as in the standard BEM. In addition, the rigid body movement can also be utilized to avoid some singular integrals. For 2D problems, BPIM with polynomial basis has no singularity problems with the moment matrix, as the boundaries are curves, and the interpolation is essentially one-dimensional (1D). Therefore, there is no reason to use MLS approximation in this case.

A detailed formulation of BPIM using polynomial basis is presented, and several numerical examples are presented to demonstrate the validity and efficiency of BPIM. A comparison study is carried out using BPIM, BNM that uses MLS shape functions, standard BEM, and analytical methods.

13.1.1 Point Interpolation on Curves

Consider a 2D domain Ω bounded by its boundary Γ, as shown in Figure 13.1. In using boundary meshfree methods, only the boundary Γ of the problem domain is represented using nodes. The point interpolants in BPIM are constructed on the 1D bounding curve Γ of

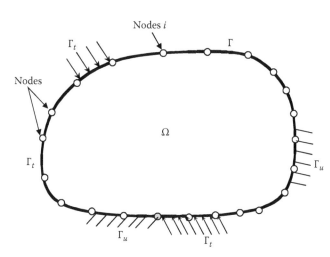

FIGURE 13.1

Domains and their boundaries: problem domain Ω bounded by Γ, including essential (displacement) boundary Γ_u and natural (force or free) boundary Γ_t. In boundary meshfree methods, only the problem boundary is represented using nodes.

the 2D domain Ω, using a set of discrete nodes on Γ. As in the conventional BEM formulation, the displacement and traction can be constructed independently using PIM shape functions. The displacement $u(s)$ and traction $t(s)$ at a point s on the boundary Γ are expressed using the surrounding nodes and polynomials:

$$u(s) = \sum_{i=1}^{n} p_i(s)a_i = \mathbf{p}^{\mathrm{T}}(s)\mathbf{a} \tag{13.1}$$

$$t(s) = \sum_{i=1}^{n} p_i(s)b_i = \mathbf{p}^{\mathrm{T}}(s)\mathbf{b} \tag{13.2}$$

where
  s is a curvilinear coordinate on Γ
  n is the number of nodes in the support domain of a point of interest at s_Q, which is often
    a quadrature point of integration
  $p_i(s)$ is a basis function of a complete polynomial with $p_1 = 1$ and $p_i = s^{i-1}$
  a_i and b_i are the coefficients that change when s_Q changes

In the matrix form, we have

$$\mathbf{a}^{\mathrm{T}} = [a_1, a_2, \ldots, a_n] \tag{13.3}$$

$$\mathbf{b}^{\mathrm{T}} = [b_1, b_2, \ldots, b_n] \tag{13.4}$$

$$\mathbf{p}^{\mathrm{T}}(s) = [1, s, s^2, \ldots, s^{n-1}] \tag{13.5}$$

The coefficients a_i and b_i can be determined by enforcing Equations 13.1 and 13.2 to be satisfied at the n nodes surrounding the point s_Q. Equation 13.1 can then be written in the following matrix form:

$$\mathbf{u}_n = \mathbf{P}_Q \mathbf{a} \tag{13.6}$$

$$\mathbf{t}_n = \mathbf{P}_Q \mathbf{b} \tag{13.7}$$

where $\mathbf{u}_n$ and $\mathbf{t}_n$ are the vectors of the nodal displacement and traction given by

$$\mathbf{u}_n = [u_1, u_2, \ldots, u_n]^{\mathrm{T}} \tag{13.8}$$

$$\mathbf{t}_n = [t_1, t_2, \ldots, t_n]^{\mathrm{T}} \quad \mathbf{t}_n = [t_1, t_2, \ldots, t_n]^{\mathrm{T}} \tag{13.9}$$

and $\mathbf{P}_Q$ is the moment matrix formed by

$$\mathbf{P}_Q^{\mathrm{T}} = [p(s_1), p(s_2), \ldots, p(s_n)] \tag{13.10}$$

Solving Equations 13.6 and 13.7 for $\mathbf{a}$ and $\mathbf{b}$, and then substituting them into Equation 13.1, we obtain

$$u(s) = \mathbf{\Phi}^{\mathrm{T}}(s)\mathbf{u}_n \tag{13.11}$$

$$t(s) = \mathbf{\Phi}^{\mathrm{T}}(s)\mathbf{t}_n \tag{13.12}$$

where the matrix of the shape function $\Phi(s)$ is defined by

$$\mathbf{\Phi}^{\mathrm{T}}(s) = \mathbf{p}^{\mathrm{T}}(s)\mathbf{P}_Q^{-1} = [\phi_1(s), \phi_2(s), \ldots, \phi_n(s)] \tag{13.13}$$

The shape function $\phi_i(s)$ obtained from the above procedure satisfies

$$\phi_i(s = s_i) = 1, \quad i = 1, n \tag{13.14}$$

$$\phi_j(s = s_i) = 0, \quad j \neq i \tag{13.15}$$

$$\sum_{i=1}^{n} \phi_i(s) = 1 \tag{13.16}$$

Therefore, the interpolation functions constructed have the Kronecker delta function property, and the boundary conditions can be easily imposed as in the standard BEM. The procedure to prove these properties can be found in Chapter 2.

The matrix $\mathbf{P}_Q$ is an $n \times n$ matrix. It needs to be invertible for the construction of the shape functions in Equation 13.13. Fortunately, in the curvilinear coordinate system, the matrix $\mathbf{P}_Q$ is, in general, reversible for 2D problems (interpolation along a 1D boundary).

It can be found that the accuracy of interpolation depends on the nodes in the support domain of a quadrature point. Therefore, a suitable support domain should be chosen to ensure a proper area of coverage for interpolation. To define the support domain for a point s_Q, a curvilinear support domain is used. The arc length of the curvilinear domain d_s is computed by

$$d_s = \alpha_s d_c \tag{13.17}$$

where
α_s is the dimensionless size of the support domain
d_c is a characteristic length that relates to the nodal spacing near the point s_Q

If the nodes are uniformly distributed, d_c is the distance between two neighboring nodes. In the case where the nodes are nonuniformly distributed, d_c can be defined as an "average" nodal spacing in the support domain of s_Q. The procedure of determining d_c can be performed following the procedure in Section 1.7.3 for the 1D case based on our current curvilinear coordinate system.

As discussed in Section 2.11, the PIM approximation could be incompatible. Similar to the domain type in PIM methods, we can also formulate nonconforming and conforming BPIMs. In using the nonconforming BPIM, the support domain is determined for each and every Gauss point. In the conforming BPIM, however, the support domain is determined for each integration cell to ensure the compatibility of the field function approximation with the cells. Because the integration cells are connected at points (not lines), the compatibility at the nodes is automatically enforced. The number of nodes, n, can be determined by counting all the nodes in the support domain. The dimensionless size of the support domain α_s should be predetermined by the analyst. Our numerical examination suggests that $\alpha_s = 2.0$–3.0 (which includes $n = 3$–6) leads to an acceptable performance for BPIM.

13.1.2 Discrete Equations of BPIMs

The well-known BIE for 2D linear elastostatics can be written as [13]

$$c_i u_i + \int_\Gamma \mathbf{ut^*}d\Gamma = \int_\Gamma \mathbf{tu^*}d\Gamma + \int_\Omega \mathbf{bu^*}d\Omega \tag{13.18}$$

where
  c_i is a coefficient dependent on the geometric shape of the boundary
  b is the body force vector
  u* and **t*** are the fundamental solution for linear elastostatics

The fundamental solution for a 2D plane strain problem is given in [13] as

$$u_{ij}^* = \frac{1}{8\pi G(1-v)}\left[(3-4v)\ln\frac{1}{r}\delta_{ij} + r_{,i}r_{,j}\right] \tag{13.19}$$

$$t_{ij}^* = \frac{-1}{4\pi(1-v)r}\left\{\frac{\partial r}{\partial n}[(1-2v)\delta_{ij} + r_{,i}r_{,j}] - (1-2v)(r_{,i}n_j - r_{,j}n_i)\right\} \tag{13.20}$$

where
  G is the shear modulus
  v is the Poisson's ratio
  δ is the Kronecker delta function
  r is the distance between the source point and the field point
  n is the normal to the boundary
  a comma designates a partial derivative with respect to the indicated spatial variable

Substituting Equations 13.11 and 13.12 into Equation 13.18 yields the BPIM system equation for all the nodes on the boundary of the problem domain:

$$\mathbf{HU} = \mathbf{GT} + \mathbf{D} \tag{13.21}$$

where

$$\mathbf{H} = \mathbf{c}_i + \int_\Gamma \mathbf{t^*\Phi^T}d\Gamma \tag{13.22}$$

$$\mathbf{G} = \int_\Gamma \mathbf{u^*\Phi^T}d\Gamma \tag{13.23}$$

$$\mathbf{D} = \int_\Omega \mathbf{bu^*}d\Omega \tag{13.24}$$

13.1.3 Implementation Issues in BPIMs

13.1.3.1 Singular Integral

To evaluate the integrals given in Equations 13.22 through 13.24, background integration cells are required. The cells should be created on the boundary of the problem domain with proper dimensions to ensure an accurate integration. From Equations 13.19 and 13.20, it can be seen that the integrands in Equations 13.22 through 13.24 consist of regular and singular functions. The regular functions can be evaluated using the usual Gaussian quadrature. Equation 13.23 for the matrix $\mathbf{G}$ contains a log singular integral. This type of a singular integral can be evaluated by log Gaussian quadrature as follows:

$$I = \int_0^1 \ln{(1/x)} f(x) \mathrm{d}x \cong \sum_{i=1}^m f(x_i) w_i \tag{13.25}$$

where the required points x_i and weights w_i can be found in [13] for conventional BEMs.

In the matrix $\mathbf{H}$ defined by Equation 13.22, $\mathbf{c}$ is a coefficient dependent on the geometric shape of the boundary that is easy to obtain for a smooth boundary. However, it is more complicated to obtain $\mathbf{c}$ for nonsmooth boundaries. In addition, $\mathbf{H}$ contains a $(1/r)$ type of singular integral. Therefore, it is a nontrivial task to directly evaluate the diagonal terms of $\mathbf{H}$. The same difficulty has been experienced in the standard BEMs. Note that shape functions of BPIM possess the delta function property; therefore, the rigid body movement can be utilized in this work to obtain the diagonal terms of $\mathbf{H}$ [13].

13.1.3.2 Application of Boundary Conditions

There are two types of boundary conditions in BPIM:

$$\mathbf{t} = \mathbf{t}_\Gamma \text{ on the natural boundary } \Gamma_t \tag{13.26}$$

$$\mathbf{u} = \mathbf{u}_\Gamma \text{ on the essential boundary } \Gamma_u \tag{13.27}$$

Because the shape functions of BPIM have the delta function property, the boundary conditions can be imposed in the same way as the standard BEM. Note that if MLS shape functions are used, proper treatments are needed [2,12].

After applying the boundary condition, the system (Equation 13.21) has $2N_B$ equations and $2N_B$ unknowns for N_B boundary nodes. The system equation can be solved using standard routines of an algebraic equation solver to obtain the displacement and traction.

13.1.3.3 Handling of Corners with Traction Discontinuities

In handling traction discontinuities in corners, special care should be taken. Double nodes and discontinuous elements at corners need to be used to overcome this problem in the BEM. Because there are no elements used in BPIM, a simple method proposed in [11] to solve this difficulty is "displacing" the nodes from the corner, meaning that we do not put a node at the corner, but the support domain for PIM interpolations stretches to and then is truncated at the corner. A small privilege of the PIM interpolation is that this can be done beyond the cells. The method is very easy to implement, is used in the following numerical examples, and is proved to be very accurate.

13.1.4 Numerical Examples

The BPIM is coded and used to solve a number of problems of mechanics. A detailed comparison study is carried out using the present BPIM, BNM, BEM, and analytical methods. In BNM, a weight function must be used for constructing MLS shape functions. The exponential weight function [2] given below is used for the examples in this section:

$$w_i(s) = \begin{cases} \dfrac{e^{-(d_i/c)^{2k}} - e^{-(d_{sQ}/c)^{2k}}}{1 - e^{-(d_{sQ}/c)^{2k}}} & d \le d_{sQ} \\ 0 & d > d_{sQ} \end{cases} \tag{13.28}$$

where
$\quad d_i$ is the arc length
$\quad d_{sQ}$ is the size of the support of the weight function w_i, which is the dimension of the support domain in BNM
$\quad k$ and c are constants

In this section, we use $k=1$ and $d_{sQ}/c = 0.75$ which were used in [2].
The following presents some examples analyzed using the nonconforming BPIM.

Example 13.1: Rectangular Cantilever

BPIM is first applied to analyze the displacement and the stress field in a rectangular cantilever, which is shown in Figure 6.4. A plane stress problem is considered. The parameters for this example are

Young's modulus for the material: $E = 3.0 \times 10^7$
Poisson's ratios for two materials: $\nu = 0.3$
Thickness of the cantilever: $t = 1$
Height of the cantilever: $D = 12$
Length of the cantilever: $L = 48$
Load: $P = 1000$

The cantilever is subjected to a parabolic traction at the free end, as shown in Figure 6.4. The analytical solution is available in Timoshenko and Goodier [14] and is listed in Equations 6.50 through 6.55.

A total of 120 uniform boundary nodes are used to discretize the boundary of the cantilever. To evaluate the integral of matrices, 120 uniform integration cells are used. The parameter α_s in Equation 13.17 for the support domains is fixed at 2.0. Therefore, three to six nodes are included in the support domain for constructing shape functions.

Figure 13.2 illustrates the comparison between the shear stress calculated analytically and that by BPIM at the section $x = L/2$. The plot shows a good agreement between the analytical and numerical results. The conventional linear BEM results of this problem are also shown in the same figure for comparison. The density of the nodes in BEM and BPIM is exactly the same. It is clearly shown that the BPIM results are more accurate than the BEM results. This is because BPIM uses more nodes for the interpolation of displacements and tractions. Therefore, the order of the interpolant in BPIM is higher than the order of the linear elements in BEM.

For a detailed error analysis, we define the following norm as an error indicator using the shear stress, as with the shear stress it is much more critical than the other stress components in the rectangular cantilever to reflect the accuracy:

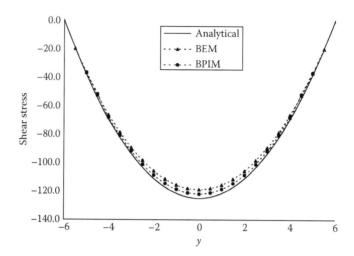

FIGURE 13.2
Shear stress τ_{xy} at the section $x = L/2$ of the beam. (From Gu, Y.T. and Liu, G.R., *Comput. Mech.*, 28, 47, 2002. With permission.)

$$e_t = \frac{1}{N} \sqrt{ \sum_{i=1}^{N} (\tau_i^{num} - \tau_i^{exact})^2 \Big/ \sum_{i=1}^{N} (\tau_i^{exact})^2 }$$

(13.29)

where
   N is the number of nodes investigated
   τ^{num} is the shear stress obtained numerically
   τ^{exact} is the analytical shear stress

The convergence for the shear stresses at the section $x = L/2$ with mesh refinement is shown in Figure 13.3. The convergences of BNM and the linear BEM are also shown in the

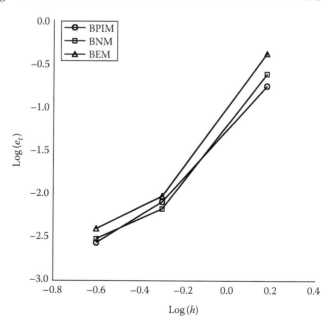

FIGURE 13.3
Convergence in the e_t norm of the error. (From Gu, Y.T. and Liu, G.R., *Comput. Mech.*, 28, 47, 2002. With permission.)

TABLE 13.1

Effects of the Dimension of the Support
Domain on the Error of the Energy Norm

Dimension of the Support Domain (α_s)	e_t^a
1.0	0.1688
2.0	0.1410
3.0	0.1775
4.0	0.1812
5.0	0.1794

a Defined by Equation 13.29.

same figure, where h is a characteristic length related to the nodal spacing. Three kinds of nodal arrangements of 40, 240, and 480 uniform boundary nodes are used. It can be observed that the accuracy of BPIM and BNM using MLS approximation is nearly the same, but both BPIM and BNM have higher accuracies than BEM. The convergence of BPIM seems to be the best among these three methods.

As mentioned above, a dimensionless size of the support domain α_s used in Equation 13.17 needs to be chosen such that a reasonable number of nodes lie in the support domain of an evaluation point. The results of e_t for different sizes of support domains are shown in Table 13.1. In this analysis, the boundary is modeled by 40 uniformly distributed nodes, and 40 uniformly spaced integration background cells are used. It is found that the accuracy of the results of BPIM changes slightly with the dimension of the support domain when the nodal density is fixed. Although the choice of the support domain may also depend a little on the type of the problem, it is found that $\alpha_s = 2.0$–3.0 works well for most of the problems investigated in this section.

It may be mentioned here that the use of a large support domain does not necessarily lead to more accurate results.

Example 13.2: Plate with a Hole

Consider now an infinite plate with a central circular hole subjected to a unidirectional tensile load of $p = 1.0$ in the x-direction. As a large finite plate can be considered a good approximation of an infinite plate, a finite square plate of 20×20 is considered. Making use of the symmetry, only the upper right quadrant of the finite plate is modeled, as shown in Figure 13.4. Plane strain condition is assumed, and the material properties are $E = 1.0 \times 10^3$ and $v = 0.3$. Symmetry conditions are imposed on the left and bottom edges, and the inner boundary of the hole is traction free. The tensile load p is imposed on the right edge in the x-direction. The exact solution for the stresses of an infinite plate is given in [14] and is listed in Equations 7.41 through 7.46.

A total of 68 nodes are used to discretize the boundary (with 10 uniformly distributed nodes on BC, CD, and AE, and 19 nonuniformly distributed nodes along AB and DE). The same number of integration background cells is used. The support domain of an evaluation point is determined, as in Equation 13.17 (with $\alpha_s = 2.0$, and the characteristic arc length: $d_Q = 1.0$ on AB, BC, CD, and DE, and $d_Q = 0.2$ along AE). If the number of nodes in the

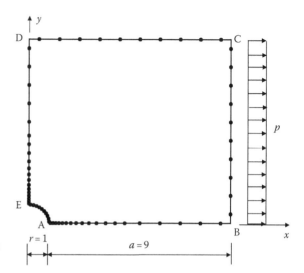

FIGURE 13.4
Nodes in a plate with a central hole subjected to a
unidirectional tensile load in the x-direction.

support domain is more than six, only six nodes with a shorter arc length to the integration
point are used in the interpolation.

As the stress is more critical in the assessment of the solution accuracy, detailed results of
the stress are presented here. The stress σ_x at $x=0$ obtained by BPIM is given in Figure 13.5
together with the analytical solution for the infinite plate. It can be observed from this
figure that BPIM gives very good results for this problem. The BNM results of this problem
are also shown in the same figure for comparison. It is clearly shown that the BPIM and
BNM results possess nearly the same accuracy.

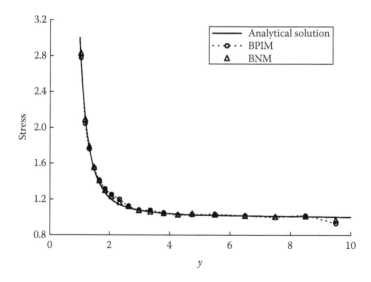

FIGURE 13.5
Stress distribution in the plate with a central hole subjected to a unidirectional tensile load (σ_x, at $x=0$). (From
Gu, Y.T. and Liu, G.R., *Comput. Mech.*, 28, 47, 2002. With permission.)

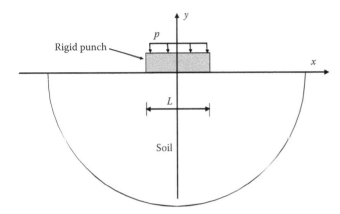

FIGURE 13.6
Rigid punch forced on a semi-infinite soil foundation. (From Gu, Y.T. and Liu, G.R., *Comput. Mech.*, 28, 47, 2002. With permission.)

Example 13.3: A Rigid Flat Punch on a Semi-Infinite Foundation

As BEM methods have a clear advantage over domain-type methods for problems with infinite domain, BPIM is used to obtain a solution for an indentation produced by a rigid flat punch in a semi-infinite soil foundation, as shown in Figure 13.6. In this case, Green's functions for a half plane are employed [13], and only the contact surface between the punch and the half-space needs to be discretized.

Consider a rigid punch of length $L = 12$ subjected to a uniform pressure $p = 100$ on the top surface. The parameters of the soil foundation are taken as $E = 3.0 \times 10^4$, and $v = 0.3$. The punch is considered to be perfectly smooth, and does not result in any fraction force in the interface between the punch and the foundation. The indentation is measured by the vertical displacement of the punch. A plane strain condition is assumed. Due to symmetry, only the right half of the contact surface is discretized by 31 distributed boundary nodes; 31 nonuniformly distributed integration background cells are used. Coordinates of these boundary nodes are obtained using the following formula:

$$x_m = \frac{6.2(m - 1)}{m} \tag{13.30}$$

where m is the node number, and $m = 1\text{--}31$.

The vertical surface displacements of the foundation are assumed to be the same as that of the punch (perfect contact). This assumption is often proved true for a rigid punch. A prescribed vertical displacement of the punch is imposed on the contact surface as a boundary constraint. The prescribed displacement of the punch can be obtained using the approximate method presented in [15], i.e., the vertical displacement of a vertically loaded rigid area in contact with the rigid punch may be approximated by the mean vertical displacement of a uniformly loaded flexible area of the same shape. The approximation is expressed as follows:

$$v_{\text{rigid}} = {}^1\!/_2(v_{\text{center}} + v_{\text{edge}})_{\text{flexible}} \tag{13.31}$$

where
v_{rigid} is the vertical displacement of the rigid area in contact with the rigid punch
v_{center} and v_{edge} are vertical displacements at the center and edge, respectively, of the contact area subjected to uniform load, when the contact area is considered flexible

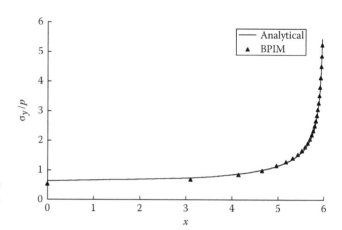

FIGURE 13.7
Contact stresses on the contact surface between the punch and the half-space. (From Gu, Y.T. and Liu, G.R., *Comput. Mech.*, 28, 47, 2002. With permission.)

The analytical solution of v_{center} and v_{edge} can be obtained in [14]. The exact solution [15] of contact stresses along the contact surface is

$$\frac{\sigma_y}{p} = \frac{2}{\pi} \frac{1}{\sqrt{1 - (2x/L)^2}} \tag{13.32}$$

BPIM is used to obtain contact stresses along the contact surface. The support domain of a quadrature point is determined by Equation 13.17 (with $\alpha_s = 2.0$, and the characteristic arc length $d_Q = 3.0$). If the number of nodes in the support domain is more than six, only six nodes with a shorter arc length to the quadrature point are used in the interpolation. When these six nodes are all on one side of the quadrature point along the boundary, one more node nearest to this evaluation point on the other side is purposely added to the support domain to avoid an extrapolation.

Figure 13.7 plots the comparison between the contact stresses calculated analytically and that by the BPIM along the contact surface. The plot shows an excellent agreement between the analytical and numerical results.

13.2 RPIM Using Radial Function Basis

For 2D problems, the boundaries of the domain are curves. Therefore, the PIM shape function using polynomial basis will have no problem, and BPIM works perfectly well without any special efforts. For 3D problems for which 2D shape functions need to be constructed, there could be an issue of singular moment matrices. One effective way is to use RPIM shape functions. This section introduces the boundary meshfree method using RPIM shape functions. This method was formulated in [10] and termed as BRPIM. Although BRPIM performs no better than BPIM for 2D problems, its full advantages are expected for 3D problems.

The formulation procedure of BRPIM is largely the same as that of BPIM, except for the formulation of the shape function. This section, therefore, focuses only on the portion of the formulation that is different from BPIM. As the radial function is used, a study on the effects of these parameters of the radial function is performed. The performance of BRPIM is discussed using example problems of 2D elastostatics.

13.2.1 Radial Basis Point Interpolation

In BRPIM, the radial basis functions $R_i(s)$ are used. The point interpolation for displacement $u(s)$ and traction $t(s)$ at a point s on the boundary Γ from the surrounding nodes uses radial basis functions, i.e.,

$$u = \sum_{i=1}^{n} R_i(s)\alpha_i = \mathbf{R}^{\mathrm{T}}(s)\boldsymbol{\alpha} \tag{13.33}$$

$$t = \sum_{i=1}^{n} R_i(s)\beta_i = \mathbf{R}^{\mathrm{T}}(s)\boldsymbol{\beta} \tag{13.34}$$

where
s is the curvilinear distance (the arc length for a 1D curve boundary) on Γ
n is the number of nodes in the support domain of a point of interest s_Q, which is usually the quadrature point
$R_i(s)$ is the radial basis function
α_i and β_i are the coefficients

In the matrix form, we have

$$\boldsymbol{\alpha}^{\mathrm{T}} = [\alpha_1, \alpha_2, \ldots, \alpha_n] \tag{13.35}$$

$$\boldsymbol{\beta}^{\mathrm{T}} = [\beta_1, \beta_2, \ldots, \beta_n] \tag{13.36}$$

The following multiquadrics (MQ) radial function is used in this section:

$$R_i(s) = (s_i^2 + C^2)^q \tag{13.37}$$

Two parameters, q and C, need to be determined in an MQ radial function. Detailed investigations of these parameters are given in the following numerical examples.

The coefficients α_i and β_I can be determined by enforcing Equations 13.33 and 13.34 to be satisfied at the n-node support domain of the point s_Q. Equations 13.33 and 13.34 can then be written in the following matrix form:

$$\mathbf{u}_n = \mathbf{B}\boldsymbol{\alpha} \tag{13.38}$$

$$\mathbf{t}_n = \mathbf{B}\boldsymbol{\beta} \tag{13.39}$$

where $\mathbf{u}_n$ and $\mathbf{t}_n$ are the vectors of nodal displacements and tractions in the support domain, given by

$$\mathbf{u}_n = [u_1, u_2, u_3, \ldots, u_n]^{\mathrm{T}} \tag{13.40}$$

$$\mathbf{t}_n = [t_1, t_2, t_3, \ldots, u_n]^{\mathrm{T}} \tag{13.41}$$

and $\mathbf{R}_Q$ is the moment matrix of radial basis functions

$$
\mathbf{R}_Q = \begin{bmatrix}
R_1(s_1) & R_2(s_1) & \cdots & R_n(s_1) \\
R_1(s_2) & R_2(s_2) & \cdots & R_n(s_2) \\
\vdots & \vdots & \ddots & \vdots \\
R_1(s_n) & R_2(s_n) & \cdots & R_n(s_n)
\end{bmatrix}
\tag{13.42}
$$

Solving Equations 13.38 and 13.39 for α and β, and then substituting them back into Equations 13.33 and 13.34, we obtain

$$
u(s) = \mathbf{\Phi}^{\mathrm{T}}(s)\, \mathbf{u}_n \tag{13.43}
$$

$$
t(s) = \mathbf{\Phi}^{\mathrm{T}}(s)\, \mathbf{t}_n \tag{13.44}
$$

where the shape function $\mathbf{\Phi}(s)$ is defined by

$$
\mathbf{\Phi}^{\mathrm{T}}(s) = \mathbf{R}^{\mathrm{T}}(s)\mathbf{R}_Q^{-1} = [\phi_1(s), \phi_2(s), \phi_3(s), \ldots, \phi_n(s)] \tag{13.45}
$$

Similar to the polynomial basis shape functions, the shape function $\phi_i(s)$ obtained through the above procedure satisfies

$$
\phi_i(s = s_i) = 1, \quad i = 1, n \tag{13.46}
$$

$$
\phi_j(s = s_i) = 0, \quad j \neq i \tag{13.47}
$$

$$
\sum_{i=1}^{n} \phi_i(s) = 1 \tag{13.48}
$$

Therefore, the RPIM shape functions constructed have the Kronecker delta function property, and the boundary conditions can be easily imposed, as in the BEM.

The moment matrix is an $n \times n$ matrix. It must be invertible for the construction of the shape functions in Equation 13.45. The existence of the inverse of $\mathbf{R}_Q$ has been proved for arbitrarily scattered nodes [16,17]. Therefore, in BRPIM, interpolation using the radial basis function is stable and flexible for arbitrarily distributed nodes on the boundary of the problem domain. These characteristics will be very beneficial when using BRPIM to solve 3D problems.

13.2.2 BRPIM Formulation

The formulation of system equations in BRPIM is exactly the same as that in BPIM, except that the PIM shape function given by Equation 13.13 is replaced by the RPIM shape function defined by Equation 13.45.

13.2.3 Comparison of BPIM, BNM, and BEM

A comparison of BPIM, BNM, and BEM is summarized concisely in Table 13.2. It can be found that BPIM, BNM, and BEM are all based on the BIE. The difference is essentially in the means of implementation.

TABLE 13.2

Comparison of BPIM, BRPIM, BNM, and BEM

	BPIM	BRPIM	BNM	BEM
Mesh	No	No	No	Yes
Approximation	Polynomial PIM	Radial PIM	MLS	Element-based polynomial
Approximation base	Distributed nodes	Distributed nodes	Distributed nodes	Element
Number of basis nodes m and interpolation nodes n	$m = n$	$m = n$	$m \neq n$	$m = n$
Overlapping of the interpolation area	Overlapping	Overlapping	Overlapping	No overlapping
Shape function	Simple	Simple	Complicated	Simple
Delta property of the shape function	Yes	Yes	No	Yes
Application of boundary conditions	Easy	Easy	More effort	Easy
Number of system equations	$2N_B$	$2N_B$	$4N_B$	$2N_B$

Source: Gu, Y.T. and Liu, G.R., *Comput. Mech.*, 28, 47–54, 2002. With permission.

13.2.3.1 BPIM vs. BEM

Both BPIM and BEM use polynomial interpolants, in which the number of monomials used in the base functions, m, is the same as the number of nodes, n, utilized. Therefore, the interpolant functions possess the Kronecker delta function property. The boundary conditions can be implemented with ease.

However, BPIM is a boundary-type meshfree method, whereas BEM is a boundary-type method based on boundary elements. As other meshfree methods (e.g., EFG, BNM, and MLPG), the interpolation procedure in BPIM is based only on a group of arbitrarily distributed nodes. The interpolation at a quadrature point in BPIM is performed over the support domain of the point, which may overlap with the support domains of other quadrature points, as shown in Figure 13.8. BEM defines the shape functions over predefined regions called elements, and there is no overlapping or gap between the elements.

13.2.3.2 BPIM vs. BNM

Both BPIM and BNM are boundary-type meshfree methods. The difference between these two methods arises from the different interpolants utilized. As discussed above, BPIM uses PIM shape functions, in which the coefficients a and b in Equation 13.1 are constant. The MLS approximations are used in BNM, in which a and b are also functions of the curvilinear coordinate s. Therefore, the shape function of BNM is more complicated than PIM. In addition, the shape function of BNM constructed using MLS approximations lacks the delta function property. It takes an extra effort to impose boundary conditions.

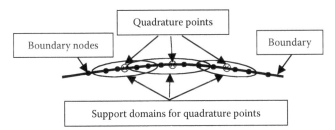

FIGURE 13.8
Interpolations in BPIM and BRPIM.

13.2.4 Numerical Examples

Example 13.4: Cantilever Beam

Example 13.1 is used here again to examine BRPIM. All the parameters are exactly the same.

13.2.4.1 Effects of Radial Function Parameters

Two parameters, α_C and q, in the MQ radial function defined in Table 2.3 are investigated and their effects on the performance of BRPIM are revealed. The characteristic length d_c is taken to be the average nodal spacing for nodes in the support domain of the quadrature point.

The parameter q is first investigated, and q is taken to be -1.5, -0.5, 0.5, and 1.5. Shear stresses for different q are obtained and compared with the analytical solution. Errors for different q are plotted in Figure 13.9a. This figure shows that $q = 0.5$ leads to a better result in the range of studies. Hence, 0.5 is used in the following studies.

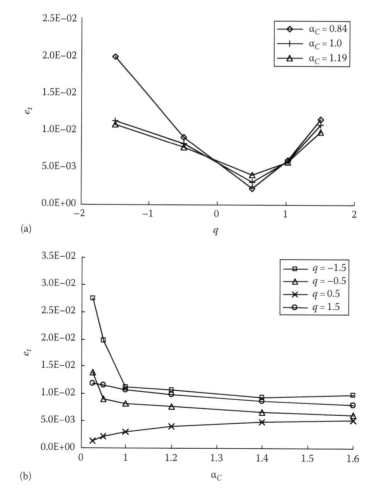

FIGURE 13.9

Effect of parameters q and α_C of the MQ radial basis function on the error of shear stress in the rectangular cantilever computed using BRPIM. e_t is defined by Equation 13.29. (a) Effect of q and (b) effect of α_C.

Errors in shear stresses for different α_C are plotted in Figure 13.9b. It is found that α_C can be chosen from a wide range, $\alpha_C = 1.0\text{--}1.6$, where steadily accurate results can be obtained. For convenience, $\alpha_C = 1.2$ is used in the following studies.

From studies for 2D interpolation, it has also been understood that α_C has a wider range of choices, but parameter q is very critical and has to be very precise to obtain good results. To determine a more precisely tuned q, more detailed study is required.

13.2.4.2 Effects of Interpolation Domain

The size of the support domain of a quadrature point is determined by the parameter α_s in Equation 13.17. Results of $\alpha_s = 1.0\text{--}5.0$ are obtained and plotted in Figure 13.10. It is found that the results obtained using a support domain with $\alpha_s = 3.0\text{--}4.5$, which covers about 6–10 nodes, are very good. Too small a support domain ($\alpha_s = 2.5$) will lead to a large error. This is because there are not enough nodes to perform interpolation for the field variable. Too big a support domain ($\alpha_s = 4.5$) will also lead to a large error. This is because there are too many nodes to perform interpolation, which results in a very complex shape function and hence a complex integrand for computing the system equations. The numerical integral error becomes very large. Therefore, $\alpha_s = 3.0\text{--}4.5$ is recommended. For convenience and consistency, $\alpha_s = 4.0$ is used in the following studies.

Comparison with the case of using BPIM, for which $\alpha_s = 2.0$ is the optimum (see Table 13.1) reveals that BRPIM requires a larger support domain and more points for interpolation.

Figure 13.11 illustrates the comparison between the shear stress calculated analytically and that by the BRPIM at the section $x = L/2$. The plot shows a good agreement between the analytical and numerical results from BRPIM. The conventional linear BEM results of this problem are also shown in the same figure for comparison. The density of the nodes in BEM and BRPIM is exactly the same. It is clearly shown that the BRPIM results are more accurate than the BEM results. This is because the BRPIM uses more nodes for the interpolation of displacements and tractions.

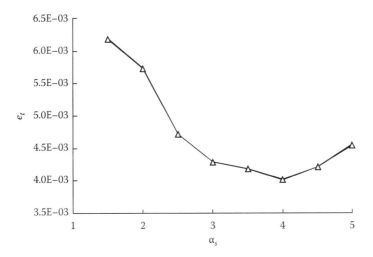

FIGURE 13.10
Influence of the parameter α_s of the interpolation domain.

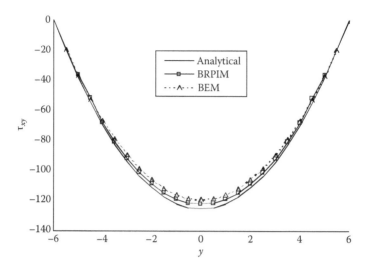

FIGURE 13.11

Shear stress at the section $x = L/2$ of the rectangular cantilever. Comparison of results obtained using three different methods.

The convergence for the shear stresses at the section $x = L/2$ with node/mesh refinement is shown in Figure 13.12, where h is a characteristic length relating the spacing of the nodes. Three kinds of nodal arrangements of 72, 240, and 480 uniform boundary nodes are used. The convergences of BNM and conventional linear BEM are also shown in the same figure. It is observed that the convergence of BRPIM is very good. It can also be observed that BRPIM has higher accuracy than BEM and BNM for this example.

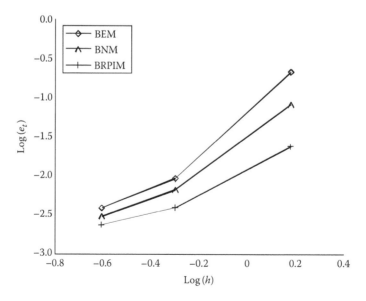

FIGURE 13.12

Convergence in the e_t norm of the error.

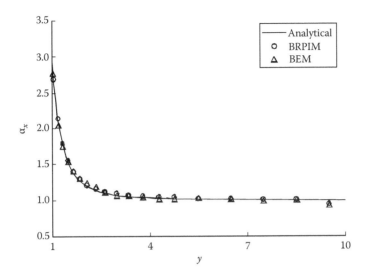

FIGURE 13.13

Stress σ_x distribution in a plate with a central hole subjected to a unidirectional tensile load. Comparison of results obtained using three different methods.

Example 13.5: Plate with a Hole

Example 13.2 is also used to examine BRPIM. All the parameters are exactly the same. The nodal distribution is shown in Figure 13.4. The stress σ_x at $x = 0$ computed using the BRPIM code is given in Figure 13.13 together with the analytical solution for the infinite plate. The BEM results of this problem are also shown in the same figure for comparison. It can be observed from this figure that BRPIM gives very good results. It is clearly shown that the BRPIM and BEM results possess nearly the same accuracy for this problem.

Example 13.6: Internally Pressurized Hollow Cylinder

A hollow cylinder under internal pressure shown in Figure 13.14 is considered. The parameters are taken as $p = 100$, $G = 8000$, and $v = 0.25$. This problem has been used by several other authors as a benchmark problem, as the analytical solution is available. Due to the symmetry of the problem, only one quarter of the cylinder needs to be modeled. The arrangement of the field nodes is shown in Figure 13.15. The boundary of this domain is discretized by 30 nodes (6 uniformly distributed nodes on *ab*, *cd*, and *ad*, and 12 uniformly distributed nodes on *bc*). The same number of background cells is used for integration. Three internal points A, B, and C are selected for examination. The polar coordinates (with the origin at the center of the cylinder) for the three internal points are A(13.75, $\pi/4$), B(17.5, $\pi/4$), and C(21.25, $\pi/4$). The dimensionless size of the support domain $\alpha_s = 2.0$ is used for all the quadrature points in the integration on the boundary.

The BRPIM and BPIM results are compared with those obtained using BNM, BEM, and the analytical solution. The radial displacements at some of the boundary nodes and the three internal points are listed in Table 13.3. The circumferential stresses σ_θ at points A, B, and C are also listed in the same table. The BRPIM and BPIM results are in a very good agreement with the analytical solution. In comparison with the conventional BEM results, the BRPIM and BPIM solutions are, in general, more accurate for both displacements and stresses.

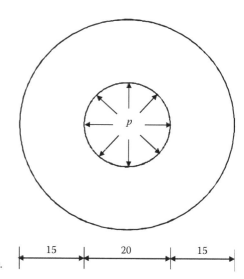

FIGURE 13.14
Hollow cylinder subjected to internal pressure.

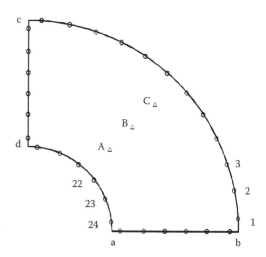

FIGURE 13.15
Arrangement of nodes for a quarter model of the hollow
cylinder.

13.3 Remarks

Boundary meshfree methods are presented in this chapter. Detailed formulations for BPIM
and BRPIM are provided for solving 2D problems of elastostatics. The BIE is discretized
using radial meshfree shape functions based on a group of arbitrarily distributed points on
the boundary of the problem domain. Numerical examples have demonstrated that
boundary meshfree methods are superior to BEM in terms of accuracy.

Compared with BNM, BPIM and BRPIM have the following advantages:

- BPIM is computationally much less expensive than BNM because of its simpler
  interpolation scheme and smaller system equation dimension. The number of
  system equations in BPIM is only half of that in BNM.

TABLE 13.3

Radial Displacements and Circumferential Stresses in the Pressurized Hollow Cylinder

Nodes	Exact	BPIM	BRPIM	BNM	BEM
Radial Displacements ($\times 10^{-2}$)					
1	0.4464	0.4465	0.4466	0.4462	0.4468
2	0.4464	0.4478	0.4475	0.4463	0.4482
3	0.4464	0.4491	0.4493	0.4498	0.4494
22	0.8036	0.8213	0.8200	0.8220	0.8266
23	0.8036	0.8214	0.8207	0.8215	0.8268
24	0.8036	0.8199	0.8215	0.8223	0.8251
A	0.6230	0.6274	0.6211	0.6256	0.6319
B	0.5294	0.5342	0.5366	0.5353	0.5374
C	0.4766	0.4809	0.4810	0.4826	0.4838
Stresses σ_θ					
A	82.0113	81.8947	82.1437	81.8513	82.0192
B	57.9226	58.1285	58.2585	58.1627	58.1691
C	45.4112	45.6471	45.6264	45.4597	45.6575

Source: Gu, Y.T. and Liu, G.R., *Comput. Mech.*, 28, 47–54, 2002. With permission.

- The imposition of boundary conditions is easy in BPIM and BRPIM because the shape functions have the Kronecker delta property.
- Rigid body movement can be used to avoid some singular integrals.

The parameters for BPIM and BRPIM should be as follows:

- In using BPIM, $\alpha_s = 2.0$–3.0 (with $n = 3$–6) yields acceptable results.
- In using BRPIM with the MQ radial function, $q = 0.5$ and $\alpha_C = 1.0$–1.6 lead to acceptable results for most problems studied. $q = 0.5$ and $\alpha_C = 1.2$ are recommended. The dimensionless size of the support domain $\alpha_s = 3.0$–4.5 should work for most problems.
- When the node distribution is very extreme, measures for preventing biased node selection and extrapolation should be taken.

BPIM and BRPIM need to be extended for 3D problems. For 2D problems, BPIM is the simplest method; it performs the best and is very stable.

References

1. Mukherjee, Y. X. and Mukherjee, S., Boundary node method for potential problems, *Int. J. Numerical Methods Eng.*, 40, 797–815, 1997.
2. Kothnur, V. S., Mukherjee, S., and Mukherjee, Y. X., Two-dimensional linear elasticity by the boundary node method, *Int. J. Solids Struct.*, 36, 1129–1147, 1999.
3. Chati, M. K., Mukherjee, S., and Mukherjee, Y. X., The boundary node method for three-dimensional linear elasticity, *Int. J. Numerical Methods Eng.*, 46, 1163–1184, 1999.

4. Chati, M. K. and Mukherjee, S., The boundary node method for three-dimensional problems in potential theory, *Int. J. Numerical Methods Eng.*, 47, 1523–1547, 2000.

5. Zhu, T. and Atluri, S. N., A modified collocation and a penalty formulation for enforcing the essential boundary conditions in the element free Galerkin method, *Comput. Mech.*, 21, 211–222, 1998.

6. Liu, G. R. and Gu, Y. T., A local point interpolation method for stress analysis of two-dimensional solids, *Struct. Eng. Mech.*, 11(2), 221–236, 2001.

7. Liu, G. R. and Gu, Y. T., On formulation and application of local point interpolation methods for computational mechanics, in *Proceedings of 1st Asia-Pacific Congress on Computational Mechanics*, Sydney, Australia, November 20–23, 2001, pp. 97–106 (invited paper).

8. Liu, G. R. and Gu, Y. T., A point interpolation method, in *Proceedings of 4th Asia-Pacific Conference on Computational Mechanics*, Singapore, December 1999, pp. 1009–1014.

9. Liu, G. R. and Gu, Y. T., A point interpolation method for two-dimensional solids, *Int. J. Numerical Methods Eng.*, 50, 937–951, 2001.

10. Gu, Y. T. and Liu, G. R., A boundary point interpolation method (BPIM) using radial function basis, in *1st MIT Conference on Computational Fluid and Solid Mechanics*, MIT, June 2001, pp. 1590–1592.

11. Gu, Y. T. and Liu, G. R., A boundary point interpolation method for stress analysis of solids, *Comput. Mech.*, 28(1), 47–54, 2002.

12. Mukherjee, Y. X. and Mukherjee, S., On boundary conditions in the element-free Galerkin method, *Comput. Mech.*, 19, 264–270, 1997.

13. Brebbia, C. A., Telles, J. C., and Wrobel, L. C., *Boundary Element Techniques*, Springer Verlag, Berlin, 1984.

14. Timoshenko, S. P. and Goodier, J. N., *Theory of Elasticity*, 3rd ed., McGraw-Hill, New York, 1970.

15. Poulos, H. G. and Davis, E. H., *Elastic Solution for Soil and Rock Mechanics*, Wiley, London, 1974.

16. Kansa, E. J., A scattered data approximation scheme with application to computational fluid-dynamics—I & II, *Comput. Math. Appl.*, 19, 127–161, 1990.

17. Kansa, E. J., Multiquadrics—A scattered data approximation scheme with applications to computational fluid dynamics, *Comput. Math. Appl.*, 19(8–9), 127–145, 1990.

14

Meshfree Methods Coupled with Other Methods

In the past few decades, the development of finite element methods (FEMs) has been accompanied by advances in boundary element methods (BEMs). The FEM is a domain discretization method, whereas the BEM is a boundary discretization method. Both methods have their strengths and weaknesses. The FEM is much more flexible for complex structures/domains with high inhomogeneity and nonlinearity but requires intensive computational resources. On the other hand, the BEM requires much less computational resources, as discretization of the structure/domain is performed only on the boundary, which leads to a much smaller discretized equation system. The BEM, however, is not efficient for inhomogeneous media/domain and nonlinear problems. Efforts to combine these two methods have been made by many (see, e.g., [1]) and have achieved remarkable results. Commercial software packages have also been developed (e.g., SYSNOISE) and used for solving a wide range of engineering problems.

In previous chapters, we presented both domain-type meshfree methods and boundary-type meshfree methods. Naturally, attempts have also been made to combine these two types of methods to take advantage of both. There is an additional motivation to couple meshfree methods that are formulated using moving least squares (MLS) shape functions and meshfree methods that are formulated using point interpolation method (PIM) shape functions or finite element (FE) shape functions. The aim is to simplify the procedure of imposing essential boundary conditions. A number of combined methods have been formulated including element-free Galerkin (EFG)/BEM [2], EFG/HBEM [3], meshless local Petrov–Galerkin (MLPG)/FEM/BEM [4,5], etc. This chapter is devoted to introducing the EFG/BEM [2] and EFG/HBEM.

14.1 Coupled EFG/BEM

This section focuses on the coupling of the EFG method with the boundary element (BE) method. Techniques for coupling the equation systems of EFG with those of BEM for continuum mechanics problems are presented in detail. This work was originally reported in [2]. The major issue was to enforce the displacement compatibility conditions on the interface boundary between the EFG domain and the BE domain. The interface elements, which are analogues of the FE interface element used in [6], are formulated and used along the interface boundary. Within the interface element the shape functions comprise the MLS and FE shape functions. Shape functions constructed in this manner satisfy both consistency and compatibility conditions on the interfaces. A number of numerical examples are presented to demonstrate the convergence, validity, and efficiency of the coupled method. It is shown that the coupled method can take full advantage of both EFG and BEM. It is very easy to implement, and very flexible for computing displacements and stresses of a desired accuracy in solids with or without infinite domains.

14.1.1 Basic Equations of Elastostatics

Consider the following two-dimensional (2D) problem of solid mechanics in domain Ω bounded by Γ:

$$\mathbf{L}_d^{\mathrm{T}}\boldsymbol{\sigma} + \mathbf{b} = \mathbf{0} \quad \text{in } \Omega \tag{14.1}$$

where

    $\boldsymbol{\sigma}$ is the stress tensor, which corresponds to the displacement field $\mathbf{u} = \{u, v\}^{\mathrm{T}}$
    $\mathbf{b}$ is the body force vector
    $\mathbf{L}_d$ is the differentiation operator defined by Equation 1.9

The boundary conditions are given as follows:

$$\mathbf{L}_n^{\mathrm{T}}\boldsymbol{\sigma} = \mathbf{t}_\Gamma \quad \text{on the natural boundary } \Gamma_t \tag{14.2}$$

$$\mathbf{u} = \mathbf{u}_\Gamma \quad \text{on the essential boundary } \Gamma_u \tag{14.3}$$

in which the superposed bar denotes the prescribed boundary values and $\mathbf{L}_n$ is the matrix of the components of the unit outward normal on Γ_t.

14.1.2 Discrete Equations of EFG

In using a coupled EFG/BEM method, one can use BEs to model the portion of the domain that includes the essential boundary and the EFG is used where there is no essential boundary. Following the procedure presented in Chapter 6, without considering the essential boundary, we have the discrete system equation of EFG for all the field nodes in the EFG domain:

$$\mathbf{K}_{\mathrm{EFG}}\mathbf{U} = \mathbf{F}_{\mathrm{EFG}} + \mathbf{P}_{\mathrm{EFG}} \tag{14.4}$$

where the subscript EFG indicates matrices obtained using a standard EFG formulation. The vector $\mathbf{F}_{\mathrm{EFG}}$ consists of the equivalent nodal forces contributed from the external force applied on the natural boundary. The nodal force can be obtained using

$$\mathbf{f}_{(\mathrm{EFG})i} = \int_{\Gamma_t} \boldsymbol{\phi}_i^H \mathbf{t}\,\mathrm{d}\Gamma \tag{14.5}$$

The force vector $\mathbf{P}_{\mathrm{EFG}}$ consists of the equivalent nodal forces contributed from the external body force in the form

$$\mathbf{P}_{(\mathrm{EFG})i} = \int_{\Omega} \boldsymbol{\phi}_i^H \mathbf{b}\,\mathrm{d}\Omega \tag{14.6}$$

Note that if EFG has to be used for the portion of the problem domain containing essential boundaries, formulations using the method of Lagrange multipliers, the penalty method, or any other method discussed in Chapter 6 must be used.

14.1.3 BE Formulation

From Equations 14.1 through 14.3, the principle of virtual displacements for linear elastic materials can be written as

$$\int_{\Omega} (\mathbf{L}_d^{\mathrm{T}} \boldsymbol{\sigma} + \mathbf{b}) \cdot \mathbf{u}^* \mathrm{d}\Omega = \int_{\Gamma_u} (\mathbf{u} - \mathbf{u}_{\Gamma}) \cdot \mathbf{t}^* \mathrm{d}\Gamma - \int_{\Gamma_t} (\mathbf{t} - \mathbf{t}_{\Gamma}) \cdot \mathbf{u}^* \mathrm{d}\Gamma \qquad (14.7)$$

where
  $\mathbf{t}$ is the surface traction
  $\mathbf{u}^*$ is the virtual displacement
  $\mathbf{t}^*$ is the virtual surface traction corresponding to $\mathbf{u}^*$

The first term on the left-hand side of Equation 14.7 can be integrated by parts to become

$$\int_{\Omega} \mathbf{b} \cdot \mathbf{u}^* \mathrm{d}\Omega + \int_{\Omega} \mathbf{L}_d^{\mathrm{T}} \boldsymbol{\sigma}^* \cdot \mathbf{u} \, \mathrm{d}\Omega = \int_{\Gamma_u} (\mathbf{u}^* \cdot \mathbf{t} - \mathbf{u}_{\Gamma} \cdot \mathbf{t}^*) \mathrm{d}\Gamma + \int_{\Gamma_t} (\mathbf{u} \cdot \mathbf{t}^* - \mathbf{u}^* \cdot \mathbf{t}_{\Gamma}) \mathrm{d}\Gamma \qquad (14.8)$$

The starting domain integral can be reduced to an integral on the boundary by finding an analytical solution that makes the second integral in Equation 14.8 vanish. The most convenient one is the fundamental solution or Green's function, which satisfies

$$\mathbf{L}_d^{\mathrm{T}} \boldsymbol{\sigma}^* + \Delta^i = \mathbf{0} \qquad (14.9)$$

where Δ^i is the Dirac delta function. Substituting Equation 14.9 into Equation 14.8, we obtain

$$c^i u^i + \int_{\Gamma} \mathbf{u} \cdot \mathbf{t}^* \mathrm{d}\Gamma = \int_{\Gamma} \mathbf{t} \cdot \mathbf{u}^* \mathrm{d}\Gamma + \int_{\Omega} \mathbf{b} \cdot \mathbf{u}^* \mathrm{d}\Omega \qquad (14.10)$$

The boundary values of $\mathbf{u}$ and $\mathbf{t}$ can now be expressed using interpolation functions and the values at the nodes of the BE on the boundary:

$$\mathbf{u} = \boldsymbol{\Phi}^{\mathrm{T}} \mathbf{u}^e \qquad (14.11)$$

$$\mathbf{t} = \boldsymbol{\Psi}^{\mathrm{T}} \mathbf{t}^e \qquad (14.12)$$

where $\boldsymbol{\Phi}^{\mathrm{T}}$ and $\boldsymbol{\Psi}^{\mathrm{T}}$ can be the conventional FE shape functions constructed based on the BEs, or the PIM shape functions constructed based on cells. $\mathbf{u}^e$ and $\mathbf{t}^e$ are the values of $\mathbf{u}$ and $\mathbf{t}$ at the boundary nodes. The resulting boundary integral (Equation 14.10) can be written in matrix form as

$$\mathbf{H}\mathbf{U} = \mathbf{B}\mathbf{T} + \mathbf{P} \qquad (14.13)$$

where $\mathbf{U}$ and $\mathbf{T}$ are vectors that collect all the nodal values of $\mathbf{u}$ and $\mathbf{t}$ at the boundary nodes, and

$$\mathbf{H} = \mathbf{c}^i + \int_\Gamma \mathbf{t}^* \mathbf{\Phi}^T d\Gamma \tag{14.14}$$

$$\mathbf{B} = \int_\Gamma \mathbf{u}^* \mathbf{\Psi}^T d\Gamma \tag{14.15}$$

$$\mathbf{P} = \int_\Omega \mathbf{b} \cdot \mathbf{u}^* d\Omega \tag{14.16}$$

The above integrals are to be carried only on boundaries, and therefore the domain need not be discretized.

To facilitate assembling the system equations of EFG and BE, the BE formulation is expressed in an equivalent form of the EFG formulation. Transforming Equation 14.13 by inverting $\mathbf{B}$ and then premultiplying the resultant by the distribution matrix $\mathbf{M}$ [7], we have

$$(\mathbf{M}\mathbf{B}^{-1}\mathbf{H})\mathbf{u} - (\mathbf{M}\mathbf{B}^{-1}\mathbf{P}) = \mathbf{M}\mathbf{T} \tag{14.17}$$

where the distribution matrix $\mathbf{M}$ is defined as

$$\mathbf{M} = \int_\Gamma \mathbf{\Phi}\mathbf{\Psi}^T d\Gamma \tag{14.18}$$

Let

$$\mathbf{K}'_{\mathrm{BE}} = \mathbf{M}\mathbf{B}^{-1}\mathbf{H} \tag{14.19}$$

$$\mathbf{P}_{\mathrm{BE}} = \mathbf{M}\mathbf{B}^{-1}\mathbf{P} \tag{14.20}$$

$$\mathbf{F}_{\mathrm{BE}} = \mathbf{M}\mathbf{T} \tag{14.21}$$

Equation 14.18 can then be written in the following equivalent form of the EFG formulation:

$$\mathbf{K}'_{\mathrm{BE}}\mathbf{U} = \mathbf{F}_{\mathrm{BE}} + \mathbf{P}_{\mathrm{BE}} \tag{14.22}$$

Note that the matrix $\mathbf{K}'_{\mathrm{BE}}$ derived from the above formulation is in general asymmetric. The asymmetry arises from the approximations involved in the discretization process and the choice of the assumed solution. In the EFG domain, however, the matrix $\mathbf{K}_{\mathrm{EFG}}$ is symmetric. If Equation 14.22 is assembled directly into the EFG matrices in Equation 14.14, the symmetry of the coefficient matrix will be destroyed, which leads to inefficiency in solving the system equations. To preserve the symmetry of the system matrix, a symmetrization operation must be performed for $\mathbf{K}'_{\mathrm{BE}}$. One simple method to perform such an operation is to minimize the squares of the errors in the asymmetric off-diagonal terms of $\mathbf{K}'_{\mathrm{BE}}$ [8]. Hence, a new symmetric equivalent BE stiffness matrix $\mathbf{K}_{\mathrm{BE}}$ can be obtained using

$$k_{\mathrm{BE}ij} = 1/2(k'_{\mathrm{BE}ij} + k'_{\mathrm{BE}ji}) \tag{14.23}$$

Equation 14.22 can be rewritten as

$$\mathbf{K}_{BE}\mathbf{U} = \mathbf{F}_{BE} + \mathbf{P}_{BE} \qquad (14.24)$$

where $\mathbf{K}_{BE}$ is now symmetric.

14.1.4 Coupling of EFG and BE System Equations

14.1.4.1 Continuity Conditions at the Interface

Consider now a problem domain consisting of two subdomains Ω_1 and Ω_2, joined by an interface boundary Γ_I. The EFG formulation is used in Ω_1 and the BE formulation is used in Ω_2, as shown in Figure 14.1. Compatibility and equilibrium conditions on Γ_I must be satisfied. Thus,

1. The nodal displacements formulated at the Γ_I for Ω_1 and that for Ω_2 should be equal, i.e.,

$$\mathbf{u}_I^{(1)} = \mathbf{u}_I^{(2)} \qquad (14.25)$$

2. The summation of the nodal force formulated on the Γ_I for Ω_1 and that for Ω_2 should be 0, i.e.,

$$\mathbf{F}_I^{(1)} + \mathbf{F}_I^{(2)} = 0 \qquad (14.26)$$

Because the MLS shape functions used in the EFG method do not possess the Kronecker delta function property, $\mathbf{u}$ in Equation 14.4 is the parameter of nodal displacement, which differs from the nodal displacement. Proper treatment is required to couple these two

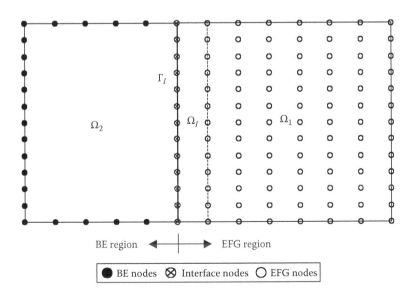

FIGURE 14.1
A problem domain divided into an EFG region and a BE region.

equation systems of EFG and BE domains along Γ_I. In these interface elements, a hybrid displacement approximation is defined so that the shape functions of the EFG domain along Γ_I possess the delta function property. Therefore, **u** in Equation 14.4 becomes the true nodal displacement on the interface. The system equations for both EFG and BE can be assembled directly.

14.1.4.2 Shape Functions for the Interface Elements

The detailed characteristics of FE interface elements can be found in [6]. Because the nodal arrangement may be irregular in the EFG domain, four to six node isoparametric interface FE elements [9] are used for the interface elements.

A detailed illustration of the interface domain is shown in Figure 14.2, where Ω_I is a layer of subdomain along the interface boundary Γ_I within the EFG domain Ω_1. The modified displacement approximation in domain Ω_1 becomes

$$u_1^h(\mathbf{x}) = \begin{cases} u^{\text{EFG}}(\mathbf{x}) + R(\mathbf{x})(u^{\text{FE}}(\mathbf{x}) - u^{\text{EFG}}(\mathbf{x})) & \mathbf{x} \in \Omega_I \\ u^{\text{EFG}}(\mathbf{x}) & \mathbf{x} \in (\Omega_1 - \Omega_I) \end{cases} \tag{14.27}$$

where
u_1^h is the displacement of a point in Ω_1
u^{EFG} is the EFG displacement given by

$$u^{\text{EFG}}(\mathbf{x}) = \sum_{i=1}^{n} \phi_i^H(\mathbf{x}) u_i \tag{14.28}$$

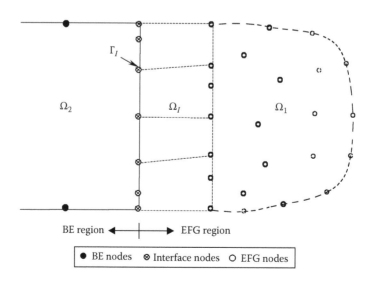

FIGURE 14.2
Interface element used in coupled EFG/BEM. (From Gu, Y.T. and Liu, G.R., *Comput. Methods Appl. Mech. Eng.*, 190, 4405, 2001. With permission.)

in which ϕ_i^H is the MLS shape function given by Equation 2.57. u^{FE} is the FE displacement defined by

$$u^{\mathrm{FE}} = \sum_{i=1}^{n_e} N_i(x) u_i, \quad n_e = 3, 4, 5, \ldots \tag{14.29}$$

where $N_i(x)$ is the FE shape function and n_e is the number of nodes in an FE interface element. The ramp function R is equal to the sum of the FE shape functions of an interface element associated with the interface element nodes that are located on the interface boundary Γ_I, i.e.,

$$R(x) = \sum_i^k N_i(x), \quad x_i \in \Gamma_I \tag{14.30}$$

where k is the number of nodes located on the interface boundary Γ_I for an interface element. According to the property of FE shape functions, R will be 1 along Γ_I and will vanish from the interface domain, i.e.,

$$R(x) = \begin{cases} 1 & x \in \Gamma_I \\ 0 & x \in \Omega_1 - \Omega_I \end{cases} \tag{14.31}$$

The new displacement approximation in the EFG domain Ω_1 can be rewritten as

$$\mathbf{u}_1^h(x) = \sum_i \tilde{\Phi}_i(x) u_i \tag{14.32}$$

where

$$\tilde{\Phi}_i(x) = \begin{cases} (1 - R(x))\Phi_i(x) + R(x)N_i(x) & x \in \Omega_I \\ \Phi_i(x) & x \in \Omega_1 - \Omega_I \end{cases} \tag{14.33}$$

The derivatives of the interface shape functions are

$$\tilde{\Phi}_{i,j} = \begin{cases} (1 - R)\Phi_{i,j} - R_{,j}\Phi_i + RN_{i,j} + R_{,j}N_i & x \in \Omega_I \\ \Phi_{i,j} & x \in \Omega_1 - \Omega_I \end{cases} \tag{14.34}$$

The approximation using the above modified shape functions will be compatible (or continuous) and reproduce the linear field exactly, which has been proved in [6].

The regular EFG and modified shape functions in one dimension are shown in Figure 14.3. It can be seen that the displacement approximation is continuous from the purely EFG domain passing to the interface domain. The derivative of it is, however, discontinuous across the boundary. These discontinuities are allowed in the weak formulation.

Using the above approximation, the shape functions of the EFG domain along Γ_I possess the Kronecker delta function property, and the system equations of the EFG domain, Equation 14.4, and the system equations for the BE domain, Equation 14.24, can be assembled together directly using the continuity condition on the interfaces of these two domains, which are defined in Equations 14.25 and 14.26.

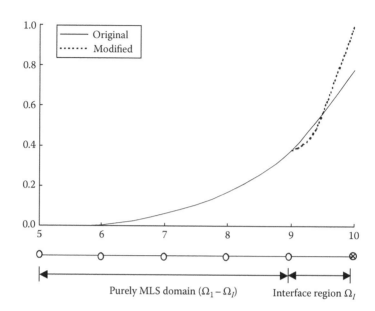

FIGURE 14.3
Comparison of original and modified shape functions in the EFG region where MLS approximation is employed;
1D case. (From Gu, Y.T. and Liu, G.R., *Comput. Methods Appl. Mech. Eng.*, 190, 4405, 2001. With permission.)

14.1.4.3 Coupling Algorithm

The flowchart of coupled EFG/BEM is given as follows:

1. Loop over in EFG domain Ω^1
   a. Determine the nodes in the support domain of point $\mathbf{x}$
   b. Compute the EFG shape functions
   c. If point $\mathbf{x}$ is in the interface element:
      Compute FE shape functions in the element, and $R(\mathbf{x})$
      Compute the interface shape functions
      End if
   d. Assemble contributions to nodes to get the stiffness matrix $\mathbf{K}_{\mathrm{EFG}}$
   e. End loop of EFG domain
2. Loop in BEs domain to obtain the matrix $\mathbf{H}$, $\mathbf{B}$
3. Compute $\mathbf{M}$, $\mathbf{K}'_{\mathrm{BE}}$ and symmetrize the $\mathbf{K}'_{\mathrm{BE}}$ to obtain $\mathbf{K}_{\mathrm{BE}}$
4. Assemble $\mathbf{K}_{\mathrm{EFG}}$ and $\mathbf{K}_{\mathrm{BE}}$ to get the global system equations
5. Solve the system equations for displacements
6. Postprocess to obtain displacement, strain, and stress

14.1.5 Numerical Results

The following examples are run to examine the coupled EFG/BEM in 2D elastostatics.
The programs are developed to combine constant, linear, and quadratic BEs with EFG.
Interface elements with four to six nodes are used.

Example 14.1: Cantilever Beam

Coupled EFG/BEM is first applied to study the benchmarking problem of the cantilever beam shown in Figure 14.4. A plane stress problem is considered. The parameters for this example are as follows:

Young's modulus for the material: $E = 3.0 \times 10^7$

Poisson's ratios for two materials: $\nu = 0.3$

Thickness of the beam: $t = 1$

Height of the beam: $D = 12$

Length of the beam: $L = 48$

Load: $P = 1000$

The beam is subjected to parabolic traction at the free end as shown in Figure 6.4. The beam is artificially divided into two parts as shown in Figure 14.4. BEs are used to model the left part of the beams in which the essential boundary is included. The EFG is used in the right part. The nodal arrangement is also shown in Figure 14.4. Background integration cells of 6×8 are used in the EFG domain. In each integration cell, a 4×4 Gauss quadrature is used to evaluate the stiffness matrix of the EFG. Linear BEs are employed in the BE domain. Rectangular elements are employed as interface elements. Only 100 nodes in total are used in the entire coupled model. The total number of nodes determines the size of the final assembled system equation, and directly affects the computation time for solving this problem.

Figure 14.5 plots the shear stress distribution on the cross section of the beam at $x = L/2$, calculated using the present coupled EFG/BEM. Results obtained using analytical formulas and FEM/BEM are also plotted in Figure 14.5 for comparison. When the FEM/BEM is used, the right portion of the beam is modeled using linear FEs instead of EFG modes. In this case, there is no need to use transition elements, as the shape functions for both FEM and BE are of the FE type,

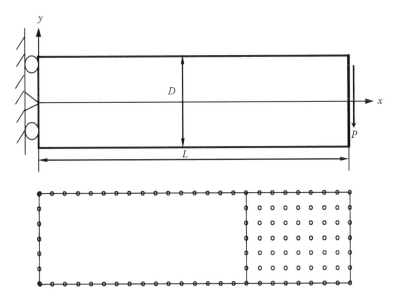

FIGURE 14.4
Nodal arrangement for the cantilever beam subjected to downward traction force on the right end of the beam. (From Gu, Y.T. and Liu, G.R., *Comput. Methods Appl. Mech. Eng.*, 190, 4405, 2001. With permission.)

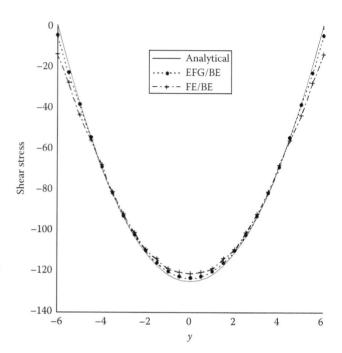

FIGURE 14.5
Shear stress τ_{xy} at the section $x = L/2$ of the cantilever beam computed using three different methods. (From Gu, Y.T. and Liu, G.R., *Comput. Methods Appl. Mech. Eng.*, 190, 4405, 2001. With permission.)

which possess the Kronecker delta function property. The plot shows an excellent agreement between the results obtained using these three methods. It can also be found that the coupled EFG/BEM yields a more accurate result than the FE/BE method. This is because the EFG performs better than the FEM of linear elements.

For quantitative error analysis, the error indicator defined in Equation 13.29 is used in the investigation. The convergence with mesh refinement is shown in Figure 14.6, where h is the nodal spacing or the element size in the FEM. It is observed that the convergence of the coupled method is very good. The convergence of the coupled FE/BE method is also shown in Figure 14.6. Figure 14.6 shows that the accuracy of the coupled EFG/BEM is a little higher than the FE/BE method because of the higher accuracy of EFG. However, the convergence rate of these two coupled methods is nearly same, and it is found to be about 2.3 for this problem. This is because the accuracy of the BEM plays a part in the convergence rate of the coupled EFG/BE and FE/BE methods.

Example 14.2: Hole in an Infinite Plate

Consider now an infinite plate with a central circular hole subjected to a unidirectional tensile load of $p = 1.0$ in the x-direction. As a large finite plate can be considered a good approximation of an infinite plate, a finite square plate of 20×20 is considered. Making use of the symmetry, only the upper right quadrant of the finite plate is modeled, as shown in Figure 14.7. A plane strain problem is considered, and the material properties are $E = 1.0 \times 10^3$, $\nu = 0.3$. Symmetry conditions are imposed on the left and bottom edges, and the inner boundary of the hole is traction free. The tensile load p is imposed on the right edge in the x-direction. The exact solution for the stresses of an infinite plate is given in the textbook [10] and is listed in Equations 7.41 through 7.46.

The plate is divided into two domains. In the area near the hole, EFG is employed. For the rest of the area of the problem domain the BEM is applied.

It is found that the numerical results obtained for displacements are almost identical to the analytical solution. As the stresses are much more critical, detailed results of stresses are presented here. The stresses σ_x at $x = 0$ obtained by the coupled method using two kinds of

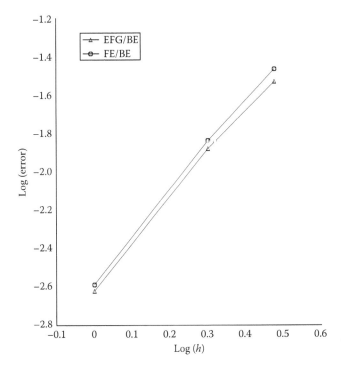

FIGURE 14.6
Convergence in energy norm of error e_t. (From Gu, Y.T. and Liu, G.R., *Comput. Methods Appl. Mech. Eng.*, 190, 4405, 2001. With permission.)

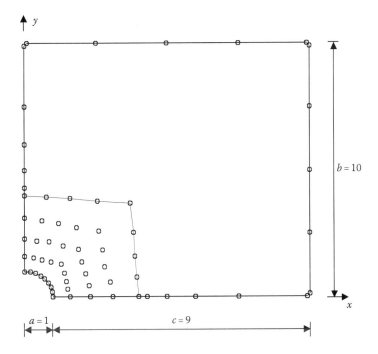

FIGURE 14.7
Nodes in a plate with a hole at its center subjected to a unidirectional tensile load in the x direction. A quarter of the plate is modeled. (From Gu, Y.T. and Liu, G.R., *Comput. Methods Appl. Mech. Eng.*, 190, 4405, 2001. With permission.)

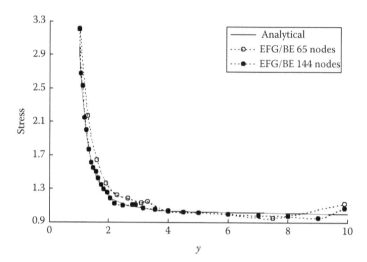

FIGURE 14.8
Stress distribution (σ_x, at $x = 0$) obtained using EFG/BEM together with analytic results for a square plate with a hole at its center. (From Gu, Y.T. and Liu, G.R., *Comput. Methods Appl. Mech. Eng.*, 190, 4405, 2001. With permission.)

nodal arrangement are given in Figure 14.8. The figure shows that the coupled method yields satisfactory results for the problem when 144 nodes are used. For comparison, the results obtained using EFG/BE, FE/BE, and EFG methods are shown in Figure 14.9. It can be found that EFG/BEM yields better results than the FE/BE method. The accuracy of EFG/BE and EFG methods is nearly the same. However, many fewer nodes are used in coupled EFG/BEM (144 nodes) than the EFG method (231 nodes).

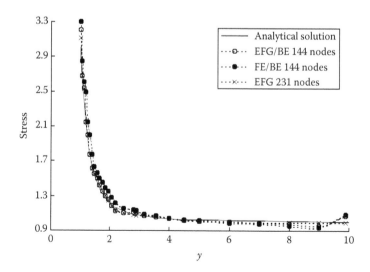

FIGURE 14.9
Stress distribution (σ_x, at $x = 0$) obtained using EFG/BEM, FE/BEM, and EFG together with analytical results for a square plate with a hole at its center. (From Gu, Y.T. and Liu, G.R., *Comput. Methods Appl. Mech. Eng.*, 190, 4405, 2001. With permission.)

There exist oscillations in the solution of the corner nodes in the BE domain, as shown in Figure 14.8. This is because the tractions are discontinuous in these corner nodes. Special care should be taken in handling traction discontinuities at the corner nodes, as discussed in Chapter 13. One method to overcome this difficulty is simply to split the corner node into two nodes with each node on one side of the corner. These two nodes are very close to the original corner node. A constant BE is then used between these two nodes. Because these two nodes belong to different sides of the corner, the discontinuity of the traction on the corner can be modeled without difficulty. The method is very simple, works very well, and is widely used in BEM. It is also used in Chapter 13 for boundary-type meshfree methods.

Example 14.3: A Structure on a Semi-Infinite Elastic Foundation

In this example, the coupled method is used to solve a foundation–structure interaction problem, illustrated schematically in Figure 14.10. A structure stands on a semi-infinite elastic foundation. The problem has been investigated using coupled FE/BEM by some researchers [11]. The infinite elastic foundation can be modeled in one of the following three ways:

1. Truncating the plane at a finite distance—approximate method
2. Using a fundamental solution corresponding to the semi-space problem rather than a full-space Green's function in BEM
3. Using an infinite element in FEM

Method 1 is used in this section because it is convenient to compare the coupled method solution with the EFG, FE, and FE/BE solutions.

As shown in Figure 14.10, Region 2 represents the semi-infinite foundation and is given a semicircular shape of very large diameter in relation to Region 1, which represents the structure. Boundary conditions to restrain rigid body movements are applied. The EFG method is used in Region 1, and BEM is used in Region 2. The nodal arrangement of coupled EFG/BEM is shown in Figure 14.11. The problem is also analyzed using FEM, EFG, and FE/BEM. The nodal arrangement of EFG is shown in Figure 14.12. Two load cases shown in Figure 14.13 are analyzed: case 1 considers five concentrated vertical loads along the top, and case 2 considers an additional horizontal load acting at the left corner.

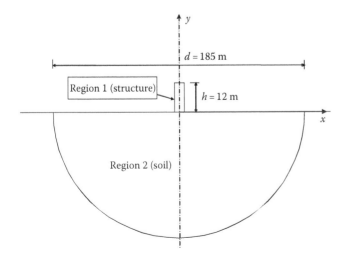

FIGURE 14.10
A structure standing on the top of a semi-infinite soil foundation. (From Gu, Y.T. and Liu, G.R., *Comput. Methods Appl. Mech. Eng.*, 190, 4405, 2001. With permission.)

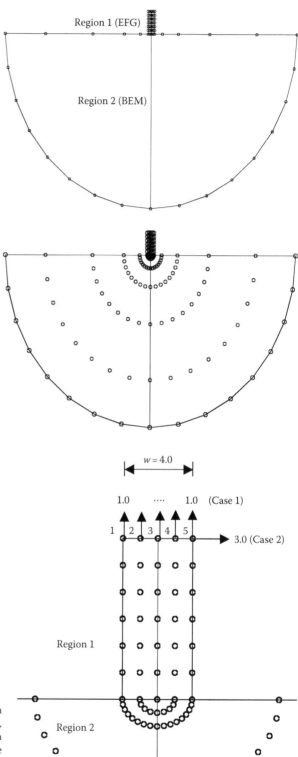

FIGURE 14.11
Nodal arrangement for the coupled EFG/ BEM model. (From Gu, Y.T. and Liu, G.R., *Comput. Methods Appl. Mech. Eng.*, 190, 4405, 2001. With permission.)

FIGURE 14.12
Nodal arrangement for the coupled EFG model. (From Gu, Y.T. and Liu, G.R., *Comput. Methods Appl. Mech. Eng.*, 190, 4405, 2001. With permission.)

FIGURE 14.13
Nodal arrangement in the structure portion where EFG is used. The structure is loaded by, case 1, a uniformly distributed normal traction in the *y*-direction or, case 2, a concentrate force in the *x*-direction at the top right corner. (From Gu, Y.T. and Liu, G.R., *Comput. Methods Appl. Mech. Eng.*, 190, 4405, 2001. With permission.)

TABLE 14.1

Vertical Displacements along the Top of the Structure

Nodes	Displacements ($\times 10^{-4}$)			
	FE	EFG	FE/BE	EFG/BE
Load case 1				
1	1.41	1.42	1.40	1.42
2	1.34	1.34	1.33	1.33
3	1.32	1.32	1.32	1.32
4	1.34	1.34	1.33	1.33
5	1.41	1.42	1.40	1.42
Load case 2				
1	−3.39	−3.43	−3.55	−3.58
2	−0.97	−1.01	−1.05	−1.04
3	1.35	1.35	1.35	1.34
4	3.61	3.67	3.70	3.68
5	6.00	6.04	6.17	6.13

Source: Gu, Y.T. and Liu, G.R., *Comput. Methods Appl. Mech. Eng.*, 190, 4405, 2001. With permission.

The results of displacement in the *y*-direction (vertical) on top of the structure are listed in Table 14.1. The results obtained using FEM, EFG, and FE/BEM are also included in Table 14.1 for comparison. The results obtained using the present EFG/BEM are in a very good agreement with those obtained using FE, EFG, and FE/BEM. However, it is interesting to note that the foundation is adequately represented using only 30 BE nodes in coupling cases as compared to 120 nodes for the EFG and FE cases.

14.2 Coupled EFG and Hybrid BEM

In Section 14.1, we demonstrated coupling of the system equations of EFG and BEM. We have seen that a symmetrization of the BE stiffness matrix must be performed before the assembly of the EFG system equations with the BE system equations. This can lead to a loss of accuracy and efficiency of the coupled method. In this section, we present an alternative approach to avoid this disadvantage in coupled EFG/BEM.

In the late 1980s, alternative BE formulations were developed based on generalized variational principles. Dumont [12] proposed a hybrid stress BE formulation based on the Hellinger–Reissner principle. DeFigueiredo and Brebbia [13,14], and Jin et al. [15] presented a hybrid displacement boundary element (HBE) formulation. The HBE formulation led to a symmetric stiffness matrix. This property of symmetry can be an added advantage in coupling the HBE with methods that produce symmetric system matrices.

This section presents a coupled EFG/HBE method for continuum mechanics problems, based on the work originally reported in [3]. The method of Lagrange multipliers is employed to impose the compatibility conditions on the interface boundary of the EFG and HBE domains. Coupled system equations are derived based on variational formulation. Several numerical examples are examined using the EFG/HBE to demonstrate the convergence, validity, and efficiency of the coupled EFG/HBE method.

Compared with coupled EFG/BEM discussed in the previous section, the present EFG/HBE method makes the following advances:

1. The coupled system equations are formulated in a different but more general manner.
2. System matrices obtained by EFG/HBE are symmetric without the need for an operation of symmetrization to the BE matrix.
3. There is no need for interface elements; therefore, mesh generation becomes much simpler and there is no special treatment needed on the interface.

The trade-off would be

1. The system matrix is larger than that of EFG/BEM.
2. The system matrix becomes nonpositive.

These drawbacks are similar to that of EFG using the method of Lagrange multipliers. Detailed formulation of the EFG/HBE is presented as follows.

14.2.1 EFG Formulation

14.2.1.1 Discrete Equations of EFG

Consider again the 2D problem of solid mechanics defined in Equations 14.1 through 14.3. The constrained functional can be written as

$$\Pi_1 = \int_\Omega \frac{1}{2}\boldsymbol{\varepsilon}^T \cdot \boldsymbol{\sigma}d\Omega - \int_\Omega \mathbf{u}^T \cdot \mathbf{b}d\Omega - \int_{\Gamma_t} \mathbf{u}^T \cdot \mathbf{t}_\Gamma d\Gamma + \int_{\Gamma_u} \boldsymbol{\lambda}^T \cdot (\mathbf{u}_\Gamma - \mathbf{u})d\Gamma \qquad (14.35)$$

where the fourth term of the integral is for the essential boundary condition, and λ is a vector of Lagrange multipliers. Following the procedure in Chapter 6, the discrete system equation of EFG for the EFG domain can be written in the form:

$$\begin{bmatrix} \mathbf{K}_{EFG} & \mathbf{G}_{EFG} \\ \mathbf{G}_{EFG}^T & 0 \end{bmatrix} \begin{Bmatrix} \mathbf{u} \\ \boldsymbol{\lambda} \end{Bmatrix} = \begin{Bmatrix} \mathbf{F}_{EFG} + \mathbf{P}_{EFG} \\ \mathbf{q}_{EFG} \end{Bmatrix} \qquad (14.36)$$

where the subscript EFG indicates matrices obtained using standard EFG formulation. The components in vectors $\mathbf{F}_{EFG}$ and $\mathbf{P}_{EFG}$ are defined in Equations 14.5 and 14.6, and the rest of the matrices have been defined in detail in Chapter 6.

14.2.2 Hybrid Displacement BE Formulation

The constrained functional for the hybrid displacement BEM can be written as

$$\Pi_2 = \int_\Omega \frac{1}{2}\boldsymbol{\varepsilon}^T \cdot \boldsymbol{\sigma}d\Omega - \int_\Omega \mathbf{u}^T \cdot \mathbf{b}d\Omega - \int_{\Gamma_t} \tilde{\mathbf{u}}^T \cdot \mathbf{t}_\Gamma d\Gamma + \int_\Gamma \boldsymbol{\lambda}^T \cdot (\tilde{\mathbf{u}} - \mathbf{u})d\Gamma \qquad (14.37)$$

where
 $\tilde{\mathbf{u}}$ is the displacement on the boundary
 $\mathbf{u}$ is the displacement in the domain

The fourth term of the integral is for the compatibility of the displacements on the boundary with that near the boundary in the domain, and $\boldsymbol{\lambda}$ is a vector of Lagrange multipliers. As the Lagrange multipliers $\boldsymbol{\lambda}$ represents the traction on the boundary, it is therefore denoted explicitly by $\tilde{\mathbf{t}}$. Hence, Equation 14.37 can be rewritten as

$$\Pi_2 = \int_\Omega \frac{1}{2}\boldsymbol{\varepsilon}^{\mathrm{T}} \cdot \boldsymbol{\sigma} \mathrm{d}\Omega - \int_\Omega \mathbf{u}^{\mathrm{T}} \cdot \mathbf{b} \mathrm{d}\Omega - \int_{\Gamma_t} \tilde{\mathbf{u}}^{\mathrm{T}} \cdot \mathbf{t}_\Gamma \mathrm{d}\Gamma + \int_{\Gamma_u} \tilde{\mathbf{t}}^{\mathrm{T}} \cdot (\tilde{\mathbf{u}} - \mathbf{u}) \mathrm{d}\Gamma \tag{14.38}$$

The first term on the right-hand side can be integrated by parts to become

$$\Pi_2 = \int_\Gamma \frac{1}{2}\mathbf{t}^{\mathrm{T}} \cdot \mathbf{u} \mathrm{d}\Gamma - \int_\Omega \mathbf{u}^{\mathrm{T}} \cdot \mathbf{b} \mathrm{d}\Omega - \int_{\Gamma_t} \tilde{\mathbf{u}}^{\mathrm{T}} \cdot \mathbf{t}_\Gamma \mathrm{d}\Gamma + \int_{\Gamma_u} \tilde{\mathbf{t}}^{\mathrm{T}} \cdot (\tilde{\mathbf{u}} - \mathbf{u}) \mathrm{d}\Gamma - \int_\Omega \frac{1}{2}\mathbf{L}_d^{\mathrm{T}} \boldsymbol{\sigma} \cdot \mathbf{u} \mathrm{d}\Omega \tag{14.39}$$

The starting domain integral in Equation 14.38 can be reduced to an integral on the boundary using the fundamental solution for Equation 14.9, which is Green's function.

The displacement vector within the domain is approximated as a series of products of $\mathbf{U}^*$, which are formed using the fundamental solutions [14] and unknown parameters $\mathbf{s}$, i.e.,

$$\mathbf{u} = \mathbf{U}^* \mathbf{s} \tag{14.40}$$

The displacement vector on the boundary is written as the product of known interpolation functions by the nodal displacement at the boundary nodes, i.e.,

$$\tilde{\mathbf{u}} = \boldsymbol{\Phi}^{\mathrm{T}} \mathbf{u}^e \tag{14.41}$$

Similarly, the traction vector is approximated as a series of products of $\mathbf{T}^*$ that are also formed using the fundamental solutions and unknown parameters $\mathbf{s}$.

$$\mathbf{t} = \mathbf{T}^* \mathbf{s} \tag{14.42}$$

The traction vector on the boundary is written as the product of known interpolation functions and the nodal traction at the boundary nodes, i.e.,

$$\tilde{\mathbf{t}} = \boldsymbol{\Psi}^{\mathrm{T}} \mathbf{t}^e \tag{14.43}$$

Substituting Green's function and Equations 14.40 through 14.43 into Equation 14.39, we can obtain

$$\Pi = -1/2\mathbf{s}^{\mathrm{T}} \mathbf{A} \mathbf{s} - \mathbf{t}^{\mathrm{T}} \mathbf{B}^{\mathrm{T}} \mathbf{s} + \mathbf{t}^{\mathrm{T}} \mathbf{L}_d \mathbf{U} - \mathbf{U}^{\mathrm{T}} \mathbf{f} - \mathbf{s}^{\mathrm{T}} \mathbf{b} \tag{14.44}$$

where

$$\mathbf{A} = \int_{\Gamma} \mathbf{U}^*\mathbf{T}^* d\Gamma \qquad (14.45)$$

$$\mathbf{B} = \int_{\Gamma} \boldsymbol{\Psi}\mathbf{U}^* d\Gamma \qquad (14.46$$

$$\mathbf{L} = \int_{\Gamma} \boldsymbol{\Psi}\boldsymbol{\Phi}^T d\Gamma \qquad (14.47)$$

$$\mathbf{f}_{\text{HBE}} = \int_{\Gamma} \boldsymbol{\Phi}\mathbf{t}_\Gamma d\Gamma \qquad (14.48)$$

$$\mathbf{g} = \int_{\Omega} \mathbf{U}^*\mathbf{b} d\Omega \qquad (14.49)$$

The stationary conditions for Π can now be found by setting its first variation of Π to zero. As this must be true for any arbitrary values of $\delta\mathbf{s}$, $\delta\mathbf{u}$, and $\delta\mathbf{t}$, one obtains

$$\mathbf{K}_{\text{HBE}}\mathbf{U} = \mathbf{F}_{\text{HBE}} + \mathbf{P}_{\text{HBE}} \qquad (14.50)$$

where

$$\mathbf{K}_{\text{HBE}} = \mathbf{R}^T\mathbf{A}\mathbf{R} \qquad (14.51)$$

$$\mathbf{R} = (\mathbf{B}^T)^{-1}\mathbf{L} \qquad (14.52)$$

$$\mathbf{P}_{\text{HBE}} = \mathbf{R}^T\mathbf{g} \qquad (14.53)$$

It can be proven that matrix $\mathbf{A}$ is symmetric; hence, matrix $\mathbf{K}$ is symmetric. Equation 14.50 shows that this hybrid displacement boundary formulation leads to an equivalent stiffness formulation. The matrix $\mathbf{K}$ may be viewed as a symmetric stiffness matrix, but the above integrals are needed to perform only on boundaries, and the domain need not be discretized.

14.2.3 Coupling of EFG and HBE

14.2.3.1 Continuity Conditions at Coupled Interfaces

Consider a problem consisting of two domains of Ω^1 and Ω^2, as schematically shown in Figure 14.14. These two domains are joined by an interface Γ_I. The EFG formulation is used in Ω^1 and the HBE formulation is used in Ω^2. Continuity conditions that must be satisfied on Γ_I are given by

$$\tilde{\mathbf{u}}_I^1 = \tilde{\mathbf{u}}_I^2 \qquad (14.54)$$

$$\mathbf{F}_I^1 + \mathbf{F}_I^2 = 0 \qquad (14.55)$$

where
 $\tilde{\mathbf{u}}_I^1$ and $\tilde{\mathbf{u}}_I^2$ are the displacements
 $\mathbf{F}_I^1$ and $\mathbf{F}_I^2$ are the forces on Γ_I for Ω^1 and Ω^2, respectively

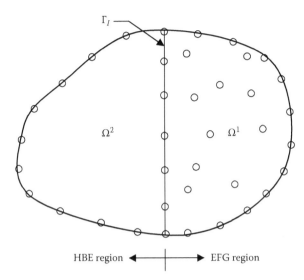

FIGURE 14.14
Domain division into EFG and HBE regions. (From Liu, G.R. and Gu, Y.T., *Comput. Mech.*, 26, 166, 2000. With permission.)

Because the shape functions of EFG are derived using MLS, the displacement vector in Equation 14.36 differs from the true nodal displacement. Proper treatments are needed to assemble these equations of EFG and HBE.

14.2.3.2 Coupling EFG with HBE via a Modified Variational Form

A subfunctional is introduced to enforce the continuity condition, Equation 14.54, by means of Lagrange multiplier λ on the interface boundary

$$\Pi_I = \int_{\Gamma_I} \boldsymbol{\gamma}^{\mathrm{T}}(\tilde{\mathbf{u}}_I^1 - \tilde{\mathbf{u}}_I^2)\mathrm{d}\Gamma = \underbrace{\int_{\Gamma_I} \boldsymbol{\gamma}^{\mathrm{T}}\tilde{\mathbf{u}}_I^1\mathrm{d}\Gamma}_{\Pi_I^1} - \underbrace{\int_{\Gamma_I} \boldsymbol{\gamma}^{\mathrm{T}}\tilde{\mathbf{u}}_I^2\mathrm{d}\Gamma}_{\Pi_I^2} = \Pi_I^1 - \Pi_I^2 \qquad (14.56)$$

In Equation 14.54, Π_I^1 and Π_I^2 are the boundary integrations along the EFG side and the HBE side, respectively. Introducing Π_I^1 and Π_I^2 separately into functions, Equations 14.35 and 14.37, generalized functional forms can be written as

$$\Pi_{\mathrm{EFG}} = \int_{\Omega} \frac{1}{2}\boldsymbol{\varepsilon}^{\mathrm{T}}\cdot\boldsymbol{\sigma}\mathrm{d}\Omega - \int_{\Omega}\mathbf{u}^{\mathrm{T}}\cdot\mathbf{b}\mathrm{d}\Omega - \int_{\Gamma_t}\mathbf{u}^{\mathrm{T}}\cdot\mathbf{t}_\Gamma\mathrm{d}\Gamma - \int_{\Gamma_u}\boldsymbol{\lambda}_{\mathrm{EFG}}^{\mathrm{T}}\cdot(\mathbf{u}-\mathbf{u}_\Gamma)\mathrm{d}\Gamma + \int_{\Gamma_I}\boldsymbol{\gamma}^{\mathrm{T}}\cdot\tilde{\mathbf{u}}_I^1\mathrm{d}\Gamma \quad (14.57)$$

$$\Pi_{\mathrm{HBE}} = \int_{\Omega} \frac{1}{2}\boldsymbol{\varepsilon}^{\mathrm{T}}\cdot\boldsymbol{\sigma}\mathrm{d}\Omega - \int_{\Omega}\mathbf{u}^{\mathrm{T}}\cdot\mathbf{b}\mathrm{d}\Omega - \int_{\Gamma_t}\tilde{\mathbf{u}}^{\mathrm{T}}\cdot\mathbf{t}_\Gamma\mathrm{d}\Gamma + \int_{\Gamma}\boldsymbol{\lambda}_{\mathrm{HBE}}^{\mathrm{T}}\cdot(\tilde{\mathbf{u}}-\mathbf{u})\mathrm{d}\Gamma - \int_{\Gamma_I}\boldsymbol{\gamma}^{\mathrm{T}}\cdot\tilde{\mathbf{u}}_I^2\mathrm{d}\Gamma \quad (14.58)$$

In these variational formulations, the domains of EFG and HBE are connected via Lagrange multiplier $\boldsymbol{\gamma}$.

In the EFG domain, $\mathbf{u}$ is given by Equation 2.61, γ is given by production of the interpolation function Λ and value of γ^I

$$\Gamma = \Lambda^T \gamma^I \tag{14.59}$$

Λ can consist of shape functions of the FE type. Substituting Equations 2.61 and 14.59 into Equation 14.57, and using the stationary condition, we can obtain the following EFG equations:

$$\begin{bmatrix} \mathbf{K}_{EFG} & \mathbf{G}_{EFG} & \mathbf{B}_{EFG} \\ \mathbf{G}_{EFG}^T & 0 & 0 \\ \mathbf{B}_{EFG}^T & 0 & 0 \end{bmatrix} \left\{ \begin{array}{c} \mathbf{U} \\ \lambda \\ \gamma \end{array} \right\} = \left\{ \begin{array}{c} \mathbf{F}_{EFG} + \mathbf{P}_{EFG} \\ \mathbf{q}_{EFG} \\ 0 \end{array} \right\} \tag{14.60}$$

where subscript EFG indicates the EFG matrices, and $\mathbf{B}$ is defined as

$$\mathbf{B}_{EFG} = \int_{\Gamma_I} \Lambda \Phi_{EFG}^T d\Gamma \tag{14.61}$$

The stationary condition of Equation 14.58 leads to the following HBE equations:

$$\begin{bmatrix} \mathbf{K}_{HBE} & -\mathbf{H}_{HBE} \\ -\mathbf{H}_{HBE}^T & 0 \end{bmatrix} \left\{ \begin{array}{c} \mathbf{U} \\ \gamma \end{array} \right\} = \left\{ \begin{array}{c} \mathbf{F}_{HBE} + \mathbf{P}_{HBE} \\ 0 \end{array} \right\} \tag{14.62}$$

where $\mathbf{K}_{HBE}$, $\mathbf{F}_{HBE}$, and $\mathbf{P}_{HBE}$ are defined by Equations 14.48, 14.50, and 14.53. $\mathbf{H}$ is defined as

$$\mathbf{H}_{HBE} = \int_{\Gamma_I} \Lambda \Phi_{HBE}^T d\Gamma \tag{14.63}$$

Because two domains are connected along the interface boundary Γ_I, assembling Equations 14.60 and 14.63 yields a linear system of

$$\begin{bmatrix} \mathbf{K}_{EFG} & 0 & \mathbf{G}_{EFG} & \mathbf{B}_{EFG} \\ 0 & \mathbf{K}_{HBE} & 0 & -\mathbf{H}_{HBE} \\ \mathbf{G}_{EFG}^T & 0 & 0 & 0 \\ \mathbf{B}_{EFG}^T & -\mathbf{H}_{HBE}^T & 0 & 0 \end{bmatrix} \left\{ \begin{array}{c} \mathbf{U}_{EFG} \\ \mathbf{U}_{HBE} \\ \lambda \\ \gamma \end{array} \right\} = \left\{ \begin{array}{c} \mathbf{F}_{EFG} + \mathbf{P}_{EFG} \\ \mathbf{F}_{HBE} + \mathbf{P}_{HBE} \\ \mathbf{q}_{EFG} \\ 0 \end{array} \right\} \tag{14.64}$$

The continuity conditions on Γ_I given in Equations 14.54 and 14.55 are satisfied via the above variational formulation. Note that the system matrix is symmetric, but enlarged and nonpositive.

14.2.4 Numerical Results

Three examples of 2D elastostatics that were examined in the previous section are reexamined using the present coupled EFG/HBE method.

Example 14.4: Cantilever Beam

All parameters and conditions are exactly the same as those in Example 14.1. The nodal arrangement is shown in Figure 14.4, and a background mesh of 6×8 is used in the EFG domain. In each integration cell, 4×4 Gauss quadrature is used to evaluate the stiffness matrix of the EFG. Only 100 nodes in total are used in the coupled method.

Figure 14.15 illustrates the comparison between the shear stress calculated analytically and using the coupled method at the section of $x = L/2$. The plot shows an excellent agreement between the analytical and numerical results. The computational result by the present coupled method with interface elements (IE) is also shown in the same figure. There is clear evidence that the accuracy of the coupled method using the modified variational formulation (MVF) is higher than that using the IE method.

The displacement along the interface boundary is shown in Table 14.2. It is shown that the continuity of the displacement is satisfied accurately using the present modified variational formulation method.

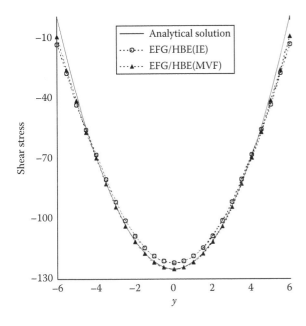

FIGURE 14.15
Shear stress τ_{xy} at the section $x = L/2$ of the beam. (From Liu, G.R. and Gu, Y.T., *Comput. Mech.*, 26, 166, 2000. With permission.)

TABLE 14.2

Vertical Displacement along the Interface Boundary (Cantilever Beam)

Node (y)	EFG/HBE (IE)[b]	EFG/HBE (MVF)[a] EFG Side	HBE Side	Exact
5.75	−4.73203E-03	−4.73090E-03	−4.73093E-03	−4.68750E-03
5.00	−4.72797E-03	−4.72617E-03	−4.72619E-03	−4.68302E-03
4.00	−4.72344E-03	−4.72050E-03	−4.72059E-03	−4.67802E-03
3.00	−4.71970E-03	−4.71664E-03	−4.71670E-03	−4.67414E-03
2.00	−4.71704E-03	−4.71419E-03	−4.71422E-03	−4.67136E-03
1.00	−4.71542E-03	−4.71257E-03	−4.71261E-03	−4.66969E-03
0.00	−4.71488E-03	−4.71199E-03	−4.71203E-03	−4.66914E-03

Source: Liu, G.R. and Gu, Y.T., *Comput. Mech.*, 26, 166, 2000. With permission.
[a] EFG/HBE (MVF): Coupled EFG/HBE method using modified variational formulation.
[b] EFG/HBE (IE): Coupled EFG/HBE method using an interface element.

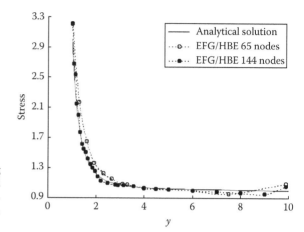

FIGURE 14.16

Stress distribution (σ_x, at $x = 0$) obtained using EFG/HBE method together with analytical results for a square plate with a hole at its center. (From Liu, G.R. and Gu, Y.T., *Comput. Mech.*, 26, 166, 2000. With permission.)

Example 14.5: Hole in an Infinite Plate

All parameters and conditions are exactly the same as those in Example 14.2. The nodal arrangement is shown in Figure 14.7. The plate is divided into two domains: in the area near the hole, EFG is employed; in the rest of the domain, the HBE method is applied.

The stresses σ_x at $x = 0$ obtained by the coupled method are plotted in Figure 14.16. The results are obtained using two kinds of nodal arrangement with 65 and 144 nodes. The nodal arrangement of 65 nodes is shown in Figure 14.7. Figure 14.16 shows that the coupled method yields satisfactory results for the problem considered. The convergence of the present method can also be observed from this figure. As the number of nodes increases, the results obtained approach the analytical solution. Compared with the EFG method, fewer nodes are needed in the present coupled method. Comparison with Figure 14.9 reveals that 231 nodes are needed in the EFG method to obtain results of the same accuracy as those obtained by the present EFG/HBE method, where only 144 nodes are required.

Example 14.6: Structure on a Semi-Infinite Foundation

All parameters and conditions are exactly the same as those in Example 14.3, which is schematically illustrated in Figure 14.10. The nodal arrangement is shown in Figures 14.11 through Figure 14.13. The only difference is that HBE is used to model the semi-infinite half space instead of BEM.

The results of displacement in the y direction on the top of the structure are given in Table 14.3. The FEM result obtained in [11] is also included in Table 14.3. The results obtained by the present method are in a very good agreement with those obtained using other methods, including FEM and EFG for the entire domain. The present method uses many fewer nodes to model the foundation. Only 30 nodes are used in the HBE method compared to 120 nodes used in EFG for the foundation.

14.3 Remarks

Methods that couple domain-type meshfree methods and boundary-type meshfree methods have been presented in this chapter. A number of benchmark examples have

TABLE 14.3

Vertical Displacements along the Top of the Structure on the Semi-Infinite Foundation

Node No.	Displacements ($\times 10^{-4}$)			
	FE	EFG	EFG/BE (IE)[a]	EFG/HBE (MVF)[b]
Load case 1				
1	1.41	1.42	1.42	1.41
2	1.34	1.34	1.33	1.33
3	1.32	1.32	1.32	1.32
4	1.34	1.34	1.33	1.33
5	1.41	1.42	1.42	1.41
Load case 2				
1	−3.39	−3.43	−3.58	−3.41
2	−0.97	−1.01	−1.04	−1.03
3	1.35	1.35	1.34	1.35
4	3.61	3.67	3.68	3.69
5	6.00	6.04	6.13	6.11

Source: Liu, G.R. and Gu, Y.T., *Comput. Mech.*, 26, 166, 2000. With permission.

[a] EFG/BE (IE): Coupled EFG/BE method using an interface element.

[b] EFG/HBE (MVF): Coupled EFG/HBE method using modified variational formulation.

demonstrated the feasibility, efficiency, and effectiveness of these coupling approaches. The following important remarks should be made before leaving this chapter:

1. The primary motivation is the same as that for coupling FEM and BEM, which is that domain discretization using FEs should be confined within the areas of inhomogeneity, nonlinearity, or complex geometry, and that boundary discretization should be used in large areas wherever Green's function is available. This kind of coupling significantly reduces the computational cost, because of the drastic reduction of the number of nodes used for modeling the problem, as well as the reduction of area integration in constructing system matrices.

2. The additional reason for performing such a coupling in meshfree methods is that the difficulty of imposing essential boundary conditions for the domain-type meshfree methods that use MLS shape functions can be overcome, by modeling the portion of the domain with essential boundaries using methods that use shape functions with the Kronecker delta function property, such as FEM, BEM, PIM, and BPIM.

3. Coupling with boundary-type meshfree methods works particularly well for problems with infinite domains.

4. BEM can be replaced by any boundary-type meshfree method, such as BNM, BPIM, radial BPIM, etc. When the boundary-type method uses MLS shape functions (e.g., BNM), special treatments of the essential boundary conditions should be performed.

5. Boundary-type methods that produce symmetric system matrices are preferred to produce a symmetric system matrix for the entire problem.

6. On the interface of different methods, there could be an issue of compatibility (not only the continuity condition for nodes), when different orders of shape functions are used on two sides of the domain. The issue can be handled simply by ensuring that the order of the shape function used on the interface is the same as that used by the other method in the domain attached to the interface. Interface or transition elements can be used if necessary.

References

1. Liu, G. R., Achenbach, J. D., Kim, J. O., and Li, Z. R., A combined finite element method/boundary element method for V(z) curves of anisotropic-layer/substrate configurations, *J. Acoust. Soc. Am. U.S.A.*, 92(5), 2734–2740, 1992.
2. Gu, Y. T. and Liu, G. R., A coupled element free Galerkin/boundary element method for stress analysis of two-dimensional solids, *Comput. Methods Appl. Mech. Eng.*, 190, 4405–4419, 2001.
3. Liu, G. R. and Gu, Y. T., Coupling of element free Galerkin and hybrid boundary element methods using modified variational formulation, *Comput. Mech.*, 26(2), 166–173, 2000.
4. Liu, G. R. and Gu, Y. T., Meshless local Petrov–Galerkin (MLPG) method in combination with finite element and boundary element approaches, *Comput. Mech.*, 26, 536–546, 2000.
5. Liu, G. R. and Gu, Y. T., Coupling of element free Galerkin method with boundary point interpolation method, in *Advances in Computational Engineering & Science*, S. N. Atluri, and F. W. Brust, eds., ICES'2K, Los Angeles, August, 2000, pp. 1427–1432.
6. Krongauz, Y. and Belytschko, T., Enforcement of essential boundary conditions in meshless approximations using finite elements, *Comput. Methods Appl. Mech. Eng.*, 131(1–2), 133–145, 1996.
7. Li, H. B. and Han, G. M., A new method for the coupling of finite element and boundary element discretized subdomains of elastic bodies, *Comput. Methods Appl. Mech. Eng.*, 54, 161–185, 1986.
8. Brebbia, C. A., Telles, J. C., and Wrobel, L. C., *Boundary Element Techniques*, Springer-Verlag, Berlin, 1984.
9. Hughes, T. J. R., *The Finite Element Method*, Prentice-Hall, Englewood Cliffs, NJ, 1987.
10. Timoshenko, S. P. and Goodier, J. N., *Theory of Elasticity*, 3rd ed, McGraw-Hill, New York, 1970.
11. Brebbia, C. A. and Georgiou, P., Combination of boundary and finite elements in elastostatics, *Appl. Math. Modeling*, 3, 212–219, 1979.
12. Dumont, N. A., The hybrid boundary element method, in *Boundary Elements*, IX, Vol. 1, C. A. Brebbia (ed.), Springer-Verlag, Berlin, 1988, pp. 117–130.
13. DeFigueiredo, T. G. B. and Brebbia, C. A., A new hybrid displacement variational formulation of BEM for elastostatics, in *Advances in Boundary Elements*, Vol. 1, C. A. Brebbia, ed., Springer-Verlag, Berlin, 1989, pp. 33–42.
14. DeFigueiredo, T. G. B., *New Boundary Element Formulation in Engineering*, Springer-Verlag, Berlin, 1991.
15. Jin, Z. L., Gu, Y. T., and Zheng, Y. L., Hybrid boundary elements for potential and elastic problem, in *Proceedings of the 7th Japan-China Symposium on Boundary Element Methods*, Japan, pp. 143–152, 1996.

15

Meshfree Methods for Adaptive Analysis

Previous chapters have described in detail the theory, formulation, procedure, and property of meshfree methods. We now discuss a very practical theme: meshfree methods for adaptive analysis for two- and three-dimensional (2D and 3D) problems. To conduct an adaptive analysis effectively we need to make our codes very efficient and robust for complicated geometries. We therefore need to discuss a few issues related to code implementation. The topics of this chapter are

- a. Triangular mesh, integration cells, and mesh automation
- b. Node numbering
- c. Fast node searching
- d. Node search for irregular boundaries
- e. Local error estimation
- f. Local adaptive refinement

Two meshfree methods will be implemented in our adaptive analysis: EFG and PIMs. For EFG, all items listed above are important, and for PIMs only items (a), (b), (e), and (f) are needed. Our discussion will focus on 2D and extend to 3D at final stage.

15.1 Triangular Mesh and Integration Cells

15.1.1 Use of Triangular Mesh

In using meshfree methods that are based on the global Galerkin method, such as the EFG method, a background mesh is required. Any mesh similar to the finite element mesh is applicable, as long as it satisfies the criteria that are stated in Chapter 6 to ensure an accurate integration. Ideally, the mesh should be adaptive: the node density should change in a automatic fashion in accordance to the field function. For this reason, the author recommends the triangular background mesh generated based on the well-established triangulation technique, such as the Delaunay triangulation [1]. The mesh vertices should be used as the field nodes and the triangular cells for integration. We understand that the integration cells can be independent of the field nodes when FEG is used, but there is really no harm in having them linked together, as long as there is no technical difficulty in doing so. It is actually helpful in adaptive analysis: Updating the density of nodes naturally leads to an update of the density of the integration cells. The advantages of using the triangular cells are as follows:

a. Matured algorithms are available in the public domain for the construction of integration cells and field nodes together.

b. Triangulation can be performed in an *automated* manner much easier for complicated two-dimensional (2D) and three-dimensional (3D) domains.

c. Smoothing domain construction based on triangular type of meshes is very straightforward.

d. Triangular mesh can be easily adapted, and hence is best suited for adaptive analyses.

One may ask now, why don't we simply use the conventional finite element method (FEM) with triangular elements, which has been fully developed? The answer is simply the well-known fact that the FEM results using triangular elements are very poor. This is due to the poor quality of the linear shape functions used with numerical operations confined within the element. In meshfree methods, however, the numerical treatments such as displacement interpolation/approximation, strain approximation, and integration are not confined within the cells/elements. Basically, it can choose "freely" and "dynamically" surrounding nodes and cells to perform these necessary operations. In addition, the demand for upper bound solutions is becoming stronger day by day for the need to provide "certified" solutions to engineering problems. Engineers will not be satisfied with a numerical solution along. They will need solutions with bounds for generally complicated problems to make reliable decisions. Meshfree methods offer a number of ways to provide solutions with bounds for complicated problems, as long as triangular cells can be built (see Chapter 8). Moreover, methods with meshfree techniques can be now much more efficient than the FEM method using linear triangular mesh (see Section 8.7).

A mesh generator called MFreePre has been developed for MFree2D$^{\copyright}$ for automatic background cells and field nodes generation. The vertices of the initially generated triangular cells serve as the initial field nodes. When new nodes are added, the triangular mesh is then modified locally and automatically. The following sections describe the techniques used in MFreePre.

In the process of background mesh generation, each node is assigned a density factor that is used to control local nodal density in the vicinity of the node. Local modification of a mesh is accomplished simply by adjusting the density factor at a node—this is particularly useful in an adaptive analysis in MFree2D$^{\copyright}$.

15.1.2 Triangular Mesh Used in EFG

In the EFG processor implemented in MFree2D$^{\copyright}$, triangular mesh is used for both integration of energy and node selection based on either influence concept or T2L-schemes. When the influence domain is opted, the triangular background mesh is used to determine the size of the influence domain for selecting nodes to be used for constructing the MLS shape functions. This ensures the automatic determination of the dimension of the influence domain for individual node during the adaptive analysis process.

A local patch for a node is first defined using the triangular cells that surround the node. The formula to compute the dimension of the influence domain of that node becomes

$$r_I = \alpha_s \sqrt{\frac{2}{n_I} \sum_{i=1}^{n_I} a_i} \tag{15.1}$$

where
  r_I is the radius of the influence domain of node I
  n_I is the number of surrounding triangular cells
  a_i is the area of the ith cell
  c is a constant scaling with the domain size

The default value is set at $\alpha_s = 2$. The size of influence domain by this approach can vary in accordance with the local nodal density and is therefore able to represent reasonably well the influence domain of a node. To ensure a successful MLS shape function construction used in EFG, a minimum number is set for the nodes contained in an influence domain. The size of the influence domain can be adjusted automatically in the code if the minimum number of nodes is not achieved.

Once the background triangular cells have been defined, the integration of the system equation for EFG can be then carried out using conventional Gauss quadrature schemes. This ensures a sufficient number of Gauss points for integrating the system matrices accurately. In addition, the density of the Gauss points is automatically tied to the density change of the nodes, thus the adequate integration is kept during the adaptive analysis process. In general, one Gauss point in a cell works well for many problems and three Gauss points in a triangular cell are sufficient for all the problems tested by our research group, which gives $\alpha_n \approx 3$ based on Equation 6.64. Hence the integration is very (probably most) economic.

15.1.3 Triangular Mesh Used in PIMs

For PIMs, the operation becomes very simple with the use of triangular background cells. All we need is to create a set of smoothing domains on top of the triangular cells, following the simple rules given in Chapter 8. Such a creation of smoothing domains can always be done without any difficulty for a given triangular mesh. When SC-PIMs are used, one may need to further divide the triangular cells, which is also trivial. The selection of nodes can always be done using a T-scheme. For strain field constructions, the integration needs to be done only along the edges of the smoothing domains and hence can be done very easily and efficiently. The energy integration either done by a simple summation over all the smoothing domains (NS- and ES-PIMs) or analytically (SC-PIMs), no numerical integration is needed. Coding of PIMs is in fact quite an easy task, and a number of PIMs have made available now in MFree2D 2.0. The PIM processors are very stable because all these issues are well controlled in a PIM setting based on weakened weak formulations.

15.2 Node Numbering: A Simple Approach

The stiffness matrix **K** is sparse and can also be banded. The bandwidth of the stiffness matrix **K** will depend on the numbering system of the field nodes. It is of little concern if a full matrix solver is used in solving the system equation. But such a full matrix solved is

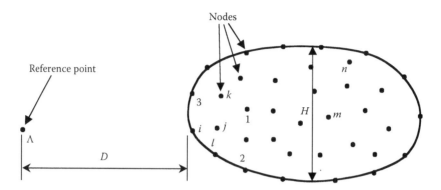

FIGURE 15.1
Reference point chosen at a large distance from the problem domain for optimizing the numbering of nodes.

never used in actual application for large systems. A much better solver is the bandwidth solver with 1D array for storage which can solve a quite large system. When a bandwidth solver is used, the CPU time for solving the system equation is proportional to the bandwidth, and hence minimization of the bandwidth of a model is of importance.

The considerations and techniques of reducing the bandwidth are similar to those in FEM. In FEM, the bandwidth depends on the largest node number difference of the elements. All one needs to do is determine which element has the largest node number difference. Optimization tools for renumbering nodes are routinely available and widely used to ensure the minimum bandwidth for different types of solvers. Most of the FEM commercial software packages include some of these optimization tools.

In the meshfree method, we do not have elements, and the bandwidth is determined by a much more complex mechanism. Here we suggest the following simplest and very robust way to minimize the bandwidth of the system matrix. The principal consideration of the "minimizer" is to minimize the maximum difference in the nodal numbers for all the nodes within a local support domain of any point in the problem domain. This simple scheme is schematically shown in Figure 15.1. The procedure is as follows:

1. Choose a reference point A that is far from the problem domain, i.e., $D \geq H$. Point A also has to be in the longitudinal direction of the problem domain.
2. Calculate the distances between point A and all the field nodes.
3. Rank all the field nodes by the shortest distance.
4. Renumber all the nodes following the rank of the nodes.

An index can be easily generated to recode the original numbering of the nodes, which may be useful in retrieving the solution originally assigned to the nodes. This simple procedure is very easy to implement and very effective. The bandwidth of the system matrix should be the minimum or at least very close to the minimum. The procedure is also implemented in MFree2D for both EFG and PIMs processors. The simplest often works best.

15.3 Bucket Algorithm for Node Searching

For meshfree methods using influence domains (including the EFG processor implemented in MFree2D©) for node selection, a procedure is required to search all the nodes that fall into the influence domain. It is very expensive (N^2 complexity) if every node in the entire problem domain has to be checked against the node, especially when the number of nodes is large. To reduce the cost of node searching, many algorithms have been developed. One very simple approach is the so-called bucket algorithm [13] that divides the problem domain into buckets, each containing nodes up to a predefined number limit. The number limit is defined according to the problem size and the maximum number of nodes allowed in an influence domain. The range of node searching can thus be reduced from the entire problem domain to a number of buckets that have overlaps with the domain of influence under construction. In numerical implementation, the structure of the bucket needs to be defined. The structure contains information about the bucket range and the nodes contained in the bucket. An example of the structure in C language is given as

```
struct bucket {
int num; //number of nodes contained in bucket
double xmin, ymin, xmax, ymax; //range of bucket
int node[BUCKETNUMLIMIT]; //domain nodes in bucket
};
```

where `BUCKETNUMLIMIT` is the predefined number limit. The bucket defined in the example above assumes a rectangular shape.

The next undertaking is to formulate all buckets in a problem domain; this is fulfilled by a recursive procedure. The procedure first sets a bucket to contain the entire domain. If the number of nodes exceeds the predefined number limit, the bucket is split at its larger dimension into two equal-sized subbuckets. The same procedure is then applied to each subbucket, and the subbuckets are further divided. The procedure terminates when the number of nodes in every bucket is less than the predefined number limit. In detail, the recursive procedure is as follows:

1. Define the maximum number of nodes that a bucket is allowed to hold.
2. Create a list to hold the buckets.
3. Set a bucket to hold the entire problem domain.
4. If the number of nodes in the bucket is less than the predefined number limit, append the bucket to the list of buckets and go to 8.
5. Split the bucket into two equal-sized buckets—`BucketOne` and `BucketTwo`.
6. Load `BucketOne` and go to 4.
7. Load `BucketTwo` and go to 4.
8. End.

An application result of this procedure is shown in Figure 15.2, where the rectangles define the bucket ranges. The list of buckets is maintained dynamically during the adaptive process. In the case of domain refinement or node exchange between adjacent buckets, the disturbed buckets are examined and adjusted accordingly.

EFG may use this algorithm if it chooses to, and an option is given in MFree2D$^©$ for the EFG processor. A PIM does not. It simply uses the connectivity provided by the triangular c, which is most efficient.

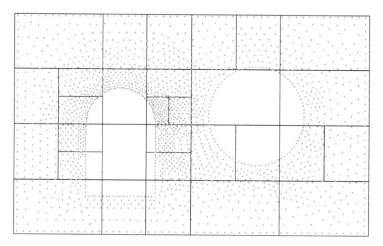

FIGURE 15.2
Problem domain represented by division into buckets. (From Liu, G. R. and Tu, Z. H., *Comput. Methods Appl. Mech. Eng.*, 191, 1923, 2002. With permission.)

15.4 Relay Model for Domains with Irregular Boundaries

The relay model is used only in EFG implemented in MFree2D$^{©}$. PIMs do not need it.

15.4.1 Problem Statement

For meshfree methods that do not use connectivity of nodes, difficulties arise in determining the influence domain of a node. The influence domain usually takes a simple shape, for example, circular or rectangular. For simple problem domains, the influence domain can be determined simply by drawing a circle of radius r_I defined by Equation 15.1. Its influence on any other point is directly computed by the distance between the node and the point via the use of the weight functions given in Chapter 2. For complex problem domains involving multiple discontinuities, such as multiple cracks, an influence domain may contain numerous irregular boundary fragments, as shown in Figure 15.3, and computation of nodal weights simply based on physical distance can be erroneous as the discontinuity of the field variables caused by the complex boundary are not accounted for. For example, in Figure 15.3, points P and Q are located at different positions within the influence domain of node O and have the same distance to source node O. A question is: Shall O impose the same influence on Q as that on P? If not, how shall the weight of influence on Q be determined, and further on an arbitrary point within the influence domain? Shall an influence domain retain the regularity in shape as in simple problem domains and, if not, how shall the profile of the influence domain be defined? These questions must be addressed properly as the determination of nodal influence is vital to the accuracy of solution using a meshfree method.

In view of the aforementioned problems, this section is specifically devoted to these aspects of defining the profile of the influence domain and computing the weight of influence in meshless approximations with irregular boundaries. By starting with a review of the existing techniques in these respects, a relay model aiming to provide a general solution for domains bounded by arbitrary boundaries is proposed. The essence of this model is to construct a hierarchical network of relay points to transmit nodal influence.

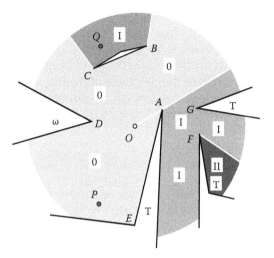

FIGURE 15.3
Influence domain containing numerous irregular boundary fragments.

Moreover, expressions derived from the circle involute curve are employed to define the profile of the portions of the influence domain that are not visible to the source node. Various numerical examples based on the EFG method are presented to verify the effectiveness of the proposed model.

A number of techniques have been reported on the construction of meshless approximations with discontinuities and nonconvex boundaries. The following sections describe three typical methods: the visibility, diffraction, and transparency methods.

15.4.2 Visibility Method

This is the earliest method used for domains with discontinuities [2]. Detailed descriptions of the method are available in papers by Belytschko et al. [3]. The essence of this approach is that the domain boundaries and any interior lines of discontinuities are treated as opaque when constructing weight functions. In this approach, the line from a point to a node is imagined to be a ray of light. If the ray encounters an opaque surface, it is terminated and the point is excluded from the domain of influence. An example of using the visibility criterion is illustrated in Figure 15.4, where the blank region is excluded from

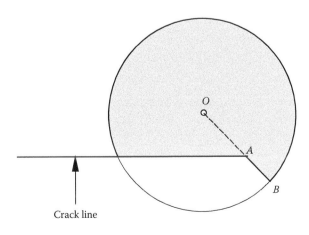

FIGURE 15.4
Domain of influence of a node in the vicinity of a crack tip determined by the visibility method (blank region is excluded by a visibility criterion).

the original influence domain of a generic node *I* in the vicinity of the tip of a straight crack that is represented by the horizontal line in Figure 15.4. This method is simple and straightforward. It is noted, however, that the weight function by the visibility criterion is discontinuous within the influence domain. Along the ray that grazes the tip of the crack, e.g., line *AB* in Figure 15.4, the weight function is nonzero to the right side of the line, but vanishes to the left. As a consequence, the shape functions constructed are also discontinuous, which is undesirable in a meshless approximation. Moreover, the visibility criterion may not be very easy to adapt to concave boundaries.

15.4.3 Diffraction Method

The diffraction method [3,4] was motivated by the way light diffracts around a sharp corner. This technique applies only to polar-type weight functions where the weights are defined as a function of a single distance-related weight parameter. The mechanism of the diffraction method is illustrated in Figure 15.5, where the straight horizontal line represents a crack. By the diffraction method, the crack line is treated as opaque, but the length of the ray is evaluated by a path that passes around the crack tip *C*. According to [3], the weight parameter *s* associated with *A* is computed by

$$s = \left(\frac{s_1 + s_2(x)}{s_0(x)}\right)^{\lambda} s_0(x) \tag{15.2}$$

where
  s_0, s_1, and s_2 are the distances between *O* and *P*, *O* and *A*, and *A* and *P*, respectively
  λ is a user-defined parameter

As a result, the domain of influence contracts around the crack tip. The diffraction method can also be applied to nonconvex boundaries. Figure 15.6 illustrates such an example, where the parameter *s* is constructed from the lengths of two line segments that just graze the boundary.

The major benefit of the diffraction method is that the weight function and shape function are continuous within the influence domain. This is, however, achieved at

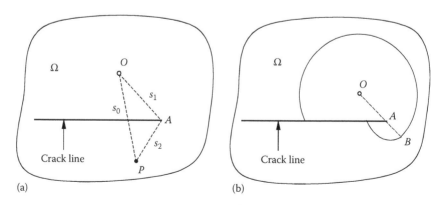

(a)　　　　　　　　　　　　　　　(b)

FIGURE 15.5
Schematic illustration of the diffraction technique to determine the influence domain. (a) Definitions of parameters s_0, s_1, and s_2; (b) domain of influence.

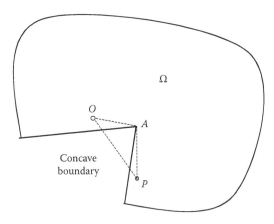

FIGURE 15.6
Diffraction technique applied to concave boundary of a domain Ω.

the cost of lower efficiency in computing the derivatives of these functions. Care should also be exercised because, when points O and P are very close, i.e., $s_0(x)$ is close to 0, the weight parameter s becomes infinite, causing difficulty in computation.

15.4.4 Transparency Method

The transparency method [4,5] was developed to smooth a meshless approximation around the tip of a discontinuity by endowing a discontinuity line with a varying degree of transparency—from completely transparent at the tip of the discontinuity to completely opaque at a distance away from the tip. The procedures to compute the weight parameter s are shown in Figure 15.7; the parameter is given by

$$s(x) = s_0(x) + s_{\max}\left(\frac{s_c(x)}{\bar{s}_c}\right)^\lambda, \quad \lambda \geq 2 \tag{15.3}$$

where
  $s_0(x)$ is the distance between O and P
  $s_{\max}$ is the radius of the nodal support
  $s_c(x)$ is the distance from the crack tip to the intersection point

The parameter S_c sets the intersection distance at which the discontinuity line is completely opaque. Care should be exercised because, for nodes very close to a boundary, the angle

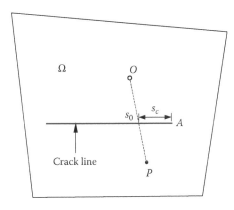

FIGURE 15.7
Schematic illustration of the transparency method for a node near a crack tip.

enclosed by the crack and the ray from the node to the crack tip is very small, causing a sharp gradient in the weight function. Therefore, it is required that all nodes have a minimum distance to the crack surface.

15.4.5 The Relay Model

The techniques described above have been applied for many problems of crack propagation. However, these methods are effective for problems with relatively simple domains (e.g., with one or two cracks, or with few nonconvex portions on the boundaries). They are not sufficient for problem domains featuring highly irregular boundaries. A highly irregular domain is a domain containing an arbitrary number of cracks, discontinuities, nonconvex boundary portions, and so forth. Further, there is no restriction to the orientations and distributions of these geometric features. In FEM, this situation causes difficulties mainly in the preprocessing work—meshing—that requires manual operation. Once the domain has been discretized into elements, shape functions are constructed readily with the aid of nodal connectivity and no further consideration is required for the geometric details in the course of computation unless refinement of the domain is needed. However, in meshfree methods, the absence of meshes requires a rather sophisticated algorithm to automatically examine the boundary details of an irregular problem domain during a computation to determine the influence domains for a node. This kind of automatic algorithm can be developed, because there is no need to provide connectivity for the nodes. All that has to be done is to determine the nodes that should be included in the influence domain. The relay model is one of such algorithms for determining the influence domain of a node in a complex problem domain. The relay model, developed in [6] is as follows.

The relay model proposed is motivated by the way a radio communication system composed of networks of relay stations works. Consider an influence domain containing a large number of irregular boundary fragments, as depicted in Figure 15.3. O is the source node and the solid lines depict the boundaries. The source node first radiates its influence in all directions equally, just as a radio signal is broadcast at a radio station, until the contained boundaries are encountered. Under the relay model, the influence from the source node is conveyed to the blocked regions via a network of relay points. The following subsections give detailed descriptions about the principles, mechanisms, and computational implementations of the relay model. The descriptions are facilitated by some definitions and notations.

15.4.5.1 Control Point and Connection

15.4.5.1.1 Control Point

A control point is a boundary point at which the angle formed by the boundary on the nonmaterial side (e.g., ω in Figure 15.3) is less than a predefined value. The predefined value is not larger than π; i.e., the boundary at the control point is nonconvex. In the discussion that follows, π is used as the predefined angle value, which means that every boundary node (not only crack tips) is qualified to be a control point if the boundary at this point is concave. (There are cases where a smaller predefined value will greatly simplify the analysis without losing the accuracy of results.) For all control points, the gradient of the boundary changes abruptly in the vicinity of their locations. In Figure 15.3, points A, B, C, D, F, and G are control points, whereas E is not. It is stressed that this definition is purely geometric and contains no physical or mechanics meaning. Therefore, it does not imply a point of singularity.

15.4.5.1.2 Connection

A connection is the shortest path that links two points within an influence domain. The path lies completely within the influence domain and is continuous without interruption by boundaries. If such a path does not exist, there is no connection between the two points within the influence domain considered. The connection is said to be direct if the path is straight; otherwise, it is indirect. To denote a connection from one point to another, an arrow is employed in the following notation. For example

$$\text{Direct connections: } P \rightarrow O, F \rightarrow A$$

$$\text{Indirect connections: } Q \xrightarrow{\{C\}} O, T \xrightarrow{\{F,A\}} O$$

The arrows for direct connections are clean, signifying that these connections require no intermediate points. For indirect connections, an assembly of the points via which the connections are established is placed over the arrow. From the left to right in the assembly, the intermediate points are from the nearest to the farthest in terms of their equivalent distances (this will be defined later) to the point on the left side of a connection. The sequence holds if the connection is viewed in reverse order.

Within the original influence domain of a node, points that have no connection with the source node are excluded from the influence domain—this is the criterion of node inclusion in the present model.

15.4.5.2 Relay Point, Relay Region, and Network of Relay Points

15.4.5.2.1 Relay Point

A relay point is a control point with only one boundary segment visible to the source point. It is so named because it is responsible for conveying the influence from the source to the blocked region. This definition is relative—whether a control point is a relay point depends on the position of this point relative to the source point. For example, to source node O, A, B, and C are relay points, whereas D is not. If at another time, P becomes the source node, D is a relay point to P. For a complex influence domain, the relay points are also ranked with the rank order assigned according to their relations with the source node. To illustrate this, the influence domain depicted in Figure 15.3 is again referenced. The relay points that have direct connections with O are the primary relay points; they are ranked level one. Specifically, the source node O is also treated as a relay point and is ranked level zero. Some relay points may possess subrelay points. For example, F is a subrelay point of A; A is called the parent (or master) of F. As F connects indirectly with O via A, F is ranked level two. Subsequently, relay points of higher order are ranked in a similar way. It should be noted that the rank order conferred to a relay point is valid only in the current influence domain; it may change if the same point is in an influence domain of a different source node.

15.4.5.2.2 Relay Region

A relay region is a fraction of the influence domain to which a relay point has an exclusive right to transmit the influence from the source node. Two conditions are set for this definition: every point within this region has a direct connection with the relay point and every point within this region has an indirect connection with the source node via the relay point. The relay point is called the master of the relay region. Relay regions are also ranked with their rank orders inherited from their respective mastering relay points. For the source

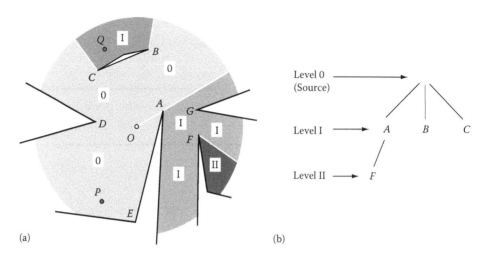

FIGURE 15.8
Relay model for an irregular influence domain. (a) Illustration of relay regions of different levels; (b) tree representation of the hierarchical network of relay points. (From Liu, G. R. and Tu, Z. H., *Comput. Methods Appl. Mech. Eng.*, 191, 1923, 2002. With permission.)

node, it governs a relay region of level zero. Figure 15.8a shows three levels of relay regions differentiated by gray colors of different degrees—the lightest corresponds to level zero, the intermediate gray to level one, and the darkest to level two. From the above definition, a relay region will have no overlap with another; the interface between two neighboring regions is generally a straight line (Figure 15.8a).

The potential boundary of a relay region is defined by the criterion that the weight function vanishes. By this definition, the potential boundary differs from the real boundary of a relay region in that the former is purely imaginary whereas the latter may comprise portions of the physical boundaries of a problem domain. For the level-zero relay region, the potential boundary coincides with the boundary of the original influence domain. For a relay region of higher level, it is assumed that the radial distance from the mastering relay point to the potential boundary of the relay region decreases with the radial line deviating from the extension of the line connecting the relay point and its parent. As shown in Figure 15.9a, the assumption leads to

$$\rho = \rho(\theta), \text{ while } \rho \downarrow \text{ with } \theta \downarrow \tag{15.4}$$

where ρ is the radial distance, decreasing with θ. The preceding expression defines the profile of a relay region that contracts around its mastering relay point (i.e., C_1 in Figure 15.9a). This concept is identical to the diffraction technique described in the preceding section. However, the present model uses expressions derived from the form of circle involute instead of Equation 15.2 that will be presented later.

15.4.5.2.3 Network of Relay Points

This is a network formed by all the multitiered relay points within an influence domain. The network is a hierarchical system and corresponds to a tree structure (Figure 15.8b). Its function is to transmit the influence from the source node to the entire range of the influence domain. This resembles the working principles of a radio communication network system except that the influence (which is the "signal" in radio communication) is

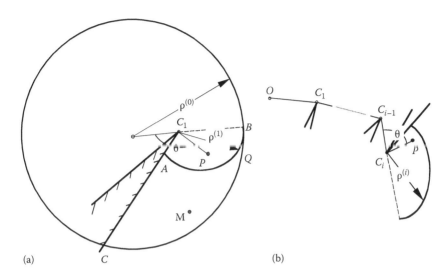

FIGURE 15.9
Profile of relay regions and schemes for computation of equivalent distance of a point using the relay model.
(a) Point in a level-one relay region; (b) point in an *i*th-level relay region.

conveyed without being enhanced at a relay point. The network formed by relay points is the core of the relay model.

15.4.5.3 Effective Relay Radius and Equivalent Distance

Effective relay radius is a parameter associated with a relay point; it defines the maximum radial distance from this relay point to the potential boundary of the relay region associated with it. In the present description, it is equal to $\rho(\pi)$. The effective relay radius is calculated by

$$r_{\text{eff}}^{(i)} = \rho^{(i-1)}(\theta) - r = \rho^{(i)}(\pi), \quad i = 1, 2, \ldots \tag{15.5}$$

where
    r_{eff} denotes the effective relay radius
    r is the distance from a relay point to its parent relay point

$\rho^{(i-1)}(\theta)$ is the radial distance defined in the relay region associated with the parent relay point, with the radial line passing the relay point considered. The superscript i in parentheses denotes the rank order. In particular, the source node, i.e., the level-zero relay point, has an effective relay radius of

$$r_{\text{eff}}^{(0)} = r_0 \tag{15.6}$$

where r_0 is the original radius of the influence domain.

Under the relay model, an influence domain is divided into numerous relay regions of different levels. Points in different relay regions have different connections with the source node. It will be unfair if the details of a connection are not accounted for in the computation of the weight parameter. This model employs the equivalent distance instead of the physical distance as the basis to measure the weight parameter. For a point in the level-zero relay region, the equivalent distance is identical to the physical distance from this point to the source node, i.e.,

$$r_{eq} = r, \text{ for points in the level-zero relay region.} \tag{15.7}$$

where r_{eq} and r are the equivalent distance and physical distance from the point to the source node (or the level-zero relay point), respectively.

For points in relay regions of higher level, the equivalent distance is evaluated in a progressive manner. To illustrate this, a point in a level-one relay region is first considered. Figure 15.9a shows a single relay point case where O is the source node, C_1 is a relay point of level one, and P is located in the relay region associated with C_1. According to Equations 15.5 and 15.7, the effective relay radius and equivalent distance associated with C_1 are

$$r_{eff}^{(1)} = \rho^{(0)} - r = r_0 - r \tag{15.8}$$

and

$$r_{eq}^{(1)} = r \tag{15.9}$$

where r is the distance between C_1 and O. Following the notation defined above, the connection between P and O is depicted as

$$P \xrightarrow{\{C_1\}} O \tag{15.10}$$

The equivalent distance of P is computed by

$$r_{eq} = \frac{r_{eq}^{(1)} + r}{r_{eq}^{(1)} + \rho^{(1)}(\theta)} r_0 \tag{15.11}$$

where

r represents the distance from P to its mastering relay point C_1
θ is the smaller angle formed by $\overrightarrow{C_1P}$ and $\overrightarrow{C_1O}$ with the direction from $\overrightarrow{C_1P}$ to $\overrightarrow{C_1O}$

If P is located in an ith-level relay region as shown in Figure 15.9b and has a connection with the source node as described by

$$P \xrightarrow{\{C_i, C_{i-1}, ..., C_1\}} O \tag{15.12}$$

the equivalent distance from P to O is computed in the same fashion, i.e.,

$$r_{eq} = \frac{r_{eq}^{(i)} + r}{r_{eq}^{(i)} + \rho^{(i)}(\theta)} r_0 \quad i = 0, 1, 2, \ldots \tag{15.13}$$

where

r_{eq}^i is the equivalent distance associated with the mastering relay point C_i
$\rho^{(i)}$ defines the potential boundary of the ith-level relay region controlled by C_i

θ is the smaller angle formed by $\overrightarrow{C_iC_{i-1}}$ and $\overrightarrow{C_iP}$; one has

$$\cos\theta = \mathbf{v}_{\overrightarrow{C_iC_{i-1}}} \cdot \mathbf{v}_{\overrightarrow{C_iP}} \tag{15.14}$$

where $\mathbf{v}_{\overrightarrow{C_iC_{i-1}}}$ and $\mathbf{v}_{\overrightarrow{C_iP}}$ are unit vectors defining the directions of $\overrightarrow{C_iC_{i-1}}$ and $\overrightarrow{C_iP}$, respectively, i.e.,

$$\mathbf{v}_{C_iC_{i-1}} = \frac{\mathbf{x}_{C_{i-1}} - \mathbf{x}_{C_i}}{|\mathbf{x}_{C_{i-1}} - \mathbf{x}_{C_i}|} = \begin{pmatrix} (x_{C_{i-1}} - x_{C_i})/r_{C_iC_{i-1}} \\ (y_{C_{i-1}} - y_{C_i})/r_{C_iC_{i-1}} \end{pmatrix} \tag{15.15}$$

$$\mathbf{v}_{C_i\vec{p}} = \frac{\mathbf{x} - \mathbf{x}_{C_i}}{|\mathbf{x} - \mathbf{x}_{C_i}|} = \begin{pmatrix} (x - x_{C_i})/r \\ (y - y_{C_i})/r \end{pmatrix} \tag{15.16}$$

Equation 15.13 presents a progressive way to compute the equivalent distance of P sequentially from $C_1, C_2, \ldots, C_i$, to P. The expression also shows that the equivalent distance of a point only needs the information of its mastering relay point; this makes the computation very easy to manage for a very complex influence domain. To save computation expense, $r_{eq}^{(i)}$ and $r_{eff}^{(i)}$ associated with all levels of relay points are computed in advance. There are occasions that a point may have various links to the source node (e.g., Q in Figure 15.8a can connect with O via either B or C). The rule practiced here is that the shortest equivalent distance among all the possible paths is used.

15.4.5.4 Weight Parameter and Its Derivatives

Weight functions play an essential role in the construction of shape functions. It rules the influence of the source node over the entire influence domain. A general polar-type weight function takes a form of

$$\widehat{W}(s) = \begin{cases} >0 & \text{for } s < 1 \\ =0 & \text{for } s \geq 1 \end{cases} \tag{15.17}$$

and its derivatives are calculated by

$$\frac{\partial \widehat{W}}{\partial x} = \frac{d\widehat{W}}{ds}\frac{\partial s}{\partial x} \tag{15.18}$$

$$\frac{\partial \widehat{W}}{\partial y} = \frac{d\widehat{W}}{ds}\frac{\partial s}{\partial y} \tag{15.19}$$

where the weight parameter $s = r/r_0$ is a normalized distance. Several commonly used weight functions are given in Chapter 5 (note that variable is now changed to s).

In the relay model, the weight parameter s is measured by the equivalent distance. For example, the weight parameter for a point in an ith-level relay region is

$$s = \frac{r_{eq}}{r_0} = \frac{r_{eq}^{(i)} + r}{r_{eq}^{(i)} + \rho^{(i)}(\theta)}, \quad i = 0, 1, 2, \ldots \tag{15.20}$$

The derivatives of s with respect to x and y are, therefore,

$$\frac{\partial s}{\partial x} = \frac{1}{r_{eq} + \rho(\theta)} \frac{\partial r}{\partial x} - \frac{r}{(r_{eq} + \rho(\theta))^2} \frac{d\rho}{d\theta} \frac{\partial \theta}{\partial x} \tag{15.21}$$

$$\frac{\partial s}{\partial y} = \frac{1}{r_{eq} + \rho(\theta)} \frac{\partial r}{\partial y} - \frac{r}{(r_{eq} + \rho(\theta))^2} \frac{d\rho}{d\theta} \frac{\partial \theta}{\partial y} \tag{15.22}$$

where the superscript (i) is dropped for simplicity of description. The derivatives of θ with respect to x and y are computed from Equation 15.14:

$$\frac{\partial \theta}{\partial x} = -\frac{y - y_{C_i}}{r^2} \tag{15.23}$$

$$\frac{\partial \theta}{\partial y} = \frac{x - x_{C_i}}{r^2} \tag{15.24}$$

if θ is counterclockwise. When the direction of θ is clockwise, the derivatives are

$$\frac{\partial \theta}{\partial x} = \frac{y - y_{C_i}}{r^2} \tag{15.25}$$

$$\frac{\partial \theta}{\partial y} = -\frac{x - x_{C_i}}{r^2} \tag{15.26}$$

15.4.5.5 Numerical Implementation

To implement the relay model, sophisticated algorithms that take into account the details of the boundary fragments are required. This necessitates, at least, two tasks: detection of all boundary fragments contained in an influence domain and construction of the network of relay points. The procedures to construct the relay model within the influence domain of an arbitrary node O are as follows:

1. Construct the influence domain associated with node O.
2. Extract all boundary fragments of the problem domain contained in the influence domain.
3. Filter out the nodes that have no connection with the source node.
4. Construct a list of all control points within the influence domain.
    a. Loop over all boundary nodes within the influence domain.
    b. Append a boundary node to the list if it is a control point.
5. Construct the network of relay points.
    a. Create a new list for relay points.
    b. Set the source node as the level-zero relay point and append it to the list.
    c. Set $L=1$ (L denotes the level of relay points to be constructed).
    d. Loop over the list of control points.
    e. Set the relay points of level L and append them to the list of relay points. This is done by checking the control points against the relay points of level $L - 1$.

   f. Update the list of control points by removing the control points that are already relay points.

   g. $L = L + 1$.

   h. Go to c.

6. Compute the effective relay radius and equivalent distance associated with each relay point.

7. End the construction of the relay model within the influence domain of O.

These procedures are complex and expensive in terms of computation time. This is the cost arising from the absence of meshes and is likely unavoidable for a very complex problem domain. The price has to be paid to relieve the labor involved in meshing. It is not difficult to imagine how much time is required to mesh a domain of complexity shown in Figure 15.3. Compared with the price for labor to perform the work, the cost of the relay model is extremely small. Moreover, if adaptive analysis is required, it is not possible to manually perform the meshing over and over again.

15.4.5.6 Profile of a Relay Region

15.4.5.6.1 Circle Involute Approach

The present model defines the profile, or potential boundary, of a relay region using forms derived from the circle involute (Figure 15.10a). By mathematical definition, the involute of a circle is a curve orthogonal to all the tangents to this circle. For a better understanding, one can imagine there is a circle with a rope wound around its perimeter. If one holds tightly to one end of the rope to unwind the rope from the circle, the circle involute is given by the track of this end. In Figure 15.10a, Q is the end of the rope and T is the point where the extracted portion of the rope makes tangential contact with the circle. The circle involute can be mathematically expressed by

$$x = a(\cos\alpha + \alpha \sin\alpha) \tag{15.27}$$

$$y = a(\sin\alpha - \alpha \cos\alpha) \tag{15.28}$$

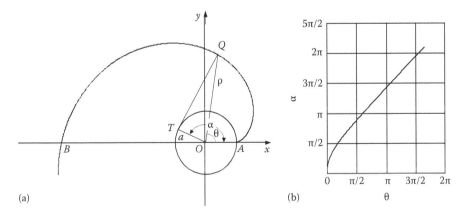

(a)　　　　　　　　　　　　　　　　(b)

FIGURE 15.10
Circle involute. (a) The involute curve; (b) relationship between α and θ.

where

    a is the radius of the circle

    α is the angle from *OA* to *OT*

Alternatively, the circle involute described in polar form is

$$\rho = a(1 + \alpha^2)^{1/2} \tag{15.29}$$

$$\theta = \alpha - \arccos\left[(1 + \alpha^2)^{-1/2}\right] \tag{15.30}$$

where (ρ, θ) is the polar position of *Q*. The preceding equations are used to define the profile of the relay region as shown in Figure 15.9a, with the value of θ confined by $(0, \pi)$.

With Equations 15.29 and 15.30, the following derivatives can be obtained:

$$\frac{d\alpha}{d\theta} = \frac{1 + \alpha^2}{\alpha^2} \tag{15.31}$$

$$\frac{d\rho}{d\alpha} = \frac{a\alpha}{(1 + \alpha^2)^{1/2}} \tag{15.32}$$

Therefore, differentiation of ρ with respect to θ yields

$$\frac{d\rho}{d\theta} = \frac{d\rho}{d\alpha}\frac{d\alpha}{d\theta} = \frac{a(1 + \alpha^2)^{1/2}}{\alpha} = \frac{\rho}{\alpha} \tag{15.33}$$

In a computation, α needs to be solved in terms of θ from the nonlinear Equation 15.30. This adds computation cost if a nonlinear equation solver is used to solve this. To avoid this, one can actually construct a table, such as Table 15.1, to store values of θ vs. α in advance. Figure 15.10b depicts the relationship between α and θ defined by Equation 15.30.

TABLE 15.1

Values of α vs. θ for the Circle Involute Approach

Θ	α	θ	α
0.0000000000E+00	0.0000000000E+00	0.1581906633E+00	0.8796459430E+00
0.8248810247E−04	0.6283185307E−01	0.1866837799E+00	0.9424777960E+00
0.6552697876E−03	0.1256637061E+00	0.2172636968E+00	0.1005309649E+01
0.2186035144E−02	0.1884955592E+00	0.2498070719E+00	0.1068141502E+01
0.5099810682E−02	0.2513274123E+00	0.2841906568E+00	0.1130973355E+01
0.9763467994E−02	0.3141592654E+00	0.4796580601E+00	0.1445132621E+01
0.1647595382E−01	0.3769911184E+00	0.7532566507E+00	0.1822123739E+01
0.2546442135E−01	0.4398229715E+00	0.1107412823E+01	0.2261946711E+01
0.3688561209E−01	0.5026548246E+00	0.1540878884E+01	0.2764601535E+01
0.5083130149E−01	0.5654866776E+00	0.2051016866E+01	0.3330088213E+01
0.6733641460E−01	0.6283185307E+00	0.2635059918E+01	0.3958406743E+01
0.8638848454E−01	0.6911503838E+00	0.3290607926E+01	0.4649557127E+01
0.1079373643E+00	0.7539822368E+00	0.4015736576E+01	0.5403539364E+01
0.1319044111E+00	0.8168140899E+00	0.4870220170E+01	0.6283185307E+01

From Table 15.1, α can be obtained by linear interpolation once θ is calculated from Equation 15.14; i.e.,

$$\alpha = \alpha_1 + \frac{\theta - \theta_1}{\theta_2 - \theta_1}(\alpha_2 - \alpha_1) \tag{15.34}$$

where
  θ falls into a range defined by two consecutive values θ_1 and θ_2 in the table
  α_2 and α_1 are values corresponding to θ_1 and θ_2, respectively

Next, parameter a is determined. This can be achieved by noting that for ρ there holds

$$\rho(0) = a \quad \text{and} \quad \rho(\pi) = r_{\text{eff}} \tag{15.35}$$

With Equation 15.29, parameter a is obtained as

$$a = r_{\text{eff}}(1 + \alpha(\pi)^2)^{-1/2} = 0.21724 r_{\text{eff}} \tag{15.36}$$

where $\alpha(\pi)$ is 4.449235 by interpolation from Table 15.1.
  By combining Equations 15.20 through 15.22, 15.29, 15.30, and 15.33, the weight parameter and its derivatives can be determined; the derived formulae are applicable to any polar-type weight function. To illustrate the effects of the circle involute approach on the weight function, three domains as shown in Figure 15.11 are used for comparison. Figure 15.11a through c represent domains containing no discontinuity, a single crack, and double cracks, respectively. The weight function used is the cubic spline. The calculated weights and derivatives over the influence domain of the source node O are depicted by the surface plots and contour plots in Figures 15.12 through 15.14, respectively.
  It is noted from Equation 15.33 that the derivative of ρ with respect to θ turns out to be infinite when α is very close to zero. This can be avoided by setting an initial value of α, e.g., $\alpha(0) = \alpha_0 > 0$.

15.4.5.6.2 *Modified Circle Involute Approach*
Figures 15.13 and 15.14 show that there are small jumps in the derivatives of the weight function across the interfaces between neighboring relay regions; i.e., the derivatives are not smooth. This is not desirable since the smoothness of weight functions is crucial to the

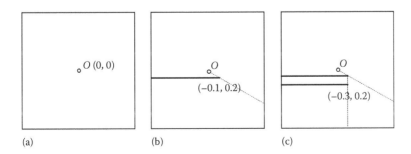

(a)   (b)   (c)

FIGURE 15.11
Three domains: (a) with no discontinuity, (b) with a single crack, and (c) with two cracks.

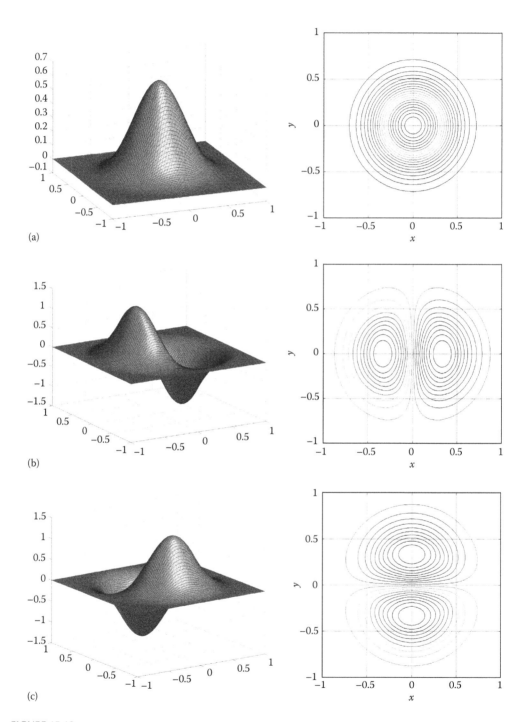

FIGURE 15.12

Cubic spline weight function and its derivatives over an influence domain with no discontinuity. Surface plots and contour plots for (a) $\hat{W}(s)$, (b) $\hat{W}_x$, and (c) $\hat{W}_y$.

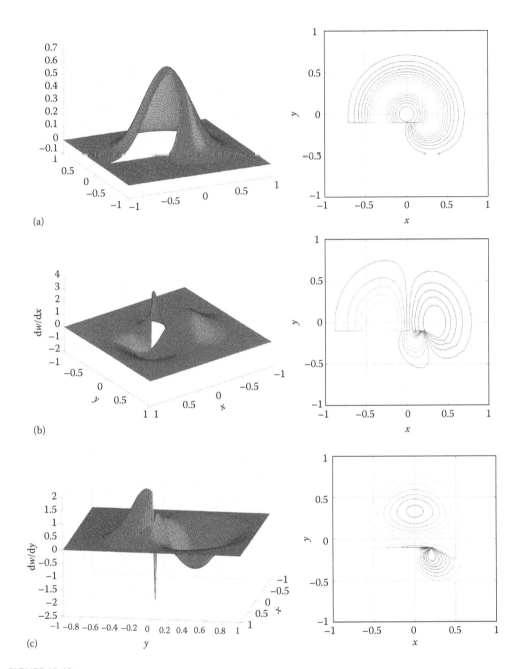

FIGURE 15.13
Cubic spline weight function and its derivatives over an influence domain with a single crack by the circle involute approach. Surface plots and contour plots for (a) $\overline{W}(s)$, (b) $\overline{W}_x$, and (c) $\overline{W}_y$.

continuity of stresses. A modified circle involute approach is therefore proposed to mitigate this problem. This approach uses a different portion of the circle involute curve as shown in Figure 15.15a, where the origin of coordinates, O, moves from the circle center to the starting point of the circle involute with the x-axis tangential to the perimeter. The resulting curve in polar form is

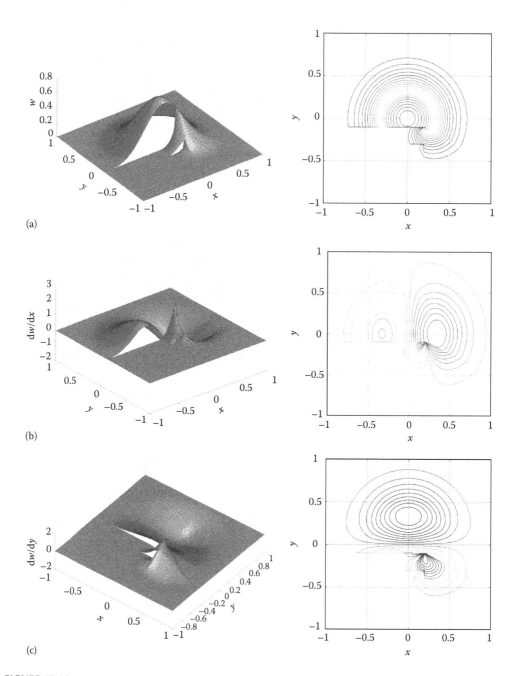

FIGURE 15.14

Cubic spline weight function and its derivatives over an influence domain with two cracks by the circle involute approach. Surface plots and contour plots for (a) $\bar{W}(s)$, (b) $\bar{W}_x$, and (c) $\bar{W}_y$. (From Liu, G. R. and Tu, Z. H., *Comput. Methods Appl. Mech. Eng.*, 191, 1923, 2002. With permission.)

$$\rho = a(2 + \alpha^2 - 2\cos\alpha - 2\alpha\sin\alpha)^{1/2} \qquad (15.37)$$

$$\cos\theta = \frac{a(\sin\alpha - \alpha\cos\alpha)}{\rho} \qquad (15.38)$$

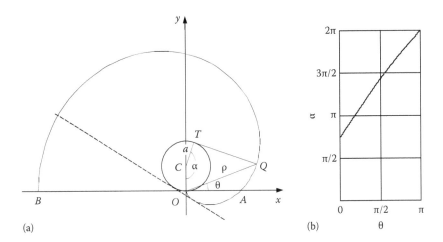

FIGURE 15.15
Modified circle involute approach. (a) The modified curve and (b) the relationship between α and θ.

The relationship of α and θ defined by Equation 15.37 is depicted in Figure 15.15b. Again, Table 15.2 is constructed to store the values of α vs. θ. For those values not directly given in the table, linear interpolations apply.

The derivative of ρ with respect to θ determined from Equations 15.37 and 15.38 is

$$\frac{d\rho}{d\theta} = \frac{\rho(1 + \sin\alpha)}{\alpha - \cos\alpha} \tag{15.39}$$

where θ is defined within $(0, \pi)$ while the bounds of α are given by $\alpha(0)$ and $\alpha(\pi)$. The constant a is

$$a = r_{\text{eff}}/2\pi \tag{15.40}$$

since $\rho = r_{\text{eff}} = 2\pi a$ at $\alpha = 2\pi$.

TABLE 15.2

Values of α vs. θ for the Modified Circle Involute Approach

θ	α	θ	α
−0.1807289880E−01	0.2304887073E+01	0.1375632706E+01	0.4241150082E+01
0.3679373644E−02	0.2336460868E+01	0.1496629531E+01	0.4398229715E+01
0.1728543851E−01	0.2356194490E+01	0.1619396195E+01	0.4555309348E+01
0.1259017101E+00	0.2513274123E+01	0.1744098373E+01	0.4712388980E+01
0.2351114085E+00	0.2670353755E+01	0.1870915404E+01	0.4869468613E+01
0.3449731099E+00	0.2827433388E+01	0.2000040107E+01	0.5026548246E+01
0.4555502106E+00	0.2984513021E+01	0.2131677850E+01	0.5183627878E+01
0.5669115049E+00	0.3141592654E+01	0.2266044530E+01	0.5340707511E+01
0.6791318229E+00	0.3298672286E+01	0.2403363059E+01	0.5497787144E+01
0.7922927290E+00	0.3455751919E+01	0.2543857832E+01	0.5654866776E+01
0.9064832817E+00	0.3612831552E+01	0.2687746603E+01	0.5811946409E+01
0.1021800854E+01	0.3769911184E+01	0.2835229208E+01	0.5969026042E+01
0.1138352007E+01	0.3926990817E+01	0.2986472681E+01	0.6126105674E+01
0.1256253398E+01	0.4084070450E+01	0.3141592654E+01	0.6283185307E+01

The cubic spline weight function and its derivatives based on the modified approach are demonstrated by surface plots and contour plots in Figures 15.16 and 15.17. The plots demonstrate that the modified approach gives smooth descriptions for the weight function and its derivatives in the level-one relay region; this can be shown easily from

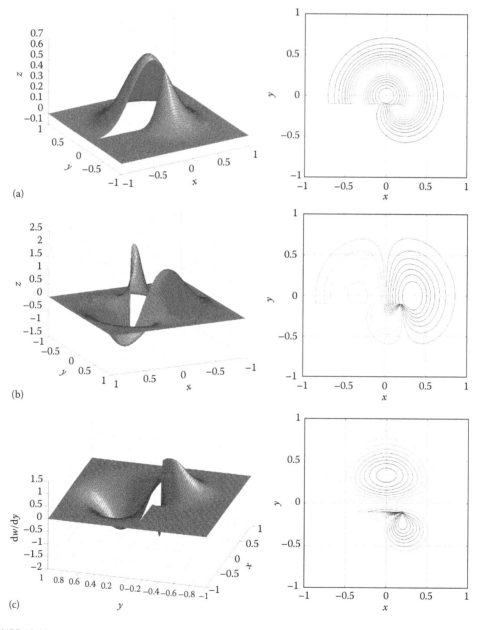

(a)

(b)

(c)

FIGURE 15.16

The cubic spline weight function and its derivatives over an influence domain with a single crack by the modified circle involute approach. Surface plots and contour plots for (a) $W(s)$, (b) W_x, and (c) W_y.

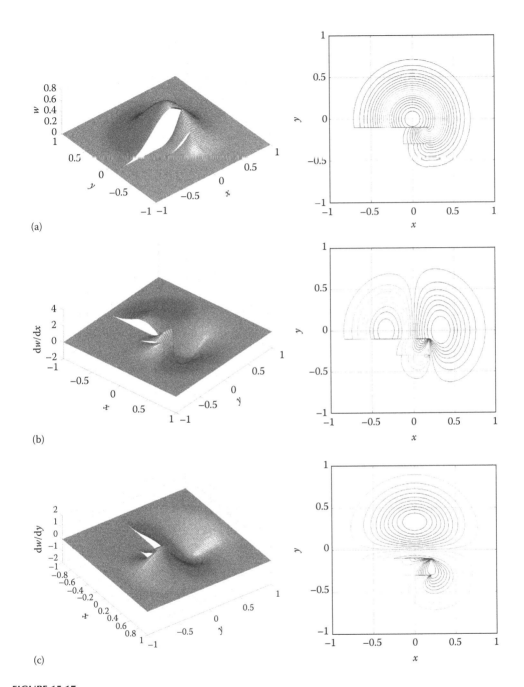

FIGURE 15.17
Cubic spline weight function and its derivatives over an influence domain with two cracks by the modified circle involute approach. Surface plots and contour plots for (a) $\widehat{W}(s)$, (b) $\widehat{W}_x$, and (c) $\widehat{W}_y$.

Equation 15.39 mathematically. However, in relay regions of higher level, there are also small jumps across the interface. A more appropriate description of the profile of the relay region is therefore needed; this unfortunately is still under investigation.

A more general form for the modified formula is expressed by

$$\rho = a(2 + \alpha^2 + 2\sin{(\alpha - \varphi)} - 2\alpha\cos{(\alpha - \varphi)})^{1/2} \tag{15.41}$$

$$\cos{\theta} = \frac{a(\cos{(\alpha - \varphi)} + \alpha\sin{(\alpha - \varphi)})}{\rho} \tag{15.42}$$

where φ is a user-defined value corresponding to an initial rotation angle. The range of φ is recommended to be $[1, \pi/2]$. The extreme case, $\varphi = \pi/2$, corresponds to Equations 15.37 and 15.38, whereas $\varphi = 1$ corresponds to the case that the x-axis makes tangential contact with the involute curve as shown by the dashed line in Figure 15.15a.

Note that PIMs do not use the relay mode, because it uses a T-scheme to select nodes.

15.5 Techniques for Adaptive Analysis

15.5.1 Issues of Adaptive Analysis

In an adaptive analysis, there are essentially two issues—*a posteriori* error estimation and domain refinement. The first requires a cheap error estimate to measure the local and global errors, whereby an adaptive procedure determines whether a refinement is required, and if it is required, which part of the domain is refined. The effectiveness and efficiency of these two operations are critical to the performance of an adaptive procedure. To conduct *a posteriori* error estimation, two values—a computed value and a reference value—are usually required. The first is the raw data from the computations for the problem, and the second is derived from the first via postprocessing (e.g., smoothing or projection). In FEM, the raw data of stresses/strains do not possess interelement continuity and have a discrepancy along element boundaries; the improved values are obtained via smoothing the interelement discontinuity on strains. The difference between the raw and improved values formulates a basis for error estimation in FEM; detailed descriptions of this approach are available in the FEM literature [7]. There are also other methods [8] used for error estimation and adaptive meshing in FEM.

In the PIMs, such as the ES-PIM and NS-PIM, we have similar stress discrepancy in the background cells, and adaptive methods will also be developed making use of this discrepancy, as will be shown later.

In some meshfree methods such as EFG, there is no interelement discontinuity of stresses and the resulting stress field is very smooth over the entire problem domain. As a result, error estimates based on stress-smoothing techniques developed for FEM cannot be used for error estimation in meshfree methods. There is a need to develop suitable error estimates for adaptive analysis for meshfree methods.

15.5.2 Existing Error Estimates

A few error estimates for meshfree methods have been reported. One is proposed by Duarte and Oden [9] for the *h-p* cloud method, which involves the computation of interior residuals and the residuals for Neumann boundary conditions. Chung and Belytschko [10] adapted the FEM stress projection technique for error analysis in EFG by computing the projected stresses from the raw stresses using a reduced domain of influence. This approach is simple and inexpensive, but its effectiveness depends on the size of the reduced influence domain. Another approach is the strain gradient method proposed by Combe and Korn [11], who make use of the fact that gradients of stresses and strains may be calculated throughout the problem domain with a high accuracy. The interpolation error is evaluated from the truncated terms in a Taylor expansion of a field variable at a point over its vicinity. This approach was demonstrated to be effective. However, it requires the computations of the second derivatives, which can be quite expensive.

15.5.3 Cell Energy Error Estimate: EFG Settings

This section presents a cell energy error (CEE) estimate that can be used for adaptive analysis in meshfree methods that use background cells for integration. The material presented here is based on the work reported in [12,13]. The work is based on the fact that many implementations of existing meshfree methods rely on a background mesh, global or local, for domain integration of governing equations. The procedure is composed of a cell-based error estimate and a local domain refinement technique. The present error estimate examines the energy error in each cell and uses it as the basis for error estimation.

15.5.3.1 Error Indicator

An error estimate for an approximation is usually constructed based on the difference between the approximate and exact solutions. For a quantity $\mathbf{q}$ defined over domain Ω and approximated by $\hat{\mathbf{q}}$, a general error indicator of an approximation is defined by

$$e = L(\mathbf{q}, \hat{\mathbf{q}}) \tag{15.43}$$

where
   L denotes a norm imposed on the exact and approximate values
   e is the error corresponding to the operation

In most problems, $\mathbf{q}$ in exact form is not available and a reference value derived from $\hat{\mathbf{q}}$ is used. In solid mechanics, the quantity can be displacement, strain, stress, or energy, depending on the needs.
   A conventional implementation of Equation 15.43 is to use the L_2 norm error, i.e.,

$$e = \left\| \mathbf{q}(x) - \hat{\mathbf{q}}(x) \right\| = (\mathbf{q}(x) - \hat{\mathbf{q}}(x))^{\mathrm{T}} \cdot (\mathbf{q}(x) - \hat{\mathbf{q}}(x))^{1/2} \tag{15.44}$$

This approach is essentially a pointwise approach, as it examines errors at individual points. Errors in local and global domains can be evaluated from pointwise error via integration of the preceding equation. To measure pointwise error, the computed and reference values at a point must be provided. A major task of an error estimate is therefore to formulate the reference value at a point. The error estimate proposed in [14] for meshfree methods follows this traditional approach, with the reference value obtained by taking a product of the computed value with the shape function constructed on a reduced influence domain. A difficulty in this approach is to minimize the size of the reduced domain while preserving the regularity of the moment matrix in the MLS approximation.

Instead of examining pointwise error, the CEE method uses the error of energy in a cell as the basic measure for error estimation in meshfree methods. The quantity $\mathbf{q}$ examined in a cell is the strain energy in the cell. For solid mechanics, the cell energy is

$$E = \int_{\text{cell}} \boldsymbol{\sigma}\boldsymbol{\varepsilon} \, d\Omega \qquad (15.45)$$

Note that we disregard the fraction of $1/2$ in front of the integration; it is immaterial in our error estimation, because it is the distribution of error that counts for adaptive analysis. In a meshfree approximation, the computed cell energy, E_{comp}, is obtained by using the same Gauss integration scheme as that used for domain integration of the weak form governing equations, i.e.,

$$E_{\text{comp}} = \sum_{i=1}^{m} c_i \boldsymbol{\sigma}_i \boldsymbol{\varepsilon}_i \qquad (15.46)$$

where
   m is the number of Gauss points used in the cell for integration
   c_i is the corresponding integration weight

The reference value, E_{ref}, is evaluated using a different Gauss integration scheme:

$$E_{\text{ref}} = \sum_{j=1,n} c_j \boldsymbol{\sigma}_j \boldsymbol{\varepsilon}_j \qquad (15.47)$$

where $n(n \neq m)$ is the number of Gauss points used for the reference value. The energy error in a cell is thus

$$e = |E_{\text{comp}} - E_{\text{ref}}| \qquad (15.48)$$

In this approach, the stresses and strains at the n Gauss points are evaluated based on the displacement field given by the original solution and therefore they have a same accuracy as those of the m Gauss points. To reduce computational cost, it is recommended that n is assumed to have a smaller value than m, i.e., $n < m$. Summation of CEE over all cells in the problem domain yields the global error, i.e.,

$$e_{\text{Global}} = \sum_{i=1}^{N_e} e_i \qquad (15.49)$$

where N_e is the total number of cells. A normalized measure of CEE is

$$\bar{e} = e/A \qquad (15.50)$$

where A is the cell area. The normalized measure is very effective in the detection of singular locations.

One prominent feature in the present approach is that it requires only one stress field: Both the computed and reference values of cell energy are computed based on values in the same field at different positions. This constitutes a major difference from the conventional pointwise error estimates where a second stress field is required. This is advantageous as the postprocessing for a second field is no longer necessary.

15.5.3.2 Error Sources

In a numerical approximation, there are mainly two sources of error—interpolation error and integration error. The first arises from the limited order of approximation to the field function, when the interpolation function is performed. For example, a kth-order interpolation results in an error of $(k+1)$th order. Compared with FEM, a meshless approximation (e.g., MLS approximation) generally has a higher order of continuity and therefore a smaller interpolation error. However, the exact order of interpolation error is difficult to determine in meshfree methods as the MLS approximation is usually implicit, in contrast to FEM approximation.

The second source of error is introduced in the numerical integration. For example, in a 1D case, m Gauss point integration causes an integration error of $(2m)$th order. The integration error can be minimized if the integration order matches the interpolation order, i.e., when $m = (k+1)/2$.

Two approaches are usually taken to improve the accuracy of the solution: One is to increase the order of interpolation function and the other is to match the integration order with the interpolation order. In an MLS approximation, the first can be implemented by choosing high-order basis functions and high-order weight functions. Implementation of the second, however, is obscure as there is no prior knowledge about the exact interpolation order. One empirical way is to use a sufficient number of Gauss points in each cell in conjunction with a sufficient number of background cells for integration. It can be expected that because the integration cells are reduced in size, the error evaluated by the proposed error estimate decreases and converges as the stress field in each cell approaches linear (or planar) variation.

The task of an error estimate is to measure the error from every source, whereby an adaptive procedure determines whether to increase the interpolation order or to use a finer background mesh, or both. In the traditional stress-smoothing approach, the smoothed value has a higher order of accuracy than the original, and therefore the interpolation error is measured by the difference. In the CEE approach, the values at the Gauss points in Equations 15.46 and 15.47 are from the same stress field and therefore are of the same order of accuracy. The order of integration accuracy in the two equations, however, is different—it is $(2m-1)$ in Equation 15.46 but $(2n-1)$ in Equation 15.47. The CEE evaluated from Equation 15.48 is therefore the integration error between the two integration schemes. It is sensitive to the order of stress field—the higher the order, the larger the energy error. From this point of view, the present approach can be viewed as a variant of the gradient approach as it reflects the gradient change in an approximation field. The major weakness of the CEE estimate is that it does not provide the accuracy of an approximation itself. As a consequence, error estimation by this strategy may be erroneous if a wrong approximation is used as the basis—this is also the case for many of the existing error estimates. Note that the CEE estimate demands that the integration cell sufficiently corresponds to the gradient of the stress/strain. If the order of the Gauss integration for computing the reference cell energy is 3 ($n = 2$), and that for computed cell energy is 1 ($m = 1$), the error estimated by the CEE will approach 0 if the cell is small enough so that the stress/strain field can be approximated by a linear function. Otherwise, a large error will be produced that may demand refinement. From this point of view, the CEE performs exactly the job that is required for adaptive analysis. In the case of using an erroneous field variable approximation, it is most likely the adaptive procedure will not converge, unless the erroneous approximation converges.

15.5.4 Cell Energy Error Estimate: NS- and ES-PIMs Settings

In ES-PIM models, two types of domains are used: triangular cell Ω_i^c and smoothing domain Ω_i^s, and they are overlaid as shown in Figure 15.18. Each triangular cell hosts three parts of smoothing domains associated with the three edges of the cell: element Ω_i^c hosts parts of Ω_1^s, Ω_2^s, and Ω_3^s. Since the smoothed strains (and stresses) are constant in each of these three smoothing domains, there will be discrepancies. Simply making use of these discrepancies, the energy error in a triangular cell can be estimated as

$$e = ((\Delta\boldsymbol{\varepsilon}_{\max})^{\mathrm{T}}\mathbf{D}(\Delta\boldsymbol{\varepsilon}_{\max}) \times A_e)^{1/2} \tag{15.51}$$

where
  A_e is the area of the triangular cell
  $\Delta\boldsymbol{\varepsilon}_{\max}$ is the maximum difference of strains between the three smoothing domains in that cell, which can be expressed as

$$\Delta\boldsymbol{\varepsilon}_{\max} = \max(|\bar{\boldsymbol{\varepsilon}}_1 - \bar{\boldsymbol{\varepsilon}}_2|, |\bar{\boldsymbol{\varepsilon}}_2 - \bar{\boldsymbol{\varepsilon}}_3|, |\bar{\boldsymbol{\varepsilon}}_1 - \bar{\boldsymbol{\varepsilon}}_3|) \tag{15.52}$$

Because the energy error in a cell is evaluated as the strain energy of maximum difference of strain among three smoothing domains, the estimated energy error is a little larger than the actual energy error. This error estimate is very simple, cheap to compute, and very easy to use. It captures well the high-error region in the problem domain, which is important for our adaptive analysis. There are clearly many alternative ways to estimate the error.

The same error estimate can also be developed for NS-PIMs, where a triangular cell hosts parts of three smoothing domains, as shown in Figure 15.18b. It is clear that the error estimation in PIMs is very simple.

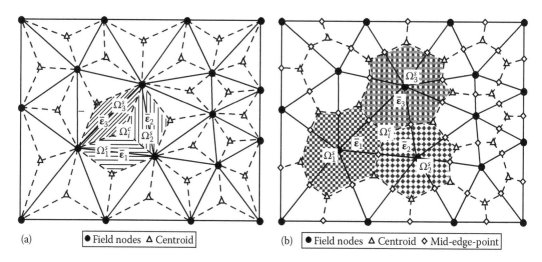

FIGURE 15.18
Strain discrepancy in a triangular cell used for error estimation: (a) ES-PIM setting; (b) NS-PIM setting.

15.5.5 Numerical Examples: Error Estimation

In the following examples, we use both EFG and PIMs processors. In EFG we use the following strategy:

- Triangular cells for background integration.
- EFG formulation with the penalty method for essential boundary conditions.
- Cell energy estimate is used. For the computed cell energy one Gauss point is used, and three Gauss points are used for the reference cell energy.
- Bucket algorithm.
- Relay model is employed.

In PIMs we use the following strategy:

- Triangular cells with smoothing domains
- T-Schemes (T3-scheme) for node selection
- Cell energy estimate for ES- or NS-PIMs

Example 15.1: Rectangular Cantilever (Error Estimation)

The cantilever described in Example 6.2 is employed for the first assessment of the error estimate. The beam is schematically drawn in Figure 6.4. For this example, we have an analytical solution (see Example 6.2), so verification can be performed easily. The parameters for this example are as follows:

Loading: $P = -1000$ N
Young's modulus for the material: $E = 3.0 \times 10^7$ N/m^2
Poisson's ratios for two materials: $\nu = 0.3$
Height of the cantilever: $D = 12.0$ m
Length of the cantilever: $L = 48.0$ m

At the left boundary $(x = 0)$ the displacements are prescribed using the analytical formulae, Equations 6.50 and 6.52, and the right boundary $(x = L)$, the traction force is computed from the analytical formula, Equations 6.53 through 6.55. The "exact" cell energy is computed based on the analytical solutions of stress and strain and integrated over the cell using three Gauss points. Note that the energy is not really exact as the integration is not exact. It can still be used to assess the accuracy of the error estimate.

The test is done first on an EFG model shown in Figure 15.19 with 737 nodes. The distribution of a normalized estimated energy error is plotted in Figure 15.19b. For the computed cell energy one Gauss point is used while three Gauss points are used for the reference cell energy. The distribution of the exact energy error is shown in Figure 15.19c. The exact error distribution is computed using the exact energy as the reference energy obtained from the analytical solution and integrated over the cells using three Gauss points. Comparison of Figure 15.19b with Figure 15.19c demonstrates a close agreement between these two error distributions. The difference in the magnitude of local error, however, is not minor: the maximum estimated value is about 2.52×10^{-4} while the exact value is 3.32×10^{-4}. This is also the case in terms of the global error—the estimated and exact values are 2.14×10^{-3} and 2.92×10^{-3}, respectively. This shows that the proposed approach is not

(a)

(b)

2.518261E−004
2.266435E−004
2.014609E−004
1.762783E−004
1.510957E−004
1.259130E−004
1.007304E−004
7.554783E−005
5.036522E−005
2.518261E−005
0.000000E+000

(c)

3.316118E−004
2.984506E−004
2.652894E−004
2.321282E−004
1.989671E−004
1.658059E−004
1.326447E−004
9.948353E−005
6.632235E−005
3.316118E−005
0.000000E+000

FIGURE 15.19
Error distribution in cantilever. (a) EFG model with 737 nodes; (b) distribution of estimated error; and (c) distribution of exact error. (From Liu, G. R. and Tu, Z. H., *Comput. Methods Appl. Mech. Eng.*, 191, 1923, 2002. With permission.)

sufficient for accurate estimation of the absolute error. It provides an indication on the error distribution sufficiently good for the purpose of adaptive analysis.

We next test on a linear ES-PIM model with 239 nodes. The distribution of estimated energy error is plotted in Figure 15.20a, and of the exact error distribution is shown in Figure 15.20b. It is seen that the estimated error distribution agrees well with the exact one. Such an agreement is sufficient for us to decide the region that needs to be refined.

Example 15.2: Infinite Plate with a Circular Hole (Error Estimation)

Example 7.4 is used here for the examination of error estimation. The geometry of the plate is plotted in Figure 7.10a. Due to the twofold symmetry, only a quarter of the plate shown in Figure 7.10b is modeled with symmetric boundary conditions applied on $x = 0$ and $y = 0$. The parameters are listed as follows:

Loading: $p = 1$ N/m
Young's modulus: $E = 1.0 \times 10^3$ N/m^2
Poisson's ratio: $\nu = 0.3$
Height of the beam: $a = 1.0$ m
Length of the beam: $b = 5$ m

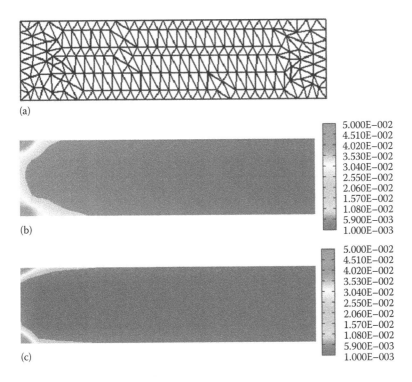

FIGURE 15.20
Error distribution in cantilever. (a) ES-PIM model with 239 nodes; (b) distribution of estimated CEE error; and (c) distribution of exact error.

The plate is subjected to a tension in the x-direction at the edge of $x = 5$. The boundary condition at $x = 5$ is $\sigma_{xx} = p$, $\sigma_{yy} = \sigma_{xy} = 0$, and the boundary condition at $y = 5$ is free of all stresses. The analytical solution of displacement and the stress fields within the plate are provided by Equations 7.41 through 7.46 in the polar coordinates (r, θ).

The test is done first on an EFG model with 204 nodes as shown in Figure 15.21a, and error distributions are plotted in Figure 15.21b and c, respectively for the estimate and the exact. Again, there is a very good correlation between the estimated and exact error distributions. Differences in the magnitude of local and global errors are also observed: the maximum estimated and exact local errors are 8.04×10^{-5} and 1.34×10^{-4} and the maximum estimated and exact global errors are 8.44×10^{-4} and 1.07×10^{-3}, respectively.

The same test is done now on the linear ES-PIM model using 87 nodes shown in Figure 15.22a, and error distributions are shown in Figure 15.22b and c. Again, the estimate predicts well the high-error region.

Example 15.3: A Square Plate Containing a Crack

A square plate containing a crack subjected to boundary conditions prescribed by the near crack-tip field solution is shown in Figure 15.23. The material properties are Young's modulus $E = 3.0 \times 10^7$ and Poisson's ratio $v = 0.3$. In the crack-tip field problem, the square plate has a side of $2a$ and the crack assumes a length of a. This corresponds to the so-called Griffith mode-I crack problem that has an analytical solution [15]

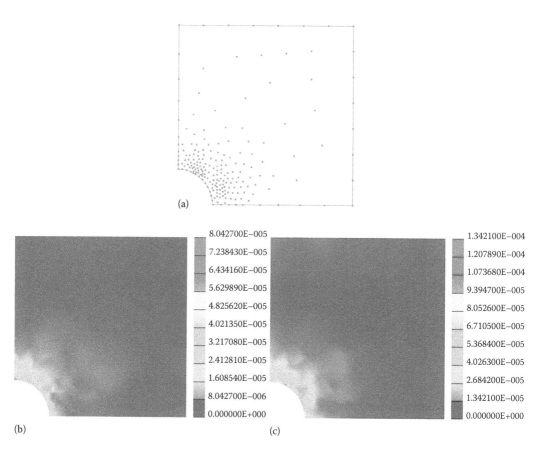

FIGURE 15.21
Error distribution for square plate with a hole problem. (a) 204 nodes used in the EFG model; (b) distribution of estimated error; and (c) distribution of exact error. (From Liu, G. R. and Tu, Z. H., *Comput. Methods Appl. Mech. Eng.*, 191, 1923, 2002. With permission.)

$$\sigma_{xx} = \frac{K_I}{\sqrt{2\pi r}} \cos\frac{\theta}{2}\left(1 - \sin\frac{\theta}{2}\sin\frac{3\theta}{2}\right)$$ (15.53)

$$\sigma_{yy} = \frac{K_I}{\sqrt{2\pi r}} \cos\frac{\theta}{2}\left(1 + \sin\frac{\theta}{2}\sin\frac{3\theta}{2}\right)$$ (15.54)

$$\sigma_{xy} = \frac{K_I}{\sqrt{2\pi r}} \sin\frac{\theta}{2} \cos\frac{\theta}{2} \cos\frac{3\theta}{2}$$ (15.55)

with the coordinate system depicted in Figure 15.23. The stress intensity factor K_I is prescribed by $K_I = p\sqrt{\pi a}$.

In the EFG model, 397 nodes are used with the majority distributed around the crack tip (Figure 15.24a) to capture the crack-tip field. The relay model is used to handle the discontinuity. As with the previous two examples, the distributions of predicted and exact errors plotted in Figure 15.24b and c shows a very good agreement.

In the ES-PIM model, 344 nodes are used. The distributions of predicted and exact errors are plotted in Figure 15.25a and b. When ES-PIM is used, there is no need for a relay model

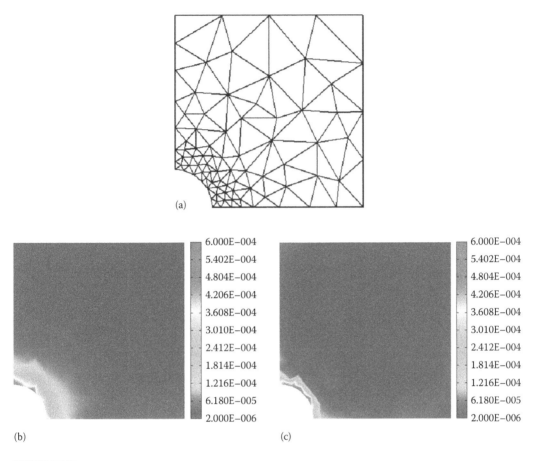

FIGURE 15.22
Error distribution for square plate with a hole problem. (a) 87 nodes used in the ES-PIM model; (b) distribution of estimated CEE error; and (c) distribution of exact error.

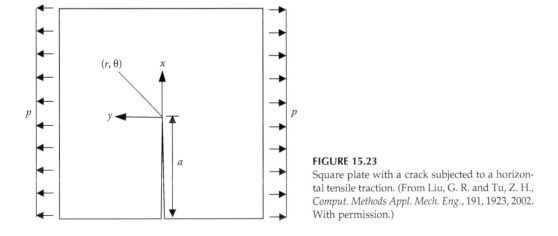

FIGURE 15.23
Square plate with a crack subjected to a horizontal tensile traction. (From Liu, G. R. and Tu, Z. H., *Comput. Methods Appl. Mech. Eng.*, 191, 1923, 2002. With permission.)

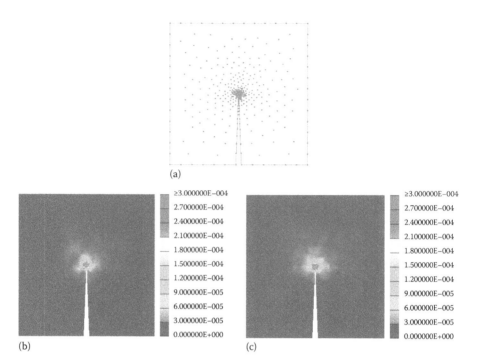

FIGURE 15.24
Error distribution for square plate with a crack problem. (a) EFG model with 397 nodes; (b) distribution of estimated error; and (c) distribution of exact error. (From Liu, G. R. and Tu, Z. H., *Comput. Methods Appl. Mech. Eng.*, 191, 1923, 2002. With permission.)

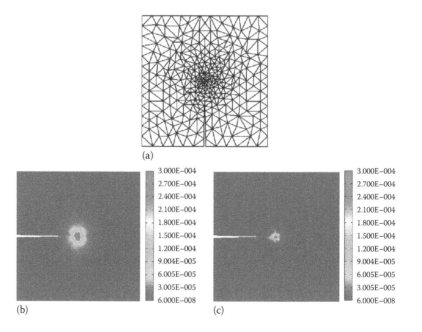

FIGURE 15.25
Error distribution for square plate with a crack problem. (a) ES-PIM model with 344 nodes; (b) distribution of estimated error; and (c) distribution of exact error.

because the node selection is based on T-scheme. Similar to the previous two examples, the stress concentration at the crack tip and high-error region can be estimated properly by our CEE estimate.

15.5.6 Strategy for Local Adaptive Refinement

In adaptive analysis, a problem domain may be refined if the desired accuracy is not achieved. It is usually undesirable to refine the entire domain, as in many cases only a few locations exhibit poor approximations. To achieve high efficiency, it is therefore required to focus on and refine those locations only. A local refinement approach is presented in this section for just this purpose. The approach is also based on a triangular mesh and uses a local Delaunay algorithm with the aid of a density factor. A detailed description of this strategy is given in the following.

15.5.7 Update of the Density Factor

The density factor at a node will be changed if refinement is required at its location. The change of density factor is based on the distribution of local error measured by the CEE estimate. This is done by converting the CEE into nodal energy error. The former is equally distributed to cell vertices and the latter is an accumulation of contributions from surrounding cells, i.e.,

$$e_{\text{nodal}} = \sum_{i=1,m} \frac{e_i}{n} \tag{15.56}$$

where
m is the number of surrounding cells associated with a node
n is the number of cell vertices ($n = 3$ for a triangular cell)
e_i is the energy error of the ith surrounding cell

A relative error measure is then defined for each node

$$R_{\text{nodal}} = \frac{e_{\text{nodal}}}{E_{\text{nodal}}} = \sum_{i=1,m} \frac{e_i}{n} \bigg/ \sum_{i=1,m} \frac{E_i}{n} \tag{15.57}$$

where
R_{nodal} is the nodal relative error
E_i is the computed energy (given by Equation 15.46) of the ith surrounding cell
E_{nodal} is the nodal energy converted from the cell energy

To determine the locations where refinement is required, a threshold nodal relative error is predefined and the relative error at each node is compared with this value. If the threshold value is exceeded, the density factor at a node will be changed to

$$S_{\text{nodal}}^* = \frac{R_{\text{threshold}}}{R_{\text{nodal}}} S_{\text{nodal}} \tag{15.58}$$

where S_{nodal} and S_{nodal}^* are the old and new values of a density factor, respectively. Refinement of the local domain is based on the change in density factor, and the degree of refinement is manageable as the density factor controls the local nodal density. Upon the change of a density factor, a local Delaunay algorithm is executed.

15.5.8 Local Delaunay Triangulation Algorithm

The Delaunay triangulation technique can be applied to an arbitrary 2D domain. Given a set of nodes and a discretized boundary that encloses the nodes, the technique can generate an optimal triangular mesh for the bounded domain based on the existing nodes. This versatility enables a local domain to be refined easily and forms the basis of our adaptive approach. To illustrate this, consider the example depicted in Figure 15.26a, where the density factor of node I changes from S_I to S_I^* ($S_I < S_I^*$). The procedures for local refinement are as follows:

1. Insert nodes. New nodes are inserted by looping the surrounding triangles: for an acute triangle a node is inserted at the center of its circumcircle whereas for an obtuse triangle, a new node is introduced at the middle of the longest edge (Figure 15.26b).

2. Formulate the local domain. This is done by drawing a circle centered at node I (Figure 15.26c) and then removing all the cell edges inside or intersecting the circle (Figure 15.26d). The circle radius dictates the block size and is used to control the range of mesh revision.

3. Triangulate the local domain using the Delaunay algorithm. This regenerates a triangular mesh for the local domain based on the existing nodes inside the block (Figure 15.26e).

4. Recalculate the density factors. Density factors for all affected nodes are updated based on the new mesh.

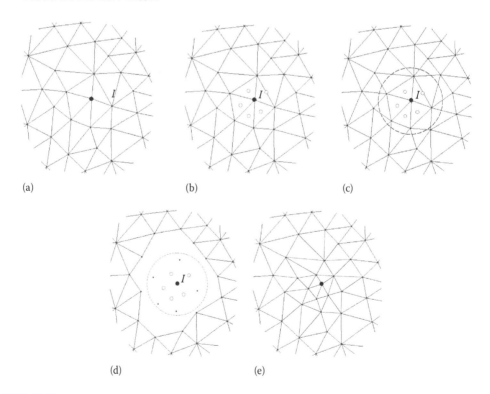

(a)   (b)   (c)

(d)   (e)

FIGURE 15.26

Procedures for local domain refinement. (a) Original mesh; (b) insertion of nodes; (c) and (d) formulation of a local block; and (e) refined mesh. (From Liu, G. R. and Tu, Z. H., *Comput. Methods Appl. Mech. Eng.*, 191, 1923–1943, 2002. With permission.)

The refinement procedures will be repeated if the updated density factor of node I is still larger than S_I^*. This local refinement approach is very efficient, especially for problems with a large number of nodes. Other variants of this strategy may be devised. In an iterative solution procedure, the variables at new nodes are evaluated based on the surrounding old nodes, thus giving a starting solution for the next iteration.

15.5.9 Numerical Examples: 2D Adaptive Analysis

We now present a number of examples of adaptive analysis using EFG, linear NS-PIM, and linear ES-PIM. All the runs are performed using the MFree2D$^{©}$ the usage of which will be detailed in Chapter 16.

Example 15.4: Infinite Plate with a Circular Hole (Adaptive Analysis)

Example 15.2 is reconsidered for adaptive analysis. The parameters are exactly the same as those in Example 15.2. The local adaptive refinement procedure is used for adaptive refinement using EFG, linear NS-PIM, and linear ES-PIM.

The results obtained using EFG are shown in Figure 15.27. The stress concentration occurs around the hole and nodes are automatically added to these locations. The result at the final step is very close to the theoretical solution.

A more detail convergence study is now carried out using linear NS-PIM and ES-PIM together with linear FEM for comparison. First, the ES-PIM is used to perform a study on the effectiveness of the adaptive analysis, and the convergence of strain energy obtained using the ES-PIM is plotted in Figure 15.28, together with that obtained using uniform refinement. It is clear that our adaptive scheme has significantly speeded up the convergence process. Figure 15.29 shows the comparison of strain energy obtained using different methods (NS-PIM, ES-PIM, and FEM) with the same meshes at various adaptive stages. The analytical solution is also plotted. It is obvious that the strain energy of ES-FEM is the closest to the analytical solution at the final stage. We also note the fact that the NS-FEM gives an upper bound solution in strain energy to the exact solution, the FEM and the ES-FEM give lower bounds. This reconfirms our findings presented in Chapter 8.

Example 15.5: Square Plate with a Square Hole (Adaptive Analysis)

Adaptive analysis of a square plate with a square hole at its center subjected to a unit tension force is performed. Making use of the symmetry, one quarter of the problem domain, as shown in Figure 15.30, is used for the analysis. Therefore, this problem is also known as the L-shaped plate, and is a classical problem for testing a refinement procedure. The refinement procedure detects that point A (see Figure 15.30) is singular and hence the location at this point is refined. With only three steps, the adaptive procedure used in the EFG yields very accurate stress distributions (Figure 15.31).

A more detailed convergence study is again carried out using linear NS-PIM and ES-PIM together with linear FEM for comparison. First, the ES-PIM is used to perform a study on the effectiveness of the adaptive analysis, and the convergence of strain energy obtained

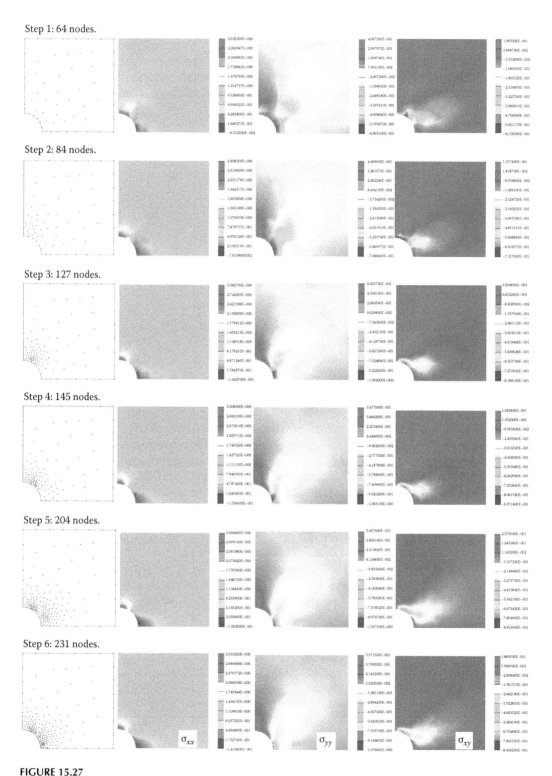

FIGURE 15.27
Adaptive FEG refinement process and distributions of stresses for the square plate with a hole subjected to a unit traction in the *x*-direction.

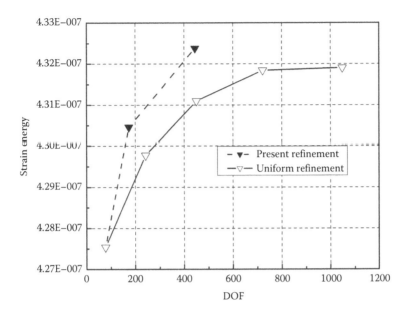

FIGURE 15.28
Infinite square plate with a hole solved using linear ES-PIM. Comparison of convergence process of solution in strain energy with and without adaptive refinement.

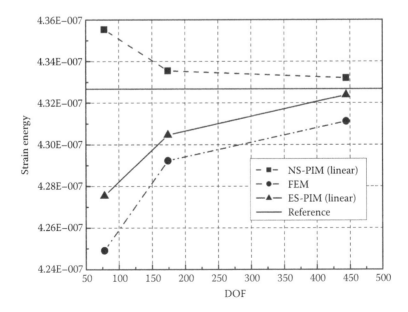

FIGURE 15.29
Infinite square plate with a hole solved using linear ES-PIM, NS-PIM, and FEM using with the same meshes. Comparison of strain energy with the exact solution.

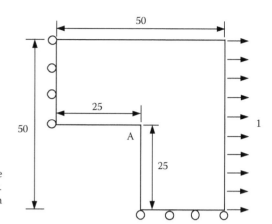

FIGURE 15.30
L-shaped plate subjected to a unit horizontal tensile traction. (From Liu, G. R. and Tu, Z. H., *Comput. Methods Appl. Mech. Eng.*, 191, 1923–1943, 2002. With permission.)

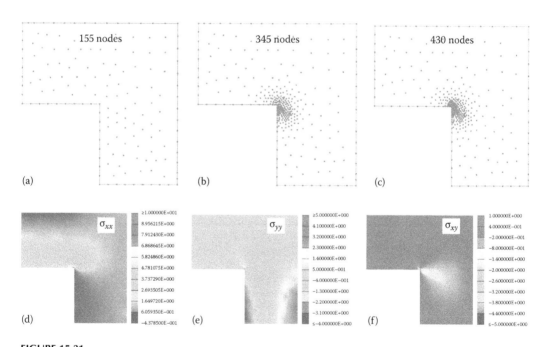

FIGURE 15.31
Adaptive EFG refinement process and the final stress distributions for L-shaped problem. (a–c) Three refinement stages and (d–f) stress distributions at the final stage. (From Liu, G. R. and Tu, Z. H., *Comput. Methods Appl. Mech. Eng.*, 191, 1923–1943, 2002. With permission.)

using the ES-PIM is plotted in Figure 15.32, together with that obtained using uniform refinement. It is clear that our adaptive scheme has significantly speeded up the convergence process. Figure 15.33 shows the comparison of strain energy obtained using NS-PIM, ES-PIM, and FEM with the same meshes at various adaptive stages, together with the reference solution (obtained using extremely fine mesh). It is again observed that the strain energy of ES-FEM is the closest to the analytical solution at all stages. In this case the ES-PIM is much more accurate than the other two. We also note the fact that the NS-FEM gives an upper bound solution in strain energy to the exact solution, the FEM and the ES-FEM give lower bounds.

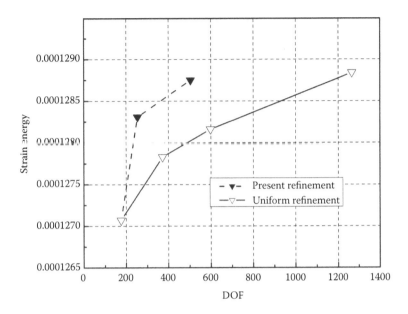

FIGURE 15.32
L-shaped plate solved using linear ES-PIM. Comparison of convergence process of solution in strain energy with and without adaptive refinement.

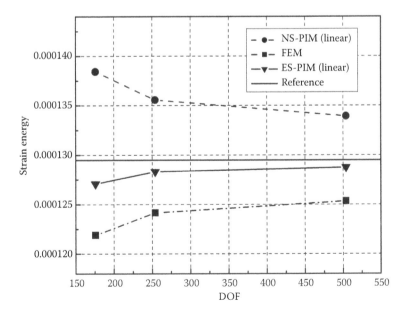

FIGURE 15.33
L-shaped plate solved using linear ES-PIM, NS-PIM, and FEM. Comparison of strain energy obtained using the same meshes and the reference solution.

Example 15.6: Rectangular Plate with a Crack (Adaptive Analysis)

Adaptive analysis of a rectangular plate with a crack loaded by horizontal unit tractions on the two vertical sides is performed. The problem domain is drawn in Figure 15.34. The domain has dimensions of 100 in. length and 50 in. height. The material parameters are Young's modulus $E = 3.0 \times 10^7$ and Poisson's ratio $v = 0.3$.

The results obtained using EFG with adaptive scheme are shown in Figure 15.35. Due to the stress singularity, the vicinity of the crack tip is mostly refined. However, this refinement will never come to an end because of the singularity at the crack tip. Therefore, the

FIGURE 15.34
Rectangular plate with a crack subjected to a unit tensile traction. (From Liu, G. R. and Tu, Z. H., *Comput. Methods Appl. Mech. Eng.*, 191, 1923–1943, 2002. With permission.)

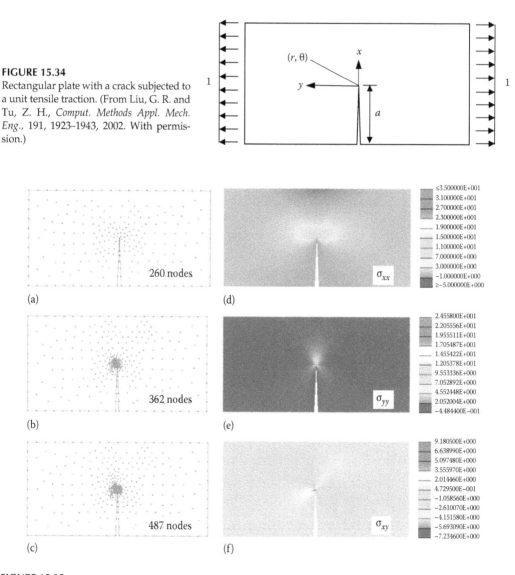

FIGURE 15.35
Adaptive EFG refinement process and the final stress distributions for the rectangular plate with a crack problem. (a–c) Three refinement stages and (d–f) stress distributions at the final stage. (From Liu, G. R. and Tu, Z. H., *Comput. Methods Appl. Mech. Eng.*, 191, 1923–1943, 2002. With permission.)

computation is terminated at the third refinement step when the crack-tip field is described with a very high resolution. The good numerical results for the three sample problems demonstrate that the proposed local adaptive refinement procedure is effective and efficient.

The ES-PIM is again used to perform a study on the effectiveness of the adaptive analysis, and the convergence of strain energy is plotted in Figure 15.36, together with that obtained using uniform refinement. It is clear that our adaptive scheme has significantly speeded up even more the convergence process, due to the presence of the singularity. Figure 15.37 shows the comparison of strain energy obtained using NS-PIM, ES-PIM, and FEM with the

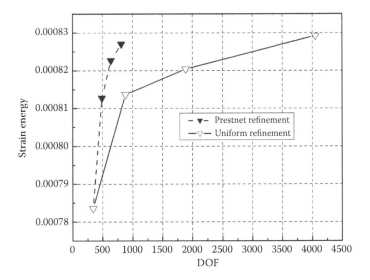

FIGURE 15.36
Rectangular plate with a crack solved using linear ES-PIM. Comparison of convergence process of solution in strain energy with and without adaptive refinement.

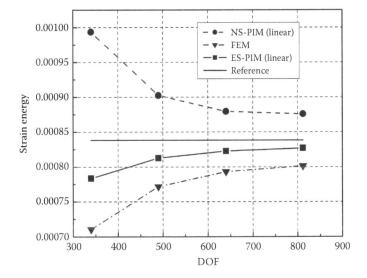

FIGURE 15.37
Rectangular plate with a crack solved using linear ES-PIM, NS-PIM, and FEM. Comparison of strain energy obtained using the same meshes and the reference solution.

same meshes at various adaptive stages, together with the reference solution (obtained using extremely fine mesh). It is again observed that the strain energy of ES-FEM is the closest to the analytical solution at all stages. In this case the ES-PIM is also much more accurate than the other two. We also note the fact that the NS-FEM gives an upper bound solution in strain energy to the exact solution, the FEM and the ES-FEM give lower bounds.

Example 15.7: Square Plate with Two Parallel Cracks (Adaptive Analysis)

Adaptive analysis of a rectangular plate with two parallel cracks loaded by horizontal unit tractions on the two vertical sides is performed. The domain has dimensions of 100 in. length and 50 in. height. The material parameters are: Young's modulus $E = 3.0 \times 10^7$ and Poisson's ratio $\nu = 0.3$. Figure 15.38a shows the meshfree model. The numerical results give very good description of the stress field around the crack tips as shown in Figure 15.38, verifying the validity of the proposed model.

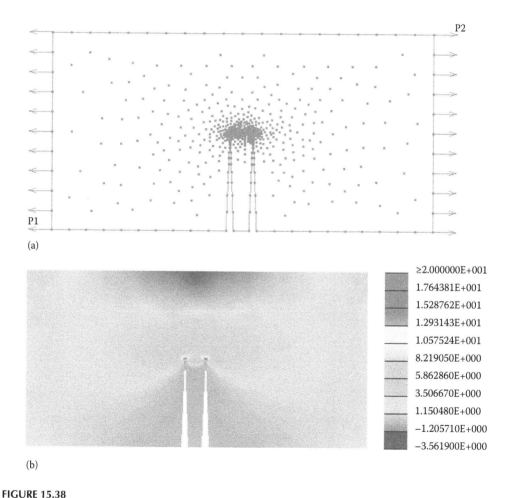

FIGURE 15.38
Interference between two cracks in a rectangular plate subject to a unit tensile traction (EFG). (a) Problem model, (b) distribution of σ_x, (c) distribution of σ_y, and (d) distribution of T_{xy}.

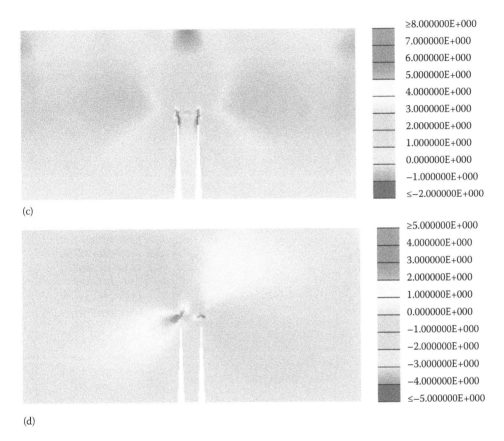

$\geq 8.000000E+000$
$7.000000E+000$
$6.000000E+000$
$5.000000E+000$
$4.000000E+000$
$3.000000E+000$
$2.000000E+000$
$1.000000E+000$
$0.000000E+000$
$-1.000000E+000$
$\leq-2.000000E+000$

(c)

$\geq 5.000000E+000$
$4.000000E+000$
$3.000000E+000$
$2.000000E+000$
$1.000000E+000$
$0.000000E+000$
$-1.000000E+000$
$-2.000000E+000$
$-3.000000E+000$
$-4.000000E+000$
$\leq-5.000000E+000$

(d)

FIGURE 15.38 (continued)
Interference between two cracks in a rectangular plate subject to a unit tensile traction (EFG). (a) Problem model, (b) distribution of σ_x, (c) distribution of σ_y, and (d) distribution of T_{xy}.

Example 15.8: Arbitrary Complex Domain (Adaptive Analysis)

A panel carved with "ACES" subjected to a tensile traction loading on the central part of the top side is investigated. The panel is fixed at all four corners. This is a multiconnected body with highly irregular boundaries. Figure 15.39a shows the meshfree model for the analysis. The material parameters are: Young's modulus $E = 3.0 \times 10^7$ and Poisson's ratio $\nu = 0.3$. The stress distributions are very well depicted, as shown in Figure 15.39, confirming that the relay model is capable of handling highly irregular problem domains.

Example 15.9: An Automotive Part: Connecting Rod

The last 2D example is a practical engineering problem of typical connecting rod used in automobiles, as shown in Figure 8.99, is studied using our adaptive linear ES-PIM code. The connecting rod is constrained along the left larger circle and subjected to a uniform unit radial pressure along half of the right circle. The problem is considered as a plane stress problem with material parameters $E = 3.0 \times 10^7$ Pa and $\nu = 0.3$.

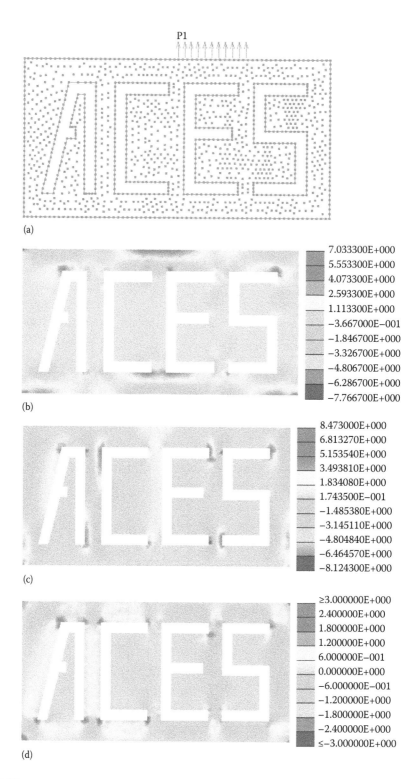

FIGURE 15.39

Panel characterized by "ACES" subjected to a unit tensile traction along the central part of the top side (EFG). (a) Problem model, (b) distribution σ_x, (c) distribution of σ_y, and (d) distribution of τ_{xy}.

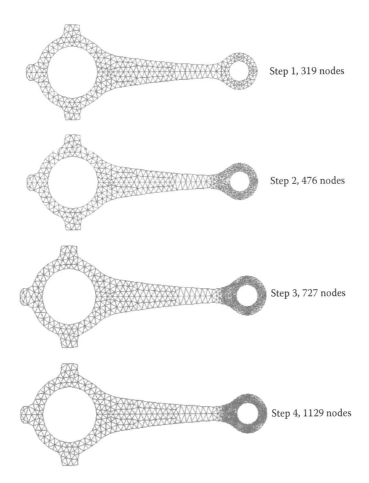

FIGURE 15.40
The adaptive ES-PIM refinement process and stress distributions for connecting rod.

The refinement process for this problem is plotted in Figure 15.40. It is found that the stress concentration occurs around the vertical radius part of the right pin hole and at the location with the transition of section, leading to an automatically refined mesh at these locations. The stress distribution at the final step agrees well with the reference solution obtained by using FEM with a very fine mesh of 16,614 nodes, as shown in Figure 15.41a and b.

For this practical problem with complicated shape, the convergence of strain energy of our adaptive scheme is found again to converge much faster than the uniform refinement, as shown in Figure 15.42. The ES-PIM produces the most accurate result that together with the exact solution is bounded by the solutions of NS-PIM and FEM (Figure 15.43). These findings again demonstrate that the proposed local adaptive refinement procedure based on ES-FEM is effective and efficient.

15.5.10 Numerical Examples: 3D Adaptive Analysis

We finally present an example of 3D adaptive analysis using linear NS-PIM. Our 3D code development is still at a very early stage, and clearly we still have a long way to go to

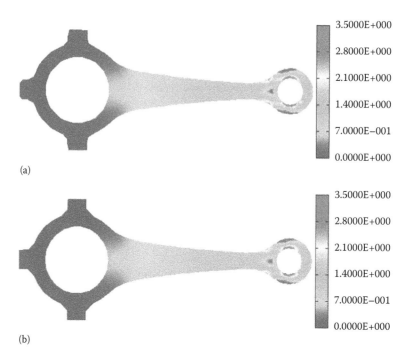

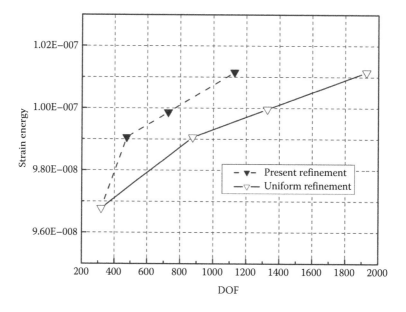

FIGURE 15.41

The adaptive ES-PIM analysis results; (a) Von Mises stress distributions at the final stage; and (b) reference Von Mises stress distributions obtained using FEM with 16,614 nodes.

FIGURE 15.42

Automotive connecting rod solved using linear ES-PIM. Comparison of convergence process of solution in strain energy with and without adaptive refinement.

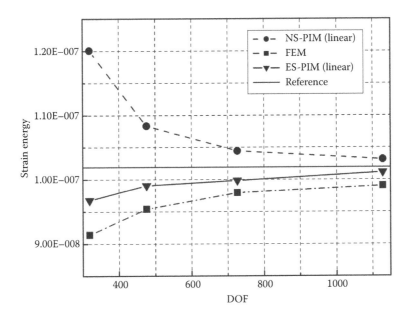

FIGURE 15.43
Automotive connecting rod solved using linear ES-PIM, NS-PIM, and FEM. Comparison of strain energy obtained using the same meshes and the reference solution.

develop an MFree3D© for adaptive analysis. Here we are able to show only one example, without much detailed description.

Example 15.10: A Elastic Cube with a Cutout

We consider here an elastic cube of $a \times a \times a$ with a 1/8 portion removed as shown in Figure 15.44. The displacement boundary conditions are imposed on faces DEFG, CDGH, and HIFG:

- On face DEFG the displacement along the x-direction is set to 0.
- On face CDGH the displacement along the y-direction is set to 0.
- On face HIFG the displacement along the z-direction is set to 0.

The cub is subjected to uniform tensile forces on the three L-shaped faces:

- On face BCHIJM $T_x = 100$ N/m^2 is applied.
- On face EFILKJ $T_y = 100$ N/m^2 is applied.
- On face ABCDEJ $T_z = 100$ N/m^2 is applied.

The parameters for the material are taken as Young's modulus $E = 3.0 \times 10^7$, Poisson's ratio $\nu = 0.3$. The dimensions are $a = 2$ m and $b = 1$ m. With this setting, we shall have strong singularity along the three lines HG, GF, and GD.

We studied this problem using our 3D linear NS-PIM code with both "uniform" and adaptive models. Four uniform refinement models with 443, 2011, 4044, and 7335 nodes

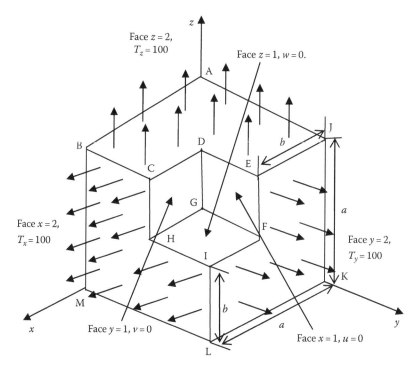

FIGURE 15.44
An elastic cube with 1/8 portion removed.

shown in Figure 15.45 are used. The adaptive analysis started from the coarse uniform mesh of 443 nodes and is performed for six steps with 5% refinement at each step. The tetrahedral cells generated by our adaptive code at each step are shown in Figure 15.46. The node distributions at these stages are plotted in Figure 15.47. The figures show that

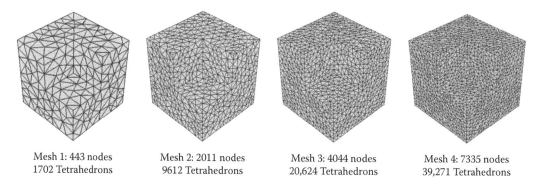

| Mesh 1: 443 nodes | Mesh 2: 2011 nodes | Mesh 3: 4044 nodes | Mesh 4: 7335 nodes |
| 1702 Tetrahedrons | 9612 Tetrahedrons | 20,624 Tetrahedrons | 39,271 Tetrahedrons |

FIGURE 15.45
Four uniform meshes used in the analysis.

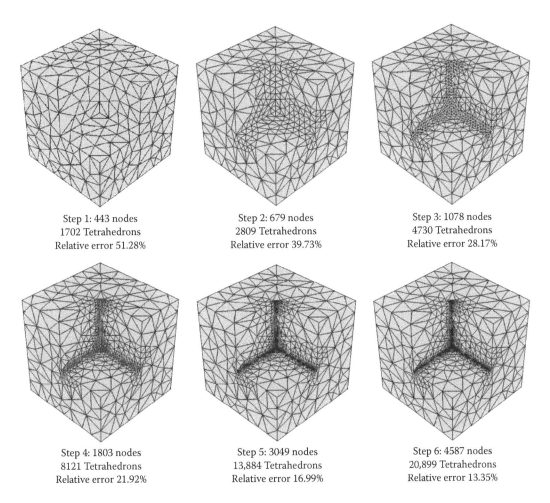

Step 1: 443 nodes 1702 Tetrahedrons Relative error 51.28%	Step 2: 679 nodes 2809 Tetrahedrons Relative error 39.73%	Step 3: 1078 nodes 4730 Tetrahedrons Relative error 28.17%
Step 4: 1803 nodes 8121 Tetrahedrons Relative error 21.92%	Step 5: 3049 nodes 13,884 Tetrahedrons Relative error 16.99%	Step 6: 4587 nodes 20,899 Tetrahedrons Relative error 13.35%

FIGURE 15.46
Tetrahedral cells generated during the adaptive analysis.

our error indicator can accurately catch the steep gradient of stresses and lead to the refinement of nodes concentrated along the three singular lines HG, GF, and GD.

Figure 15.48 shows the comparisons on convergence of strain energy between the results obtained using the uniform meshes and adaptive analysis. It is clear that when uniform mesh is used, the solution converges very slowly. Our adaptive analysis converges much faster than that using the uniform mesh.

Figure 15.49a plots the contours of the Von Mises stress at the fifth adaptive step (3,049 nodes and 13,884 tetrahedrons). For comparison a FEM reference result of a very fine mesh (50,354 nodes and 283,963 tetrahedrons) is plotted in Figure 15.49b. It clearly shows that the Von Mises stress obtained with adaptive mesh is in good agreement with the reference results. We confirmed also the NS-PIM solution is an upper bound and the FEM solution is a lower bound for this problem.

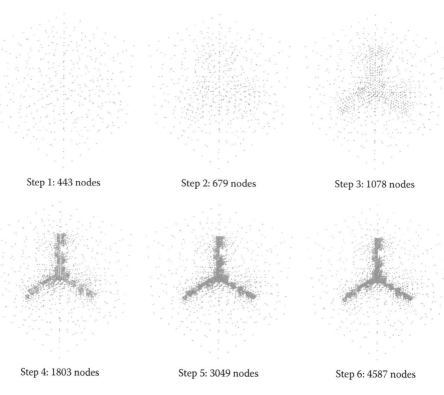

Step 1: 443 nodes   Step 2: 679 nodes   Step 3: 1078 nodes

Step 4: 1803 nodes   Step 5: 3049 nodes   Step 6: 4587 nodes

FIGURE 15.47
Nodes generated during the adaptive analysis.

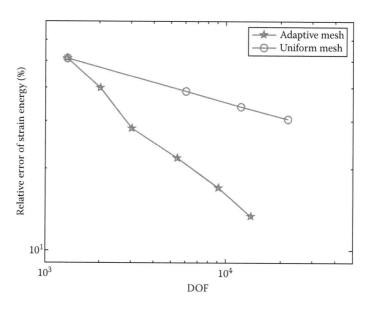

FIGURE 15.48
Comparison of the convergence of the solution in strain energy.

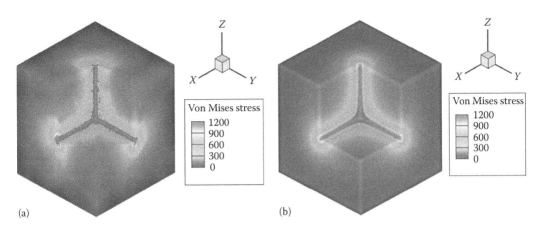

FIGURE 15.49
Comparison of the solution in Von Mises stress. (a) 3D NS-PIM with 3,049 nodes and 13,884 tetrahedrons; (b) FEM with very fine mesh of 50,354 nodes and 283,963 tetrahedrons.

15.6 Concluding Remarks

For practical applications, meshfree methods offer good choices for adaptive analysis. Based on the study in this chapter we note the following:

a. Triangular types of meshes need to, and can be used. We have no reason to reject them.

b. In our implementation, the EFG worked quite well, but is a little complicated for adaptive analysis.

c. PIMs are much simpler, accurate, robust, very efficient, and offer solutions of special properties. Using PIMs for more challenging problems like crack propagation can be developed and it is currently an ongoing project at ACES.

d. 3D adaptive analysis is very challenging. The challenge is not too much on technical issues, rather on implementation and coding issues. As a start, the linear NS-PIM has shown some promising results, and we are making some progress. We should be able to report more results in the near future.

References

1. Delaunay, B., Sur la sphere vide, *Bull. Acad. Sci. URSS Class. Sci. Nat.*, 793–800, 1934.
2. Belytschko, T., Lu, Y. Y., and Gu, L., Element-free Galerkin methods, *Int. J. Numer. Methods Eng.*, 37, 229–256, 1994.
3. Belytschko, T., Krongauz, Y., Fleming, M., Organ, D., and Liu, W. K., Smoothing and accelerated computations in the element free Galerkin method, *J. Comput. Appl. Math.*, 74, 111–126, 1996.
4. Organ, D., Fleming, M., Terry, T., and Belytschko, T., Continuous meshless approximations for nonconvex bodies by diffraction and transparency, *Comput. Mech.*, 18(3), 225–235, 1996.

5. Belytschko, T., Krongauz, Y., Organ, D., Fleming, M., and Krysl, P., Meshless method: An overview and recent developments, *Comput. Methods Appl. Mech. Eng.*, 139, 3–47, 1996.
6. Liu, G. R. and Chen, X. L., A mesh-free method for static and free vibration analyses of thin plates of complicated shape, *J. Sound Vib.*, 241(5), 839–855, 2001.
7. Zienkiewicz, O. C. and Taylor, R. L., *The Finite Element Method*, 5th edn., Butterworth Heimemann, Oxford, 2000.
8. Radovitzky, R. M. O., Error estimation and adaptive meshing in strongly nonlinear dynamic problems, *Comput. Methods Appl. Mech. Eng.*, 172, 203–240, 1999.
9. Duarte, C. A. and Oden, J. T., An *hp* adaptive method using clouds, *Comput. Methods Appl. Mech. Eng.*, 139, 237–262, 1996.
10. Chung, H. J. and Belytschko, T., An error estimate in the EFG method, *Comput. Mech.*, 21, 91–100, 1998.
11. Combe, U. H. and Korn, C., An adaptive approach with the element-free-Galerkin method, *Comput. Methods Appl. Mech. Eng.*, 162, 203–222, 1998.
12. Tu, Z. H. and Liu, G. R., An error estimate based on background cells for meshless methods, in *Advances in Computational Engineering and Sciences*, S. N. Atluri and F. W. Brust, eds., Tech Science Press, Palmdale, CA, pp. 1487–1492, 2000.
13. Liu, G. R. and Tu, Z. H., An adaptive procedure based on background cells for meshless methods, *Comput. Method Appl. Mech. Eng.*, 191, 1923–1942, 2002.
14. Chung, H. J. and Belytschko, T., An error estimate in the EFG method, *Comput. Mech.*, 21, 91–100, 1998.
15. Anderson, T. L., *Fracture Mechanics: Fundamentals and Applications*, CRC Press, Boca Raton, FL, 1991.

16

MFree2D$^{©}$

16.1 Overview

MFree2D$^{©}$ is a software package being developed by a team at the Centre for Advanced Computations in Engineering Science (ACES) at the National University of Singapore, based on meshfree technologies. It consists of three components—*MFreePre*, *MFreeApp*, and *MFreePost*. *MFreePre* is a preprocessor used to formulate the input required by *MFreeApp*. The latter performs computations and yields the output, which is then fed to *MFreePost* for postprocessing. These three processors work together seamlessly, and users are not usually aware that they are different processors. However, these three integrated processors can also be separated to perform their work independently. The current *MFree2D* 2.0 is limited to two-dimensional (2D) elastostatics. It runs on a PC in a Window environment. MFree2D was first showcased at the 4th International Asia-Pacific Conference on Computational Mechanics, which was held in Singapore in 1999. It is available for download at: http://www.crcpress.com/e_products/downloads/download.asp?cat_no = 1238.

MFree2D is currently a freeware with a user guide built in, but without technical support. Interested readers should make their own arrangements with ACES. Because it is new and still in the development phase, the functions are changing quite frequently and, naturally, there are bugs. ACES is constantly trying to update the new developments and changes on its Web site. All users who are interested in trying or using this package are welcome to do so at their own risk, but are currently required by ACES to register and to agree on the terms and conditions set by ACES.

Main features of *MFree2D*
- The problem domain is discretized using scattered nodes, and a triangular mesh is used. The domain discretization process is fully automatic.
- Automatic adaptive refinement techniques are implemented (for EFG and ES-PIM processors) to ensure results of desired accuracy.
- Operations are performed on graphic, manual-based user-friendly interfaces.

Advantages over FEM packages
- Cost of labor for meshing is reduced significantly (for both EFG and PIMs)
- Higher accuracy—results are as accurate as desired (for both EFG and PIMs)
- There is no problem relating to mesh distortion (for both EFG and PIMs)

- Upper bound solutions (for node-based smoothed point interpolation methods [NS-PIMs])
- Significant computational efficiency (for linear ES-PIM [edge-based smoothed point interpolation method], see Section 8.7)

16.2 Techniques Used in MFree2D

The tasks involved in a meshfree analysis are choice of the size of the influence domain; search of the nodes for the influence domain; computation of nodal weight, which is a coefficient that controls the dimension of the influence domain or cells locally; generation of the background cells/mesh; and selection of the integration scheme. MFree2D performs all these tasks. The techniques used in MFree2D 2.0 for different processors are listed in Table 16.1, with references to the relevant chapters of this book for detailed technical explanations on each technique.

16.3 Preprocessing in MFree2D

MFreePre is used to define and model a problem for analysis. It creates a geometric model, creates nodes and background triangular cells, defines material properties and initial and boundary conditions, and performs solution control. One salient feature is that troublesome and time-consuming manual mesh generations are no longer necessary.

TABLE 16.1

Techniques Used in MFree2D 2.0 and Ongoing Development

Techniques	EFG Processor	PIM Processors
Shape function	MLS (Chapter 2)	PIMs (Chapter 2)
Weak form	Galerkin weak form (Chapters 3, 4, and 6)	GS-Galerkin weakened-weak form (Chapters 3, 4, and 8)
Essential boundary condition	Penalty method (Chapters 5 and 6)	Direct approach as in FEM
Background integration	Triangular cells + Gauss quadrature (Chapter 15)	Smoothing domains built on triangular cells
		No Gauss integration for energy
Node searching (influence domain)	Bucket algorithm (Chapter 15)	T-schemes (Chapter 1)
	Relay model (Chapter 15)	
	T2L-scheme (Chapter 1)	
Error estimation	Cell energy estimate via different Gauss integration schemes (Chapter 15)	Cell energy estimate using strain discrepancy in cells
Refinement strategy	Density factor control and Delaunay triangulation (Chapter 15)	Density factor control and Delaunay triangulation (Chapter 15)

In discretizing a problem domain, users do not need to work on the geometric model part by part. One needs only to define the boundaries of the problem domain by specifying them. MFreePre automatically identifies the geometry and models the domain using scattered nodes and background cells, simply using the Delaunay triangulation algorithm. This saves significant labor cost in mesh creation by engineers. MFreePre allows an analysis to be customized with its open environmental settings, while providing default settings for new users. MFree2D requires only the minimum from users, such as geometry, boundary conditions, and material properties needed to define the problem. All issues related to the mesh can be handled automatically. To ease the effort involved in inputting material properties, these properties can be stored in a database for future use.

A brief user guide is provided in the following sections. Note that there may be discrepancies between the brief guide shown here and the current version of the code downloaded from the Web site. The guide here serves just to describe how MFreePre works, rather than describing detailed operations.

16.3.1 Main Windows

MFreePre is developed in the Microsoft Windows environment. It is driven by an interactive graphical user interface. All geometry data, material properties, and boundary conditions can be input by a mouse or keyed in with an interactive window. The main working interface (Figure 16.1) consists of six parts: main menu, toolbars, tree view, text view and command view, graphical display area, and status bar. Note that the window has minimal items, because we do not need to create sophisticated element meshes. All we need is geometry creation and boundary, loading, and material definition.

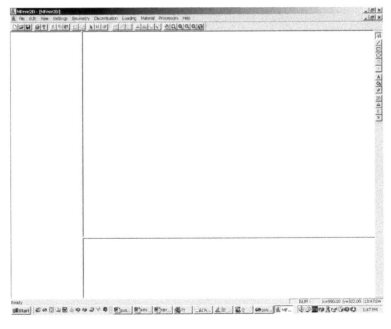

FIGURE 16.1
Main window of MFree2D.

Main menu: The main menu contains all input items and operation facilities. Most of the items are also available as buttons in the toolbars.

Toolbars: Toolbars contain buttons for actions that are related to creating the geometry model, adding loads, defining the boundary conditions, editing, and so on.

Tree view: Tree view is used to show the boundary and to edit its properties.

Text view: Text view is used to show the commands and the keyed-in data.

Graphical display area: The graphical display area is where the geometry model is created. All operations for the creation of geometry or postprocessing are mainly done by means of a mouse. When a mouse input does not give the desired accuracy, a direct keyboard input is available.

Status bar: The status bar contains three indicators: a line indicator showing the operation of the selected menu item, a cursor position indicator giving the current position of the mouse cursor on the screen, and a time indicator providing the current time.

16.3.2 Geometry Creation

The generation of an analysis model begins with the creation of a geometry model. A geometry model can be formed using points, lines, squares, arcs, and circles. All these entities can be inputted or edited by means of a mouse or a keyboard.

Seed Node

This icon is used to create a *seed node* to control the density of discrete nodes when representing the problem domain with nodes. It is particularly useful when you intend to increase or decrease the density of nodes in the area surrounding the seed node. Use it only when necessary, as for most cases the default works just fine. This item can be selected from the drawing submenu, *source point*, or directly from the toolbar. The point position can be selected with a mouse or keyed in from the keyboard. When the point position has been input, a dialog window (Figure 16.2) will prompt you to key in the values of weight and effect radius. The weight value is used to control the density of discrete nodes, i.e., the average nodal spacing around the seed node. Figure 16.3 shows an example of using a seed node to increase the nodal density at a particular point.

Line

This input icon is used to create a *geometric line*. This item can be selected from the geometry menu as well as directly from the toolbar. You can pick two points using a mouse on the graphical display area to generate a line or key in the data (Figure 16.4) for

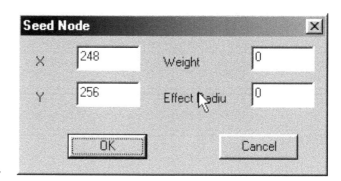

FIGURE 16.2
Edit dialog box for seed nodes.

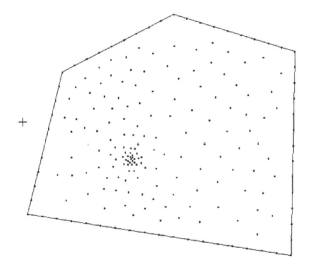

FIGURE 16.3
Example of the effect of a seed node with large density weight.

FIGURE 16.4
Example of generating a line using the keyboard input.

the desired accuracy. After the first point has been input, a line will be dragged by moving the mouse, and a right click will fix the ending point. Exact coordinates for the ending point can be also specified using the dialog box that is prompted by pressing any of the 10 digit keys.

Square
This icon is used to create a *square geometry*. This can be done using a keyboard input for specifying the coordinates for the two diagonal points. The square can also be created by dragging the mouse while pressing the left button.

Circle
This icon is used to generate a *circular geometry*. The operation can be carried out by selecting this item from the geometry menu, or directly from the toolbar. First, the center point of the circle can be selected in the graphical display area, using a mouse or a direct input from the keyboard. Next, the radius for generating the circle can be keyed in or specified by dragging the mouse to a desired radius of the circle. The key-in boxes are shown in Figure 16.5. When the mouse is used to generate a circle, the snap mode can assist you to select particular points conveniently.

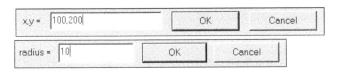

FIGURE 16.5
Example of generating a circle with the keyboard input.

Arc

This icon is used to generate an *arc geometry*. This item can be selected from the geometry menu or directly from the toolbar. In this command, a three-point input method is used to generate an arc. The first point determines the center of the arc, and the second point is used to determine both the start angle and the radius of the arc. The latter value is calculated by simply taking the distance between the first point, the center of the arc, and the second point. The third point is used only to determine the end angle of the arc. All the point inputs can be done by selecting in the graphical display area with a mouse or by keying in from the keyboard. The second and third points follow the right-hand rule related to the center point of the arc; i.e., the arc should be specified in a counterclockwise direction.

16.3.3 Boundary Conditions and Loads

Boundary conditions are special conditions that are imposed on geometric entities to constrain the displacements. Boundary conditions come in two variants, intensive (point) and distributed constraints.

The user can define two types of loads: intensive (concentrated) forces and distributed tractions. Concentrated forces are applied to geometric points that lie in the geometry. Distributed tractions are applied to geometric entities.

Point Displacement Constraint (Intensive Displacement)

This icon is used to define a *point displacement constraint* on a certain point. This can be performed from the loading menu, boundary condition → *intensive displacement*, or directly from the toolbar. The user can adjust the point location with a dialog box, as shown in Figure 16.6, which will prompt after the position of the point is defined.

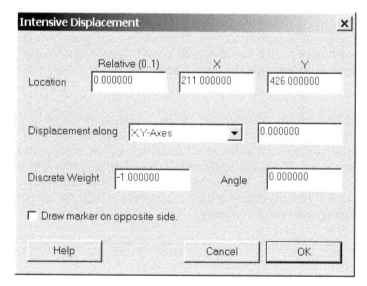

FIGURE 16.6
Dialog box for specifying a point displacement constraint.

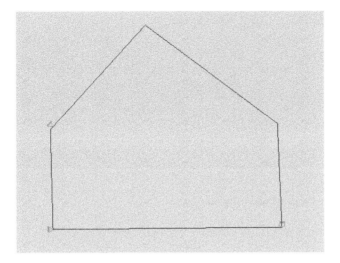

FIGURE 16.7
Example of specified point constraints.

There are seven items in the dialog box. *Location* shows the coordinates of the point. *Displacement* is used to prescribe the fashion of the constraint. The combo box contains five items for selection—*Constraint along X, Constraint along Y, Constraint along X and Y, Surface Normal,* and *Surface Tangent*—which are used to define the direction of the displacement. The last box is used for the prescribed value of this displacement constraint. If the user selects *Constraint along X and Y* and no zero is set for *displacement,* the angle of the displacement related to the x-direction must be defined. *Discrete Weight* is used to control the density of nodes around the point. The value of *Discrete Weight* is the average nodal spacing around the point load. Figure 16.7 shows an example of specified point constraints along x, y, and x and y.

Distributed Displacement

This icon is used to define distributed constraints along a certain line. This can be performed from the loading menu, boundary condition $\rightarrow$ *distribution displacement,* or directly from the toolbar. The user can define a linear distribution displacementalong a line, circle, and arc, and can input coordinates of the start point and the end point by a mouse. Figure 16.8 shows the dialog box of distribution displacement. Figure 16.9 shows an example of specified distributed displacement constraints.

Intensive Force (Point Force)

This icon is used to define *intensive (point) force.* Intensive force can be selected from the preprocess menu, *loading $\rightarrow$ intensive force,* or by clicking the icon button on the toolbar. The position of the point on which the force will be specified can be input by a mouse. The load value and direction can be input in the dialog box, Intensive Force, shown in Figure 16.10. In the dialog box, the location shows the coordinates of the point on which the force is to be imposed. The loading value is the magnitude of the force. The angle to the x-axis is the force direction related to the x-axis. *Discrete weight* is the average nodal spacing around the point of force. Within the direction box, there are two radio

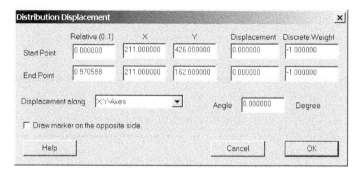

FIGURE 16.8
Dialog box for specifying the distribution displacements.

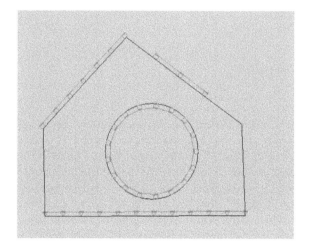

FIGURE 16.9
Example of specified distributed displacement constraints.

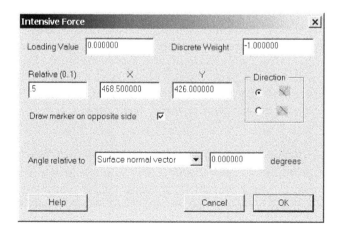

FIGURE 16.10
Dialog box for specifying an intensive force.

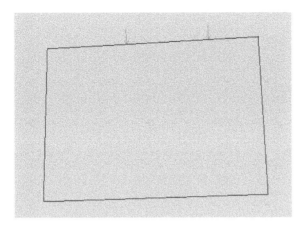

FIGURE 16.11
Example of specified point forces.

buttons that are used to specify the direction of the force. Figure 16.11 shows an example of specified intensive forces.

Traction
This icon is used to define distributed *traction* on geometric entities. This item can be selected from the preprocess menu, *loading → traction*, or by clicking the icon button on the toolbar. The distributed traction represents a distributed force applied on a segment of lines, circles, or arcs. The positions of the start and end points of the traction effect can be selected using a mouse. The user can define the linear traction. Figure 16.12 shows a dialog box for specifying a distributed traction. The meaning of all items in this dialog box is the same as that in the intensive force dialog box. Figure 16.13 shows an example of a specified traction applied to a structure.

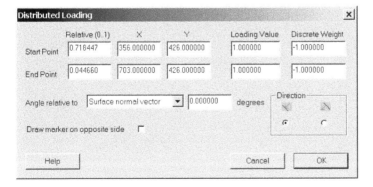

FIGURE 16.12
Dialog box for specifying a distributed traction.

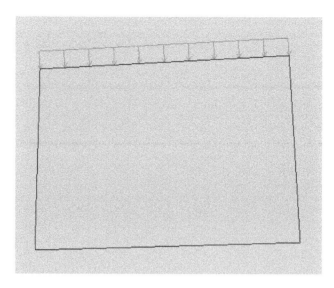

FIGURE 16.13
Example of a specified distributed traction.

16.3.4 Modify and Delete Boundary Conditions and Loads

E

This icon is used to edit (*modify* and *delete*) boundary conditions and loads. The user can select the boundary constraint or load applied to the selected entity by double-clicking the entity in the Tree view. After a dialog window similar to Figure 16.14 appears, the user can edit and delete selected boundary conditions and loads.

16.3.5 Node Generation

In MFree2D, the domain is represented using scattered nodes. All the processes for the nodal generation are fully automated based on the boundary of the geometry model

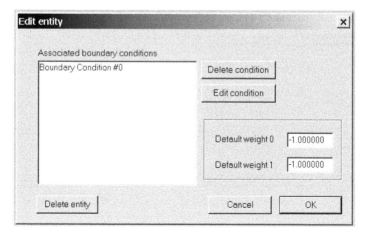

FIGURE 16.14
Example of a dialog box for editing boundary and loading conditions.

created at the initial stage. The density of nodes in the domain can be controlled by the user by setting the *discrete weights* on the end points of each boundary item. Users can conveniently change the local density of nodes by adding seed nodes to obtain a desired density of nodal distribution at specified locations.

Boundary Specification

This icon is used to *specify the boundary* of the geometry of the model. There are two types of boundaries. The first is an external boundary and the second is an internal boundary. The external boundary is the profile that covers the whole domain; internal boundaries are the edges that form holes within the domain of the model. MFreePre will automatically decide which boundary is the external one, and which are the internal ones. The user can specify a boundary by double-clicking the geometric entity that is part of the boundary. All boundaries that form the geometry of the model will be shown in the Tree view. Figure 16.15 shows an example of specified boundaries (green lines on the screen) for a model.

Modification of Discrete Nodal Density on a Boundary

The user can double-click the item after selecting it on the tree window (the left window) that needs to be modified, and a dialog box will be prompted to show the discrete weight on the item. The user can then modify the parameter in the dialog box as needed, by simply changing the value. Figure 16.16 shows an example of the nodal distribution on the boundaries of an MFree model.

The user can also perform the modification by selecting from the menu *settings → options* and then modifying the weights on the dialog window. Figure 16.17 shows a dialog box for altering some of the options.

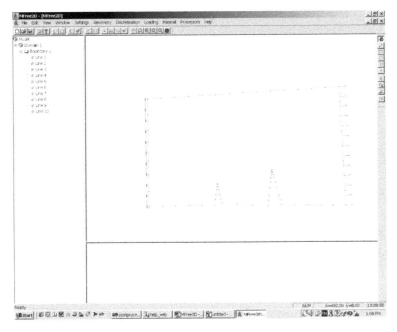

FIGURE 16.15
Example of specified boundaries (shown as green lines on the screen) for a model.

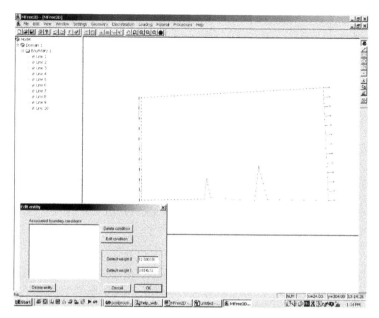

FIGURE 16.16
Example of nodal distribution on the boundaries of an MFree model.

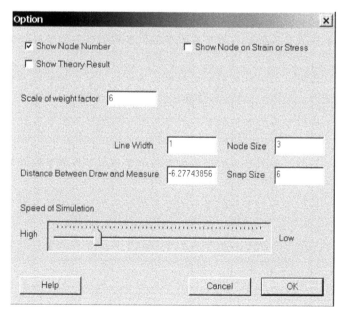

FIGURE 16.17
Dialog box for altering options.

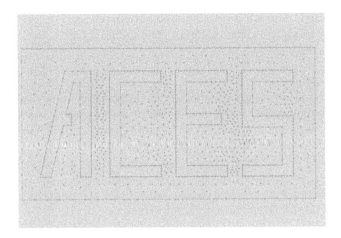

FIGURE 16.18
Example of nodal distribution generated by MFreePre for a very complex domain.

Node Generation
This icon is used to generate nodes in the domain. It can be selected from the preprocess menu, *discretization*, or by clicking the icon button on the toolbar. This command implements the task of generating nodes in the domain, based on the density of nodes specified (or given by default) on the boundaries and at the seed nodes. Figure 16.18 shows an example of the nodal distribution generated by MFreePre for a very complex domain. It is not too difficult to imagine what is required if one has created an FEM model for a geometry of this complexity.

16.3.6 Materials Property Input

The parameters of materials can be input using dialog boxes.

Create Material
This item can be selected from *create* on the *Material* menu. It can be used to create a new material for the material database built together with MFree2D. Figure 16.19 shows a

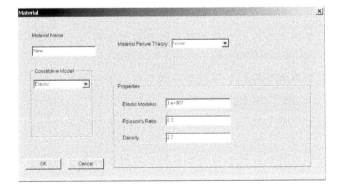

FIGURE 16.19
Dialog box for inputting material properties.

dialog box for inputting material properties. The current MFree2D version released to the public handles only solids of elastic materials. The material created will be stored in the database for future use.

Show and Update Material

This item can be selected from *show & update* on the *Material* menu. It can be used to show, modify, and delete a material from the material database. Figure 16.20 shows a dialog box to display the properties of materials created. Modification is done by simply changing the values. Deleting is done by clicking the delete button.

Material Selection

This item can be selected from *selection* on the *Material* menu. Figure 16.21 shows a dialog box for selecting a material from the material database.

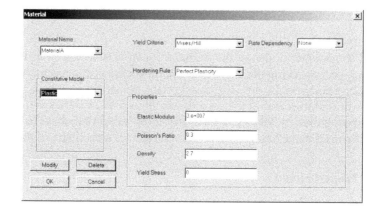

FIGURE 16.20
Dialog box displaying the material properties created. Modification is done by simply changing the values. Deletion is done by clicking the delete button.

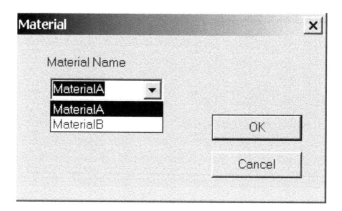

FIGURE 16.21
Dialog box for selecting a material from the material database.

16.3.7 Miscellaneous

MFreePre provides some auxiliary functions to help users view or draw geometry models and change the color of the background or lines.

Scale of Graphical Display Area

This icon is used to change the scale of the graphical display area. It can be selected by clicking the icon button on the toolbar. The default coefficient of the scale is 1. Users can change it in the following dialog box. Figure 16.22 shows a dialog box for changing the display scaling factor.

Pan Graphical View

This icon is used to pan the graphical view. Users can press the left button of the mouse and drag the drawing to shift it.

Clear

This icon is used to clear the screen. This does not delete the model that you are creating.

Zoom In

This icon is used to zoom in the view in the display region with a zooming rate of 200%.

Zoom Out

This icon is used to zoom out the view in the display region with a zooming rate of 50%.

Zoom Region

To view a detail of a drawing, users can also use this function to magnify a selected area with a window.

Global View

This icon is used to restore the entire view of the model within the screen.

Background Color

This icon is used to change the background color. When this icon is selected, a standard color dialog box (Figure 16.23a) will appear. You can directly select a color from the dialog box or define a custom color for the background.

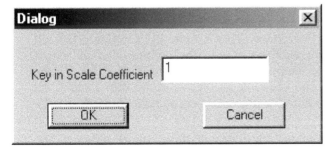

FIGURE 16.22
Dialog box for changing the display scaling factor.

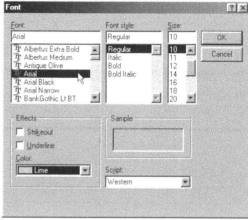

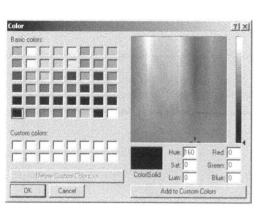

(a)   (b)

FIGURE 16.23
(a) Color editing dialog box; and (b) dialog box for changing fonts.

Entity Color

This icon is used to change the color of entities that will be drawn on the screen. When this icon is selected, a standard color dialog box will appear. You can directly select a color from the dialog box or define a custom color for the selected entities.

Font

This icon is used to change the font, font size, font style, and font color. When this icon is selected, a standard font dialog box (shown in Figure 16.23b) will prompt. Users can change the font properties to meet their requirements.

16.3.8 Adaptive Parameter Setting

MFreePre provides parameter setting for adaptive refinement operations in MFree2D. Default settings work fine for problems that we have tested. Changes to the settings can be done by selecting *Adaptive parameter* on the *settings* menu. Figure 16.24 shows a dialog box for setting adaptive parameters.

The adaptive parameters contain two categories: refinement criterion for local mesh refinement and stop criterion for computation termination. There are three items in the dialog box for the refinement criterion.

- *The percentage of nodes to be refined* is used to prescribe the percentage of the nodes in the whole domain to be refined. This is done by computing the cell energy error (CEE) indicator for all the nodes in the problem domain. For a node the CEE is obtained by a simple average of the (relative) CEEs for the surrounding cells of the node. The CEEs for all the nodes are then ranked. All the nodes at the top of the list within the prescribed percentage are marked to be refined later.

- *Update of scaling factors for the refined nodes* is done with three possible schemes. The first scheme is to use the CEE values of the nodes to be refined. The scaling factors

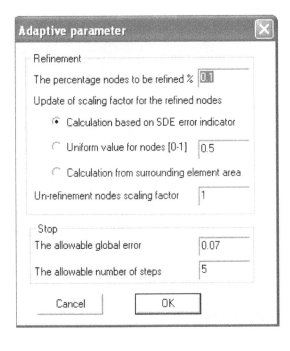

FIGURE 16.24
Adaptive parameter setting dialog box.

for these nodes are updated in relation to their CEE values. The second is to input one coefficient in [0, 1] manually to scale down the original scaling factors for all the nodes to be refined. The last scheme is to simply recompute the scaling factors for refined nodes, based on the newly updated local triangular mesh in the process of mesh updating.

- *Unrefinement nodes scaling factor* is used to define the scaling factor of the unrefinement nodes. Normally, the scaling factors of these unrefinement nodes do not need change, and the default input value is 1.0. Unless there is a special need, there is no need to change the default value.

Two items are included for the stop criterion. *The allowable global error* and *the allowable number of steps* are used to predefine the desired accuracy in terms of the global relative error and the maximum allowable number of adaptive iterations.

16.3.9 Numerical Methods/Processor Selection

The numerical methods or processors in the current MFree2D 2.0 version released to the public are five smoothed PIMs, EFG, and Linear FEM.

Since ES-PIM (linear) is found to have the highest computational efficiency (least computation time for the same accuracy) among all the smoothed PIMs for the same background mesh of triangular elements, it is chosen for adaptive analysis. NS-PIM provides the upper bound solutions with respect to the exact solution in an energy norm, and the FEM produces the lower bound solutions in an energy norm (for force-driving problems). Hence, it may prove useful to use both of them for the fine mesh at the final stage of the analysis to produce quality bounds for the exact solution.

EFG is equipped with two node selection schemes: influence domain-based and surrounding elements (T2L-scheme). This option can be chosen by selecting *adaptive* and then

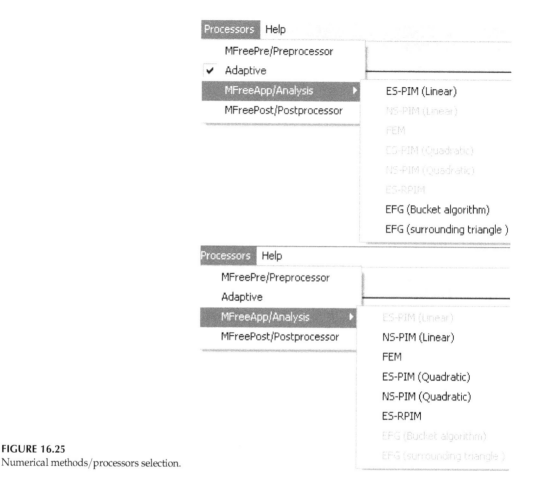

FIGURE 16.25
Numerical methods/processors selection.

MFreeApp/Analysis on the *Processors* menu. Figure 16.25 shows the pop-up menu for numerical methods selection.

16.4 Postprocessing in MFree2D

MFreePost provides a convenient graphical user interface for visualizing numerical solutions of MFree2D, e.g., initial and deformed domain displaying, field contouring, vector viewing, section projecting, and surface and curve plotting. In addition, it allows animation of the converging and nodal-refining processes of the adaptive analysis.

16.4.1 Start of MFreePost

To start MFreePost, choose "Post processor/MFreePost" from "Processors" in the integrated main window shown in Figure 16.1. You can also start MFreePost as an executable file.

16.4.2 Window of MFreePost

Figure 16.26 shows the graphical user interface of the MFreePost. It consists of a title bar, a menu bar, a toolbar, an entity pane (left part of the window), a display region (right area of the window), and a status bar at the bottom of the window.

Main Menu
The menu consists of the following menu items: File, Edit, View, Settings, Field, Section, Animation, Export, and Help.

File
Figure 16.27a shows the submenu in the "File" menu. The submenu items are as follows:

Open: To open an MFree2D result file. The file has an extension of ".out" and is generated by MFreeApp.

Save Image: To save the image displayed in the display region. The file format is Bitmap.

Save Image As: To save the image in the display region with a given file name. The file format is Bitmap.

Print: To print the image in the display region.

Print Preview: To preview the printing effect of the image in the display region.

Print Setup: To set up the printer for use.

List of Recent Files: To display the files most recently used. Maximally four files are displayed.

Exit: To end the MFreePost program.

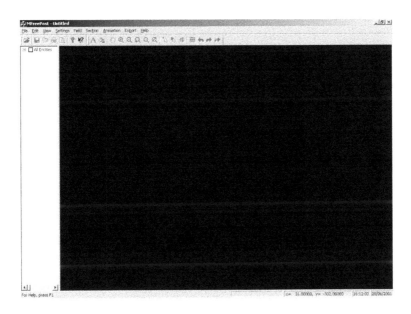

FIGURE 16.26
Graphical user interface of MFreePost.

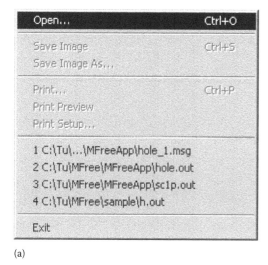

(a)

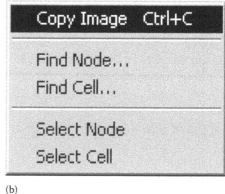

(b)

FIGURE 16.27
(a) File submenu in MFreePost; and (b) Edit submenu in MFreePost.

Edit
Figure 16.27b shows the submenu in the "Edit" item. The submenu items and their functions are as follows:

Copy Image: To copy the image in the display region to the clipboard. This enables the image to be pasted in many word/image-processing applications.

Find Node: To find the location of a specific node by giving the node ID in the dialog box shown in Figure 16.28.

Find Cell: To find the location of a cell by giving the cell ID in the dialog box shown in Figure 16.29. The cell will be highlighted when it is found. If the cell is out of the display region, it will be shifted to the central location of that region.

Select Node: To enable the node selection mode. By turning on this mode, the information (e.g., ID, coordinates, displacement, strain, stress, velocity, acceleration, etc.) of every point within the problem domain can be displayed simply by clicking the left mouse button. Figure 16.30 shows the output information of a point.

Select Cell: To enable the cell selection mode. By turning on this mode, the information (e.g., cell ID, vertices, cell area, etc.) of every cell in the problem domain can be displayed simply by clicking the left mouse button. The output information related to the clicked cell will be shown as in Figure 16.31.

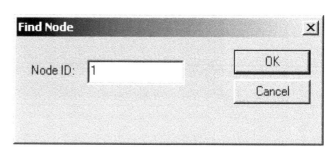

FIGURE 16.28
Find Node dialog box prompts for a node ID.

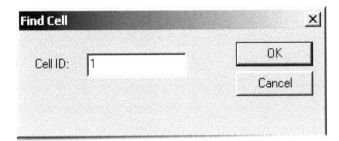

FIGURE 16.29
Find Cell dialog box requires a cell ID

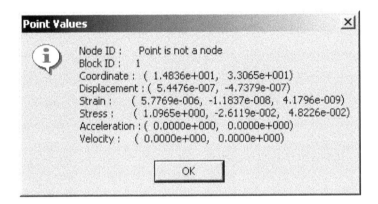

FIGURE 16.30
Output information of a point.

View

The submenu of the "View" item is shown in Figure 16.32. The functions of the submenu items are as follows:

Toolbar: To show or hide the toolbar.

Status Bar: To show or hide the status bar.

Visualization Bar: To show or hide the visualization bar.

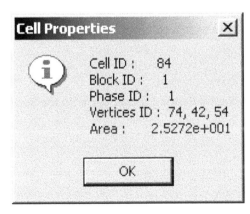

FIGURE 16.31
Output information of a cell.

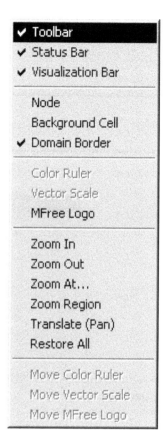

FIGURE 16.32
View submenu in MFreePost.

Node: To show or hide the domain nodes.

Background Cell: To show or hide the background cells.

Domain Border: To show or hide the domain border.

Color Ruler: To show or hide the color ruler.

Vector Scale: To show or hide the vector scale.

MFree Logo: To show or hide the MFree logo.

Zoom In: To zoom in the view in the display region with a zooming rate of 50%.

Zoom Out: To zoom out the view in the display region with a zooming rate of 200%.

Zoom At: To zoom at the view in the display region with a given zooming rate. Figure 16.33 shows a dialog box that requires a magnification factor.

Zoom Region: To zoom the view into a selected rectangular region. The region is selected by dragging the mouse and is projected to the full display region.

Translate (Pan): To shift the view in the display region by moving the mouse with the left button pressed.

Restore All: To restore the image in the display region to its original size and location.

Move Color Ruler: To move the color ruler in the display region.

Move Vector Scale: To move the vector scale in the display region.

Move MFree Logo: To move the MFree logo in the display region.

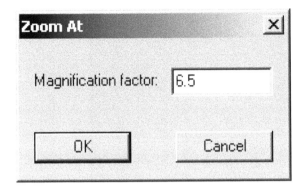

FIGURE 16.33
The Zoom At dialog box requires a magnification factor.

Settings

The menu item "Settings" is responsible for all view settings in the display region. Figure 16.34 shows its submenu in MFreePost. The functions of the submenu items are as follows:

Font: To define the font size, color, and type for the view in the display region (as shown in Figure 16.34).

Colors: To define the background color and drawing color for the display region. The color is selected using the color dialog shown in Figure 16.23.

Color Ruler: To define the style, mode, size, location, and other properties of the color ruler. All the settings related to the color ruler are done using the Color Ruler Settings dialog box (Figure 16.35).

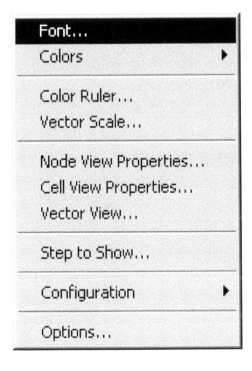

FIGURE 16.34
Submenu of settings in MFreePost.

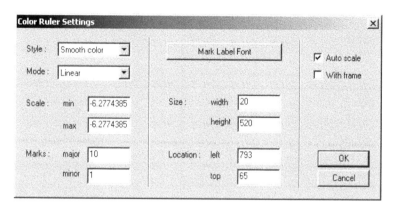

FIGURE 16.35
Dialog box for Color Ruler Settings.

Style: To define Line only, Leaped color, and Smooth color. Line only is a style that shows contour lines, Leaped color style shows the contour regions and fills each region using a unique color, and Smooth color style fills the entire problem domain with smoothly changed colors.

Mode: To define Linear and Logarithmic. The Linear Mode defines that the color along the color scale changes linearly with the value it represents and the Logarithmic Mode defines a logarithmic relationship between the colors and the values.

Scale: To define the minimum and maximum values measured by the color ruler.

Marks: To define the major divisions of the color ruler and the minor divisions in a major division.

Mark Label Font: To define the label font for the color ruler.

Size: To define the size (in pixels) of the color ruler.

Location: To define the location (in pixels) of the color ruler.

Auto Scale: To mark the box to set the scale of the color ruler automatically.

With Frame: To mark the box to bound the color ruler with a frame.

Vector Scale: To define the properties of the vector scale. The dialog box for vector scale settings is shown in Figure 16.36.

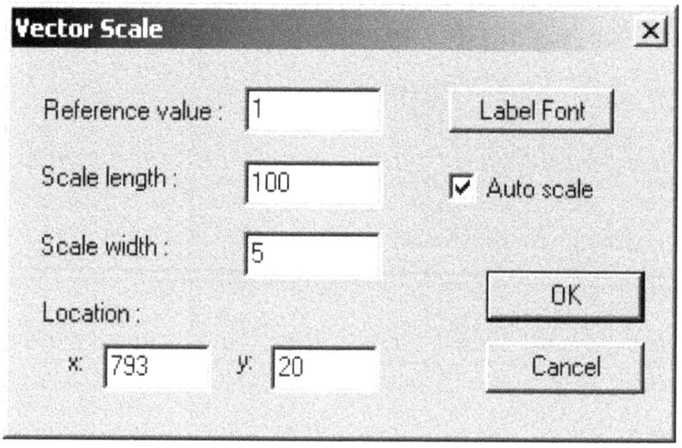

FIGURE 16.36
Dialog box for Vector Scale settings.

Reference Value: To define the value represented by the length of the vector scale.

Scale Length: To define the length (in pixels) of the vector scale.

Scale Width: To define the thickness of the vector scale in pixels.

Location: To define the coordinates (in pixels) of the left origin of the vector scale.

Label Font: To define the label font of the vector scale.

Auto Scale: To mark the box to set the reference value automatically.

Node View Properties: To define the viewing properties for the domain nodes. The dialog box for setting the node view is shown in Figure 16.37.

Show Node Symbol: To mark the box to show the node symbol.

Show Node ID: To mark the box to show the node ID.

Shape: To choose the nodal shape. The options are none, circle, and rectangle.

Foreground Color: To define the drawing color for the node symbol.

Fill Color: To define the color to fill the node symbol.

Size: To define the size of the node symbol in pixels.

Line Width: To define the line width for drawing the node symbol.

Distance from Nodal Position: To define the relative location of the node label to the nodal position.

Font: To define the node label font.

Cell View Properties: To define the cell view properties. The dialog box for setting the cell view properties is shown in Figure 16.38.

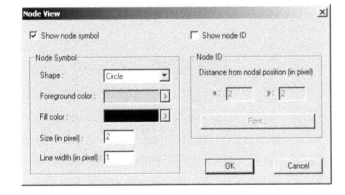

FIGURE 16.37
Dialog box for setting Node View properties.

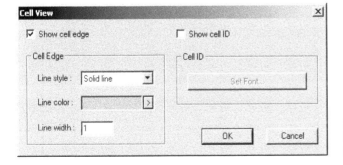

FIGURE 16.38
Dialog box for setting Cell View properties.

Show Cell Edge: To mark the box to show the cell edge.

Show Cell ID: To mark the box to show the cell ID label.

Line Style: To define the line style for drawing the cell edge. The options are solid line, dash line, dot line, and dashdot line.

Line Color: To define the edge color.

Line Width: To define the edge thickness in pixels.

Set Font: To set the label font for the cell ID.

Vector View: To define the vector properties for a vector field. The dialog box for setting the vector properties is shown in Figure 16.39.

Style: To define the line style of the vector arrow. The options are solid line, dot line, and dash line.

Color: To define the vector arrow color.

Width: To define the thickness of the vector arrow in pixels.

Arrowhead Length: Fixed: to define an arrow with a fixed head length. Proportional: to define an arrow with the head length proportional to the arrow length. The length is measured in pixels.

Step to Show: To define the step to be shown in the display region. The dialog box is shown in Figure 16.40.

Show Mesh Refinement Step: To show a specified mesh refinement step.

Show Load Increment Step: To show a specified load increment step.

Configuration: To set the configuration of the problem domain. There are two options of configurations: initial configuration and current configuration (Figure 16.41). The former is based on the original undeformed problem domain whereas the latter uses the deformed problem domain at the current step of iteration.

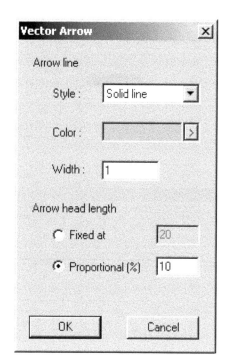

FIGURE 16.39
Dialog box to define Vector Arrow properties.

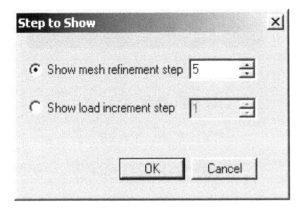

FIGURE 16.40
Dialog box to define Step to Show.

FIGURE 16.41
Submenu in the Configuration item.

Options: To define other general options.

Field
This item consists of functions related to field variables. The submenu is shown in Figure 16.42. The functions of the submenu items are as follows:

Strain: To display the strain field. The strain components consist of strains in the x and y directions (strain_xx and strain_yy), the shear strain (strain_xy), principal strains, the equivalent strain, etc. Figure 16.43 is the submenu of Strain.

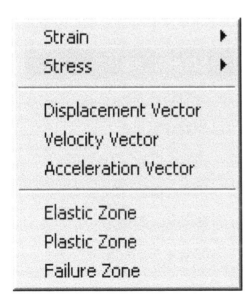

FIGURE 16.42
Submenu of field.

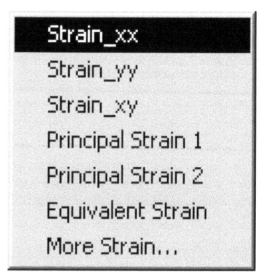

FIGURE 16.43
Strain components that can be displayed.

Stress: To display the stress field. The stress components consist of stresses in the *x*- and *y*-directions (stress_xx and stress_yy), the shear stress (strain_xy), principal stresses, Mises stress, etc. Figure 16.44 is the submenu of Stress.

Displacement Vector: To show the displacement field using vector arrows. A vector arrow starts at the location of a node and points to the displacement direction. The arrow length represents the magnitude of the nodal displacement.

Velocity Vector: To show the velocity field using vector arrows. A vector arrow starts at the location of a node and points to the velocity direction. The arrow length represents the magnitude of the nodal velocity.

Acceleration Vector: To show the acceleration field using vector arrows. A vector arrow starts at the location of a node and points to the acceleration direction. The arrow length represents the magnitude of the nodal acceleration.

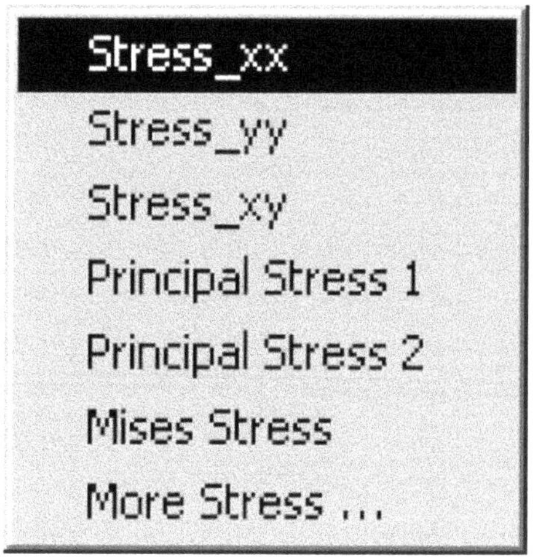

FIGURE 16.44
Stress components that can be displayed.

Elastic Zone: To paint the elastic region in the problem domain with a specified color (green by default).

Plastic Zone: To paint the plastic region in the problem domain with a specified color (yellow by default).

Failure Zone: To paint the failure region in the problem domain with a specified color (red by default).

Section

This is a collection of commands related to the section view. The submenu items are shown in Figure 16.45.

Create Section: To create a section inside the problem domain. This is done by moving the cursor to the starting point, pressing and holding the left mouse button, dragging the cursor across the problem domain to the ending point, and releasing the button. A section line is then shown on the screen. Note that only one section line can be created at a time and the creation of a new section will replace the existing one.

Remove Section: To remove the existing section if any.

Section Properties: To define the starting and ending points of a section precisely. The dialog box to set section properties is shown in Figure 16.46.

Show Section Line: To show or hide a section line.

Show Curves: To show curves of field variables along the section line. The choice of this item initiates a separate Section View window (Figure 16.47) where the curves are displayed.

Export Data: To export variable values along a section to a text file.

Animation: To define animation. This comprises all commands related to animation. The submenu of animation is shown in Figure 16.48.

Set Up Animation: To set up an animation using a dialog box shown in Figure 16.49. There are two panels in the dialog box: the source panel is for defining the source data for an animation and the advance panel defines the animation advancing mode.

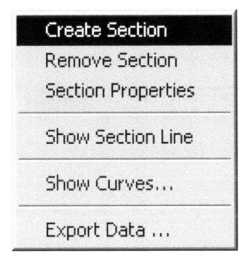

FIGURE 16.45
Submenu of Create Section.

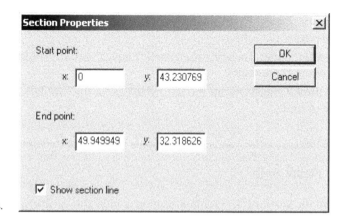

FIGURE 16.46
Dialog box for setting Section Properties.

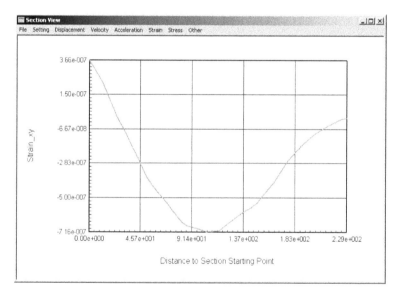

FIGURE 16.47
Section View window to display curves along a section.

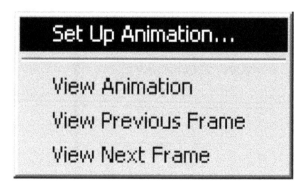

FIGURE 16.48
Submenu of Animation.

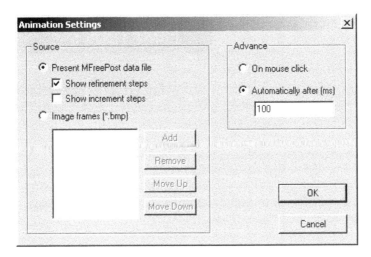

FIGURE 16.49
Dialog box for setting up an animation.

Present MFreePost Data File: To define the source data from the currently opened data file. Further, whether to show refinement steps or to show increment steps can also be defined.

Image Frames: To define the source data from a list of image frames. The list can be maintained using four commands: Add, Remove, Move Up, and Move Down.

On Mouse Click: Advancing mode is made by clicking the left mouse button and frame by frame.

Automatically After: The advancing mode is automatic with a specified interval time.

View Animation: To display an animation with a specified advancing mode.

View Previous Frame: To show the previous frame in an animation.

View Next Frame: To show the next frame in an animation.

Export
This item is specifically used for exporting values to text files. The submenu is shown in Figure 16.50.

Node Value: To export the values of a specified node to a text file. The values include coordinates, displacement, velocity, acceleration, strain, stress, and deformation status at all steps.

FIGURE 16.50
Submenu of Export.

FIGURE 16.51
Submenu of Help Topics.

Section Value: To export the variable values along a specified section to a text file.

Help
This item comprises help contents and description of the software package. The submenu is shown in Figure 16.51.

Help Topics: To search for a specific help topic.
About MFreePost: To describe the credit and the copyright information of MFreePost.

Index

A

ABAQUS, 260, 263, 267, 270, 321, 610
Adaptive analysis, 31, 156, 323, 379, 386, 432,
        435, 596, 733–734
    arbitrary complex domain, 707
    bucket algorithm, node searching, 664–666
    cell energy error (CEE) estimate
        error indicator, 687–689
        error sources, 689–690
        NS- and ES-PIM settings, 690
    connecting rod, 707–709
    density factor, 697
    elastic cube, 711–715
    error estimation, 686–687
    infinite plate, circular hole
        error estimation, 692–693
        NS-PIM and ES-PIM, 699
    inviscid flow, 442–445
    local adaptive refinement, 697
    local Delaunay triangulation algorithm,
        698–699
    node numbering, 663–664
    Poisson problems, 440–442
    rectangular cantilever, 691–692
    rectangular plate, 704–706
    relay model, domains
        circle involute approach, 677–686
        control point and connection,
            670–671
        diffraction method, 668–669
        effective relay radius and equivalent
            distance, 673–675
        numerical implementation, 676–677
        problem statement, 666–667
        relay point, 671–673
        relay region, 671–672
        transparency method, 669–670
        visibility method, 667–668
        weight parameter and derivatives,
            675–676
    square plate
        crack, 693–697, 706–707
        square hole, 699–703
    triangular mesh and integration cells
        EFG, 662–663
        PIMs, 663
        uses, 661–662

Adaptive gradient smoothing method (A-GSM)
    advancing front technique, 438
    directional error indicator, 435–437
    inviscid flow, 442–445
    meshing parameters, 437–438
    Poisson problems, 440–442
    procedure, 438–440
Adaptive parameter setting, 732–733
Adaptive refinement, 701, 703, 710, 717, 732
    local, 661, 697, 699, 705, 709
Admissible conditions, 112, 135–138, 149,
        172–173, 175–177, 179–180, 188
Admissible displacement, 9, 99, 145, 147, 176
Admissible field function, 171, 176
Amplitude, 270, 343, 362, 464, 494, 571
Analytical solution, 214, 216, 228–229, 231, 252,
        255, 260, 285, 293, 300, 304, 311,
        318–321, 328, 332, 334, 336, 488, 496,
        498, 510, 532–534, 537, 553, 555–556,
        576, 581, 609, 621, 624, 626, 630, 633,
        639, 646, 658, 691, 693, 699, 702, 706
    Blasius analytical solution, 430–431
    Poisson problem
        adaptive analysis, 440–441
        benchmarking, 419–420
Anisotropic material, 4, 480–481
Approximation function, 54
    boundary flux, 141–142, 434
    characters, 13
    field approximation, 92, 210, 564
    field variable approximation, 14, 37, 221,
        223, 690
    function, 19, 73, 74, 89, 90, 140, 153–156,
        214, 618
    surface, 570
Arc geometry, 722
Artificial compressibility, 389, 396–398, 443
Artificial viscosity, 395–396

B

Background cell, 11, 14–15, 17, 18, 26–28, 39, 79,
        89–91, 104, 106–107, 136, 147, 164, 169,
        178, 193, 201, 216, 218, 237, 276–277,
        316, 329, 338, 341, 345, 348, 351,
        366–367, 369–370, 408, 414, 420–421,
        434–435, 531, 542, 596, 623, 625, 633,
        662–663, 686–687, 689, 718–719, 738

Milton Keynes UK
Ingram Content Group UK Ltd.
UKHW051927141024
449569UK00027B/1389